Handbook of Capillary
Electrophoresis Applications

JOIN US ON THE INTERNET VIA WWW, GOPHER, FTP OR EMAIL:

WWW: http://www.thomson.com
GOPHER: gopher.thomson.com
FTP: ftp.thomson.com
EMAIL: findit@kiosk.thomson.com

A service of I(T)P®

Handbook of Capillary Electrophoresis Applications

Edited by

H. SHINTANI
Section Director
National Institute of Health Sciences
Tokyo
Japan

and

J. POLONSKÝ
Professor of Analytical Chemistry
Slovak Technical University
Bratislava
Slovakia

BLACKIE ACADEMIC & PROFESSIONAL
An Imprint of Chapman & Hall
London · Weinheim · New York · Tokyo · Melbourne · Madras

Published by Blackie Academic and Professional, an imprint of
Chapman & Hall, 2–6 Boundary Row, London SE1 8HN, UK

Chapman & Hall, 2–6 Boundary Row, London SE1 8HN, UK

Chapman & Hall GmbH, Pappelallee 3, 69469 Weinheim, Germany

Chapman & Hall USA, 115 Fifth Avenue, New York, NY 10003, USA

Chapman & Hall Japan, ITP-Japan, Kyowa Building, 3F, 2-2-1 Hirakawacho, Chiyoda-ku, Tokyo 102, Japan

DA Book (Aust.) Pty Ltd, 648 Whitehorse Road, Mitcham 3132, Victoria, Australia

Chapman & Hall India, R. Seshadri, 32 Second Main Road, CIT East, Madras 600 035, India

First edition 1997

© 1997 Chapman & Hall

Typeset in 10/12 Times By Best-set Typesetter Ltd, Hong Kong
Printed in Great Britain by Hartnolls Ltd, Bodmin, Cornwall

ISBN 0 7514 0359 8

Apart from any fair dealing for the purposes of research or private study, or criticism or review, as permitted under the UK Copyright Designs and Patents Act, 1988, this publication may not be reproduced, stored, or transmitted, in any form or by any means, without the prior permission in writing of the publishers, or in the case of reprographic reproduction only in accordance with the terms of the licences issued by the Copyright Licensing Agency in the UK, or in accordance with the terms of licences issued by the appropriate Reproduction Rights Organization outside the UK. Enquiries concerning reproduction outside the terms stated here should be sent to the publishers at the London address printed on this page.

The publisher makes no representation, express or implied, with regard to the accuracy of the information contained in this book and cannot accept any legal responsibility or liability for any errors or omissions that may be made.

A catalogue record for this book is available from the British Library

Library of Congress Catalog Card Number: 96–70902

∞ Printed on acid-free text paper, manufactured in accordance with ANSI/NISO Z39.48-1992 (Permanence of Paper)

Contents

List of contributors xvii

Foreword xxi

Preface xxiii

Introduction 1
K. KITAGISHI

I.1	Electroosmotic flow (EOF) and analyte mobilization	1
I.2	Separation modes and samples	4
	I.2.1 Electrophoretic mode	4
	I.2.2 Electrokinetic mode	6
	I.2.3 Polymer sieving mode	8
	I.2.4 Isoelectric mode	12
	I.2.5 Other separation modes	12
References		13

Part One   Equipment Systems

1 Sample injection 17
K. KITAGISHI

1.1	Modes of sample injection	17
	1.1.1 Hydrodynamic injection	17
	1.1.2 Electrokinetic injection	18
1.2	Injection reproducibility	18
1.3	Sample preconcentration	20
1.4	Other microhandling techniques for sample injection	22
References		23

2 Quantitative analysis, precision and reproducibility 25
K. KITAGISHI

2.1	Quantitation of sample mass	25
2.2	Integration techniques	26
2.3	Quantitative information derived from migration time	29
	2.3.1 Ionization constant, pK_a in free zone capillary electrophoresis	29
	2.3.2 MEKC partition coefficient	29
	2.3.3 Molecular weight in linear polymer sieving and capillary gel electrophoresis (CGE)	31
	2.3.4 Isoelectric point (pI) in capillary isoelectric focusing	31
	2.3.5 Binding constant, K_b, in affinity CE	31
	2.3.6 Relative chiral separation factor/percent chiral separation in chiral CE	32
2.4	Peak broadening	32
2.5	Temperature effects in CE	33
2.6	Temperature controlled device	34

	2.7	Attempts to improve reproducibility and accuracy	34
	References		37

3 Precision and quantitation in capillary electrophoresis   41
G.A. ROSS

	3.1	Introduction	41
	3.2	Precision	41
		3.2.1 Migration time precision	41
		3.2.2 Peak area and height precision	45
	3.3	Quantitation	50
		3.3.1 Quantitation of impurities and chiral analyses	51
		3.3.2 Internal standards	51
		3.3.3 Calibration	53
		3.3.4 Accuracy	53
	3.4	Conclusions	54
	References		54

4 Column technology   56
K. KITAGISHI

4.1	Electroosmotic control	56
4.2	Techniques for coating the capillary surface	57
4.3	Gel-filled capillaries	61
4.4	Column connection	63
4.5	Novel column designs	63
References		66

5 Electrolyte systems   71
K. KITAGISHI

5.1	Optimization of electrolytes with respect to separation characteristics	71
5.2	Electrolytes in free zone capillary electrophoresis	73
5.3	Electrolytes in MEKC and related EKC	74
5.4	Electrolytes in polymer sieving	76
5.5	Electrolytes for chiral separation	78
5.6	Additives to suppress the wall adsorption of basic compounds	78
References		79

6 Detection   84
K. KITAGISHI

6.1	On-column and off-column detection	84
6.2	Absorbance	86
6.3	Fluorescence	87
6.4	Chemiluminescence	89
6.5	Indirect detection	90
6.6	Conductivity	90
6.7	Potentiometry	92
6.8	Amperometry	92
6.9	Mass spectroscopy	95
6.10	Refractive index	96
6.11	Other detection techniques	97
References		97

CONTENTS

7 Recent advances in fluorescence detection technology — 104
M. SAITO and Y. KUROSU

7.1	Introduction	104
7.2	Inherent complexity in fluorescence detection	104
7.3	Recent advances in fluorescence detection technology	107
	7.3.1 Flowcell arrangements	107
	7.3.2 Laser-induced fluorescence detection (LIFD)	109
	7.3.3 Simultaneous detection with UV absorption and fluorescence	114
7.4	Conclusion	114
	References	114

8 Other systems — 118
K. KITAGISHI

8.1	Fraction collection	118
8.2	Field modulation	119
8.3	Computer simulation	119
	References	120

9 Advanced systems — 122
K. KITAGISHI

9.1	Capillary electrophoresis on a glass chip	122
9.2	Imaging detector	123
9.3	Automated analysis	123
9.4	DNA sequencing and diagnosis	124
9.5	Other new approaches in CE	125
	References	126

10 Automated fraction collection in capillary electrophoresis — 128
R. GRIMM

10.1	Introduction	128
10.2	Principle of fraction collection	129
10.3	Fraction collection of peptides using capillary zone electrophoresis	130
10.4	Peptide analysis by micellar electrokinetic chromatography (MEKC)	130
10.5	Protein and oligonucleotide analysis by capillary gel electrophoresis (CGE)	133
10.6	Protein mixture analysis by capillary isoelectric focusing (CIEF)	133
10.7	Conclusion	133
	References	136

11 Comparison with other analytical methods — 137
K. KITAGISHI

11.1	Comparison with high performance liquid chromatography	137
	11.1.1 Separation capability	137
	11.1.2 Detection	137
	11.1.3 Sample preparation	138
	11.1.4 Reliable quantitation	138
	11.1.5 Resources utilization and environmental friendliness	139
	11.1.6 Assay time, sample amount and throughput	140
11.2	Comparison with gel electrophoresis	140
	11.2.1 Preparation of separation medium	141
	11.2.2 Sample preparation and injection	141

	11.2.3	Detection		142
	11.2.4	Reliable quantitation and reproducibility	142
	11.2.5	Assay time, throughput, sample amount and cost	143
11.3	Comparison with ion chromatography	143
	11.3.1	Separation capability	143
	11.3.2	Sample preparation	144
	11.3.3	Detection, quantitation and reproducibility	144
	11.3.4	Assay time and running cost	144
References		145

Part Two Biochemistry Applications

12 Amino acids — 149
L.J. BRUNNER

12.1	Introduction	149
12.2	Application of CE to amino acid analysis	149
12.3	Conclusion	156
References		156

13 Sodium dodecyl sulfate (SDS)–non-gel molecular sieving capillary electrophoresis for proteins — 158
Y. KUROSU and M. SAITO

13.1	Introduction	158
13.2	Principles of molecular sieving CE	159
13.3	Molecular sieving CE in the absence of sodium dodecyl sulfate (SDS)	159
13.4	Fundamental studies of the use of molecular sieving media (gels and non-gels) for proteins and DNA	160
13.5	SDS–non-gel molecular sieving electrophoresis for proteins	162
13.6	Conclusions	164
References		170

14 Protein analysis by capillary electrophoresis — 173
F.-T.A. CHEN and R.A. EVANGELISTA

14.1	Introduction	173
14.2	Capillary zone electrophoresis	175
	14.2.1 Untreated fused-silica capillary	175
	14.2.2 Chemically modified fused-silica capillary	179
14.3	Sodium dodecyl sulfate capillary gel electrophoresis	180
14.4	Micellar electrokinetic chromatography	183
14.5	Capillary isoelectric focusing	183
References		194

15 Analysis of erythropoietin by capillary electrophoresis — 198
F.-T.A. CHEN

15.1	Introduction	198
15.2	Native protein	199
15.3	Peptide mapping and glycopeptides analysis	200
15.4	Carbohydrate analysis	202
15.5	Future trends in CE analysis of r-EPO	204
References		205

16 Assay of enzymes by capillary electrophoresis — 207
F.-T.A. CHEN and R.A. EVANGELISTA

16.1	Introduction	207
16.2	Off-line analysis of enzyme catalysis and applications	208
16.3	On-line immobilized enzymes and their applications	213
16.4	Postcolumn enzyme assay	213
16.5	On-line analysis of enzyme activity and applications	214
16.6	Future trends in CE analysis of enzymes	215
	References	216

17 Capillary electrophoresis-based immunoassays — 219
R.A. EVANGELISTA and F.-T.A. CHEN

17.1	Introduction	219
17.2	Immunoassays by CE with UV detection	220
17.3	Immunoassays by CE with laser-induced fluorescence (LIF) detection	221
17.4	Multianalyte immunoassays	229
17.5	Labeling considerations	231
17.6	Assay considerations	233
17.7	Summary	237
	References	237

18 Analysis of antibodies by capillary electrophoresis — 240
T. PRITCHETT

18.1	Introduction	240
18.2	Analysis of antibodies using capillary zone electrophoresis	240
18.3	Analysis of antibodies using capillary isoelectric focusing	245
18.4	Analysis of antibodies using SDS–capillary gel electrophoresis	245
18.5	Analysis of antibodies using micellar electrokinetic capillary chromatography	247
18.6	Quantitative analysis of antibodies	248
	References	252

19 Analysis of peptide hormones and model peptides by capillary electrophoresis — 255
J.C. OSBORNE

19.1	Introduction	255
19.2	Peptides from the pituitary gland	255
19.3	Peptides from the hypothalamus	263
19.4	Peptides from the pancreas	264
19.5	Peptides from the gastrointestinal tract	264
19.6	Polypeptides from the placenta	265
19.7	Polypeptides from the kidney	265
	References	265

20 Affinity analysis by capillary electrophoresis — 270
J.E. WIKTOROWICZ

20.1	Introduction		270
20.2	Size/mobility-based affinity CE		271
20.3	General applications		273
	20.3.1	Measurement of the binding constant	273
	20.3.2	Measurement of the kinetic constants	275

	20.3.3 Measurement of the binding stoichiometry	276
20.4	Immobilized affinity analysis	278
20.5	Conclusions	280
References		281

21 Analysis of nucleic acids by capillary electrophoresis 283
K. KITAGISHI

21.1	Introduction	283
21.2	Bases, nucleosides, nucleotides and their related compounds	283
21.3	Oligonucleotides	291
21.4	Molecular sieving separation of DNA restriction fragments and polymerase chain reaction (PCR) products	291
21.5	DNA sequencing	293
21.6	Possible application to RNA analysis	294
21.7	Conclusion	295
References		295

Part Three Pharmaceutical Science Applications

22 Analysis of pharmaceuticals by capillary electrophoresis 301
T.J. O'SHEA

22.1	Introduction	301
22.2	Method validation	301
22.3	Impurity analysis	308
22.4	Antiasthmatics	309
22.5	Antibiotics	309
22.6	Anticancer drugs	311
22.7	Antiulcer drugs	312
22.8	Antidepressants	312
22.9	Antimigraine drugs	312
22.10	Analgesics	313
22.11	Cardiovascular drugs	313
22.12	Chinese herbal drugs	314
22.13	Drugs of abuse	315
22.14	Miotic agents	316
22.15	Protein/peptide drugs	316
22.16	Miscellaneous	318
References		319

23 Analysis of medicinal plants: comparison of capillary electrophoresis with high performance liquid chromatography 324
P.G. PIETTA

23.1	Introduction	324
23.2	Analysis of medicinal plants	325
23.3	Conclusions	331
References		332

24 Quantitative applications of the resolution of enantiomers by capillary electrophoresis 334
K.D. ALTRIA

24.1	Introduction	334
24.2	Method validation	334

	24.2.1	Detection limits	334
	24.2.2	Precision	338
	24.2.3	Linearity	338
	24.2.4	Selectivity	339
	24.2.5	Recovery	339
	24.2.6	Cross validation	339
	24.2.7	Freedom from interference	339
	24.2.8	Robustness evaluation	340
	24.2.9	Method transfer	340
24.3	Quantitation		340
24.4	Quantitative application areas		341
	24.4.1	Enantiomeric purity testing	341
	24.4.2	Reaction rate monitoring	341
	24.4.3	Formulation stability testing	341
	24.4.4	Clinical and forensic analysis	342
24.5	Conclusions		342
References			342

25 Applications of chiral capillary electrophoresis (cyclodextrin–capillary zone electrophoresis; CD–CZE)   345
A. AUMATELL

25.1	Introduction		345
25.2	Amino acids		345
25.3	Sympathomimetics		347
	25.3.1	β-Agonists	347
	25.3.2	β-Antagonists	351
	25.3.3	Agenergic agents	352
25.4	Stimulants		352
25.5	Anesthetics and sedatives		353
25.6	Anticoagulants		355
25.7	Antidepressants		355
25.8	Non-steroidal anti-inflammatory drugs		355
25.9	Polycyclic aromatic compounds		356
25.10	Cationic drugs and solutes		357
25.11	Anionic solutes		360
25.12	Chiral method development		361
	Abbreviations		361
References			362

26 Separation of enantiomeric compounds by micelle-mediated capillary electrokinetic chromatography   366
M.E. SWARTZ, J.R. MAZZEO and E.R. GROVER

26.1	Introduction	366
26.2	Indirect separation of enantiomers by MEKC	368
26.3	Direct separation of enantiomers by cyclodextrin-modified MEKC	369
26.4	Direct separation of enantiomers by MEKC with natural product chiral micelles formed from bile salt surfactants	373
26.5	Direct separation of enantiomers by MEKC with synthetic chiral surfactants	376
26.6	Conclusion	379
References		382

27 Therapeutic drug monitoring by capillary electrophoresis   386
Z.K. SHIHABI

27.1	Introduction	386

		27.1.1	History	386
		27.1.2	Role of CE in therapeutic drug monitoring	388
	27.2	Practical aspects		390
		27.2.1	Sample preparation for CE	390
		27.2.2	General guidelines for improving separation in CE	391
		27.2.3	Basic principles of TDM	392
	27.3	Drug analysis by CE		392
		27.3.1	Common drugs	392
		27.3.2	Miscellaneous non-common drugs	399
		27.3.3	Therapeutic drug monitoring by combination of CE and immunoassay	401
		27.3.4	Comparison of CE with HPLC	401
	27.4	Chiral separation		401
	27.5	Quantitation and reproducibility		402
	27.6	Drug screening for drugs of abuse and forensic applications		403
	27.7	Concluding remarks		404
	References			405

Part Four Bioscience Applications

28 Biomedical applications and biological systems   411
P.L. WEBER

28.1	Introduction		411
28.2	Tissue analysis		411
	28.2.1	Brain tissue	411
	28.2.2	Other tissues	417
28.3	Single cell analysis		418
	28.3.1	Erythrocyte	418
	28.3.2	Other cells	419
28.4	Sampling of tissues by microdialysis		420
28.5	Conclusion		421
References			422

29 Analysis of selected common clinical tests by capillary electrophoresis   425
Z.K. SHIHABI

29.1	Introduction		425
29.2	Serum proteins		427
	29.2.1	Clinical significance	427
	29.2.2	Present methods of analysis	427
	29.2.3	CE separation	427
	29.2.4	Capillary isoelectric focusing (CIEF)	430
	29.2.5	Specific serum proteins	430
	29.2.6	Immunofixation (electrophoresis)	430
	29.2.7	Cryoglobulins	431
29.3	Urinary proteins		431
29.4	Cerebrospinal fluid (CSF) proteins		434
29.5	Serum lipoproteins		436
29.6	Enzymes		436
	29.6.1	General	436
	29.6.2	Proteolytic enzymes	438
	29.6.3	Isoenzymes	438
29.7	Hemoglobin variants, A_{1c} and globins		440
29.8	Miscellaneous tests		442

		CONTRIBUTORS	xiii

	29.9	Practical aspects	443
		29.9.1 General method for serum protein analysis	443
		29.9.2 Method for urine clean-up	443
		29.9.3 Method for analysis of cathepsin D	443
		29.9.4 Recommended guidelines for CE in clinical analysis	444
		29.9.5 Quantitation and reproducibility	444
	29.10	Concluding remarks	445
	References		445

30 Gene analysis and nucleic acid sequencing   449
H.-M. WENZ and J.E. WIKTOROWICZ

	30.1	Introduction	449
	30.2	Fragment and gene analysis	451
		30.2.1 Detection of sequence-induced anomalous migration of dsDNA	451
		30.2.2 Mutation detection	451
		30.2.3 Determination of length variations for mutation detection, mapping of heritable diseases, carrier detection and forensic identification	453
		30.2.4 Quantitative assays	455
	30.3	Sequence analysis	456
		30.3.1 Instrumentation	456
		30.3.2 Separation chemistry	457
	30.4	Conclusions	458
	Acknowledgments		463
	References		463

31 Gene mutation and DNA sequencing   466
K. KITAGISHI

	31.1	Introduction	466
	31.2	Polymerase chain reaction (PCR) product analysis	466
		31.2.1 Size determination and quantitation of PCR-amplified DNA fragments	466
		31.2.2 DNA mutational analysis	476
	31.3	DNA sequencing	478
		31.3.1 DNA sequencing in CE with gel-filled capillaries	478
		31.3.2 DNA sequencing in CE with soluble linear polymers	480
	31.4	Enhancement of throughput in gene analysis in CE	480
		31.4.1 Parallel runs in an array of capillaries	480
		31.4.2 High speed separations on a micromachined CE chip	482
	31.5	Conclusion	483
	References		483

32 Analysis of body fluids: urine, blood, saliva and tears   486
L.A. COLON

	32.1	Introduction	486
	32.2	Capillary electrophoresis of urine samples	487
	32.3	Capillary electrophoresis of blood samples	488
	32.4	Capillary electrophoresis of saliva and tear samples	493
	32.5	Concluding remarks	494
	References		494

33 Toxins associated mainly with uremia and cancer   499
H. SHINTANI

    33.1   Introduction   499
    33.2   Analysis of toxic compounds from uremia or cancer patients   501
    33.3   Conclusion   503
    References   504

Part Five   Ion Analysis Applications

34 Capillary electrophoresis of metal complexes   509
W. BUCHBERGER

    34.1   General strategies for metal ion analysis after complexation   509
    34.2   Precapillary and on-capillary complexation of metal ions   510
    34.3   Modes for CE separation of metal complexes   511
    34.4   Applications of metal ion analysis after complexation   515
    References   515

35 Metal chelation   517
F.B. ERIM

    35.1   Introduction   517
    35.2   Metal chelation in capillary electrophoresis   517
    35.3   Conclusion   528
    References   528

36 Inorganic ions   531
W. BUCHBERGER

    36.1   Introduction   531
    36.2   Requirements for inorganic ion analysis by CZE   531
    36.3   Carrier electrolytes for separation of inorganic anions   533
    36.4   Carrier electrolytes for separation of inorganic cations   534
    36.5   Applications of capillary zone electrophoresis in inorganic ion analysis   535
    36.6   Conclusions   543
    References   544

37 Organic acids and organic ions   550
F.S. STOVER

    37.1   Introduction   550
    37.2   Aspects of CE separations of organic ions and organic acids   551
    37.3   Applications of CE to organic acids and organic ions   553
          37.3.1   Organic acids and anions   553
          37.3.2   Organic bases and cations   560
    References   567

Part Six   Food Analysis Applications

38 Food analysis by capillary electrophoresis   583
P.F. CANCALON

    38.1   Introduction   583
    38.2   Analysis of food components   589
          38.2.1   Small ions and organic acid analysis   590
          38.2.2   Saccharides   592
          38.2.3   Amino acids, peptides and proteins   593

		38.2.4	Lipids and hormones	595

           38.2.4   Lipids and hormones   595
           38.2.5   Vitamins   595
           38.2.6   Natural compounds   596
           38.2.7   Drugs, additives, toxins and degradation products   597
    38.3   Multiple analyses   598
    References   599

39 Capillary zone electrophoresis analysis of additives in food samples   607
R. SCHUSTER

    39.1   Introduction   607
    39.2   Artificial sweeteners   607
    39.3   Natural and synthetic dyes   609
    39.4   Concrete CZE analytical procedure   610
           39.4.1   Sweeteners   610
           39.4.2   Synthetic dyes   611
    39.5   Application   612
           39.5.1   Sweeteners   612
           39.5.2   Dyes   612
    39.6   Reproducibility and linearity   616
    39.7   Conclusion   616

40 Analysis of underivatized carbohydrates by capillary electrophoresis and electrochemical detection   617
R.P. BALDWIN

    40.1   Introduction   617
    40.2   Electrochemical detection of carbohydrates   619
    40.3   CE separations at high pH   619
    40.4   Applications   622
    40.5   Conclusions   625
    References   626

Part Seven   Environmental Science Applications

41 Hydrocarbons   629
E. DABEK-ZLOTORZYNSKA

    41.1   Introduction   629
    41.2   Application of micellar electrokinetic capillary chromatography to
           analysis of hydrocarbons   629
    References   636

42 Environmental pollutants   639
E. DABEK-ZLOTORZYNSKA

    42.1   Introduction   639
    42.2   Phenol and its derivatives   639
    42.3   Aromatic and aliphatic amines   645
    42.4   Nitrosoamines   646
    42.5   Carbonyl compounds   646
    42.6   Phthalate esters   647
    42.7   Explosive compounds   647
    42.8   Surfactants   647
    42.9   Dyes and their hydrolyzed compounds   648
    42.10 Pesticides and herbicides   649

	42.11 Organic heavy metal compounds	651
	References	652

43 Analysis of synthetic polymers by capillary electrophoretic techniques — 657
J.A. BULLOCK

43.1	Introduction	657
43.2	Analysis of impurities and residual monomers in polymers	664
43.3	Analysis of polymeric particulates	665
43.4	Separation of oligomers of homopolymers and copolymers	666
43.5	Capillary electrophoresis of high molecular weight polymers	668
43.6	Capillary electrophoresis of inorganic polymers	669
43.7	Polymer composition analysis by capillary electrophoresis	669
43.8	Investigation of polymerization kinetics by capillary electrophoresis	670
43.9	Conclusions	670
	References	671

Appendix A Commercially available instrumentation for capillary electrophoresis — 673

Appendix A1 Overview of commercial instruments — 673
T. WEHR and M. ZHU

A1.1	Introduction	673
A1.2	Power supply	673
A1.3	Injection	678
A1.4	Capillary temperature control	679
A1.5	Liquid handling	680
A1.6	Autosamplers	681
A1.7	Detectors	681
	A1.7.1 UV–vis absorbance	681
	A1.7.2 Fluorescence	683
	A1.7.3 Other detectors	684
	A1.7.4 On-line coupling with mass spectrometry (MS)	684
A1.8	Fraction collection	685

Appendix A2 The CAPI series of instruments — 686
K. KITAGISHI

A2.1	CAPI-3100	686
	A2.1.1 Specifications	686
	A2.1.2 Features	688
A2.2	CAPI-1000	688
	A2.2.1 Specifications	688
	A2.2.2 Features	689

Appendix B Troubleshooting — 691
K. KITAGISHI

Appendix C Commercially available buffer reagents — 695
H. SHINTANI

Index — 697

Contributors

Kevin D. Altria Pharmaceutical Analysis Division, Glaxo Group Research and Development, Park Road, Ware, Hertfordshire SG12 0DP, UK

Anthony Aumatell 9 Oxford Street, Lidcombe, New South Wales 2141, Australia

Richard P. Baldwin Department of Chemistry, University of Louisville, Louisville KY 40292, USA

Lane J. Brunner College of Pharmacy, PHR 4.214E, The University of Texas at Austin, Austin TX 78712-1074, USA

Wolfgang Buchberger Johannes Kepler University, Department of Analytical Chemistry, Altenbergerstrasse 69, A-4040 Linz, Austria

John A. Bullock Proclinical Inc., 300 Kimberton Road, Phoenixville PA 19460, USA

Paul F. Cancalon Florida Department of Citrus, 700 Experiment Station Road, Lake Alfred FL 33850-2299, USA

Fu-Tai A. Chen Beckman Instruments Inc., 2500 Harbour Boulevard, D-20A, Fullerton CA 92634, USA

Luis A. Colon State University of New York at Buffalo, Department of Chemistry, Natural Sciences and Mathematics Complex, Buffalo NY 14260-3000, USA

Ewa Dabek-Zlotorzynska Chemistry Division, Environmental Technology Centre, Environment Canada, 3439 River Road, Ottawa ON K1A 0H3, Canada

F. Bedia Erim Department of Chemistry, Technical University of Istanbul, Maslak 80626, Istanbul, Turkey

Ramon A. Evangelista	Beckman Instruments Inc., 2500 Harbour Boulevard, D-20A, Fullerton CA 92634, USA
R. Grimm	Hewlett-Packard GmbH, Analytical Division, Hewlett-Packard Strasse 8, W-7517 Waldbronn, Germany
Edward R. Grover	Waters Corporation, 34 Maple Street, Department TG, Milford MA 01757, USA
Keiko Kitagishi	Otsuka Electronics Co., 3-26-3, Shodai-Tajika, Hirakata, Osaka, Japan 573
Yasuyuki Kurosu	Liquid Chromatography Division, JASCO Corporation, 2967-5, Ishikawa-cho, Hachioji, Tokyo, Japan 192
Jeffrey R. Mazzeo	Waters Corporation, 34 Maple Street, Department TG, Milford MA 01757, USA
James C. Osborne	Beckman Instruments Inc., 2500 Harbour Boulevard, Box 3100, Fullerton CA 92634-3100, USA
Thomas J. O'Shea,	G.D. Searle & Co., Research and Development, 4901 Searle Parkway, Slokie IL 60077, USA
P.G. Pietta	Universita of Milan, Department of Science and Technology, Via Celoria 2, 20133 Milan, Italy
Tom Pritchett	Beckman Instruments Inc., 2500 Harbour Boulevard, D-20A, Fullerton CA 92634, USA
Gordon A. Ross	Hewlett-Packard GmbH, Analytical Division, Hewlett-Packard Strasse 8, W-7517 Waldbronn, Germany
Muneo Saito	Liquid Chromatography Division, JASCO Corporation, 2967-5, Ishikawa-cho, Hachioji, Tokyo, Japan 192
R. Schuster	Hewlett-Packard GmbH, Analytical Division, Hewlett-Packard Strasse 8, W–7517 Waldbronn, Germany

Zak K. Shihabi	Pathology Department, Bowman Gray School of Medicine, Wake Forest University, Winston-Salem NC 27157, USA
Hideharu Shintani	National Institute of Health Sciences, Ministry of Health & Welfare, 1-18-1, Kamiyoga, Setagaya, Tokyo, Japan
Frederick S. Stover	Performance Materials Technology, Monsanto Chemical Company, 800 North Lindbergh Boulevard, St. Louis MO 63167, USA
Michael E. Swartz	Waters Corporation, 34 Maple Street, Department TG, Milford MA 01757, USA
Paul L. Weber	Briar Cliff College, Department of Chemistry, 3303 Rebecca, P.O. Box 2100, Sioux City IA 51104-2100, USA
Tim Wehr	Bio-Rad Laboratories, 2000 Alfred Nobel Drive, Hercules CA 94547, USA
H. Michael Wenz	Perkin Elmer, Applied Biosystems Division, 850 Lincoln Center Drive, Foster City CA 94404, USA
John E. Wiktorowicz	Perkin Elmer, Applied Biosystems Division, 850 Lincoln Center Drive, Foster City CA 94404, USA
Mingde Zhu	Bio-Rad Laboratories, 2000 Alfred Nobel Drive, Hercules CA 94547, USA

Foreword

Few areas in the history of the analytical sciences have enjoyed as explosive a growth as capillary electrophoresis. The lure of virtually infinite separation efficiency, the choice of a large number of separation, preconcentration and detection modes, extremely small sample requirement and mass detection limits that are frequently below femtomole quantities are too much for most researchers to resist. Further, the technique is intrinsically simple in principle. Indeed, electroosmosis or electroendosmosis were described by Reuss as early as 1809 and many steps that we perform today in the practice of separating components in a conduit by the application of an electric field are qualitatively little different from those conducted half a century ago. The great leap in performance that capillary electrophoresis enjoys today is derived as much from technological advances as conceptual breakthroughs.

This is not to say that all, or even the majority of, the problems that are associated with the routine analytical practice of capillary electrophoresis have been solved. Detractors often claim therefore that there are few real applications of capillary electrophoresis. While there are many books on the general subject area, the present volume is the first to offer a convincing rebuttal to such criticism – capillary electrophoresis is alive and thriving in real-life applications. The compilation of such a huge volume is a monumental task. Further, the pressures induced by the rapid growth of the subject virtually guarantee the presence of errors or omissions. Nevertheless, the practitioner will find this volume a handy reference.

Purnendu K. Dasgupta

Preface

In 1988, the first commercial high performance capillary electrophoresis (HPCE) instrument became available on the market. Since then inter- and intralaboratory data comparison using identical equipment and analytical conditions has been possible. Currently HPCE is a developing instrument with status identical to that achieved by high performance liquid chromatography (HPLC) in 1970. Thus compared with HPLC today, HPCE is somewhat inferior in areas such as sensitivity, selectivity, reproducibility, selection of analytical optimization conditions, and especially when considering introducing the use of HPCE into clinical and medical analysis for diagnosis. Therefore this handbook will be useful in describing reference analytical procedures. Analytical conditions will be revised, depending on advances in instrument hardware. Reproducible and reliable data will be required for clinical and medical analysis for diagnosis using HPCE. In this area HPLC, especially fast HPLC or capillary HPLC, has been successful, so HPLC is officially approved in hospitals for routine analysis. HPCE is superior in terms of the smaller sample needed compared with HPLC. Additionally HPCE has the advantage of providing simultaneous analysis of polar and non-polar compounds.

When the inferiorities associated with HPCE can be overcome, this technique will be superior to HPLC. HPLC–MS–MS is a significant tool for trace analysis as well as in identification of chemical structure even though the problem of non-polar compound analysis has yet to be resolved. HPCE–MS–MS has identical problems to HPLC–MS–MS. In addition to this, HPCE has several other problems to be resolved. For example, difficulties concerning sample injection techniques and the interface between HPCE and MS must be resolved. Currently only a limited number of companies supply HPCE–MS instruments because of the difficulties to be conquered. The pressurized sample injection CE system is capable of being connected successfully with MS; making the connection between CE and MS is an important factor to overcome. My own experience of the use of HPLC–MS–MS has caused me to conclude that HPLC–MS–MS as currently available is not reliable for analysis due to the unreliability of MS detection of non-polar compounds especially in the case of atomic pressure ionization (API).

HPCE is developing day-by-day by innovation and that it should prove very useful in the future is the hope of all analytical chemists. We have written this preface, keeping these hopes in mind.

This book has significant contributions from HPCE experts from all over the world. We are grateful for their prompt and valuable contributions. H. Shintani thanks his wife, Miharu Shintani for her patient understanding during the publication of this book. We sincerely thank Dr Paul Sayer and Ms Nina Kapoor for giving us the opportunity of publishing this book and helping us overcome several difficulties.

It is our hope that the format of this book will be a helpful reference for HPCE researchers enabling them to find the appropriate analytical conditions with ease.

November 1996

Hideharu Shintani
Jozef Polonský

Introduction
K. KITAGISHI

Capillary electrophoresis (CE) has recently become an important technique in the separation analysis field. In 1967, Hjertén successfully achieved free-zone electrophoresis in a narrow-bore tube of 3 mm inner diameter [1]. His pioneering work was followed by Virtanen [2] and Mikkers *et al.* [3]. Jorgenson and Lukacs have improved the technique substantially with capillaries of inner diameter less than 100 μm [4–6]. They demonstrated the characteristics of modern CE, i.e. the sufficient dissipation of heat resulting from high applied voltage, on-column detection and the high-efficiency separation.

Over the last few years, commercially available CE instruments have been developed and distributed on the market. CE is now expanding its fields of application. Some excellent texts and reviews have been published since 1992 [7–14]. Detailed information about the principles and theory of CE can be obtained from these publications. The outline of the CE separation method is described in this Introduction.

I.1 Electroosmotic flow (EOF) and analyte mobilization

Narrow-bore hollow tubes made from glass, fused-silica, fluorohydrocarbon resin, etc., can be employed as CE columns. Fused silica capillaries, inner diameter between 20 and 100 μm and 10–100 cm long, are commonly used. An external polyimide coating on fused-silica capillaries protects the thin fragile wall. Although the separation efficiency of electrophoresis in free solution is limited by thermal diffusion, narrow bore capillaries improve the rapid thermal dissipation due to the high surface-to-volume ratio. A small cross-sectional area of the capillary permits a relatively low current even with a high voltage. The high electrical field, up to 30 kV, gives a relatively quick separation for a short period. Because of the extremely small volume of capillary column, in the order of microliters, several nanoliters of sample were injected into the column for separation and analysis.

The basic instrumental setup of CE is shown in Figure I.1. A capillary is filled with electrolyte and both ends are immersed in buffer reservoirs. A high voltage power supply applies the electrophoretic field with the electrodes between two buffer reservoirs. The sample solution is injected from

INTRODUCTION

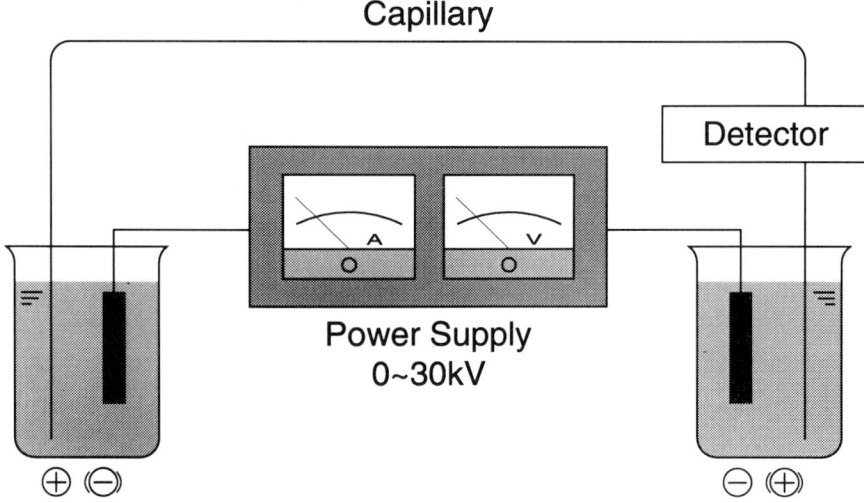

Figure I.1 Block diagram of basic capillary electrophoresis system.

one end of the capillary (inlet end) and the on-line detector is generally placed near the outlet end. Absorbance in UV–VIS region is the most common form of on-line detection. Other detection techniques are fluorescence, amperometry, conductivity, potentiometry, mass spectroscopy (MS), etc. Currently available commercial CE systems are equipped with capillary temperature control function and an autosampler in order to improve reproducibility.

Fused silica capillaries for CE have silanol groups on the inner wall. In the pH region above 3, the negative charges on the silanol groups form a double layer of counterions and the mobile cations from a diffuse layer generates electroosmotic flow (EOF) towards the cathode as shown in Figure I.2. EOF, a driving force both for ionic and non-ionic species, corresponds to the pumping pressure in high performance liquid chromatography (HPLC). An advantage of EOF in CE is the flat profile which does not disperse the solute zone, although a laminar or parabolic flow can be produced by the pump-driven system as HPLC broadens the band. A flat flow profile in electrokinetic flow and a parabolic flow profile in hydrodynamic flow were confirmed by imaging based on microscope optics and a charge-coupled device (CCD) camera [15].

Under conditions where EOF is ignored, the electrophoretic migration velocity of an ion, v, is proportional to the electrophoretic mobility, μ_{ep}, and electric field, E

$$v = \mu_{ep} E = \mu_{ep} V/L \tag{I.1}$$

where V is the applied voltage across the capillary and L is the capillary length. The electrophoretic mobility is a constant for the ion in the medium and is defined by the electric force of the ion and the frictional force through the medium

$$\mu_{ep} = q/6\pi\eta r \qquad (I.2)$$

where q, η and r represent the ion charge, solution viscosity and ion radius, respectively. When EOF exists, the migration velocity of a given species can be expressed as the vector sum of electrophoretic and EOF mobilities (Figure I.3);

$$v = (\mu_{ep} + \mu_{eo})E = (\mu_{ep} + \mu_{eo})V/L \qquad (I.3)$$

where μ_{eo} is the mobility of EOF.

The EOF must be lowered in some cases, especially when the adsorption of analytes onto the capillary wall significantly reduces the separation efficiency. Examples of analyses are given for a protein mixture by capillary gel electrophoresis (CGE), linear polymer sieving (LPS) and capillary isoelectric focusing (CIEF) and separations of cationic substances in capillary zone electrophoresis (CZE). Otherwise, EOF is a useful factor in optimizing the separation for capillary zone electrophoresis and micellar electrokinetic capillary chromatography (MEKC).

Much effort must be put into controlling EOF. First the migration velocity of separands must be controlled and second the sample adsorption onto capillary inner wall must be suppressed. Different methods have been developed including pH adjustment, the addition of basic organic compounds (quaternary ammonium salts, linear amines, morpholines, etc.) or zwitterions to the buffer, and modifying the covalent bonding of the silica

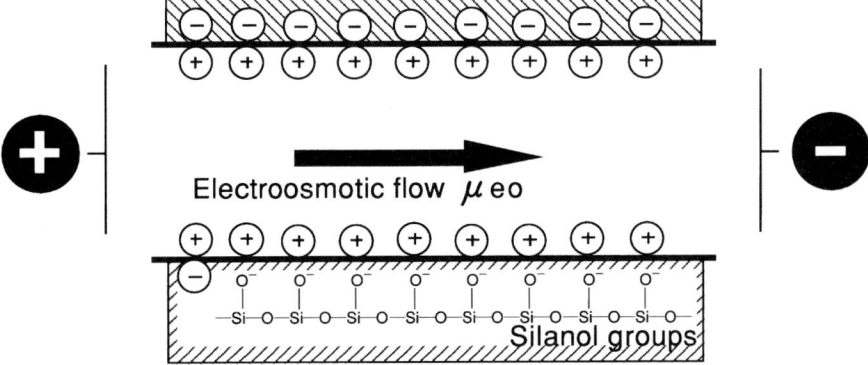

Figure I.2 Schematic diagram of capillary surface and generation of electroosmotic flow.

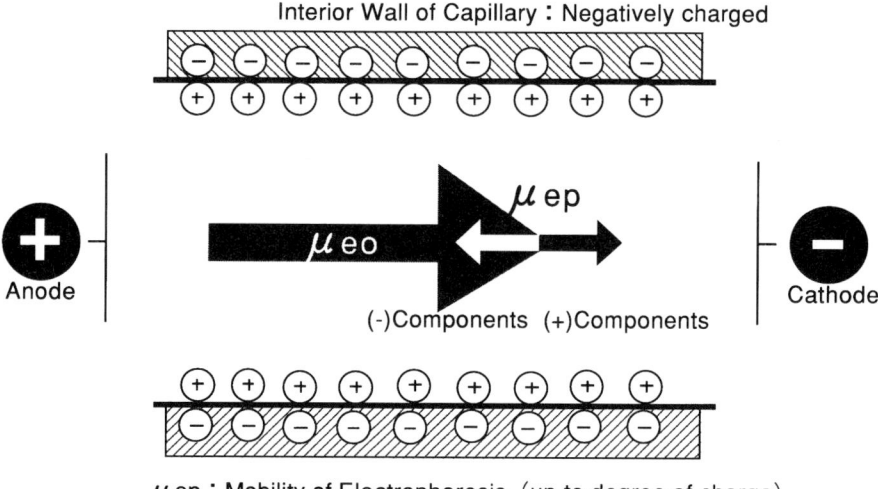

Figure I.3 Migration of analyte in the presence of electroosmotic flow. The migration velocity of a given species can be expressed as a vector sum of electrophoretic velocity, v_{ep}, and electroosmotic flow velocity, v_{eo}.

surface of the capillaries. Details are described in the Chapter 5 on Electrolyte Systems and in Chapter 4 on Column Technology.

I.2 Separation modes and samples

The separation in CE is based on the difference in the electric charge which various substances possess. It is apparent that separation by CE is restricted to ionic substances. An additive to the electrolyte which interacts with the sample molecules through electrostatic force, the coordinate bond, the hydrophobic bond, etc., expands the separation mode of CE and some additives enable us to separate non-ionic substances by CE. The four main separation modes in CE are listed in Table I.1.

I.2.1 Electrophoretic mode

The electrophoretic mode is the most general and basic state in CE as described by equations I.1–I.3. CE in this mode is called capillary zone electrophoresis (CZE) or free zone capillary electrophoresis (FZE). Solutes migrate in the buffer inside the capillary depending on their own electric mobility. Separation of a mixture of different solutes is affected by several factors, e.g. ionic strength of the solution, pH, net charge density of the sample, conductivity, temperature, electric field, cross-sectional area

Table I.1 Principal separation modes in capillary electrophoresis

Separation mode	Separation technique	Additive used	Description
Electrophoretic mode	Free zone capillary electrophoresis (FZE)	No additive (buffer only); Zwitterion/cation; UV absorbent (indirect UV); Chelate agent (indirect/direct UV); Coated capillary	A difference in the velocity of migration of a substance for its own electric charge results in its separation
Electrokinetic mode	Micellar electrokinetic chromatography (MEKC);	Surfactant	A difference in the partition equilibrium of a substance between surfactant and water (or including equilibrium of a substance in CD) results in its separation
	Cyclodextrin electrokinetic chromatography (CD/EKC);	Cyclodextrin	
	Cyclodextrin micellar electrokinetic chromatography (CD/MEKC)	Cyclodextrin, surfactant and organic solvent	
Polymer sieving mode	Capillary gel electrophoresis (CGE);	Polyacrylamide gel	Components of a substance are separated due to the sieving mechanism of a polymer produced by a capillary filled with gel or polymer solution
	Linear polymer sieving (LPS)	Linear polymer	
Isoelectric mode	Capillary isoelectric focusing electrophoresis (CIEF)	Ampholyte	pH gradient is created in a capillary, the components of a sample are separated at a pH where their isoelectric points are zeroed

and length of the capillary, additives such as surfactants or organic solvents, etc. Most of them affect both EOF and sample migration velocities. Highly efficient separation can be achieved by the optimization of these factors. Computer simulations to optimize the separation have been reported, and are described in Chapter 5 on Electrolyte Systems and Chapter 8 on Other Systems. However, it should be recognized that computer simulation is only a simulation, indicating that experiment is required to confirm simulation results, especially when handling complicated matrix compounds such as blood, soil, etc. In most of these cases, the result of computer simulation is not effective and is not useful according to the editor's experience.

Several small anionic and cationic compounds cannot be detected by the on-line absorbance mode because they have no characteristic absorption in the detectable wavelength region with the exception of some metal cations and organic/inorganic anions. Bromide, nitrate, nitrite and most organic acids are monitored directly by their own absorbances. When a UV absorbent is added to the electrolyte, the anions which do not have sufficient absorbance for detection can be followed; this is called 'indirect UV detection'. Chromate is the most useful UV absorbent due to the broad UV absorption spectrum throughout the entire UV range. Indirect UV detection can be applied to organic acids and often shows higher sensitivity than direct UV detection depending on the sort of compounds used. Phthalate and benzoate are added as UV absorbents for anions with lower migration velocities like organic acids, alkyl sulfonates, etc.

For indirect UV detection of cations, creatine, benzylamine or imidazole are used as UV absorbent. Chelating agents which can interact with cations, e.g. α-hydroxyisobutyrate, lactate, succinate and crown ethers are often added to improve the separation. Direct detection of inorganic cations is performed upon the addition of ethylene diaminetetraacetic acid (EDTA), cycloethylene diaminetetraacetic acid (CyDTA) or others as a UV absorbing and chelating agent. Cationic surfactant is often added to suppress the adsorption of cations to the capillary inner surface, causing the reversion of EOF.

1.2.2 Electrokinetic mode

The electrokinetic mode is based on the difference in the incorporation of substances to an additive in the buffer. The additive is a surfactant like sodium dodecyl sulfate (SDS) in micellar electrokinetic chromatography (MEKC), cyclodextrin (CD) in CD/electrokinetic chromatography (CD/EKC), or both in CD/MEKC. MEKC has been developed as the combination of electrophoretic and chromatographic modes [16]. When more surfactant is added to the buffer than the critical micelle concentration, the formation of micelles is observed. Micelles consisting of surfactant molecules migrate at different velocities to water molecules due to the negative or positive charges of micelles. As hydrophobic substances congregate inside the micelle while hydrophilic ones do not, the analyte has a migration velocity dependent on its hydrophobicity in the presence of ionic micelles. This means that the separation of MEKC is based on the partition of the analyte between water and micelle. The mobility of the analyte depends on the original electrophoretic mobility due to its own charge, the partition coefficient between water and micelle and the mobilities of water (EOF) and micelles (Figure I.4).

Chiral separation in CE has been successfully performed by using CDs

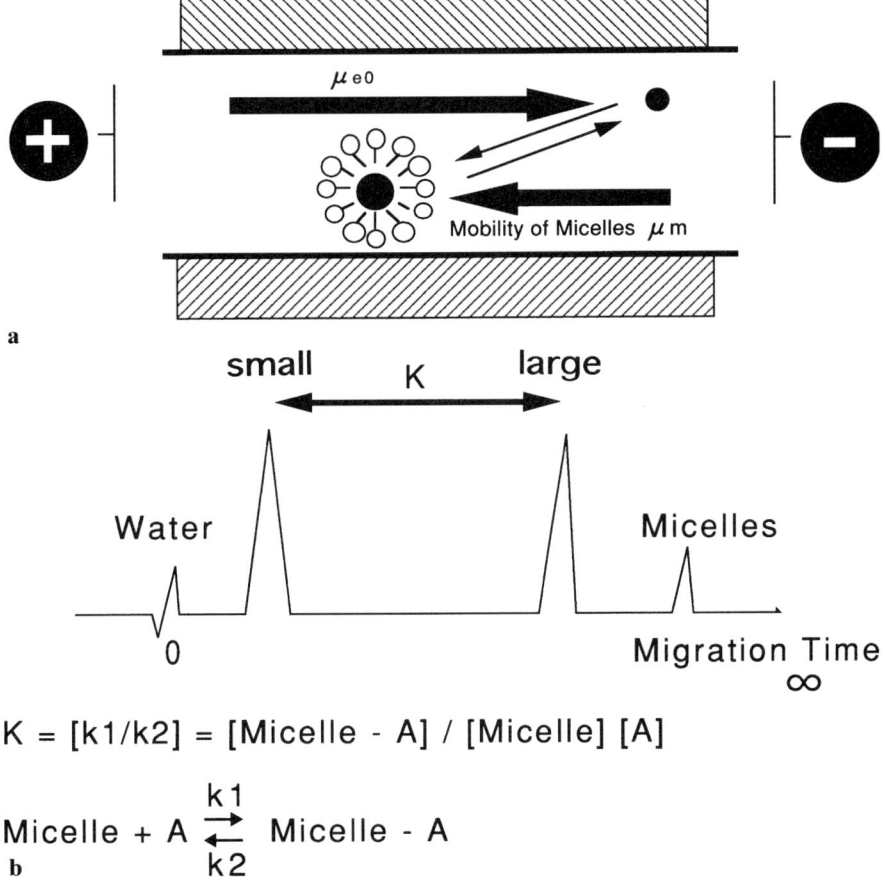

$K = [k_1/k_2] = [\text{Micelle} - A] / [\text{Micelle}] [A]$

$$\text{Micelle} + A \underset{k_2}{\overset{k_1}{\rightleftarrows}} \text{Micelle} - A$$

b

Figure I.4 (a) Micellar electrokinetic chromatography (MEKC). The migration velocity of a given species depends on its own charge, electroosmotic flow velocity, micelle velocity and partition of the substance between water and micelle. (b) Schematic electropherogram in MEKC.

as chiral additives. CDs are commercially available cyclic non-reducing oligosaccharides with six to eight glucopyranose units through α-(1,4) linkages, which have a cone shape with a relatively hydrophobic interior cavity (Figure I.5). CDs are ideal as chiral selectors in CE due to their ability to form host–guest complexes with the stability dependent on the size and spatial configuration of the analyte molecules. The difference in host–guest inclusion complex stabilities for chiral compounds yields chiral separation in CE. CD/EKC represents the chiral separation with CDs as chiral selectors (Figure I.6). Under conditions where the CD and analyte molecules migrate at different velocities and the host–guest complex stabilities are varied for chiral molecules, CE separation of chiral compounds is performed. The combination of chiral recognition by CDs with incorporation into SDS micelles is widely used as a chiral separation for compounds with low solubility, called CD/MEKC (Figure I.7) [14, 17–19].

I.2.3 Polymer sieving mode

Polymer sieving mode includes capillary gel electrophoresis (CGE). Analytes migrate at different velocities through polymer networks dependent on their molecular sizes. Agarose or polyacrylamide gel-filled capillaries

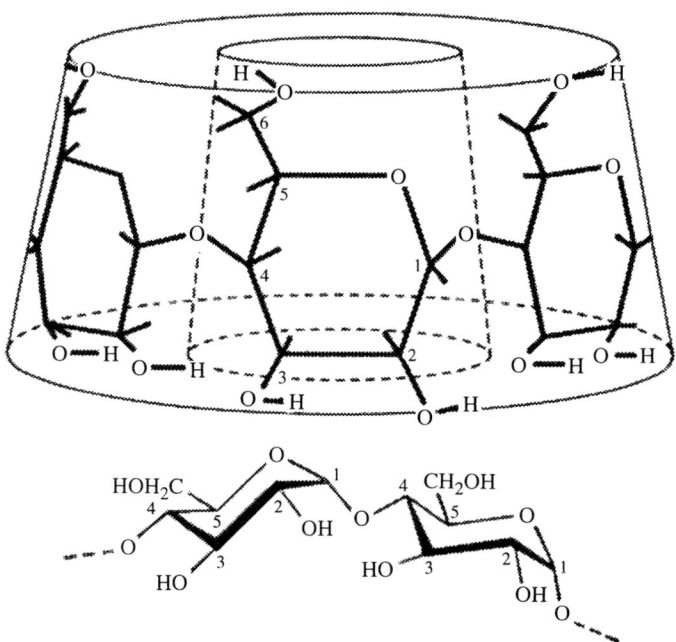

Figure I.5 Structure of cyclodextrin.

SEPARATION MODES

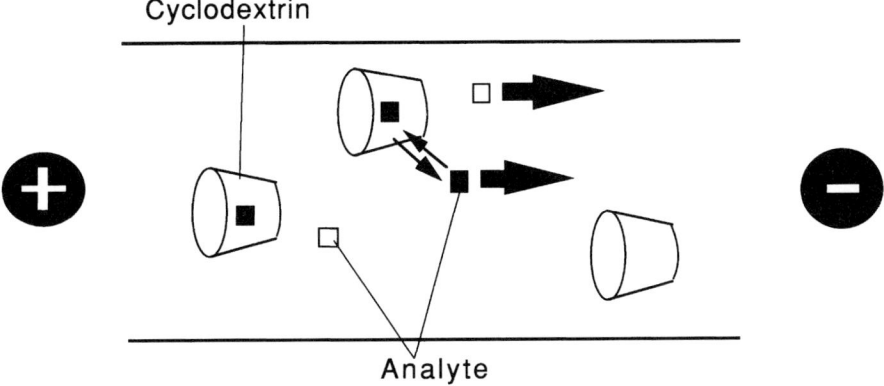

Figure I.6 Cyclodextrin electrokinetic chromatography (CD/EKC).

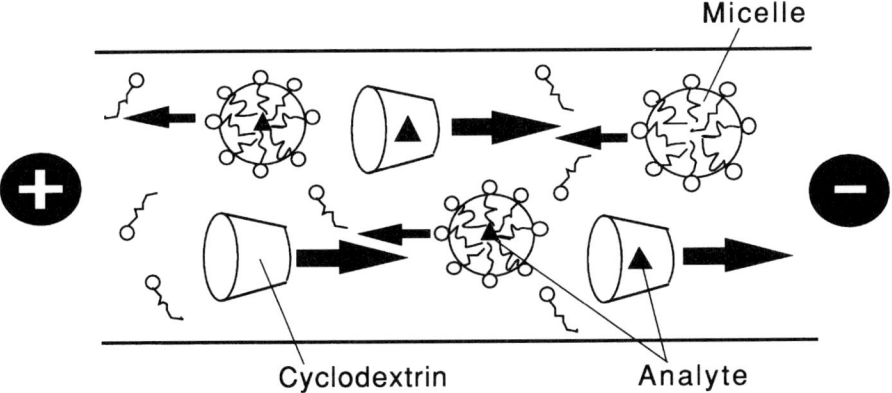

Figure I.7 Cyclodextrin micellar electrokinetic chromatography (CD/MEKC). Cyclodextrin molecules migrate at the same velocity as electroosmotic flow.

are widely used as a column. Recently a replaceable polymer solution has been applied as a sieving medium, e.g. uncrosslinked polyacrylamide, dextran or cellulose derivative solution, in what has been called linear polymer sieving (LPS) [20, 21]. Although polymer solutions are relatively viscous, they can be replaced automatically using a hydrodynamic pump attached to the CE instrument as a washing facility. Durability of the polymer solution for a number of experiments is not required in LPS, because the solution is easily replaced before each measurement. Although LPS with uncrosslinked polyacrylamide or other viscous polymer solutions is defined as CGE in some publications, these are considered to be LPS, not CGE, here and in Part One (Equipment Systems) of this book, in order to avoid confusion. The analytes used in polymer sieving mode are biopolymers including DNA, nucleotides, polysaccharides, SDS–protein complexes [22] and industrial polymers.

Table I.2 Samples and analytical methods in capillary electrophoresis

Class L	Code	Class M	Class S (Samples)	Separation mode	Additive used in FZE techniques
Ions	A	Inorganic anion	Cl, F, Br, NO$_3$, NO$_2$, SO$_4$, SO$_3$, PO$_4$, ClO$_4$	FZE	Indirect UV/reversible EOF
	B	Organic anion	Organic acid (acetic acid, formic acid, lactic acid, malic acid, maleic acid, citric acid, etc.)	FZE	Indirect UV/reversible EOF
	C	Inorganic cation	Metallic cation (Ca, Mg, Ba, Li, Na, K, Fe, Cu)	FZE	Indirect UV/chelate agent
	D	Organic cation	NH$_4$, aliphatic amine	FZE	Indirect UV/chelate agent
Biological materials	E	Protein	Serum protein, enzyme, collagen, antibody	FZE CGE/LPS CIEF	No addition/zwitterion/ reversible EOF/coating
	F	Peptide	Synthetic peptide, glutathione, pseudopeptide	FZE MEKC CIEF	No addition/zwitterion/ reversible EOF/coating
	G	Amino acid and its derivatives	Amino acid, creatine, low-molecular weight hormone, dopamine, histamine	FZE MEKC CD-EKC CIEF	No addition/zwitterion/ reversible EOF
	H	Nucleic acid	DNA, RNA, PCR product, nucleoside, nucleotide	FZE CGE/LPS	No addition
	I	Saccharide	Monosaccharide, oligosaccharide, polysaccharide	FZE MEKC CGE	No addition

Group	Code	Subcategory	Examples	Method	Notes
	J	Lipid	Fat, phospholipid, glycolipid, fatty acid	FZE/MEKC	Indirect UV
	K	Steroid	Steroidal hormone, cholesterol	MEKC/CD-EKC	No addition
	L	Vitamin, coenzyme	Vitamin derivative, nicotine amide	FZE/MEKC	No addition
	M	Others	Porphyrin compounds	FZE/MEKC	
Drugs	N	Drugs	Plant extract, antibiotics, crude drugs	FZE/MEKC	Reversible EOF
Chemical materials	O	Surfactants	SDS, LAS, POE, alkyl ether	FZE/MEKC	No addition/reversible EOF
	P	Polymer	Emulsion, polystyrene derivative, polyacrylamide	FZE/MEKC/CGE/LPS	No addition
	Q	Aromatics	Phenol, aromatic carbonic acid, aromatic amine	MEKC/CD-EKC/CD-MEKC	
	R	Pigment/dye		MEKC	
	S	Others		MEKC	
Materials requiring special separation techniques	X-1	Hardly soluble compound		MEKC/organic solvent	
	X-2	Optical isomers	DL amino acid	CD-EKC/CD-MEKC	
	X-3	Drug in blood/urine	Anticancer drug	MEKC	

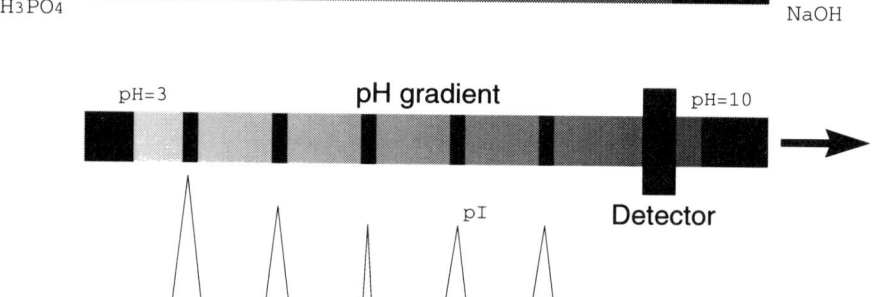

Figure I.8 Capillary isoelectric focusing electrophoresis (CIEF). Analytes are focused corresponding to their own isoelectric points on the pH gradient formed in the capillary, after which the zones move to one end of the capillary.

I.2.4 Isoelectric mode

The isolelectric mode represents capillary isoelectric focusing electrophoresis (CIEF) [23, 24]. A mixture of electrolytes, ampholytes and analytes are introduced into the capillary. A pH gradient is formed and analytes are focused corresponding to their own isoelectric points in an electric field. The anode and cathode are placed into an acidic and an alkaline solution, respectively. After focusing, zones move to one end of the capillary with exchange of the anodic or cathodic solution, or with the pressure of continuous focusing (Figure I.8). A coating on the capillary to suppress EOF and protein adsorption is required. One of the advantages of CIEF is the concentration effect of the analytes. A dilute protein solution even in a concentration below the detection limit can be concentrated into a condensed zone by focusing and can consequently be detected.

I.2.5 Other separation modes

Separation modes which are not described involve capillary isotachophoresis (CITP) and capillary electrochromatography (CEC), both of which are often excluded from CE separation modes. In CITP separation, continuous analyte zones are formed between leading and terminating constituents [25]. Zones are detected not with peaks but with a stepwise increase in conductivity. In CEC, a packed capillary-like chromatographic column must be prepared and EOF is used at the point of pressure exerted by the pump in HPLC [26–29].

Samples which can analyzed by CE are covered over a wide field of applications as shown in Table I.2, which summarizes the analytical methods, used for different samples.

References

1. Hjertén, S. (1967) Free zone electrophoresis. *Chromatogr. Rev.* **9**, 122–219.
2. Virtanen, R. (1974) Zone electrophoresis in a narrow-bore tube employing potentiometric detection. A theoretical and experimental study. *Acta Polytech. Scand.* **123**, 1–67.
3. Mikkers, F.E.P.; Everaerts, F.M.; Verheggen, T.P.E.M. (1979) High-performance zone electrophoresis. *J. Chromatogr.* **169**, 11–20.
4. Jorgenson, J.W.; Lukacs, K.D. (1981) Zone electrophoresis in open-tubular glass capillaries. *Anal. Chem.* **53**, 1298–1302.
5. Jorgenson, J.W.; Lukacs, K.D. (1981) High-resolution separations based on electrophoresis and electroosmosis. *J. Chromatogr.* **218**, 209–216.
6. Green, J.S.; Jorgenson, J.W. (1984) High speed zone electrophoresis in open-tubular fused silica capillaries. *J. High Resolut. Chromatogr.* **7**, 529–531.
7. Li, S.F.Y. (1992) *Capillary Electrophoresis: Principles, Practice, and Applications*; Elsevier, Amsterdam.
8. Jandik, P.; Bonn, G. (1993) *Capillary Electrophoresis of Small Molecules and Ions*; VCH Publishers, New York.
9. Grossman, P.D.; Colburn, J.C. (eds) (1992) *Capillary Electrophoresis: Theory and Practice*; Academic Press, San Diego.
10. Kuhn, R.; Hoffstetter-Kuhn, S. (1993) *Capillary Electrophoresis: Principles and Practice*; Springer-Verlag, Berlin.
11. Monnig, C.A.; Kennedy, R.T. (1994) Capillary electrophoresis. *Anal. Chem.* **66**, 280R–314R.
12. Marina, M.L.; Torre, M. (1994) Capillary electrophoresis. *Talanta* **41**, 1411–1433.
13. Knox, J.H. (1994) Terminology and nomenclature in capillary electroseparation systems. *J. Chromatogr.* **680**, 3–13.
14. Terabe, S.; Otsuka, K.; Nishi, H. (1994) Separation of enantiomers by capillary electrophoretic techniques. *J. Chromatogr.* **666**, 295–319.
15. Taylor, J.A.; Yeung, E.S. (1993) Imaging of hydrodynamic and electrokinetic flow profiles in capillaries. *Anal. Chem.* **65**, 2928–2932.
16. Terabe, S.; Otsuka, K.; Ichikawa, K.; Tsuchiya, A.; Ando, T. (1984) Electrokinetic separations with micellar solutions and open-tubular capillaries. *Anal. Chem.* **56**, 111–113.
17. Ward, T.J. (1994) Chiral media for capillary electrophoresis. *Anal. Chem.* **66**, 633A–640A.
18. Novotny, M.; Soini, H.; Stefansson, M. (1994) Chiral separation through capillary electromigration methods. *Anal. Chem.* **66**, 646A–655A.
19. Bereuter, T.L. (1994) Enantioseparation by capillary electrophoresis. *LC–GC* **12**, 748–766.
20. Zhu, M.D.; Hansen, D.L.; Burd, S.; Gannon, F. (1989) Factors affecting free zone electrophoresis and isoelectric focusing in capillary electrophoresis. *J. Chromatogr.* **480**, 311–319.
21. Bae, Y.C.; Soane, D. (1993) Polymeric separation media for electrophoresis: cross-linked systems or entangled solutions. *J. Chromatogr.* **652**, 17–22.
22. Dolník, V. (1994) Capillary gel electrophoresis. *J. Microcol. Sep.* **6**, 315–330.
23. Hjertén, S.; Liao, J.-L.; Yao, K. (1987) Theoretical and experimental study of high-performance electrophoretic mobilization of isoelectrically focused protein zones. *J. Chromatogr.* **387**, 127–138.
24. Zhu, M.; Rodriguez, R.; Wehr, T. (1991) Optimizing separation parameters in capillary isoelectric focusing. *J. Chromatogr.* **559**, 479–488.
25. Oefner, P.; Haefele, R.; Bartsch, G.; Gunther, B. (1990) Isotachophoretic separation of organic acids in biological fluids. *J. Chromatogr.* **516**, 251–262.
26. Kitagawa, S.; Tsuda, T. (1994) Effects of pH and organic solvent on chromatographic behavior in capillary electrochromatography. *J. Microcol. Sep.* **6**, 91–96.
27. Gordon, D.B.; Lord, G.A.; Jones, D.S. (1994) Development of packed capillary column electrochromatography/mass spectrometry. *Rapid Commun. Mass Spectrom.* **8**, 544–548.
28. Yan, C.; Schaufelberger, D.; Erni, F. (1994) Electrochromatography and micro high-

performance liquid chromatography with 320 µm I.D. packed columns. *J. Chromatogr.* **670**, 15–23.
29 Smith, N.W.; Evans, M.B. (1994) The analysis of pharmaceutical compounds using electrochromatography. *Chromatographia* **38**, 649–657.

Part One

Equipment Systems

1 Sample injection

K. KITAGISHI

In capillary electrophoresis, a small amount of sample solution is injected from the inlet end of a capillary and a high voltage is applied between the two ends of the capillary, both of which are immersed in electrolyte. Satisfactory reproducibility of the injection volume is required for quantitative analysis.

In order to exhibit the high separation efficiency of capillary electrophoresis (CE), the sample solution should be injected as a narrow plug. When the injection volume is in excess, broadening of the migration peaks affects the separation adversely. On the other hand, the analyte concentration zone should be relatively greater than that in conventional separation techniques such as high performance liquid chromatography (HPLC) because the short light-path of the capillary cell derives low limits of CE detection. Many studies have reported on the sample preconcentration on column or precolumn, described later.

1.1 Modes of sample injection

The most common modes of sample injection in CE is generally classified into two categories, hydrodynamic injection and electrokinetic injection.

1.1.1 Hydrodynamic injection

Hydrodynamic injection is accomplished by application of pressure at one end of the capillary. It involves injection by gravity flow, pressure and vacuum suction. The advantage of hydrodynamic injection is that the quantity of individual analytes injected is constant, independent of their own mobility, in contrast to electrokinetic injection.

When the inlet end of the capillary immersed in the sample solution is placed at a higher position than the outlet end in the container of buffer, gravity flow generates sample injection. In gravity flow injection, the injection volume, V_i, is regulated by the difference in height between the two vessels, Δh, and by the injection time, t. It is dependent on the inner radius of the capillary, r, the capillary length, L, the solution density, ρ, the acceleration due to gravity, g and the solution viscosity, η

$$V_i = \rho g \pi r^4 \Delta h t / (8 \eta L) \qquad (1.1)$$

This mode is mechanically the simplest one and adequate for samples with a low viscosity.

When a regulated gas pressure is applied to the sealed sample vessel at the inlet end (pressure injection) or when the electrolyte vessel at the outlet side is under reduced pressure (vacuum injection), the sample solution penetrates into the capillary. Pressure injection is suitable for viscous samples. Vacuum injection is often unfavorable because of poor reproducibility accompanied by the difficulty in regulating reduced pressure. Pressure and vacuum injection both depend on pressure difference across the capillary, ΔP, the injection period, t, inner radius of the capillary, the capillary length and solution viscosity.

$$V_i = \Delta P \pi r^4 t / (8 \eta L) \tag{1.2}$$

1.1.2 Electrokinetic injection

No special exclusive devices or regulation mechanisms are required for electrokinetic injection. The inlet end of a capillary is placed in a sample vessel instead of a buffer and a high voltage is applied, that is for electromigration to take place. Electrokinetic injection is applicable for capillary columns filled with gel in capillary gel electrophoresis (CGE). A unique characteristic, often a disadvantage, is that the quantity of each analyte injected in the sample solution depends on its own mobility. This means that the amount injected is not defined as a specific sample volume.

1.2 Injection reproducibility

In both of the above methods, the injection reproducibility decreases as the injection amount decreases, especially for capillaries with a large inner diameter. When an injection period varies at a constant height differential in the gravity flow mode, the plot of injection volume versus injection period is linear (not through the origin) with a positive intercept (Figure 1.1). This phenomenon was first reported by Grushka and MaCormick [1] as an ubiquitous injection. It is also termed spontaneous extraneous injection. The effects of diffusion and inadvertent hydrodynamic flow upon the sample injection were simulated by Dose and Guiochon [2]. They concluded that analytes enter the capillary by diffusion even at a low concentration of the analyte especially for small for small molecules and that the difference between the liquid levels in the two reservoirs mainly generates an incidental hydrodynamic flow.

A sample solution with high visocosity shows satisfactory injection repro-

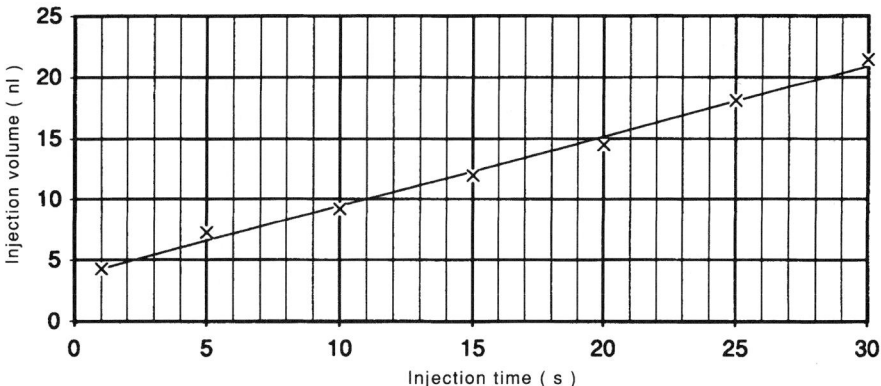

Figure 1.1 Injection volume of sample solution plotted versus injection time in gravity flow injection. Differences in height, capillary inner diameter and capillary length are 2.5 cm, 75 μm and 50 cm, respectively. Sample and electrolyte solutions are aqueous at low viscosity.

ducibility. The plot of injection volume versus injection time originates near the origin. This means that spontaneous extraneous injection diminishes as the viscosity of the sample solution increases.

Zare and his group observed this phenomenon with a video-speed charge-coupled device (CCD) camera [3]. They concluded that spontaneous injection is primarily caused by an interfacial pressure difference formed at the inlet of the capillary. It is recommended that extraneous injection is reduced by etching the capillary inlet or by using a thin-walled capillary [3]. The shape of the edge of the capillary inlet end affects the separation efficiency as well. Straight-edge capillaries show more efficient pherograms with more symmetrical peaks than slanted-edge capillaries [4].

Most commercial CE instruments are equipped with an autosampler making sample injections automatically for successive electrophoreses. The automatically controlled insertion into and exit from the capillary of the sample solution improves the reproducibility of sample injection, derived not only from the reproducible operation of injection time, height and voltage, but also from the small deviation in the amount of spontaneous extraneous injection [5]. In order to improve the reproducibility of sample injection in pressure mode, continuous pressure control during the injection process has been applied to some commercial systems.

However, capillary-to-capillary differences cannot be corrected by using either an autosampler or continuous pressure-controlled injection.

Devices to inject a constant volume of sample solution have been demonstrated. A microdrop injector permits the injection of a small droplet [6] and a slide-type injector enables constant injection of 2 nl [7]. The former requires automatic micropositioning for alignment for general use but the latter presents a problem with respect to microfabrication of the injector. An injection method based on spontaneous fluid displacement has been proposed by Fishman et al. [8]. The droplet of sample solution left on the inlet end of the capillary when it is removed from the sample solution can be injected into the capillary by the interfacial pressure difference across the curved surface of the droplet [8]. Improvement of injection reproducibility is expected by automating and controlling the environment at the point of injection and by manufacturing a flat edged capillary.

Electrokinetic injection generally shows less reproducibility than hydrodynamic injection. Lee and Yeung postulated that the integration of the current during electrokinetic injection could correct the deviation in sample injection [9]. Temporal fluctuations in the analyte migration velocity are similarly corrected by monitoring the current during separation [9].

In capillary gel electrophoresis (CGE), the peak height of analytes decreases as the number of consecutive electrokinetic injections increases [10]. This phenomenon, explained by the pH change of sample solution during the electrokinetic sample introduction process, is solved simply with an electrokinetic injection of water (water preinjection) prior to the sample injection. It was also reported that the actual physical shape of the capillary inlet affects the resolution and that of the migration peaks.

Although injection reproducibility is essential for quantitative reproducibility, the final reproducibility is expressed as peak height, peak area or other parameters to be evaluated such as migration time. One of the primitive solutions to quantitative reproducibility involving final output and capillary-to-capillary correction is the addition of internal standards to the sample solution. Deviation of injection volume can be corrected based on addition of internal standards. Examples of recent publications are listed in Chapter 2 on Quantitative Analysis, Precision and Reproducibility (Table 2.2).

1.3 Sample preconcentration

For relatively low detection limits of CE, a large volume of injection fluid is required for samples at low concentration. Since a narrow concentrated band is necessary to obtain a high resolution and adequate detection, sample preconcentration is a requisite technique in CE.

A simple method of sample preconcentration is sample stacking, which is achieved with the diluted buffer solution or water-containing analytes. Analyte molecules dissolved in water migrate with much higher velocity than in electrolyte because the higher resistance of diluted sample solution induces a higher electric field across the sample zone. When rapidly moving analyte molecules enter the electrolyte across the boundary of the sample solution and electrolyte, the migration velocity is slowed down considerably due to the lower electric field. Consequently, analyte molecules are stacking and condensed.

Combination of isotachophoresis (ITP) and capillary zone electrophoresis (CZE) provides a powerful tool for sample preconcentration. Analytes can be focused during ITP with a discontinuous buffer system containing leading and terminating buffer and be separated by CZE. The ITP plus CZE preconcentration can be carried out with a single capillary or a coupled column capillary. This technique improves the detectability in CE–MS (mass spectrometry) [11, 12]. The leading, terminating and running electrolytes should be selected to be appropriate for the analytes [13].

By using surface-modified or solid-phase packed capillaries connected to the separation capillary, on-line preconcentration can be held as chromatographic preconcentration. Many papers have been published since 1992 [24–41] and are listed in Table 1.1.

Hjertén and co-workers introduced new approaches to sample concentration and desalting of proteins and peptides. The sample solution is concentrated by electrophoresis towards a steep pH gradient, a small-pore gel, a piece of dialysis tubing, a gradient in conductivity or by a combination of displacement electrophoresis and a counterflow [14, 15].

Zone sharpening of neutral solutes in micellar electrokinetic chromatography (MEKC) has been achieved with electrokinetic injection when the effective electrophoretic velocity of the neutral analytes and the electroosmotic flow velocity are in the same direction during injection. Cationic mixed micelles should be added in sample solutions for the

Table 1.1 Recent publications on sample preconcentration

Preconcentration mode	Analyte	Ref.
Stacking	Anion standard	24
Stacking	Acidic/basic amino acids	25, 26
Stacking	Iso-α-acids in beer	27
Stacking	Creatinine, clenbuterol	28
ITP+CZE	Benzoic acid derivatives	29
ITP+CZE	Ionic mixtures in urine	30
ITP+CZE	Amino acids	31
ITP+CZE	Anthracyclines	11
ITP+CZE	Antimuscarinic drugs	32, 33
ITP+CZE	Enzyme inhibitor of plasma extracts	34
ITP+CZE	Protein mixture	12, 35
ITP+CZE	Tryptic digest of β-casein	36
Chromatographic	Triazine herbicides	37
Chromatographic	Propranolol in serum	38
Chromatographic	Doxepin and propranolol in urine	39
Chromatographic	Papaverine	40
Chromatographic	Carbonic anhydrase	41

running buffer containing only sodium dodecyl sulfate (SDS) micelles [16].

1.4 Other microhandling techniques for sample injection

In order to control the injection amount of sample matrices independently of their own mobility in electrokinetic injection, microinjection can be performed by fracturing the capillary and applying an electric field between the fracture and outlet end. EOF pulls the sample into a separation capillary without the biased introduction of analytes. This technique enables reproducible and quantitative injection [17].

Although the injection volume in CE is usually from 10 to 50 nl, commercial CE equipment uses sample vials larger than 10 μl. A multisample holder on a silicon wafer with a volume of 118 nl has been made by Jansson et al. for electrokinetic injection from a small sample volume [18]. No additional band broadening was observed compared with injection from conventional sample vials.

Interposition of narrow-bore tubular membranes in the capillary permits sample injection by diffusion and/or permeation. This kind of membrane interface has been applied to the injection of gaseous samples, wastewater and low molecular analytes in human blood plasma [19].

By the connection of an enzyme-modified capillary reactor and an electrophoretic separation capillary, on-line digestion of protein and analysis for mapping were made possible in an extremely small sample volume [20].

When an analyte and its analytical reagent are injected as two different zones with different migration rates, the chemical reaction proceeds as long as the two zones overlap. This technique, termed electrophoretically mediated microanalysis (EMMA), can be illustrated by the enzymatic oxidation by alcohol dehydrogenase as a chemical analysis with ultramicrosamples [21].

An on-line coupling of microdialysis sampling to capillary electrophoresis can be realized by using microcolumn separation, which has conquered the sample volume restriction. The interface was designed to derive a 60 nl sample plug from a continuous microdialysis flow and to inject a portion of the sample plug into a separation capillary. This allows the on-line pharmacokinetic analysis *in vivo* with a 90 s interval [22].

Moore and Jorgenson reported an optical gating injection, a unique system which achieves a higher speed separation CZE than conventional CZE [23]. It is based on the photodecomposition of fluorescein isothiocyanate (FITC) induced by an argon ion laser beam. Fluorescent-labeled amino acids could be analyzed within 2–3 s with a performance better than 80% of the expected separation efficiency.

References

1. Grushka, E.M.; McCormick, R.M. (1989) Zone broadening due to sample injection in capillary zone electrophoresis. *J. Chromatogr.* **471**, 421–428.
2. Dose, E.V.; Guiochon, G. (1992) Problems of quantitative injection in capillary zone electrophoresis. *Anal. Chem.* **64**, 123–128.
3. Fishman, H.A.; Amudi, N.M.; Lee, T.T.; Scheller, R.H.; Zare, R.N. (1994) Spontaneous injection in microcolumn separations. *Anal. Chem.* **66**, 2318–2329.
4. Cohen, N.; Grushka, E. (1994) Influence of the capillary edge on the separation efficiency in capillary electrophoresis. *J. Chromatogr.* **684**, 323–328.
5. Rose, D.J.; Jorgenson, J.W. (1988) Characterization and automation of sample introduction methods for capillary zone electrophoresis. *Anal. Chem.* **60**, 642–648.
6. Sziele, D.; Brüggemann, O.; Döring, M.; Freitag, R.; Schügerl, K. (1994) Adaptation of microdrop injector to sampling in capillary electrophoresis. *J. Chromatogr.* **669**, 254–258.
7. Hanai, T.; Tsuruta, H. (1994) 2 nl Injector for capillary electrophoresis. *Instrum. Sci. Tech.* **22**, 151–155.
8. Fishman, H.A.; Scheller, R.H.; Zare, R.N. (1994) Microcolumn sample injection by spontaneous fluid displacement. *J. Chromatogr.* **680**, 99–107.
9. Lee, T.T.; Yeung, E.S. (1992) Compensating for instrumental and sampling biases accompanying electrokinetic injection in capillary zone electrophoresis. *Anal. Chem.* **64**, 1226–1231.
10. Guttman, A.; Schwartz, H.E. (1995) Artifacts related to sample introduction in capillary gel electrophoresis affecting separation performance and quantitation. *Anal. Chem.* **67**, 2279–2283.
11. Reinhoud, N.J.; Tinke, A.P.; Tjaden, U.R.; Niessen, W.M.A.; van der Greef, J. (1992) Capillary isotachophoretic analyte focusing for capillary electrophoresis with mass spectrometric detection using electrospray ionization. *J. Chromatogr.* **627**, 263–271.
12. Thompson, T.J.; Foret, F.; Vouros, P.; Karger, B.L. (1993) Capillary electrophoresis/ electrospray ionization mass spectrometry: Improvement of protein detection limits using on-column transient isotachophoretic sample preconcentration. *Anal. Chem.* **65**, 900–906.
13. Schwer, C.; Gaš, B.; Lottspeich, F.; Kenndler, E. (1993) Computer simulation and experimental evaluation of on-column sample preconcentration in capillary zone electrophoresis by discontinuous buffer systems. *Anal. Chem.* **65**, 2108–2115.
14. Hjertén, S.; Liao, J.-L.; Zhang, R. (1994) New approaches to concentration on a microliter scale of dilute samples, particularly biopolymers with special reference to analysis of peptides and proteins by capillary electrophoresis. I. Theory. *J. Chromatogr.* **676**, 409–420.
15. Liao, J.-L.; Zhang, R.; Hjertén, S. (1994) New approaches to concentration on a microliter scale of dilute samples, particularly biopolymers with special reference to analysis of peptides and proteins by capillary electrophoresis. II. Applications. *J. Chromatogr.* **676**, 421–430.
16. Nielsen, K.R.; Foley, J.P. (1994) Zone sharpening of neutral solutes in micellar electrokinetic chromatography with electrokinetic injection. *J. Chromatogr.* **686**, 283–291.
17. Linhares, M.C.; Kissinger, P.T. (1991) Use of an on-column fracture in capillary zone electrophoresis for sample introduction. *Anal. Chem.* **63**, 2076–2078.
18. Jansson, M.; Emmer, A.; Roeraade, J.; Lindberg, U.; Hök, B. (1992) Micro vials on a silicon wafer for sample introduction in capillary electrophoresis. *J. Chromatogr.* **626**, 310–314.
19. Bao, L.; Dasgupta, P.K. (1992) Membrane interfaces for sample introduction in capillary zone electrophoresis. *Anal. Chem.* **64**, 991–996.
20. Amankwa, L.N.; Kuhr, W.G. (1993) On-line peptide mapping by capillary zone electrophoresis. *Anal. Chem.* **65**, 2693–2697.
21. Harmon, B.J.; Patterson, D.H.; Regnier, F.E. (1993) Mathematical treatment of electrophoretically mediated microanalysis. *Anal. Chem.* **65**, 2655–2662.
22. Hogan, B.L.; Lunte, S.M.; Stobaugh, J.F.; Lunte, C.E. (1994) On-line coupling of in vivo microdialysis sampling with capillary electrophoresis. *Anal. Chem.* **66**, 596–602.
23. Moore, Jr., A.W.; Jorgenson, J.W. (1993) Study of zone broadening in optically gated high-speed capillary electrophoresis. *Anal. Chem.* **65**, 3550–3560.

24 Burgi, D.S. (1993) Large volume stacking of anions in capillary electrophoresis using an electroosmotic flow modifier as a pump. *Anal. Chem.* **65**, 3726–3729.
25 Chien, R.L.; Burgi, D.S. (1992) Sample stacking of an extremely large injection volume in high-performance capillary electrophoresis. *Anal. Chem.* **64**, 1046–1050.
26 Burgi, D.S.; Chien, R.L. (1992) Improvement in the method of sample stacking for gravity injection in capillary zone electrophoresis. *Anal. Biochem.* **202**, 306–309.
27 Szücs, R.; Vindevogel, J.; Sandra, P.; Verhagen, L.C. (1993) Sample stacking effects and large injection volumes in micellar electrokinetic chromatography of ionic compounds: Direct determination of iso-α-acids in beer. *Chromatographia* **36**, 323–329.
28 Beckers, J.L.; Ackermans, M.T. (1993) Effect of sample stacking on resolution, calibration graphs and pH in capillary zone electrophoresis. *J. Chromatogr.* **629**, 371–378.
29 Hirokawa, T.; Ohmori, A.; Kiso, Y. (1993) Analysis of a dilute sample by capillary zone electrophoresis with isotachophoretic preconcentration. *J. Chromatogr.* **634**, 101–106.
30 Kaniansky, D.; Marák, J.; Madajová, V.; Šimuničová, E. (1993) Capillary zone electrophoresis of complex ionic mixtures with on-line isotachophoretic sample pretreatment. *J. Chromatogr.* **638**, 137–146.
31 Reinhoud, N.J.; Tjaden, U.R.; van der Greef, J. (1993) Automated isotachophoretic analyte focusing for capillary zone electrophoresis in a single capillary using hydrodynamic back-pressure programming. *J. Chromatogr.* **641**, 155–162.
32 Reinhoud, N.J.; Tjaden, U.R.; van der Greef, J. (1993) Strategy for setting up single-capillary isotachophoresis-zone electrophoresis. *J. Chromatogr.* **653**, 303–312.
33 Mazereeuw, M.; Tjaden, U.R.; van der Greef, J. (1994) In-line isotachophoretic focusing of very large injection volumes for capillary zone electrophoresis using a hydrodynamic counterflow. *J. Chromatogr.* **677**, 151–157.
34 Larsson, M.; Någård, S. (1994) On-column ITP focusing in CZE for the quantitation of low-molecular-weight basic solutes in biological samples. *J. Microcol. Sep.* **6**, 107–113.
35 Foret, F.; Szoko, E.; Karger, B.L. (1992) On-column transient and coupled column isotachophoretic preconcentration of protein samples in capillary zone electrophoresis. *J. Chromatogr.* **608**, 3–12.
36 Schwer, C.; Lottspeich, F. (1992) Analytical and micropreparative separation of peptides by capillary zone electrophoresis using discontinuous buffer systems. *J. Chromatogr.* **623**, 345–355.
37 Cai, J.; El Rassi, Z. (1992) On-line preconcentration of triazine herbicides with tandem octadecyl capillaries–capillary zone electrophoresis. *J. Liq. Chromatogr.* **15**, 1179–1192.
38 Morita, I.; Sawada, J. (1993) Capillary electrophoresis with on-line sample pretreatment for the analysis of biological samples with direct injection. *J. Chromatogr.* **641**, 375–381.
39 Swartz, M.E.; Merion, M. (1993) On-line sample preconcentration on a packed-inlet capillary for improving the sensitivity of capillary electrophoretic analysis of pharmaceuticals. *J. Chromatogr.* **632**, 209–213.
40 Debets, A.J.J.; Mazereeuw, M.; Voogt, W.H.; van Iperen, D.J.; Lingeman, H.; Hupe, K.-P.; Brinkman, U.A.T. (1992) Switching valve with internal micro precolumn for on-line sample enrichment in capillary zone electrophoresis. *J. Chromatogr.* **608**, 151–158.
41 Cai, J.; El Rassi, Z. (1993) Selective on-line preconcentration of proteins by tandem metal chelate capillaries–capillary zone electrophoresis. *J. Liq. Chromatogr.* **16**, 2007–2024.

2 Quantitative analysis, precision and reproducibility

K. KITAGISHI

In capillary electrophoresis (CE) analysis, principally two quantitative values are obtained. One is a peak area or height as an index of concentration in the detection zone which corresponds to the sample mass in the objective zone. The other is the migration time which generally represents the qualitative characteristics of the sample. Quantitative indices can be derived from the migration time in certain measuring systems, dissociation/association constants in affinity CE, the molecular weight in molecular sieving mode, isoelectric point (pI) in capillary isoelectric focusing (CIEF), etc. These two pieces of information, the peak area/height and the migration time, are described in this chapter.

2.1 Quantitation of sample mass

The peak area in a CE electropherogram is proportional to the sample mass when the detectors have an electronic linear function. One of the problems of quantitative analysis in CE is that sample molecules do not pass through the detection window at a constant velocity, in contrast to other chromatographic analyses, e.g. high performance liquid chromatography (HPLC). Molecules with a smaller migration velocity remain for longer periods at the detection window than those with a larger velocity. Therefore, the integrated peak area from the sample zone where the migration velocity was slow shows a comparatively large value, indicating that the peak area does not correlate exactly with amount of sample. In order to eliminate the effects of migration velocity on peak width, the peak width observed as a function of time, w_t, can be corrected by multiplying the observed peak width by the zone velocity at the detection window. Since zone velocity is defined as the value of capillary length to detector, l, divided by the migration time of the zone, t_m. The peak width independent of the migration velocity, w, is calculated as follows;

$$w = w_t(l/t_m) \qquad (2.1)$$

The spatial width of the sample migration, w_s, is defined when the spatial width of the detection window is w_d [1];

$$w_s = w_t(l/t_m) - w_d \qquad (2.2)$$

Conveniently, the correction for the quantitation can be achieved by the division of integrated peak area by the migration time [2–7]. The corrected peak areas are proportional to the sample concentration even if the mobility of an analyte changes from run to run.

Shear *et al.* [8] attempted in field programming CE to make molecules pass through the detection window at the same migration velocity by changing the applied electric field as a function of the separation time. This would be useful for the separation and quantitation of the mixture of derivatives labeled with the same chromophore.

Under conditions of overloading, such as when the sample injection is in excess, the peak shape is distorted. However, the integrated peak area still represents the sample mass, since peak area versus analyte concentration remains linear [9]. Although the peak height is sometimes used in place of peak area to give information about the sample mass, peak height is a less significant index since the value is responsive to the peak shape.

2.2 Integration techniques

Most modern commercial CE systems include data processing reporting devices, e.g. inexpensive integrators or a personal computer with diverse software packages. Whatever the case, an electric analog signal from the CE detectors is converted into a digital signal for quantitation. Much analytical software has been designed on the principle that a peak should consist of a minimum of 10 data points for accurate estimates to be made [10]. This means that the interval between data points should be less than 0.1 of the full peak width. The signal-to-noise ratio (S/N) of the electropherogram increases if more data points are used.

Suppose there are two Gaussian peaks at a certain interval with random noise and calculate the peak area by integration in order to find the critical factor which results in deviation of the integrated results under certain frequently used analytical conditions (Figure 2.1). Five different sources of random noise are added to a theoretical electropherogram which consists of two peaks with the same peak areas. The calculated areas should be unity, when any disturbing factors are corrected. The results are shown in Table 2.1. The magnitude of random noise mainly affects the deviation in the of calculated peak area. When the S/N ratio is less than 4, the deviation mostly exceeds ±5% and the average varies from unity. Data smoothing improves the deviation but is not very effective. Improvement in the S/N ratio of raw data is a determining factor in quantitation accuracy. Although the S/N ratio is partially improved by data processing, e.g. adaptive smoothing [11], improvement in the design and/or arrangement of the detectors is more beneficial than data processing.

The detection window is another important filtering device for the detec-

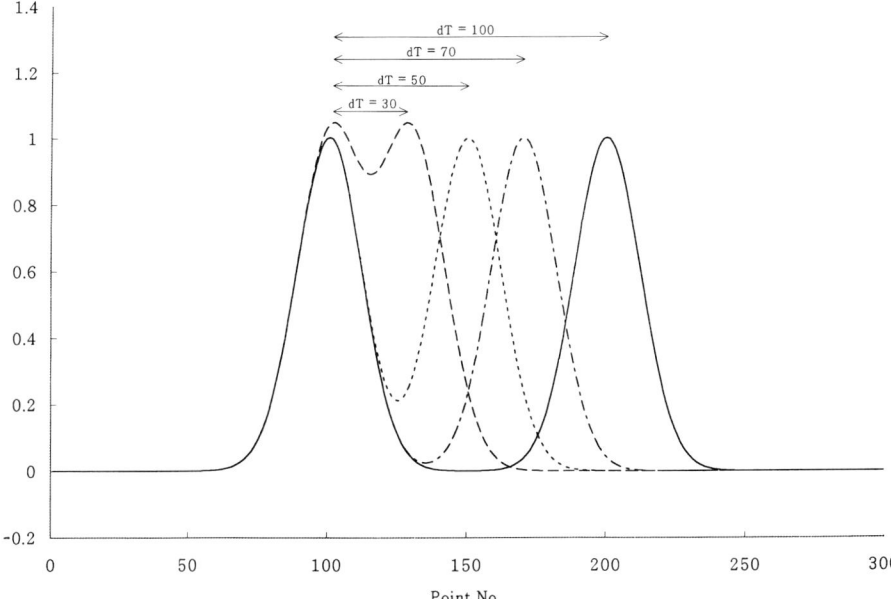

Figure 2.1 Theoretical overlay of two Gaussian peaks of the same width and the same height at various intervals between the two. —, near 0; — · — ·, near 30; ———, near 50; ----, near 70. A Gaussian peak consists of 100 points and the peakwidth at a half of the peak height is 29 points. The values of dT represent point numbers at intervals between the two peaks.

Table 2.1 Peak areas calculated by integration for simulated electropherograms[a]

S/N	dT			
	100	75	50	30
50	1.0000 ± 0.0095	0.9975 ± 0.0040	1.0024 ± 0.0054	0.9888 ± 0.0241
32	0.9934 ± 0.0176	0.9950 ± 0.0165	0.9912 ± 0.0078	1.0018 ± 0.0253
16	1.0050 ± 0.0280	0.9975 ± 0.0267	1.0061 ± 0.0187	0.9903 ± 0.0362
8	1.0131 ± 0.0487	1.0234 ± 0.0247	1.0386 ± 0.0294	0.9838 ± 0.0318
4	1.0590 ± 0.0333	1.0528 ± 0.0501	1.0286 ± 0.0569	1.0316 ± 0.0537
2	1.0681 ± 0.0976	1.0554 ± 0.1180	0.9979 ± 0.0861	0.9794 ± 0.1031

[a] Electropherograms where random noises are added to the theoretical peaks as in Figure 2.1. If the noise does not affect the integration, the peak area should be unity. Since electropherograms of two peaks with five different noises are simulated, 10 peak areas are calculated at each point.
The values in the table represent average ± standard deviation.

tor signal. Shorter capillaries or more slowly migrating analytes are more sensitive to the spatial width of the detection window. The signal is continuously monitored to allow data point smoothing for 0.5s for an analyte detected after 10 min; the length of capillary to the detector is 30 cm and the detector window is 0.25 mm wide. Figure 2.2 shows electropherograms for

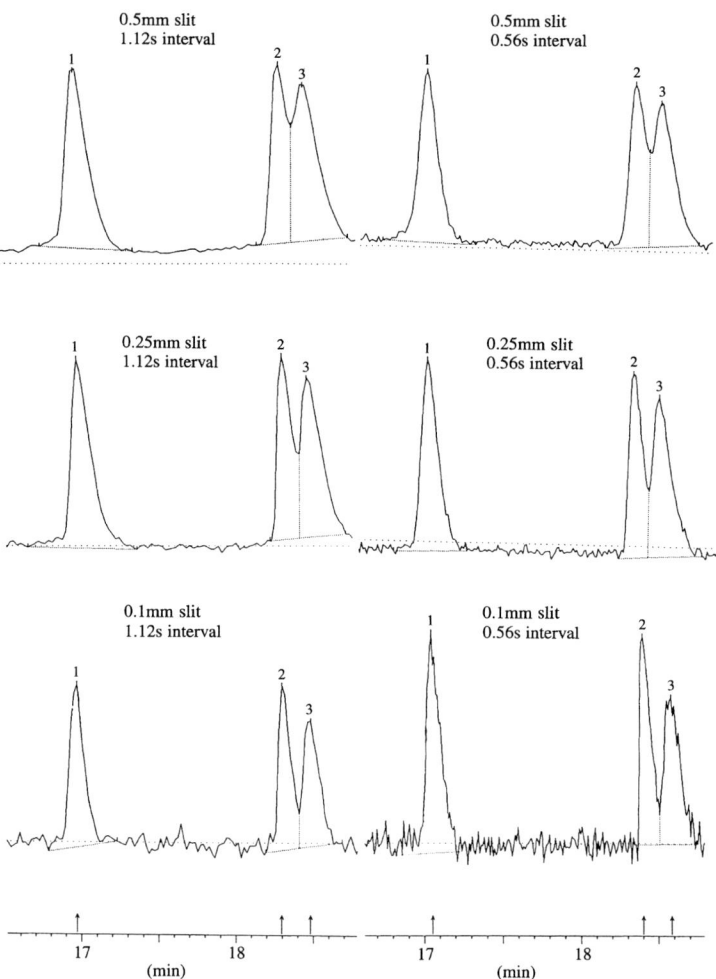

Figure 2.2 Electropherograms of DNA fragments in linear polymer sieving (LPS) mode. Detection windows had spacial widths 0.1, 0.25 and 0.5 mm at sampling intervals of 0.56 and 1.12 s. The slit width is the spatial width of the detection window. Capillary: uncoated, 50 μm inner diameter, 500 mm length. Sample: DNA molecular weight marker; pBR322/Msp I (Nippon Gene Co.) Electrolyte: 0.1 M Tris-borate buffer, pH 8.5, 2 mM EDTA containing hydroxypropylcellulose as sieving polymer. Migration: 12 kV applied potential. Detection: on-line UV absorption at 260 nm. The detector is placed 375 mm from inlet end of the capillary. The electropherograms of DNA fragments of (1) 309 base pairs, (2) 242 base pairs and (3) 238 base pairs are shown.

DNA fragments migrating under linear polymer sieving (LPS) with three kinds of detector windows at sampling intervals of 0.56 s and 1.12 s. Similar effects are noticed for each detector window with respect to sampling interval for the signal filtering. An incompatibility can be seen between improvement in *S/N* ratio and enhancement of separation efficiency.

2.3 Quantitative information derived from migration time

Migration time represents the qualitative aspect of the analyte inducing quantitative index. The six useful quantitative indices derived from migration time are described below.

2.3.1 Ionization constant, pK_a in free zone capillary electrophoresis

The migration velocity of ionizable molecules in free zone capillary electrophoresis (FZE) is dependent on the pK_a of the compound. Ionization equilibrium is represented as a function of pH.

$AH = A^- + H^+$ AH: weak acid
$BH^+ = B + H^+$ B: weak base

The pK_a is estimated by the measurement of the migration velocity over the appropriate pH range.

An example is shown in Figure 2.3 which is used to evaluate the ionization equilibrium of HClO, a strong bleaching agent. The electrophoretic mobility of the analyte, μ_{ep}, is estimated using the migration time of the analyte, t_m, the migration time of electroosmotic flow (EOF), t_{eo}, the capillary length to the detector, l, the total capillary length, L and the applied potential for separation, V;

$$\mu_{ep} = (1/t_{eo} - 1/t_m)(l \times L/V) \qquad (2.3)$$

As the pH dependence of electrophoretic mobility corrected by EOF between pH 5 and 9 is similar to the pH ionization curve, this analytical method by CE is thought to be appropriate for estimation of the ratio of HClO and ClO$^-$. Applications for weak acids and bases have been published [12–14]. The pK_a values determined by CE are reasonably close to those in the literature. When the ionized and neutral forms of a solute give different spectra, the pK_a values can be determined by the peak area in a series of buffer solutions at different pHs [15].

2.3.2 MEKC partition coefficient

A migration velocity in micellar electrokinetic chromatography (MEKC) is expressed by means of capacity factor, k [16, 17]. The value of k is repre-

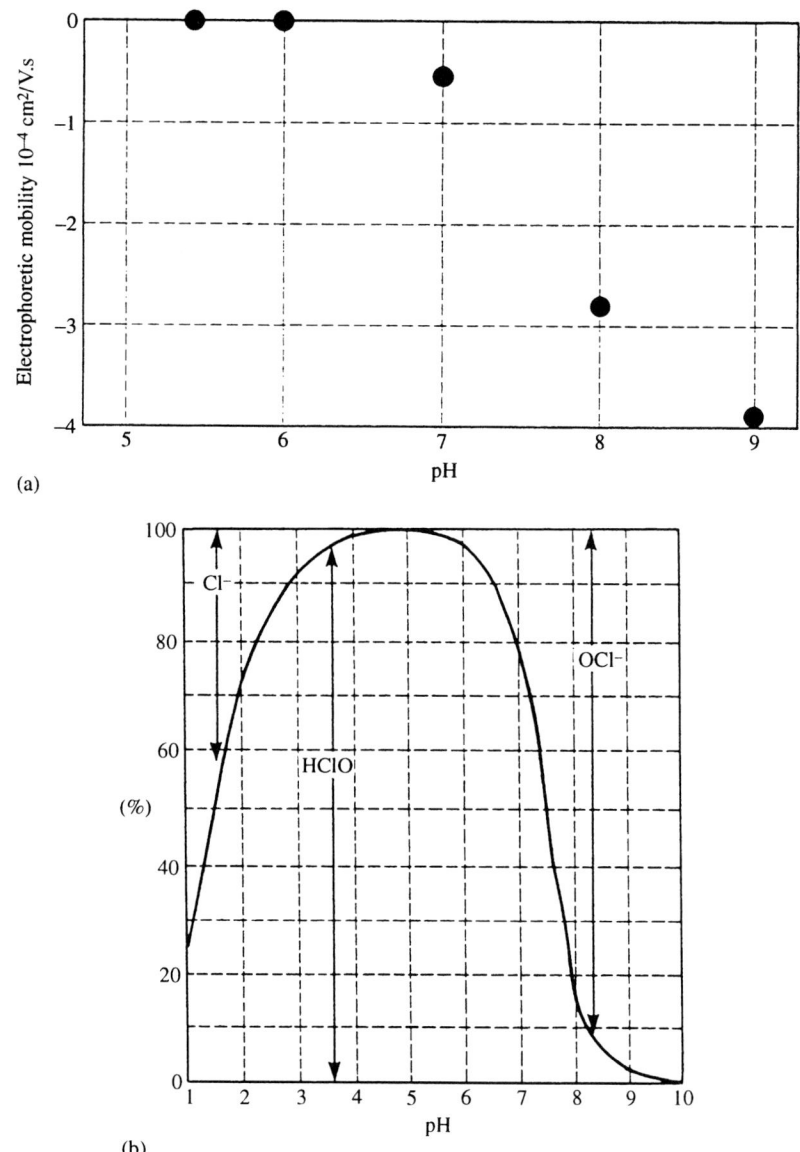

Figure 2.3 pH dependence (a) upon electrophoretic mobility of HClO and (b) upon the ratio of HClO and ClO$^-$ by a conventional technique. In (a), the pH range was between 5 and 9 at 25°C. Sample: NaHClO in running buffer (diluted 500 times). Electrolyte: 100 mM borate with NaOH to adjust pH. Capillary: 75 μm inner diameter, 72.5 cm length. Sample injection: gravity flow injection with the difference in height of 25 mm for 30 s. Migration: 15 kV of applied potential. Detection: on-line UV absorption at 290 nm.

sented with the migration time of the sample molecules, water and micelles at t_m, t_0 and t_{mc}, respectively;

$$k = (t_m - t_0) / \{t_0(1 - t_m/t_{mc})\} \tag{2.4}$$

A retention index scale in MEKC has been derived which is analogous to the retention index in gas chromatography (GC), based on a homologous series with an increasing number of methylene groups, and which has already reached the text books. Muijselaar *et al.* [18] have evaluated the retention index for a homologous series of alkylbenzenes and alkylaryl ketones. A linear relationship is obtained between the retention index in MEKC and the logarithm of the octanol–water partition coefficient measured by UV spectroscopy, a conventional physical properties–structure relationship. This result can be speculated in advance before the experiment is performed as the mechanisms of GC, HPLC and MEKC are identical. This speculation was confirmed experimentally.

2.3.3 Molecular weight in linear polymer sieving and capillary gel electrophoresis (CGE)

Sodium dodecylsulfate (SDS)–protein complexes or double stranded DNAs have the same migration velocity in FZE, since both have the same charge density independent of the molecular weight [19, 20]. In a sieving medium like polyacrylamide gel, smaller molecules migrate faster than larger ones in the absence of EOF, since the sieving medium affects the migration of an analyte dependent upon the size of the analyte molecules. When the molecular weight (or DNA size) dependence of a standard mixture upon migration time or electrophoretic mobility has been obtained [21], the unknown molecular weight can be estimated [22] in a similar way to GPC (gel permeation chromatography).

Note that the mobility of the SDS–protein complex is not constant for proteins of molecular weights less than 10 000 [23].

2.3.4 Isoelectric point (pI) in capillary isoelectric focusing

The pI of an unknown protein or polypeptide can be estimated by comparing the detection time of the unknown with those of standard protein mixtures in CIEF.

2.3.5 Binding constant, K_b, in affinity CE

Affinity CE employs changes in the electrophoretic mobility of a protein on complexation with a charged ligand present in the buffer. Comparison of the migration velocity of the protein of interest with those of the internal standards by the change in the concentration of the ligand yields the binding constant of the ligand to proteins. The binding constant between carbonic anhydrase B and charged benzenesufonamides can be estimated by this strategy [24].

2.3.6 Relative chiral separation factor/percent chiral separation in chiral CE

The relative chiral separation factor (RCS) and percent chiral separation (%CS) have been introduced as a quantitative expression of chiral separation efficiency in CE. They are useful in evaluating the performance through method development [25].

2.4 Peak broadening

In considering the separation, quantitation and reproducibility in CE, it is essential to understand the factors controlling the observed peak width. Huang *et al.* [1] have identified three major parameters for peak width in capillary zone electrophoresis (CZE): (1) the sample injection length, (2) longitudinal diffusion during electromigration, (3) the analyte–wall interaction. The peak width is primarily defined as the sample injection length when other factors, which are called as peak dispersion factors, are negligible, for example, in the case of a large injection length and a weak wall interaction. Some models of peak dispersion have been proposed in CE. A model of the Van Deemter equation in chromatography has been developed by Huang *et al.* [1]. Delinger and Davis [26] reported satisfactory agreement between experimental peak profiles and theoretical ones under conditions where any dispersion factors are practically eliminated excluding longitudinal diffusion [26]. Kenndler and co-workers expressed peak broadening, based on the Einstein–Nernst equation, independently of the diffusion coefficients of the analytes, although longitudinal diffusion is the only process responsible for peak broadening [27, 28]. This model is applicable to FZE in the presence [27] and in the absence [28] of EOF as well as the separation of oligonucleotides with gel-filled capillaries [29]. The peak dispersion coefficient and the diffusion coefficient of a solute can be evaluated from peak profiles based on the Taylor–Aris dispersion theory [30]. Bandwidths of DNA by CGE were estimated by injection length, diffusion, thermal gradient and detection volume as a function of DNA fragment length. It was concluded that longitudinal diffusion was an important factor for bandwidth in contrast to the insignificance of radial temperature gradient in CGE [31].

The analyte–wall interaction often gives poor reproducibility in continuous measurements because the adsorption of the analyte onto the capillary wall is not reproducible in a large number of cases. Theoretical models were proposed for resolution loss caused by wall adsorption, adhesion or retention. The dependence of capillary diameter on the resolution loss was considered [32]. The deviation in wall adsorption makes it impossible to predict selectivity and efficiency with sufficient accuracy [33]. When it is assumed that the zeta potential and the concentration of buffer at the capillary wall are independent of the longitudinal coordinate, the dependence of the efficiency and resolution of the separation on analyte–wall interactions including thermal effects can be investigated mathematically [34].

The analytical system (the choice of column, buffer, electric field, temperature, etc.) should be designed to eliminate the analyte–wall interaction for reproducible measurement.

2.5 Temperature effects in CE

A relatively high electric field in CE results in self heating of the capillary depending on the capillary diameter, ionic strength of buffer, cooling efficiency and other conditions. The undesired temperature rise often alters the viscosity and resistance of the buffer. These cause variation in the migration velocity of the analyte and consequently reduce the reproducibility. The effects of column temperature on migration time due to a temperature induced viscosity change in water were estimated in FZE of peptides [35, 36]. Knox and McCormack [37] investigated zone broadening by decreasing the viscosity with self heating of the capillary and predicted it by calculating the internal column temperature and using the actual diffusion coefficient of the temperature. Fluctuation in the electric current dependent on the thermal coefficient of electric conductivity of the buffer results in variation in the reproducibility of migration time. A statistical algorithm, software package and experimental validation were studied for these effects [38]. In contrast to the estimate of internal capillary temperature using electroosmotic velocity [37], electric current [37, 38] or electrophoretic mobility of samples [37], the local temperature inside a capillary is measured non-invasively with Raman thermometry [39]. Raman thermometry with a microprobe enables estimation of the local temperature with a resolution of 1 to 5μm, as well as estimation of transient temperature gradient [40]. The transient thermal gradient could be mathematically simulated based on some physical assumptions [41–43]. Effective thermostated cooling improves the reproducibility which is lowered by thermal effects.

2.6 Temperature controlled device

Commercial CE systems, except the economy models, have forced cooling functions in capillary cartridges to reduce thermal fluctuations when a high voltage is applied. They are classified either as forced air or forced liquid thermostatted capillaries. Liquid cooling is generally considered to be preferable for high performance [44], although often little difference is recognized between the two [45]. The analytical conditions probably determine the thermal gradient in the capillary. The cooling efficiency of the cooling system is decisive by its design, assuming a high thermostatic efficiency for jacketing the outer surface of capillary compared with air cooling [46].

An example shows how the thermostatic effect on the capillary relates to the observed migration data. A sample solution of adenosine is injected in hydrodynamic mode and migrates with an applied potential of 15 kV in 20 mM phosphate buffer, pH 6.0. Electropherograms of ten runs are superimposed on each other without and with controlling the capillary temperature by forced-air thermostatting and are shown in Figure 2.4(a) and (b), respectively. Even the use of pure phosphate buffer, which reveals less repeatability of EOF than tris(hydroxymethyl)aminomethane- or ethylenediamine-containing buffers [47], indicated that a remarkable improvement in the migration reproducibility could be recognized with a capillary temperature-controlled device.

2.7 Attempts to improve reproducibility and accuracy

Reproducibility of CE should be discussed from two aspects. One is peak area/height reproducibility and the other is migration time. The former is related to injection reproducibility, mainly concerning how to inject a appropriate amount of sample. The latter is regulated by the EOF and electrophoretic mobility of the analyte.

The EOF is influenced by a number of factors, including the dielectric constant and viscosity of the electrolyte and the zeta potential at the shear plane in the electrical double layer. The nature of running buffer plays an important role in the reproducibility of electroosmotic mobility, although the method of manufacture of the capillary columns and their pretreatment are less important [47]. Dependence of electroosmotic mobility upon the applied electric field varies with surrounding buffer, implying the participation of the properties of the solution–wall interaction. The reproducibility of EOF is related to the cation penetration into the capillary wall [48]. Trace cationic constituents of the buffer adsorb onto the capillary wall and induce poor reproducibility of EOF [49]. Grushka's group found that addition of amines to the running buffer stabilized electroosmotic mobility [50]. A

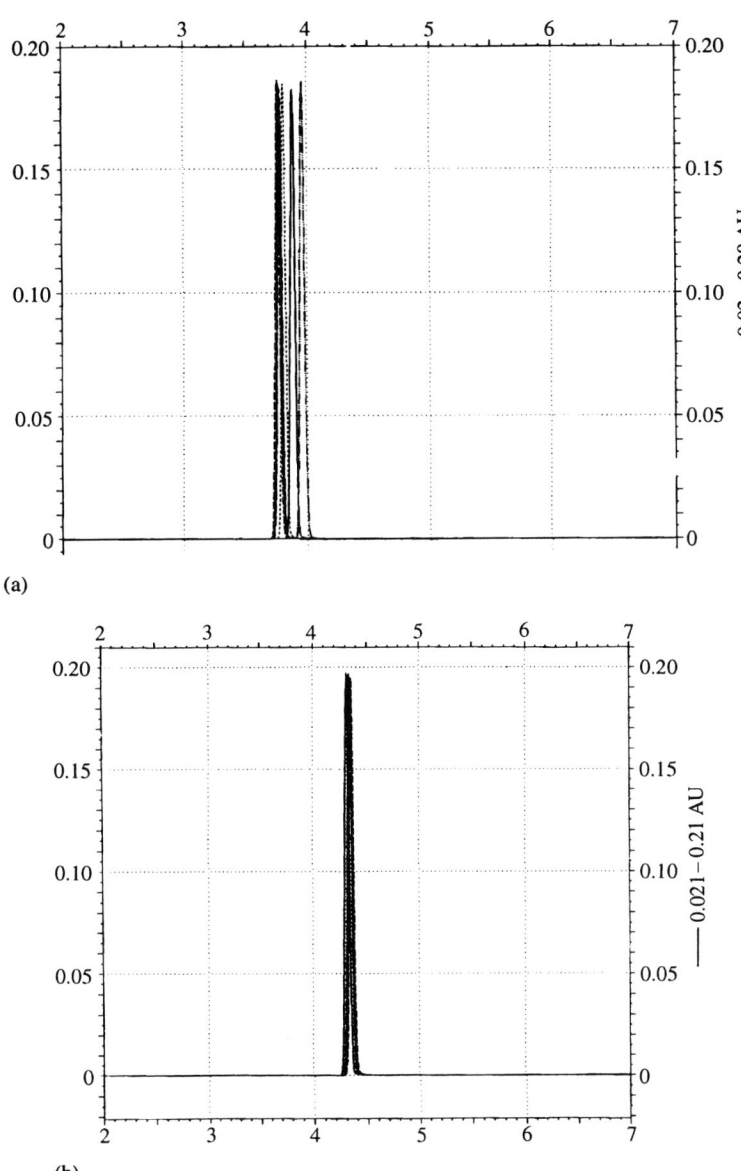

Figure 2.4 Electrophoretic migration (a) without controlling the temperature of the capillary and (b) controlling the temperature of the capillary at 25°C over ten runs superimposed upon one another. Sample: 0.5 mg ml^{-1} adenosine in running buffer. Electrolyte: 20 mM phosphate buffer, pH 6.0. Capillary: 75 μm inner diameter, 50 cm length. Sample injection: gravity flow injection with the a difference in height of 2.5 cm for 30 s. Migration: 15 kV of applied potential. Detection: on-line UV absorption at 260 nm. The detector is placed 375 mm from inlet end of the capillary.

method has been developed for real time measurements of EOF by Lee *et al.* [51]. This technique is applicable to accurate determination of the electrophoretic mobility of the analyte.

Reproducibility is sometimes subject to considerable control by the capillary washing process. Only flushing the running buffer essentially shows the most stable run-to-run reproducibility, if wall adsorption of solutes is negligible. A washing process with a 1M HCl acidic or 1M NaOH alkaline solution followed by the refill of running buffer between each run provides a stable measurement by removing adsorbing molecules from the capillary wall [52, 53]. High reproducibility can be obtained by controlling delicate injection factors, such as minimizing the period in which the capillary ends are not immersed in buffer or sample solution, minimizing the numbers of assays for each buffer vial or maintaining a constant fluid level for vials [54]. Insufficient buffer capacity lowers the reproducibility of the migration time due to pH drift during long term continuous CE measurements [55]. This influence is suppressed by selection of an appropriate buffer which has a sufficiently larger buffer capacity allowed by frequent replacement of the electrolyte in the buffer reservoirs and by increasing the buffer volume.

In order to improve the injection precision and normalize EOF, many quantitative studies have been done using internal standards (IDs). IDs contribute to the improvement in the reproducibilities of both peak area/height and migration time. This method has been applied not only in CE with the hydrodynamic injection [7, 56, 57], but also in electrokinetic injection which has a poor injection–injection reproducibility [3] and CE–ion spray MS [58]. Some examples of applications of IDs are listed in Table 2.2. Significant improvements could be shown in most cases.

Table 2.2 Examples of applications of internal standards to improve reproducibility

Sample	Internal standard	CE mode	Mass/rate[a]	Ref.
Drug counterions	Counter ions	FZE	Mass	56
Neuropeptides	Hemoglobin	FZE	Mass	9
Alanylglutamine	Triglycine	FZE	Mass	57
Isoquinoline alkaloids	Tetrahydroberberine	FZE–MS	Mass	58
Anthraquinone sulfonates	Analog of sample	FZE	Mass	5
Oligonucleotides	Oligonucleotide	CGE	Mass	3
Vitamin B	Paracetamol	FZE/MEKC	Mass/rate	60
Potassium ion	Sodium ion	FZE	Mass	61
Carboxylic acids	Benzoic acid	FZE	Mass	62
Heroin/amphetamine	Crystal violet	MEKC	Mass/rate	63
Dipeptides	Benzyl alcohol	FZE	Rate	64

[a] Mass and rate indicate that internal standards are applied to the correction of peak area/height and migration time, respectively.

Reproducibility in CE mostly refers to run-to-run or day-to-day variations and rarely refers to intercompany cross validation [59]. When the analytical conditions are satisfactory established, reproducible and accurate validation in CE should be accomplished independent of the manufacturer [5].

References

1. Huang, X.; Coleman, W.F.; Zare, R.N. (1989) Analysis of factors causing peak broadening in capillary zone electrophoresis. *J. Chromatogr.* **480**, 95–110.
2. Nielen, M.W.F. (1993) Chiral separation of basic drugs using cyclodextrin-modified capillary zone electrophoresis. *Anal. Chem.* **65**, 885–893.
3. Srivatsa, G.S.; Batt, M.; Schuette, J.; Carlson, R.H.; Fitchett, J.; Lee, C.; Cole, D.L. (1994) Quantitative capillary gel electrophoresis assay of phosphorothioate oligonucleotides in pharmaceutical formulations. *J. Chromatogr.* **680**, 469–477.
4. Wätzig, H.; Dette, C.; Uhl, H. (1993) Quantification of the important pharmaceutical intermediate pyridine-2,4-dicarboxylic acid using capillary electrophoresis (CE). *Pharmazie* **48**, 527–531.
5. Williams, S.J.; Goodall, D.M.; Evans, K.P. (1993) Analysis of anthraquinone sulphonates: comparison of capillary electrophoresis with high-performance liquid chromatography. *J. Chromatogr.* **629**, 379–384.
6. Altria, K.D. (1993) Essential peak area normalisation for quantitative impurity content determination by capillary electrophoresis. *Chromatographia* **35**, 177–182.
7. Altria, K.D.; Goodall, D.M.; Rogan, M.M. (1994) Quantitative applications and validation of the resolution of enantiomers by capillary electrophoresis. *Electrophoresis* **15**, 824–827.
8. Shear, J.B.; Dadoo, R.; Zare, R.N. (1994) Field programming to achieve uniform sensitivity for on-line detection in electrophoresis. *Electrophoresis* **15**, 225–227.
9. Lee, H.G.; Desiderio, D.M. (1994) Capillary zone electrophoresis of two synthetic neuropeptides: examination of detectability and resolution as a function of peptide concentration and buffer concentration. *J. Chromatogr.* **655**, 9–19.
10. Matthews, D.E.; Hayes, J.M. (1976) Systematic errors in gas chromatography–mass spectrometry isotope ratio measurements. *Anal. Chem.* **48**, 1375–1382.
11. Kawata, S.; Minami, S. (1984) Adaptive smoothing of spectroscopic data by a linear mean-square estimation. *Appl. Spectrosc.* **38**, 49–58.
12. Cai, J.; Smith, J.T.; El Rassi, Z. (1992) Determination of the ionization constants of weak electrolytes by capillary zone electrophoresis. *J. High Resolut. Chromatogr.* **15**, 30–32.
13. Cleveland, Jr., J.A.; Benko, M.H.; Gluck, S.J.; Walbroehl, Y.M. (1993) Automated pK_a determination at low solute concentrations by capillary electrophoresis. *J. Chromatogr.* **652**, 301–308.
14. Gluck, S.J.; Cleveland, Jr., J.A. (1994) Investigation of experimental approaches to the determination of pK_a values by capillary electrophoresis. *J. Chromatogr.* **680**, 49–56.
15. Cleveland, Jr, J.A.; Martin, C.L.; Gluck, S.J. (1994) Spectrophotometric determination of ionization constants by capillary zone electrophoresis. *J. Chromatogr.* **679**, 167–171.
16. Terabe, S.; Otsuka, K.; Ichikawa, K.; Tsuchiya, A.; Ando, T. (1984) Electrokinetic separations with micellar solutions and open-tubular capillaries. *Anal. Chem.* **56**, 111–113.
17. Terabe, S.; Otsuka, K.; Ando, T. (1985) Electrokinetic chromatography with micellar solution and open-tubular capillary. *Anal. Chem.* **57**, 834–841.
18. Muijselaar, P.G.H.M.; Claessens, H.A.; Cramers, C.A. (1994) Application of the retention index concept in micellar electrokinetic capillary chromatography. *Anal. Chem.* **66**, 635–644.
19. Grossman, P.D.; Soane, D.S. (1991) Capillary electrophoresis of DNA in entangled polymer solutions. *J. Chromatogr.* **559**, 257–266.

20 Ganzler, K.; Greve, K.S.; Cohen, A.S.; Karger, B.L.; Guttman, A.; Cooke, N.C. (1992) High-performance capillary electrophoresis of SDS–protein complexes using UV-transparent polymer networks. *Anal. Chem.* **64**, 2665–2671.

21 Guttman, A.; Shieh, P.; Hoang, D.; Horváth, J.; Cooke, N. (1994) Effect of operational variables on the separation of proteins by capillary sodium dodecyl sulfate-gel electrophoresis. *Electrophoresis* **15**, 221–224.

22 Guttman, A.; Shieh, P.; Lindahl, J.; Cooke, N. (1994) Capillary sodium dodecyl sulfate gel electrophoresis of proteins II. On the Ferguson method in polyethylene oxide gels. *J. Chromatogr.* **676**, 227–231.

23 Karim, M.R.; Shinagawa, S.; Takagi, T. (1994) Electrophoretic mobilities of the complexes between sodium dodecyl sulfate and various peptides or proteins determined by free solution electrophoresis using coated capillaries. *Electrophoresis* **15**, 1141–1146.

24 Gomez, F.A.; Avila, L.Z.; Chu, Y.-H.; Whitesides, G.M. (1994) Determination of binding constants of ligands to proteins by affinity capillary electrophoresis: compensation for electroosmotic flow. *Anal. Chem.* **66**, 1785–1791.

25 D'Hulst, A.; Verbeke, N. (1994) Quantitation in chiral capillary electrophoresis: theoretical and practical considerations. *Electrophoresis* **15**, 854–863.

26 Delinger, S.L.; Davis, J.M. (1992) Influence of analyte plug width on plate number in capillary electrophoresis. *Anal. Chem.* **64**, 1947–1959.

27 Schwer, C.; Kenndler, E. (1992) Peak broadening in capillary zone electrophoresis with electro-osmotic flow: dependence of plate number and resolution on charge number. *Chromatographia* **33**, 331–335.

28 Friedl, W.; Kenndler, E. (1993) Resolution as a function of the pH of the buffer based on the analyte charge number for multivalent ions in capillary zone electrophoresis without electroosmotic flow: theoretical prediction and experimental evaluation. *Anal. Chem.* **65**, 2003–2009.

29 Kenndler, E.; Schwer, C. (1992) Peak dispersion and separation efficiency in high-performance zone electrophoresis with gel-filled capillaries. *J. Chromatogr.* **595**, 313–318.

30 Bello, M.S.; Rezzonico, R.; Righetti, P.G. (1994) Use of Taylor–Aris dispersion for measurement of a solute diffusion coefficient in thin capillaries. *Science* **266**, 773–776.

31 Luckey, J.A.; Norris, T.B.; Smith, L.M. (1993) Analysis of resolution in DNA sequencing by capillary gel electrophoresis. *J. Phys. Chem.* **97**, 3067–3075.

32 Schure, M.K.; Lenhoff, A.M. (1993) Consequences of wall adsorption in capillary electrophoresis: theory and simulation. *Anal. Chem.* **65**, 3024–3037.

33 Friedl, W.; Kenndler, E. (1994) Limitations of the optimization of the resolution by the buffer pH in capillary zone electrophoresis. *Fresenius J. Anal. Chem.* **348**, 576–582.

34 Andreev, V.P.; Lisin, E.E. (1993) On the mathematical model of capillary electrophoresis. *Chromatographia* **37**, 202–210.

35 Zhang, Y.K.; Chen, N.; Wang, L. (1993) The effect of column temperature on the migration time of peptides in free-solution capillary electrophoresis. *J. Liq. Chromatogr.* **16**, 3689–3697.

36 Chen, N.; Wang, L.; Zhang, Y. (1993) Influence of column temperature and physicochemical properties on the electrophoretic behaviour of polyglycine peptides in free-solution capillary electrophoresis. *J. Chromatogr.* **644**, 175–182.

37 Knox, J.H.; McCormack, K.A. (1994) Temperature effects in capillary electrophoresis. 1: Internal capillary temperature and effect upon performance. *Chromatographia* **38**, 207–214.

38 Bello, M.S.; Levin, E.I.; Righetti, P.G. (1993) Computer-assisted determination of the inner temperature and peak correction for capillary electrophoresis. *J. Chromatogr.* **652**, 329–336.

39 Davis, K.L.; Liu, K.-L.K.; Lanan, M.; Morris, M.D. (1993) Spatially resolved temperature measurements in electrophoresis capillaries by Raman thermometry. *Anal. Chem.* **65**, 293–298.

40 Liu, K.-L.K.; Davis, K.L.; Morris, M.D. (1994) Raman spectroscopic measurement of spatial and temporal temperature gradients in operating electrophoresis capillaries. *Anal. Chem.* **66**, 3744–3750.

REFERENCES

41 Bello, M.S.; Righetti, P.G. (1992) Unsteady heat transfer in capillary zone electrophoresis I. A mathematical model. *J. Chromatogr.* **606**, 95–102.
42 Bello, M.S.; Righetti, P.G. (1992) Unsteady heat transfer in capillary zone electrophoresis II. Computer simulations. *J. Chromatogr.* **606**, 103–111.
43 Dose, E.V.; Guiochon, G. (1993) Timescales of transient processes in capillary electrophoresis. *J. Chromatogr.* **652**, 263–275.
44 Knox, J.H.; McCormack, K.A. (1994) Temperature effects in capillary electrophoresis. 2: Some theoretical calculations and predictions. *Chromatographia* **38**, 215–221.
45 Landers, J.P.; Oda, R.P.; Madden, B.; Sismelich, T.P. Spelsberg, T.C. (1992) Reproducibility of sample separation using liquid or forced air convection thermostatted high performance capillary electrophoresis systems. *J. High Resolut. Chromatogr.* **15**, 517–525.
46 Bello, M.S.; de Besi, P.; Righetti, P.G. (1993) Thermally induced fluctuations of electric current and baseline in capillary electrophoresis. *J. Chromatogr.* **652**, 317–327.
47 Coufal, P.; Štulík, K.; Claessens, H.A.; Cramers, C.A. (1994) The magnitude and reproducibility of the electroosmotic flow in silica capillary tubes. *J. High Resolut. Chromatogr.* **17**, 325–334.
48 Bello, M.S.; Capelli, L.; Righetti, P.G. (1994) Dependence of the elctroosmotic mobility on the applied electric field and its reproducibility in capillary electrophoresis. *J. Chromatogr.* **684**, 311–322.
49 Gassner, B.; Friedl, W.; Kenndler, E. (1994) Wall adsorption of small anions in capillary zone electrophoresis induced by cationic trace constituents of the buffer. *J. Chromatogr.* **680**, 25–31.
50 Cohen, N.; Grushka, E. (1994) Controlling electroosmotic flow in capillary zone electrophoresis. *J. Chromatogr.* **678**, 167–175.
51 Lee, T.T.; Dadoo, R.; Zare, R.N. (1994) Real-time measurement of electroosmotic flow in capillary zone electrophoresis. *Anal. Chem.* **66**, 2694–2700.
52 Shieh, P.C.H.; Hoang, D.; Guttman, A.; Cooke, N. (1994) Capillary sodium dodecyl sulfate gel electrophoresis of proteins I. Reproducibility and stability. *J. Chromatogr.* **676**, 219–226.
53 Barron, A.E.; Blanch, H.W.; Soane, D.S. (1994) A transient entanglement coupling mechanism for DNA separation by capillary electrophoresis in ultradilute polymer solutions. *Electrophoresis* **15**, 597–615.
54 Thomas, B.R.; Fang, X.G.; Chen, R.J.; Tyrrell, R.J.; Ghodbane, S. (1994) Validated micellar electrokinetic capillary chromatography method for quality control of the drug substances hydrochlorothiazide and chlorothiazide. *J. Chromatogr.* **657**, 383–394.
55 Zhu, T.; Sun, Y.; Zhang, C.; Ling, D.; Sun, Z. (1994) Variation of the pH of the background electrolyte as a result of electrolysis in capillary electrophoresis. *J. High Resolut. Chromatogr.* **17**, 563–564.
56 Altria, K.D.; Goodall, D.M.; Rogan, M.M. (1994) Quantitative determination of drug counter-ion stoichiometry by capillary electrophoresis. *Chromatographia* **38**, 637–642.
57 Jones, D.; Scarborough, A.; Tier, C.M. (1994) Quantitative determination of a dipeptide in personal wash liquid by capillary electrophoresis. *J. Chromatogr.* **661**, 1–6.
58 Henion, J.D.; Mordehai, A.V.; Cai, J. (1994) Quantitative capillary electrophoresis–ion spray mass spectrometry on a benchtop ion trap for the determination of isoquinoline alkaloids. *Anal. Chem.* **66**, 2103–2109.
59 Altria, K.D.; Clayton, N.G.; Hart, M.; Harden, R.C.; Hevizi, J.; Makwana, J.V.; Portsmouth, M.J. (1994) An inter-company cross-validation exercise on capillary electrophoresis testing of dose uniformity of paracetamol content in formulations. *Chromatographia* **39**, 180–184.
60 Boonkerd, S.; Detaevernier, M.R.; Michotte, Y. (1994) Use of capillary electrophoresis for the determination of vitamins of the B group in pharmaceutical preparations. *J. Chromatogr.* **670**, 209–214.
61 Altria, K.D.; Wood, T.; Kitscha, R.; Roberts-McIntosh, A. (1995) Validation of a capillary electrophoresis method for the determination of potassium counter-ion levels in an acidic drug salt. *J. Pharm. Biomed. Anal.* **13**, 33–38.
62 Leube, J.; Roeckel, O. (1994) Quantification in capillary zone electrophoresis for samples differing in composition from the electrophoresis buffer. *Anal. Chem.* **66**, 1090–1096.

63 Krogh, M.; Brekke, S.; Tønnesen, F.; Rasmussen, K.E. (1994) Analysis of drug seizures of heroin and amphetamine by capillary electrophoresis. *J. Chromatogr.* **674**, 235–240.
64 Chen, N.; Wang, L.; Zhang, Y. (1993) Electrophoretic selectivity as a function of operating parameters in free-solution capillary electrophoretic separation of dipeptides. *J. Liq. Chromatogr.* **16**, 3609–3622.

3 Precision and quantitation in capillary electrophoresis
G.A. ROSS

3.1 Introduction

Capillary electrophoresis (CE) describes a variety of electrophoretic separations which may be used to provide different data. In all cases, however, it is the peak migration time and peak dimensions which are used to derive the quantitative and qualitative data. Capillary gel electrophoresis (CGE) and capillary isoelectric focusing (CIEF) modes may be used to determine qualitative data but the most generally used CE techniques for performing quantitative analysis of analytes are capillary zone electrophoresis (CZE) and micellar electrokinetic chromatography (MEKC).

3.2 Precision

CE is increasingly being used in quantitative analyses. Initially CE was considered less reproducible than other separation techniques, e.g. high performance liquid chromatography (HPLC) and gas chromatography (GC). However, with the maturing of the technique and increasing familiarity of workers with its operation an increasing number of reports show that acceptable precision can be readily achieved. Precision of the peak parameters determines the precision of the analytical data derived from these. Migration time precision is important for peak identification and for assessing the stability of the analytical method while peak area precision is of great importance in the quantitative assay. Recent studies have demonstrated that not only is CE capable of exhibiting good precision in a quantitative assay [1–3], but that this precision can be preserved when methods are transferred between laboratories and between instruments [4, 5].

3.2.1 Migration time precision

Table 3.1 describes the various factors which can affect migration time precision, the sources and effects of these factors and some solutions to preserving migration time precision. These are discussed in more detail in the following section.

In CZE the migration time of a compound is dependent upon its appar-

Table 3.1 Reproducibility of mobility and migration time

Factor	Cause	Effect	Solution
Temperature changes	• Changes in viscosity	• Changes in EOF • Changes in mobility	• Stable capillary thermostatting
Capillary wall adsorption	• Electrostatic adsorption of analyte to capillary wall	• Changes in EOF	• Control pH to effect electrostatic wall repulsion • Use coated capillaries
Hysteresis of wall charge	• Conditioning capillary at high (or low) pH and operation at low (or high) pH	• Unstable or irreproducible EOF	• Avoid pH differences between cleaning/conditioning solutions and operating buffer • Allow sufficient equilibration time
Changes in buffer composition	• pH changes due to electrolysis • Buffer evaporation • Conditioning waste flushed into buffer outlet vial	• Changes in pH • Changes in buffer concentration • Changes in organic modifier concentration • Changes in buffer ionic strength	• Replenish buffer reservoirs • Cap buffer vials and cool sample tray • Use separate reservoir for wash waste • Wash capillary end between washes
Variation in silica batches	• Variation in silanol content between batches of silica	• Variable EOF between capillaries	• Measure EOF and normalize if necessary
Variations in applied voltage	• Unstable voltage applied	• Changes in applied electric field (E) proportional changes in migration time	• Instrument dependent (not user accessible)

ent mobility (μ_{app}) which is the sum of its own mobility (μ_{ep}) plus that of the electroosmotic flow (EOF) (μ_{eo}). Consequently those factors which influence both the EOF and the electrophoretic mobility of an ion must be kept constant in order to realize precise migration times. In MEKC analyses of neutral compounds the EOF is important in maintaining a constant mobility of the neutral compounds as they interact with the pseudostationary micellar phase. The mobility of the pseudostationary phase must also be preserved.

3.2.1.1 Electroosmotic flow (EOF) stabilization. With bare fused silica capillaries, there is an inherent EOF which is pH dependent and also varies depending on the pH history of the capillary [6, 7] (Figure 3.1). In order to stabilize the EOF, the capillary should be appropriately preconditioned and should have sufficient time to equilibrate at the operating pH prior to analysis. It is also important to match the preconditioning and cleaning of a capillary to the operating conditions. This means the determination of appropriate pretreatment regimens which may be assay specific. An example is the separation of tryptic digests of proteins to determine sequence variations. Such peptide separations are routinely performed using a low pH phosphate buffer and the appropriate wash regime includes a capillary pretreatment with H_3PO_4 prior to a buffer wash [8, 9]. In this way the capillary wall chemistry is stabilized and the EOF minimized. EOF is less reproducible at higher pHs where its magnitude is greater [10].

Silica capillaries supplied by different suppliers can have different silanol concentrations such that their associated EOF can vary. Different batches of silica from a single manufacturer can have different silanol concentrations even when using the same wash regimen [11]. Table 3.2 shows the EOF associated with seven different batches of silica supplied by Polymicro Technologies with and without a pre-use wash method. In each case the precision of the EOF was improved, sometimes quite dramatically by using the indicated prewash method. It should, however, be noted that even after using a prewash method there is still a degree of variation in the measured EOF between silica batches.

Coufal *et al.* [12] investigated intensively the effects of various preconditioning techniques on capillaries from different manufacturers. They conclude that the run buffer plays a more significant role in determining the precision and magnitude of the EOF than the manufacturing process or the capillary pretreatment. Furthermore, they suggest the use of amine additives to stabilize the capillary wall charge and ensure EOF precision. Using

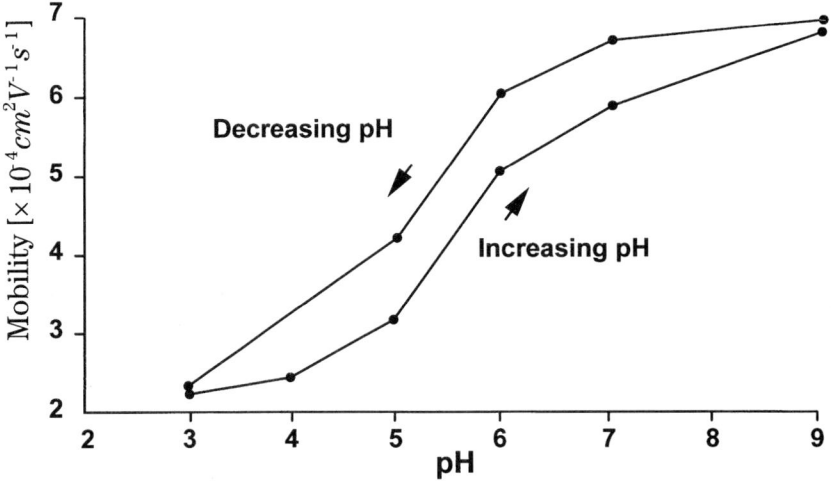

Figure 3.1 Hysteresis of electroosmotic flow.

Table 3.2 EOF variations with pretreatment and silica batch

Silica batch	No pretreatment		Pretreated: 5 min MeOH, water, 1N NaOH, water 20 min run buffer	
	EOF (mm s^{-1}) mean	%RSD ($n = 10$)	EOF (mm s^{-1}) mean	%RSD ($n = 10$)
QQT11A	0.2307	1.31	0.2837	0.86
QNR01	0.2353	1.18	0.2828	0.78
MNZ04B	0.2303	5.95	0.2811	1.48
KZGO1A	0.1886	23.97	0.2580	1.71
MSZ01	0.1538	11.32	0.2195	0.95
KYL12	0.2185	13.80	0.2517	0.42
KYL05	0.1406	19.82	0.2900	0.97
	0.1997	19.65	0.2667	9.47

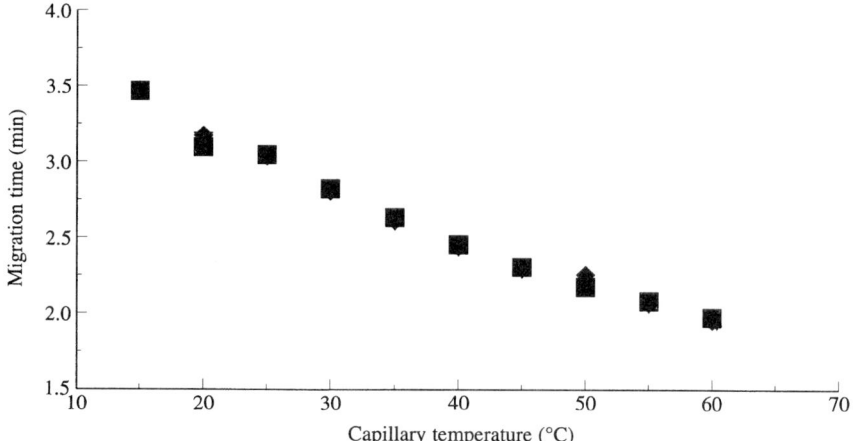

Figure 3.2 The effect of increasing capillary temperature on migration time of *p*-hydroxyacetophenone. Other conditions: buffer, 20 mM borate pH 9.2; capillary, 48 cm (40 cm effective) × 50 μm i.d.; detection, 192 nm (4 nm bandwidth); injection, 250 mbar s^{-1}; run, 30 kV.

buffer additives [13] or voltage preconditioning [14] can also stabilize the EOF.

3.2.1.2 Temperature changes. Temperature control has been identified as one of the major factors in achieving precision in migration times and in peak areas [15, 16]. Temperature changes affect the run buffer viscosity and therefore the mobility of the EOF and of analyte ions which will change by approximately 2% per °C. Figure 3.2 shows the effects of increasing capillary temperature on peak migration time. In order to achieve good migration time reproducibility it is important that the capillary temperature is kept stable. Stability is of more importance than the absolute accuracy of the capillary temperature.

3.2.1.3 Wall adsorption. Changes to the capillary wall caused by adsorption of analyte or sample matrix components have a detrimental effect on EOF. This is a particular problem in the analysis of proteins. Irreversible adsorption of analyte onto the capillary wall causes a progressive decrease in EOF. In order to counter this, higher concentrations of run buffer may be used or buffer additives [12, 13, 17]. If the EOF is not critical for the separation, then a coated capillary may be employed.

3.2.1.4 Buffer composition. During electrophoresis the run buffer will undergo electrolysis [18]. When using higher buffer concentrations (>100 mM), it can result in detrimental Joule heating and therefore a buffer should be selected so that it is well suited to the operating pH. A buffer has an optimal buffering capacity where pH = $pK_a \pm 1$.

However, if the buffer is outside this optimal pH buffering range or has insufficient capacity then its pH will change over time. This will affect both the EOF and the mobility of the ion and consequently the analyte migration time and precision.

In some cases a buffer is chosen for properties other than its buffering capacity. In indirect detection techniques a buffer is chosen for its UV absorption and mobility properties rather than its pH. Subsequently the buffer frequently has a low buffering capacity. In both cases regular buffer replacement or frequent replenishment is sufficient to ensure that the buffer properties are kept constant to preserve the migration time precision.

3.2.1.5 Concentration-dependent peak shapes. When observing for precision of migration time there are some analytical methods which may give rise to apparent variation in migration time although the system itself is very stable. In particular, in the analysis of small ions the peak shape is dependent upon the concentration of the peak [31]. In these cases it is normal for a peak to have a pronounced triangular shape with the apex varying according to the analyte concentration. This has various ramifications for peak identification. It may therefore be of more use to use the peak center of gravity as the indicative time point for migration time rather than the more variable peak apex.

3.2.2 Peak area and height precision

Peak area and peak height are dependent upon the amount of sample injected into the capillary. Most instruments are capable of performing hydrodynamic and electrokinetic injection although hydrodynamic injections are better suited to quantitative analyses. Electrokinetic injection exploits the charged nature of the analytes and/or the presence of an EOF to inject sample.

In CE peak areas are linearly related to sample concentration over a broader range than peak heights [19]. For this reason peak areas are used as the basis for quantitative analysis. Table 3.3 describes some factors which can lead to imprecision in peak areas, some causes and effects and solutions to achieving peak area precision.

3.2.2.1 Temperature effects. Since temperature changes affect the viscosity of a solution, subsequent pressure-mediated injection will also be affected. It is interesting to note that the capillary temperature has a much greater effect on the amount loaded and peak area (Figure 3.3a) than the sample tray temperature (Figure 3.3b). For precise injection volumes, therefore, the capillary thermostatting is of more importance than the sample tray thermostatting.

3.2.2.2 Sample considerations.
Sample concentration. Figure 3.4 shows how the %RSD (relative standard deviation) of peak corrected area varies inversely with the analyte concentration. This is mainly due to variations in integration of reported peak areas. This is observed with signals with very low signal to noise. Although the injected amount may be reproducible, if the reported peak area is not, then precision is lost. This is of particular importance where small volumes or low concentrations are being introduced [15, 16]. The simple solution to this, where appropriate, is to increase the analyte concentration or the sample load so that a signal to noise ratio of at least 40 is achieved.

Matrix evaporation. Evaporation of the sample leads to an increase in analyte concentration. Figure 3.5 shows the effects of repeated injection from a single vial containing analytes in 80% MeOH, over the course of 4 h. It should be noted that the vial was capped but had to be pierced for the first injection. In comparison with a subsequent injection from a vial which had remained capped over the course of the experiment it is clear that capping is vital to preserving sample concentration. From these data it is important that samples with a high organic content should remain capped over the course of an analysis. Further sample tray cooling should be used to reduce the evaporation of such sample matrices. The effect will be less pronounced with aqueous sample solutions although small volume samples (ca. 5–10 μl) will be affected to a greater degree than larger volume samples (>100 μl).

3.2.2.3 Injection artifact.
Spontaneous injection. Sample can be injected onto a capillary by the simple act of inserting the capillary into a sample solution [20, 21]. The effect is exaggerated by the time taken between injection of sample and

Table 3.3 Factors affecting peak reproducibility

Factor	Cause	Effect	Solution
• Temperature changes	• Viscosity changes	• Changes in injected amount	• Capillary thermostatting • Sample tray thermostatting
• Sample matrix evaporation	• XXX sample tray temperatures • Uncapped vials	• Increases in sample concentration	• Cap sample vials • Cool sample tray
• Sample carry-over	• Extraneous injection	• Peak tailing and loss in resolution • Irreproducible integration of peak area	• Use capillary with flat smooth injection end • Remove polyimide from injection end • Dip capillary end in appropriate solution prior to run
• Zero injection caused by dipping capillary in sample solution	• Caused by dipping capillary end in sample solution	• Increased area	• Cannot be eliminated but can be incorporated in quantitation by control of timing
• Low signal to noise ratio		• Integration errors	• Optimize integration parameters • Increase sample concentration • Use peak height
• Sudden application of high voltage	• Lack of voltage ramp capability	• Heating, thermal expansion of buffer and/or sample and expulsion of sample	• Use voltage ramp during run • Inject buffer post injection
• Concentration-dependent peak shapes		• Peak heights become non-linear	• Use peak areas for quantitation
• Intercapillary peak area changes	• Changes in capillary i.d. or alignment	• Changes in peak area between capillary changes	• Careful alignment procedures • Recalibrate after capillary change
• Hydrodynamic injection	• Changes in applied pressure • Changes capillary length and i.d.	• Changes in volume injected	• Validate method on capillary dimensions within instrument • Validate method on pressure applied
• Electrokinetic injection		• Variation in injected amount due to variations in sample matrix	• Careful matching of calibration and sample solutions • Use hydrodynamic injection

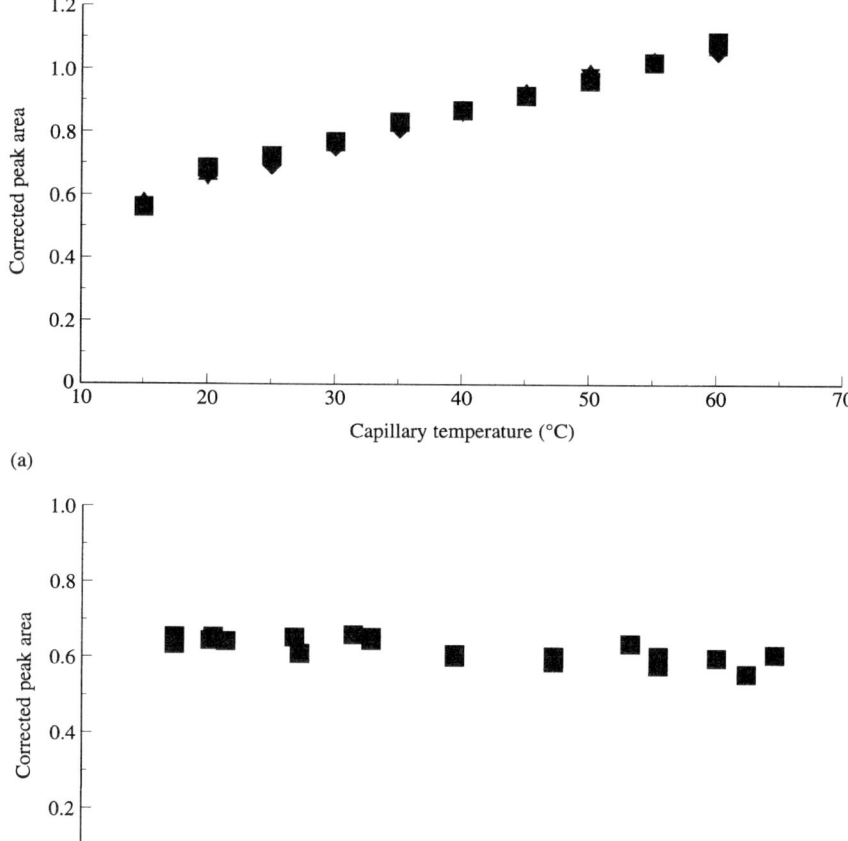

Figure 3.3 (a) The effect of increasing capillary temperature on corrected peak area from 250 mbar s^{-1} injection of p-hydroxyacetophenone. Other conditions were identical to Figure 3.2. (b) The effect of increasing sample tray temperature on corrected peak area from 250 mbar s^{-1} injection of p-hydroxyacetophenone. Other conditions: buffer, 20 mM borate pH 9.2; capillary, 48 cm (40 cm effective) × 50 μm i.d.; detection, 192 nm (4 nm bandwidth); injection, 250 mbar s^{-1}; run, 30 kV, capillary temperature 20°C.

insertion of capillary into the run buffer. The effect is of greater importance in ultramicrosampling than in routine quantitative assays performed using CE. Reports of the significance of this phenomenon vary in the literature [16, 22] and the effect may be affected by buffer properties. The use of automated injection will also help to reduce this phenomenon.

Sample carry-over. Irreproducible amounts of sample may also be introduced by contamination of the external surface of the capillary. This second factor can also result in peak tailing and baseline shifts leading to integrator errors and loss of resolution. This can be avoided simply by dipping the capillary injection end in a suitable solvent between injection and run to wash the capillary end. It should be noted that this dip solution should be able to solvate the analytes of interest.

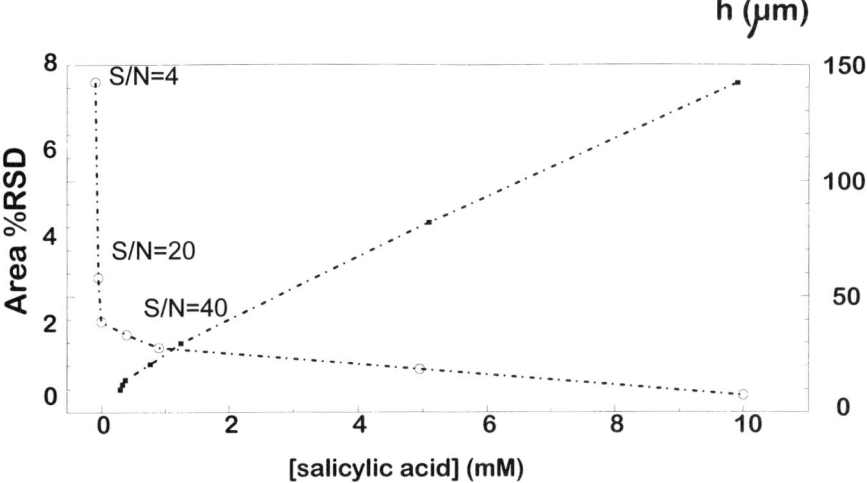

Figure 3.4 The effect of low concentration injections of sample on peak area precision and peak efficiency. Concentration range: 0.01–10 mg ml^{-1}.

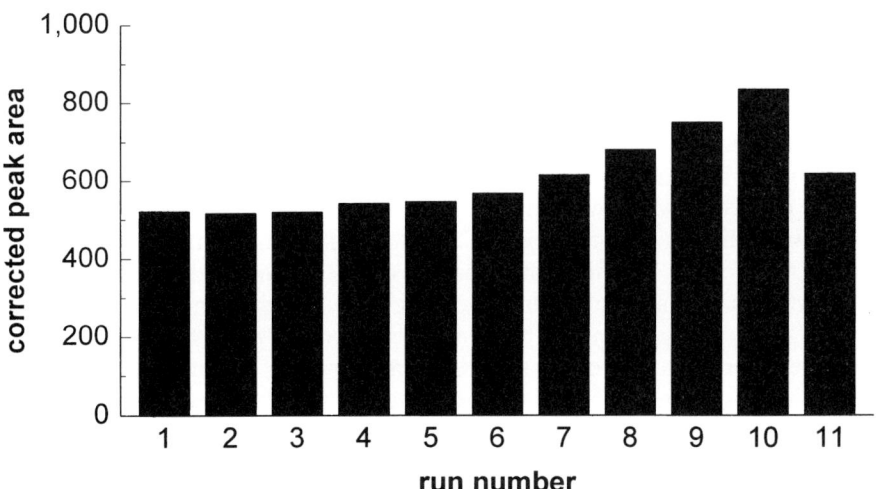

Figure 3.5 The effects of sample matrix evaporation on peak corrected area.

Thermal effects. The sudden application of a high voltage to a sample of low conductivity can lead to its thermal expansion and expulsion from the capillary end [23]. In order to circumvent this, the voltage may be applied over a period of time (voltage ramp) or a plug of buffer can be injected after the sample zone. The same effect may also result in sample precipitation and subsequent capillary blockage [9].

Capillary-end geometry. The geometry of the capillary end is also of extreme importance in preserving injected volume precision and eliminating carry-over. Rough or ragged capillary ends can lead to irreproducible injected volumes and losses in resolution and peak efficiency. Any capillary used in a quantitative assay should have a flat, smooth-cut injection end in order to avoid these problems [7].

3.2.2.4 Internal standards. Using internal standards can also help to improve peak area precision [5, 24, 25]. In many cases the %RSD has been reported at <1%, for peak areas. Many manufacturers quote around 2% RSD for corrected areas, but with an experienced operator this can be improved to below 1%.

3.3 Quantitation

Quantitative analysis in CE may be approached similarly to high performance liquid chromatography (HPLC). Quantitative aspects of the application of CE to pharmaceutical analysis have been reviewed by Altria [2] and specific concerns for precision and calibration have been discussed by Watzig [19, 26].

Quantitative data for an unknown sample from peak parameters are derived by determining the peak response (e.g. UV absorbance) for a known concentration and relating this to the peak response in the unknown. In CZE and MEKC peaks are separated by virtue of their different mobilities and therefore travel at different velocities within the detection window. In this case a slow moving peak will have a longer residence time in the detection window than a faster moving peak. So the slower moving peak will elicit a larger response. This well known phenomenon can be corrected for by dividing the peak area by its migration time. When analyzing the linearity of an analytical method it is apparent that while peak heights become non-linear the peak areas remain linearly related to the sample concentration over a wider concentration range [19]. Although integration software for HPLC may be used in CE, subsequent derivation of quantitative data should be performed using CE specific software capable of using corrected peak areas.

Quantitative data are reported as absolute concentrations for the analyses of components in a sample, e.g. drugs in biofluids or main components in pharmaceutical preparations. In the analysis of impurities in chemical formulations or in reporting the amount of excess enantiomer in chiral analyses, data may be reported as absolute concentration or as percentage area/area.

3.3.1 Quantitation of impurities and chiral analyses

In quantitation in these assays the reported peak areas must always be corrected for migration time. This is of particular importance when reporting quantitative results as chiral excess.

In pharmaceutical analysis a detection level of 0.1% area/area is accepted as a minimum requirement for a related impurities assay. In the chemical industry the acceptable impurities level depends upon the nature of the impurities and the subsequent use to which the chemical will be put.

In CE the injection volume can be easily altered usually via software control. Altria [2] has reported the use of high–low injection to determine impurity levels. In this method a first injection is made to determine the response of the main component (low injection). The injection volume is then increased to inject sufficient quantities to detect the impurities with the main component peak being off-scale (high injection). The main component peak area is then multiplied by the increased injection factor and the resultant impurity peak is then reported as percent area/area of impurity peak area/(main component peak area × injection parameter ratio). Although this obviates the need for sample dilution, it should be accompanied by determination of the linearity of the injection mechanism. In this case the linearity of injection range should also be determined using a concentration of analyte which is at the lower end of the linearity range so that detector non-linearity does not compromise determination of the injector linearity.

Where an adequate linear range is available, direct determination from a single injection may be performed. The sum of the corrected areas is used to generate data as percent area/area. Usually the peak responses are determined at single wavelengths. Figure 3.6 shows the analysis of terephthalic acid containing four minor impurities at the ppm level.

The principal area for application of chiral analysis is in enatiomeric purity testing of drug formulations. In this case peak areas must always be corrected for migration time. The lower costs associated with chiral CE analyses and the acceptable chiral excess levels have promoted their use in analysis of pharmaceutical formulations and in determinations of enantiomers in biofluids [27].

3.3.2 Internal standards

With the use of internal standards in an extracted assay, the recovery of the internal standards should be determined independently. Although 100%

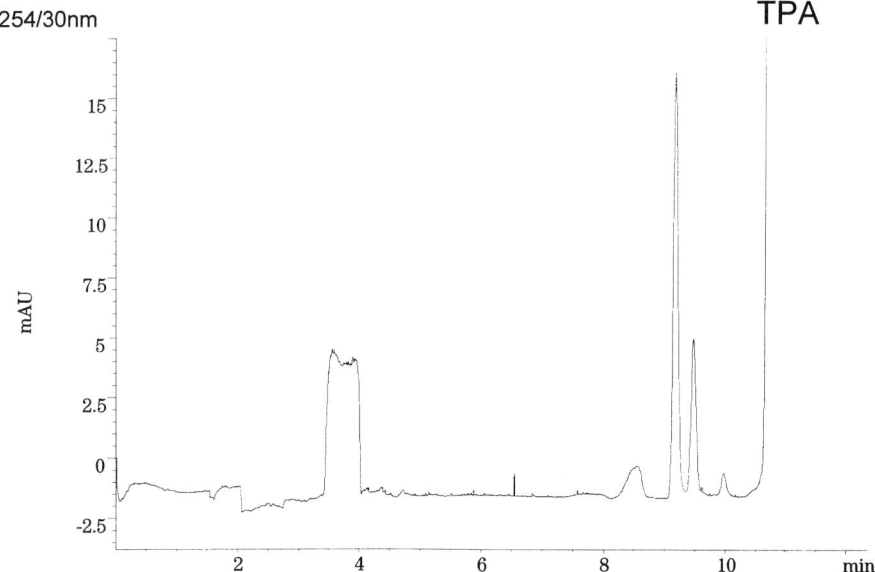

Figure 3.6 Analysis of impurities in terephthalic acid using large volume injections and extended lightpath capillaries. Buffer, 14 g l^{-1} hexane sulfonic acid; capillary, 56 cm eff × 75 μm i.d. (extended lightpath); injection, 2500 mbar s; temperature, 20°C; voltage, 25 kV; sample, 16 mg ml^{-1} TPA.

recovery is desirable, limits of not less than 50, 80 and 90% have all been used. The most important is the satisfactory reproducibility of the recovery rate.

For the internal standard method a known amount of substance is added as soon as possible in the analytical process to enable correction for sample loss during the assay. An internal standard should be well resolved from other peaks in the electropherogram. With main component assay this is easier to accomplish than in the analysis of components in a more complex sample matrix (e.g. biofluids). The internal standard should be as similar as possible to the analyte(s) of interest in terms of its chemical and physical properties. Consequently the detector response is similar to the analyte being determined. For fixed wavelength detectors the wavelength used for detection and quantification should also be chosen such that this lies on a plateau of UV absorption for both analyte and internal standard, although this is of greater importance with double beam variable wavelength UV detectors with moving gratings than with diode array detectors where the optical bench is fixed. With the use of a photodiode array detector where multiple wavelengths can be monitored, the sample detection wavelength may be different from that used for the internal standard. This provides a greater flexibility in choice of internal standard and in choice of detection wavelength to provide maximum absorption. Internal standards have been used in CE analysis to improve precision from the 1–2% RSD level to %RSD levels of 0.2% [1, 4, 5, 20, 26, 28].

3.3.3 Calibration

The linearity of an analytical method is its ability to produce test results which are directly, or by means of a well defined mathematical transformation, proportional to the concentrations of analytes within a given range. Usually it is important to demonstrate that a linear relationship exists between detector response and analyte concentration. Linearity should be determined by injection of about six standards of varying concentrations which cover 50 to 150% of the expected working range of the assay. In CE using direct detection methods, a linear range covering three orders of magnitude is obtainable, however, in indirect detection modes, linear ranges of two orders of magnitude are common. A linear regression analysis applied to the results should have an intercept not significantly different from zero. Errors associated with each concentration level should be determined by multiple injection at each concentration level. Baumann and Watzig [26] investigated the use of linear regression analysis of concentration ranges used to calculate analyte concentration. According to their study the standard deviation of detector response was dependent on concentration of compounds of interest, therefore ordinary least squares analysis should be used where the concentration range differs by a factor of 5. Where the concentration range differs by more than a factor of 50 then weighted least squares regression must be used and is advised down to a concentration range of 5.

3.3.3.1 Calibration samples. When using hydrodynamic injection, it is an important premise that the calibration samples and unknown samples have the same viscosity [7]. This can be of particular importance when directly injecting biofluids such as saliva or plasma whose viscosity may differ from that of aqueous calibration solutions [3]. When electrokinetic injection is used, the calibration and unknown samples should be carefully matched not only to the unknown sample matrix but also to the run buffer [7, 24, 29] and two internal standards may be necessary [20]. In these cases it is advisable to use a blank sample matrix to construct the calibration solutions.

3.3.4 Accuracy

The accuracy of the data reflects the degree of agreement between test results generated by the assay and the true value. The results can be compared to a reference method although this assumes no error in the reference method used. Frequently, correlation between two methods is used to determined the accuracy of an assay although this in fact measures only the strength of a relationship between two variables and not the agreement between them.

More correctly, orthogonal correlation or correlation of the difference between means and the mean ratio should be determined [30]. Alternatively spiking the sample matrix of interest with a known amount of analyte and comparing the results with the true value can be used to assess the accuracy of a system. Accuracy may also be determined for pharmaceutical formulations by comparison of results with the label claim.

Where sample extraction is used, recovery experiments should be performed to assess the loss of sample during the extraction process. The efficiency of the extraction may be expressed as percentage of response to a pure standard which had not been exposed to the sample extraction process and is defined as absolute recovery. Relative recovery is measured by comparing the response of the analyte matrix (e.g. plasma) to the response of the analyte in pure solvent (e.g. water) [3, 7].

3.4 Conclusions

CE is a quantitative separation technique. CE may be used to derive quantitative data from a number of separation modes. Migration time precision in CE can be optimized using appropriate procedures to give %RSD (relative standard deviation) less than 1%. In some cases if internal standards are used, this can be reduced to <0.5% RSD. CE may be used for the quantitative determination of analytes in a variety of sample matrices.

References

1 Altria, K.D. (1993) Quantitative aspects of the application of capillary electrophoresis to the analysis of pharmaceuticals and drug related impurities. *J. Chromatogr.* **646**, 245–257.
2 Altria, K.D. (1993) Determination of salbutamol related impurities by capillary electrophoresis. *J. Chromatogr.* **634**, 323–328.
3 Perrett, D.; Ross, G.A. (1995) Rapid determination of drugs in biofluids by capillary electrophoresis. Measurement of antipyrine in saliva for pharmacokinetic studies. *J. Chromatogr.* **700**, 179–185.
4 Altria, K.D.; Harden, R.C.; Hart, M.; Hevizi, J.; Hailey, P.A.; Makwana, J.V.; Portsmouth, M.J. (1993) Inter-company cross-validation exercise on capillary electrophoresis. I Chiral analysis of clenbuterol. *J. Chromatogr.* **641**, 147–153.
5 Altria, K.D.; Clayton, N.G.; Hart, M.; Harden, R.C.; Hevizi, J.; Makwana, J.V.; Portsmouth, M.J. (1994) An inter-company cross-validation exercise on capillary electrophoresis testing of dose uniformity of paracetamol content in formulations. *Chromatographia* **39**, 180–184.
6 Lambert, W.J.; Middleton, D.L. (1990) pH hysteresis effect with silica capillaries in capillary zone electrophoresis. *Anal. Chem.* **62**, 1585–1587.
7 Ross, G.A. (1994) Biomedical applications of capillary electrophoresis including the analysis of structural haemoglobinopathies. *PhD Thesis*, University of London.

REFERENCES

8. McCormick, R.M. (1988) Capillary zone electrophoretic separation of peptides and proteins using low pH buffers in modified silica capillaries. *Anal. Chem.* **60**, 2322–2328.
9. Herold, M.; Ross, G.A.; Grimm, R.; Heiger, D.N. (1966) Separation of peptides and protein digests by capillary electrophoresis. In *Methods in Molecular Biology, Vol. XX: Capillary Electrophoresis*, Altria, K.D. (ed.), Human Press, Totoiwa, N.J.
10. Bello, M.S.; Capelli, L.; Righetti, P.G. (1994) Dependence of the electroosmotic mobility on the applied electric field and its reproducibility in capillary electrophoresis. *J. Chromatogr.* **684**, 311–322.
11. Wood, P. (1996) Personal communication.
12. Coufal, P.; Stulik, K.; Claessens, H.A.; Cramers, C.A. (1994) The magnitude and reproducibility of the electroosmotic flow in silica capillary tubes. *J. High Resolut. Chromatogr.* **17**, 325–334.
13. Cohen, N.; Grushka, E. (1994) Controlling electroosmotic flow in capillary electrophoresis. *J. Chromatogr.* **678**, 167–175.
14. Ross, G.A. (1996) Voltage pre-conditioning technique for optimisation of migration time reproducibility in capillary electrophoresis. *J. Chromatogr.* **718**, 444–447.
15. Watzig, H.; Dette, C. (1993) Precise quantitative capillary electrophoresis (CE): Methodological and instrumental aspects. *J. Chromatogr.* **636**, 31–38.
16. Altria, K.D.; Fabre, H. (1995) Approaches to optimisation of precision in capillary electrophoresis. *Chromatographia* **40**, 313–320.
17. Swedberg, S.A.; Miller, C.A. (1994) Protein analysis via phosphate-deactivated fused silica capillaries using salt additives. Hewlett-Packard Application Note 12-5962-9965E.
18. Zhu, T.; Sun, Y.; Zhang, C.; Ling, D.; Sun, Z. (1994) Variation of the pH of the background electrolyte and a result of electrolysis in capillary electrophoresis. *J. High Resolut. Chromatogr.* **17**, 563–564.
19. Watzig, H. (1995) Appropriate calibration functions for capillary electrophoresis I. Precision and sensitivity using peak areas and heights. *J. Chromatogr.* **700**, 1–8.
20. Dose, E.V.; Guiochon, G. (1992) Problems of quantitative injection in capillary zone electrophoresis. *Anal. Chem.* **64**, 123–128.
21. Fishman, H.A.; Amudi, N.M.; Lee, T.T.; Scheller, R.H.; Zare, R.N. (1994) Spontaneous injection in microcolumn separations. *Anal. Chem.* **66**, 2318–2329.
22. Thomas, B.R.; Fang, X.G.; Chen, X.; Tyrrell, R.J.; Ghodbane, S. (1994) Validated micellar electrokinetic capillary chromatography method for quality control of the drug substances hydrochlorothiazide and chlorothiazide. *J. Chromatogr.* **657**, 383–394.
23. Knox, J.H.; McCormack, K. (1994) Temperature effects in capillary electrophoresis. 1. Internal capillary temperature and effect upon performance. *Chromatographia* **38**, 207–221.
24. Dose, E.V.; Guiochon, G. (1991) Internal standardisation technique for capillary zone electrophoresis. *Anal. Chem.* **63**, 1154–1158.
25. Altria, K.D.; Filbey, S.D. (1993) *J. Liq. Chromatogr.* **633**, 221–225.
26. Baumann, K.; Watzig, H. (1995) Appropriate calibration functions for capillary electrophoresis II. Heteroscedasticity and its consequences. *J. Chromatogr.* **700**, 9–20.
27. Prunonosa, J.; Obach, R.; Diez-Cascon, A.; Gouesclou, L. (1992) Determination of cicletanine enantiomers in plasma by high-performance capillary electrophoresis. *J. Chromatogr.* **574**, 127–133.
28. Altria, K.D.; Goodall, D.M.; Rogan, M.M. (1994) Quantitative determination of drug counter-ion stoichiometry by capillary electrophoresis. *Chromatographia* **38**, 637–642.
29. Srivatsa, G.S.; Batt, M.; Schuette, J.; Carlson, R.H.; Fitchett, J.; Lee, C.; Cole, D.L. (1994) Quantitative capillary gel electrophoresis assay of phosphorothioate oligonucleotides in pharmaceutical formulations. *J. Chromatogr.* **680**, 469–477.
30. Bland, J.M.; Altman, D.G. (1986) Statistical methods for assessing agreement between two methods of clinical measurement. *Lancet* **i**, 307–310.
31. Nielen, M.W.F. (1991) Quantitative aspects of indirect UV detection in capillary zone electrophoresis. *J. Chromatogr.* **588**, 321–326.

4 Column technology
K. KITAGISHI

A capillary plays two main roles in capillary electrophoresis. One is as a column, its internal volume providing a separation space in electrophoresis. The other is as a pump, the silanol groups on its inner surface yielding electroosmotic flow (EOF). Remarkable progress has been made in making improvements in capillaries to achieve high quality and uniformity in CE. However, more improvements in several properties of capillaries are for further advancement. In this chapter, recent developments in column technology are described including how to control EOF, the method of coating the surface of inner wall of the capillary, gel-filled columns, connection of capillaries and design of capillary shape.

4.1 Electroosmotic control

EOF is a well-known concept in CE, because most commercial CE systems are equipped with fused-silica capillaries yielding EOF. Although fused-silica capillaries have many advantages, i.e. durability, easy handling, diversity of manipulation, stable supply at low prices, etc., fluctuation in EOF due to surface conditions causes poor reproducibility. Andreev and Lisin reported a mathematical model based on the concept of a double electrical layer that could explain the dependence of the EOF profile on buffer concentrations and capillary diameters [1]. A porous gel model of the silica–solution interface was applied to evaluate the pH dependence of EOF mobility [2]. For measurement of EOF velocity, neutral markers are generally used. Typical organic solvents could be employed both for the dissolution of analytes and for the estimation of electroosmotic velocity in micellar electrokinetic chromatography (MEKC) [3]. Direct monitoring devices of EOF have been developed using a conductivity cell [4] and weighing the amount of liquid transferred [5].

The electroosmotic properties of capillaries made from other polymers were evaluated. A hollow polypropylene (PP) fiber has a lower EOF than fused silica however PP easily adsorbs surface modifiers such as non-ionic surfactants due to its strong hydrophobicity [6]. The EOF generated in capillaries of polyfluorocarbon, polyethylene, poly(vinylchloride) [7], polytetrafluoroethylene and polytetrafluoroethylene–polyhexafluoropropylene copolymer [5] was measured as a function of pH.

Direct control of EOF using an additional electric field applied from outside of the capillary was studied by several researchers. The control of EOF by a radial potential gradient demonstrated by Lee and co-workers increased the separation efficiency of peptides and proteins due to reduction in their adsorption onto the capillary wall [8] as well as increasing the separation efficiency in MEKC of phenylthiohydantoin amino acids [9]. An attempt was made to simplify the instrumental design by using ionized air as the conductive medium for applying external electric field instead of the buffer solution [10]. Adsorbed tin ions and dodecyltrimethylammonium bromide at the silica surface caused a shielding effect of radial potential gradient on the direct control of EOF [11]. Ewing and co-workers demonstrated EOF control by application of radial voltage with a conductive sheath onto the polyimide outer surface of the capillary. Both the Nefion ionomer film formed by painting and drying solution over a maximum available surface area [12, 13] and a small portion of dried silver paint [14] gave sufficient control of EOF. The usefulness of EOF control was demonstrated by the remarkable separation of a peptide mixture at low pH as well as by the reduction in the analytical time [13].

4.2 Techniques for coating the capillary surface

In order to suppress EOF or diminish protein adsorption, methods of coating the capillary surface with surfactants and polymers have been developed. Belder and Schomburg classified capillary coating methods into dynamic coating and permanent coating [15]. The former implies that modifiers are attached to the capillary surface through adsorption or electrostatic interaction by prewashing with buffer containing the modifier. The latter indicates that the coating is effected either through covalent bonds between modifier and silanol group on the capillary wall or by adsorption of modifier followed by the immobilization. Other capillary surface coatings are either covalently bonded or non-covalently bonded.

Surface modifiers are linked to the capillary wall by covalent bonds in several coating methods to give a permanent covalently bonded coating. When separation is achieved using a capillary conditioned by prewashing with cationic surfactant or amphipathic polymer solution and electrolytes are used in which there may be or may not be modifiers, the coating method is called dynamic non-covalent bonding. Permanent, non-covalently bonded capillaries are those to which modifiers are immobilized by heating or forced drying following adsorption by prewashing of the modifier solution.

Examples of applications of capillary coating are listed in Table 4.1. For other publications on dynamic and non-covalently bonded coating with cationic surfactants refer to Chapter 5 on Electrolyte Systems. The linear

Table 4.1 Recent publications on capillary surface coating

Coating reagent	Coating method	Samples used for evaluation (separation mode)	Remarks	Ref.
Cationic surfactant FC135	Dynamic	Protein standard mixture (FZE)	KOH prewash, FC135 in washing buffer and electrolyte	57
FC135	Dynamic	Antagonist G (peptide) (FZE)	FC135 in washing buffer and electrolyte	58
Methylcellulose (MC)	Dynamic	Human hemoglobin variants (CIEF)	MC is added to the catholyte	59
Cellulose acetate; cellulose triacetate; crosslinked hydroxypropyl cellulose	Permanent and non-covalently bonded	Protein mixture (FZE)	Drying immobilization under He stream following the flush with coating reagent; stable and reproducible in the pH range of 2 to 7.5	60
Poly(ethylene oxide) (PEO)	Dynamic	DNA sequencing (LPS)	Prewash and molecular sieving with PEO	61
Chitosan	Dynamic	Basic proteins (FZE)	Chitosan in preconditioning buffer and electrolyte	62
Chitosan	Permanent and non-covalently bonded	Basic proteins (FZE)	Drying immobilization by forced air	63
Polybren	Dynamic	Chiral basic drugs; chiral organic acids (CD-EKC)	Modifiers in washing buffer but not in electrolyte	15
Poly(vinylalcohol) (PVA)	Permanent and non-covalently bonded	Chiral basic drugs; chiral organic acids (CD-EKC)	Thermal immobilization	15
Poly(vinylalcohol) (PVA)	Dynamic	Protein standards (FZE)	Including the results of hydroxyethyl cellulose (HEC), hydroxypropyl methyl cellulose (HPMC) and dextran as modifiers	64
Poly(vinylalcohol) (PVA)	Permanent and non-covalently bonded	Protein standards (FZE)	Thermal immobilization	64

Poly(vinylalcohol) (PVA)	Dynamic	Basic proteins (FZE)	PVA in electrolyte; comparable performance with permanently coated capillaries	65
2-Acrylamido-2-methyl-1-propanesulfonic acid (AMPSA)	Covalently bonded	Organic acids, phenols (MEKC), amino acids (CD-EKC)	Coating by siloxane bonding; pH-independent EOF	66, 67
Polyethylene; poly(ethylene glycol); poly(ethylene propylene glycol)	Permanent and non-covalently bonded	Basic proteins (FZE)	Thermal immobilization; poor resolution for polyethylene-coated capillary	68
Polyacrylamide; poly(hydroxypropyl-cellulose)	Covalently bonded	Basic proteins (FZE)	Coating by siloxane bonding	68
Vinyl group	Covalently bonded	Benzylalcohol (FZE)	Coating by siloxane bonding; comparison of capillaries from different suppliers	69
Polyacrylamide (PAA)	Covalently bonded	Proteins; peptides; DNA molecules (FZE)	PAA coating to the highly crosslinked polysiloxane–diol sublayer	70
Methylcellulose	Covalently bonded	Proteins (FZE, CIEF)	Coating by siloxane bonding; highly stable using α-glycidoxypropyltrimethoxysilane	71
Epoxy-diol layer with oxiranes	Covalently bonded	Basic proteins (FZE)	Epoxypolymer bonded through glyceryl propyl silane	72
Polyethyleneglycol; polyethyleneimine	Permanent and non-covalently bonded	Proteins (FZE)	Thermal immobilization; evaluation of coating by streaming potential and frontal chromatography; poor performance for polyethyleneimine coating	73
Cyclodextrin	Covalently bonded	Epinephrine enantiomers (FZE with chiral recognized capillary)	Immobilization of cyclodextrin on polyacrylamide coated capillary	74

Table 4.1 *Continued*

Coating reagent	Coating method	Samples used for evaluation (separation mode)	Remarks	Ref.
Polyethylene oxide–polypropylene oxide–polyethylene oxide triblock copolymer	Covalently bonded	Protein mixture (FZE)	Coating by siloxane bonding	75
Multilayered coating with hydroxypropylcellulose, polyethyleneglycol and polyethyleneimine	Covalently bonded	Proteins and peptides; oligosaccharides; nucleotides (FZE)	Coating by siloxane bonding; capillaries with no EOF, weak anodal EOF and strong anodal EOF	76
Polyacrylamide	Covalently bonded	Oligonucleotides (LPS)	Through Si—C linkage	77
Polyethyleneglycol	Covalently bonded	Methanol (MEKC)	Coating by siloxane bonding; Carbowax 20M is used for coating with polyethyleneglycol	78
Polymethylhydroxysilane; polymethyloctadecyl-silane; polydimethylsilane C-1; C-18	Covalently bonded	Methanol (FZE)	H; C-1; C-18 groups on a siloxane backbone layer	78
C-1; C-18	Covalently bonded	Lanthanide (FZE)	Trimethylsilane and dimethyloctadecylsilane are used for bonding reaction of C-1 and C-18 capillaries, respectively	79
C-18 + surfactants or polymers	Covalently bonded (C-18), dynamic (surfactants or polymers)	Protein mixture (FZE, CIEF)	Octadecyltrichlorosilane in toluene is used for bonding reaction; electroosmotic flow manipulation	80
Multilayered coating of positively charged quaternary ammonium group and polyethylene glycol	Covalently bonded	Nucleosides; basic proteins; oligosaccharides (FZE)	Coating by siloxane bonding	81
Titanium oxide; aluminum oxide	Covalently bonded	Lysozyme; myoglobin (FZE)	Coating by Ti—O—Si and Al—O—Si; resistant to alkaline solution	82

acrylamide coating onto the capillary surface through the siloxane bond was firstly developed by Hjertén in 1985 [16]. Most coating procedures in which modifiers are covalently bound to the capillary wall via the siloxane bond are based on the ideas of Hjertén. Polyacrylamide layer coating on the capillary inner surface by the method of Hjertén [16] is known to be stable under acidic conditions, but easily hydrolyzed under alkaline conditions. The coating by Si—C bonding developed by Cobb et al. is more stable, but the processes are less mild [17]. Several commercial coated capillaries are now available, e.g. neutral hydrophilic coated capillaries [18–20] and polyamine-coated capillaries [21].

4.3 Gel-filled capillaries

Capillary gel electrophoresis (CGE) is undoubtedly one of the fields of greatest potential in CE because an extremely high separation efficiency is expected due to less diffusion in the gel than in the solution. Moreover, its fields of application, analyses of DNA and sodium dodecyl sulfate (SDS)–protein mixtures are key technologies in the biochemical and biomedical fields. For conventional gel electrophoresis, polyacrylamide (PAA) and agarose have been used as a gel. In CGE, PAA gel is mostly applied because of its high resolution and its probable capacity to control the polymerization of acrylamide monomers. Generally, PAA gel-filled capillaries are prepared by polymerization of acrylamide monomers with N,N'-methylenebis (acrylamide) (BIS) as a crosslinker, similar to the conventional procedures for preparation of PAA slab gels. Appropriate amounts of acrylamide and BIS are dissolved in the buffer solution which is degassed carefully. An aliquot of freshly prepared solution containing initiator and catalyst, ammonium peroxydisulfate and N,N,N',N'-tetramethylethylenediamine (TEMED), respectively, is added to acrylamide solution to initiate polymerization. The polymerizing solution quickly fills the capillary and the polymerization is completed in a few hours. In order to stabilize the gel, a bifunctional reagent, e.g. 3-methacryloxypropyltrimethoxysilane, is often employed to attach the gel to the capillary inner wall. The ends of the gel-filled capillaries are trimmed prior to the run.

Although PAA gel has high separation efficiency, it also has several disadvantages, i.e. air bubble formation, instability under alkali conditions, poor durability, and so on. Preparation of PAA gel-filled capillaries was improved to give stable, reproducible measurements by using special equipment to inject a polymerizing solution of acrylamide into the capillaries [22]. Swerdlow et al. tried to improve gel instability, aiming at automated sequencing of DNA [23]. They prepared a sample solution which was treated with uracyl DNA glycosylase, replaced the denaturant urea with formamide and added acrylamide monomers and formamide to the running buffer. PAA gel-filled capillaries containing urea for DNA denaturation are available commercially.

Righetti and co-workers [24–31] investigated the polymerization of PAA gels by capillary zone electrophoresis (CZE). Since the redox couple of peroxydisulfate and TEMED produces N-oxides by oxidizing the amino groups in the polymerizing solution and since the N-oxides attack the –SH residues transforming them into S—S bridges, the procedures used to polymerize acrylamide described above are not suitable for protein analysis. Conditions for photopolymerization of acrylamide must be established which have a high efficiency. The kinetics of acrylamide polymerization by riboflavin [24–26] or methylene blue [27–29] instead of peroxydisulfate were studied, as functions of temperature, oxygen partial pressure, catalyst concentration and pH. Moreover, PAA gel formation with substituted acrylamide monomers and other crosslinkers was investigated in order to change the pore size or hydrophobicity of gels [29–31]. The polymerization of a PAA gel could be followed by using Raman microprobe spectroscopy [32], while Righetti and co-workers monitored it by UV detection in CE.

A highly condensed PAA gel capillary of more than 15% PAA is much more difficult to prepare than the more common capillary with a PAA concentration of less than 10%. Chen *et al.* studied optimization of polymerization for highly condensed PAA gel-filled capillaries by applying a slight pressure and selecting the buffer components [33].

Coiling of capillaries is supposed to affect the column efficiency in CGE more than free zone capillary electrophoresis (FZE) due to the immobilized structure of the matrix. The influence of capillary coiling in CGE was examined as a function of the content of gel [34].

Some separation conditions were defined to optimize CGE. For DNA sequencing, the resolution was optimized by functions of the electric field strength [35, 36] and capillary length [36]. Luckey and Smith proposed a modified model of the Ogston sieving theory for the migration behavior of single-stranded DNA in a gel-filled capillary [37].

To overcome the lower detection limit in the UV region due to the strong UV absorption of PAA gel, PAA gel-filled capillaries with UV transparent buffer at the detection window have been designed [38]. These capillaries are prepared by partial or step gradient filling of gel and by coupling a gel-filled capillary with a buffer-filled capillary.

In capillary affinity gel electrophoresis, a new field in CE, receptor molecules are mostly bound to the gel. Selective electrophoresis of drugs was possible with imprinted polymers prepared by template-dispersion polymerization [39].

4.4 Column connection

Two capillaries with different properties are connected to obtain a new capability for separation.

An uncoated capillary was coupled to polyethylene glycol coated capillary in order to control EOF for rapid protein analysis with high separation efficiency [40, 41]. When a separation capillary was connected to a second, auxiliary capillary to which high voltage was independently applied, the migration velocity in the separation capillary was controlled by the applied voltage to the second capillary. This technique gives improvements in the electrostacked band profile and the separation efficiency for charged solutes [42]. Fraction collection of separated analytes is achieved by connection of a separation capillary to several capillaries with different EOF mobilities by a multiport sliding valve [43].

Combination of a stationary phase-packed column and an open-tubular capillary offers enhanced selectivity in CE. Separation of polar nitrosamines was improved by using diol, aminocyano, C-18 and silica as the packing materials used for supports [44]. A solid phase-packed column was used for preconcentration prior to analyte separation by CE. The coupled solid phase column can alter the EOF mobility and the peptide separation properties in the separation capillary [45].

Two-dimensional separation of peptides and proteins was studied by developing a coupled column for size exclusion chromatography (SEC) and capillary zone electrophoresis (CZE) [46]. A transverse flow gating interface was designed to suppress extra band broadening with sample injection [47]. Reversed phase liquid chromatography (LC) was coupled to CZE, in a similar way to SEC–CZE coupling. This instrumentation provides a unique two-dimensional map with great resolving power [48].

When a separation capillary in FZE was coupled with immobilized capillary enzyme reactors, a mixture of oligonucleotides which were not resolved in FZE could be separated. On the capillary inner wall of the enzyme reactors, ribonuclease T1, hexokinase and adenosine deaminase were immobilized through Shiff base covalent linkage using conjugation between glutaraldehyde and a terminal amine [49]. This immobilization method is a conventional procedure described in the literature.

4.5 Novel column designs

Rectangular capillaries should become available commercially. Higher speed analysis is possible with a rectangular column than with a cylindrical one, deriving from better heat dissipation [50]. The EOF profile in a rectan-

COLUMN TECHNOLOGY

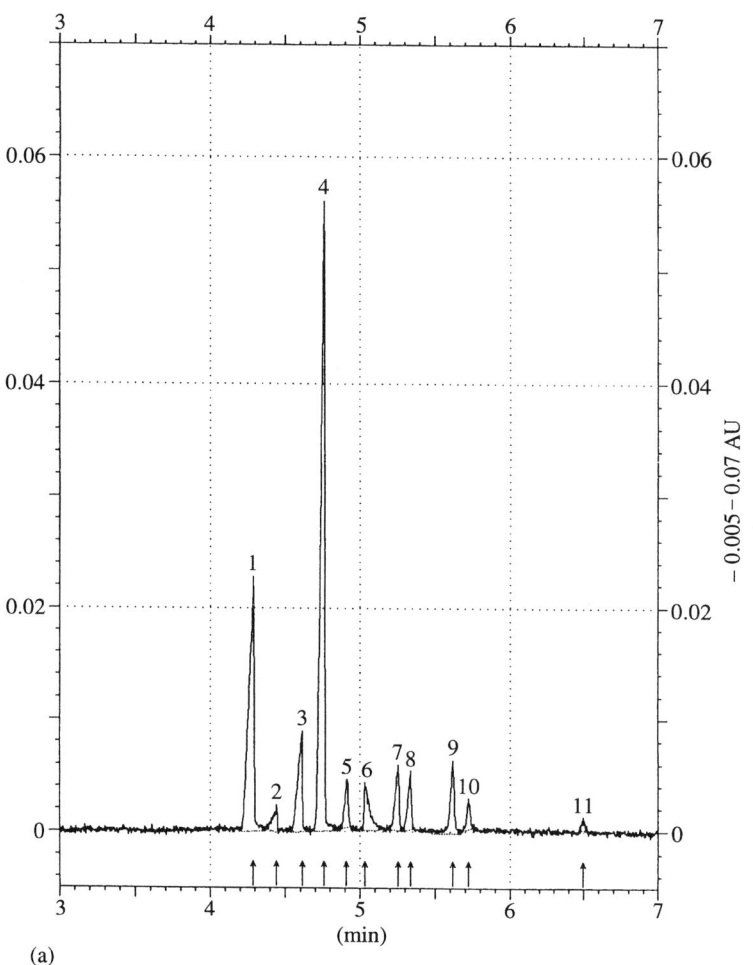

CAPI-3100	Integrator	(peak)	correct		Otsuka Electronics
6MBK1011.DT3	06/09/95	13:09:59		Sample name	10mix 50ppm
Y-scale			0.075 AU/FS	Paper speed	30 mm/min
Sampling time			90 ms * 2	Baseline	3–7 min
Sense			normal	Resolution	5 nm
Time range			3–7 min	Interval	0.18 s
Smoothing			5 points	Slope	0.005 AU/min
Drift			0.001 AU/min	Height	0.002 AU
Width			0.001 min	Min. area	0.00015 AU*min
Time double			30 min	Minus peak	OFF

(a)

Figure 4.1 Electropherograms of organic acids in FZE at 25°C. (a) A normal capillary and (b) a capillary with bubble-shaped detection window (unpublished data). Peaks are identified as (1) oxalic acid, (2) formic acid, (3) malonic acid, (4) maleic acid, (5) L-malic acid, (6) citric acid, (7) acetic acid, (8) glycolic acid, (9) propionic acid and (10), (11) lactic acid. Sample: 50 ppm each in distilled water. Electrolyte: 50 mM sodium tetraborate including 0.1 mM

NOVEL COLUMN DESIGNS

CAPI-3100 Integrator (peak) correct			Otsuka Electronics
COOH1011.DT3	06/22/95 19:37:13	Sample name	10mix 50ppm
Y-scale	0.075 AU/FS	Paper speed	30 mm/min
Sampling time	55 ms * 3	Baseline	3–7 min
Sense	normal	Resolution	2 nm
Time range	3–7 min	Interval	0.16 s
Smoothing	5 points	Slope	0.005 AU/min
Drift	0.001 AU/min	Height	0.0005 AU
Width	0.001 min	Min. area	0.0001 AU*min
Time double	30 min	Minus peak	OFF

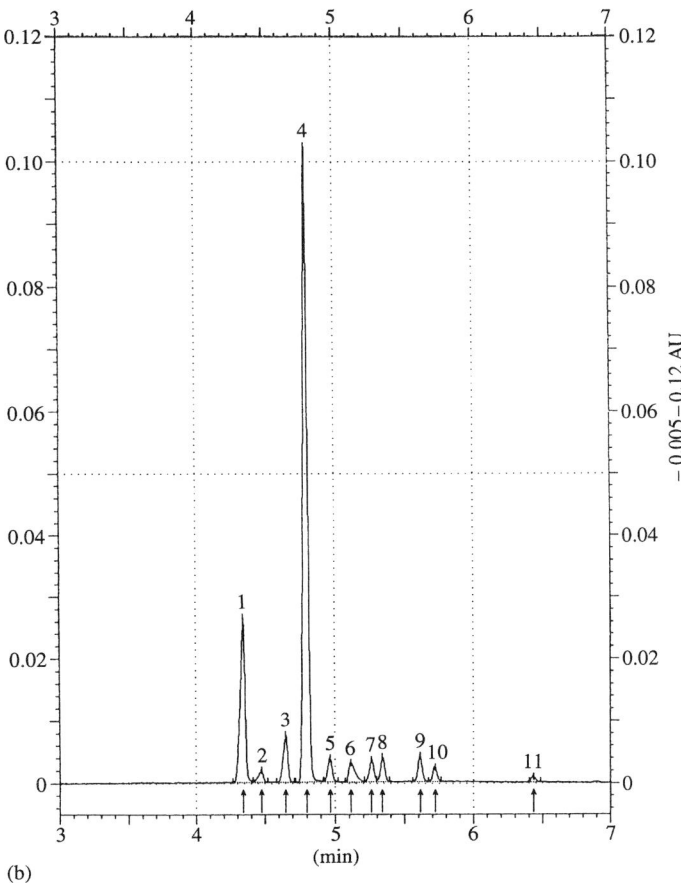

Figure 4.1 *Continued.*

hexadecyltrimethylammonium sulfate and 10% polyethylene glycol. Capillary: 75 µm inner diameter, 60 cm length. Sample injection: gravity flow injection with a difference in height of 20 mm for (a) 120 s or (b) 80 s. Migration: −15 kV applied potential. Detection: on-line UV absorption at 195 nm. The detector is placed at 47.5 cm from the inlet end of the capillary.

gular capillary was observed with a microscope charge coupled device camera–video system [51].

A capillary column which has an extended optical path compensates for one disadvantage of CE, i.e. poor concentration detection limits due to a short light path. A Z-shaped cell with a ball lens could give more than tenfold sensitivity enhancement [52]. A postcolumn cell with an extended optical path was developed and confirmed for enhanced sensitivity for UV–vis detection [53]. A bubble-shaped cell capillary could also improve the detection limit [54]. Although signal enhancement with a bubble-shaped capillary is less than ten times that of a Z-shaped cell capillary, the resolution loss due to a change in geometry is smaller. An example of improvement in detection sensitivity accompanied by noise reduction is shown in Figure 4.1 for a bubble-shaped capillary. Some capillaries with an extended optical path are available commercially.

A theoretical model of electrofocusing in a tapered capillary has been presented and the feasibility of application of this technique has been suggested [55].

A microconcentric column with a thin separation chamber of ca. 95 µm has been developed for CE. The column was prepared by using a smaller diameter capillary to insert a line of synthetic resin into a larger capillary. It has been suggested that micropreparative CE will become feasible with the use of a microconcentric column [56].

References

1. Andreev, V.P.; Lisin, E.E. (1992) Investigation of the electroosmotic flow effect on the efficiency of capillary electrophoresis. *Electrophoresis* **13**, 832–837.
2. Huang, T.-L. (1993) On the pH hysteresis of electroosmotic mobility with capillary zone electrophoresis in silica capillary. *Chromatographia* **35**, 395–398.
3. Ahuja, E.S.; Little, E.L.; Foley, J.P. (1992) Selected organic solvents as electroosmotic velocity markers in micellar electrokinetic capillary chromatography. *J. Liq. Chromatogr.* **15**, 1099–1113.
4. Wanders, B.J.; van de Goor, T.A.A.M.; Everaerts, F.M. (1993) On-line measurement of electroosmosis in capillary electrophoresis using a conductivity cell. *J. Chromatogr.* **652**, 291–294.
5. Rohlíček, V.; Deyl, Z.; Mikšík, I. (1994) Determination of the isoelectric point of the capillary wall in capillary electrophoresis. Application to plastic capillaries. *J. Chromatogr.* **662**, 369–373.
6. Nielen, M.W.F. (1993) Capillary zone electrophoresis using a hollow polypropylene fiber. *J. High Resolut. Chromatogr.* **16**, 62–64.
7. Schützner, W.; Kenndler, E. (1992) Electrophoresis in synthetic organic polymer capillaries: Variation of electroosmotic velocity and ζ potential with pH and solvent composition. *Anal. Chem.* **64**, 1991–1995.
8. Wu, C.-T.; Lopes, T.; Patel, B.; Lee, C.S. (1992) Effect of direct control of electroosmosis on peptide and protein separations in capillary electrophoresis. *Anal. Chem.* **64**, 886–891.
9. Tsai, P.; Patel, B.; Lee, C.S. (1993) Direct control of electroosmosis and retention window in micellar electrokinetic capillary chromatography. *Anal. Chem.* **65**, 1439–1442.

REFERENCES

10 Wu, C.-T.; Lee, C.S.; Miller, C.J. (1992) Ionized air for applying radial potential gradient in capillary electrophoresis. *Anal. Chem.* **64**, 2310–2311.
11 Huang, T.-L.; Tsai, P.; Wu, C.-T.; Lee, C.S. (1993) Mechanistic studies of electroosmotic control at the capillary–solution interface. *Anal. Chem.* **65**, 2887–2893.
12 Hayes, M.A.; Ewing, A.G. (1992) Electroosmotic flow control and monitoring with an applied radial voltage for capillary zone electrophoresis. *Anal. Chem.* **64**, 512–516.
13 Hayes, M.A.; Kheterpal, I.; Ewing, A.G. (1993) Effects of buffer pH on electroosmotic flow control by an applied radial voltage for capillary zone electrophoresis. *Anal. Chem.* **65**, 27–31.
14 Hayes, M.A.; Kheterpal, I.; Ewing, A.G. (1993) Electroosmotic flow control and surface conductance in capillary zone electrophoresis. *Anal. Chem.* **65**, 2010–2013.
15 Belder, D.; Schomburg, G. (1994) Chiral separations of basic and acidic compounds in modified capillaries using cyclodextrin-modified capillary zone electrophoresis. *J. Chromatogr.* **666**, 351–365.
16 Hjertén, S. (1985) High-performance electrophoresis. Elimination of electroendosmosis and solute adsorption. *J. Chromatogr.* **347**, 191–198.
17 Cobb, K.A.; Dolnik, V.; Novotny, M. (1990) Electrophoretic separations of proteins in capillaries with hydrolytically stable surface structures. *Anal. Chem.* **62**, 2478–2483.
18 Huang, T.-L.; Shieh, P.C.H.; Koh, E.V.; Cooke, N. (1994) Evaluation of a neutral hydrophilic coated capillary for capillary zone electrophoretic separation of proteins. *J. Chromatogr.* **685**, 313–320.
19 Piccoli, G.; Fiorani, M.; Biagiarelli, B.; Palma, F.; Potenza, L.; Amicucci, A.; Stocchi, V. (1994) Simultaneous high-performance capillary electrophoretic determination of reduced and oxidized glutathione in red blood cells in the femtomole range. *J. Chromatogr.* **676**, 239–246.
20 Cole, R.B.; Varghese, J.; McCormick, R.M.; Kadlecek, D. (1994) Evaluation of a novel hydrophilic derivatized capillary for protein analysis by capillary electrophoresis–electrospray mass spectrometry. *J. Chromatogr.* **680**, 363–373.
21 Richards, M.P. (1994) Application of a polyamine-coated capillary to the separation of metallothionein isoforms by capillary zone electrophoresis. *J. Chromatogr.* **657**, 345–355.
22 Baba, Y.; Matsuura, T.; Wakamoto, K.; Morita, Y.; Nishitsu, Y.; Tsuhako, M. (1992) Preparation of polyacrylamide gel filled capillaries for ultrahigh resolution of polynucleotides by capillary gel electrophoresis. *Anal. Chem.* **64**, 1221–1225.
23 Swerdlow, H.; Dew-Jager, K.E.; Brady, K.; Grey, R.; Dovichi, N.J.; Gesteland, R. (1992) Stability of capillary gels for automated sequencing of DNA. *Electrophoresis* **13**, 475–483.
24 Gelfi, C.; De Besi, P.; Alloni, A.; Righetti, P.G.; Lyubimova, T.; Briskman, V.A. (1992) Kinetics of acrylamide photopolymerization as investigated by capillary zone electrophoresis. *J. Chromatogr.* **598**, 277–285.
25 Chiari, M.; Micheletti, C.; Righetti, P.G.; Poli, G. (1992) Polyacrylamide gel polymerization under non-oxidizing conditions, as monitored by capillary zone electrophoresis. *J. Chromatogr.* **598**, 287–297.
26 Caglio, S.; Righetti, P.G. (1993) On the pH dependence of polymerization efficiency, as investigated by capillary zone electrophoresis. *Electrophoresis* **14**, 554–558.
27 Righetti, P.G.; Caglio, S. (1993) On the kinetics of monomer incorporation into polyacrylamide gels, as investigated by capillary zone electrophoresis. *Electrophoresis* **14**, 573–582.
28 Caglio, S.; Righetti, P.G. (1993) On the efficiency of methylene blue versus persulfate catalysis of polyacrylamide gels, as investigated by capillary zone electrophoresis. *Electrophoresis* **14**, 997–1003.
29 Righetti, P.G.; Chiari, M.; Nesi, M.; Caglio, S. (1993) Towards new formulations for polyacrylamide matrices, as investigated by capillary zone electrophoresis. *J. Chromatogr.* **638**, 165–178.
30 Gelfi, C.; De Besi, P.; Alloni, A.; Righetti, P.G. (1992) Investigation of the properties of novel acrylamido monomers by capillary zone electrophoresis. *J. Chromatogr.* **608**, 333–341.
31 Gelfi, C.; Alloni, A.; De Besi, P.; Righetti, P.G. (1992) Investigation of the properties of acrylamide bifunctional monomers (crosslinkers) by capillary zone electrophoresis. *J. Chromatogr.* **608**, 343–348.

32. Rapp, T.L.; Kowalchyk, W.K.; Davis, K.L.; Todd, E.A.; Liu, K.-L.; Morris, M.D. (1992) Acrylamide polymerization kinetics in gel electrophoresis capillaries. A Raman microprobe study. *Anal. Chem.* **64**, 2434–2437.
33. Chen, Y.; Höltje, J.-V.; Schwarz, U. (1994) Preparation of highly condensed polyacrylamide gel-filled capillaries. *J. Chromatogr.* **680**, 63–71.
34. Wicar, S.; Vilenchik, M.; Belenkii, A.; Cohen, A.S.; Karger, B.L. (1992) Influence of coiling on performance in capillary electrophoresis using open tubular and polymer network columns. *J. Microcol. Sep.* **4**, 339–348.
35. Luckey, J.A.; Smith, L.M. (1993) Optimization of electric field strength for DNA sequencing in capillary gel electrophoresis. *Soc. Photooptic. Instrum. Eng.* **1891**, 21–26.
36. Nishikawa, T.; Kambara, H. (1994) Separation of long DNA fragments by capillary gel electrophoresis with laser-induced fluorescence detection. *Electrophoresis* **15**, 215–220.
37. Luckey, J.A.; Smith, L.M. (1993) A model for the mobility of single-stranded DNA in capillary gel electrophoresis. *Electrophoresis* **14**, 492–501.
38. Chen, Y.; Höltje, J.-V.; Schwarz, U. (1994) Preparation of highly condensed polyacrylamide gel-filled capillaries with low detection background. *J. Chromatogr.* **685**, 121–129.
39. Nilsson, K.; Lindell, J.; Norrlöw, O.; Sellergren, B. (1994) Imprinted polymers as antibody mimetics and new affinity gels for selective separations in capillary electrophoresis. *J. Chromatogr.* **680**, 57–61.
40. Nashabeh, W.; El Rassi, Z. (1992) Coupled fused silica capillaries for rapid capillary zone electrophoresis of proteins. *J. High Resolut. Chromatogr.* **15**, 289–292.
41. Nashabeh, W.; El Rassi, Z. (1993) Fundamental and practical aspects of coupled capillaries for the control of electroosmotic flow in capillary zone electrophoresis of proteins. *J. Chromatogr.* **632**, 157–164.
42. Dasgupta, P.K.; Liu, S. (1994) Auxiliary electroosmotic pumping in capillary electrophoresis. *Anal. Chem.* **66**, 3060–3065.
43. Nashabeh, W.; Smith, J.T.; El Rassi, Z. (1993) Studies in capillary zone electrophoresis with a post-column multiple capillary device for fraction collection and stepwise increase in electroosmotic flow during analysis. *Electrophoresis* **14**, 407–416.
44. Ng, C.L.; Ong, C.P.; Lee, H.K.; Li, S.F.Y. (1994) Capillary electrophoretic separation of nitrosamines using combined open-tubular and packed capillary columns. *J. Chromatogr. Sci.* **32**, 121–125.
45. Tomlinson, A.J.; Benson, L.M.; Oda, R.P.; Braddock, W.D.; Strausbauch, M.A.; Wettstein, P.J.; Naylor, S. (1994) Modification of electroosmotic flow in preconcentration–capillary electrophoresis (PC–CE). *J. High Resolut. Chromatogr.* **17**, 669–671.
46. Lemmo, A.V.; Jorgenson, J.W. (1993) Two-dimensional protein separation by microcolumn size-exclusion chromatography–capillary zone electrophoresis. *J. Chromatogr.* **633**, 213–220.
47. Lemmo, A.V.; Jorgenson, J.W. (1993) Transverse flow gating interface for the coupling of microcolumn LC with CZE in a comprehensive two-dimensional system. *Anal. Chem.* **65**, 1576–1581.
48. Larmann, Jr., J.P.; Lemmo, A.V.; Moore, Jr., A.W.; Jorgenson, J.W. (1993) Two-dimensional separations of peptides and proteins by comprehensive liquid chromatography–capillary electrophoresis. *Electrophoresis* **14**, 439–447.
49. Nashabeh, W.; El Rassi, Z. (1992) Enzymophoresis of nucleic acids by tandem capillary enzyme reactor–capillary zone electrophoresis. *J. Chromatogr.* **596**, 251–264.
50. Cifuentes, A.; Poppe, H. (1994) Rectangular capillary electrophoresis: Some theoretical considerations. *Chromatographia* **39**, 391–404.
51. Tsuda, T.; Ikedo, M.; Jones, G.; Dadoo, R.; Zare, R.N. (1993) Observation of flow profiles in electroosmosis in a rectangular capillary. *J. Chromatogr.* **632**, 201–207.
52. Moring, S.E.; Reel, R.T.; van Soest, R.E.J. (1993) Optical improvements of a Z-shaped cell for high-sensitivity UV absorbance detection in capillary electrophoresis. *Anal. Chem.* **65**, 3454–3459.
53. Kim, S.; Kim, W.; Hahn, J.H. (1994) Extended path length post-column flow cell for UV-visible absorbance detection in capillary electrophoresis. *J. Chromatogr.* **680**, 109–116.
54. Xue, Y.; Yeung, E.S. (1994) Characterization of band broadening in capillary electrophoresis due to nonuniform capillary geometries. *Anal. Chem.* **66**, 3575–3580.

55 Šlais, K. (1994) Model of electrophoretic focusing in a natural pH gradient moving in a tapered capillary. *J. Chromatogr.* **684**, 149–161.
56 Fujimoto, C.; Matsui, H.; Sawada, H.; Jinno, K. (1994) The use of a microconcentric column in capillary electrophoresis. *J. Chromatogr.* **680**, 33–42.
57 Muijselaar, W.G.H.M.; de Bruijn, C.H.M.M.; Everaerts, F.M. (1992) Capillary zone electrophoresis of proteins with a dynamic surfactant coating. Influence of a voltage gradient on the separation efficiency. *J. Chromatogr.* **605**, 115–123.
58 Reubsaet, J.L.E.; Beijnen, J.H.; Bult, A.; Teeuwsen, J.; Koster, E.H.M.; Waterval, J.C.M.; Underberg, W.J.M. (1994) Reversed-phase high-performance liquid chromatography and capillary electrophoresis in the stability study of the neuropeptide growth factor antagonist [Arg^6,D-Trp^{7,9},MePhe^8]-Substance P{6-11}: A comparative study. *Anal. Biochem.* **220**, 98–102.
59 Molteni, S.; Frischknecht, H.; Thormann, W. (1994) Application of dynamic capillary isoelectric focusing to the analysis of human hemoglobin variants. *Electrophoresis* **15**, 22–30.
60 Busch, M.H.A.; Kraak, J.C.; Poppe, H. (1995) Cellulose acetate-coated fused-silica capillaries for the separation of proteins by capillary zone electrophoresis. *J. Chromatogr.* **695**, 287–296.
61 Fung, E.N.; Yeung, E.S. (1995) High-speed DNA sequencing by using mixed poly(ethylene oxide) solutions in uncoated capillary columns. *Anal. Chem.* **67**, 1913–1919.
62 Yao, Y.J.; Li, S.F.Y. (1994) Capillary zone electrophoresis of basic proteins with chitosan as a capillary modifier. *J. Chromatogr.* **663**, 97–104.
63 Sun, P.; Landman, A.; Hartwick, R.A. (1994) Chitosan coated capillary with reversed electroosmotic flow in capillary electrophoresis for the separation of basic drugs and proteins. *J. Microcol. Sep.* **6**, 403–407.
64 Gilges, M.; Kleemiss, M.H.; Schomburg, G. (1994) Capillary zone electrophoresis separations of basic and acidic proteins using poly(vinyl alcohol) coatings in fused silica capillaries. *Anal. Chem.* **66**, 2038–2046.
65 Gilges, M.; Husmann, H.; Kleemiss, M.-H.; Motsch, S.R.; Schomburg, G. (1992) CZE separations of basic proteins at low pH in fused silica capillaries with surfaces modified by silane derivatization and/or adsorption of polar polymers. *J. High Resolut. Chromatogr.* **15**, 452–457.
66 Sun, P.; Landman, A.; Barker, G.E.; Hartwick, R.A. (1994) Synthesis and evaluation of anionic polymer-coated capillaries with pH-independent electroosmotic flows for capillary electrophoresis. *J. Chromatogr.* **685**, 303–312.
67 Landman, A.; Sun, P.; Hartwick, R.A. (1994) Enhanced micellar electrokinetic capillary chromatography separations on anionic polymer-coated capillary with pH-independent electroosmotic flow. *J. Chromatogr.* **669**, 259–262.
68 Zhao, Z.; Malik, A.; Lee, M.L. (1993) Solute adsorption on polymer-coated fused-silica capillary electrophoresis columns using selected protein and peptide standards. *Anal. Chem.* **65**, 2747–2752.
69 Kohr, J.; Engelhardt, H. (1993) Characterization of quartz capillaries for capillary electrophoresis. *J. Chromatogr.* **652**, 309–316.
70 Schmalzing, D.; Piggee, C.A.; Foret, F.; Carrilho, E.; Karger, B.L. (1993) Characterization and performance of a neutral hydrophilic coating for the capillary electrophoretic separation of biopolymers. *J. Chromatogr.* **652**, 149–159.
71 Liao, J.L.; Abramson, J.; Hjertén, S. (1995) A highly stable methyl cellulose coating for capillary electrophoresis. *J. Capillary Electrophoresis* **2**, 191–196.
72 Towns, J.K.; Bao, J.; Regnier, F.E. (1992) Synthesis and evaluation of epoxy polymer coatings for the analysis of proteins by capillary zone electrophoresis. *J. Chromatogr.* **599**, 227–237.
73 Wang, T.; Hartwick, R.A. (1992) Capillary modification and evaluation using streaming potential and frontal chromatography for protein analysis in capillary electrophoresis. *J. Chromatogr.* **594**, 325–334.
74 Szemán, J.; Ganzler, K. (1994) Use of cyclodextrins and cyclodextrin derivatives in high-performance liquid chromatography and capillary electrophoresis. *J. Chromatogr.* **668**, 509–517.
75 Ng, C.L.; Lee, H.K.; Li, S.F.Y. (1994) Prevention of protein adsorption on surfaces by

polyethylene oxide–polypropylene oxide–polyethylene oxide triblock copolymers in capillary electrophoresis. *J. Chromatogr.* **659**, 427–434.
76. Smith, J.T.; El Rassi, Z. (1993) Capillary zone electrophoresis of biological substances with fused silica capillaries having zero or constant electroosmotic flow. *Electrophoresis* **14**, 396–406.
77. Nakatani, M.; Shibukawa, A.; Nakagawa, T. (1994) Preparation and characterization of a stable polyacrylamide sieving matrix-filled capillary for high-performance capillary electrophoresis. *J. Chromatogr.* **661**, 315–321.
78. Wu, Q.; Claessens, H.A.; Cramers, C.A. (1992) The influence of surface treatments on the electroosmotic flow in micellar electrokinetic capillary chromatography. *Chromatographia* **33**, 303–308.
79. Chen, M.; Cassidy, R.M. (1992) Bonded-phase capillaries and the separation of inorganic ions by high-voltage capillary electrophoresis. *J. Chromatogr.* **602**, 227–234.
80. Yao, X.-W.; Wu, D.; Regnier, F.E. (1993) Manipulation of electroosmotic flow in capillary electrophoresis. *J. Chromatogr.* **636**, 21–29.
81. Smith, J.T.; El Rassi, Z. (1992) Capillary zone electrophoresis of biological substances with surface-modified fused silica capillaries with switchable electroosmotic flow. *J. High Resolut. Chromatogr.* **15**, 573–578.
82. Tsai, P.; Wu, C.-T.; Lee, C.S. (1994) Electrokinetic studies of inorganic coated capillaries. *J. Chromatogr.* **657**, 285–290.

5 Electrolyte systems
K. KITAGISHI

The selection of an electrolyte in capillary electrophoresis is of strategic importance to the success or otherwise of the separation. The electrolyte in CE corresponds both to the stationary phase and to the eluting solution in high performance liquid chromatography (HPLC). Features worthy of note for electrolytes in CE are as follows; (1) sufficient buffering capacity to suppress pH shift caused by migration of ions in the electrolyte, (2) low electric conductivity to avoid heat generation caused by high current, (3) low absorbance in the wavelength of detection not to disturb the on-column UV detection of the analyte zone.

The multifunctional nature of the electrolytes in CE often requires many kinds of additives to the electrolytes. Examples of additives are surfactants to control EOF and to form micelles in micellar electrokinetic chromatography (MEKC), zwitterions to eliminate the wall adsorption of proteins, soluble linear polymers for molecular sieving, cyclodextrins (CDs) as chiral selectors, organic solvents to change the characteristics of MEKC, UV-absorbing probes for indirect UV detection, urea to suppress hydrogen bonds, and so on. In this chapter, attempts to optimize electrolytes to provide satisfactory separation conditions and recommended additives which have been recently developed will be presented.

5.1 Optimization of electrolytes with respect to separation characteristics

The characteristics of an electrolyte, i.e. pH, viscosity, electric conductivity and interaction with analyte, depend on its composition. The migration behavior of the analyte can be simulated by varying these factors. Simulation suggests a composition of electrolyte which will provide the best separation in the shortest analysis time. Such a process is called optimization.

The separation of a mixture of monovalent ions, phenols and benzoic acids has been optimized as a function of the pH of the electrolyte [1]. Khaledi and co-workers [2, 3] simulated the separation of phenols with known pK_a values by one parameter modeling as a function of pH of the running buffer in free zone capillary electrophoresis (FZE) [2] and by two-parameter modeling with pH and micelle concentration in MEKC [3].

In MEKC, optimization has been reported to be dependent on micellar concentration, applied voltage and temperature [4], and on pH and micellar concentration [5–7]. Discrepancies between theoretical and experimental migration behavior were found in low micelle concentrations [5]. When the sample concentration is not negligible compared to the micelle concentration and the sample is extremely hydrophobic and incorporates micelles strongly, the theory that partition of analytes between water and micelle similar to HPLC becomes invalid.

Separation of metal cations in CE with indirect UV detection was optimized by two electrolyte factors, the pH and the concentration of the complexing agent. The optimal conditions predicted from modeling are very close to observed values [8]. Optimization of indirect UV detection of inorganic anions has been reported. The variable parameters were type and concentration of background electrolytes, capillary diameter, additive polymer and pH. Under the optimized conditions, sub-fmol concentrations of anions could be detected [9]. Indirect UV detection of inorganic anions with carrier electrolytes, chromate, phthalate, benzoate, trimellitate and pyromellitate has been evaluated and optimized. Not only separation efficiency, but also detection sensitivity should be optimized since carrier electrolyte also acts as a detection probe. For high mobility anions, chromate is considered to be the best carrier electrolyte [10, 11]. 2,6-Naphthalenedicarboxylate is a suitable carrier electrolyte for the analysis of aliphatic acids, providing detection at concentrations of 0.025 to 1000 mg l^{-1} of aliphatic acids under optimized condition [12].

Several studies have been reported of chiral separation in CE. Wren and co-workers investigated chiral separation by using a simple model for the separation of propranolol enantiomers with β-CD (cyclodextrin) and methyl β-CD as chiral selectors [13]. This study has been applied to CE in the presence of organic solvent [14] and to CE separation of other β-blockers [15]. The model shows good agreement with the observed data. The dependence of pH and CD concentration upon chiral selectivity and peak resolution of weak acid enantiomers has been studied using a theoretical equilibrium model [16, 17]. A systematic and experimental approach to optimization of chiral separation was investigated by Goodall and co-workers [18]. They designed a series of experiments with five variables, i.e. pH, CD concentration, electrolyte ionic strength, methanol concentration and injection time [18]. Optimization to determine the chiral purity of a drug was studied with the primary factors described above. The robustness of the assay was also investigated with respect to secondary factors like hydroxypropyl-β-CD source, lot and degree of substitution of CD [19].

5.2 Electrolytes in free zone capillary electrophoresis

The pH and viscosity of buffers are known to vary as temperature changes. Temperature-induced pH and viscosity changes affect the migration behavior of the analytes. A temperature program based on this phenomena in FZE improves the separation efficiency of weak acids by simple modification of the apparatus [20]. Joule heat is generated when an electrolyte at a high ionic strength fills the capillary and a high voltage is applied to the capillary, and pH shift by Joule heating changes the degree of ionization of analytes. A step change in operating voltage from 15 to 25 kV produces a pH gradient in the capillary which results in better separation of phenols than a constant voltage [21].

Tsuda [22] has developed a pH gradient system for CE using a solvent delivery program. The ratio of Na_2HPO_4 and NaH_2PO_4 is changed in his solvent delivery system. The pH can be estimated from the current through the capillary [22].

Non-aqueous media have been considered to have some advantages in classical electrophoresis, particularly for the sample matrices which are insoluble or indicate identical mobilities with others in water. The use of non-aqueous media has not been explored in depth in CE. It has been demonstrated recently that formamide with a higher dielectric constant and higher viscosity than water could be a candidate solvent [23].

For the separation and quantitation of cations or anions using indirect UV detection, an appropriate background electrolyte should be chosen which is UV absorbent with a mobility close to that of the analyte. Theoretical analysis of anion mobilities in binary buffers predicts that when the mobility of the analyte is between the mobilities of two buffer components, the analyte will displace the buffer component with the mobility closest to that of the analyte [24]. Cation analysis with indirect UV detection has generally used creatine, benzylamine or imidazole as UV absorbent and a chelating agent, while chromate, phthalate and benzoate are commonly added for indirect UV separation of anions. Additional developments and modifications of the additives for indirect UV detection have continued in order to improve the separation efficiency and enhance the application. The aluminum ion is difficult to analyze simultaneously with other metal cations in CE but in ion chromatography (IC), analysis of aluminum ion is not a problem. Ephedrine was selected for the analysis of aluminum ions in solutions of multiple cations [25]. Simultaneous separation of alkali, alkaline earth and transition metal ions has been developed using indirect UV detection. A complete separation of 17 metal ions can be achieved using pyridine or imidazole as background absorbent and glycolic acid as complexing agent [26]. Ammonium and 12 metal ions can be separated simultaneously in an electrolyte composed of imidazole, 2-

hydroxyisobutyric acid, 18-crown-6 and methanol in 12 min. This separation was applied to metal ions in Chinese tea [27] and methanol was found to be the dominating factor affecting EOF. 15-crown-5, 18-crown-6 or cryptand 2.2.2 were reported to be effective EOF modifiers in the separation of inorganic anions both in direct and indirect UV detection modes [28, 29]. The addition of phytic acid to an electrolyte has been used to lower the EOF and consequently to improve the resolution of CE separation of inorganic acids using direct UV detection [30]. Cationic polymers used as electrolyte additives for the separation of inorganic anions alter the direction of EOF, the separation selectivity and the resolution in the presence of chromate as a background chromophore. They may be able to control the separation of a series of inorganic anions [31].

Seven amino acids were separated using 5 mM salicylate as a background electrolyte at pH 11 [32]. CE separation of a mixture of 20 amino acids was carried out with UV indirect detection. Seventeen to 19 peaks can be observed using p-aminosalicylic acid and 4-(N,N-dimethyl)aminobenzoic acid as background electrolytes in the pH range 10.3 to 11.2. Metal cations and cationic surfactants used as electrolyte additives improve the resolution [33]. Inositol phosphates in fermentation broth could be determined by FZE with 1-naphthol-3,6-disulfonic acid as the UV-absorbent [34].

Indirect detection is widely used for UV absorbance and can be applied to fluorescence. Xue and Yeung demonstrated indirect fluorescence detection with fluorescein for the determination of lactate and pyruvate in single erythrocytes. Limits of detection are roughly estimated to be of the order of 10^{-16} mol [35]. This method is applicable to DNA fragment separation. Detection sensitivity is, however, still the same order as for the direct absorption [36].

Metal ions can be analyzed in FZE and MEKC with direct UV detection by using 8-hydroxyquinoline-5-sulfonic acid [37] and 2-(5-nitro-2-pyridylazo)-5-(N-propyl-N-sulfopropylamino) phenol as chelating agents [38].

5.3 Electrolytes in MEKC and related EKC

In MEKC, analytes are separated dependent mainly upon their hydrophilicity/lipophilicity, even when they have no electric charge, by surfactants like the micelles in sodium dodecyl sulfate (SDS). In order to improve the separation efficiency and resolution, novel surfactants and additives have been developed.

One of the most common surfactants for MEKC is bile salt [39]. Bile salts which have several advantages over SDS are used as pseudostationary phases. Since micelles of bile salts are more stable in the presence of organic solvents and show more polarity than those of SDS, the MEKC separation is possible for more hydrophobic compounds with better resolution. Chiral separations of some enantiomers were successful using bile salts [39].

Disodium-5,12-bis(dodecyloxymethyl)-4,7,10,13-tetraoxa-1,16-hexadecanedisulfonate was synthesized as a surfactant with two lipophilic chains and two ionized groups. It has a wider migration time window and lower CMC (critical micelle concentration) than conventional surfactants such as SDS [40].

When various tetraalkylammonium ions are added to SDS micelle solutions, selectivity and resolution in MEKC can be modified. Systematic MEKC separation of neutral mixtures suggests that the major influences are due to change in micelle properties derived from the interaction of tetraalkylammonium ions with the micelles [41].

El Rassi and co-workers [42–44] developed micelles with an adjustable surface charge in order to have micelles with adjustable retention windows in MEKC. The micelles are obtained by complexation of neutral surfactants with anionic borates through the association of polyhydroxy groups with borate ions under alkali conditions. They introduced octylglucosides [42], N-gluco-N-methylalkanamides [43] and alkyldisaccharides [44] as neutral surfactants. A wide range of solutes, especially hydrophobic compounds, could be separated in MEKC due to the balanced hydrophilic–lipophilic character of the new neutral surfactant–borate micelles.

A few studies have been done on neutral surfactants. Micelles of a neutral, non-ionic surfactant, Tween-20, could separate closely related peptides [45] or dansyl amino acids [46] under acidic conditions via the weak interaction between the analytes and the micelles. Non-ionic surfactants have the advantage of insignificant Joule heating because they do not increase the electric conductivity of electrolyte in contrast to cationic or anionic surfactants such as SDS.

Like cationic surfactants, tetraalkylammonium ions are widely used. Crosby and El Rassi investigated the properties of cationic surfactants for the separation of urea herbicides, alkylbenzenes and phenylalkylalcohols by using a series of alkyltrimethylammonium chlorides and bromides [47]. Separation of phenolic carboxylic derivatives can be achieved in MEKC with cetyltrimethylammonium bromide. Conditions of temperature, applied voltage, pH, electrolyte and surfactant concentration were arranged to obtain an adequate performance. Phenolic carboxylic acids in standard solutions and in samples prepared from plant materials can be efficiently separated [48].

Since conventional MEKC separations with ionic surfactants have problems of limited elution range and poor selectivity for extremely hydrophobic analytes, addition of organic solvents to the electrolyte was examined by some research groups. A difficulty to be overcome is that widely used micelles like SDS are significantly unstable in the presence of organic solvent. Bile salts solve the problem reasonably as already described. Shi and Fritz [49] reported that very hydrophobic compounds such as cyclic aromatic hydrocarbons as well as hydrophilic compounds could be separated by using tetraheptylammonium salt as an additive in aqueous–acetonitrile electrolyte. The mechanism of separation was not based on MEKC, because surfactants are not supposed to form micelles but on the interaction between hydrophobic analytes and the hydrocarbon chains of quaternary ammonium salts [49]. Oligomerization of sodium 10-undecylenate provides a stable micelle-like pseudostationary phase for successful electrokinetic chromatography in the presence of organic solvents [50]. Starburst dendrimers, poly(amidoamines) were useful as carriers in electrokinetic chromatography in aqueous and aqueous–organic solvent electrolytes. The size of the poly(amidoamines), pH and the content of organic solvent affect the separation [51]. Microemulsion (oil-in-water), prepared by mixing oil, surfactant, cosurfactant and water, heptane–SDS–butanol buffer (pH 7), enabled the extension of electrokinetic chromatography to more hydrophobic analytes [52]. However, more systematic investigations are required for reproducible performance of this technique.

5.4 Electrolytes in polymer sieving

Molecular sieving gels which have been used in conventional slab or disk gel electrophoresis can be applied to CE as capillary gel electrophoresis (CGE). However, CGE has several disadvantages. Filling gel into the capillary should be achieved with much caution to avoid air bubbles getting into the gel. Careful manipulation is required so as not to induce shrinkage and drying of the gel. Replacement of the gel with polymer solution in CE solves these problems. Polymer solutions can be easily prepared, filled into the capillary without any additional aids, handled in a similar way to general electrolytes and replaced with no difficulty.

Three types of sieving polymer which can be used commonly are linear polyacrylamide (PAA), polysaccharides and poly (ethylene glycol) or poly (ethylene oxide). Linear polyacrylamide can be prepared by polymerization of acrylamide monomers without crosslinking reagents. A double-stranded DNA mixture has been separated from the linear polyacrylamide solution with high resolution [53–57]. The addition of intercalating agents such as ethidium bromide induces higher resolution [54] already reported for the separation of DNA fragments from cellulose derivatives [58, 59].

One of the advantages of polysaccharides and poly(ethylene glycol/oxide) is their low UV absorption in contrast to the high UV absorption of PAA. SDS–protein complexes have been resolved based on their molecular weight using dextran or poly(ethylene glycol) [60]. The temperature dependence of the separation was investigated by Guttman et al. [61]. Cellulose derivatives, i.e. hydroxyethylcellulose, hydroxypropylcellulose, hydroxypropylmethylcellulose and methylcellulose are frequently used as sieving polymers for the separation of DNA fragments [58, 59, 62–67] as well as a natural polysaccharide, glucomannan [68]. The addition of hydroxyethylcellulose to electrolytes resulted in the separation of poly(styrenesulfonates) over a wide range of molecular weight with high resolution [69].

The mechanism of molecular sieving of polymer solutions was explained as mesh formation by polymer entanglement similar to a gel matrix [62]. The mesh size could be determined from the viscosity [62] or dynamic light scattering of polymer solutions [70]. Guidelines were presented showing that a polymer solution at higher concentration can separate smaller DNA molecules [65] and that a polymer solution of larger molecular weight is appropriate to the separation of larger analytes [55]. On the other hand, sufficient molecular sieving could occur under conditions where polymer molecules cannot entangle one another. Separation of DNA fragments at a hydroxyethylcellulose concentration below the entanglement threshold was shown by Barron and co-workers [67, 71, 72]. Karim et al. [73] reported that dextran solutions with a molecular weight less than 10 000 Da could separate SDS–protein mixtures at similar concentrations to those conventionally adopted for dextrans of larger molecular weights. A slight conformation change in DNA molecules could be recognized by video microscopy of DNA migration behavior in dilute hydroxyethylcellulose solution, possibly caused by the entanglement of the DNA molecule with hydroxyethylcellulose molecule or by a hydroxyethylcellulose cluster [74]. Evidence by Shi et al. [74] indicates agreement with speculation following from the results of electrophoretic migration by Barron et al. [71].

Poly(ethylene glycol) and poly(ethylene oxide) are often used as sieving polymers, since purified polymers of poly(ethylene glycol) and poly(ethylene oxide) in various molecular weight ranges are available commercially. SDS–protein complexes can be separated by poly(ethylene glycol) or poly(ethylene oxide) with performances similar to those of polyacrylamide gel in CE [75–77]. The separation efficiency of DNA using poly(ethylene glycol) or poly(ethylene oxide) solutions is also comparable to that of polyacrylamide gel [78]. Poly(ethylene glycol) was found to provide a new separation mode in CE. The separation of benzoic acid derivatives was achieved through hydrogen bonding with poly(ethylene glycol) [79].

Low melting point agarose, i.e. methoxylated agarose [80] and poly(n-acryloylaminoethoxyethanol) [81] have been introduced as new replaceable polymers.

5.5 Electrolytes for chiral separation

CD derivatives have been conventionally used as chiral selectors for chiral separation. Enantiomeric separation with CD derivatives is not ubiquitous, because the cyclic structure can recognize analytes of certain dimensions. In order to extend chiral recognition, chemical modification of CD and other chiral selectors as mobile phases have been developed.

The addition of linear maltodextrins, oligosaccharides obtained by the hydrolysis of amylose to electrolytes, causes enantiomeric recognition of some pharmaceuticals under various conditions [82, 83]. Several enantiomers of antimalarial drugs and antihistamines can be separated with heparin, a polydisperse, linear glycosaminoglycan as a chiral selector [84]. Yang and Hage reported the separation of warfarin and tryptophan enantiomers using bovine serum albumin as an electrolyte additive [85]. Ansamycins, such as rifamycin B, a class of macrocyclic antibiotics, can be chiral selectors of racemic amino alcohols. Since rifamycin B absorbs in both the UV and visible spectral regions, analyte zones can be detected easily by indirect spectroscopy [86]. Dobashi and co-workers reported that chiral selection of amino acid derivatives could be achieved by the addition of (S)-N-dodecanoylvaline, a chiral surfactant [87, 88]. Enantiomer mixtures such as norephedrine, atenolol, propranolol, benzoin, etc. can be separated with (R)- and (S)-N-dodecoxycarbonylvaline by MEKC [89].

5.6 Additives to suppress the wall adsorption of basic compounds

In order to analyze basic compounds in CE, the fact that adsorption of the analytes onto the capillary inner wall causes peak distortion and poor recovery needs to be resolved. Towns and Regnier [90] attempted to quantitate the adsorption effect of cationic polymers such as basic proteins upon transport velocity, peak symmetry and efficiency. Cationic polymers are adsorbed preferably near the inlet end of a capillary and produce a non-uniform zeta potential across the length of the capillary, which significantly reduces peak symmetry and separation efficiency [90]. Several techniques have been performed to suppress this undesirable adsorption; (1) permanent or semi-permanent coating on the surface of the inner capillary as described in Chapter 4 on Column Technology, (2) lowering the electrolyte pH to decrease the net charge on the capillary surface, (3) elevation of the electrolyte pH to invert the net charge of the zwitterionic analytes as basic proteins resulting in electrostatic repulsion between the analytes and the capillary wall, (4) shielding of silanol groups on the capillary wall with cationic surfactants or zwitterions. The following focuses on the fourth strategy.

A cationic surfactant, cetyltrimethylammonium bromide (CTAB), is the most popular additive to control EOF and reduce protein adsorption. Molecules of CTAB are adsorbed on the capillary inner surface by dynamic electrostatic interaction between the positively charged tertiary ammonium group of CTAB and the negatively charged silanol groups on the capillary surface. The magnitude of EOF controlled by CTAB depends on the nature of the capillary surface, the pH and the concentration of CTAB. The direction of EOF is reversed at a CTAB concentration of 2.5×10^{-4} M [91] from positive mobility to negative mobility. For separation of pyridinecarboxylic isomers, the addition of CTAB at a concentration of 30 mM in 10 mM phosphate buffer at pH 2.7 improved the separation remarkably, while optimum resolution for hydroxy-substituted pyridinecarboxylic isomers was observed at pH 7.5 with 30 mM CTAB [91].

Chang and Yeung reported an EOF gradient produced by changing the buffer reservoir containing a low CTAB concentration at the anodic end for an electrolyte with a higher CTAB concentration, which suppresses the EOF prior to the run [92]. This technique proved useful in improving the resolution in the separation of organic acids. The concept of a CTAB stationary front that is shifted slightly with the interaction between capillary surface and sample matrix was introduced by Chang and Yeung. Reproducibility in the migration time was improved on the EOF gradient, by investigating conditions at the constant stationary front [93].

Amines which suppress EOF as an additive ameliorate the separation of basic proteins. The performance of ethylenediamine, 1,3-diaminopropane and 1,4-diaminobutane as cationic additives have been investigated [94].

Combined use of zwitterion and cationic fluorosurfactant provided high efficiency separation of basic proteins. Separation selectivity could be tuned by changing the concentration ratio between zwitterion and cationic fluorosurfactant [95].

References

1. Kenndler, E.; Friedl, W. (1992) Adjustment of resolution and analysis time in capillary zone electrophoresis by varying the pH of the buffer. *J. Chromatogr.* **608**, 161–170.
2. Smith, S.C.; Khaledi, M.G. (1993) Optimization of pH for the separation of organic acids in capillary zone electrophoresis. *Anal. Chem.* **65**, 193–198.
3. Sahota, R.S.; Khaledi, M.G. (1994) Target factor modeling of migration behavior in capillary electrophoresis, *Anal. Chem.* **66**, 2374–2381.
4. Hayashi, Y.; Matsuda, R.; Terabe, S. (1993) Optimization of precision and throughput in micellar electrokinetic chromatography. *Chromatographia* **37**, 149–155.
5. Smith, S.C.; Khaledi, M.G. (1993) Prediction of the migration behavior of organic acids in micellar electrokinetic chromatography. *J. Chromatogr.* **632**, 177–184.
6. Ng, C.L.; Ong, C.P.; Lee, H.K.; Li, S.F.Y. (1992) Systematic optimization of micellar electrokinetic chromatographic separation of flavonoids. *Chromatographia* **34**, 166–172.
7. Quang, C.; Strasters, J.K.; Khaledi, M.G. (1994) Computer-assisted modeling, prediction,

and multifactor optimization in micellar electrokinetic chromatography of ionizable compounds. *Anal. Chem.* **66**, 1646–1653.
8. Quang, C.; Khaledi, M.G. (1994) Prediction and optimization of the separation of metal cations by capillary electrophoresis with indirect UV detection. *J. Chromatogr.* **659**, 459–466.
9. Ma, Y.; Zhang, R. (1992) Optimization of indirect photometric detection of anions in high-performance capillary electrophoresis. *J. Chromatogr.* **625**, 341–348.
10. Cousins, S.M.; Haddad, P.R.; Buchberger, W. (1994) Evaluation of carrier electrolytes for capillary zone electrophoresis of low-molecular-mass anions with indirect UV detection. *J. Chromatogr.* **671**, 397–402.
11. Buchberger, W.; Cousins, S.M.; Haddad, P.R. (1994) Optimisation of indirect UV detection in capillary zone electrophoresis of low-molecular-mass anions. *Trends Anal. Chem.* **13**, 313–319.
12. Tindall, G.W.; Wilder, D.R.; Perry, R.L. (1993) Optimizing dynamic range for the analysis of small ions by capillary zone electrophoresis. *J. Chromatogr.* **641**, 163–167.
13. Wren, S.A.C.; Rowe, R.C. (1992) Theoretical aspects of chiral separation in capillary electrophoresis. I. Initial evaluation of a model. *J. Chromatogr.* **603**, 235–241.
14. Wren, S.A.C.; Rowe, R.C. (1992) Theoretical aspects of chiral separation in capillary electrophoresis. II. The role of organic solvent. *J. Chromatogr.* **609**, 363–367.
15. Wren, S.A.C. (1993) Theory of chiral separation in capillary electrophoresis. *J. Chromatogr.* **636**, 57–62.
16. Rawjee, Y.Y.; Staerk, D.U.; Vigh, G. (1993) Capillary electrophoretic chiral separations with cyclodextrin additives. I. Acids: chiral selectivity as a function of pH and the concentration of β-cyclodextrin for fenoprofen and ibuprofen. *J. Chromatogr.* **635**, 291–306.
17. Rawjee, Y.Y.; Vigh, G. (1994) A peak resolution model for the capillary electrophoretic separation of the enantiomers of weak acids with hydroxypropyl β-cyclodextrin-containing background electrolytes. *Anal. Chem.* **66**, 619–627.
18. Rogan, M.M.; Altria, K.D.; Goodall, D.M. (1994) Plackett–Burman experimental design in chiral analysis using capillary electrophoresis. *Chromatographia* **38**, 723–729.
19. Rickard, E.C.; Bopp, R.J. (1994) Optimization of a capillary electrophoresis method to determine the chiral purity of a drug. *J. Chromatogr.* **680**, 609–621.
20. Whang, C.-W.; Yeung, E.S. (1992) Temperature programming in capillary zone electrophoresis. *Anal. Chem.* **64**, 502–506.
21. Chang, H.-T.; Yeung, E.S. (1993) Voltage programming in capillary zone electrophoresis. *J. Chromatogr.* **632**, 149–155.
22. Tsuda, T. (1992) pH gradient capillary zone electrophoresis using a solvent program delivery system. *Anal. Chem.* **64**, 386–390.
23. Sahota, R.S.; Khaledi, M.G. (1994) Nonaqueous capillary electrophoresis. *Anal. Chem.* **66**, 1141–1146.
24. Wang, T.; Hartwick, R.A. (1992) Binary buffers for indirect absorption detection in capillary zone electrophoresis. *J. Chromatogr.* **589**, 307–313.
25. Barger, W.R.; Mowery, R.L.; Wyatt, J.R. (1994) Separation and indirect detection by capillary zone electrophoresis of ppb (w/w) levels of aluminum ions in solutions of multiple cations. *J. Chromatogr.* **680**, 659–665.
26. Lee, Y.-H.; Lin, T.-I. (1994) Determination of metal cations by capillary electrophoresis. Effect of background carrier and complexing agents. *J. Chromatogr.* **675**, 227–236.
27. Yang, Q.; Smeyers-Verbeke, J.; Wu, W.; Khots, M.S.; Massart, D.L. (1994) Simultaneous separation of ammonium and alkali, alkaline earth and transition metal ions in aqueous–organic media by capillary ion analysis. *J. Chromatogr.* **688**, 339–349.
28. Reid, R.H.P. (1994) Electrophoretic behaviour of a group of organic anions of biochemical interest in a functionally coherent series of buffers. *J. Chromatogr.* **669**, 151–183.
29. Lamb, J.D.; Edwards, B.R.; Smith, R.G.; Garrick, R. (1995) The use of macrocycles as electroosmotic flow modifiers in the separation of inorganic anions by capillary electrophoresis. *Talanta* **42**, 109–117.
30. Birrell, H.C.; Camilleri, P.; Okafo, G.N. (1994) Phytic acid can greatly enhance resolution in capillary electrophoresis. *J. Chem. Soc., Chem. Commun.* 43–44.
31. Stathakis, C.; Cassidy, R.M. (1994) Cationic polymers for selectivity control in the capillary electrophoretic separation of inorganic anions. *Anal. Chem.* **66**, 2110–2115.

REFERENCES

32 Bruin, G.J.M.; van Asten, A.C.; Xu, X.; Poppe, H. (1992) Theoretical and experimental aspects of indirect detection in capillary electrophoresis. *J. Chromatogr.* **608**, 97–107.
33 Lee, Y.-H.; Lin, T.-I. (1994) Capillary electrophoretic determination of amino acids with indirect absorbance detection. *J. Chromatogr.* **680**, 287–297.
34 Buscher, B.A.P.; Irth, H.; Andersson, E.; Tjaden, U.R.; van der Greef, J. (1994) Determination of inositol phosphates in fermentation broth using capillary zone electrophoresis with indirect UV detection. *J. Chromatogr.* **678**, 145–150.
35 Xue, Q.; Yeung, E.S. (1994) Indirect fluorescence determination of lactate and pyruvate in single erythrocytes by capillary electrophoresis. *J. Chromatogr.* **661**, 287–295.
36 Chan, K.C.; Whang, C.-W.; Yeung, E.S. (1993) Separation of DNA restriction fragments using capillary electrophoresis. *J. Liq. Chromatogr.* **16**, 1941–1962.
37 Timerbaev, A.; Semenova, O.; Bonn, G. (1993) Metal ion capillary zone electrophoresis with direct UV detection: Comparison of different migration modes for negatively charged chelates. *Chromatographia* **37**, 497–500.
38 Motomizu, S.; Mori, N.; Kuwabara, M.; Oshima, M. (1994) Separation and sensitive determination of metal ions by capillary zone electrophoresis with 2-(5-nitro-2-pyridylazo)-5-(N-propyl N-sulfopropylamino) phenol. *Anal. Sci.* **10**, 101–103.
39 Cole, R.O.; Sepaniak, M.J. (1992) The use of bile salt surfactants in micellar electrokinetic capillary chromatography. *LC-GC* **10**, 380–385.
40 Tanaka, M.; Ishida, T.; Araki, T.; Masuyama, A.; Nakatsuji, Y.; Okahara, M.; Terabe, S. (1993) Double-chain surfactant as a new and useful micelle-forming reagent for micellar electrokinetic chromatography. *J. Chromatogr.* **648**, 469–473.
41 Nielsen, K.R.; Foley, J.P. (1994) Effect of the dodecyl sulfate counterion on selectivity and resolution in micellar electrokinetic chromatography: II. Organic counterions. *J. Microcol. Sep.* **6**, 139–149.
42 Cai, J.; El Rassi, Z. (1992) Micellar electrokinetic capillary chromatography of neutral solutes with micelles of adjustable surface charge density. *J. Chromatogr.* **608**, 31–45.
43 Smith, J.T.; Nashabeh, W.; El Rassi, Z. (1994) Micellar electrokinetic capillary chromatography with *in situ* charged micelles. 1. Evaluation of N-D-gluco-N-methylalkanamide surfactants as anionic borate complexes. *Anal. Chem.* **66**, 1119–1133.
44 Smith, J.T.; El Rassi, Z. (1994) Micellar electrokinetic capillary chromatography with *in situ* charged micelles: II. Evaluation and comparison of octylmaltoside and octylsucrose surfactants as anionic borate complexes in the separation of herbicides. *J. Microcol. Sep.* **6**, 127–138.
45 Matsubara, N.; Terabe, S. (1992) Separation of closely related peptides by capillary electrophoresis with a nonionic surfactant. *Chromatographia* **34**, 493–496.
46 Matsubara, N.; Terabe, S. (1994) Separation of 24 dansylamino acids by capillary electrophoresis with a non-ionic surfactant. *J. Chromatogr.* **680**, 311–315.
47 Crosby, D.; El Rassi, Z. (1993) Micellar electrokinetic capillary chromatography with cationic surfactants. *J. Liq. Chromatogr.* **16**, 2161–2187.
48 Bjergegaard, C.; Michaelsen, S.; Sørensen, H. (1992) Determination of phenolic carboxylic acids by micellar electrokinetic capillary chromatography and evaluation of factors affecting the method. *J. Chromatogr.* **608**, 403–411.
49 Shi, Y.; Fritz, J.S. (1994) Capillary zone electrophoresis of neutral organic molecules in organic-aqueous solution. *J. High Resolut. Chromatogr.* **17**, 713–718.
50 Palmer, C.P.; Khaled, M.Y.; McNair, H.M. (1992) A monomolecular pseudostationary phase for micellar electrokinetic capillary chromatography. *J. High Resolut. Chromatogr.* **15**, 756–762.
51 Tanaka, N.; Tanigawa, T.; Hosoya, K.; Kimata, K.; Araki, T.; Terabe, S. (1992) Starburst dendrimers as carriers in electrokinetic chromatography. *Chem. Lett.* 959–962.
52 Terabe, S.; Matsubara, N.; Ishihama, Y.; Okada, Y. (1992) Microemulsion electrokinetic chromatography: comparison with micellar electrokinetic chromatography. *J. Chromatogr.* **608**, 23–29.
53 Chiari, M.; Nesi, M.; Fazio, M.; Righetti, P.G. (1992) Capillary electrophoresis of macromolecules in 'syrupy' solutions: Facts and misfacts. *Electrophoresis* **13**, 690–697.
54 Pariat, Y.F.; Berka, J.; Heiger, D.N.; Schmitt, T.; Vilenchik, M.; Cohen, A.S.; Foret, F.; Karger, B.L. (1993) Separation of DNA fragments by capillary electrophoresis using replaceable linear polyacrylamide matrices. *J. Chromatogr.* **652**, 57–66.

55. Chiari, M.; Nesi, M.; Righetti, P.G. (1993) Movement of DNA fragments during capillary zone electrophoresis in liquid polyacrylamide. *J. Chromatogr.* **652**, 31–39.
56. Manabe, T.; Chen, N.; Terabe, S.; Yohda, M.; Endo, I. (1994) Effects of linear polyacrylamide concentrations and applied voltages on the separation of oligonucleotides and DNA sequencing fragments by capillary electrophoresis. *Anal. Chem.* **66**, 4243–4252.
57. Heller, C.; Viovy, J.L. (1994) Electrophoretic separation of oligonucleotides in replenishable polyacrylamide-filled capillaries. *Appl. Theor. Electrophoresis* **4**, 39–41.
58. Ulfelder, K.J.; Schwartz, H.E.; Hall, J.M.; Sunzeri, F.J. (1992) Restriction fragment length polymorphism analysis of ERBB2 oncogene by capillary electrophoresis. *Anal. Biochem.* **200**, 260–267.
59. Zhu, H.; Clark, S.M.; Benson, S.C.; Rye, H.S.; Glazer, A.N.; Mathies, R.A. (1994) High-sensitivity capillary electrophoresis of double-stranded DNA fragments using monomeric and dimeric fluorescent intercalating dyes. *Anal. Chem.* **66**, 1941–1948.
60. Ganzler, K.; Greve, K.S.; Cohen, A.S.; Karger, B.L.; Guttman, A.; Cooke, N.C. (1992) High-performance capillary electrophoresis of SDS–protein complexes using UV-transparent polymer networks. *Anal. Chem.* **64**, 2665–2671.
61. Guttman, A.; Horváth, J.; Cooke, N. (1993) Influence of temperature on the sieving effect of different polymer matrices in capillary SDS gel electrophoresis of proteins. *Anal. Chem.* **65**, 199–203.
62. Grossman, P.D.; Soane, D.S. (1991) Capillary electrophoresis of DNA in entangled polymer solutions. *J. Chromatogr.* **559**, 257–266.
63. Singhal, R.P.; Xian, J. (1993) Separation of DNA restriction fragments by polymer-solution capillary zone electrophoresis. Influence of polymer concentration and ion-pairing reagents. *J. Chromatogr.* **652**, 47–56.
64. McGregor, D.A.; Yeung, E.S. (1993) Optimization of capillary electrophoretic separation of DNA fragments based on polymer filled capillaries. *J. Chromatogr.* **652**, 67–73.
65. Baba, Y.; Ishimaru, N.; Samata, K.; Tsuhako, M. (1993). High-resolution separation of DNA restriction fragments by capillary electrophoresis in cellulose derivative solutions. *J. Chromatogr.* **653**, 329–335.
66. McCord, B.R.; Jung, J.M.; Holleran, E.A. (1993) High resolution capillary electrophoresis of forensic DNA using a non-gel sieving buffer. *J. Liq. Chromatogr.* **16**, 1963–1981.
67. Barron, A.E.; Soane, D.S.; Blanch, H.W. (1993) Capillary electrophoresis of DNA in uncross-linked polymer solutions. *J. Chromatogr.* **652**, 3–16.
68. Izumi, T.; Yamaguchi, M.; Yoneda, K.; Isobe, T.; Okuyama, T.; Shinoda, T. (1993) Use of glucomannan for the separation of DNA fragments by capillary electrophoresis. *J. Chromatogr.* **652**, 41–46.
69. Poli, J.B.; Schure, M.R. (1992) Separation of poly(styrenesulfonates) by capillary electrophoresis with polymeric additives. *Anal. Chem.* **64**, 896–904.
70. Grossman, P.D.; Hino, T.; Soane, D.S. (1992) Dynamic light-scattering studies of hydroxyethyl cellulose solutions used as sieving media for electrophoretic separations. *J. Chromatogr.* **608**, 79–83.
71. Barron, A.E.; Blanch, H.W.; Soane, D.S. (1994) A transient entanglement coupling mechanism for DNA separation by capillary electrophoresis in ultradilute polymer solutions. *Electrophoresis* **15**, 597–615.
72. Barron, A.E.; Sunada, W.M.; Blanch, H.W. (1995) The use of coated and uncoated capillaries for the electrophoretic separation of DNA in dilute polymer solutions. *Electrophoresis* **16**, 64–74.
73. Karim, M.R.; Janson, J.-C.; Takagi, T. (1994) Size-dependent separation of proteins in the presence of sodium dodecyl sulfate and dextran in capillary electrophoresis: Effect of molecular weight of dextran. *Electrophoresis* **15**, 1531–1534.
74. Shi, X.; Hammond, R.W.; Morris, M.D. (1995) DNA conformational dynamics in polymer solutions above and below the entanglement limit. *Anal. Chem.* **67**, 1132–1138.
75. Cifuentes, A.; de Frutos, M.; Diet-Masa, J.C. (1994) Polymeric networks vs cross-linked polyacrylamide bonded gels for CE separations of whey proteins. *Am. Lab.* **26**, 46–51.
76. Benedek, K.; Guttman, A. (1994) Ultra-fast high-performance capillary sodium dodecyl sulfate gel electrophoresis of proteins. *J. Chromatogr.* **680**, 375–381.
77. Benedek, K.; Thiede, S. (1994) High-performance capillary electrophoresis of proteins using sodium dodecyl sulfate–poly(ethylene oxide). *J. Chromatogr.* **676**, 209–217.

78 Fung, E.N.; Yeung, E.S. (1995) High-speed DNA sequencing by using mixed poly(ethylene oxide) solutions in uncoated capillary columns. *Anal. Chem.* **67**, 1913–1919.
79 Esaka, Y.; Yamaguchi, Y.; Kano, K.; Goto, M.; Haraguchi, H.; Takahashi, J. (1994) Separation of hydrogen-bonding donors in capillary electrophoresis using polyethers as matrix. *Anal. Chem.* **66**, 2441–2445.
80 Hjertén, S.; Srichaiyo, T.; Palm, A. (1994) UV-transparent, replaceable agarose gels for molecular-sieve (capillary) electrophoresis of proteins and nucleic acids. *Biomed. Chromatogr.* **8**, 73–76.
81 Chiari, M.; Nesi, M.; Righetti, P.G. (1994) Capillary zone electrophoresis of DNA fragments in a novel polymer network: Poly (*N*-acryloylaminoethoxyethanol). *Electrophoresis* **15**, 616–622.
82 D'Hulst, A.; Verbeke, N. (1992) Chiral separation by capillary electrophoresis with oligosaccharides. *J. Chromatogr.* **608**, 275–287.
83 Soini, H.; Stefansson, M.; Riekkola, M.-L.; Novotny, M.V. (1994) Maltooligosaccharides as chiral selectors for the separation of pharmaceuticals by capillary electrophoresis. *Anal. Chem.* **66**, 3477–3484.
84 Stalcup, A.M.; Agyei, N.M. (1994) Heparin: A chiral mobile-phase additive for capillary zone electrophoresis. *Anal. Chem.* **66**, 3054–3059.
85 Yang, J.; Hage, D.S. (1994) Chiral separations in capillary electrophoresis using human serum albumin as a buffer additive. *Anal. Chem.* **66**, 2719–2725.
86 Armstrong, D.W.; Rundlett, K.; Reid III, G.L. (1994) Use of a macrocyclic antibiotic, rifamycin B, and indirect detection for the resolution of racemic amino alcohols by CE. *Anal. Chem.* **66**, 1690–1695.
87 Dobashi, A.; Ono, T.; Hara, S.; Yamaguchi, J. (1989) Optical resolution of enantiomers with chiral mixed micelles by electrokinetic chromatography. *Anal. Chem.* **61**, 1984–1986.
88 Dobashi, A.; Ono, T.; Hara, S.; Yamaguchi, J. (1989) Enantioselective hydrophobic entanglement of enantiometric solutes with chiral functionalized micelles by electrokinetic chromatography. *J. Chromatogr.* **480**, 413–420.
89 Mazzeo, J.R.; Grover, E.R.; Swartz, M.E.; Petersen, J.S. (1994) Novel chiral surfactant for the separation of enantiomers by micellar electrokinetic capillary chromatography. *J. Chromatogr.* **680**, 125–135.
90 Towns, J.K.; Regnier, F.E. (1992) Impact of polycation adsorption on efficiency and electroosmotically driven transport in capillary electrophoresis. *Anal. Chem.* **64**, 2473–2478.
91 Janini, G.M.; Chan, K.C.; Barnes, J.A.; Muschik, G.M.; Issaq, H.J. (1993) Separation of pyridinecarboxylic acid isomers and related compounds by capillary zone electrophoresis. *J. Chromatogr.* **653**, 321–327.
92 Chang, H.-T.; Yeung, E.S. (1992) Optimization of selectivity in capillary zone electrophoresis via dynamic pH gradient and dynamic flow gradient. *J. Chromatogr.* **608**, 65–72.
93 Chang, H.-T.; Yeung, E.S. (1993) Self-regulating dynamic control of electroosmotic flow in capillary electrophoresis. *Anal. Chem.* **65**, 650–652.
94 Song, L.; Ou, Q.; Yu, W. (1994) Study on phosphate of ethylenediamine, 1,3-diaminopropane and 1,4-diaminobutane as carrying electrolyte in open-tubular capillary electrophoresis. *J. Liq. Chromatogr.* **17**, 1953–1969.
95 Emmer, Å.; Roeraade, J. (1994) Performance of zwitterionic and cationic fluorosurfactants as buffer additives for capillary electrophoresis of proteins. *J. Liq. Chromatogr.* **17**, 3831–3846.

6 Detection
K. KITAGISHI

Although analysis in capillary electrophoresis has the advantage of speed with high separation efficiency, its concentration sensitivity is unfavorably low. Since the capillary diameter in CE is less than 100 μm and the entire volume of the capillary in most cases is less than 10 μl, the optical path should be short enough to detect analyte zones without loss of separation efficiency. Consequently, the detection limit is between 1 and 0.1 ppm with conventional on-line UV detection. Several studies have been carried out to enhance the detection sensitivity, e.g. sample preconcentration, choice of detection system including strong light source, a detector with high sensitivity and efficient optical alignment, and derivatization of analytes [1].

The connection between mass spectroscopy (MS) and CE has been particularly noted since 1990. Detection with MS is a powerful tool to obtain structural information on analytes.

In this chapter, the outline of various detection systems will be described. The characteristics of the four primary detection modes in CE which are available in commercial products are shown in Table 6.1.

6.1 On-column and off-column detection

Detection in CE is usually achieved on-column by techniques such as UV-vis absorbance, fluorescence, chemiluminescence, refractive index (RI), etc. When commercially available fused-silica capillaries are used as a column, part of the polyimide coating on the fused-silica capillary should be thermally peeled away and submitted to the detection cell for on-column detection.

More sensitive detection, like electrochemical detection (ECD) is difficult on-column excepting when using an ultramicroelectrode, because the application of high voltage for CE separation interferes with the on-column ECD. Off-column detection in some cases is arranged by the connection of two columns, one for separation and one for detection. Postcolumn derivatization of analytes and detection with MS requires an off-column system. The obvious disadvantage of off-column detection is zone broadening which is generated by laminar flow in the second capillary. In addition, the back pressure from the second pressure disturbs the EOF in the separation capillary. Kok [2] proposed application of compensating pressure at

Table 6.1 Characteristics of the four main detection modes in capillary electrophoresis

Detection mode	Principal components	Advantages	Disadvantages
Absorbance	Light source (D_2; I_2; Xe); detector (photodiode; photodiode array); slit	Many commercial systems available; low cost; easy maintenance; high stability; good reproducibility; detectable for most kinds of molecules (by direct and indirect detection)	Relatively low sensitivity Low selectivity
Fluorescence	Light source (gas-phase laser; semiconductor laser; laser diode; Xe; I_2; etc.) Detector (photomultiplier; photodiode); filter; microscope objective	(General) high selectivity; stable baseline; high sensitivity (High energy laser) very high sensitivity; easy to focus (Low energy laser) low cost (Semiconductor laser/laser diode) low cost; easy maintenance; compact size; easy operation (Xenon; I_2) tunable excitation wavelength; low cost	Fluorescence chromophore is requested Expensive light source; excitation wavelength restricted Excitation wavelength restricted Temperature dependence of wavelength emitted by light source; excitation wavelength restricted Difficult to focus
Electrochemical (amperometry; conductivity; potentiometry)	Microelectrodes; (field decoupler)	High sensitivity; easy evaluation	Fragile; difficult to align electrode; high noise and drift; low reproducibility
Mass spectroscopy	Interface of CE-MS; MS	(General) useful for identification (On-line) easy to be ionized (Off-line) structural information can be obtained; wider applicable conditions	Expensive Restriction of separation conditions Fraction collection requested; difficult to set up

the front end of the separation capillary during electrophoresis, balancing the back pressure in order to maintain plug-like EOF. Cassidy *et al.* [3] evaluated the introduction of auxiliary EOF for end-column detection across a short detection capillary with postreaction reagent aligned at the end of the separation capillary. The experimental optimization of applied potential for auxiliary EOF, separation capillary diameter, distance between the two capillaries and capillary displacement allows off-column ECD with a small peak broadening from 420 000 to 400 000 of the theoretical plate number [3]. Other attempts at off-column detection will be described later in section 6.6 on conductivity, section 6.7 on potentiometry and section 6.8 on amperometry.

6.2 Absorbance

Absorption detection is most popular due to its wide application, convenience, low cost, flexibility, etc., in spite of relatively poor sensitivity.

A deuterium lamp is frequently used as a light source because it emits a stable light over a wide range of wavelengths in the UV region where many analytes have a high molar absorptivity. A halogen or xenon lamp which emits light in the visible region is rarely selected because only limited chromophores can be absorbed in the visible region with low molecular absorptivity. Coherent UV light based on an argon ion laser improved the limit of detection four-fold using an instrumental setup of double-beam detection with two photodiodes or phototubes [4]. When UV lasers can be incorporated into the CE system by light-controlled feedback or by frequency doubling of diode lasers, sensitivity will be improved remarkably.

Capillaries exhibit the optical effect of a lens in CE owing to their cylindrical hollow tube shape. Maystre and Bruno [5] investigated the geometry of the capillary in an attempt to suppress the lens effect with a laser beam, allowing the development of more reliable detectors with enhanced detection sensitivity, especially in fluorescence detection [5]. A novel design in UV-absorption optics in CE has been reported by Flint *et al.* [6]. It consisted of deuterium lamp, mirrors, nickel pinhole, Perspex bracket, and so on and offers high sensitivity, variable wavelength and a single path UV detector [6].

A photodiode array detector was applied to UV/vis absorbance detection in CE to obtain spectral information concerning the analyte zones. Spectral information is useful not only for identification and purification by comparing the accumulated spectra library [7], but also to reduce white noise by subtracting/dividing pherograms of two wavelengths.

A two-dimensional imaging system along the capillary axis is a powerful tool for capillary isoelectric focusing, since it enables separand zone detection without zone movement following the focusing step. Imaging detectors have been developed using a photodiode array [8] or a capillary scanning device [9].

The derivatization of proteins, peptides and amino acids with fluorescamine, which is commonly used as a fluorogenic reagent, could improve the detection sensitivity even in the UV absorbance region. Furthermore, a fluorescamine-derivatized sample was separated with better resolution than an underivatized one [10].

6.3 Fluorescence

Fluorescence has been widely used in the field of spectroscopy or high performance liquid chromatography (HPLC), because it is in general considered to be more sensitive and selective in detection than absorbance. Although many commercial instruments like the fluorospectrometer or fluorescence detector in HPLC are available, not many applications of fluorescence detection in CE have been reported. One of the main reasons is that scattering of incident light from the capillary wall together with stray light give a large background which is difficult to eliminate. If these difficulties are overcome, CE combined with fluorescence may prove popular due to its advantage in terms of higher sensitivity.

Various types of lasers are equipped as light sources, since the incident light is focused into a small volume inside the capillary to excite the analyte of interest with high efficiency and to reduce the scattering effect of the capillary wall.

A simple instrumental setup for laser-induced fluorescence (LIF) is on-column detection with an argon ion laser which is a stable strong laser with little tunability in the excitation wavelength and microscope objectives, mirrors, filters and a photomultiplier. A He–Cd laser or a He–Ne laser can be replaced as the light source, dependent on the excitation wavelength. For example, a system build-up in the laboratory indicated a limit of detection of 3 pM solution of fluorescein [11]. Toulas and Hernandez presented another instrumental design for LIF detection by using collinear rather than orthogonal geometry. They obtained a limit of detection of 10^{-10} to 10^{-13} M [12]. The LIF system from Beckman Co. has collinear geometry with an optical fiber and an ellipsoidal mirror. Taylor and Yeung demonstrated an axial beam excitation scheme, in which an optical fiber was inserted into the separation capillary and a photomultiplier was placed perpendicular to the capillary as a detector [13]. Their technique can be applied to a bundle of capillaries using a charge-coupled device (CCD) camera as a detector. Simultaneous monitoring of 10 to 12 capillaries is possible and the feasibility of DNA sequencing with multicapillaries was suggested [14]. A laser excited, confocal fluorescence scanner with four capillaries on a translation stage was developed for DNA sequencing by Huang et al. [15].

Recently, Lee and Yeung [16, 17] demonstrated UV laser induced fluorescence with an argon ion laser operating at 275 nm which has lower power than at 488 nm and is accompanied by a higher background. Nevertheless, high sensitivity detection has been achieved for native tryptophan- or tyrosine-containing proteins [16, 17]. A similar detection was applied to the separation of polycyclic aromatic hydrocarbons [18]. Chan et al. reported that a relatively economical, compact, pulsed UV laser could replace the expensive, large argon ion laser for monitoring native peptides and proteins with a limit of detection in the low nanomoles [19].

Imasaka and co-workers [20, 21] emphasized the advantages of a semiconductor laser or laser diode with respect to low price, compact size and ease of operation [20, 21], in spite of lower directivity and less coherence than a gas laser such as argon ion, He–Ne, He–Cd lasers, etc. Excitation wavelengths for semiconductor lasers/laser diodes are in the visible blue region like the second harmonic emission of the near infrared [20, 22] and the deep-red region [21, 23]. The blue semiconductor laser developed recently was evaluated as a light source for fluorescence detection in CE [24]. A He–Cd laser and a blue semiconductor laser utilized for the separation of polycyclic aromatic hydrocarbons indicate a limit of detection of 10^{-7} M in fluorescence compared with 10^{-5} M in absorbance detection [24]. CE in fluorescence detection has been developed recently with a blue laser diode (Nichia Chemicals Co., NLPB-500) as a light source and a photomultiplier (Hamamatsu Photonics, H5784) as a detector. A linear relationship between the fluorescence signal and the sample concentration is obtained in the range of 5×10^{-9} to 5×10^{-5} M fluorescein isothiocyanate (FITC) (unpublished data). A prototype diode-pumped, frequency-tripled, Nd:YLF laser, a solid laser, could be used for time-resolved fluorescence detection in CE. This laser produces 2.5 ns pulses at 349 nm with kHz repetition rates [25].

Cheng and Dovichi [26] developed a sheath-flow cuvette where a buffer stream of sheath analytes is eluted from the capillary. Off-column fluorescence detection can be achieved, reducing the light scatter [26]. The sheath-flow cuvette was applied to analyses with various light sources; subattomole detection of amino acid derivatives with a deep-red semiconductor laser [23], one order-of-magnitude improvement in detection sensitivity over absorbance of DNA fragments with an argon ion laser operating in the UV region [27], 5×10^{-23} mol detection of rhodamine 6G with a 1 mW He–Ne laser [28], 100 molecules detection of enzyme substrate and product with a 1 mW He–Ne laser [29] and 250 to 17 molecules detection of sulforhodamine 101 with an 8 mW He–Ne laser [30].

Shear et al. [31] proposed the optimization of fluorescence detection by the correction of the velocity in each migration zone. They showed that the signal-to-noise ratio in fluorescence was drastically affected by the velocity. The control of the data digitization and the excitation intensity could be improved by the signal-to-noise ratio of analyte depending on the velocity [31]. An algorithm was developed for automated velocity programming in order to increase the detection sensitivity. The separation field is reduced automatically when the analyte bands pass through the detection zone and the signal-to-noise ratio increased notably [32].

Derivatization of analytes with a suitable fluorescent reagent for the excitation wavelength is important for sensitive detection of biomolecules with strong visible gas lasers. Fluorescein isothiocyanate (FITC), the most popular fluorescent reagent, and 3-(4-carboxybenzoyl)-2-quinoline-carboxaldehyde (CBQCA) reacted with primary amines [33, 34] are appropriate for excitation at 488 nm by an argon ion laser. Chloroacetaldehyde selectively bound to adenine [35] and oligosaccharides labeled with 8-aminonaphthalene-1,3,6-trisulfonic acid [36] can be detected with a He–Cd laser at 325 nm.

Tunable fluorescence detection can be achieved with a xenon arc lamp with a bandpass filter. Arriaga *et al.* reported a limit of detection of 2×10^{-20} mol of fluorescein with the system [37]. A similar design was demonstrated with a high pressure mercury lamp, which could monitor rhodamine B at 2.1×10^{-18} mol [38].

6.4 Chemiluminescence

Chemiluminescence is a highly sensitive detection technique without a light source. Chemiluminescence detection is performed in a postcapillary reactor with a reaction tee (T joint) to mix the migration and catalyst solutions. Dadoo *et al.* [39] adapted on-line chemiluminescence detection to CE with a postcapillary reactor. Chemiluminescence from a luminol-labeled compound was detected with a photomultiplier connected to a photon counting system. This detection can enhance the sensitivity by two to three orders of magnitude over the absorption mode [39]. A similar setup is demonstrated by Ruberto and Grayeski [40], Wu and Huie [41] and Hara and co-workers [42, 43]. An acridinium chemiluminescence detection interface with a postcolumn reactor in a sheath-flow profile was designed by Ruberto and Grayeski [40]. Peroxyoxalate chemiluminescence could be performed by elution under pressure, following switching off the CE power supply (dynamic elution). The limit of detection is about 1.2 fmol of dansylated amino acids [41]. The eosin–bis(2,4,6-trichlorophenyl) oxalate (TCPO)–H_2O_2 system provided a limit of detection of 1.7 fmol by measuring the chemiluminescence of an eosin–bovine serum albumin complex [42]. Protein detection with a high sensitivity could be performed with chemiluminescence by using rhodamine B isothiocyanate or tetramethylrhodamine isothiocyanate isomer R [43].

Electrogenerated chemiluminescence has been developed based on the reaction of luminol and H_2O_2 with a cylindrical electrode [44]. End-column chemiluminescence detection was designed by Zare's group [45]. The signal is generated at the column outlet as with luminol chemiluminescence of firefly luciferase bioluminescence. The limits of detection are lower than those of absorption detection by the approximately three orders of magnitude [45].

6.5 Indirect detection

The selection of background absorbent to optimize separation was described in Chapter 5 on Electrolyte Systems. The signal-to-noise ratio in direct UV detection is mainly controlled by two independent factors, the background absorbent and the detector. A new concept of 'noise coefficient' was introduced as a ratio of the concentration fluctuation to the concentration of the absorbing agent, and the limit of detection can be evaluated based on this concept [46].

Williams et al. [47] developed a diode laser-based detector with a double-beam arrangement for near infrared indirect absorbance. With a 670 nm laser diode and rhodamine 700 as an absorbing agent, 2×10^{-5} M tetrabutylammonium ion could be detected [47]. A double-beam detector equipped with a 10 mW He–Ne laser (632.8 nm) improved the limit of detection of 1×10^{-7} M of pyruvate by using 0.5 mM bromocresol green as a background absorbent for anions. Malachite green (0.5 mM) was added to the separation buffer for cations [48].

Time-resolved luminescence of terbium(III)–acetyl acetone chelating agent can be adopted as an indirect detection method. The dynamic quenching by selected anions indicates higher sensitivity of the limit of detection, calculated to be 3×10^{-9} M nitrite [49].

6.6 Conductivity

Electrochemical detection (ECD) is commonly used in the field of HPLC because of its advantages, i.e. simple design, higher selectivity, higher sensitivity and easier operation. Miniaturization of ECD has yet to be solved in CE. As with ECD in HPLC, setting up a detector in CE first involves the connection of a separation capillary and an electrochemical cell to avoid interference from a high applied electric field. This type of design may cause loss of separation efficiency. ECD recently has concerned on-column or end-column detection in CE [50]. ECD in CE can be classified in three different modes, amperometry, conductivity and potentiometry.

Zare and co-workers present different types of conductivity detectors, for both on-column and end-column detections [51–53] (Figure 6.1(a) and (b)). Detection of conductivity in CE suffers from high background noise from the high applied separation voltage and poor sensitivity resulting from interference by high conductivity in the running buffer. Suppressed conductometric CE was designed to suppress the conductivity of the running buffer by a short ion-exchange membrane placed between the separation capillary and the detector cell for the end-column detection [54, 55]. Inorganic ions at 1 to 10 ppb can be demonstrated with this system [54].

Kar *et al.* reported a new design for the conductivity cell based on bifilar

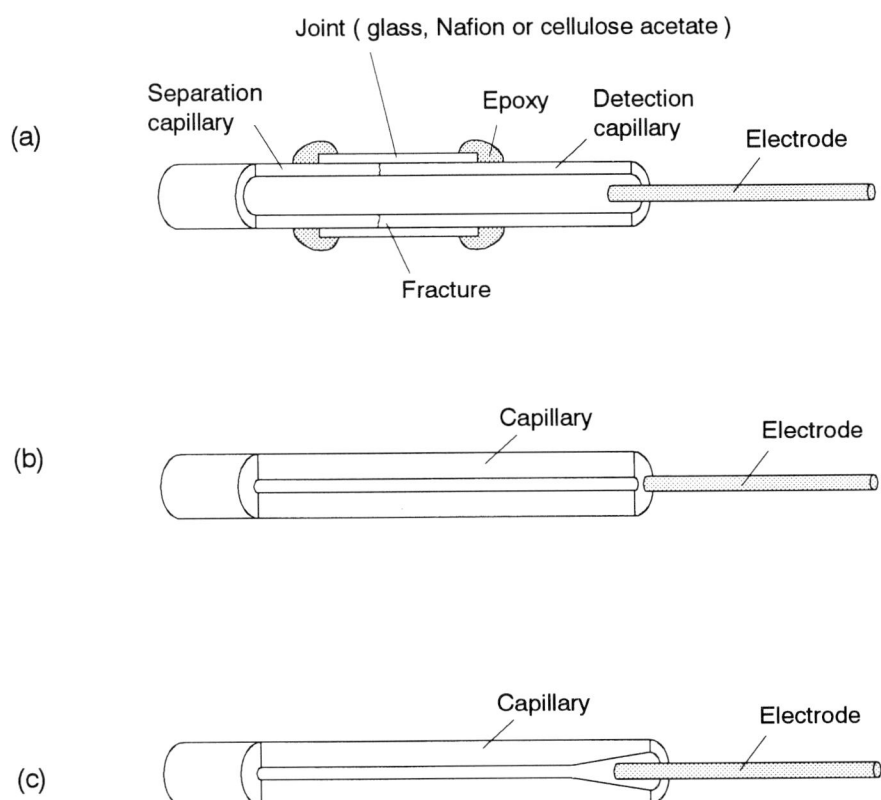

Figure 6.1 Electrochemical detectors. (a) Off-column detection; the separation voltage is decoupled with a joint over the fracture of the capillary. The electrode is inserted into the detection capillary. (b) End-column detection. (c) Optimized end-column detection by etching the detector end of the capillary into a conical aperture.

wire electrodes. Since these electrodes are put directly at the end of the capillary, little diffusion of analytes occurs because there is no flow of fluid between the electrodes [56].

6.7 Potentiometry

On-column potentiometric detection with ion-selective electrodes suffers from drift and noise caused by the high level of applied electrophoretic field. Nann and Simon reported a drastic reduction in drift and noise by etching the detector end of the capillary into a conical aperture [57] (Figure 6.1(c)). This capillary modification enabled on-column potentiometry for K^+, Na^+, Rb^+, Ca^{2+}, dopamine and histamine in the range of 10^{-5} to 10^{-3} M [58] and for Br^-, I^-, NO_3^-, ClO_4^-, SCN^- and salicylate with a limit of ClO_4^- detection of 5×10^{-8} M (approximately 5 ppb) [59]. Hauser et al. demonstrated similar potentiometric detection of Cl^-, NO_3^-, NO_2^-, Br^-, I^- and ClO_4^- with a sensitivity limit for ClO_4^- at 10 ppb [60].

6.8 Amperometry

Amperometry is the most generally used in ECD. Wallingford and Ewing developed ECD to decouple the electrode from the separation potential [61]. The capillary has a fracture covered with porous glass in a grounded electrolyte container. The separation potential across the capillary is dropped to zero at the position of the fracture (Figure 6.1(a)). Consequently end-capillary amperometric detection could be performed by inserting the microelectrodes from the end of the capillary at the potential at zero with limits of detection of 10^{-8} M for several catechol compounds [61, 62]. The implement has a glass coupler which is fragile and difficult to maintain. Porous glass as an on-column joint could be replaced by Nafion tubing [63] or a thin layer of cellulose acetate film [64, 65]. A robust and easy-to-handle field decoupler has been constructed with a palladium metal joint. The disadvantages of the palladium decoupler are the limitation of buffer composition and zone broadening [66].

End-column amperometric detection does not require a field decoupler (Figure 6.1(b)). Capillaries with a narrow inner diameter are preferable for end-column detection because the current across the capillary is negligible and it does not significantly affect the amperometric detection. Huang et al. demonstrated end-column amperometric detection of neurotransmitters with a capillary of 5 μm inner diameter [52]. Amperometric detection sensitivity was improved by etching the detector end of the capillary (Figure 6.1(c)). This design facilitates the alignment of the electrode to the capillary bore [67]. End-column detection with ultramicroelectrodes of 10 μm carbon fiber was achieved for CE separation with 25 μm capillaries by compensat-

Table 6.2 Recent publications on modifications of electrodes and applications in amperometric detection in CE

Sample	Electrode	Mode of detection	Capillary diameter (μm)	Limit of detection	Remarks	Ref.
Metal ions	Pt; Au; carbon fiber; mercury film	End-column	25	10 amol to 50 fmol	Under constant- and pulsed-voltage operations	110
Anthraquinone-2-carboxylic acid (*1); γ-aminobutyric acid(*2); mitomycin C in human serum	Carbon fiber	With a field decoupler	50	1.3 fmol (*1) 1.6 fmol (*2)	Connection with Nafion joint	111
Phenols; phenols in industrial waste water	Carbon fiber	With a field decoupler	50	0.03 ppm for 2,4-DMP	Connection with cellulose acetate joint	112
Thiols	Au/Hg	With a field decoupler	75	0.53 fmol for glutathione	Connection with Nafion joint	113
Thiols; disulfides	Mixed-valence ruthenium cyanide-modified carbon fiber array	End-column	20	2.5 fmol for glutathione disulfide; 1.3 fmol for cystine; 1.1 fmol for homocystine	Separation at pH 2.8; applied to the detection of cystine in the urine of a patient with kidney stones	114
Cysteine(*3); cystine(*4)	Au/Hg	End-column with a field decoupler	75	5 μmol l^{-1} (*3); 100 μmol l^{-1} (*4)	Dual-electrode detection; applied to human urine samples	115
Cysteine(*5); glutathione(*6); glucose	Cobalt phthalo-cyanine modified (for cysteine and glutathione); glucose oxidase immobilized (for glucose)	End-column with a field decoupler	25	31 nmol l^{-1} (*5); 0.3 μmol l^{-1} (*6)	Connection with Nafion joint	116

Table 6.2 Continued

Sample	Electrode	Mode of detection	Capillary diameter (μm)	Limit of detection	Remarks	Ref.
Amino acid derivatives	Carbon fiber	With a field decoupler	50	80 nmol l^{-1} for aspartate; 100 nmol l^{-1} for glutamate	Connection with Nafion joint; derivatization with naphthalenedialdehyde-cyanide prior to analysis; applied to *in vivo* sample	117
Amino acids; peptides	Cu	End-column in a wall-jet configuration	25	0.8 fmol for Trp and Tyr (min); 640 fmol for Phe (max)	Applied to the detection of amino acids in human urine, dipeptides in soft drinks and pentapeptides synthesized with solid phase	118
L-dopa and its metabolites	Carbon fiber	With a field decoupler	50	98 amol for L-dopa	Connection with Nafion joint; applied to microdialysis samples	119
Carbohydrates	Cu	End-column	50	Below 50 fmol	Separation at strong alkaline pH	120
Carbohydrates; sugar acids; alditols	Cu	End-column in a wall-jet configuration	25	At or below fmol level	Separation at strong alkaline pH	121
p-Aminophenol (microassay of alkaline phosphatase)	Pt	Detector assembly consists of two compartments screwed together	50	1.2 pmol for alkaline phosphatase	*p*-Aminophenol is the product of alkaline phosphatase	122
L-dopa; glutathione; cysteine; carbohydrates; amino acids; peptides	Carbon fiber; Au/Hg; Co phthalocyanine modified carbon; Au (in detail, see the original paper)	With a field decoupler	25; 50; 75	See original paper	Connection with Nafion joint; examples of pharmaceutical and biomedical applications	123

ing the interfering separation currents by a high impedance circuit. Comparable limits of detection for the narrower capillaries were obtained at 1.1 amol catechol with careful optimization of the electrode alignment [68].

More robust and easier-to-construct amperometric detectors have been developed for reproducible measurements. An ECD cell was designed for micro-liquid chromatography and CE, which has an effective cell volume of less than 1 nl. The detector cell in CE with a field decoupler shows the limit of detection less than 1 fmol catecholamines [69]. Ye and Baldwin demonstrated end-column detection with normal size electrodes in a wall-jet-like configuration [70].

Pulsed amperometric detection, a well established technique in HPLC, has been applied in CE. The pulsed operation reduces the fouling of the electrode and reconditions its surface. Pulsed amperometric detection for carbohydrates at noble metal electrodes was investigated [71–73].

Modifications of electrodes and applications in amperometric detection for CE were reported by several groups (Table 6.2).

6.9 Mass spectroscopy

Mass spectroscopy (MS) is commonly used as a useful detection technique in combination with gas chromatography (GC) or liquid chromatography (LC), because it provides chemical structure information on the analytes including molecular weight. Analytes should be ionized for MS detection. This should be easy, but depending upon the type of interface between CE and MS (there are several interface methods), it can present a problem especially for neutral compounds for which MEKC is applicable. In CE, analytes migrate in the capillary with a high separation efficiency and in most cases elute from the end of the capillary as ionized species. Therefore, an electrospray ionizer connected to the end of separation capillary is considered to be reasonable for the interface of CE and MS (CE–MS) developed by Smith and co-workers [74–76].

The CE–MS interface is mainly based on the electrospray ionization (EI) by vaporization of the solvent and concentration, using either coaxial sheath flow or a liquid–junction interface. A serious constraint currently for CE–MS is the limit of sensitivity although the analyte is partially concentrated during EI. The use of smaller inner diameter capillaries from 100 μm to 10 μm or from 50 μm to 5 μm improved the limit of detection by 25 to 50 times to below the femtomole level for peptide and protein mixtures [77, 78]. Fast CE separation combined with EI–MS detection was demonstrated for the separation of eight sulfonylureas by laboratory-made CE with short capillaries. While a simple CE–MS system with an electrospray interface provides the molecular weight of analyte, Garcia and Henion additionally demonstrated that CE–MS under tandem mass spectrometric conditions (CE–MS–MS) produces a full-scan collision-induced dissociation spectrum which represents the structural information [79]. The feasibility of MS detection following the CE separation of inorganic cations was suggested for an ion spray–sheath flow interface. Although the selected ion monitoring provides a picogram level for the limit of detection for Br$^-$, SCN$^-$ and HSeO$_4^-$, the separation of several cations has poor efficiency suffering from background interference of the CE buffer additives [80]. The addition of cetyltrimethyl ammonium chloride to the separation electrolyte facilitates the analysis of various cationic molecules by CE–MS. The surfactant decreases EOF and induces the sample flow to the CE–MS interface [81]. The CE–MS system is accompanied by background dependent on electrolyte composition, pH, flow rates, interface condition, etc. A combination of broadband collisional activation and resonance ejection reduced the background and improved the signal-to-noise ratio [82]. Garcia and Henion [83] found that gel-filled capillaries containing high concentrations of urea and Tris–borate buffer did not affect the performance of MS with an electrospray interface. This means that CE separation is decoupled from MS detection by the gel in the capillary [83]. Hofstadler *et al.* reported that on-line CE with Fourier transform ion cyclotron resonance MS based on EI probably enables higher resolution of separation, detection and sensitivity [84]. Conversion of a Finnigan-MAT TSQ-70 thermospray ionization interface to EI was investigated for connection to the CE system [85]. Other applications of CE–MS with an electrospray interface are reported [86–88].

Continuous flow fast atom bombardment can provide an interface for CE to MS often providing the structural information. The gas phase ions for detection are generated by bombardment with a xenon gun. Two types of fast atom bombardment interface were designed one with an coaxial flow chamber by Moseley *et al.* [89] and the other with liquid junction coupling by Reinhoud *et al.* [90]. Nichols *et al.* reported on the performance of the CE–MS interface with liquid junction coupling for industrial applications, comparing the results with the electrospray interface [91].

Although electrospray and flow atom bombardment ionizations are suitable for on-line coupling, the separation conditions are restricted. Off-line coupling with matrix-assisted laser adsorption ionization or ^{252}Cf plasma desorption ionization can be applied to wider separation conditions. It may become a powerful tool especially for the evaluation of the molecular weight and structure of proteins and peptides. Recent applications have been reported by several groups [92–97].

6.10 Refractive index

Refractive index (RI) is not yet an established detection method in CE, in spite of its universal use as an HPLC detector. The principal problem for RI detection in CE is the lower sensitivity while preserving higher separation efficiency. Krattiger and co-workers demonstrated the RI detector for CE consisting of a He–Ne laser, capillaries immersed in RI matching fluid and photodiode. Immersion of capillaries in RI matching fluid simplifies the light scattering pattern of the capillaries and reduces the thermal noise due to Joule heating in CE [98, 99]. Krattiger and co-workers developed another design for a RI detector for an off-axis model. Coherent light from a laser diode is partially deflected by the holographic optical element and the capillary is glued between two parallel glass plates with RI matching glue. Probing and reference beams are monitored with a photodiode array [100]. Wu and Pawliszyn [101] reported a monitoring device for a moving boundary. The direction of the beam is deflected when it passes through the RI gradient produced by the sample zone. This detector consists of a low power He–Ne laser or a laser diode and a photodiode position sensor and is applied to various separation modes in capillary zone electrophoresis (CZE), capillary isotachophoresis (CITP) and capillary isoelectric focusing (CIEF) [101].

Thermo-optical absorbance is one of the applications of RI detection with higher sensitivity. The subtle temperature rise caused by the excimer laser can be detected as a change in RI of the heated sample by taking advantage of the high temperature sensitivity of RI. Waldron and Dovichi [102] demonstrated that 20 phenylthiohydantoin (PTH)–amino acids were separated in MEKC and detected thermo-optically with a low energy krypton fluoride laser as an excimer and a 3 mW He–Ne laser as a probe beam. The limit of detection is 0.6 nl of 9×10^{-7} M PTH–glycine, corresponding to 0.5 fmol [102].

6.11 Other detection techniques

Development of many CE detectors based on other detection techniques is currently in progress. Details of some of examples can be referred to the original papers as follows;

1. radiodetection [103, 104];
2. laser-induced vibration [105];
3. multipoint detection [106];
4. frame photometric detection [107];
5. inductively coupled plasma atomic emission [108];
6. ^{1}H-NMR [109].

References

1 Szulc, M.E.; Krull, I.S. (1994) Improved detection and derivatization in capillary electrophoresis. *J. Chromatogr.* **659**, 231–245.

2 Kok, W.T. (1993) Off-column detection with pressure compensation in capillary electrophoresis. *Anal. Chem.* **65**, 1853–1860.
3 Cassidy, R.M.; Lu, W.; Tse, V.-P. (1994) Auxiliary electroosmotic flow for postcapillary reaction detection in capillary electrophoresis. *Anal. Chem.* **66**, 2578–2583.
4 Xue, Y.; Yeung, E.S. (1994) Laser-based ultraviolet absorption detection in capillary electrophoresis. *Appl. Spectrosc.* **48**, 502–506.
5 Maystre, F.; Bruno, A.E. (1992) Laser beam probing in capillary tubes. *Anal. Chem.* **64**, 2885–2887.
6 Flint, C.D.; Grochowicz, P.R.; Simpson, C.E. (1994) Design, construction and evaluation of an ultraviolet absorbance detector for capillary electrophoresis. *Anal. Proc. Anal. Commun.* **31**, 117–121.
7 Beck, W.; van Hoek, R.; Engelhardt, H. (1993) Application of a diode-array detector in capillary electrophoresis. *Electrophoresis* **14**, 540–546.
8 Wu, J.; Pawliszyn, J. (1992) Universal detection for capillary isoelectric focusing without mobilization using a concentration gradient imaging system. *Anal. Chem.* **64**, 224–227.
9 Wang, T.; Hartwick, R.A. (1992) Whole column absorbance detection in capillary isoelectric focusing. *Anal. Chem.* **64**, 1745–1747.
10 Guzman, N.A.; Moschera, J.; Bailey, C.A.; Iqbal, K.; Malick, A.W. (1992) Assay of protein drug substances present in solution mixtures by fluorescamine derivatization and capillary electrophoresis. *J. Chromatogr.* **598**, 123–131.
11 Yeung, E.S.; Wang, P.; Li, W.; Giese, R.W. (1992) Laser fluorescence detector for capillary electrophoresis. *J. Chromatogr.* **608**, 73–77.
12 Toulas, C.; Hernandez, L. (1992) Applications of a laser-induced fluorescence detector for capillary electrophoresis to measure attomolar and zeptomolar amounts of compounds. *LC-GC* **10**, 471–476.
13 Taylor, J.A.; Yeung, E.S. (1992) Axial-beam laser-excited fluorescence detection in capillary electrophoresis. *Anal. Chem.* **64**, 1741–1744.
14 Taylor, J.A.; Yeung, E.S. (1993) Multiplexed fluorescence detector for capillary electrophoresis using axial optical fiber illumination. *Anal. Chem.* **65**, 956–960.
15 Huang, X.C.; Quesada, M.A.; Mathies, R.A. (1992) Capillary array electrophoresis using laser-excited confocal fluorescence detection. *Anal. Chem.* **64**, 967–972.
16 Lee, T.T.; Yeung, E.S. (1992) High-sensitivity laser-induced fluorescence detection of native proteins in capillary electrophoresis. *J. Chromatogr.* **595**, 319–325.
17 Lee, T.T.; Yeung, E.S. (1992) Quantitative determination of native proteins in individual human erythrocytes by capillary zone electrophoresis with laser-induced fluorescence detection. *Anal. Chem.* **64**, 3045–3051.
18 Nie, S.; Dadoo, R.; Zare, R.N. (1993) Ultrasensitive fluorescence detection of polycyclic aromatic hydrocarbons in capillary electrophoresis. *Anal. Chem.* **65**, 3571–3575.
19 Chan, K.C.; Janini, G.M.; Muschik, G.M.; Issaq, H.J. (1993) Pulsed-UV laser-induced fluorescence detection of native peptides and proteins in capillary electrophoresis. *J. Liq. Chromatogr.* **16**, 1877–1890.
20 Higashijima, T.; Fuchigami, T.; Imasaka, T.; Ishibashi, N. (1992) Determination of amino acids by capillary zone electrophoresis based on semiconductor laser fluorescence detection. *Anal. Chem.* **64**, 711–714.
21 Fuchigami, T.; Imasaka, T.; Shiga, M. (1993) Ultratrace analysis of biological substances by capillary electrophoresis/semiconductor laser fluorometry. *SPIE* **1885**, 435–438.
22 Jansson, M.; Roeraade, J.; Laurell, F. (1993) Laser-induced fluorescence detection in capillary electrophoresis with blue light from a frequency-doubled diode laser. *Anal. Chem.* **65**, 2766–2769.
23 Fuchigami, T.; Imasaka, T.; Shiga, M. (1993) Subattomole detection of amino acids by capillary electrophoresis based on semiconductor laser fluorescence detection. *Anal. Chim. Acta* **282**, 209–213.
24 Kaneta, T.; Yamashita, T.; Imasaka, T. (1995) Separation of polycyclic aromatic hydrocarbons by micellar electrokinetic chromatography with laser fluorescence detection. *Anal. Chim. Acta* **299**, 371–375.
25 Miller, K.J.; Lytle, F.E. (1993) Capillary zone electrophoresis with time-resolved fluorescence detection using a diode-pumped solid-state laser. *J. Chromatogr.* **648**, 245–250.
26 Cheng, Y.F.; Dovichi, N.J. (1988) Subattomole amino acid analysis by capillary zone electrophoresis and laser-induced fluorescence. *Science* **242**, 562–564.

27 McGregor, D.A.; Yeung, E.S. (1994) Detection of DNA fragments separated by capillary electrophoresis based on their native fluorescence inside a sheath flow. *J. Chromatogr.* **680**, 491–496.
28 Chen, D.Y.; Dovichi, N.J. (1994) Yoctomole detection limit by laser-induced fluorescence in capillary electrophoresis. *J. Chromatogr.* **657**, 265–269.
29 Zhao, J.Y.; Dovichi, N.J.; Hindsgaul, O.; Gosselin, S.; Palcic, M.M. (1994) Detection of 100 molecules of product formed in a fucosyl tranferase reaction. *Glycobiology* **4**, 239–242.
30 Chen, D.Y.; Adelhelm, K.; Cheng, X.L.; Dovichi, N.J. (1994) A simple laser-induced fluorescence detector for sulforhodamine 101 in a capillary electrophoresis system: Detection limits of 10 yoctomoles or six molecules. *Analyst* **119**, 349–352.
31 Shear, J.B.; Dadoo, R.; Fishman, H.A.; Scheller, R.H.; Zare, R.N. (1993) Optimizing fluorescence detection in chemical separations for analyte bands traveling at different velocities. *Anal. Chem.* **65**, 2977–2982.
32 Shear, J.B.; Colón, L.A.; Zare, R.N. (1993) Automated velocity programming for increased detection zone residence times in capillary electrophoresis. *Anal. Chem.* **65**, 3708–3712.
33 Liu, J.; Hsieh, Y.-Z.; Wiesler, D.; Novotny, M. (1991) Design of 3-(4-carboxybenzoyl)-2-quinolinecarboxaldehyde as a reagent for ultrasensitive determination of primary amines by capillary electrophoresis using laser fluorescence detection. *Anal. Chem.* **63**, 408–412.
34 Bergquist, J.; Gilman, S.D.; Ewing, A.G.; Ekman, R. (1994) Analysis of human cerebrospinal fluid by capillary electrophoresis with laser-induced fluorescence detection. *Anal. Chem.* **66**, 3512–3518.
35 Tseng, H.C.; Dadoo, R.; Zare, R.N. (1994) Selective determination of adenine-containing compounds by capillary electrophoresis with laser-induced fluorescence detection. *Anal. Biochem.* **222**, 55–58.
36 Klockow, A.; Widmer, H.M.; Amadò, R.; Paulus, A. (1994) Capillary electrophoresis of ANTS labelled oligosaccharide ladders and complex carbohydrates. *Fresenius J. Anal. Chem.* **350**, 415–425.
37 Arriaga, E.; Chen, D.Y.; Cheng, X.L.; Dovichi, N.J. (1993) High-efficiency filter fluorometer for capillary electrophoresis and its application to fluorescein thiocarbamyl amino acids. *J. Chromatogr.* **652**, 347–353.
38 Chan, K.T.; Yao, Y.J.; Ng, C.L.; Li, S.F.Y. (1994) Fluorescence detection in capillary electrophoresis using a variable wavelength epi-fluorescence microscope. *Fresenius J. Anal. Chem.* **349**, 487–490.
39 Dadoo, R.; Colón, L.A.; Zare, R.N. (1992) Chemiluminescence detection in capillary electrophoresis. *J. High Resolut. Chromatogr.* **15**, 133–135.
40 Ruberto, M.A.; Grayeski, M.L. (1992) Acridinium chemiluminescence detection with capillary electrophoresis. *Anal. Chem.* **64**, 2758–2762.
41 Wu, N.; Huie, C.W. (1993) Peroxyoxalate chemiluminescence detection in capillary electrophoresis. *J. Chromatogr.* **634**, 309–315.
42 Hara, T.; Kayama, S.; Nishida, H.; Nakajima, R. (1994) Simple apparatus for on-line chemiluminescence detection of proteins separated by capillary zone electrophoresis. *Anal. Sci.* **10**, 223–225.
43 Hara, T.; Nishida, H.; Nakajima, T. (1994) Highly sensitive detection of proteins separated by capillary zone electrophoresis using on-line chemiluminescence detection. *Anal. Sci.* **10**, 823–825.
44 Gilman, S.D.; Silverman, C.E.; Ewing, A.G. (1994) Electrogenerated chemiluminescence detection for capillary electrophoresis. *J. Microcol. Sep.* **6**, 97–106.
45 Dadoo, R.; Seto, A.G.; Colón, L.A.; Zare, R.N. (1994) End-column chemiluminescence detector for capillary electrophoresis. *Anal. Chem.* **66**, 303–306.
46 Wang, T.; Hartwick, R.A. (1992) Noise and detection limits of indirect absorption detection in capillary zone electrophoresis. *J. Chromatogr.* **607**, 119–125.
47 Williams, S.J.; Bergström, E.T.; Goodall, D.M.; Kawazumi, H.; Evans, K.P. (1993) Diode laser-based indirect absorbance detector for capillary electrophoresis. *J. Chromatogr.* **636**, 39–45.
48 Xue, Y.; Yeung, E.S. (1993) Double-beam laser indirect absorption detection in capillary electrophoresis. *Anal. Chem.* **65**, 2923–2927.

49 Nielen, M.W.F. (1992) Indirect time-resolved luminescence detection in capillary zone electrophoresis. *J. Chromatogr.* **608**, 85–92.
50 Ewing, A.G.; Mesaros, J.M.; Gavin, P.F. (1994) Electrochemical detection in microcolumn separations. *Anal. Chem.* **66**, 527A–537A.
51 Huang, X.; Pang, T.-K.J.; Gordon, M.J.; Zare, R.N. (1987) On-column conductivity detector for capillary zone electrophoresis. *Anal. Chem.* **59**, 2747–2749.
52 Huang, X.; Zare, R.N.; Sloss, S.; Ewing, A.G. (1991) End-column detection for capillary zone electrophoresis. *Anal. Chem.* **63**, 189–192.
53 Huang, X.; Zare, R.N. (1991) Improved end-column conductivity detector for capillary zone electrophoresis. *Anal. Chem.* **63**, 2193–2196.
54 Avdalovic, N.; Phol, C.A.; Rocklin, R.D.; Stillian, J.R. (1993) Determination of cations and anions by capillary electrophoresis combined with suppressed conductivity detection. *Anal. Chem.* **65**, 1470–1475.
55 Dasgupta, P.K.; Bao, L. (1993) Suppressed conductometric capillary electrophoresis separation systems. *Anal. Chem.* **65**, 1003–1011.
56 Kar, S.; Dasgupta, P.K.; Liu, H.; Hwang, H. (1994) Computer-interfaced bipolar pulse conductivity detector for capillary systems. *Anal. Chem.* **66**, 2537–2543.
57 Nann, A.; Simon, W. (1993) On-column detection in capillary zone electrophoresis with ion-selective microelectrodes in conical capillary apertures. *J. Chromatogr.* **633**, 207–211.
58 Nann, A.; Silvestri, I.; Simon, W. (1993) Quantitative analysis in capillary zone electrophoresis using ion-selective microelectrodes as on-column detectors. *Anal. Chem.* **65**, 1662–1667.
59 Nann, A.; Pretsch, E. (1994) Potentiometric detection of amions separated by capillary electrophoresis using an ion-selective microelectrode. *J. Chromatogr.* **676**, 437–442.
60 Hauser, P.C.; Renner, N.D.; Hong, A.P.C. (1994) Anion detection in capillary electrophoresis with ion-selective microelectrodes. *Anal. Chim. Acta* **295**, 181–186.
61 Wallingford, R.A.; Ewing, A.G. (1987) Capillary zone electrophoresis with electrochemical detection. *Anal. Chem.* **59**, 1762–1766.
62 Wallingford, R.A.; Ewing, A.G. (1989) Separation of serotonin from catechols by capillary zone electrophoresis with electrochemical detection. *Anal. Chem.* **61**, 98–100.
63 O'Shea, T.J.; Greenhagen, R.D.; Lunte, S.M.; Lunte, C.E.; Smyth, M.R.; Radzik, D.M.; Watanabe, N. (1992) Capillary electrophoresis with electrochemical detection employing an on-column Nafion joint. *J. Chromatogr.* **593**, 305–312.
64 Whang, C.-W.; Chen, I.-C. (1992) Cellulose acetate-coated porous polymer joint for capillary zone electrophoresis. *Anal. Chem.* **64**, 2461–2464.
65 Chen, I-C.; Whang, C.-W. (1993) Capillary electrophoresis with amperometric detection using a porous cellulose acetate joint. *J. Chromatogr.* **644**, 208–212.
66 Kok, W.T.; Sahin, Y. (1993) Solid-state field decoupler for off-column detection in capillary electrophoresis. *Anal. Chem.* **65**, 2497–2501.
67 Sloss, S.; Ewing, A.G. (1993) Improved method for end-column amperometric detection for capillary electrophoresis. *Anal. Chem.* **65**, 577–581.
68 Lu, W.; Cassidy, R.M.; Baranski, A.S. (1993) End-column electrochemical detection for inorganic and organic species in high-voltage capillary electrophoresis. *J. Chromatogr.* **640**, 433–440.
69 Tüdös, A.J.; Van Dyck, M.M.C.; Poppe, H.; Kok, W.T. (1993) An electrochemical detector cell for open tubular liquid chromatography and capillary electrophoresis. *Chromatographia* **37**, 79–85.
70 Ye, J.; Baldwin, R.P. (1993) Amperometric detection in capillary electrophoresis with normal size electrodes. *Anal. Chem.* **65**, 3525–3527.
71 O'Shea, T.J.; Lunte, S.M.; LaCourse, W.R. (1993) Detection of carbohydrates by capillary electrophoresis with pulsed amperometric detection. *Anal. Chem.* **65**, 948–951.
72 Lu, W.; Cassidy, R.M. (1993) Pulsed amperometric detection of carbohydrates separated by capillary electrophoresis. *Anal. Chem.* **65**, 2878–2881.
73 Roberts, R.E.; Johnson, D.C. (1994) Fast-pulsed electrochemical detection at noble metal electrodes: The frequency-dependent response at gold electrodes for chromatographically separated carbohydrates. *Electroanalysis* **6**, 269–273.
74 Smith, R.D.; Olivares, J.A.; Nguyen, N.T.; Udseth, H.R. (1988) Capillary zone electrophoresis–mass spectrometry using an electrospray ionization interface. *Anal. Chem.* **60**, 436–441.

75 Smith, R.D.; Loo, J.A.; Edmonds, C.G.; Barinaga, C.J.; Udseth, H.R. (1990) New developments in biochemical mass spectrometry: Electrospray ionization. *Anal. Chem.* **62**, 882–899.
76 Smith, R.D.; Fields, S.M.; Loo, J.A.; Barinaga, C.J.; Udseth, H.R.; Edmonts, C.G. (1990) Capillary isotachophoresis with UV and tandem mass spectrometric detection for peptides and proteins. *Electrophoresis* **11**, 709–717.
77 Wahl, J.H.; Goodlett, D.R.; Udseth, H.R.; Smith, R.D. (1992) Attomole level capillary electrophoresis–mass spectrometric protein analysis using 5-μm-i.d. capillaries. *Anal. Chem.* **64**, 3194–3196.
78 Wahl, J.H.; Goodlett, D.R.; Udseth, H.R.; Smith, R.D. (1993) Use of small-diameter capillaries for increasing peptide and protein detection sensitivity in capillary electrophoresis–mass spectrometry. *Electrophoresis* **14**, 448–457.
79 Garcia, F.; Henion, J. (1992) Fast capillary electrophoresis–ion spray mass spectrometric determination of sulfonylureas. *J. Chromatoger.* **606**, 237–247.
80 Huggins, T.G.; Henion, J.D. (1993) Capillary electrophoresis/mass spectrometry determination of inorganic ions using an ion spray–sheath flow interface. *Electrophoresis* **14**, 531–539.
81 Varghese, J.; Cole, R.B. (1993) Cetyltrimethylammonium chloride as a surfactant buffer additive for reversed-polarity capillary electrophoresis–electrospray mass spectormetry. *J. Chromatogr.* **652**, 369–376.
82 Ramsey, R.S.; Goeringer, D.E.; McLuckey, S.A. (1993) Active chemical background and noise reduction in capillary electrophoresis/ion trap mass spectrometry. *Anal. Chem.* **65**, 3521–3524.
83 Garcia, F.; Henion, J.D. (1992) Gel-filled capillary electrophoresis/mass spectrometry using a liquid junction–ion spray interface. *Anal. Chem.* **64**, 985–990.
84 Hofstadler, S.A.; Wahl, J.H.; Bruce, J.E.; Smith, R.D. (1993) On-line capillary electrophoresis with Fourier transform ion cyclotron resonance mass spectrometry. *J. Am. Chem. Soc.* **115**, 6983–6984.
85 Jackett, S.; Moini, M. (1994) Conversion of the Finnigan-MAT TSQ-70 thermospray ionization interface to an electrospray ionization interface. *Rev. Sci. Instrum.* **65**, 591–596.
86 Hsieh, F.Y.L.; Cai, J.; Henion, J. (1994) Determination of trace impurities of peptides and alkaloids by capillary electrophoresis–ion spray mass spectrometry. *J. Chromatogr.* **679**, 206–211.
87 Tomlinson, A.J.; Benson, L.M.; Johnson, K.L.; Naylor, S. (1994) Investigation of drug metabolism using capillary electrophoresis with photodiode array detection and on-line mass spectrometry equipped with an array detector. *Electrophoresis* **15**, 62–71.
88 Tomlinson, A.J.; Benson, L.M.; Gorrod, J.W.; Naylor, S. (1994) Investigation of the *in vitro* metabolism of the H_2-antagonist mifentidine by on-line capillary electrophoresis–mass spectrometry using non-aqueous separation conditions. *J. Chromatogr.* **657**, 373–381.
89 Moseley, M.A.; Deterding, L.J.; Tomer, K.B.; Jorgenson, J.W. (1989) Coupling of capillary zone electrophoresis and capillary liquid chromatography with coaxial continuous-flow fast atom bombardment tandem sector mass spectrometry. *J. Chromatogr.* **480**, 197–209.
90 Reinhoud, N.J.; Niessen, W.M.A.; Tjaden, U.R.; Gramberg, L.G.; Verheij, E.R.; van der Greef, J. (1989) Performance of a liquid-junction interface for capillary electrophoresis mass spectrometry using continuous-flow fast-atom bombardment. *Rapid Commun. Mass Spectrom.* **3**, 348–351.
91 Nichols, W.; Zweigenbaum, J.; Garcia, F.; Johansson, M.; Henion, J. (1992) CE–MS for industrial applications using a liquid junction with ion-spray and CF–FAB mass spectrometry. *LC–GC* **10**, 676–686.
92 Keough, T.; Takigiku, R.; Lacey, M.P.; Purdon, M. (1992) Matrix-assisted laser desorption mass spectrometry of proteins isolated by capillary zone electrophoresis. *Anal. Chem.* **64**, 1594–1600.
93 van Veelen, P.A.; Tjaden, U.R.; van der Greef, J.; Ingendoh, A.; Hillenkamp, F. (1993) Off-line coupling of capillary electrophoresis with matrix-assisted laser desoprtion mass spectrometry. *J. Chromatogr.* **647**, 367–374.
94 Castoro, J.A.; Chiu, R.W.; Monnig, C.A.; Wilkins, C.L. (1992) Matrix-assisted laser

desorption/ionization of capillary electrophoresis effluents by Fourier transform mass spectrometry. *J. Am. Chem. Soc.* **114**, 7571–7572.
95 Weinmann, W.; Parker, C.E.; Deterding, L.J.; Papac, D.I.; Hoyes, J.; Przybylski, M.; Tomer, K.B. (1994) Capillary electrophoresis–matrix-assisted laser-desorption ionization mass spectrometry of proteins. *J. Chromatogr.* **680**, 353–361.
96 Weinmann, W.; Baumeister, K.; Kaufmann, I.; Przybylski, M. (1993) Structural characterization of polypeptides and proteins by combination of capillary electrophoresis and ^{252}Cf plasma desorption mass spectrometry. *J. Chromatogr.* **628**, 111–121.
97 Weinmann, W.; Parker, C.E.; Baumeister, K.; Maier, C.; Tomer, K.B.; Przybylski, M. (1994) Capillary electrophoresis combined with ^{252}Cf plasma desorption and electrospray mass spectrometry for the structural characterization of hydrophobic polypeptides using organic solvents. *Electorphoresis* **15**, 228–233.
98 Bruno, A.E.; Krattiger, B.; Maystre, F.; Widmer, H.M. (1991) On-column laser-based refractive index detector for capillary electrophoresis. *Anal. Chem.* **63**, 2689–2697.
99 Krattiger, B.; Bruno, A.E.; Widmer, H.M.; Geiser, M.; Dändliker, R. (1993) Laser-based refractive-index detection for capillary electrophoresis: ray-tracing interference theory. *Appl. Optics* **32**, 956–965.
100 Krattiger, B.; Bruin, G.J.M.; Bruno, A.E. (1994) Hologram-based refractive index detector for capillary electrophoresis: Separation of metal ions. *Anal. Chem.* **66**, 1–8.
101 Wu, J.; Pawliszyn, J. (1992) Multi-purpose capillary electrophoresis system with concentration gradient detection. *Talanta* **39**, 1281–1288.
102 Waldron, K.C.; Dovichi, N.J. (1992) Sub-femtomole determination of phenylthiohydantoin-amino acids: Capillary electrophoresis and thermooptical detection. *Anal. Chem.* **64**, 1396–1399.
103 Westerberg, G.; Lundqvist, H.; Kilár, F.; Långström, B. (1993) β$^+$-Selective radiodetector for capillary electrophoresis. *J. Chromatogr.* **645**, 319–325.
104 Tracht, S.; Toma, V.; Sweedler, J.V. (1994) Postcolumn radionuclide detection of low-energy β emitters in capillary electrophoresis. *Anal. Chem.* **66**, 2382–2389.
105 Odake, T.; Kitamori, T.; Sawada, S. (1992) Direct detection of laser-induced capillary vibration by a piezoelectric transducer. *Anal. Chem.* **64**, 2870–2871.
106 Srichaiyo, T.; Hjertén, S. (1992) Simple multi-point detection method for high-performance capillary electrophoresis. *J. Chromatogr.* **604**, 85–89.
107 Sänger-van de Griend, C.E.; Kientz, C.E.; Brinkman, U.A.T. (1994) Capillary electrophoresis coupled on-line with flame photometric detection. *J. Chromatogr.* **673**, 299–302.
108 Tarr, M.A.; Zhu, G.; Browner, R.F. (1993) Microflow ultrasonic nebulizer for inductively coupled plasma atomic emission spectrometry. *Anal. Chem.* **65**, 1689–1695.
109 Wu, N.; Peck, T.L.; Webb, A.G.; Magin, R.L.; Sweedler, J.V. (1994) Nanoliter volume sample cells for ^{1}H NMR: Application to on-line detection in capillary electrophoresis. *J. Am. Chem. Soc.* **116**, 7929–7930.
110 Lu, W.; Cassidy, R.M. (1993) Evaluation of ultramicroelectrodes for the detection of metal ions separated by capillary electrophoresis. *Anal. Chem.* **65**, 1649–1653.
111 Malone, M.A.; Weber, P.L.; Smyth, M.R.; Lunte, S.M. (1994) Reductive electrochemical detection for capillary electrophoresis. *Anal. Chem.* **66**, 3782–3787.
112 Chen, I-C.; Whang, C.-W. (1994) Capillary zone electrophoresis of eleven priority phenols with amperometric detection. *J. Chin. Chem. Soc.* **41**, 419–424.
113 O'Shea, T.J.; Lunte, S.M. (1993) Selective detection of free thiols by capillary electrophoresis-electrochemistry using a gold/mercury amalgam microelectrode. *Anal. Chem.* **65**, 247–250.
114 Zhou, J.; O'Shea, T.J.; Lunte, S.M. (1994) Simultaneous detection of thiols and disulfides by capillary electrophoresis–electrochemical detection using a mixed-valence ruthenium cyanide-modified microelectrode. *J. Chromatogr.* **680**, 271–277.
115 Lin, B.L.; Colón, L.A.; Zare, R.N. (1994) Dual electrochemical detection of cysteine and cystine in capillary zone electrophoresis. *J. Chromatogr.* **680**, 263–270.
116 O'Shea, T.J.; Lunte, S.M. (1994) Chemically modified microelectrodes for capillary electrophoresis/electrochemistry. *Anal. Chem.* **66**, 307–311.
117 O'Shea, T.J.; Weber, P.L.; Bammel, B.P.; Lunte, C.E.; Lunte, S.M.; Smyth, M.R. (1992) Monitoring excitatory amino acid release *in vivo* by microdialysis with capillary electrophoresis–electrochemistry. *J. Chromatogr.* **608**, 189–195.

118 Ye, J.; Baldwin, R.P. (1994) Determination of amino acids and peptides by capillary electrophoresis and electrochemical detection at a copper electrode. *Anal. Chem.* **66**, 2669–2674.
119 O'Shea, T.J.; Telting-Diaz, M.W.; Lunte, S.M.; Lunte, C.E.; Smyth, M.R. (1992) Capillary electrophoresis–electrochemistry of microdialysis samples for pharmacokinetic studies. *Electroanalysis* **4**, 463–468.
120 Colón, L.A.; Dadoo, R.; Zare, R.N. (1993) Determination of carbohydrates by capillary zone electrophoresis with amperometric detection at a copper microelectrode. *Anal. Chem.* **65**, 476–481.
121 Ye, J.; Baldwin, R.P. (1994) Determination of carbohydrates, sugar acids and alditols by capillary electrophoresis and electrochemical detection at a copper electrode. *J. Chromatogr.* **687**, 141–148.
122 Wu, D.; Regnier, F.E.; Linhares, M.C. (1994) Electrophoretically mediated micro-assay of alkaline phosphatase using electrochemical and spectrophotometric detection in capillary electrophoresis. *J. Chromatogr.* **657**, 357–363.
123 Lunte, S.M.; O'Shea, T.J. (1994) Pharmaceutical and biomedical applications of capillary electrophoresis/electrochemistry. *Electrophoresis* **15**, 79–86.

7 Recent advances in fluorescence detection technology
M. SAITO AND Y. KUROSU

7.1 Introduction

Capillary electrophoresis (CE) is the most rapidly expanding area in separation science. In 1989 when Hernadez *et al.* reviewed the total number of articles on CE, only 120 had been published. They stated 'During the last decade the number of scientific papers published in this field has grown exponentially' [1]. Expansion in publication has continued. Now, only five years after their review, the total number of the articles published in 1994 was nearly 1000. The number published in 1994 was approximately 50 times that in 1985 (Figure 7.1).

Detection in CE has mainly been by ultraviolet light absorption (UV absorption) owing to instrumental simplicity; part of the (fused silica) capillary separation tube can serve as the flowcell. Fluorescence (FL) detection has also been used; however, the number of reports have been fewer compared with UV detection. More than 90% of reports using CE have been carried out using UV detection, though the extremely short path length of the flowcell (internal diameter of the capillary of 50–75 µm) restricts the achievable sensitivity. This ratio between the frequency of use of UV and FL seems to have not changed since 1985 (Figure 7.1).

7.2 Inherent complexity in fluorescence detection

There are a few points which must be overcome in order to utilize FL detection efficiently in CE. The small internal diameter of the capillary, only 50 µm, also limits the size of the flowcell in FL with the consequence that the proportion of light scattered at the cell becomes very high, resulting in too high a background signal and hence noise.

In general, a CE flowcell includes boundaries of air-to-silica, silica-to-liquid, liquid-to-silica and silica-to-air. These boundaries are the causes of light scattering that makes FL detection more complex than UV detection. Light is reflected at any optical boundary where a refractive index (RI) change occurs. When light falls obliquely on such a boundary some of the light is reflected and the reflectance varies with the incident angle. Polarization and phase changes also take place but these can be neglected because

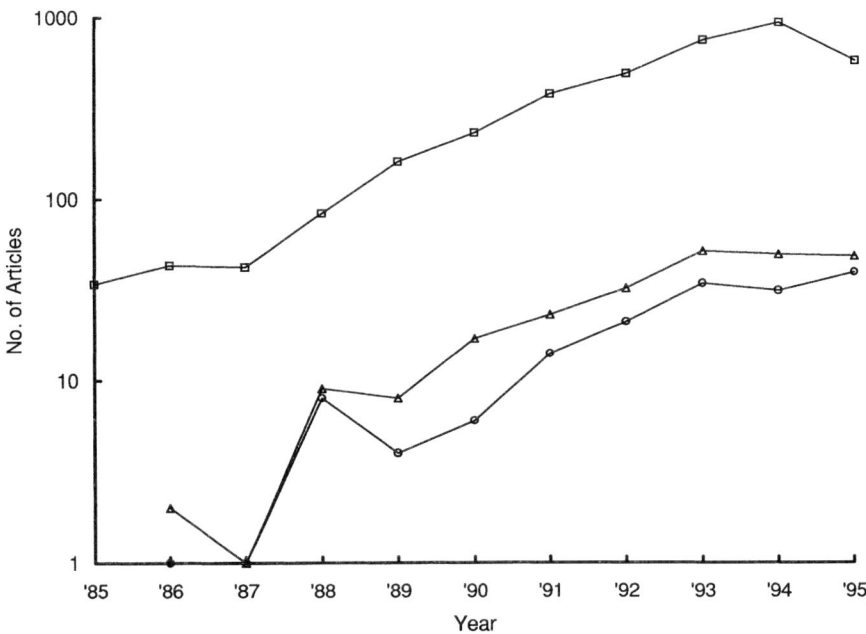

Figure 7.1 The number of articles on capillary electrophoresis published each year since 1985. Note that the vertical axis is logarithmic and the numbers in 1995 include only those up to the third quarter. Total (□): total number of articles with 'electrophoresis' and 'capillary' as the key words. FL total (△): total number of articles that include 'fluorescence' in the 'Total'. LIF (○): total number of articles that include 'laser' in the 'FL' total.

only the overall reflectance is taken into account in the case of FL detection. At the boundary of media with refractive indices n_1 and n_2, the overall reflectance is given by the following equation (Fresnel's law):

$$R = \frac{1}{2}\left(\frac{\sin^2(\phi - \phi')}{\sin^2(\phi + \phi')} + \frac{\tan^2(\phi - \phi')}{\tan^2(\phi + \phi')}\right) \quad (7.1)$$

where
$$\frac{\sin\phi}{\sin\phi'} = \frac{n_2}{n_1}.$$

Four of these boundaries occur in a liquid-filled fused-silica capillary mentioned above. Scattering by total and multiple reflection is greatest when light is incident on a boundary from high to low RI at an angle greater than the critical angle (defined as critical angle = $\sin^{-1}(n_1/n_2)$). Two such boundaries (silica-to-liquid (electrolyte) and silica-to-air) cause internal reflection in a CE flowcell, leading to high background noise and poor sensitivity [2].

An ordinary FL cell or flowcell has a rectangular cross-section in order to minimize total and multiple reflections. The excitation light beam strikes one of the cell walls at a right angle (which means the incident angle is 0) and the FL emission is collected through the wall adjacent to the one used for the excitation beam, i.e. emission is collected at right angles to the excitation beam. By this arrangement, the amount of scattered light arising from total and multiple reflection is significantly reduced. For CE, however, it is extremely difficult to make a flowcell with the above configuration and a volume small enough to match the diameter of the capillary separation tube and it is, therefore, necessary to utilize part of the separation tube as a flowcell by removing the polymer coating in the same manner as for a UV detection cell, or other techniques will be required.

When a FL flowcell similar to an UV detection cell is employed, the cell has a cylindrical cross-section. Figure 7.2 illustrates schematically the scattering of the excitation beam by a cylindrical flowcell. The flow direction is along the axis of the cylindrical cell, i.e. perpendicular to the plane of the paper. Part of the excitation light beam is reflected on the outer surface of the cell, which is an air-to-silica boundary. The amount of light reflected at this point may, however, be less than a few tenths of the incident light because the boundary is from low-to-high refractive index and the incident angle is close to 0 (except for the beam grazing the silica surface). Most of the light passes through the boundary and hits the internal surface of the cell, a silica-to-liquid boundary. Though the RI difference is small, this boundary is normally from high-to-low RI and, therefore, total reflection

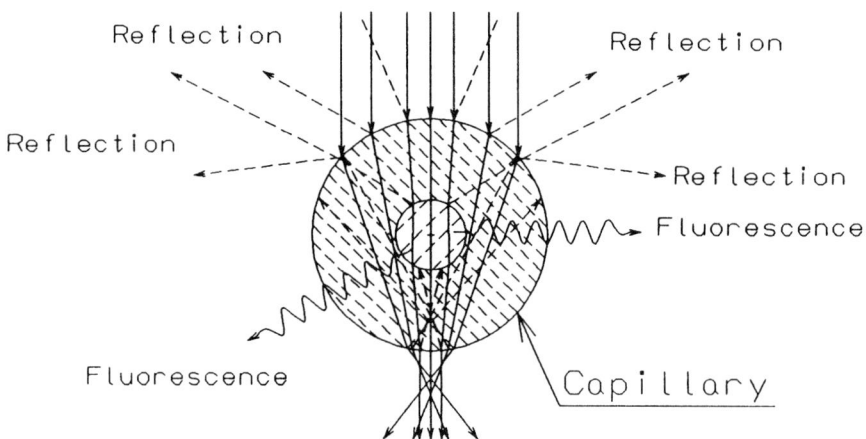

Figure 7.2 Scattering of excitation beam by a cylindrical flowcell.

can occur for light passing far from the center of the beam (where the incident angle approaches 90°). When this totally reflected light hits the silica-to-air boundary, it is efficiently reflected by the same mechanism and directed back towards the silica-to-liquid boundary. In this way, multiple reflection takes place. Whenever light strikes a boundary at an incident angle smaller than the critical angle, part of the light passes through the outer wall of the cell; this results in scattering of the excitation light beam in every direction. In order to avoid these unwanted effects, it is essential to irradiate the cell or capillary wall over as small a spot as possible.

7.3 Recent advances in fluorescence detection technology

Several techniques to reduce the scattering and to increase the detection sensitivity have been proposed. Such techniques involve flowcell arrangements, optical systems, selection of excitation light sources and fluorescent derivatization.

7.3.1 Flowcell arrangements

Since a flowcell configuration is very important for obtaining higher sensitivity in CE–FL, there are many reports on various types of flowcell arrangements; a microscope, a sheath flowcell and an immersed flowcell.

7.3.1.1 FL microscopy. Hernandez et al. proposed an epillumination FL microscope with collinear or confocal optical configuration [3]. In this optical setup, light from a 50 W high pressure mercury lamp is directed to the capillary wall by means of a chromatic beam splitter through an objective lens, and then the FL emission is collected by the same objective lens and is passed to the beam splitter along the same optical axis. Emission light that has a longer wavelength than the excitation light passes through the beam splitter to a photomultiplier. Excitation light above 405 nm was removed using a filter and a chromatic beam splitter allows emission above 420 nm. Later, the excitation source was replaced with an argon ion laser (488 nm) [4].

7.3.1.2 Sheath-flow cuvette flow chamber. Dovichi and co-workers demonstrated the usefulness of a sheath-flow cuvette flow chamber [5–10]. A schematic diagram of a typical sheath-flow curvette is shown in Figure 7.3. This arrangement is used to reduce scattering of excitation light by reducing the number of boundaries where changes in refractive indices take place. The chamber they used was typically 200 μm square in cross-section, 20 mm long and had 2 mm thick windows. The 190 μm outer diameter capillary was placed within this chamber and effluent from the capillary was swept by a

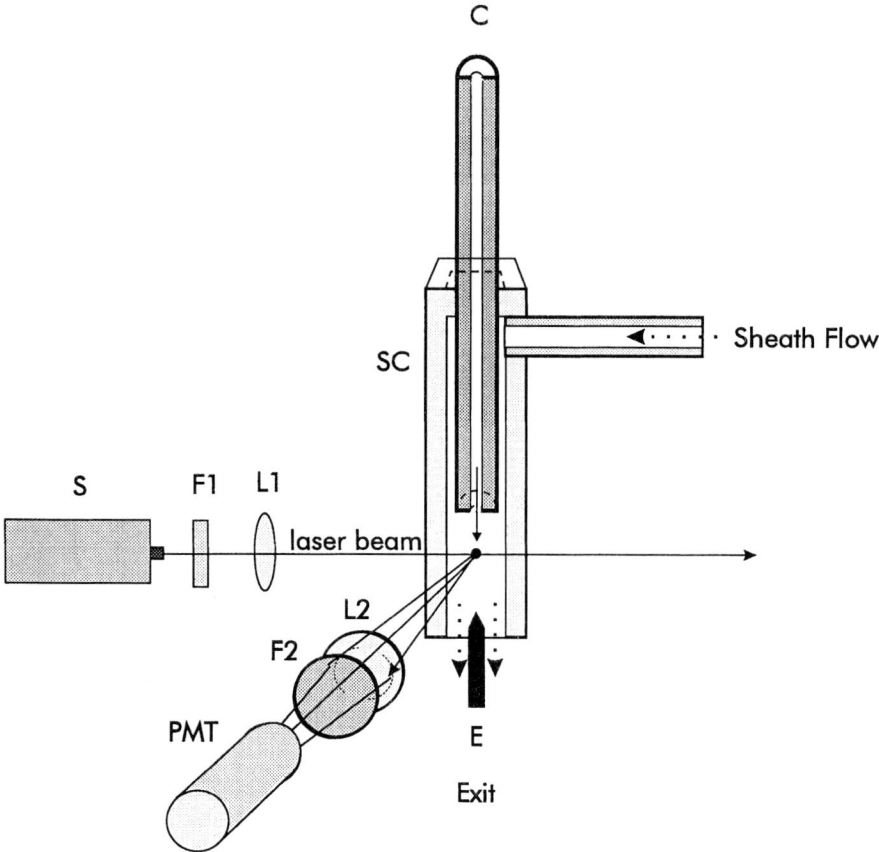

Figure 7.3 A typical sheath-flow cuvette. Components: S = laser, F1 = optical filter for excitation beam, L1 = beam condenser lens, SC = sheath-flow cuvette, C = separtion capillary, E = electrode, L2 = objective lens, F2 = optical filter for fluorescence emission, PMT = photomultiplier.

sheath stream which was usually composed of the same buffer as is used for the separation. By this configuration, rectangular walls can be formed which enclose liquid containing analytes, eliminating the most unwanted boundary of small circular cross-section; this significantly reduces scattering caused by internal and multiple reflection in the quartz material.

Dovichi and co-workers stated that in their zone electrophoresis detector, the best detection limit (3σ) was 50 yoctomol (1 yoctomol = 10^{-24} mol) of rhodamine 6G with a 1 mW He–Ne laser (543.5 nm) as the excitation source [10].

7.3.1.3 Immersed flowcell. Kurosu *et al.* proposed an immersed flowcell to reduce scattering in FL detection [2]. They directed the capillary tube through an ordinary rectangular HPLC FL flowcell and filled the space between the tube and the flowcell with a liquid that has an appropriate RI. In this manner, the flowcell is immersed as shown in Figure 7.4.

By this arrangement, though the circular boundary still exists, the amount of scattered light could be significantly reduced by minimizing the differences between the refractive indices of the boundary materials [2]. It is noted that as the RI of the immersing liquid approaches that of quartz, the background signal rapidly decreases.

Kurosu *et al.* employed an HPLC FL detector consisting of a 150 W Xe lamp as the excitation source, excitation and emission monochromators, and demonstrated the emission spectrum of riboflavine (vitamin B2) [2].

7.3.2 Laser-induced fluorescence detection (LIFD)

The history of laser-induced FL detection (LIFD) in CE can be traced back to 1985 when Gassmann *et al.* used a He–Cd laser as an excitation source at 325 nm for the detection of dansylated amino acids in chiral separation [11]. They guided the laser beam through an optical filter and a lens to the capillary wall by using optical fiber, and the FL emission was collected at right angles through a lens and directed to a monochromator. The effective cell volume was approximately 0.5 nl. Using this arrangement, detection of femtomole amounts of dansylated amino acids was achieved. Although reported in the mid 1980s, LIFD began attracting interest rapidly in 1990 as shown in Figure 7.1. Since then, 70–80% of FL detection has been per-

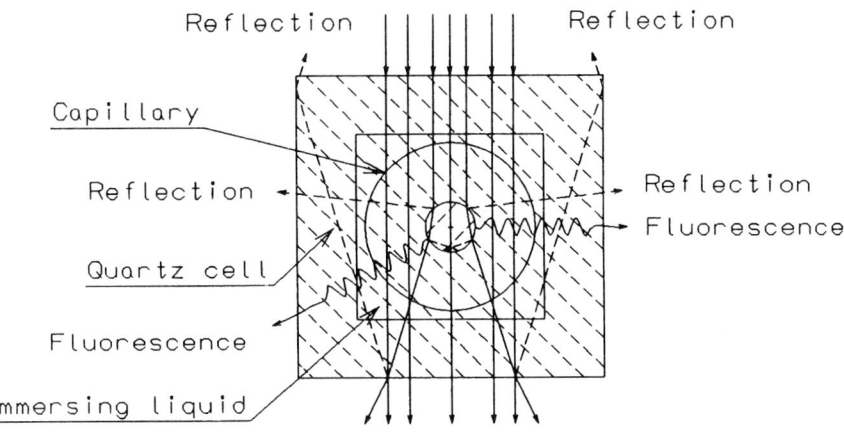

Figure 7.4 Cross-sectional view of immersed flowcell.

formed by employing a laser as an excitation source. A coherent laser beam is easily condensed and effectively focused as a tiny spot on the fused-silica capillary, thus matching the small internal diameter and irradiated volume. However, available wavelengths are limited, e.g. a helium–cadmium laser at 325 nm, an argon ion laser at 488 and 514 nm, a helium–neon green laser at 543 and 594 nm and a diode laser at 670 nm. Yeung reviewed available wavelengths, types of laser, costs and other characteristics such as stability and focusing [12].

7.3.2.1 Considerations in optics. Although the laser beam can be focused as a small enough spot at circular boundaries between air–silica and silica–liquid, scattering is still a major cause of noise and several methods have been proposed to reduce the noise resulting from such scattering.

Gassmann *et al.* utilized optical fiber to guide the laser beam to the capillary wall [11]. Taylor and Yeung demonstrated axial-beam laser excitation by inserting a 51 μm diameter optical fiber into a 75 μm internal diameter capillary [13]. They suggested the possibility of whole column detection. FL emission was collected by using a 1 mm diameter optical fiber placed perpendicular to the capillary wall. Dovichi and co-workers introduced the sheath flow cuvette for LIFD [6]. Takahashi *et al.* used the sheath flow system for multiple capillary detection [14]. Hernandez and co-workers utilized an epillumination FL microscope with confocal optical configuration for LIFD [4, 15]. Similar types of optical configuration were studied [16–18]. Beale and Sudmeier utilized the epillumination confocal optics for spatial scanning LIFD [19]. In their system, the capillary mounting table can be moved to allow scanning of the detection point along the entire length of the capillary.

7.3.2.2 Native FL detection. Many organic compounds are fluorescent. However, most of such compounds emit FL only when they are excited with UV light. At present, wavelengths available by lasers are limited and prices of such lasers which emit light beams of wavelengths shorter than 600 nm increase exponentially. This restricts wide use of LIFD of native FL and the number of reports is limited. However, laser excitation of native FL greatly facilitates sample pretreatment and allows extremely small quantities of samples to be handled.

Yeung and co-workers [20, 21] and Belenky *et al.* [22] utilized argon ion laser emission at 275 nm for the native FL detection of DNA fragments in sequencing. Liu *et al.* reported the usefulness of argon ion laser emission (488 nm) for the detection of anticancer drug doxorubicin, which exhibits FL emission at 550 nm when excited at 490 nm, in order to obtain information regarding the site of drug attachment on each of conjugated fragments [23]. Wu *et al.* also utilized a 488 nm emission of an argon ion laser to detect porphyrin acids by their native FL [24]. The limits of detection (LOD) were in the range of 1–10 nM. Nie *et al.* demonstrated ultrasensitive detection of polycyclic aromatic hydrocarbons at the level of 30–150 zeptmol (1 zeptmol = 10^{-21} mol) [25].

7.3.2.3 Indirect FL detection. There are many compounds which have poor chromophores, thus causing optical absorption or FL detection to be difficult. In order to detect these compounds, indirect UV or FL detection has been used in electrophoresis as well as in liquid chromatography (LC) since the late 1960s, especially in the detection of organic and inorganic anions. A well-known example of the latter is ion chromatography analysis combined with indirect UV detection using phthalate eluent as an example.

In 1988, Kuhr and Yeung first demonstrated the usefulness of indirect FL detection in native amino acids [26], they also showed the optimization of sensitivity and separation in the detection method [27]. Later, Yeung and co-workers extended the application to analyses of sugars [28], high molecular weight polysaccharides [29] and free cyanide and related compounds [30]. Baechmann *et al.* reported simultaneous determination of inorganic cations and anions with indirect FL detection [31]. Chao and Whang applied the indirect detection to the analysis of 11 priority phenols [32]. Andersson *et al.* compared direct UV absorption, indirect UV absorption detection and indirect LIF detection for the determination of isoprenyl pyrophosphates [33]. The LOD by direct UV was ca. 50 µM, whereas that by indirect LIF was ca. 0.5 µM, indicating that the latter is more sensitive.

7.3.2.4 Use of diode laser. A UV laser costs about 100 times as much as a visible or near-infrared laser. A typical visible laser is of He–Ne. Recently, a diode laser that offers a very stable visible or near-infrared wavelength emission has become available inexpensively, and research work on application of such a laser has been increasing. Few compounds absorb light with a wavelength longer than 670 nm, which is a wavelength typical of a diode laser emission. To overcome this weakness and to utilize a diode laser successfully there are two major techniques: frequency doubling to obtain shorter wavelengths and fluorescent labeling. These techniques are often used in combination.

In 1988 Imasaka *et al.* described the use of the second harmonic emission of a near-infrared diode laser [34]. They applied this technique to the detection of native FL of polycyclic aromatic hydrocarbons by micellar electrokinetic chromatography (MEKC) [35]. The original diode laser emission was 830 nm with 40 mW power and the wavelength utilized was 415 nm with 50 µW power. The LOD achieved [35] were 1×10^{-8} M and 1×10^{-6} M for 1-aminoanthracene and benzo[*a*]pyrene, respectively. Jansson *et al.* reported a high power frequency doubled diode laser source that offered emission at 424 nm with 0.5 mW power [36]. They used this system for the detection of naphthalene-2,3-dicarboxaldehyde (NDA) derivatized amino acids. The mass LOD of 0.9 amol was obtained ($S/N = 3$) corresponding to a concentration detection limit of 5.9 nM injected.

Extensive research work on the use of diode lasers with emissions at 670 nm and 415 nm (frequency doubled), respectively, was undertaken by Imasaka and co-workers [37]. They used a 670 nm laser to detect the native FL of chlorophyll in the deep-red region. They also examined indirect FL detection by using methylene blue as a chromophore for the detection of alkylsilane-derivatized amino acids. The LOD obtained was ca. 1 pmol. In their study on fluorescent labeling, fluoresceinisothiocyanate (FITC) and 7-(diethylamino)coumarin-3-carboxylic acid succinimidyl ester (DCCS) were compared using a blue diode laser (415 nm). The LOD of FL labeled amino acids with FITC and DCCS were ca. 2 fmol and ca. 100 amol, respectively. Later they proposed a pyronin succinimidyl ester as a suitable FL labeling reagent, and the LOD obtained was 800 zmol for glycine [38]. Mank and Yeung also reported derivatized amino acid detection with a diode laser-based LIF [39]. They employed dicarbocyanine succinimidyl ester, and the limit of detection achieved was 0.1 amol for glycine.

7.3.2.5 Derivatization. In the previous section FL derivatization in diode laser-based LIFD was partly reviewed. There are many reports of FL derivatization detection with conventional lasers. Pentoney et al. demonstrated o-phthaldialdehyde (OPA) derivatization of amino acid in CE with LIF using a 325 nm emission from a He–Cd laser [40]. The LOD obtained was 800 amol for L-histidine. They employed on-column (or postseparation) derivatization as being identical to the conventional OPA postcolumn derivatization detection in HPLC. Normally, derivatization of analytes to fluorescent derivatives is carried out before the analytes are subjected to CE, i.e. prederivatization. Wu and Dovichi [5] utilized FITC as a fluorogenic reagent for amino acid analysis using a 50 mW argon ion laser (488 nm)-based LIF with sheath flow cuvette, and achieved a LOD of 1.7×10^{-21} mol for FITC-labeled arginine. Amankwa et al. employed 2,3-naphthalenedialdehyde (NDA) as FL labeling reagent to detect poly(oxyalkylene)diamine polymers in CE with LIF using a 442 nm He–Cd laser [41]. Novotny and co-workers designed 3-(4-carboxybenzoyl)-2-quinolinecarboxaldehyde (CBQCA) as a fluorogenic reagent for primary amines in capillary electrophoresis with LIFD using a 442 nm He–Cd laser. They demonstrated detection of amino acids, peptides and amino sugars [42, 43]. The typical LOD were low attomole levels (10^{-18} mol). Albin et al. evaluated derivatizing reagents and techniques for FL detection [44].

More recently, highly sensitive amino acid detection by LIF has excited much interest. As described in the previous section, NDA, FITC, DCCS and other fluorogenic reagents have been studied. Mattusch and Dittrich

Table 7.1 Various fluorescence detection modes and their limits of detection (LOD)

Ref.	Year	Excitation source	Wavelength (nm)	LOD	Analyte	Remarks
40	1988	He–Cd	325	800 amol	OPA-L-histidine	Derivatized
5	1989	Ar ion	488	1.7×10^{-21} mol	FITC-arginine	Derivatized
35	1991	Blue diode	415	10^{-8} M	1-Aminoanthracene	Native
42,43	1991	He–Cd	442	Low amol	CBQCA-amino acids	Derivatized
37	1992	Blue diode	415	2 fmol	FITC-amino acids	Derivatized
37	1992	Blue diode	415	100 amol	DCCS-amino acids, peptides, amino sugars polycyclic aromatics	Derivatized
25	1993			30–150 amol		Native
36	1993	Blue diode	424	0.9 amol	NDA-amino acids	Derivatized
10	1994	He–Ne	543.5	50 ymol	Rhodamine 6G	Native
24	1995	Ar ion	488	1–10 nM	Porphyrin acids	Native
33	1995	He–Cd	325	0.5 μM	Isoprenyl pyrophosphates	Indirect
39	1995	Diode	670	0.1 amol	Derivatized amino acids	Dicarbocyanine succinimidyl ester

Note that the detection limits are presented in the same form as they appear in the original paper.

improved separation and long term stability of (FITC I)-derivatized amino acids. They used a liquid ion-exchange resin LA-2 to extract excessive FITC I from the sample solution after derivatization [45]. Dawson *et al.* applied NDA derivatization in the analysis of glutamate in striatal microdialysates using a He–Cd laser-based LIF. The LOD of their system was at the attomole level [46]. Williams and Soper successfully used a near-IR fluorescent dye in DNA sequencing applications [47]. A universal M13 sequencing primer was labeled on the 5' end with a near-IR dye containing isothiocyanate group, and the detection was performed by using a 10 mW excitation beam (795 nm) from a Ti–sapphire laser. They reported that the LOD was 3.4×10^{-20} mol [47].

7.3.3 Simultaneous detection with UV absorption and fluorescence

Although LIF offers extremely high sensitivity for many compounds, there are several intrinsic drawbacks to the use of a laser as an excitation light source in FL detection for CE. These drawbacks may include unavailability of continuous variation of excitation wavelength, high cost and the requirement of having a large system in comparison with a CE separation system itself. In order to overcome these limitations, Caslavska *et al.* described modification of a fast scanning multiwavelength UV detector for the simultaneous absorbance and FL detection in CE [48]. The system allowed changing the excitation wavelength in the range of 190–800 nm using a light source from deuterium and tungsten lamps. For the detection of emission lights, bandpath filters of 340, 366, 405, 450 and 520 nm were used as required for the compounds of interest. The emission light was collected at a right angle to the excitation light beam by using a fused-silica optical fiber. LOD ($S/N = 3$) obtained by UV absorbance and FL detections for fluorescein were 6.0 and 0.4 µg ml^{-1}, respectively. The system was used for profiling of body fluids for drugs and endogenous compounds, utilizing the native FL of the compounds.

7.4 Conclusion

There have been several methods for FL detection and the major means of FL detection in current CE is LIFD. Reports of use of this detection mode are rapidly increasing as shown in Figure 7.1. In Table 7.1, various FL detection modes and their LOD are summarized.

References

1 Hernandez, L.; Narahari, J.; Verdeguer, P.; Guzman, N.A. (1993) Laser-induced FL detection for capillary electrophoresis: a powerful analytical tool for the separation and

REFERENCES

detection of trace amounts of analytes. In *Capillary Electrophoresis Technology*, Guzman, N.A. (ed.), Chromatogr. Sci. Ser. Vol. 64, Marcel Dekker, New York, p. 605.

2. Kurosu, Y.; Sasaki, T.; Saito M. (1991) Fluorescence detection with an immersed flow cell in capillary electrophoresis. *J. High Resolut. Chromatogr.* **14**, 186.
3. Hernandez, L.; Marquina, R.; Escalona, J.; Guzman, N.A. (1990) Detection and quantification of capillary electrophoresis zones by fluorescence microscopy. *J. Chromatogr.* **502**, 247.
4. Hernandez, L.; Escalona, J.; Guzman, N.A. (1991) Laser-induced fluorescence and fluorescence microscopy for capillary electrophoresis zone detection. *J. Chromatogr.* **559**, 183.
5. Wu, S.; Dovichi, N. (1989) High-sensitivity fluorescence detector for fluorescein isothiocyanate derivatives of amino acids separated by capillary zone electrophoresis. *J. Chromatogr.* **480**, 141.
6. Swerdlow, H.; Wu, S.; Harke, H.; Dovichi, N. (1990) Capillary gel electrophoresis for DNA sequencing: laser-induced fluorescence detection with the sheath flow cuvette. *J. Chromatogr.* **516**, 61.
7. Zhang, J.; Chen, D.Y.; Harke, H.R.; Dovichi, N. (1993) Laser-induced fluorescence detection for capillary electrophoresis: laser-induced fluorescence detection. In *Capillary Electrophoresis Technology*, Guzman, N.A. (ed.), Chromatogr. Sci. Ser. Vol. 64, Marcel Dekker, New York, p. 631.
8. Cheng, Y.; Wu, S.; Chen, D.Y.; Dovichi, N.J. (1990) Interaction of capillary zone electrophoresis with a sheath flow cuvette detector. *Anal. Chem.* **62**, 496.
9. Swerdlow, H.; Wu, S.; Harke, H.; Dovichi, N.J. (1990) Capillary gel electrophoresis for DNA sequencing: laser-induced fluorescence detection with the sheath low cuvette. *J. Chromatogr.* **516**, 61.
10. Chen, D.Y.; Dovichi, N.J. (1994) Yoctomole detection limit by laser-induced fluorescence in capillary electrophoresis. *J. Chromatogr.* **657**, 265.
11. Gassmann, E.; Kuo, J.E.; Zare, R.N. (1985) Electrokinetic separation of chiral compounds. *Science* **230**, 813.
12. Yeung, E.S. (1993) Optical detection schemes for capillary electrophoresis. In *Capillary Electrophoresis Technology*, Guzman, N.A. (ed.), Chromatogr. Sci. Ser. Vol. 64, Marcel Dekker, New York, p. 605.
13. Taylor, J.A.; Yeung, E.S. (1992) Axial-beam laser-excited fluorescence detection in capillary electrophoresis. *Anal. Chem.* **64**, 1741.
14. Takahashi, S.; Murakami, K.; Anazawa, T.; Kambara, H. (1994) Multiple sheath-flow gel capillary-array electrophoresis for multicolor fluorescent DNA detection. *Anal. Chem.* **66**, 1021.
15. Hernandez, L.; Joshi, N.; Murzi, E.; Verdeguer, P.; Mifsud, J.C.; Guzman, N. (1993) Collinear laser-induced fluorescence detector for capillary electrophoresis. Analysis of glutamic acid in brain dialyzates. *J. Chromatogr.* **652**, 399.
16. Huang, X.C.; Quesada, M.A.; Mathies, R.A. (1992) Capillary array electrophoresis using laser-excited confocal fluorescence detection. *Anal. Chem.* **64**, 967.
17. Lee, T.T.; Yeung, E.S. (1992) Quantitative determination of native proteins in individual human erythrocytes by capillary zone electrophoresis with laser-induced fluorescence detection. *Anal. Chem.* **64**, 3045.
18. Lee, T.T.; Yeung, E.S. (1992) High-sensitivity laser-induced fluorescence detection of native proteins in capillary electrophoresis. *J. Chromatogr.* **595**, 319.
19. Beale, S.C.; Sudmeier, S.J. (1995) Spatial-scanning laser fluorescence detection for capillary electrophoresis. *Anal. Chem.* **67**, 3367.
20. Milofsky, R.E.; Yeung, E.S. (1993) Native fluorescence detection of nucleic acids and DNA restriction fragments in capillary electrophoresis. *Anal. Chem.* **65**, 153.
21. McGregor, D.A.; Yeung, E.S. (1994) Detection of DNA fragments separated by capillary electrophoresis based on their native fluorescence inside a sheath flow. *J. Chromatogr.* **680**, 491.
22. Belenky, A.; Smisek, D.L.; Cohen, A.S. (1995) Sequencing of antisense DNA analogues by capillary gel electrophoresis with laser-induced fluorescence detection. *J. Chromatogr.* **700**, 137.
23. Liu, J.; Abid, S.; Lee, M.S. (1995) Analysis of monoclonal antibody chimeric BR96-doxorubicin immunoconjugate by sodium dodecyl sulfate–capillary gel electrophoresis with ultraviolet and laser-induced fluorescence detection. *Anal. Biol. Chem.* **229**, 221.

24. Wu, N.; Li, B.; Sweedler, J.V. (1994) Recent developments in porphyrin separations using capillary electrophoresis with native fluorescence detection. *J. Liq. Chromatogr.* **17**, 1917.
25. Nie, S.; Dadoo, R.; Zare, R.N. (1993) Ultrasensitive fluorescence detection of polycyclic aromatic hydrocarbons in capillary electrophoresis. *Anal. Chem.* **65**, 3571.
26. Kuhr, W.G.; Yeung, E.S. (1988) Indirect fluorescence detection of native amino acids in capillary zone electrophoresis. *Anal. Chem.* **60**, 1832.
27. Kuhr, W.G.; Yeung, E.S. (1988) Optimization of sensitivity and separation in capillary zone electrophoresis with indirect fluorescence detection. *Anal. Chem.* **60**, 2642.
28. Garner, T.; Yeung, E.S. (1990) Indirect fluorescence detection of sugars separated by capillary zone electrophoresis with visible laser excitation. *J. Chromatogr.* **515**, 639.
29. Richmond, M.D.; Yeung, E.S. (1993) Development of laser-excited indirect fluorescence detection for high-molecular-weight polysaccharides in capillary electrophoresis. *Anal. Biochem.* **210**, 245.
30. Marti, V.; Aguilar, M.; Yeung, E.S. (1995) Indirect fluorescence detection of free cyanide and related compounds by capillary electrophoresis. *J. Chromatogr.* **709**, 367.
31. Baechmann, K.; Haumann, I.; Groh, T. (1992) Simultaneous determination of inorganic cations and anions in capillary zone electrophoresis (CZE) with indirect fluorescence detection. *Fresenius J. Anal. Chem.* **343**, 901.
32. Chao, Y.-C.; Whang, C-W. (1994) Capillary zone electrophoresis of eleven priority phenols with indirect fluorescence detection. *J. Chromatogr.* **663**, 229.
33. Andersson, P.E.; Pfeffer, W.D.; Blomberg, L.G. (1995) Indirect detection in capillary electrophoresis; Comparison between indirect UV and indirect laser-induced fluorescence detection for the determination of isoprenyl pyrophosphates. *J. Chromatogr.* **699**, 323.
34. Imasaka, T.; Okazaki, T.; Ishibashi, N. (1988) Semiconductor laser fluorometry for enzyme and enzymatic assays. *Anal. Chim. Acta* **208**, 327.
35. Imasaka, T.; Nishitani, K.; Ishibashi, N. (1991) Cyclodextrin-modified micellar electrokinetic chromatography combined with semiconductor laser fluorimetry. *Analyst*, **116**, 1407.
36. Jansson, M.; Roeraade, J.; Laurell, F. (1993) Laser-induced fluorescence detection in capillary electrophoresis with blue light from a frequency-doubled diode laser. *Anal. Chem.* **65**, 2766.
37. Higashijima, T.; Fuchigami, T.; Imasaka, T.; Ishibashi, N. (1992) Determination of amino acids by capillary zone electrophoresis based on semiconductor laser fluorescence detection. *Anal. Chem.* **64**, 711.
38. Fuchigami, T.; Imasaka, T.; Shiga, M. (1993) Subattomole detection of amino acids by capillary electrophoresis based on semiconductor laser fluorescence. *Anal. Chim. Acta* **282**, 209.
39. Mank, A.J.G.; Yeung, E.S. (1995) Diode laser-induced fluorescence detection in capillary electrophoresis after pre-column derivatization of amino acids and small peptides. *J. Chromatogr.* **708**, 309.
40. Pentoney, Jr, S.L.; Huang, X.; Burgi, D.S.; Zare, R.N. (1988) On-line connector for microcolumns: application to the on-column o-phthaldialdehyde derivatization of amino acids separated by capillary zone electrophoresis. *Anal. Chem.* **60**, 2625.
41. Amankwa, L.N.; Scholl, J.; Kuhr, W.G. (1990) Characterization of the oligomeric dispersion of poly(oxyalkylene)diamine polymers by precolumn derivatization and capillary zone electrophoresis with fluorescence detection. *Anal. Chem.* **62**, 2189.
42. Liu, J.; Hsieh, Y.-Z; Wiesler, D.; Novotny, M. (1991) Design of 3-(4-carboxybenzoyl)-2-quinolinecarboxaldehyde as a reagent for ultrasensitive determination of primary amines by capillary electrophoresis using laser fluorescence detection. *Anal. Chem.* **63**, 408.
43. Liu, J.; Shirota, O.; Wiesler, D.; Novotny, M. (1991) Capillary electrophoresis of amino sugars with laser-induced fluorescence detection. *Anal. Chem.* **63**, 413.
44. Albin, M.; Weinberger, R.; Sapp, E.; Moring, S. (1991) Fluorescence detection in capillary electrophoresis: evaluation of derivatizing reagents and techniques. *Anal. Chem.* **63**, 417.
45. Mattusch, J.; Dittrich, K. (1994) Improvement of laser-induced fluorescence detection of amino acids in capillary zone electrophoresis. *J. Chromatogr.* **680**, 279.
46. Dawson, L.A.; Stow, J.M.; Dourish, C.T.; Routledge, C. (1995) Analysis of glutamate in striatal microdialysates using electrophoresis and laser-induced fluorescence detection. *J. Chromatogr.* **700**, 81.

47 Williams, D.C.; Soper, S.V. (1995) Ultrasensitive near-IR fluorescence detection for capillary gel electrophoresis and DNA sequencing applications. *Anal. Chem.* **67**, 3427.
48 Caslavska, J.; Gassmann, E.; Thormann, W. (1995) Modification of a tunable UV–visible capillary electrophoresis detector for simultaneous absorbance and fluorescence detection: profiling of body fluids for drugs and endogenous compounds. *J. Chromatogr.* **709**, 147.

8 Other systems
K. KITAGISHI

8.1 Fraction collection

The collection of analyte zones in capillary electrophoresis which are separated with high efficiency allows the identification and characterization of the analytes by analytical techniques which cannot be directly connected to CE. One common design of a fraction collector in CE is based on the replacement of the outlet buffer vial with collection vials, which can be achieved by modification of commercial instruments. This arrangement often gives inexact collection and, in the cases where the collection vials contain a small volume of buffer, gives dilution of the fraction. The first problem is solved by calculating the migration time for fraction collection corresponding to the related migration behavior [1].

The second approach for CE fraction collection includes the use of a membrane [2] or drum which moves during migration. It has the advantage that the collection is achieved with no loss of the sample but the disadvantage that it is difficult to achieve a satisfactory rate of recovery of the collected sample from the membrane or drum. This method of collection is useful when restricted to postanalysis techniques being with samples adsorbed on the solid surface.

Other procedures utilize a field decoupler prior to fraction collection, which has the greatest possibility of being developed as a common device for general applications. Fujimoto et al. [3] designed a fraction collector with an on-column fracture assembly which is immersed in a buffer reservoir connected to high voltage. The analyte zones are detected after fracture with laser induced fluorescence (LIF) and collected directly from outlet end of the capillary [3]. Precise fraction collection with a sheath flow connection was developed by Muller et al. [4]. This device consists of a fiber optic-based UV detection cell close to the sheath flow unit to be grounded and a fraction collector with a support for collection capillaries operated by a stepper motor and controlled by computer software which calculates the exit time for the zones of interest according to the migration time at the detection window. All DNA fragments of ΦX-174/Hae III ranging from 72 to 1356 base pairs could be successfully collected into separate capillaries [4].

8.2 Field modulation

In conventional gel electrophoresis, very large biopolymers could not be separated sufficiently because analyte molecules larger than the mesh size migrate and are squeezed like 'snakes'. Where symmetrical square wave or sine wave signals are applied as a potential field, a pulsed field technique is effective in overcoming this separation difficulty. Pulsed field CE is expected to enhance the separation range applicable for large polymers in polymer sieving.

The separation of large DNA fragments of 4363 and 7253 base pairs was improved by pulsed field gel CE by optimizing the waveform frequency between 0.1 and 1000 Hz [5]. Pulsed field CE could be satisfactorily applied for the separation of polysaccharides with molecular weights of 8 to 2000 kDa [6] and sulfonated polystyrenes of 400 to 1132 kDa [7] by using a soluble polymer solution as the sieving medium.

Unexpected distortion of the waveform of a pulsed electric field was reported, resulting from the significant resistance–capacitance (RC) time constants with narrow-bore capillaries and/or high viscosity solution. It is necessary to understand the factors which control the RC rise time for optimization of the separation of large molecules in pulsed field CE [8].

Another fundamental study has been done on the influence of field modulation on electroosmotic flow (EOF) profile. The flow profile varies between laminar flow and plug flow dependent on the modulation depth and induces a radial movement of analyte molecules [9].

8.3 Computer simulation

A number of variables affect migration behavior in CE. These include instrumental variables, physicochemical properties of electrolytes and samples and factors deriving from injection, detection, heat generation, electroosmosis, electromigration, etc. Heinrich and Wagner have developed a high speed simulation program for optimization of electrophoretic separation in free zone capillary electrophoresis (FZE) containing up to 15 components [10]. Reijenga and Kenndler developed a simulation program for graphic illustration of electropherograms as functions of the variables [11, 12]. Simulation was achieved on the influence of (1) the length of injection zone on peak broadening, (2) the low zeta potential of capillary inner surface on migration times, (3) the pH of the electrolyte on the resolution, (4) the ionic strength of the electrolyte on the selectivity and (5) the co-ion mobility on the migration dispersion. The experimental electropherograms on benzene derivatives and amino acids show satisfactory agreement with the simulated ones based on the theoretical model introduced [12].

Numerical algorithms applied to mass transport equations in electrokinetic processes have been proposed by three research groups. Ermakov *et al.* compared them and attempted to predict the shape and structure of the boundaries [13].

When thermal expansion of the liquid in the capillary caused by initial self heating is more rapid than the electromigration velocity of analytes, a part of the injected sample is lost from the inlet end of the capillary. The limiting condition to avoid sample loss was calculated by Knox and McCormack. They recommended instrumental incorporation of ramp-up voltage control [14].

The development of computer simulation as described above will provide a compass to optimize the separation in CE; however the reader should note that simulation is only a simulation, and therefore the simulation result must be evaluated experimentally as the exact experiment will be affected by many unexpected factors including the matrix factor, which is impossible to predict by computer simulation. In a typical example, separation conditions using standard sample will not be applicable to the analysis of the compounds in blood by computer simulation due to unexpected interference by certain compounds. So, computer simulation is restricted and can be applicable to the preliminary test, but this should not be considered under the final CE conditions.

References

1 Lee, H.G.; Desiderio, D.M. (1994) Preparative capillary zone electrophoresis of synthetic peptides. Conversion of an autosampler into a fraction collector. *J. Chromatogr.* **686**, 309–317.
2 Cheng, Y-F.; Fuchs, M.; Andrews, D.; Carson, W. (1992) Membrane fraction collection for capillary electrophoresis. *J. Chromatogr.* **608**, 109–116.
3 Fujimoto, C.; Fujikawa, T.; Jinno, K. (1992) Sample collection by a capillary zone electrophoretic system with an on-column fracture. *J. High Resolut. Chromatogr.* **15**, 201–203.
4 Müller, O.; Foret, F.; Karger, B.L. (1995) Design of a high-precision fraction collector for capillary electrophoresis. *Anal. Chem.* **67**, 2974–2980.
5 Heiger, D.N.; Cohen, A.S.; Karger, B.L. (1990) Separation of DNA restriction fragments by high performance capillary electrophoresis with low and zero crosslinked polyacrylamide using continuous and pulsed electric fields. *J. Chromatogr.* **516**, 33–48.
6 Sudor, J.; Novotny, M. (1993) Electromigration behavior of polysaccharides in capillary electrophoresis under pulsed-field conditions. *Proc. Natl Acad. Sci. USA* **90**, 9451–9455.
7 Sudor, J.; Novotny, M.V. (1994) Pulsed-field capillary electrophoresis: Optimizing separation parameters with model mixtures of sulfonated polystyrenes. *Anal. Chem.* **66**, 2139–2147.
8 Heiger, D.N.; Carson, S.M.; Cohen, A.S.; Karger, B.L. (1992) Wave form fidelity in pulsed-field capillary electrophoresis. *Anal. Chem.* **64**, 192–199.
9 Demana, T.; Guhathakurta, U.; Morris, M.D. (1992) Effects of analyte velocity modulation on the electroosmotic flow in capillary electrophoresis. *Anal. Chem.* **64**, 390–394.
10 Heinrich, J.; Wagner, H. (1992) High speed electrophoresis simulation for optimization of continuous flow electrophoresis and high performance capillary techniques: Part I. Computer model. *Electrophoresis* **13**, 44–49.

11 Reijenga, J.C.; Kenndler, E. (1994) Computational simulation of migration and dispersion in free capillary zone electrophoresis: I. Description of the theoretical model. *J. Chromatogr.* **659**, 403–415.
12 Reijenga, J.C.; Kenndler, E. (1994) Computational simulation of migration and dispersion in free capillary zone electrophoresis: II. Results of simulation and comparison with measurements. *J. Chromatogr.* **659**, 417–426.
13 Ermakov, S.V.; Bello, M.S.; Righetti, P.G. (1994) Numerical algorithms for capillary electrophoresis. *J. Chromatogr.* **661**, 265–278.
14 Knox, J.H.; McCormack, K.A. (1994) Volume expansion and loss of sample due to initial self-heating in capillary electroseparation (CES) systems. *Chromatographia* **38**, 279–282.

9 Advanced systems
K. KITAGISHI

9.1 Capillary electrophoresis on a glass chip

The benefits of miniaturization have been realized in the field of electrophoresis by using capillaries instead of cylindrical hollow tubing or slab gels. An additional miniaturization has been demonstrated by the use of photolithographically fabricated microstructure by Ciba Geigy [1, 2]. Photolithographic patterning on the silicon wafer has become a well-known technique in microelectronics and has been applied to gas and liquid chromatography. The process generally consists of the exposure of silicon wafer coated with photoresist to light through a metal mask followed by etching the exposed glass with a HF-based solution. Capillaries can be formed by covering the fabricated glass with other glass plate [1]. Capillary electrophoresis on a planar chip with a highly flexibile design is feasible by the fabrication of complicated microstructures. Manz and co-workers reported preliminary micro-CE on a 15 cm × 4 cm chip, which provides separation on a minute scale [1, 3, 4]. When the scale was decreased to 2 cm × 1 cm fluorescent labeled amino acids could be separated with up to 75 000 theoretical plates in 15 s [2] and with up to 6800 theoretical plates in 4 s [5]. Fast and efficient separation of fluorescent labeled amino acids was presented by changing the separation length of the capillary to between 5 to 50 mm and the applied potential with an integrated sample injection of 100 pl, which resulted in an analysis from a few seconds to a few tens of seconds with theoretical plate numbers of 5800 to 16 000, respectively [6–8]. Synchronized cyclic CE on a chip was demonstrated to increase the theoretical plate number remarkably [8]. Ramsey and co-workers evaluated a high-speed microchip electrophoresis with a separation length of 0.9 to 11.1 mm [9].

Gel electrophoresis on micromachined glass chips were reported for the separation of oligonucleotide mixtures. Oligonucleotide monomers from 10 to 25 base pairs in length could be separated in one minute [10]. Linear polymer sieving with hydroxyethyl cellulose on a glass chip for DNA fragments was demonstrated by Wooley and Mathies. The reproducible separation of DNA mixture ranging from 72 to 1353 base pairs was shown in 120 s [11].

A microchip with a postcolumn reactor was fabricated. Postcolumn derivatization of amino acids with *o*-phthaldialdehyde (OPA) was performed with this device [12].

Selective isolation of fraction zones after separation was achieved on a microchip. A 90 pl volume of injected sample plug was diluted by a factor of only 3 after separation [13].

The bottlenecks in CE on a microchip are reliable fabrication, separation efficiency and sensitivity. A valveless flow control at the intersection point causes analyte diffusion. Zone dilution due to diffusion makes detection difficult, although laser-induced fluorescence with an argon ion laser, a detection mode with an extremely high sensitivity is generally applied to microchip CE. In spite of these problems, it is certain that high-speed analysis based on micromachining will be one of the principal techniques in electrophoresis in the near future.

9.2 Imaging detector

The detector in CE is commonly placed at a single point on a separation capillary in on-column absorbance or fluorescence detection. The analytes migrate inside the capillary and pass through the detection window. In capillary isoelectric focusing (CIEF), the analytes are focused on the pH gradient along the capillary according to their isoelectric point (pI). Zone mobilization should follow for the detection of hydrodynamic flow with applied pressure or by the addition of salt to either of the electrolytes. During the mobilization, the pH gradient is distorted. Imaging detection is favorable in CIEF for fast analysis with a high efficiency. Wu and Pawliszyn [14–17] developed three kinds of imaging detectors for CIEF with a short capillary of 4 cm length by using a charge-coupled device (CCD). The first is refractive index (RI) gradient, which can be achieved with 670 nm diode laser [14, 15] or an argon ion laser [16] as a light source and 1024 pixel linear CCD as the detector. The second one absorption imaging with an argon ion laser and a 1-D CCD [16, 17] or with a halogen lamp and a 2-D CCD sensor [14]. The third is a highly sensitive fluorescence detection using an argon ion laser and a cooled CCD camera [14]. All types of imaging detector allow CIEF in 2–4 min. RI gradient imaging with a diode laser and a linear CCD provides high speed separation of proteins with high efficiency [15]. The feasibility of peptide mapping was suggested with RI gradient imaging by using a He–Ne laser and a linear CCD even for the peptides containing no tyrosine and tryptophan residues [18].

9.3 Automated analysis

CE is suitable for automated analysis because of its compact size and simple configuration. Some commercial CE systems are equipped as an autosampler both at the inlet and the outlet ends of the capillary and can be controlled by programming several operation processes.

Ruyters and van der Wal demonstrated full automated analysis in the determination of chiral impurities by on-column derivatization following separation with a commercial CE system. The on-column derivatization of amino acids with 4-fluoro-7-nitrobenz-2,1,3-oxadiazole gave a limit of detection of 0.014% L-phenylalanine in D-phenylalanine using laser-induced fluorescence (LIF) detection [19].

Combination of CE separation with postcolumn enzymatic assay was designed for two enzymes, glucose-6-phosphate dehydrogenase and 6-phosphogluconate hydrogenase. Separation and reaction capillaries are connected with a tee (T) joint, and substrate flow is driven by pressure [20] or by a high applied voltage [21]. The later shows the lower limit of detection of glucose-6-phosphate dehydrogenase to be 4×10^{-9} M. This technique will be applicable to clinical examination and enzymatic engineering.

Miller and Lytle [22] immobilized yeast cells on a capillary with a frit for enzyme profiling of aminopeptidase using different amino acid β-naphthylamides. Profiling can be achieved with only 500 cells and the use of autosampler enables the automatic switching of different substrates. The CE-based method shows a profile similar to the result by the conventional filter/cuvette method [22].

A ultradilute solution of glutathion and amino acids at nM concentration was attainable by CE with LIF detection. On-column fluorescence derivatization during isotachophoresis (ITP) preconcentration was automatically followed by electrophoretic separation in a miniaturized scale of reagents [23].

Gilman and Ewing [24] demonstrated quantitative analysis of single mammalian cells with CE separation and LIF detection following on-column derivatizations. Cell injection and on-column derivatization reduce the dilution of cell contents compared to derivatization in microvials. Dopamine and five amino acids in individual cells could be quantitatively determined in the range from 180 ± 110 amol/cell to 5.1 ± 1.5 fmol/cell [24].

9.4 DNA sequencing and diagnosis

Automatic, high-speed DNA analysis with a high sensitivity is one of the most sought after techniques in biomedical science. CE has the advantage over conventional electrophoresis of being capable of automated and miniaturized analysis. In order for CE to be applied to DNA sequencing or diagnosis, the problem of highly sensitive analysis with an increase in sample throughput needs to be solved urgently.

Parallel separation of an array of capillaries with simultaneous monitoring was proposed to enhance the sample throughput of CE. Huang et al. designed a confocal fluorescence capillary array scanner which is continuously tracing the optical system at 2 cm s^{-1} and obtained an electropherogram image of DNA sample by scanning a 24-capillary array [25].

Takahashi et al. [26] developed capillary array electrophoresis with 20 capillaries for DNA sequencing. A multiple sheath-flow cell was used in order to eliminate considerable background light scattered at the capillary surface. The detection system consists of an image-splitting prism and a cooled CCD camera coupled with a cooled image intensifier. The migrated zones of all capillaries are irradiated simultaneously with two lasers (Ar laser 488 nm and YAG laser 532 nm) for excitation. Fluorescence labeled DNA with four different fluorophores could be detected at a minimum concentration of 2×10^{-12} M with the base reading speed of 200 bases/h [26].

Ueno and Yeung [27] constructed an array of 100 capillaries with a CCD camera for image detection. Fluorescence signals excited with an argon ion laser can be detected simultaneously for all capillaries by a CCD camera at the rate of 0.6 frame s^{-1} with negligible crosstalk [27].

Lifetime measurement for different fluorophores are useful to distinguish the fluorescent markers. Two fluorescent dyes, rhodamine 800 and JA 32, could be recognized with a pulse laser by collecting 800 photons. This technique suggests the possibility of applying DNA sequencing with high sensitivity and simple compact configuration [28].

9.5 Other new approaches in CE

Continuous zone electrophoretic separation was developed by Mesaros et al. [29]. Two-dimensional continuous electrophoresis in a narrow rectangular channel is performed, coupled to a small-bore capillary which can be moved across the entrance of the channel by the stepper motor. Detection by a photodiode array coupled to a fiber optic array provides a three-dimensional image of electrophoretic separation. This system has a large sample capacity due to efficient heat dissipation over the rectangular shape of the channel [29].

Culbertson and Jorgenson [30] proposed flow counterbalanced CE to improve efficiency and resolution. The cycle of movement of analytes back-and-forth through the detection window is repeated in this method, where one cycle is carried out by migrating the analyte through the detection window followed by pushing back the analyte by applied pressure through the window. A theoretical plate number of 17.3 million was obtained for the separation of amino acid derivatives after an 8 h separation [30].

References

1. Manz, A.; Harrison, D.J.; Verpoorte, E.M.J.; Fettinger, J.C.; Paulus, A.; Lüdi, H.; Widmer, H.M. (1992) Planar chips technology for miniaturization and integration of separation techniques into monitoring systems. Capillary electrophoresis on a chip. *J. Chromatogr.* **593**, 253–258.
2. Harrison, D.J.; Fluri, K.; Seiler, K.; Fan, Z.; Effenhauser, C.S.; Manz, A. (1993) Micromachining a miniaturized capillary electrophoresis-based chemical analysis system on a chip. *Science* **261**, 895–897.
3. Harrison, D.J.; Manz, A.; Fan, Z.; Lüdi, H.; Widmer, H.M. (1992) Capillary electrophoresis and sample injection systems integrated on a planar glass chip. *Anal. Chem.* **64**, 1926–1932.
4. Seiler, K.; Harrison, D.J.; Manz, A. (1993) Planar glass chips for capillary electrophoresis: Repetitive sample injection, quantitation, and separation efficiency. *Anal. Chem.* **65**, 1481–1488.
5. Fan, Z.H.; Harrison, D.J. (1994) Micromachining of capillary electrophoresis injectors and separators on glass chips and evaluation of flow at capillary intersections. *Anal. Chem.* **66**, 177–184.
6. Effenhauser, C.S.; Manz, A.; Widmer, H.M. (1993) Glass chips for high-speed capillary electrophoresis separations with submicrometer plate heights. *Anal. Chem.* **65**, 2637–2642.
7. Manz, A.; Verpoorte, E.; Effenhauser, C.S.; Burggraf, N.; Raymond, D.E.; Harrison, D.J.; Widmer, H.M. (1993) Miniaturization of separation techniques using planar chip technology. *J. High Resolut. Chromatogr.* **16**, 433–436.
8. Manz, A.; Verpoorte, E.; Effenhauser, C.S.; Burggraf, N.; Raymond, D.E.; Widmer, H.M. (1994) Planar chip technology for capillary electrophoresis. *Fresenius J. Anal. Chem.* **348**, 567–571.
9. Jacobson, S.C.; Hergenröder, R.; Koutny, L.B.; Ramsey, J.M. (1994) High-speed separations on a microchip. *Anal. Chem.* **66**, 1114–1118.
10. Effenhauser, C.S.; Paulus, A.; Manz, A.; Widmer, H.M. (1994) High-speed separation of antisense oligonucleotides on a micromachined capillary electrophoresis device. *Anal. Chem.* **66**, 2949–2953.
11. Woolley, A.T.; Mathies, R.A. (1994) Ultra-high-speed DNA fragment separations using microfabricated capillary array electrophoresis chips. *Proc. Natl Acad. Sci. USA* **91**, 11348–11352.
12. Jacobson, S.C.; Koutny, L.B.; Hergenröder, R.; Moore, Jr., A.W.; Ramsey, J.M. (1994) Microchip capillary electrophoresis with an integrated postcolumn reactor. *Anal. Chem.* **66**, 3472–3476.
13. Effenhauser, C.S.; Manz, A.; Widmer, H.M. (1995) Manipulation of sample fractions on a capillary electrophoresis chip. *Anal. Chem.* **67**, 2284–2287.
14. Wu, J.; Pawliszyn, J. (1994) Imaging detection methods for capillary isoelectric focusing. *Am. Lab.* **26**, 48–52.
15. Wu, J.; Pawliszyn, J. (1995) Diode laser-based concentration gradient imaging detector for capillary isoelectric focusing. *Anal. Chim. Acta* **299**, 337–342.
16. Wu, J.; Pawliszyn, J. (1994) Dual detection for capillary isoelectric focusing with refractive index gradient and absorption imaging detectors. *Anal. Chem.* **66**, 867–873.
17. Wu, J.; Pawliszyn, J. (1994) Application of capillary isoelectric focusing with absorption imaging detection to the analysis of proteins. *J. Chromatogr.* **657**, 327–332.
18. Vonguyen, L.; Wu, J.; Pawliszyn, J. (1994) Peptide mapping of bovine and chicken cytochrome c by capillary isoelectric focusing with universal concentration gradient imaging. *J. Chromatogr.* **657**, 333–338.
19. Ruyters, H.; van der Wal, S. (1994) Fully automated analysis of amino acid enantiomers by derivatization and chiral separation on a capillary electrophoresis instrument. *J. Liq. Chromatogr.* **17**, 1883–1897.
20. Emmer, Å.; Roeraade, J. (1994) Capillary electrophoresis, combined with an on-line micro post-column enzyme assay. *J. Chromatogr.* **662**, 375–381.
21. Emmer, Å.; Roeraade, J. (1994) Micro enzymatic assay coupled to capillary electrophoresis via liquid junction. *Chromatographia* **39**, 271–278.

REFERENCES

22 Miller, K.J.; Lytle, F.E. (1994) Enzymatic profiling of immobilized cells using CZE. *Anal. Chem.* **66**, 2420–2423.
23 Reinhoud, N.J.; Tjaden, U.R.; van der Greef, J. (1994) Automated on-capillary isotachophoretic reaction cell for fluorescence derivatization of small sample volumes at low concentrations followed by capillary zone electrophoresis. *J. Chromatogr.* **673**, 255–266.
24 Gilman, S.D.; Ewing, A.G. (1995) Analysis of single cells by capillary electrophoresis with on-column derivatization and laser-induced fluorescence detection. *Anal. Chem.* **67**, 58–64.
25 Huang, X.C.; Quesada, M.A.; Mathies, R.A. (1992) Capillary array electrophoresis using laser-excited confocal fluorescence detection. *Anal. Chem.* **64**, 967–972.
26 Takahashi, S.; Murakami, K.; Anazawa, T.; Kambara, H. (1994) Multiple sheath-flow gel capillary-array electrophoresis for multicolor fluorescent DNA detection. *Anal. Chem.* **66**, 1021–1026.
27 Ueno, K.; Yeung, E.S. (1994) Simultaneous monitoring of DNA fragments separated by electrophoresis in a multiplexed array of 100 capillaries. *Anal. Chem.* **66**, 1424–1431.
28 Bachteler, G.; Drexhage, K.-H.; Arden-Jacob, J.; Han, K.-T.; Köllner, M.; Müller, R.; Sauer, M.; Seeger, S.; Wolfrum, J. (1994) Sensitive fluorescence detection in capillary electrophoresis using laser diodes and multiplex dyes. *J. Lumin.* **62**, 101–108.
29 Mesaros, J.M.; Luo, G.; Roeraade, J.; Ewing, A.G. (1993) Continuous electrophoretic separations in narrow channels coupled to small-bore capillaries. *Anal. Chem.* **65**, 3313–3319.
30 Culbertson, C.T.; Jorgenson, J.W. (1994) Flow counterbalanced capillary electrophoresis. *Anal. Chem.* **66**, 955–962.

10 Automated fraction collection in capillary electrophoresis
R. GRIMM

10.1 Introduction

In the last few years capillary electrophoresis (CE) has become established alongside high performance liquid chromatography (HPLC) as a complementary, powerful separation technique for peptides and proteins. Along with the high speed, low sample requirement and overall lower running costs, a major advantage of CE is its flexibility. On one hand, typical HPLC separation modes, like reversed phase HPLC using differences in the hydrophobicity of components or ion-exchange chromatography using differences in the net charge of components can also be performed by micellar electrokinetic chromatography (MEKC) or capillary zone electrophoresis (CZE), respectively. On the other hand, typical slab gel electrophoresis separation modes like SDS (sodium dodecyl sulfate)–polyacrylamide gel electrophoresis (SDS–PAGE) or isoelectric focusing (IEF) can also be carried out by CE as capillary gel electrophoresis (CGE) or capillary isoelectric focusing (CIEF).

Although CE was very rapidly accepted as a powerful analytical tool, it has hardly been applied for microgreparative purposes for various reasons. One major obstacle is the inner diameter of the capillaries. Typically capillaries with an inner diameter of 50 to 75 µm are used. The total volume of a capillary with a length of 50–100 cm is between 1–5 µl. Therefore, sample loading is quite limited and very often 5–30 micropreparative separations had to be performed to collect sufficient material for further structural identification methods such as amino acid sequencing [1–3]. The micropreparative usage of CE also suffered from the lack of automation which means full computer control of fraction collection.

This chapter mainly focuses on the latest developments in automation of fraction collection. Examples of analysis of peptides or proteins will be shown for the fraction collection from a single run followed by a further characterization of the collected fraction by amino acid sequencing and by MALDI–TOF/MS (matrix assisted laser desorption ionization–time of flight/mass spectrometry) from four major separation modes of CZE, MEKC (or MECC), CGE and CIEF. These modes are summarized in Table 10.1.

Table 10.1 CE modes and quantitative data

CE mode	Peak parameters	Quantitative data
CZE	Migration time and peak dimension	Analyte concentration
MECC	Migration time and peak dimensions	Analyte concentration
CGE	Migration time (or relative to an internal standard)	ssDNA base number dsDNA base pair number Protein/peptide molecular weight
CIEF	Migration time	Protein isoelectric point (pI)

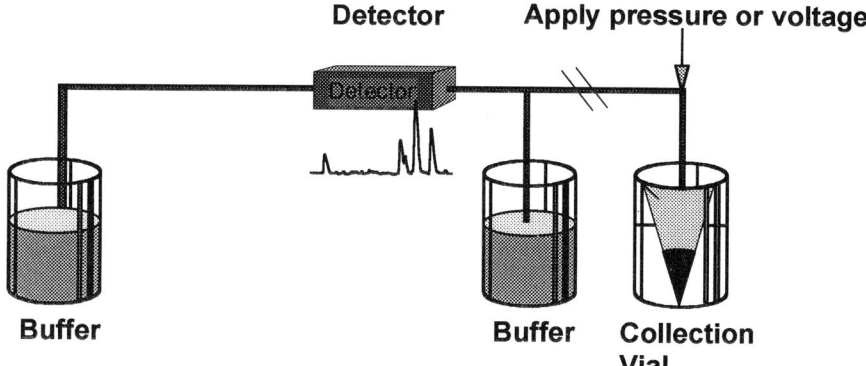

Figure 10.1 How fraction collection works in capillary electrophoresis.

10.2 Principle of fraction collection

The principle of fraction collection function is illustrated in Figure 10.1. Due to on-capillary detection of components in CE, the time taken for a peak to migrate from the point of detection to the capillary end has to be calculated very accurately, otherwise a loss of collection will result. Peak sensing performed by a diode-array detector together with the fixed length of the capillary between detector and capillary outlet allows the computer to calculate exactly the residual migration time of a peak as well as the collection of peaks which exceed a certain peak height. When a peak arrives at the outlet side, the voltage is dropped immediately and the outlet vial containing electrolyte solution is automatically exchanged with a microvial containing typically 5–10 µl of an appropriate solution (refer to Table 10.2). Components are then collected into the microvial either by the application of pressure or by the application of voltage for a certain period (Figure 10.1). After collection the microvial is automatically exchanged with the electrolyte vial again and the separation continues.

Modern software not only allows automated time calculations and auto-

Table 10.2 Solutions routinely used for fraction collection with pressure or voltage using various separation modes

Separation mode	Pressure collection	Voltage collection
CZE, ITP	2% Acetic acid, electrolyte	Electrolyte
MECC	Electrolyte, 5 mM NaOH	Electrolyte
CGE	–	Electrolyte, water
CIEF	Carrier ampholyte solution, 20 mM NaOH	–

mated exchanges of electrolyte and collection microvials, but also provides further possibilities, such as:

- collection of all peaks above a certain peak height which can be set before starting a run;
- peaks only within a certain timeframe of the entire separation;
- several individual peaks;
- collection of peaks either by pressure, voltage or both together.

The sample tray also serves as the fraction collector allowing unattended collection of several components from one single run or from multiple runs.

10.3 Fraction collection of peptides using capillary zone electrophoresis

Automated fraction collection of peptides from a peptide mixture separated using CZE is shown in Figure 10.2(a). All four peptides were pressure collected from a single run into 10 µl of 2% acetic acid. Successful collection of the peptides can be confirmed either by reinjection of the fraction into a capillary or by subjecting an aliquot of the fraction to MALDI–TOF/MS as shown in Figure 10.2(b). The advantage of the latter method is that besides confirmation of successful fraction collection and purity testing, collected components can be quickly characterized and identified by their individual mass. Another micropreparative CZE separation with single run fraction collection of a peptide mapping was recently described [4].

10.4 Peptide analysis by micellar electrokinetic chromatography (MEKC)

The micropreparative MEKC (or MECC) separation of a crude peptide synthesis preparation is shown in Figure 10.3(a). The major synthesis product could be well separated from all synthesis by-products. After collection of the largest peak from a single run, the peptide could be identified

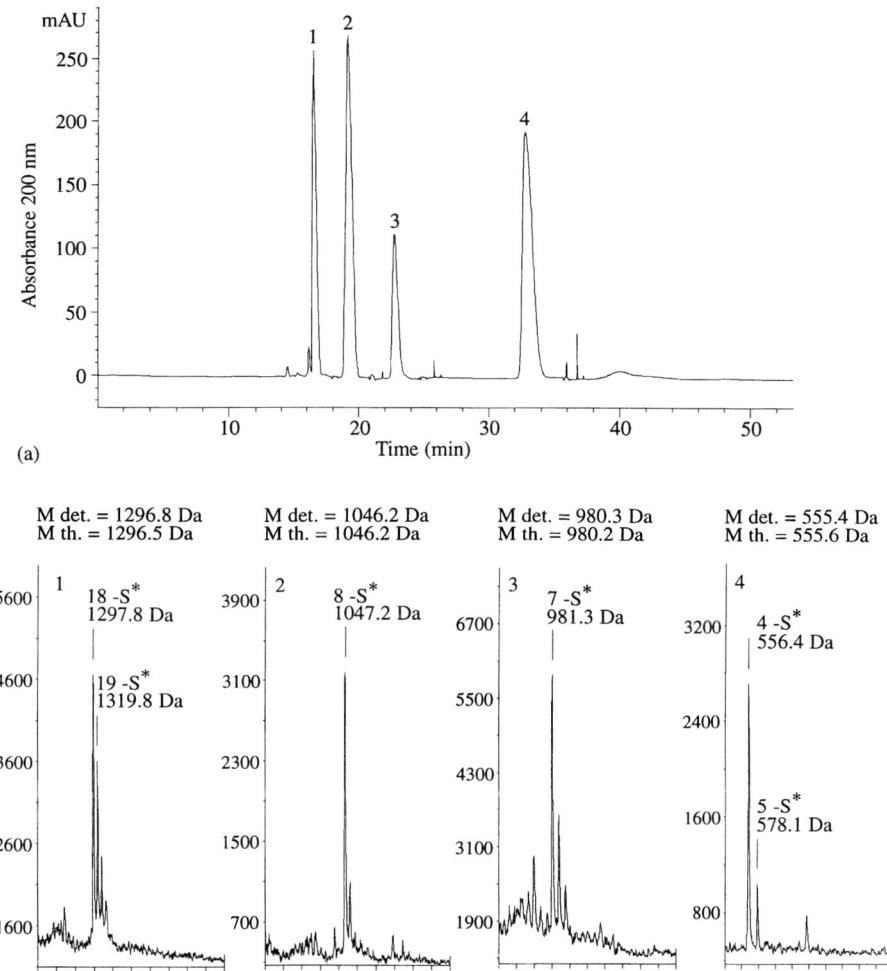

Figure 10.2 (a) CZE separation of a mixture of synthetic peptides. (b) MALDI–TOF/MS analysis of all four collected peptide fractions. Conditions: effective capillary length = 88 cm, i.d. = 100 μm, electrolyte: 100 mM sodium phosphate pH 2.5, 20 kV. Detection: 200 nm, temperature: 30°C. For (b) sinapic acid was used as matrix. Others were identical to (a).

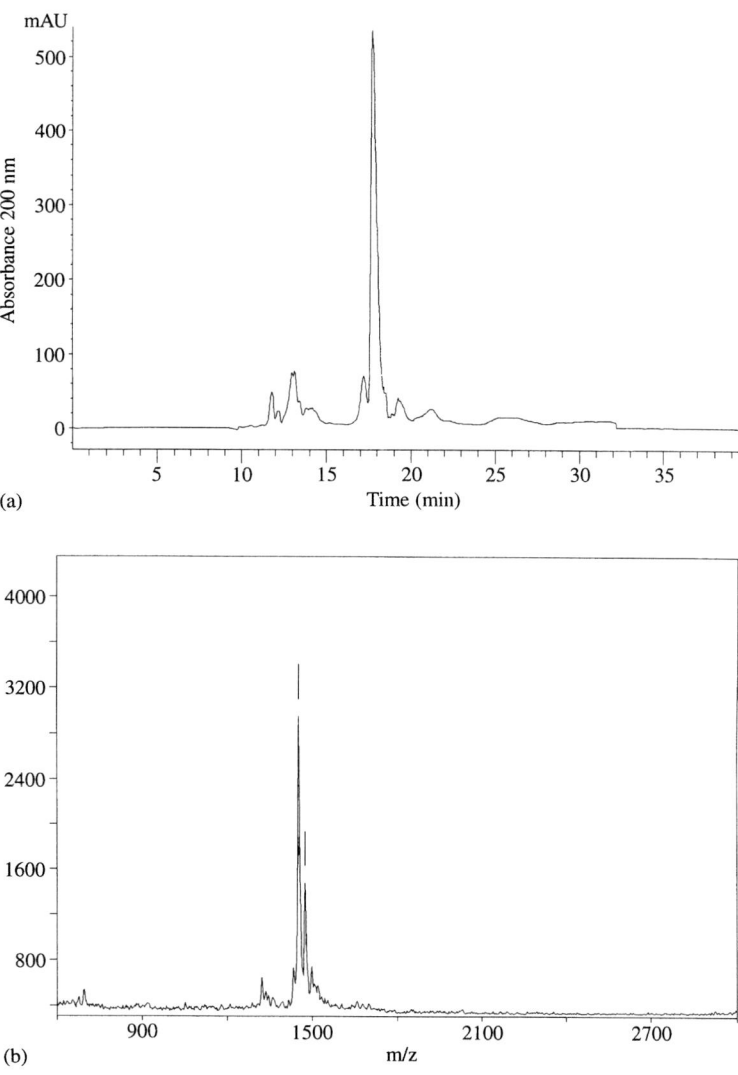

Figure 10.3 (a) MEKC separation of a crude peptide synthesis preparation. (b) MALDI–TOF/MS analysis of the collected major peptide fraction. Conditions: effective capillary length = 88 nm, i.d = 100 μm, electrolyte: 50 mM SDS, 100 mM borate pH 9.0, 22 kV. Detection: 200 nm, temperature: 35°C. For (b) sinapic acid was used as matrix. Others were identical to (a).

as the desired product, again by MALDI–TOF/MS analysis (Figure 10.3(b)).

10.5 Protein and oligonucleotide analysis by capillary gel electrophoresis (CGE)

CGE is typically applied either for protein size determination or for the separation of oligonucleotides. Figure 10.4(a) shows the analytical separation of a synthetic 20 mer oligonucleotide. As can be seen, just after the 20-mer a 21-mer oligonucleotide is present in the synthetic product. With narrow settings for the fraction collection peak width, it was possible to collect selectively only the 20-mer oligonucleotide in a micropreparative separation (Figure 10.4(b)). Successful fraction collection and identification of the 20 mer oligonucleotide was simply and quickly confirmed by analyzing the crude oligomer preparation and the CGE fraction collected using MALDI–TOF/MS (Figure 10.4(c)).

10.6 Protein mixture analysis by capillary isoelectric focusing (CIEF)

Finally the micropreparative separation of a protein mixture by CIEF is described. A mixture of four proteins with different isoelectric point (pI) values was separated and all the proteins collected from one single run into a carrier ampholyte solution (Figure 10.5(a)). After vortexing the collected fractions, these were further analyzed by refocusing in a very short capillary. As shown in Figure 10.5(b) all proteins could be successfully fractionated and purified to homogeneity. Further identification of the proteins was then performed by subjecting them directly to protein sequencing without prior removal of the carrier ampholytes [5]. The first protein fraction collected could be N-terminally sequenced for 45 amino acids (Figure 10.5(c)).

10.7 Conclusion

Even though the amount of sample in CE is quite tiny, chemical structure determination is attainable using a combination of several modes of CE analysis, automated fractionation collection and MALDI–TOF/MS. It is essential that an accurate sampling time must be determined on the premise that CE separation is reproducible, otherwise resulting in loss of collection. In this sense, reproducibility of elution time of CE is essential for quantitative and qualitative determination as well as automated fractionation.

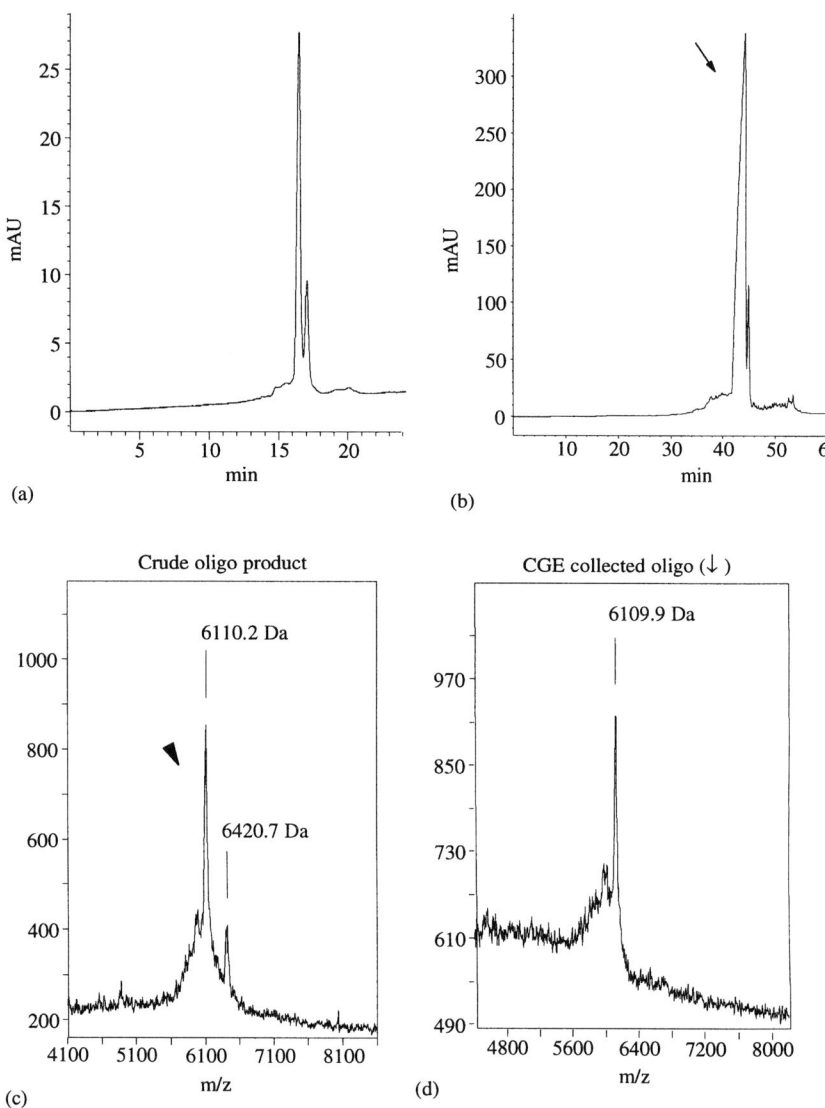

Figure 10.4 (a) Analytical CGE separation of a synthetic 20 mer oligonucleotide preparation using the Beckman eCAP ssDNA 100 kit. Conditions: effective capillary length = 40 cm, i.d = 100 μm, −20 kV. Detection: 256 nm, temperature: 30°C. (b) Micropreparative CGE separation. Conditions were identical to (a) excepting applied voltage of −7 kV. (c) MALDI–TOF/MS analysis of the crude oligomer preparation. CGE collected was 20-mer oligonucleotide using dihydroxyacetophenone as matrix.

CONCLUSION

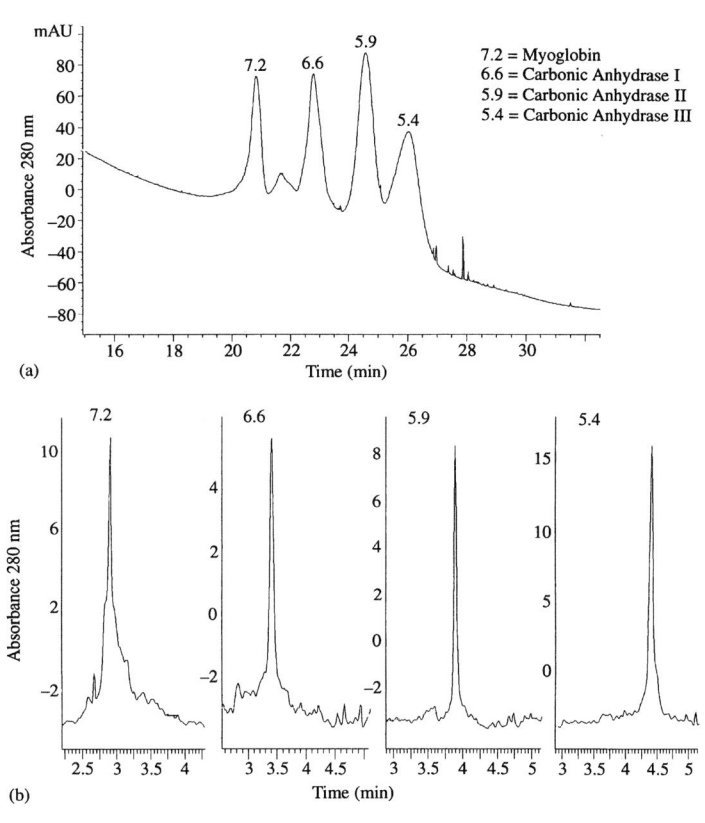

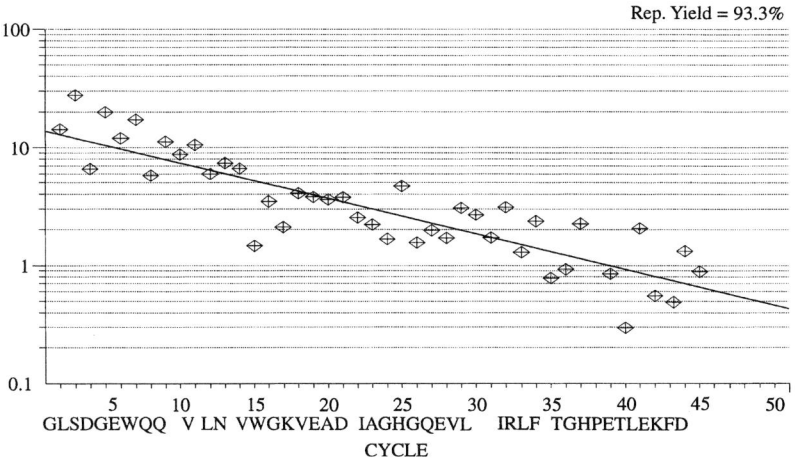

References

1. Albin, M.; Chen, S.M.; Louie, A.; Pairaud, C.; Colburn, J.; Wiktorowicz, J. (1992) The use of capillary electrophoresis in a micropreparative mode: methods and applications. *Anal. Biochem.* **205**, 382–388.
2. Herold, M.; Wu, S.-L. (1994) Automated peptide fraction collection in CE. *LC–GC* **12**, 531–533.
3. Boss, H.J.; Rohde, M.F.; Rush, R.S. (1995) Multiple sequential fraction collection of peptides and glycopeptides by high-performance capillary electrophoresis. *Anal. Biochem.* **203**, 123–129.
4. Grimm, R.; Herold, M. (1994) Micropreparative single run fraction collection of peptides separated by CZE for protein sequencing. *J. Capillary Electrophoresis*, **1**, 74–77.
5. Grimm, R. (1995) Micropreparative capillary isoelectric focusing of protein and peptide samples followed by protein sequencing. *J. Capillary Electrophoresis*, **2**, 111–115.

Figure 10.5 (a) Micropreparative capillary isoelectric focusing of a protein mixture. Conditions: effective capillary length = 88 cm, i.d = 100 µm, carrier ampholytes: 1.25% Servalyt 3–10 in 0.4% hydroxymethylcellulose, anolyte: 100 mM phosphoric acid, catholyte: 20 mM NaOH, 30 kV. Detection: 280 nm, focusing time: 15 min, pressure mobilization. (b) Refocusing of the collected protein fractions in a 25 cm effective length capillary. Further conditions deviating from (a): capillary i.d: 50 µm, focusing time: 2.2 min. (c) Repetitive yield (93.3%) plot of the protein sequence analysis of the first protein fraction (myoglobin).

11 Comparison with other analytical methods
K. KITAGISHI

11.1 Comparison with high performance liquid chromatography

High performance liquid chromatography (HPLC) is a most important technique for general purposes and has a potential market in the field of analytical chemistry. The principles and theory of HPLC can be partially applied to CE. The comparison of CE with HPLC provides a useful index in considering a future development plan in CE.

11.1.1 Separation capability

Since CE shows higher efficiency for theoretical plate numbers than HPLC by an order of one or two [1, 2], CE is essentially considered to be suitable for samples which contain a large number of components with a subtle difference in physicochemical properties [3]. In HPLC, a wide range of separation selectivity can be provided by using a diverse combination of solid phases and eluents. CE allows separation with many kinds of buffers and additives as the electrolyte, although probably still less than HPLC at present. Both polar and non-polar compounds can be analyzed by HPLC. Although CE shows superior separation for charged compounds, its capability for non-polar, uncharged compounds is restricted. The reasons are that solubility in aqueous solutions is often significantly low and that there is no difference in electric charge to induce the same migration velocity. MEKC partly solves these problems [4], but very non-polar compounds cannot be used even in MEKC due to their insolubility.

CE is very convenient for chiral separation using cyclodextrin (CD) solution instead of expensive chiral support of HPLC. However, the chiral selectivity of CD is more restricted than chiral support HPLC.

11.1.2 Detection

A variety of detectors are commercially available for HPLC. For CE, an absorbance model is commercially common, and a limited number of laser-induced fluorescence (LIF), electrochemical detection (ECD) and mass spectroscopy (MS) systems are available. Other modes are still home-made in the laboratory and do not always attain a satisfactory evaluation.

The MS detection sensitivity achievable in CE is impressive owing to the small dimensions of the capillary and the low sample loading capacity [2]. However, the concentration sensitivity of CE is generally lower than that of HPLC when UV absorbance detection is employed in both techniques. This is due to the short path length imposed by the small diameter in CE.

To improve the concentration sensitivity of CE, a more specialized detection system, such as laser-induced fluorescence and ECD, may be employed. Alternatively, sample preconcentration may be performed off-column or on-column as discussed in the next section.

11.1.3 Sample preparation

The sample preparation methods employed for CE are similar to or simpler than those for HPLC [5]. Typically, simple filtration is used to remove particles for both separations. Some other interfering substances should be eliminated before analysis in HPLC, which is often expensive. The interfering substances may mask peaks of interest and/or damage permanently the HPLC support. Such pretreatments are not always required in CE. The electrolyte can be replaced easily between each run, preventing damage of the column in CE. Moreover, when the interfering substances have different migration velocities from analyte ions, the injection amount can be reduced by the control of EOF.

In both HPLC and CE, direct injection of the sample can be performed when the concentrations of analytes lie in the range of detection sensitivity. When the concentrations of the analytes are above the linear range of the detection system, the sample can be simply diluted with a suitable solvent or the running buffer. The running buffer is more appropriate. When the concentrations of the analytes are below the detection limits, solvent extraction or solid phase extraction (SPE) is often used to preconcentrate or extract the desired components from the sample matrix. Automated on-line sample preconcentration systems are available commercially for HPLC, whereas in CE on-column preconcentration can be readily achieved either by coupling isotachophoresis (ITP) with other modes of CE or by field amplified injection like sample stacking.

An example of a field amplified injection is as follows. When a dilute solution of analyte ions is introduced into the capillary, stacking of sample ions at the interface between the dilute solution and the operating buffer occurs because of the rapid migration of the ions under the greater potential gradient in dilute solutions. Preconcentration factors of as much as 100 to 1000 times can be obtained by the stacking.

11.1.4 Reliable quantitation

Although highly automated instruments are available for both HPLC and CE, the accuracy of an assay may still depend to some extent on the skill of the analyst. There should be little difference in the quantitative capabilities of HPLC and CE, especially when the analyses are performed by experienced operators. Reproducible retention times (less than 0.5% RSD (relative standard deviation)) and peak area/height (less than 1% RSD) can be obtained in HPLC, provided that the peaks are reasonably well separated and there is negligible peak tailing. In CE, the amount introduced into the column is often three orders smaller than HPLC (in the range of nanoliters instead of microliters). Therefore, syringe or valve type of injection methods are not frequently used in CE owing to practical difficulties encountered when handling minute amounts of samples. Instead, hydrodynamic or electrokinetic injection is generally used for CE sample introduction. Similar reproducibilities to HPLC systems have been obtained with the currently available commercial CE instruments [6, 7]. Electrokinetic injection or the small amount for injection of low viscosity sample solutions is less reproducible. In all cases, further improvements in quantitation or reproducibility of the data can be obtained by precisely controlling the temperature of the capillary and/or by using internal standards [3]. A potential problem that may affect reproducibility in CE is irreversible adsorption. Usually adsorbed compounds can be removed by washing the capillary. However, for strongly adsorbed components such as proteins other methods to reduce the undesirable effects of irreversible adsorption may be utilized. Various approaches have been developed to reduce adsorption, including the use of a coated capillary, the use of buffer additives, the use of high ionic strength buffers and the use of either high- or low-pH buffers. In the areas of validation and ruggedness, HPLC has been more widely tested than CE. Nevertheless, there is now increasing attention being paid to validation and ruggedness testing in CE methodologies, and the initial results obtained have been favorable for CE.

11.1.5 *Resources utilization and environmental friendliness*

In terms of resource utilization and environmental friendliness, CE possesses many advantages over HPLC. Although the costs of commercial HPLC and CE instruments are comparable, the cost of maintenance for CE is expected to be lower, since replacement items, such as injection valves and pump seals, are not required. In addition, the cost of HPLC columns tends to be higher than those for CE. CE utilizes low cost fused silica capillaries [7]. Furthermore, solvent and daily reagent consumption is much smaller for CE than for HPLC. CE does not generate significant quantities of waste, mainly because organic solvent are not required for the analysis, and only small amounts of aqueous buffer are normally required. Consequently, there should be considerable potential for cost savings in terms of column, solvents and waste disposal by performing CE instead of HPLC in routine analyses. This applies to conventional HPLC. Fast HPLC or capillary HPLC will not be considered. Current HPLC utilizes an even sized silica particle with a diameter size of 2 μm, so that 4 mm i.d. × 2 cm is enough for fast HPLC.

11.1.6 Assay time, sample amount and throughput

Both HPLC and CE typically require between several minutes and an hour per analysis. The assay time by CE is often shorter than that by HPLC [2]. Although CE is capable of extremely fast separations (e.g. several milliseconds per run), there has been little practical implementation of such rapid CE separation methods.

A long preconditioning process is required only for HPLC. Flushing with running buffer is sufficient in CE. Stabilization of the system for a half day is often desirable for good reproducibility in ion pair HPLC [6].

In terms of sample requirement, CE has the advantage that the amount of sample required is minimal at several nanoliters per analysis [7]. This feature makes CE a noteworthy technique for samples in limited supply or for valuable samples [1]. On the other hand, the sample capacity of CE is generally low and, therefore, it would be difficult to use CE for preparative scale separations.

A recent development in CE is the use of multiple capillaries. The use of open tubular capillaries with very small amounts of reagent allows implementation of a simultaneous multicolumn CE system with a simple and economical configuration. In addition, a significant increase in sample throughput can be achieved. Further developments in this respect should provide significant gains in throughput in CE analyses.

Judging from the advances made to date, the prospect for CE appears excellent. The widespread use of CE has been accelerated by the emergence of numerous commercially available CE instruments on the market. The main features of these commercial instruments are temperature control, autosampling, computer-controlled analysis, polarity switching, ease of operation, several sorts of detectors, etc. These instruments provide reproducible qualitative and quantitative analyses. At the current stage of development, CE is considered by many researchers to be either an attractive complement or an alternative to HPLC.

11.2 Comparison with gel electrophoresis

Conventional gel electrophoresis (GE) is routinely used for confirmation and identification of biopolymers such as proteins and nucleic acids in the fields of biochemistry and biomedicine. Polymer sieving and isoelectric focusing in CE can be a substitute for GE. Polymer sieving for SDS–

proteins and DNA provides information about their molecular weight. Isoelectric focusing is a separation technique based on the isoelectric points (pI) of proteins or peptides.

11.2.1 Preparation of separation medium

Polyacrylamide and agarose commonly serve as sieving media for GE. For polyacrylamide gel, acrylamide monomers which are highly toxic and N,N'-methylenebis(acrylamide) are dissolved and polymerized in the presence of catalysts, e.g. ammonium peroxydisulfate and N,N,N',N'-tetramethylethylenediamine. Agarose gel is formed by lowering the temperature of hot agarose solution. In both cases, air bubbles should be eliminated because they reduce the separation efficiency considerably. Gel preparation consists of time consuming, laborious and unreproducible procedures.

Polyacrylamide and agarose gels can be applied to CE as well. Furthermore, solutions containing soluble linear polymers, i.e. cellulose derivatives, dextran, linear polyacrylamide, poly(ethylene glycol) and poly(ethylene oxide), have been recently developed as molecular sieving media. The larger the molecular weight of a component, the larger the resistance from the sieving medium. One of the advantages of molecular sieving with linear polymer solutions is simple reproducible procedures with no special preparative skills for the sieving medium. Earlier polymer solutions were used to show that the separation resolution on DNA is similar to agarose gel but is inferior to polyacrylamide gel. Later on the resolution for polymer solutions was improved significantly. Very high resolution close to polyacrylamide gel is often reported.

Isoelectric focusing is achieved on a slab gel containing an ampholyte zwitterion mixture, which forms a favorable pH gradient before analysis by prefocusing. In CE, preparation procedures are extremely simple so that dilute ampholyte solutions can be filled into the capillary. Viscous polymers are often included to suppress hydrodynamic flow during electrophoresis. The analytes, peptides and proteins are sometimes injected into the capillary in a mixture with ampholyte.

11.2.2 Sample preparation and injection

A similar sample is used for molecular sieving in CE as is used in GE. For the analysis of SDS–proteins, SDS is added to the solution of proteins or peptides of interest and the mixed solution is kept in a hot water bath to form stable SDS–protein complexes. Nucleic acid samples are served in the buffer as in conventional GE.

The sample of protein or peptide in capillary isoelectric focusing (CIEF) is premixed with ampholyte before injection as described above or injected following the filling of ampholyte into the capillary. In contrast in isoelectric focusing with a slab gel it is common to place a piece of paper soaked in the sample solution onto the prefocused slab gel with the pH gradient.

11.2.3 Detection

Conventional GE generally requires a staining process for detection. In GE for SDS–protein and isoelectric focusing, the gel is soaked in dye solution such as Coomassie brilliant blue to be stained followed by washout of the extra dye on or in the gel. Densitometry is frequently used for detection. DNA separated by GE is detected by Ag staining, or by fluorescing with intercalating dye.

In CE, detection can be achieved on-column, either absorbance or fluorescence. While the limit of mass for detection in CE is relatively lower than that in GE for molecular sieving, the concentration limit of detection is comparable between the two. CIEF shows very high detection sensitivity due to the concentration effect during focusing.

11.2.4 Reliable quantitation and reproducibility

Both in conventional slab gel electrophoresis and in polymer sieving in CE, the injection accuracy of sample mass has not been considered much. Relative quantitation of the amount of analyte often provides important information. Examples are determination of the content of isozymes, or impurities and quantitation of competitive polymerase chain reaction (PCR) product [8]. The reproducibility of migration velocity is more meaningful than that of peak area/height because migration velocities reflect quantitative indices, molecular weights or pI of peptides or proteins. Samples are electrophoretically migrated in parallel with standards for a direct visual comparison of the results in slab gel analysis. With CE analysis, run-to-run reproducibility is crucial, as the samples are run separately from the standards. The use of internal standards both compared with standards for quantitation and compared with unknown compounds facilitates the direct comparison [9–11]. Adsorption of proteins onto the capillary inner wall makes the reproducibility worse. The coating on the capillary surface which minimizes protein adsorption much improves the reproducibility.

Molecular weights of proteins estimated by CE with SDS–polymer are not significantly different from those with SDS–PAGE [10, 12] and are, moreover, anticipated to be more precise values [10]. For the analysis of double-stranded DNA ranging from 179 to 203 base pairs, an average standard deviation of 0.5 base pairs was obtained for eight replicate runs with appropriate internal standards, comparable to the results by polyacrylamide slab-gel electrophoresis [9].

11.2.5 Assay time, throughput, sample amount and cost

The analysis time for one sample in CE is much shorter than in GE, e.g. within 10 min for electrophoresis with the corresponding period for rinsing and purging for SDS–protein [12] and DNA [8, 9]. For a number of samples, however, the total separation time in GE is shorter than in CE, because samples in slab gel electrophoresis can be run in parallel, while CE measures samples sequentially. Nevertheless additional processes for gel preparation and detection, which sometimes includes staining of analyte zones and computer analysis are time-consuming and laborious in GE [8]. Developments in multicapillaries in CE aimed at DNA sequencing will possibly improve the limit of throughput in CE with a single capillary electrophoretic lane.

The sample amount for CE is tiny. Efficient separation of PCR products can be obtained with minimum of sample preparation [11].

The price of commercial CE equipment is greater than that of a GE assembly. However, the former has the advantage of automated operation and no requirement for an experienced analytical technician. CE has a much lower running cost than GE, i.e. particularly tiny amounts of reagent. CE shows considerable power for routine analysis with reproducible quantitation.

11.3 Comparison with ion chromatography

Ion chromatography (IC) is commonly used for routine analysis of cations or anions. In several reports, excellent separation of ions even for complex sample matrices by IC as well as CE has been demonstrated. CE might be an alternative for IC due to its diverse applicability.

11.3.1 Separation capability

The separation capability of IC depends on the column and the eluent. Anions and cations require different columns for separation. Furthermore, separate individual columns are often required for analysis of monovalent and divalent cations. Samples consisting of complex matrices can be analyzed in IC by replacing columns or by coupling columns. The selectivity in CE is determined by the components of the electrolyte. The CE system is capable of simultaneous analysis of anionic and cationic samples by changing the buffer, for both monovalent and divalent cations.

A large number of theoretical plates in CE indicates the separation of anions and cations, for example, separation of 36 anions in 3 min [13]. The theoretical plate number in this case is calculated to be 1×10^6 at maximum, larger than that of IC by 100 times. The mode and mechanism of analysis by

IC and CE differ markedly, so comparison of the theoretical plate number is not always an appropriate marker; however it is certain that the column with the greater theoretical plate number gives better resolution over an identical analytical period.

11.3.2 Sample preparation

In IC, counterions or interfering organic compounds are eluted before the ions of interest. Disturbances by these substances produce strong shifts in the migration time, decreases in peak resolution and often reduce the lifetime of the columns. Consequently, pretreatment of sample is required for removal of the counterions and interfering compounds in IC. Such a treatment is dispensable for CE, because the target ions are allowed to migrate first with the remainder flowing out. Filtration and, if necessary, dilution of sample solution are only recommended before sample injection.

Analysis of compounds of interest in a complicated matrix such as blood, food, soil etc. will require pretreatment with SPE, dialysis or ultrafiltration to avoid matrix interference.

11.3.3 Detection, quantitation and reproducibility

In respect of detection sensitivity, IC is superior to CE. Although ions which have characteristic absorbance in the UV or visible region can be detected directly in CE, indirect UV detection is applied for non-absorbing ions, which is identical to IC. In IC, conductivity and UV detectors are equipped normally. The concentration detection sensitivity in IC is one order higher than in CE [14, 15]. On the other hand, CE is two orders of magnitude more sensitive than IC on a mass basis, since the injection volume in CE is very small [16]. Sample preconcentration by stacking deriving from electrokinetic injection of dilute sample solution or by isotachophoretic zone sharpening with a discontinuous buffer system can improve the concentration detection sensitivity in CE.

The linearities of peak area to sample concentration are comparable between CE and IC [14, 16], while the range of dynamic detection is relatively smaller in CE [16].

Quantitative estimation of concentrations in sample matrices by CE indicates no significant difference from the results by IC [14, 15]. Run-to-run and day-to-day reproducibilities are similar for IC and CE.

11.3.4 Assay time and running cost

CE is superior to IC for analysis time. For example, IC requires 5 to 30 min for separation of anions, while 2 to 8 min is taken for the similar anion analysis in CE. Analysis time in CE depends on ion exchange capacity and shorter is not always better. Shorter times with significant baseline separation and satisfactory reproducibility are essential in analytical chemistry. Reproducible measurements can be achieved only by flushing the capillary after each run in CE. In contrast, IC needs extensive cleaning and regeneration of the column.

The running cost in CE is lower than in IC. IC columns are expensive and their regeneration produces a large amount of hazardous waste. An eluent flows at a rate of the order of ml/min. The CE system with capillaries uses a small volume of rinsing solution and electrolyte.

Automated, simultaneous analysis for anions and cations can be achieved in one CE system easily with an autosampler and a programming controlled inversion of applied potential polarity. For IC, often at least two or more kinds of columns or two systems may be required for a similar analysis. CE is more appropriate for simultaneous separation of many compounds.

References

1. Sun, P.; Mariano, G.J.; Barker, G.; Hartwick, R.A. (1994) Comparison of micellar electrokinetic capillary chromatography and high-performance liquid chromatography on the separation and determination of caffeine and its analogues in pharmaceutical tablets. *Anal. Lett.* **27**, 927–937.
2. Loregian, A.; Scremin, C.; Schiavon, M.; Marcello, A.; Palù, G. (1994) Quantitative analysis of ribonucleotide triphosphates in cell extracts by high-performance liquid chromatography and micellar electrokinetic capillary chromatography: A comparative study. *Anal. Chem.* **66**, 2981–2984.
3. Klockow, A.; Paulus, A.; Figueiredo, V.; Amadò, R.; Widmer, H.M. (1994) Determination of carbohydrates in fruit juices by capillary electrophoresis and high-performance liquid chromatography. *J. Chromatogr.* **680**, 187–200.
4. Jinno, K.; Sawada, Y. (1995) Comparison of cyclodextrin modified micellar electrokinetic capillary chromatography and reversed-phase liquid chromatography for separation of polycyclic aromatic hydrocarbons. *J. Capillary Electrophoresis* **2**, 151–155.
5. Lagoutte, D.; Lombard, S.; Nisseron, S.; Papet, M.P.; Saint-Jalm, Y. (1994) Determination of organic acids in cigarette smoke by high-performance liquid chromatography and capillary electrophoresis. *J. Chromatogr.* **684**, 251–257.
6. Carneiro, M.C.; Puignou, L.; Galceran, M.T. (1994) Comparison of capillary electrophoresis and reversed-phase ion-pair high-performance liquid chromatography for the determination of paraquat, diquat and difenzoquat. *J. Chromatogr.* **669**, 217–224.
7. Issaq, H.J. (1994) Comparison of chiral separations in capillary zone electrophoresis with other methods. *Instrum. Sci. Tech.* **22**, 119–149.
8. Kuypers, A.W.H.M.; Meijerink, J.P.P.; Smetsers, T.F.C.M.; Linssen, P.C.M.; Mensink, E.J.B.M. (1994) Quantitative analysis of DNA aberrations amplified by competitive polymerase chain reaction using capillary electrophoresis. *J. Chromatogr.* **660**, 271–277.
9. Butler, J.M.; McCord, B.R.; Jung, J.M.; Allen, R.O. (1994) Rapid analysis of the short tandem repeat HUMTH01 by capillary electrophoresis. *BioTechniques* **17**, 1062–1070.
10. Guttman, A.; Nolan, J. (1994) Comparison of the separation of proteins by sodium dodecyl sulfate–slab gel electrophoresis and capillary sodium dodecyl sulfate–gel electrophoresis. *Anal. Biochem.* **221**, 285–289.
11. Rossomando, E.F.; White, L.; Ulfelder, K.J. (1994) Capillary electrophoresis: separation and quantitation of reverse transcriptase polymerase chain reaction products from polio virus. *J. Chromatogr.* **656**, 159–168.

12 Tsuji, K. (1994) Sodium dodecyl sulfate polyacrylamide gel- and replaceable polymer-filled capillary electrophoresis for molecular mass determination of proteins of pharmaceutical interest. *J. Chromatogr.* **662**, 291–299.
13 Jones, W.R.; Jandik, P. (1992) Various approaches to analysis of difficult sample matrices of anions using capillary ion electrophoresis. *J. Chromatogr.* **608**, 385–393.
14 Cammann, K.; Kleiböhmer, W.; Mussenbrock, E.; Roß, B.; Zuther, F. (1994) Fast chromatographic separation techniques as competitors to chemical and biochemical sensor systems. *Fresenius J. Anal. Chem.* **349**, 338–345.
15 Stahl, R. (1994) Routine determination of anions by capillary electrophoresis and ion chromatography. *J. Chromatogr.* **686**, 143–148.
16 Hepler, R.W.; Yu Ip, C.C. (1994) Application of capillary ion electrophoresis and ion chromatography for the determination of O-acetate groups in bacterial polysaccharides. *J. Chromatogr.* **680**, 201–208.

Part Two

Biochemistry Applications

12 Amino acids
L.J. BRUNNER

12.1 Introduction

Several analytical methods have been reported for standard or derivatized amino acids using combinations of several capillary electrophoresis modes. This chapter will present several of these current methods and explain exactly how this technique is applied. These applications are summarized in Table 12.1.

12.2 Application of CE to amino acid analysis

Gilman and Ewing were able to analyze single mammalian cells quantitatively using on-column derivatization with laser-induced fluorescence (LIF) detection [1]. The front end of the capillary was used as a derivatization chamber and analytes were separated by capillary electrophoresis. This method afforded the ability to analyse the 2,3-naphthalenedialdehyde (NDA)-derivatized contents of a single cell in less than 30 min. Femtomole to attomole levels of alanine, aspartic acid, glutamic acid and glycine were reported [1].

Weber et al. determined the distribution of oligosaccharide linkage positions in O-linked glycoproteins using alkaline sulfite treatment, followed by hydrolysis, derivatization by NDA and UV detection [2]. Separations were done in a sodium borate buffer, however, the addition of sodium dodecycl sulfate (SDS) was needed to quantify the absolute amount of threonine in the samples.

Matsubara and Terabe used the non-ionic surfactant, Tween-20, in a phosphate buffer to separate dansyl-derivatized amino acids in solution [3]. Due to the increased Joule heating found with the high SDS concentrations needed for amino acid separation, Tween-20 was chosen as the buffer additive. Since Tween-20 is electrically neutral, increased concentrations did not raise the conductivity of the separation buffer. Furthermore, the use of Tween-20 did not decrease the separation resolution.

O'Shea et al. used capillary zone electrophoresis (CZE) with electrochemical detection (ECD) to measure the extracellular levels of aspartate, glutamate and alanine from the frontoparietal cortex of rats implanted with microdialysis probes [4]. Samples were NDA-derivatized and separated with a pH 9.0 borate buffer. The use of CE as an analytical tool for microdialysis sample analysis is noteworthy, due to the small sample volumes often obtained from microdialysis experiments. In the work of O'Shea et al., 3 µl samples were used for the derivatization. The volume of samples analysed were in the nanoliter range. Hernandez et al. used LIF to measure glutamic acid in the corpus striatum of concious rats through the use of microdialysis sampling probes [5]. Samples were fluorescein isothiocyanate (FITC)-derivatized and separated in a carbonate buffer. A low sample volume, 120 pl, was analyzed, and samples were resolved in less than 3 min.

Ma et al. used indirect photometric detection to measure arginine, histidine and lysine in cell culture media [6]. Analytes were separated in less than 10 min in the presence of quinine sulfate and hydroxypropylmethyl cellulose (HPMC). In addition, amino acids in the cell culture media were separated from polyamines extracted from the rat pheochromocytoma cell line, PC12.

Chan et al. used LIF to detect N-(9-fluorenylmethoxycarbonyl) (FMOC)-derivatized amino acids which were separated in a borate buffer, with or without SDS [7]. Seventeen amino acids were detected in about 16 min. The derivativization and detection methods were sensitive and provided a limit of detection for alanine in the order of 5×10^{-10} M.

Tracht et al. used postcolumn radionuclide detection to measure six radiolabeled amino acids [8]. Following separation in an alkaline borate buffer, analytes are directed onto peptide binding membranes which have been coated with a solid scintillator. Light emission from the membrane scintillation is imaged onto a charge-coupled device by the use of camera lenses. Detection limits are in the femtomole and attomole range for ^{3}H-labeled and ^{35}S-labeled analytes, respectively.

Yang and Hage used human serum albumin (HSA) as a buffer additive to separate tryptophan enantiomers [9]. While adsorbed proteins may have an important role in the separation of certain compounds, these researchers demonstrated that for the separation of tryptophan enantiomers, it was the human serum albumin dissolved in the acidic phosphate buffer that determined separation. It was further shown that field strength was a significant factor in achieving adequate separations.

Bergquist et al. used LIF to measure amino acids in human cerebrospinal fluid samples [10]. Samples were labeled with 3-(4-carboxybenzoyl)-2-quinoline carboxaldehyde (CBQCA) prior to analysis and separated in a borate–SDS buffer containing dimethylsulfoxide (DMSO). Samples were resolved in less than 35 min. Samples were analyzed from patients diagnosed with Alzheimer's dementia, childhood neurological disorders and control patients. The differences in cerebrospinal fluid levels of amino acids found between patient groups appeared to be related to the age and disease. This study demonstrates how CE may be used as a rapid and sensitive analytical tool for clinical diagnosis.

Table 12.1 Summary of applications in amino acid analysis

Compound	Method	Separation and detection conditions	Reference
Alanine (Ala)	CZE	100 mM borate; pH 9.5, LIF	1
	CZE	20 mM borate; pH 9.0, NDA-derivatization, UV 254	2
	MECC[a]	25 mM phosphate, 100 mM Tween-20; pH 2.4, dansyl-derivatization, UV 214	3
	CZE	20 mM borate; pH 9.0, NDA-derivatization, electrochemical detection	4
	MECC	20 mM borate, 25 mM SDS; pH 9.2, FMOC-derivatized, LIF	7
	CZE	20 mM borate; pH 9.2, FMOC-derivatized, LIF	7
	MECC	50 mM borate, 30 mM SDS, 20% DMSO; pH 9.0, CBQCA-derivatized, LIF	10
Arginine (Arg)	CZE	50 mM phosphate; pH 2.5, UV 200 nm	14
	MECC	25 mM phosphate, 100 mM Tween-20; pH 2.4, dansyl-derivatization, UV 214	3
	CZE	8 mM quinine sulfate in 20% ethanol, 0.5% HPMC; pH 5.9, UV 236 nm	6
	MECC	20 mM borate, 25 mM SDS; pH 9.2, FMOC-derivatized, LIF	7
	MECC	50 mM borate, 30 mM SDS, 20% DMSO; pH 9.0, CBQCA-derivatized, LIF	10
Asparagine (Asp)	MECC	25 mM phosphate, 100 mM Tween-20; pH 2.4, dansyl-derivatization, UV 214	3
Aspartic acid (Asp)	CZE	100 mM borate; pH 9.5, LIF	1
	CZE	20 mM borate; pH 9.0, NDA-derivatization, UV 254	2
	CZE	50 mM phosphate; pH 2.5, UV 200 nm	14
	MECC	25 mM phosphate, 100 mM Tween-20; pH 2.4, dansyl-derivatization, UV 214	3
	CZE	20 mM borate; pH 9.0, NDA-derivatization, electrochemical detection	4

Table 12.1 *Continued*

Compound	Method	Separation and detection conditions	Reference
	MECC	20 mM borate, 25 mM SDS; pH 9.2, FMOC-derivatized, LIF	7
	MECC	50 mM borate, 30 mM SDS, 20% DMSO; pH 9.0, CBQCA-derivatized, LIF	10
	CZE	10 mM borate; pH 9.5, NDA-derivatization, LIF	11
Cysteine (Cys)	MECC	25 mM phosphate, 100 mM Tween-20; pH 2.4, dansyl-derivatization, UV 214	3
	MECC	20 mM borate, 25 mM SDS; pH 9.2, FMOC-derivatized, LIF	7
Glutamine (Gln)	MECC	25 mM phosphate, 100 mM Tween-20; pH 2.4, dansyl-derivatization, UV 214	3
	MECC	50 mM borate, 30 mM SDS, 20% DMSO; pH 9.0, CBQCA-derivatized, LIF	10
Glutamic acid (Glu)	CZE	100 mM borate; pH 9.5, LIF	1
	CZE	20 mM borate; pH 9.0, NDA-derivatization, UV 254	2
	MECC	25 mM phosphate, 100 mM Tween-20; pH 2.4, dansyl-derivatization, UV 214	3
	CZE	20 mM borate; pH 9.0, NDA-derivatization, electrochemical detection	4
	CZE	20 mM carbonate; pH 9.5, FITC-derivatized, LIF	5
	MECC	20 mM borate, 25 mM SDS; pH 9.2, FMOC-derivatized, LIF	7
	MECC	50 mM borate, 30 mM SDS, 20% DMSO; pH 9.0, CBQCA-derivatized, LIF	10
	CZE	10 mM borate; pH 9.5, NDA-derivatization, LIF	11
Glycine (Gly)	CZE	100 mM borate; pH 9.5, LIF	1
	CZE	50 mM phosphate; pH 2.5, UV 200 nm	14

AMINO ACID ANALYSIS

Table 12.1 *Continued*

Compound	Method	Separation and detection conditions	Reference
	CZE	20 mM borate; pH 9.0, NDA-derivatization, UV 254	2
	MECC	25 mM phosphate, 100 mM Tween-20; pH 2.4, dansyl-derivatization, UV 214	3
	MECC	20 mM borate, 25 mM SDS; pH 9.2, FMOC-derivatized, LIF	7
	MECC	50 mM borate, 30 mM SDS, 20% DMSO; pH 9.0, CBQCA-derivatized, LIF	10
Histidine (His)	CZE	50 mM MOPS[b] + Tris; pH 7.75, UV 210 nm	15
	CZE	8 mM quinine sulfate in 20% ethanol, 0.5% HPMC; pH 5.9, UV 236 nm	6
	MECC	20 mM borate, 25 mM SDS; pH 9.2, FMOC-derivatized, LIF	7
Isoleucine (Ile)	CZE	50 mM phosphate, pH 2.5, UV 200 nm	14
	MECC	25 mM phosphate, 100 mM Tween-20; pH 2.4, dansyl-derivatization, UV 214	3
	MECC	20 mM borate, 25 mM SDS; pH 9.2, FMOC-derivatized, LIF	7
Leucine (Leu)	CZE	50 mM phosphate; pH 2.5, UV 200 nm	14
	MECC	25 mM phosphate, 100 mM Tween-20; pH 2.4, dansyl-derivatization, UV 214	3
	MECC	20 mM borate, 25 mM SDS; pH 9.2, FMOC-derivatized, LIF	7
	CZE	50 mM borate; pH 10.0, radiochemical	8
Lysine (Lys)	MECC	25 mM phosphate, 100 mM Tween-20; pH 2.4, dansyl-derivatization, UV 214	3
	CZE	8 mM quinine sulfate in 20% ethanol, 0.5% HPMC; pH 5.9, UV 236 nm	6
	MECC	20 mM borate, 25 mM SDS; pH 9.2, FMOC-derivatized, LIF	7
	CZE	50 mM borate; pH 10.0, radiochemical	8
Methionine (Met)	MECC	20 mM borate, 25 mM SDS; pH 9.2, FMOC-derivatized, LIF	7

Table 12.1 Continued

Compound	Method	Separation and detection conditions	Reference
	CZE	50 mM borate; pH 10.0,	8
	MECC	75 mM SDGlu^c, 50 mM SDS, radiochemical	16
Phenylalanine (Phe)	CZE	50 mM phosphate; pH 6.52, DNB-derivatized, UV 214 nm	17
	CZE	50 mM phosphate; pH 2.5, UV 200 nm	14
	CZE	25 mM borate; pH 10, UV 214 nm	12
	MECC	25 mM phosphate, 100 mM Tween-20; pH 2.4, dansyl-derivatization, UV 214 nm	3
	CZE	7.84 mM β-cyclodextran, pH 6.52, DNB-derivatized, UV 214 nm	17
	MECC	20 mM borate, 25 mM SDS; pH 9.2, FMOC-derivatized, LIF	7
	CZE	50 mM borate; pH 10.0, radiochemical	8
Proline (Pro)	MECC	25 mM phosphate, 100 mM Tween-20; pH 2.4, dansyl-derivatization, UV 214 nm	3
	MECC	20 mM borate, 25 mM SDS; pH 9.2, FMOC-derivatized, LIF	7
	CZE	20 mM borate; pH 9.2, FMOC-derivatized, LIF	7
	CZE	50 mM borate; pH 10.0, radiochemical	8
Serine (Ser)	MECC	25 mM phosphate, 100 mM Tween-20; pH 2.4, dansyl-derivatization, UV 214 nm	3
	MECC	20 mM borate, 25 mM SDS; pH 9.2, FMOC-derivatized, LIF	7
	MECC	50 mM borate, 30 mM SDS, 20% DMSO; pH 9.0, CBQCA-derivatized, LIF	10
Threonine (Thr)	MECC	20 mM borate, 50 mM SDS; pH 9.0, NDA-derivatization, UV 254 nm	21
	MECC	25 mM phosphate, 100 mM Tween-20; pH 2.4, dansyl-derivatization, UV 214 nm	3
	MECC	20 mM borate, 25 mM SDS; pH 9.2, FMOC-derivatized, LIF	7

Table 12.1 Continued

Compound	Method	Separation and detection conditions	Reference
	MECC	50 mM borate, 30 mM SDS, 20% DMSO; pH 9.0, CBQCA-derivatized, LIF	10
Tryptophan (Trp)	CZE	25 mM borate; pH 10, UV 214 nm	12
	CZE	5 mM phosphate; pH 10.2, LIF	13
	MECC	25 mM phosphate, 100 mM Tween-20; pH 2.4, dansyl-derivatization, UV 214 nm	3
	CZE	25 mM phosphate, 10 µM HSA[d]; pH 7.4, UV 280 nm	9
		1 M urea, 10% methanol; pH 9.0, PTH-derivatization[e], UV	16
Tyrosine (Tyr)	CZE	50 mM phosphate; pH 2.5, UV 200 nm	14
	CZE	25 mM borate; pH 10, UV 214 nm	12
	MECC	20 mM borate, 25 mM SDS; pH 9.2, FMOC-derivatized, LIF	7
	CZE	50 mM borate; pH 10.0, radiochemical	8
Valine (Val)	CZE	20 mM borate; pH 9.0, NDA-derivatization, UV 254 nm	2
	MECC	25 mM phosphate, 100 mM Tween-20; pH 2.4, dansyl-derivatization, UV 214 nm	3
	MECC	20 mM borate, 25 mM SDS; pH 9.2, FMOC-derivatized, LIF	7
	MECC	50 mM borate, 30 mM SDS, 20% DMSO; pH 9.0, CBQCA-derivatized, LIF	10
	MECC	50 mM digitonin, 50 mM STDC,[f] 1 M urea; pH 2.5, PTH-derivatization, UV	16

[a] MECC = micellar electrophoretic capillary chromatography; [b] MOPS = 4-morpholine propanesulfonic acid; [c] SDGlu = N-dodecanoyl-L-glutamate; [d] HSA = human serum albumin; [e] PTH = phenylthiohydantoin; [f] STDC = sodium taurodeoxycholate.

Hernandez *et al.* used a collinear LIF detector to measure glutamic and aspartic acids in brain dialysates [11]. Microdialysis props were placed into the striatum of the rat and perfused with artificial cerebrospinal fluid. *In vivo* measurement of the release of these amino acid neurotransmitters was performed through the use of narrow-bore capillaries (15 μm). Despite the release and uptake of these neurotransmitters occuring over milliseconds, dialysate volumes were collected over 72 ms. Thus, this method theoretically provides the basis for monitoring the release of amino acid neurotransmitters in the rat brain.

Tagliaro et al. used CZE to measure serum concentrations of phenylalanine [12]. This method was used in the diagnosis of patients with the degenerative genetic disease, phenylketonuria. Patients with phenylketonuria have a deficit in phenylalanine hydroxylase, which results in significantly elevated levels of serum phenylalanine. Samples analyzed from patients with confirmed phenylketonuria showed large peaks corresponding to phenylalanine. The assays demonstrate many of the advantages of the clinical application of CZE; high resolution, rapid analysis times, decreased reagent costs and automation.

Gahm and Stalcup [17] used N-(3,5-dinitrobenzoyl) DNB-derivitization and UV detection to measure phenylalanine in solution. Phenylalanine enantiomers were separated by electrophoresis in a phosphate buffer. The addition of millimolar concentrations of β-cyclodextran improved chiral separation.

Lee and Yeung applied LIF following CZE separation to measure tryptophan in solution [13]. Detection of tryptophan was rapid, with the peak migrating in less than 4 min. The limit of detection was in the nanomolar range.

12.3 Conclusion

Innovative of technology for CE application to amino acids analysis is significant. Several techniques described in this paper have been used. For instance, the microdialysis method is applicable to real time *in vivo* determination of amino acids in moving animals. Isotope labeled amino acids are also significant for selective and sensitive analysis, especially when combined with the radioisotope detector currently available on the market. Post-CE derivatization will be required and know-how at the tee joint. Currently, these methods are mostly restricted to laboratory research. Science continues developing and these methods will be innovated further. Development of an easy operating method will be desired for routine analysis.

References

1 Gilman, S.D.; Ewing, A.G. (1995) Analysis of single cells by capillary electrophoresis with on-column derivatization and laser-induced fluorescence detection. *Anal. Chem.* **67**, 58–64.

2. Weber, P.L.; Bramich, C.J.; Lunte, S.M. (1994) Determination of the number and distribution of oligosaccharide linkage positions in O-linked glycoproteins by capillary electrophoresis. *J. Chromatogr.* **680**, 225–232.
3. Matsubara, N.; Terabe, S. (1994) Separation of 24 dansylamino acids by capillary electrophoresis with a non-ionic surfactant. *J. Chromatogr.* **680**, 311–315.
4. O'Shea, T.J.; Weber, P.L.; Bammel, B.P.; Lunte, C.E.; Lunte, S.M. (1992) Monitoring excitatory amino acid release *in vivo* by microdialysis with capillary electrophoresis–electrochemistry. *J. Chromatogr.* **608**, 189–195.
5. Hernandez, L.; Tucci, S.; Guzman, N.; Paez, X. (1993) *In vivo* monitoring of glutamate in the brain by microdialysis and capillary electrophoresis with laser-induced fluorescence detection. *J. Chromatogr.* **652**, 393–398.
6. Ma, Y.; Ahang, R.; Cooper, C.L. (1992) Indirect photometric detection of polyamines in biological samples separated by high-performance capillary electrophoresis. *J. Chromatogr.* **608**, 93–96.
7. Chan, K.C.; Janini, G.M.; Muschik, G.M.; Issaq, H.J. (1993) Laser-induced fluorescence detection of 9-fluorenylmethyl chloroformate derivatized amino acids in capillary electrophoresis. *J. Chromatogr.* **653**, 93–97.
8. Tracht, S.; Toma, V.; Sweedler, J.V. (1994) Postcolumn radionuclide detection of low-energy β emitters in capillary electrophoresis. *Anal. Chem.* **66**, 2382–2389.
9. Yang, J.; Hage, D.S. (1994) Chiral separations in capillary electrophoresis using human serum albumin as a buffer additive. *Anal. Chem.* **66**, 2719–2725.
10. Bergquist, J.; Gilman, S.D.; Ewing, A.G.; Ekman, R. (1994) Analysis of human cerebrospinal fluid by capillary electrophoresis with laser-induced fluorescence detection. *Anal. Chem.* **66**, 3512–3518.
11. Hernandez, L.; Joshi, N.; Murzi, E.; Verdeguer, P.; Mifsud, J.C.; Guzman, N. (1993) Collinear laser-induced fluorescence detector for capillary electrophoresis. *J. Chromatogr.* **652**, 399–405.
12. Tagliaro, F.; Moretto, S.; Valentini, R.; Gambaro, G.; Antonioli, C.; Moffa, M.; Tató, L. (1994) Capillary zone electrophoresis determination of phenylalanine in serum: A rapid, inexpensive and simple method for the diagnosis of phenylketonuria. *Electrophoresis* **15**, 94–97.
13. Lee, T.T.; Yeung, E.S. (1992) High-sensitivity laser-induced fluorescence detection of native proteins in capillary electrophoresis. *J. Chromatogr.* **595**, 319–325.
14. Agerberth, B.; Lee, J.; Bergman, T.; Carlquist, M.; Boman, H.G.; Mutt, V.; Jörnvball, H. (1991) Amino acid sequence of PR-39: isolation from pig intestine of a new member of the family of proline–arginine-rich antibacterial peptides. *Eur. J. Biochem.* **202**, 489–854.
15. Klepárnik, K.; Slais, K.; Bocek, P. (1993) Determination of the isoelectric points of low and high molecular mass ampholytes by capillary electrophoresis. *Electrophoresis* **14**, 475–479.
16. Otsuka, K.; Kashihara, M.; Kawaguchi, Y.; Koike, R.; Hisamitsu, T.; Terabe, S. (1993) Optical resolution by high-performance capillary electrophoresis. Micellar electrokinetic chromatography with sodium N-dodecanoyl-L-glutamate and digitonin. *J. Chromatogr.* **652**, 253–257.
17. Gahm, K.; Stalcup, A.M. (1995) Capillary zone electrophoresis study of naphthylethylcarbamoylated beta-cyclodextrins. *Anal. Chem.* **67**, 19–25.

13 Sodium dodecyl sulfate (SDS)–non-gel molecular sieving capillary electrophoresis for proteins

Y. KUROSU AND M. SAITO

13.1 Introduction

The long monopoly enjoyed by polyacrylamide slab gel electrophoresis in protein separation by molecular size is finally coming to the end as the novel method of capillary electrophoresis comes to the fore [1, 2]. There are three reasons why gel packed in a glass tube or between glass plates has long been used for separation:

1. heat convection can be suppressed;
2. electroosmosis can be suppressed;
3. separation by molecular size can be realized.

In the case of CE, it is well known that even without the gel, (1) and (2) are realized due to the smaller size of the capillary tube. Therefore, the gel is necessary only for (3), separation by molecular size. A gel-packed capillary offers another benefit, that is, applicability of a higher voltage because of good heat dissipation, which shortens the analysis time.

In CE, there are two methods of realizing the separation by gel filtration. One is to form gel, which is similar to the conventional gel, in the capillary as the sieving medium. The method is often referred to as 'capillary gel electrophoresis' (CGE). The other is to pack non-crosslinked medium in the capillary instead of conventional gel to perform gel filtration. In this method, non-branched or branched macromolecule solutions such as non-crosslinked polyacrylamide, cellulose derivatives and dextran are used as sieving media. It seems that there are no standardized terms for this method and the sieving media yet. At present, the media are referred to by various terms such as 'gel', 'non-gel', 'polymer gel', 'polymer solution', 'linear', 'liquid', 'fluidified', 'non-crosslinked', 'uncrosslinked', 'crosslink-free' and 'entangled'. In this chapter, we will use the term 'gel' for the media which have a three-dimensional structure formed by chemical covalent bonds and the term 'polymer solution (non-gel)' for the media which do not have this structure. We will also use the term 'molecular sieving capillary electrophoresis' for the method that uses both types of the media.

Use of polymer solutions (non-gel) for CE as sieving media offers the following advantages:

1. The medium can be changed easily for every run according to analytical requirements.
2. There is wide selection of media because there is no need for the media to be able to form gels.
3. Well customized media for the compounds of interest can be used.

Considering the above features, it is expected that CE using polymer solution (non-gel) as sieving media will be more commonly used in the future.

13.2 Principles of molecular sieving CE

There are two well-known theories for the migration of a flexible macromolecule through a polymer network, the Ogston model and the reptation model.

The Ogston model assumes that the matrix consists of a random network of interconnected pores having an average pore size. The migrating solute behaves as an undeformable particle. In this model, smaller molecules migrate faster because they have access to a larger fraction of the available pores. This model ceases to be applicable when the radius of the particle exceeds the pore size. The reptation model assumes that the migrating solute moves as an unraveled coil from head to tail. This model can elucidate a long and flexible chain molecule, such as DNA, that can migrate even when the radius is much larger than a pore size [3–6].

Bae and Soane [4] state that it is necessary to introduce the simple scaling law of De Gennes in the case of polymer solution. Three types of polymer solution can be imaged; dilute solutions, semi-dilute solutions and concentrated solutions. Only dilute and semi-dilute solutions are important for separation. Their calculation shows that it would be difficult to achieve high resolution for high molecular masses using polymer solutions that are neither concentrated nor extremely viscous like the sieving media [4].

13.3 Molecular sieving CE in the absence of sodium dodecyl sulfate (SDS)

Molecular sieving capillary electrophoresis, which does not use SDS corresponding to conventional disc electrophoresis, may be useful for some applications. However, it is, at present, not so popular. The reason is that proteins having isoelectric points (pI) similar to the pH value of the electrolyte do not migrate and tend to remain in the capillary. Accordingly, these proteins do not leave the capillary and cannot be detected.

Wu and Regnier [7] utilized acrylamide polymer solution for native protein separation in the absence of SDS. Polymer with a concentration range of 3.5–5% did not cause separation by size for native proteins with molecular weights from 20 000 to 47 000 [7].

Nucleic acids are always negatively charged in a commonly used electrolyte with pH (3–10), because the pK_a of the functional group of phosphoric acid is approximately 1. Therefore, nucleic acids always migrate in the capillary and reach the detector. Due to this simple migration behavior in CE, polynucleotide mixtures are the preferred materials for the study of the separation mechanism as well as evaluation and investigation of sieving media themselves.

13.4 Fundamental studies of the use of molecular sieving media (gels and non-gels) for proteins and DNA

Many fundamental studies on the use of molecular sieving media for proteins and DNA have been reported lately. Effects of factors such as column temperature, internal diameter and column length on the performance of SDS–acrylamide polymer solution capillary electrophoresis were evaluated for the analysis of protein. An increase in column temperature resulted in an exponential decrease in separation efficiency. A linear relationship existed between the length of the capillary column and the peak migration time and/or theoretical plates [8].

Separation of DNA restriction fragments by CE with low and zero crosslinked polyacrylamide using continuous and pulsed electric fields was reported [9]. Three different polyacrylamide-based media were investigated by chemically synthesized oligodeoxyribonucleotides [10]. Righetti et al. [12] investigated a series of crosslinkers as potential candidates for a novel class of polyacrylamide matrices, exhibiting high hydrophilicity, high resistance to hydrolysis and larger pore sizes than conventional polyacrylamide gels [11–14]. To achieve sensitive and fast separations, polymer solution CE was examined.

A systematic study of the influence of the linear polymer concentration on the separation of restriction fragments indicated that larger DNA fragments were resolved with greater efficiency by using a lower concentration of polymer, while the smaller ones were better resolved with a higher polymer concentration when added to the buffer [15]. The relative amounts of DNA fragments in a mixture injected into the capillary by electromigration or hydrodynamically by pressure were compared. The experiments were performed in capillaries filled with a solution of liquified agarose, a replaceable sieving medium [16]. The kinetics of riboflavin-mediated photopolymerization of acrylamide were re-examined by CE [17]. The acrylamide polymerization in gel electrophoresis capillaries was monitored by Raman microprobe spectroscopy and found to follow second-order kinetics [18].

To study the effect of temperature on the separation of SDS–protein complexes in the molecular weight range of 14 400–97 400, branched (dextran) and linear (poly(ethylene oxide)), were used in high performance capillary gel electrophoresis [19]. A stable sieving medium polyacrylamide filled in a capillary was developed. The inner wall of a fused silica capillary was covalently bonded with a linear polyacrylamide through Si—C linkages, in which polyacrylamide gel or non-gel was filled [20, 21]. Glucomannan, a natural polysaccharide was useful as a sieving medium [22].

Dynamic light-scattering experiments have been performed on semi-dilute solutions of hydroxyethyl cellulose which have been shown to be suitable as molecular sieving media for electrophoretic separation in a capillary format. The relationship between polymer concentration and the mesh size of the entangled network was found to be in good agreement with predictions based on the scaling theory of De Gennes and intrinsic viscosity measurements [23].

A specially optimized and reproducible procedure improved the performance and stability of capillary gel electrophoresis. Bubble formation during gel polymerization can be avoided [24]. Single-base resolution of double-strand DNA between 123 and 124 base pairs can be achieved by the use of homogeneous media prepared from poly(ethylene oxide) (2.5%, molecular wt 8 000 000), and even better separation is achieved by using mixed polymer media with formamide and urea [25]. Similarly, the performance and the efficiency of several cellulose derivatives as molecular sieving media for the capillary electrophoretic separation of DNA restriction fragments or DNA amplified by the polymerase chain reaction were investigated. The migration time and the resolution of DNA fragments were manipulated by varying several parameters, such as the size (viscosity) and the concentration of cellulose derivatives and the applied field strength [26–30].

Two injection-related artifacts with capillary gel electrophoresis were investigated. The first occurs with consecutive injections from the same, low volume aqueous sample. The second artifact observed is related to the physical shape of the inlet of the gel-filled capillary [31].

An acrylamide polymer solution prepared by different methods was evaluated. Standard polyacrylamide polymer as prepared with typical levels of catalyst (1 µl pure N,N,N',N'-tetramethylethylenediamine and 4 µl 10% peroxodisulfate per ml of solution) at room temperature, have extremely high weight-average molecular mass values (in excess of 2×10^6 Da) and can be injected or extruded from a capillary at a concentration above 6% only with great difficulty [32]. Careful examination of the polymerization reaction of acrylamide non-gel has led to methodology that has proved to be reproducible for obtaining DNA sequencing information M13mp 18 phage for 350 nucleotides in close to 30 min [33]. Loss of resolution with replaceable acrylamide non-gels was observed when increasing the length of the sample plug with pressure injection. This allows for compensation for any possible migration time variation caused by high ionic strength sample matrices [34].

Best *et al.* demonstrated DNA sequencing with polyacrylamide solution at an electric field of 200 V cm^{-1} and at room temperature. Resolution was observed to decrease exponentially with fragment length [35]. The use of low percentage (1.5–6% T) replaceable polyacrylamide network matrices for rapid separation of double-strand DNA fragments was explored. Separations of fragments ranging from 20 to 23 000 base pairs (bp) were readily achieved [36]. Cheng *et al.* have observed that addition of glycerol to a polymer solution containing (hydroxypropyl)methylcellulose in Tris–borate buffer markedly improved the separation of double-strand DNA ranging from less than 100 bp to about 1 kbp [37].

The performance of the separation, the relationship between resolution and analysis time, has been examined using poly(dT)16–500 by changing the acrylamide polymer concentration, capillary length and electric field strength. It was found that for large DNA fragments, the migration time interval between bands decreases linearly as DNA fragment size increases [38]. A systematic study of the separation of double-stranded DNA in hydroxypropylcellulose (HPC) with a molecular mass of 10^6 Da was undertaken, using a variety of concentrations and different electric fields. The data showed that at high polymer concentrations and low fields, the separation mechanism was similar to that occurring in gels. The results were in good agreement with theoretical models. For more dilute solutions and higher fields, however, the separation pattern cannot be explained by existing theories [39].

13.5 SDS–non-gel molecular sieving electrophoresis for proteins

Many groups investigated the use of polymer solutions (non-gel) as media for CE separations of proteins. The separation of proteins according to their molecular masses between 17 800 and 77 000 Da, was demonstrated in a buffer containing an acrylamide polymer as a sieving medium [40]. Wu and Regnier utilized non-crosslinked linear polyacrylamide solution to perform size-based separations of SDS–protein complexes with masses from 2500 to 205 000 Da. Separation with model proteins indicated that baseline resolution between protein species that vary by 10% in molecular mass can be achieved [41]. In another report on the sieving of non-crosslinked linear polyacrylamide solution electrophoresis, proteins differing by as little as

4% in molecular weight could be resolved. The logarithm of the standard molecular weight of proteins correlated linearly with the relative mobility of the denatured proteins over the molecular weight range 14 000 to 205 000 [42]. In polyethylene oxide, a similar investigation was performed. Larger errors were possible, however, when special groups such as glycoproteins or lipoproteins were present due to a different ratio of binding SDS molecules [43]. Rapid separations of proteins in the moleculer mass range of 20 000–200 000 Da by SDS–gel CE (non-gel may be used) was demonstrated with excellent linearity and intra- and inter-day reproducibility of the migration time. Ultrafast analysis of proteins, based on SDS-mediated non-gel electrophoresis, was developed in which protein molecular mass standards ranging from 14 200 to 94 700 Da were separated in 3 min. A 50 µm diameter uncoated fused-silica capillary and a high field strength (888 V cm^{-1}) were used [44]. Monomer–dimer forms of the recombinant human ciliary neurotrophic factory were examined by the use of this method under reducing and non-reducing conditions [45].

The rapid separation of SDS–protein complexes according to their molecular masses by CE was described. Standard proteins with masses in the range 29 000–97 400 Da were resolved to the baseline in less than 2 min by utilizing a column with a separation distance of 7 cm [46]. UV-transparent polymer networks of dextran and poly(ethylene glycol) (PEG) were substituted for polyacrylamide with successful molecular weight sieving of SDS–protein complexes at 214 nm. Due to their moderately lower viscosity, these networks could be routinely replaced leading to the possibility of hundreds of injections with a single column. Migration time reproducibilities of 0.5% RSD (relative standard deviation) or less were found with replacement of polymer between each run. Using dextran, calibration plots of peak area vs. concentration of standard protein were linear over the range of 0.5 µg ml^{-1} up to at least 0.25 mg ml^{-1} [47]. Significant separation of standard SDS–protein complexes in dextran polymer gel are shown [47]. Branched (dextran) and linear (poly(ethylene oxide)) polymers were used to study the effect of temperature on the separation of SDS–protein complexes in the molecular mass range 14 000–97 400 Da [48]. A novel polymer network was investigated for efficient sieving of SDS–protein complexes: poly(vinyl alcohol) [49]. Takagi and Karim [2] provided an overview of the separation of SDS–protein complexes by SDS–polymer solution molecular sieving CE. The use of polymer solution in the separation of proteins was discussed [2]. Separation of collagen type I chain polymers was achieved by SDS–acrylamide non-gel CE [50]. The molecular mass of antithrombin III was investigated by SDS–dextran solution. The results showed an excellent correlation with those achieved with conventional slab gels [51]. The separation of the four major whey proteins by SDS–acrylamide non-gel CE was reported. The SDS–PAGE (polyacrylamide gel electrophoresis) and HPLC methods and results were compared [52]. SDS–acrylamide non-gel CE was examined as an alternative to the high-performance size-exclusion chromatographic method for analysis of recombinant bovine growth hormone [53].

13.6 Conclusions

Studies of molecular sieving CE have historically been performed mainly for DNA, using polyacrylamide gel. Applications to proteins have been

Table 13.1 Summary of applications for proteins

Separation conditions Sieving media Coating Voltage, current Capillary length with i.d. Detection	Proteins (molecular weight)	Reference
Antibody 100 mM Tris + 250 mM boric acid, 0.1% SDS, 7M urea 2% acrylamide non-gel Bonded linear acrylamide −12 kV 45 cm (25 cm to detector), 75 μm i.d. 230 nm	Bovine IgG (partial reduction) 22 000–75 000	24, 41
Commercial (ProSort kit) 0.2% SDS, pH 7.0 Acrylamide non-gel Untreated −12 kV 42 cm (22 cm to detector), 55 μm i.d. 215 nm	Bovine γ-globulin (unreduced and reduced) 30 000–135 000	42
Commercial (eCAP SDS 14–200 kit) PEO Treated −300 V cm^{-1}, 25–30 μA 27 cm (20 cm to detector), 100 μm i.d. 214 nm	Human IgG (reduced)	43
0.1 M Tris–CHES, 0.1% SDS, pH 8.6 10% dextran (mol. wt 2 000 000) 4% linear acrylamide coated −740 V cm^{-1} 200 nm	Human IgG (reduced)	46
60 mM AMPD–CACO 0.1% SDS, pH 8.8 10% dextran (mol. wt 2 000 000) 6% linear acrylamide + dextran coated −300 V cm^{-1}, 20 μA (30 cm to detector), 75, 100 μm i.d. 214, 280 nm	Human IgG (reduced)	47

CONCLUSIONS 165

Table 13.1 *Continued*

Separation conditions Sieving media Coating Voltage, current Capillary length with i.d. Detection	Proteins (molecular weight)	Reference
100 mM Tris–CHES 0.1% SDS, pH 8.8 3% PEG (mol.wt. 100 000) 6% linear acrylamide + dextran coated −300 V cm^{-1} (40 cm to detector), 100 µm i.d. 214 nm	Monoclonal antibody (Fab fragments)	47
Antithrombin III 100 mM Tris–CHES 0.1% SDS, pH 8.6 10% dextran (mol. wt 2 000 000) Linear acrylamide coated or dextran coated 27 cm (7 cm to detector), 100 µm i.d. 200, 254, 280 nm	Antithrombin III (AT) thrombin (TH) AT–TH complex	51
Collagen 50 mM Tris–glycine 0.1% SDS, pH 8.8 4% acrylamide non-gel Untreated −10 kV 45 cm (35 cm to detector), 75 µm i.d. 220 nm	Collagen type I	50
E. coli *crude extract* 100 mM Tris–CHES 0.1% SDS, pH 8.8 3% PEG (mol. wt 100 000) 6% linear acrylamide + dextran coated −200 V cm^{-1} 100 µm i.d. 214 nm		47
Fetuin Commercial (ProSort kit) 0.2% SDS, pH 7.0 Acrylamide non-gel Untreated −285 V cm^{-1} 42 cm (22 cm to detection), 55 µm i.d. 215 nm	(Native and deglycosylated)	42
Growth hormone 120 mM Tris–120 mM histidine 0.1% SDS, pH 8.8 8% acrylamide non-gel 6% linear acrylamide coated −350 V cm^{-1}, 30 µA 12 cm, 75, 100 µm i.d. 214 nm	Recombinant human growth hormone (monomer and dimer)	47

Table 13.1 *Continued*

Separation conditions Sieving media Coating Voltage, current Capillary length with i.d. Detection	Proteins (molecular weight)	Reference
Tris–ethylene glycol 0.1% SDS, pH 8.8 Acrylamide non-gel Linear acrylamide coated −14.1 kV, −300 V cm^{-1}, 24 μA 40 cm to detector, 100 μm i.d. 214 nm	Somatotropin (oligomers) 20 000–80 000	53
Salivary proteins 100 mM Tris + 250 mM boric acid 0.1% SDS, 7 M urea 4% acrylamide non-gel Bonded linear acrylamide −12 kV 45 cm (25 cm to detector), 75 μm i.d. 230 nm	Human salivary proteins 14 000–94 000	24, 41
Standard mixture 100 mM Tris–CHE 0.1% SDS, pH 8.8 8% dextran (av. mol. wt 2 000 000) Linear acrylamide coated −20 kV, 10 μA 50.5 cm (38.3 cm to detector), 50 μm i.d. 214 nm	Carbonic anhydrase ovalbumin β-galactosidase lysozyme phosphorylase b BSA 14 400–96 000	2
50 mM phosphate 0.5% SDS, pH 5.5 10% acrylamide non-gel Untreated −20 kV, 120–160 μA 57 cm (50 cm to detector), 100 μm i.d. 254 nm	Myoglobin ovalbumin BSA conalbumin 17 800–77 000	40
100 mM Tris + 250 mM boric acid 0.1% SDS, 7 M urea 3–8% acrylamide non-gel Untreated and bonded Linear acrylamide −12 kV 45 cm (25 cm to detector), 75 μm i.d. 230 nm	Myoglobin and its fragments aprotinin cytochrome c trypsin inhibitor trypsinogen carbonic anhydrase glyceraldehyde-3-phosphate dehydrogenase ovalbumin BSA conalbumin phosphorylase b β-galactosidase myosin 2500–205 000	24, 41

Table 13.1 *Continued*

Separation conditions Sieving media Coating Voltage, current Capillary length with i.d. Detection	Proteins (molecular weight)	Reference
Commercial (ProSort kit) 0.2% SDS, pH 7.0 Acrylamide non-gel Untreated −12 kV 42 cm (22 cm to detector), 55 μm i.d. 215 nm	α-Lactalbumin carbonic anhydrase ovalbumin BSA phosphorylase b β-galactosidase myosin 14 000–205 000	42
Commercial (eCAP SDS 14–200 kit) PEO Treated −300 V cm^{-1}, 25–30 μA 27 cm (20 cm to detector), 100 μm i.d. 214 nm	α-Lactalbumin carbonic anhydrase ovalbumin BSA phosphorylase b β-galactosidase myosin 14 000–205 000	43
Commercial (eCAP SDS 14–200 kit) PEO Treated −633, −888 V cm^{-1} 27 cm (20 cm to detector), 50 or 100 μm i.d. 214 nm	α-Lactalbumin trypsin inhibitor carbonic anhydrase ovalbumin BSA phosphorylase b 14 200–94 700	44
Commercial (eCAP SDS- 200 kit) PEO Treated −300 V cm^{-1}, 25–30 μA 47 cm (40 cm to detector), 100 μm i.d. 214 nm	Carbonic anhydrase ovalbumin BSA phosphorylase b β-galactosidase myosin 20 000–200 000	45
0.1 M Tris–CHES 0.1% SDS, pH 8.6 10% dextran (mol. wt 2 000 000) 4% linear acrylamide coated −740 V cm^{-1} 27 cm (20 cm to detector), 50, 100 μm i.d. 200 nm	Carbonic anhydrase ovalbumin BSA phosphorylase b 29 000–97 400	46
120 mM Tris–120 mM histidine 0.1% SDS, pH 8.8 8–16% acrylamide non-gel 6% linear acrylamide coated −560 V cm^{-1}, 44 μA (15 cm to detector), 75, 100 μm i.d. 280 nm	Lysozyme carbonic anhydrase ovalbumin BSA phosphorylase b 14 000–96 000	47

Table 13.1 *Continued*

Separation conditions Sieving media Coating Voltage, current Capillary length with i.d. Detection	Proteins (molecular weight)	Reference
60 mM AMPD–CACO 0.1% SDS, pH 8.8 10% dextran (mol. wt 2 000 000) 6% linear acrylamide + dextran coated −400 V cm^{-1}, 30 µA (18 cm to detector), 75, 100 µm i.d.	Myoglobin carbonic anhydrase ovalbumin BSA β-galactosidase myosin 14 000–205 000	47
100 mM Tris–CHES 0.1% SDS, pH 8.2 10% dextran (mol. wt 2 000 000) 6% linear acrylamide + dextran coated −300 V cm^{-1}, 15 µA, (30 cm to detector), 75, 100 µm i.d. 214, 280 nm	α-Lactalbumin trypsinogen carbonic anhydrase glyceraldehyde-3-phosphate dehydrogenase ovalbumin BSA 14 000–96 000	47
100 mM Tris–CHES 0.1% SDS, pH 8.8 3% PEG (mol. wt 100 000) 6% linear acrylamide + dextran coated −200 V cm^{-1} (40 cm to detector), 100 µm i.d. 214 nm	α-Lactalbumin trypsin inhibitor carbonic anhydrase ovalbumin BSA phosphorylase b 14 000–96 000	47
100 mM Tris–CHES 0.1% SDS, pH 8.8 15% dextran (mol. wt 72 000) Linear acrylamide–dextran coated −300 V cm^{-1}, 11–18 µA 47 cm (40 cm to detector), 100 µm i.d. 214 nm	Lysozyme trypsin inhibitor carbonic anhydrase ovalbumin BSA phosphorylase b 14 000–97 400	48
100 mM Tris–CHES 0.1% SDS, pH 8.8 3% PEO (mol. wt 100 000) Untreated −300 V cm^{-1}, 18–30 µA 47 cm (40 cm to detector), 100 µm i.d. 214 nm	Lysozyme trypsin inhibitor carbonic anhydrase ovalbumin BSA phosphorylase b 14 000–97 400	48
60 mM AMPD–CACO 0.1% SDS, pH 8.8 10% dextran (mol. wt 2 000 000) Acrylamide–dextran coated −8 kV, 11.5 µA 34 cm, 75 µm i.d. 214 nm	α-Lactalbumin trypsin inhibitor carbonic anhydrase ovalbumin BSA phosphorylase b 14 000–94 000	49

CONCLUSIONS

Table 13.1 *Continued*

Separation conditions Sieving media Coating Voltage, current Capillary length with i.d. Detection	Proteins (molecular weight)	Reference
60 mM AMPD–CACO 0.1% SDS, pH 8.8 3–8% PVA AAEE coated −8, −12, −16 kV 34 cm, 25, 50, 75 µm i.d. 214 nm	α-Lactalbumin trypsin inhibitor carbonic anhydrase ovalbumin BSA phosphorylase b 14 000–94 000	49
375 mM Tris–2.5 M ethylene glycol 0.1% SDS, pH 8.8 Acrylamide non-gel Linear acrylamide coated −300 V cm^{-1}, 24 µA (40 cm to detector), 100 µm i.d. 214 nm	Lysozyme trypsin inhibitor carbonic anhydrase ovalbumin BSA phosphorylase b 14 000–97 000	53
Plasma proteins 60 mM AMPD–CACO 0.1% SDS, pH 8.8 10% dextran (mol. wt 2 000 000) 6% linear acrylamide–dextran coated −300 V cm^{-1}, 20 µA (30 cm to detector), 75, 100 µm i.d. 214 nm	Rat plasma	47
Whey proteins Commercial (ProSort SDS–protein analysis kit)	β-Lactoglobulin α-lactalbumin BSA IgG lactoferrin proteose–peptone	52

AAEE, *N*-acryloylaminoethoxyethanol; AMPD, 2-amino-2-methyl-1-3-propanediol; BSA, bovine serum albumin; CACO, cacodylic acid; CHE, 2-(cyclohexyl amino)ethanesulfonic acid; CHES, 2-(*N*-cyclohexylamino)ethanesulfonic acid; HPC, hydroxypropylcellulose; IgG, immunoglobulin G; PEG, poly(ethylene glycol); PEO, poly(ethylene oxide); PVA, poly(vinyl alcohol); SDS, sodium dodecyl sulfate; Tris, tris(hydroxymethyl)aminomethane.

reported since 1987 in the search to obtain higher performance than SDS–PAGE [1, 54]. However, it is not easy to prepare an acrylamide gel-filled capillary, because once a bubble is generated in the capillary, it becomes unusable. Therefore, polyacrylamide without crosslinking or polymer solution (non-gel) without polymerization, such as cellulose derivatives, dextran, PEG, poly(ethylene oxide) (PEO) etc., is preferred for the molecular mass-based separation of proteins with the molecular mass range of 10 000 to 200 000 Da. A polymer solution-filled capillary can offer higher reproducibility and easier handling than those with a gel filled one, because the sieving medium can be replaced at each run.

However, at present, there are still several problems to be solved:

1. The effect of treatment on the surface of the internal wall of the capillary has not yet been revealed. If it is necessary, some standardization is required for comparison of data obtained in different laboratories.
2. There is experimentally low resolution for proteins in the region of low molecular mass due to difficulty in using higher concentration polymer solution and thus higher viscosity. Calculation shows that it would be difficult to achieve higher resolution for high molecular masses [4].
3. Study of the principles of molecular sieving CE has not yet been successfully attained to elucidate data obtained by different types of media, e.g. differences in molecular mass and in concentration.

In future, resolution in SDS–non-gel molecular sieving CE will certainly exceed by far that in SDS–PAGE. In addition, a mass spectrometer for detection will be required to determine correctly the mass of proteins because the retention of SDS–protein complexes does not reflect the correct mass of proteins.

A summary of applications of SDS–non-gel molecular sieving electrophoresis for proteins is shown in Table 13.1.

References

1. Heller, C. (1995) Capillary electrophoresis of proteins and nucleic acids in gels and entangled polymer solution. *J. Chromatogr.* **698**, 19–31.
2. Takagi, T.; Karim, M.R. (1994) SDS–polymer solution capillary electrophoresis. *Tanpakushitu Kakusan Koso.* **39**, 930–940.
3. Grossman, P.D.; Soane, D.S. (1991) Capillary electrophoresis of DNA in entangled polymer solutions. *J. Chromatogr.* **559**, 257–266.
4. Bae, Y.C.; Soane, D. (1993) Polymeric separation media for electrophoresis: cross-linked systems or entangled solutions. *J. Chromatogr.* **652**, 17–22.
5. Liu, J.; Dolink, V.; Hsieh, Y.; Novotny, M. (1992) Experimental evaluation of the separation efficiency in capillary electrophoresis using open tubular and gel-filled columns. *Anal. Chem.* **64**, 1328–1336.
6. Cohen, A.S.; Karger, B.L. (1987) High-performance sodium dodecyl sulfate polyacrylamide gel capillary electrophoresis of peptides and proteins. *J. Chromatogr.* **397**, 409–417.
7. Wu, D.; Regnier, E. (1993) Native protein separations and enzyme microassays by capillary zone and gel electrophoresis. *Anal. Chem.* **65**, 2029–2035.
8. Tsuji, K. (1994) Factors affecting the performance of sodium dodecyl sulfate gel-filled capillary electrophoresis. *J. Chromatogr.* **661**, 257–264.
9. Heiger, D.N.; Cohen, A.S.; Karger, B.L. (1990) Separation of DNA restriction fragments by high performance capillary electrophoresis with low and zero crosslinked polyacrylamide using continuous and pulsed electric fields. *J. Chromatogr.* **516**, 33–48.
10. Cordier, Y.; Roch, O.; Cordier, P.; Bischoff, R. (1994) Capillary gel electrophoresis of oligonucleotides: prediction of migration times using base-specific migration coefficients. *J. Chromatogr.* **680**, 479–489.

REFERENCES

11 Gelfi, C.; Alloni, A.; Besi, P.; Righetti, P.G. (1992) Investigation of the properties of acrylamide bifunctional monomers (cross-linkers) by capillary zone electrophoresis. *J. Chromatogr.* **608**, 343–348.
12 Righetti, P.G.; Chiari, M.; Nesi, M.; Caglio, S. (1993) Towards new formulations for polyacrylamide matrices, as investigated by capillary zone electrophoresis. *J. Chromatogr.* **638**, 165–178.
13 Gelfi, C.; Besi, P.; Alloni, A.; Righetti, P.G. (1992) Investigation of the properties of novel acrylamide monomers by capillary zone electrophoresis. *J. Chromatogr.* **608**, 333–341.
14 Chiari, M.; Nesi, M.; Righetti, P.G. (1993) Movement of DNA fragments during capillary zone electrophoresis in liquid polyacrylamide. *J. Chromatogr.* **652**, 31–39.
15 Singhal, R.P.; Xian, J. (1993) Separation of DNA restriction fragments by polymer-solution capillary zone electrophoresis. *J. Chromatogr.* **652**, 47–56.
16 Kleparnik, K.; Garner, M.; Bocek, P. (1995) Injection bias of DNA fragments in capillary electrophoresis with sieving. *J. Chromatogr.* **698**, 375–383.
17 Gelfi, C.; Besi, P.D.; Alloni, A.; Righetti, P.G.; Lyubimova, T.; Briskman, V.A. (1992) Kinetics of acrylamide photopolymerization as investigated by capillary zone electrophoresis. *J. Chromatogr.* **598**, 277–285.
18 Rapp, T.L.; Kowalchyk, W.K.; Davis, K.L.; Todd, E.A.; Liu, K.L.; Morris, M.D. (1992) Acrylamide polymerization kinetics in gel electrophoresis capillaries. A Raman microprobe study. *Anal. Chem.* **64**, 2434–2437.
19 Guttman, A.; Horvath, J.; Cooke, N. (1993) Influence of temperature on the sieving effect of different polymer matrices in capillary SDS gel electrophoresis of proteins. *Anal. Chem.* **65**, 199–203.
20 Nakatani, M.; Shibukawa, A.; Nakagawa, T. (1994) Preparation and characterization of a stable polyacrylamide sieving matrix-filled capillary for high-performance capillary electrophoresis. *J. Chromatogr.* **661**, 315–321.
21 Cobb, K.A.; Dolnik, V.; Novotny, M. (1990) Electrophoretic separations of proteins in capillaries with hydrolytically stable surface structures. *Anal. Chem.* **62**, 2478–2483.
22 Izumi, T.; Yamaguchi, M.; Yoneda, K.; Isobe, T.; Okuyama, T.; Shinoda, T. (1993) Use of glucomannan for the separation of DNA fragments by capillary electrophoresis. *J. Chromatogr.* **652**, 41–46.
23 Grossman, P.D.; Hino, T.; Soane, D.S. (1992) Dynamic light-scattering studies of hydroxyethyl cellulose solutions used as sieving media for electrophoretic separations. *J. Chromatogr.* **608**, 79–83.
24 Yin, H.F.; Lux, J.A.; Schomburg, G. (1990) Production of polyacrylamide gel filled capillaries for capillary gel electrophoresis (CGE): Influence of capillary surface pretreatment on performance and stability. *J. High Resolut. Chromatogr.* **13**, 624–627.
25 Chang, H.T.; Yeung, E.S. (1995) Poly(ethylene oxide) for high-resolution and high-speed separation of DNA by capillary electrophoresis. *J. Chromatogr.* **669**, 113–123.
26 Baba, Y.; Ishimaru, N.; Samata, K.; Tsuhako, M. (1993) High-resolution separation of DNA restriction fragments by capillary electrophoresis in cellulose derivative solutions. *J. Chromatogr.* **653**, 329–335.
27 McCord, B.R.; Jung, J.M.; Holleran, E.A. (1993) High resolution capillary electrophoresis of forensic DNA using a non-gel sieving buffer. *J. Liq. Chromatogr.* **16**, 1963–1981.
28 Chan, K.C.; Whang, C.W.; Yeung, E.S. (1993) Separation of DNA restriction fragments using capillary electrophoresis. *J. Liq. Chromatogr.* **16**, 1941–1962.
29 Barron, A.E.; Soane, D.S.; Blanch, H.W. (1993) Capillary electrophoresis of DNA in uncross-linked polymer solutions. *J. Chromatogr.* **652**, 3–16.
30 McGregor, D.A.; Yeung, E.S. (1993) Optimization of capillary electrophoretic separation of DNA fragments based on polymer filled capillaries. *J. Chromatogr.* **652**, 67–73.
31 Guttman, A.; Schwartz, H.E. (1995) Artifacts related to sample introduction in capillary gel electrophoresis affecting separation performance and quantitation. *Anal. Chem.* **67**, 2279–2283.
32 Gelfi, C.; Orsi, A.; Leoncini, F.; Righetti, P.G. (1995) Fluidified polyacrylamides as molecular sieves in capillary zone electrophoresis of DNA fragments. *J. Chromatogr.* **689**, 97–105.
33 Martinez, M.C.R.; Berka, J.; Belenkii, A.; Foret, F.; Miller, A.W.; Karger, B.L. (1993) DNA sequencing by capillary electrophoresis with replaceable linear polyacrylamide and laser-induced fluorescence detection. *Anal. Chem.* **65**, 2851–2858.

34 Schans, M.J.; Allen, K.A.; Wanders, B.J.; Guttman, A. (1994) Effects of sample matrix and injection plug on dsDNA migration in capillary gel electrophoresis. *J. Chromatogr.* **680**, 511–516.
35 Best, N.; Arriaga, E.; Chen, D.Y.; Dovichi, N. (1994) Separation of fragments up to 570 bases in length by use of 6% T non-cross-linked polyacrylamide for DNA sequencing in capillary electrophoresis. *Anal. Chem.* **66**, 4063–4067.
36 Pariat, Y.F.; Berka, J.; Heiger, D.N.; Schmit, T.; Vilenchik, M.; Cohen, A.S.; Foret, F.; Karger, B.L. (1993) Separation of DNA fragments by capillary electrophoresis using replaceable linear polyacrylamide matrices. *J. Chromatogr.* **652**, 57–66.
37 Cheng, J.; Mitchelson, K.R. (1994) Glycerol-enhanced separation of DNA fragments in entangled solution capillary electrophoresis. *Anal. Chem.* **66**, 4210–4214.
38 Manabe, T.; Chen, N.; Terabe, S.; Yohda, M.; Endo, I. (1994) Effects of linear polyacrylamide concentrations and applied voltages on the separation of oligonucleotides and DNA sequencing fragments by capillary electrophoresis. *Anal. Chem.* **66**, 4243–4252.
39 Mitnik, L.; Salome, L.; Viovy, J.L.; Heller, C. (1995) Systematic study of field and concentration effects in capillary electrophoresis of DNA in polymer solutions. *J. Chromatogr.* **710**, 309–321.
40 Widhalm, A.; Schwer, C.; Blaas, D.; Kenndler, E. (1991) Capillary zone electrophoresis with a linear, non-cross-linked polyacrylamide gel: separation of proteins according to molecular mass. *J. Chromatogr.* **549**, 446–451.
41 Wu, D.; Regnier, F.E. (1992) Sodium dodecyl sulfate–capillary gel electrophoresis of proteins using non-cross-linked polyacrylamide. *J. Chromatogr.* **604**, 349–356.
42 Werner, W.E.; Demorest, D.M.; Stevens, J.; Wiktorowicz, J.E. (1993) Size-dependent separation of proteins denatured in SDS by capillary electrophoresis using a replaceable sieving matrix. *Anal. Biochem.* **212**, 253–258.
43 Guttman, A.; Shieh, P.; Lindahl, J.; Cooke, N. (1994) Capillary sodium dodecyl sulfate gel electrophoresis of proteins II. On the Ferguson method in polyethylene oxide gels. *J. Chromatogr.* **676**, 227–231.
44 Benedek, K.; Guttman, A. (1994) Ultra-fast high-performance capillary sodium dodecyl sulfate gel electrophoresis of proteins. *J. Chromatogr.* **680**, 375–381.
45 Guttman, A.; Nolan, A.N.; Cooke, N. (1993) Capillary sodium dodecyl sulfate gel electrophoresis of proteins. *J. Chromatogr.* **632**, 171–175.
46 Lauch, T.; Scheper, T.; Reif, O.-W.; Fleisher, J.; Freitag, R. (1993) Rapid capillary gel electrophoresis of proteins. *J. Chromatogr.* **654**, 190–195.
47 Ganzler, K.; Greve, K.S.; Cohen, A.S.; Karger, B.L.; Guttmann, A.; Cooke, N.C. (1992) High-performance capillary electrophoresis of SDS–protein complexes using UV-transparent polymer networks. *Anal. Chem.* **64**, 2665–2671.
48 Guttman, A.; Horvath, J.; Cooke, N. (1993) Influence of temperature on the sieving effect of different polymer matrices in capillary SDS gel electrophoresis of proteins. *Anal. Chem.* **65**, 199–203.
49 Alfonso, E.S.; Conti, M.; Gelgi, C.; Righetti, P.G. (1995) Sodium dodecyl sulfate capillary electrophoresis of proteins in entangled solutions of poly(vinyl alcohol). *J. Chromatogr.* **689**, 85–96.
50 Deyl, Z.; Miksik, I. (1995) Separation of collagen type I chain polymers by electrophoresis in non-cross-linked polyacrylamide-filled capillaries. *J. Chromatogr.* **698**, 369–373.
51 Reif, O.W.; Freitag, R. (1994) Control of the cultivation process of antithrombin II and its characterization by capillary electrophoresis. *J. Chromatogr.* **680**, 383–394.
52 Kinghorn, N.M.; Norris, C.S.; Paterson, G.R.; Otter, D.E. (1995) Comparison of capillary electrophoresis with traditional methods to analyse bovine whey proteins. *J. Chromatogr.* **700**, 111–123.
53 Tsuji, K. (1993) Evaluation of sodium dodecyl sulfate non-acrylamide polymer gel-filled capillary electrophoresis for molecular size separation of recombinant bovine somatotropin. *J. Chromatogr.* **652**, 139–147.
54 Manabe, T. (1994) Capillary electrophoresis: application for biopolymers. *Bunseki*, 467–472.

14 Protein analysis by capillary electrophoresis
F.-T.A. CHEN AND R.A. EVANGELISTA

14.1 Introduction

Zone electrophoresis performed on supporting media such as agarose and polyacrylamide gels provides a powerful and indispensable tool for the separation and characterization of proteins. The analysis of proteins in the resulting electrophoresis gels requires staining with Coomassie Blue or silver stain to reveal the protein species of interest, unless the proteins are required to be prelabeled with radioisotopes or fluorophores. The procedures involved in casting the gel, sample application and staining are labor-intensive, skill-dependent, tedious and the entire process requires several hours.

Protein analysis by zone electrophoresis can be performed in capillaries with distinct advantages over slab gel electrophoresis in terms of speed, automation, higher resolution, small sample requirement, low reagent consumption and real-time data analysis. Proteins are separated at shorter times than in slab gels because the voltage gradient can be increased for greater electrophoretic mobility without much Joule heating due to more effective heat dissipation in narrow capillaries. Moreover, the on-line UV absorbance detection provides a great improvement in quantitation compared to estimating the amount of protein in a band on a slab gel by densitometer. The electropherograms are amenable to automated data storage and manipulation just as an high performance liquid chromatography (HPLC) chromatogram. A densitometer can also be used, but the reproducibility of the determined amount is poorer than with other methods. UV absorbance detection has a much wider linear range than the staining method which is dependent on the variable dye-to-protein binding.

Earlier attempts by Jorgenson and Lukacs [1] to electrophorese proteins in untreated fused-silica capillaries resulted in broad peaks and irreproducible migration of the sample zones. The fused-silica surface contains weakly acidic silanol groups. The pK_a value of the silanols on the fused-silica was estimated to be between 4.3 to 5.0 by Tsuda [2] based on data from Hayes and Ewing [3] and Towns and Regnier [4]. Thus, the silica surface is essentially negatively charged above pH 2.0, and the degree of ionization increases with increasing buffer pH. Proteins are polyelectrolytes, consisting of both positively and negatively charged moieties. The pK_a values of the guanidinium and ε-NH_2 groups of the arginine and lysine residues in proteins are 10.5 and 12, respectively. These groups comprise most of the positively charged moieties in proteins, other than the α-NH_2 terminal which has a pK_a of between 7.5 and 9, and the imidazole moiety of histidine with a pK_a of 6.0. At buffer pH above 2.0, the fused-silica surface can act as a cation exchanger. The protein–silica surface interaction can be viewed as an ion-exchange phenomenon [5, 6]. Such interactions severely degraded the separation efficiency and reproducibility in capillary electrophoresis separation of proteins in untreated fused-silica capillaries. To avoid or minimize the protein–silica interaction, CE separation of proteins in untreated fused-silica capillaries have been previously performed in buffers with a pH either substantially higher than the isoelectric points of the sample proteins [5], or at a very acidic pH [7], or in a relatively basic buffer with a high salt concentration to avoid adhesion or absorption of proteins onto bare fused silica [6].

Methods to overcome the problems in protein analysis in a bare fused-silica capillary have focused on the modification of the silica surface. Thus by using an appropriate chemical coating, or by physical masking using additives, the zeta potential of the silica surface can be minimized or altered for specific applications. Chemical coating with a neutral functional group reduces or eliminates the interaction of silanoate (Si—O~) with the positively charged moieties of proteins. Thus, by using a silica capillary covalently bonded with polyacrylamide, Hjertén [8] was able to perform protein separations at neutral and acidic pH. Cobb *et al.* [9] extended the protein separation capability of Hjertén's coating to the basic pH range by introducing a stable Si—C bond for further attachment with linear polyacrylamide. Several other chemical modification procedures including cellulose and dextran-based coating procedures resulted in a substantial improvement in column technology for protein separations by CE to allow operation in the pH range 1.5 to 12 in the presence of 5% SDS (sodium dodecyl sulfate) [10, 11].

The conventional slab gel-based electrophoretic separation of proteins by agarose, SDS–polyacrylamide (SDS–PAGE (polyacrylamide gel electrophoresis)) and isoelectric focusing (IEF) methods can be performed by CE procedures with little or no modification in procedures, and in many instances produce substantially better resolution. The following review of the practical applications of protein analysis by CE is categorized into four sections: capillary zone electrophoresis (CZE) deals with the separation of proteins in free solution, capillary gel electrophoresis (CGE) in the presence of SDS for the characterization of molecular size, micellar electrokinetic capillary chromatography (MECC) and capillary isoelectric focusing (CIEF) for the determination of isoelectric point of proteins.

14.2 Capillary zone electrophoresis

14.2.1 Untreated fused-silica capillary

High resolution CE separation of proteins in untreated fused-silica capillary was first demonstrated in 1986 [5]. The sharp sample zones in CE were reported with a theoretical plate number of approaching $1\,000\,000\,m^{-1}$, using buffer pH substantially higher than the isoelectric point (pI) of the model proteins. Green and Jorgenson [6] reported the use of alkaline metal salt to minimize protein adsorption for CE separation of proteins. Well-resolved electropherograms with lysozyme, trypsinogen, myoglobin and β-lactoglobulin A and B mixtures by CE were achieved in 0.1 M 2-(cyclohexylamino)ethane sulfonic acid (CHES) buffer at pH 9.0 with an addition of 0.25 M potassium sulfate. Both lysozyme and trypsinogen which contain a significant region of the positively charged moieties at pH 9.0 showed appreciable adsorption in the case of no addition of salt. The addition of K_2SO_4 in the buffer system had profound effect in the CE separation resulting in well-resolved peaks.

McCormick [7] demonstrated that phosphate buffers at low pH were extremely efficient in separating many different peptides and protein species. In one example showing separation of cytochrome c from six different species (pI ~ 10.5) by CE, a well-resolved pattern in the electropherogram was obtained in 0.15 M phosphate buffer at pH 5.0, despite the fact that the silica surface is partially ionized (pK_a of silica is between 4 and 5) at pH 5.0. Tran et al. [12] have reported the separation of isoforms of recombinant erythropoietin on a phosphate-modified fused-silica surface. It is recognized that phosphate buffer interacted with the silica surface to form a dynamic equilibrium preventing protein adsorption. As the phosphate buffer pH increases from 5.0 to 5.25, the CE separation of cytochrome c showed a drastic deterioration in resolution, presumably due to the increment of negative charges on the silica surface. A phosphate concentration at 0.15 M simply does not provide enough interaction with the silica surface to prevent the adsorption of proteins on a more negatively charged silica surface at pH 5.25 than that at pH 5.0 or below. Chen and co-workers [13, 39] extended the application of phosphate buffers to pH ranges of 4 to 10 and considerably higher concentration at 0.4 to 0.5 M for separating proteins with broad pI ranges. The use of high ionic strength buffer in CE results in larger currents that produce Joule heat. Rush et al. [14] reported the effect of column temperature on the conformational transition of proteins by CE, and pointed out the need for temperature control in CE for reproducibility. A smaller diameter capillary reduces the current and enhances separation efficiency. Figure 14.1 showed the CE separation of proteins in serum (a) and ascites fluid (b) in a 20 μm × 25 cm capillary with

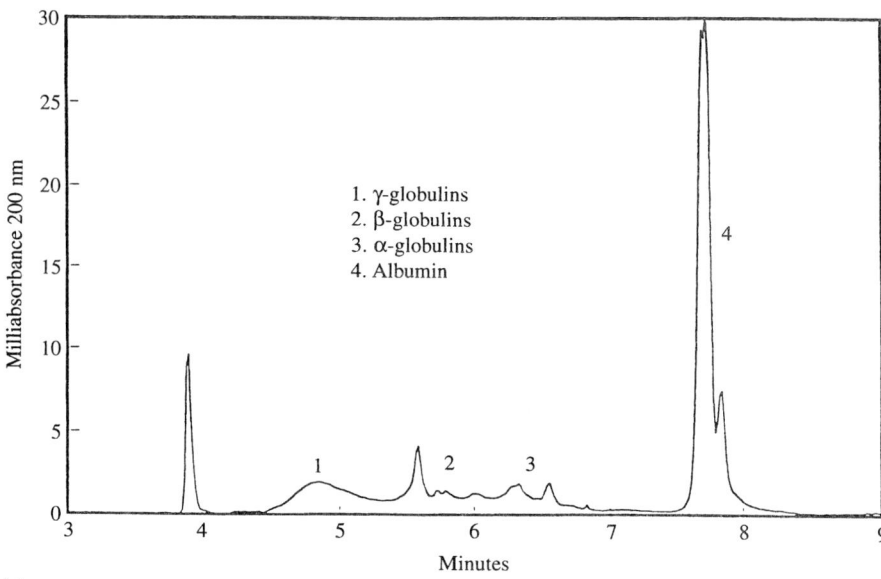

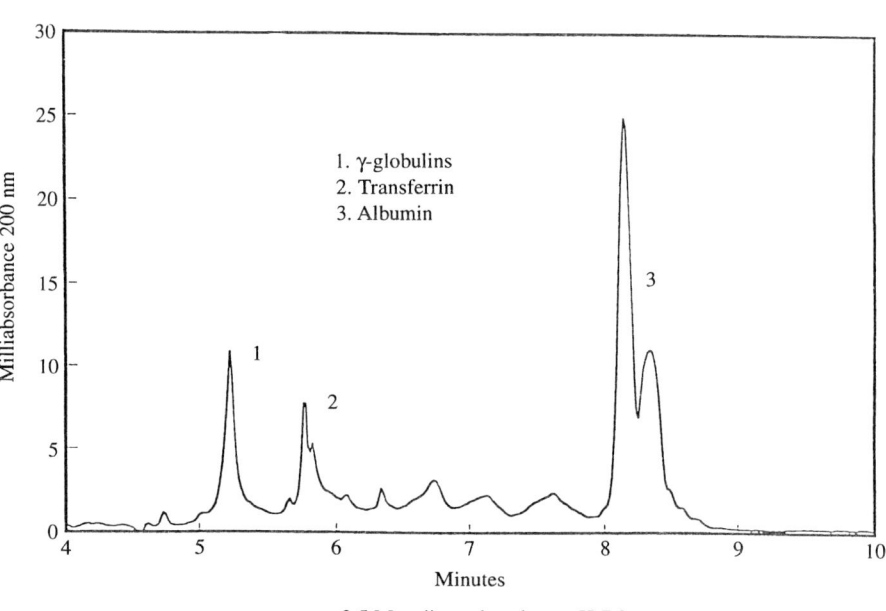

Figure 14.1 Electropherograms of proteins, in (a) normal human serum, (b) mouse ascites fluid (GO 144). Conditions: untreated fused-silica capillary, 20 μm × 25 cm; UV: 200 nm; buffer: 0.5 M sodium phosphate pH 7.0; 10 kV/62 mA [15].

0.5 M phosphate at pH 7.0 [15]. The electropherogram in Figure 14.1(a) shows an extremely broad γ-globulin peak, indicative of the heterogeneity of the polyclonal antibody. In the mouse ascites fluid however, the γ-globulin region shown in Figure 14.1(b) is a considerably sharper peak due to the monoclonal nature of the antibody. Electrophoresis of serum and ascites in phosphate buffer at a concentration of 0.4 M or less resulted in a rather broad peak with longer migration time, consistent with the reduction of electroosmotic flow (EOF) due to protein adsorption.

Protein analysis by CE using both high salt buffer and high pH conditions provides a rapid separation without compromising resolution. Borate buffer at pH 10.0 or higher was reported for CE separation of proteins in human serum [16, 17], cerebrospinal fluid and urine [17]. A rapid serum protein analysis by CE in borate buffer with resolution comparable to that obtained by the conventional agarose gel electrophoresis was achieved in 90 s [18]. High resolution protein analysis by CE was achieved with the use of borate at a concentration of 200 mM or higher [18]. The reduced EOF rate resulting from higher borate buffer salt concentration allows for a longer electrophoretic separation. Higher salt concentration also decreases protein–protein interactions [19]. Hemoglobin variants A, F, S and C were resolved by CE in 200 mM barbital buffer [17]. Figure 14.2 shows the electropherogram of a whole blood sample from a heterozygous hemoglobin C trait patient. Sample was lysed in water and directly injected

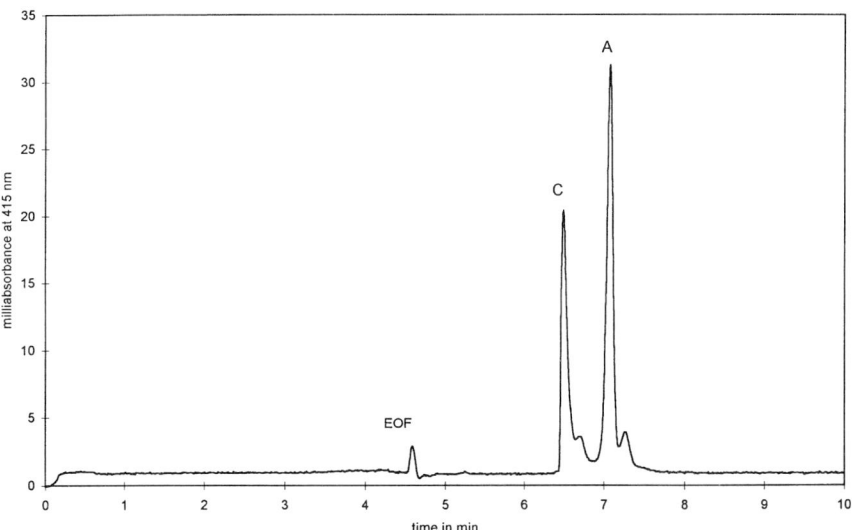

Figure 14.2 Electropherogram of lysed whole blood of a hemoglobin variant C patient sample. Conditions: untreated fused-silica capillary, 50 μm × 25 cm; visible: 415 nm; buffer: 200 mM barbital buffer, pH 8.6; 5 kV/83 mA [15].

and analyzed by CE with a 415 nm detection system. Only hemoglobin and related species were detected as shown in Figure 14.2.

The effect of buffer additives such as zwitterions at high molar concentration in reducing protein adsorption for CE separation was first demonstrated by Bushey and Jorgenson [20]. Fang et al. [21] studied the structural requirement of zwitterions effectively to prevent adsorption of proteins and concluded that $(CH_3)_3N^+CH_2CH_2CH_2CH_2SO_3^-$, trimethylammonium butanesufonate, is the most effective zwitterion. Cationic surfactants [22, 23] and diaminoalkane derivatives [24–29] at appropriate concentrations were shown to be very effective for CE separation of basic proteins and glycoproteins. Watson and Yao [29] reported the use of 1,4-diaminobutane for CE separation of human recombinant erythropoietin (r-EPO) with nearly baseline resolution of six species. The CE separation pattern of r-EPO was comparable to the results obtained by the conventional gel-based isoelectric focusing (IEF) technique [29]. The resolution of six species by IEF was much better, but no concrete comparison of efficiency was attained. The addition of 1,4-diaminobutane resulted in the reduction of EOF, while urea was added to the buffer to prevent aggregation of r-EPO by cleavage hydrogen bonding with urea. Similarly, the addition of phytic acid (inositol hexaphosphate) in a borate-based buffer system appeared to improve the separation efficiency of proteins [30]. High resolution CE separation of histones was studied extensively by Lindner et al. [31] using phosphate buffer with various counterions and hydroxylpropylmethylcellulose at pH 2.0. Banks and Paquette [32] studied the migration behavior of fluorescein-labeled myoglobin by CE. Addition of 10% dimethylformamide (DMF) in borate-based buffer was found to enhance the resolution of labeled species with different molar ratios of fluorescein to myoglobin. Whey proteins were well resolved in a buffer system containing non-ionic surfactant [33]. Kalman and co-workers [34, 35] studied the the electrophoretic mobilities of nuclease A and closely related mutants produced by site-directed mutagenesis. Selected aliphatic di-, tri- and tetraamine salts of phosphate and/or trifluoroacetate buffers were formulated effectively to reduce the silica surface charge and protein adsorption. Such buffer systems were very efficient for CE separation of minute differences in size and charge of the nuclease A variants result in small changes in tertiary structure. The migration order of each mutant species could be predicted from the protein structure as determined by X-ray crystallography and NMR measurements.

A method related to the use of additives to modify the functional behavior of the capillary was explored by Yao et al. [36] using polybrene, a polyquaternary amine polymer. Capillary coating was prepared by simply rinsing with buffer containing diluted polybrene. Excess polybrene solution was flushed with electrophoresis buffer prior to each sample analysis. Between runs, capillary coating and rinsing procedures were repeated. Using polybrene coated capillary and a combination of buffer systems, CE analysis of the mobilities of proteins vs. pH were plotted and the pH value corresponding to zero mobility should yield the isoelectric point of each protein. The pI values for both acidic and basic proteins obtained by the above method were reported to correlate well with the literature values.

It should be noted that the addition of ionic surfactants, diaminoalkanes, polycations or polyanions have a profound effect on electroosmosis and the conductivity of the buffer system. Polycationic species such as polybrene produce a positively charged capillary surface with reversed EOF. Diaminoalkanes and related compounds or other divalent cationic species bind and neutralize the silica surface resulting in reduction or reverse electroosmosis. Anionic surfactants and polyanions such as inositol-hexaphosphate reduce EOF. In CE separations using these additives, the potential interaction of the additives with proteins should always be taken into consideration.

14.2.2 Chemically modified fused-silica capillary

Jorgenson and Lukacs [1] reported a considerable decrease in protein adsorption using glycol-modified fused-silica capillaries. Separation patterns for human serum proteins by conventional agarose gel electrophoresis could be reproduced by the CE procedure in the coated capillary. Neutral hydrophilic surface with non-crosslinked polyacrylamide [8, 9] on the fused-silica capillary was by far the most reliable and reproducible coating for protein analysis. The use of Si—C bond-based vinyl coating resulted in a stable polyacrylamide attachment to the silica surface which is resistent towards hydrolysis at high buffer pH (pH 9.5) [9]. Cobb et al. [9] showed the electropherograms of proteins in the coated and uncoated fused-silica capillary. The resolution of proteins using coated silica capillary was better than that using uncoated capillary. The order of migration of proteins toward the anode in a coated capillary is exactly reverse that of the uncoated capillary with a cathodic outlet [9]. Cobb et al. [9] used running buffer at pH 9.5, while the pI values of all seven proteins are ≤7.6.

Improvements to coating procedures were made by Hjertén and Kubo [10] by copolymerization of silica surface activated methacryl residue with allyl dextran or methylcellulose. Huang et al. [11] applied a similar procedure using an activated silica surface with a hydrophobic olefinic residue. Crosslinking and copolymerization were initiated by the addition of methylene bisacrylamide along with hydroxypropylcellulose and hydroxyethylmethacrylate. Both methods produced coatings which were stable at high pH buffer containing 5% SDS. Surface-bound dextran coating was also explored by Mechref and El Rassi [37] for protein analysis. Swedberg [38] reported a pentafluorobenzoyl (PFB) coated capillary that performed a separation with excellent efficiency for proteins having a broad range of isoelectric points (from 5 to 11) using 200 mM phosphate buffer containing 100 mM KCl at a pH of 7.0. Despite the coating procedure, the PFB-derivatized silica surface exhibited significant EOF, thus a significant concentration of buffer salts was required to prevent protein adsorption.

Several other covalent-bond coating procedures for silica surface [40–43] were explored with a limited operational pH range of ≤7.0. A simple physical coating on fused-silica column was reported by Busch et al. [43] and appeared to be extremely effective for CE separation of proteins. A capillary was filled with cellulose acetate in acetone, followed by flushing the capillary with helium gas to create a thin film of cellulose acetate on the silica surface. The resulting coating appeared to mask the silanol groups effectively for CE separation of basic proteins. The CE separation efficiency close to 10^6 plates per meter was achieved for ribonuclease A. Such a simple coating procedure can potentially be adapted in the automated CE system for on-line capillary coating for specific applications.

A few comercially available coated capillaries are available for CE separation of proteins, and most of them are operative at buffer pH below 7.0. Quantitative CE analysis of milk proteins using citrate buffer at pH 2–3 containing 0.05% methylhydroxyethylcellulose and 6.0 M urea was elegantly demonstrated by de Jong et al. [44] using a hydrophilically coated silica capillary. Under the urea-denatured condition, proteins from the milk of goats, sheep and cows showed unique patterns for each species, and could potentially be used to detect adulteration. In spite of the use of a coated capillary, the CE analysis of milk proteins showed a minimum ionic strength requirement to suppress protein adsorption even at pH 3.0 or below.

In the coated capillaries with reduced electroosmosis, the electrophoretic mobility of the protein is the driving force toward the detection window. In a neutral buffer at pH 7.0 with a detection window at the cathodic side, only the species in the sample with pI > 7.0 is detected, while the species in the same sample with pI ≤ 7.0 migrates toward the anode, the injection side, and never reaches the detection window. To avoid this shortcoming, protein analysis by CE in coated capillaries is often performed in acidic buffer pH.

14.3 Sodium dodecyl sulfate capillary gel electrophoresis

Polyacrylamide gel electrophoresis in the presence of sodium dodecyl sulfate (SDS–PAGE) is a widely accepted method for characterization of proteins for purity assessment and molecular weight determination. In SDS–PAGE, the protein analytes are usually treated with a thiol reducing agent such as mercaptoethanol or dithiothreitol to remove disulfide linkages which may connect the constituent sub-units. During the electrophoretic separation, the anionic detergent present in the electrolyte is bound to the hydrophobic region of the protein at a constant surfactant/ mass ratio resulting in a uniform charge density. The proteins, which assume an extended conformation surrounded by negative charges, migrate through the gel which acts as a sieving medium to separate the proteins according to their molecular weights. After migration a certain distance along the gel, the proteins are stained and their migration distance is compared to that of a standard composed of a mixture of proteins of known molecular weights.

Protein analysis by gel electrophoresis in the presence of SDS in capillaries offers substantial advantages over slab gel electrophoresis. Proteins are separated within shorter times in capillaries than in slab gels because the voltages can be increased for greater electrophoretic mobility without much Joule heating. Other advantages have been described in the Introduction (section 14.1).

The first reported methods of SDS–protein analysis by capillary gel electrophoresis (CGE) utilized crosslinked gels prepared by polymerization inside the capillary of acrylamide and bisacrylamide mixture, the same components used in slab gel preparation. These methods employed capillaries whose interior surfaces were treated wih silane reagents to modify silanol groups to eliminate EOF. Cohen and Karger [45] reported the first CGE/SDS method which utilized a gel-filled capillary. Excellent linearity between log molecular weight and mobilities was obtained. The separation efficiencies were in the order of 40000 theoretical plates per column which were much superior to those obtained by slab gel methods. A more detailed study of the mobilities as a function of gel concentration showed excellent Ferguson plots demonstrating that the separation is based solely on molecular size.

Tsuji [46] used a silane reagent to produce acrylic groups on the inner capillary surface which were then copolymerized with the acrylamide monomers in the bulk of the solution to form a fixed gel inside the capillary. The results showed good linear relationship between migration times and molecular weights ($r > 0.999$) for proteins typically used as molecular weight standards in slab gel SDS–PAGE. Analysis of a recombinant biotechnology-derived protein showed that the peak area is a linear function of the protein concentration over two orders of magnitude (20–2000 $\mu g\,ml^{-1}$) and the method can be used for detection of an impurity with a lower molecular weight.

Ganzler et al. [47] first reported the use of non-crosslinked hydrophilic polymers as sieving media for separation of protein–SDS complexes based on molecular weights. It was found that the linear polyacrylamide not only provided comparable separation resolution to crosslinked polyacrylmide with, but also produced greater column lifetime due to the flexibility of the polymer chains. Dextran and poly(ethylene glycol) (PEG) provided more significant advantages as sieving media, namely UV transparency at 214 nm and lower viscosity which allows the separation medium to be replaced by pressure rinsing, making possible hundreds of runs per column. The calibration plots of peak area vs. protein concentration were linear in the range 0.5 mg ml^{-1} to 0.25 mg ml^{-1} and the migration time RSDs were 0.5% or less.

Subsequent reports on CGE analysis of SDS–proteins describe the use of other non-crosslinked soluble polymers as sieving media such as polyethylene oxide [49, 59–63], poly(vinyl alcohol) [54, 58], pullulan [52], hydroxyethylcellulose [54], hydroxypropyl methylcellulose [54] and methoxylated agarose [51]. Aqueous solutions of these UV-transparent hydrophilic polymers also have sufficiently low viscosity such that some were used as replaceable separation media. The concentration of the polymer in the separation medium is usually optimized to suit the particular molecular weight range of the proteins being separated. Most of these separations were performed in capillaries whose inner walls have been coated with hydrophilic polymers to reduce endoosmotic flow and protein adhesion. In many cases, these lower viscosity sieving media produced resolutions comparable to those obtained with the fixed crosslinked gels.

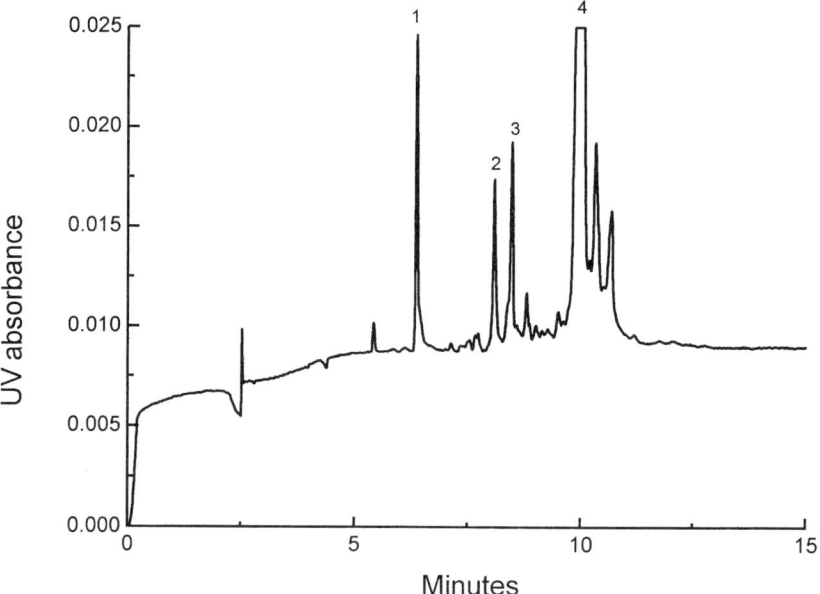

Figure 14.3 CGE/SDS analysis of bovine milk proteins. Conditions: Beckman eCAP SDS 14–200 kit with 100 μm × 27 cm coated capillary and replaceable polymer buffer (PEG/PEO mixture); 300 V cm^{-1}. Peak i.d.: 1 = orange G reference marker; 2 = α-lactalbumin; 3 = β-lactoglobulin; 4 = casein. Reprinted with permission of Elsevier Science from Ref. [59].

Capillary gel electrophoresis of SDS–protein complexes has been applied to the analysis of proteins from plasma [47], milk [59], egg white [54, 59], whey [64] and various biotechnology-derived products such as soluble CD_4 [46], recombinant bovine somatotropin [56] and polyclonal antithymocyte equine immune globulin [56]. The electropherogram of milk proteins analyzed by CGE/SDS is presented in Figure 14.3 [59].

14.4 Micellar electrokinetic chromatography

A few reports have dealt with the use of MECC for analysis of proteins. Primarily developed for separation of neutral species which are difficult to separate by capillary zone electrophoresis (CZE), separation by MECC is mostly attributed to the partitioning of molecules between the pseudostationary hydrophobic micelle phase and the aqueous phase, which has an identical mechanism in reversed-phase liquid chromatography (LC). For charged species, however, analyte separation is based on a combined effect of the differences in electrophoretic mobility and the degree of partitioning with the negatively charged micelles. In MECC, the separation medium contains a surfactant, most commonly SDS, at a concentration above its critical micellar concentration.

Pedersen et al. [69] showed that MECC produced better resolution than CZE for the separation of three *Serratia marcescens* nuclease isoforms. The effect of various additives such as methanol, hydroxypropylmethylcellulose, sodium chloride, magnesium chloride and l-butanol were investigated but the best separation was obtained with the simple tris (hydroxymethyl)aminomethane (Tris)–SDS mixture. Similarly, Beattie and Richards [71] found MECC to be suitable as a separation method for isoforms of metallothionein in sheep liver extracts and found that capillaries with a hydrophobic internal coating did not produce any advantage over uncoated capillaries. Wu et al. [66] employed laser-induced fluorescence (LIF) detection using the 457.9 nm line of the argon ion laser as excitation source for the MECC analysis of serum biliprotein species in normal sera without sample pretreatment and obtained detection limits which are two orders of magnitude lower than with absorption detection.

14.5 Capillary isoelectric focusing

Isoelectric focusing (IEF), another electrophoretic technique which is traditionally performed on slab gels, is also a method of great importance for characterization of proteins. In IEF, the proteins are separated based on their isoelectric points (pIs) in a medium with a pH gradient formed by the migration of ampholytes in an electric field. During electrophoresis, a protein migrates along the pH gradient until its pI is equal to the pH of the surrounding medium. As in SDS–PAGE, the detection is performed by staining; thus the entire procedure is time consuming, as described in the Introduction (section 14.1), and so there was great motivation to apply this method to a capillary format.

In the development of a CE-based IEF method, two factors that were not of any consideration in slab gels had to be dealt with: first, reduction of EOF and, second, mobilization of the focused analyte zones for detection by the UV absorbance detector. Most of the reports on CIEF describe the use of capillaries coated with hydrophilic polymers which eliminate or reduce EOF and prevent protein adhesion. The most common polymers used for capillary wall coating are methylcellulose and polyacrylamide.

Almost all the published CIEF methods used phosphoric acid as anolyte and sodium hydroxide solution as catholyte. The samples were mixed first with commercially available ampholytes typically used for slab gel IEF. The samples are electrophoresed until the monitored current decreases to less than 1 µA, which is the main evidence of completion of focusing. The focused protein zones are then mobilized to move them past the stationary UV detector to produce peaks in an HPLC-like electropherogram. The absorbance monitoring is usually done at 280 nm because the ampholytes have substantial absorbance at shorter wavelengths. Thus, the proteins are detected due mainly to their tyrosine and tryptophan residues. Salt mobilization and pressure mobilization procedures have been developed for postelectrophoresis detection. The process of salt mobilization involves changing the cathode reservoir to an NaOH/NaCl solution (cathodic mobilization) or changing the H_3PO_4 reservoir to a H_3PO_4/NaCl solution (anodic mobilization) and then applying high voltage. Pressure mobilization is accomplished by pushing the entire liquid in the capillary containing the focused protein zones with pressure exerted by an inert gas.

CIEF analysis of an unknown protein or separation of several proteins with a wide range of pIs can be performed with a wide range ampholyte pH such as 3–10. The use of an ampholyte within a narrower range such as pH 6–8 allows higher resolution as the case of hemoglobin analysis.

Hjertén and Zhu [72] reported the first application of capillary IEF for separation of hemoglobin and transferrin components. The capillary inner surface was coated with methylcellulose to eliminate zone distortion caused by EOF. Both hemoglobin and transferrin exhibited multiple components which the authors attributed to variation in iron content. A subsequent more detailed study of transferrin by CIEF in combination with enzymatic desialylation showed that the presence of multiple components was due to variation in sialic acid content in the carbohydrate portion of the glycoprotein [74].

Mazzeo and Krull [77] demonstrated that complete elimination of EOF is not an absolute requirement to produce stable focused zones in CIEF.

These authors used additives such as methylcellulose in uncoated fused silica capillaries which reduced EOF sufficiently to allow formation of a stable pH gradient and shielded the capillary wall against protein adsorption. It was found that addition of TEMED (N,N,N',N'-tetramethylethylenediamine) to the ampholytes had beneficial effects for the separation of basic proteins by acting as a blocking agent, focusing in the region past the detection window and extending the pH gradient to 12. Quantitative treatment of the data showed that the migration time percentage RSDs were about 2% and the peak area RSDs were all below 8%.

Huang et al. [81] reported results of a detailed quantitative treatment using CIEF separation of model proteins over the wide pI range of 2.75–9.5. The focusing was performed in a neutral hydrophilic coated capillary with methylcellulose added to the catholyte and ampholyte and then mobilization was achieved by pushing the liquid containing the focused zones with inert gas pressure. Aside from its role as an EOF suppressor,

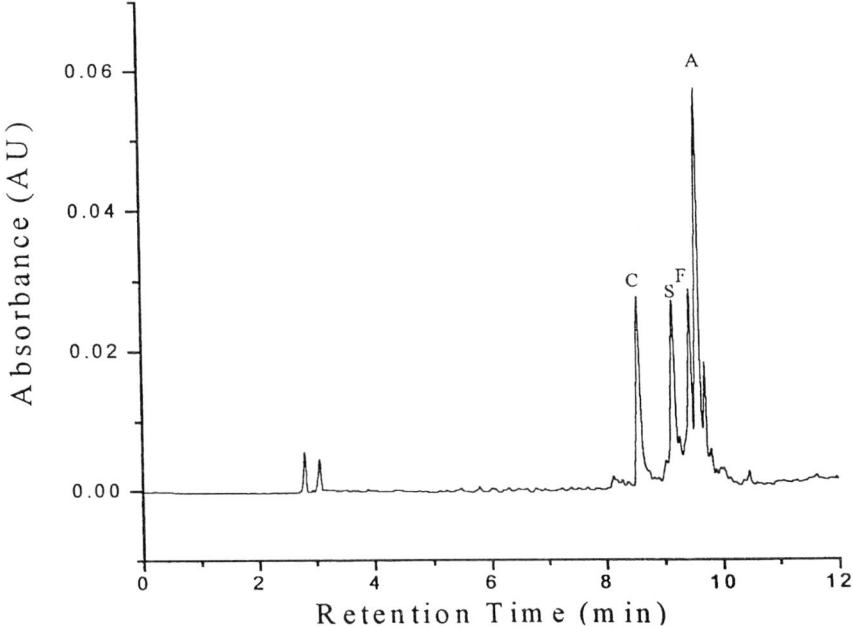

Figure 14.4 CIEF analysis of hemoglobin variants. Concentration of hemoglobin mixture = $0.16 \, mg \, ml^{-1}$ in the ampholyte solution (0.5% Servalyte in 0.3% methylcellulose). Conditions: focusing, $700 \, V \, cm^{-1}$; pressure mobilization, $900 \, V \, cm^{-1}$: $50 \, \mu m \times 27 \, cm$ neutral hydrophilic coated capillary from Beckman eCAP™ neutral capillary methods development kit/proteins; catholyte = 20 mM NaOH, 0.4% methylcellulose; anolyte = 25 mM H_3PO_4, 0.4% methylcellulose; UV detection at 280 nm. Reprinted with permission of A.A. Weis Vieweg from Ref. [81].

Table 14.1 Protein analysis by capillary electrophoresis

Separation mode Separation medium/buffer Voltage/current Capillary Detection	Analytes	Matrix	Ref.
CZE 100 mM Tris–HCl, pH 8.5 10 kV 75 µm × 50 cm, glycol-coated 230 nm	Proteins in serum	Serum	1
CZE 10 mM phosphate, 0.001% (w/w) Brij 35, pH 7.0 300 V cm^{-1}, 26 µA 75 µm × 50 cm (30 cm), alkylsilane coated 200 nm	Lysozyme, cytochrome c, ribonuclease, myoglobin, α-chymotrypsinogen		4
CZE 25 mM sodium borate, pH 8.25 297 V cm^{-1}, 38 mA 52 mm × 101 cm (55 cm) 200 nm	Myoglobin, carbonic anhydrase A and B, β-lactoglobulin A and B		5
CZE 0.1 M CHES, 0.25 M K$_2$SO$_4$, 1 mM EDTA, pH 9.0 5 kV 25 µm × 100 cm (63 cm) 193 nm	Lysozyme, myoglobin, trypsinogen, β-lactoglobulin A and B		6
CZE 0.15 M sodium phosphate, pH 1.5 2.5–30 kV linear program 300 s 53 µm × 110 cm (75 cm) 30 kV constant 190 nm	Lysozyme, cytochrome c myoglobin, β-lactoglobulin A, paralbumin		7
CZE 50 mM glutamine–triethylamine, pH 9.5 10 kV/15 µA 50 µm × 60 cm (50 cm), Si–C coated 214 nm	Insulin chain A and B, albumin ovalbumin, insulin, β-casein, α-lactalbumin		9
CZE 25 mM Tris/HCl, 2% 1-propanol, pH 3.0 3 kV/9 µA 75 µm × 55 cm (35 cm), coated 214 nm	Ribonuclease A and glycoforms of ribonuclease B		11
CZE 100 mM acetate–phosphate, pH 4.0, equilibrate for >2 h 20 kV/18 µA; 25°C 75 µm × 27 cm (20 cm) 200 nm	Human r-EPO (CHO)		12
CZE 0.5 M sodium phosphate, pH 4.5 to 10, 0.2–0.4 M borate, pH 8.5–10 10 kV; 23°C 20 µm × 27 cm (20 cm) 200 nm	Serum proteins and model proteins	Serum and PBS	13, 39

Table 14.1 *Continued*

Separation mode Separation medium/buffer Voltage/current Capillary Detection	Analytes	Matrix	Ref.
CZE 0.4 M sodium phosphate, pH 4.0 10 kV/62 μA; 23°C 20 μm × 25 cm (18 cm) 200 nm	Serum proteins	Serum	15
CZE 80 to 300 mM sodium borate, pH 10.0 10 to 20 kV; 23°C 20 μm × 25 cm (18 cm) 200 nm	Human serum proteins and model proteins		18
CZE 100 mM sodium phosphate, 1.0 M TMAB, pH 6.9; TMAB: zwitterionic species 190 V cm^{-1} 75 μm × 70 cm (50 cm) 214 nm	Cytochrome c, β-lactoglobulin A and B		21
CZE 20 mM phosphate, 200 μg ml^{-1} zwitterionic fluorosurfactant, pH 3 to 5 20 kV; 25°C 50 μm × 60 cm (50 cm) 210 nm	Lysozyme, ribonuclease, myoglobin, trypsinogen		22, 23
CZE 30 mM putrescine, 20 mM K$_2$SO$_4$, pH 7.0 20 kV; 20°C 50 μm × 57 cm (50 cm) 200 nm	Ricin, a toxic glycoprotein		24
CZE 40 mM tetramethyl 1,3-butanediamine pH 4.0–6.5 adjusted with phosphoric acid 10 kV; 28°C 75 μm × 37 cm (30 cm) 214 nm	Cytochrome c, lysozyme, ribonuclease A, α-chymotrypsinogen		25
CZE 25 mM borate–phosphate, 0.5 mM spermidine, pH 9.2 11 kV/28 μA; 25°C 75 μm × 69 cm (63 cm) 200 nm	Arginase isoforms		26
CZE 10 mM Tricine, 10 mM NaCl 2.5 mM 1,4-diaminobutane, pH 6.2 10 kV 75 μm × 57 cm (50 cm) 214 nm	r-EPO		29
CZE 150 mM sodium borate, 15 mM phytic acid, pH 9.5 10 kV; 25°C 50 μm × 27 cm (20 cm) 214 nm	Lysozyme, myoglobin, β-lactoglobulin A and B		30

Table 14.1 *Continued*

Separation mode Separation medium/buffer Voltage/current Capillary Detection	Analytes	Matrix	Ref.
CZE 50 mM phosphate, 90 mM perchlorate–TEA, pH 2.0 with 0.02% HPMC 12 kV; 23°C 50 μm × 60 cm (53 cm) 200 nm	Histones		31
CZE 0.1 M borate, pH 9.28 with 10% DMF 24 kV/44 μA 50 μm × 60 cm 200 nm	FITC-labeled myoglobin		32
CZE 50 mM MES, 0.1% Tween-20, 0.1% ethanolamine, pH 8.0 20 kV 50 μm × 72 cm (50 cm) 200 nm	β-Lactoglobulin A, B and C variants		33
CZE 12.5 mM triethylenetetraamine–TFA, pH 6.8 20 kV/65 μA: 25°C 75 μm × 47 cm (40 cm) 214 nm	Nuclease A and its genetic variants		34, 35
CZE 20 mM citrate/MES, pH 6.0 23.5 kV; 25°C 50 μm × 47 cm (40 cm), dextran–PEG coated 214 nm	Basic proteins: lysozyme, cytochrome c, ribonuclease A, α-chymotrypsinogen A		37
CZE 0.2 M phosphate, 0.1 M KCl, pH 7.0 250 V cm^{-1} 20 μm × (100 cm), PFB coated 220 nm	Lysozyme, myoglobin carbonic anhydrase ribonuclease, myoglobin trypsinogen		38
CZE 25 mM potassium phosphate, pH 2.6 25 kV; 30°C 50 μm × 57 cm (50 cm), coated 200 nm	Model proteins and peptides mixture		41
CZE 50 mM Tris–HCl, pH 4.5 cellulose acetate coated; 20 kV 50 μm × 78.8 cm (49.4 cm) 210 nm	Lysozyme, trypsinogen cytochrome c, ribonuclease A α-chymotrypsinogen A		43
CZE 10 mM sodium citrate, 6 M urea, 0.05% MHEC, pH 2.45 25 kV; 45°C 50 μm × 57 cm (50 cm), coated 214 nm	Milk proteins from genetic variants of cows	milk	44
CGE/SDS 10% T, 3.3% C polyacrylamide, 90 mM Tris–phosphate pH 8.6, 0.1% SDS, 8 M urea 400 V cm^{-1}, 36 μA 75 μm × 20 cm bifunctional reagent treated UV detection	α-Lactalbumin, β-lactoglobulin trypsinogen, pepsin		45

Table 14.1 *Continued*

Separation mode Separation medium/buffer Voltage/current Capillary Detection	Analytes	Matrix	Ref.
CGE/SDS 5% T, 3% C polyacrylamide, 375 mM Tris 3.2 mM SDS, 2.35 M ethylene glycol, pH 8.8 83 V cm^{-1}, 12 μA 50 μm × 24 cm acrylic silane-treated 214 nm	Lysozyme, trypsin inhibitor, carbonic anhydrase, ovalbumin, BSA phosphorylase b		46
CGE/SDS 3% PEG 100 000, 0.1 M Tris–CHES, 0.1% SDS 300 V cm^{-1} 100 μm × 40 cm polyacrylamide coated UV 214 cm	α-Lactalbumin, trypsin inhibitor, carbonic anhydrase, ovalbumin, BSA phosphorylase b		47
CGE/SDS 10% w/v dextran, 0.1% SDS, 100 mM Tris–CHES pH 8.6 370 V cm^{-1} 100 μm × 27 cm polyacrylamide coated 200 nm	Carbonic anhydrase, ovalbumin, phosphorylase b, BSA		48
CGE/SDS 15% Dextran (72 kDa) or 3% PEO (100 kDa) 100 mM Tris–CHES pH 8.8, 0.1% SDS 300 V cm^{-1} 100 μm × 47 cm coated 214 nm	Lysozyme, trypsin inhibitor, carbonic anhydrase, ovalbumin, BSA phosphorylase b		49
CGE/SDS Beckman eCAP SDS-200 kit 300 V cm^{-1}, 25–30 μA 100 μm × 47 cm coated 214 nm	Carbonic anhydrase, ovalbumin, BSA, β-galactosidase, phosphorylase b		50
CGE/SDS 1% Methoxylated agarose, 2 mM EDTA, 1% SDS in 40 mM Tris–acetate pH 8.3 2.5 kV/13 μA 75 μm × 240 mm 210 nm	Albumin monomers, dimers, trimers, etc.		51
CGE/SDS 7% Pullulan, 100 mM Tris–CHES, 0.1% SDS pH 8.7 15 kV/15 μA 75 μm × 50 cm polyacrylamide coated 214 nm	α-lactalbumin, trypsin inhibitor carbonic anhydrase, ovalbumin, BSA, phosphorylase b, β-galactosidase		52
CGE/SDS 1–3% PEO, 100 mM Tris–CHES pH 8.5, 20% SDS 300 V cm^{-1} 100 μm × 27 cm uncoated fused silica UV 214 nm	α-lactalbumin, trypsin inhibitor, carbonic anhydrase, ovalbumin, BSA, phosphorylase b, β-galactosidase		53
CGE 0.05% PVA (50 kDa), 70 mM phosphate, pH 3.0 429 V cm^{-1}, 33 μA 50 μm × 70 cm, PVA dynamically coated 214 nm	Cytochrome c, lysozyme, trypsinogen, trypsin, α-chymotrypsinogen		54

Table 14.1 *Continued*

Separation mode Separation medium/buffer Voltage/current Capillary Detection	Analytes	Matrix	Ref.
CGE/SDS 10% w/v dextran in 100 mM CHES pH 8.8, 0.1% SDS 400 V cm^{-1}, 48–50 μA 100 μm × 50 cm, polyacrylamide coated UV 214 nm	Myoglobin, carbonic anhydrase, BSA, phosphorylase b		55
CGE/SDS 5% T, 2% C polyacrylamide with ethylene glycol 375 mM Tris pH 8.8, 0.1% SDS, 2.5 M ethylene glycol 83 V cm^{-1} 75 μm × 27 cm, acryloxypropylsilane treated 214 nm	Lysozyme, trypsin inhibitor carbonic anhydrase, ovalbumin, BSA, phosphorylase b		56
CGE/SDS Beckman eCAP SDS-200 gel 14.1 kV 100 μm × 47 cm polyacrylamide coated 200 nm	HDL, apo A-1, A-II		57
CGE/SDS 3–8% PVA, 60 mM AMPD–CACO, pH 8.8, 0.1% SDS 8 kV/11.5 μA 75 μm × 34 cm poly(N-acryloylaminoethoxyethanol) coated 214 nm	α-Lactalbumin, trypsin inhibitor, carbonic anhydrase, ovalbumin, BSA, phosphorylase b		58
CGE/SDS Beckman eCAP SDS 14–200 kit (replaceable PEG/PEO) 300 V cm^{-1} 100 μm × 27 cm polyacrylamide coated 214 nm	α-Lactalbumin, ovalbumin, BSA, phosphorylase b, myosin, carbonic anhydrase, β-galactosidase		59, 60
CGE/SDS Beckman eCAP SDS 60 kit (replaceable PEO with SDS) 300 V cm^{-1}, 35–45 μA 100 μm × 27 cm polyacrylamide coated 214 nm	α-Lactalbumin, trypsin inhibitor, carbonic anhydrase, ovalbumin, BSA		61
CGE/SDS Beckman eCAP SDS 200 kit (replaceable PEO with SDS) 300 V cm^{-1}, 25–30 μA 100 μm × 47 cm polyacrylamide coated UV 214	Carbonic anhydrase, BSA, phosphorylase b, myosin, β-galactosidase, ovalbumin		62
CGE/SDS 0.25% to 4% PEO (100–900 kDa), 100 mM Tris–CHES pH 8.8, 0.1% SDS 300 V cm^{-1}, 25–30 mA 100 μm × 47 cm uncoated 214 nm	α-Lactalbumin, trypsin inhibitor, carbonic anhydrase, ovalbumin phosphorylase b, BSA		63
CGE/SDS Applied Biosystems Pro-Sort SDS–protein Analysis Kit	Bovine whey proteins		64

CAPILLARY ISOELECTRIC FOCUSING 191

Table 14.1 *Continued*

Separation mode Separation medium/buffer Voltage/current Capillary Detection	Analytes	Matrix	Ref.
CZE/detergent 30 mM borate pH 9, 10 mM DOC, 3.5 mM SDS, 3.5 mM Triton X-100 50 μm × 80 cm polyacrylamide coated 220 nm	Plasma apolipoproteins		65

CZE/detergent
40 mM SDS, 12 mM BSA
  10 mM borate/phosphate pH 9.0, 5% MeOH,
  5% CH_3CN
18 kV/65 μA
75 μm × 57 cm uncoated
LIF: 457.9 Ar^+ ion ex/520 em
	Serum bilirubins, biliprotein		66

CZE/detergent 20 mM borate pH 8.0, 50 mM SDS, 0–25% CH_3CN 25 kV 50 μm × 57 cm uncoated 200 nm	rDNA-derived protein in *E. coli* fermentation broth		67
CZE/detergent Borate–phosphate pH 7.2, 34.7 or 50 mM SDS 25 kV 50 μm × 72 cm uncoated Diode array scanning at 200 nm	Catalase subunits		68
MECC 40 mM Tris pH 7.5, 25 mM SDS 15 kV 50 μm × 52 cm 214 nm	*S. marcescens* nuclease isoforms		69
MECC 12 mM borate pH 9.4, 25 mM SDS 600 V cm^{-1} 75 μm × 57 cm uncoated 214 nm	BR96 antibody isoforms		70
MECC 300 mM borate pH 8.4, 85 mM SDS 10 kV 75 μm × 57 cm uncoated Diode array monitoring 200 nm, scanning 190–340 nm	Metallothionen isoforms		71
CIEF Catholyte 20 mM NaOH; anolyte 20 mM H_3PO_4 ampholyte: Pharmalyte 3–10 Focusing: 1 kV for 7 min; 3 kV for 13 min 200 μm × 120 mm methylcellulose coated 280 nm	Hemoglobin, transferrin		72

CIEF
Catholyte 20 mM NaOH; anolyte 20 mM H_3PO_4;
  ampholyte: 1% v/v Biolyte (pH 6–8)
Focusing: 1 kV, 7 min; 3 kV, 13 min;
  anodic mobilization
200 μm × 130 mm methylcellulose coated
280 nm
	Hemoglobin A_{1c} fraction		73

Table 14.1 *Continued*

Separation mode Separation medium/buffer Voltage/current Capillary Detection	Analytes	Matrix	Ref.
CIEF Catholyte 20 mM NaOH; anolyte 20 mM H_3PO_4; ampholyte: 2% v/v Biolyte 5 7 Focusing: 6 kV; anodic mobilization 100 µm × 185 mm methylcellulose coated 280 nm	Transferrin isoforms		74
CIEF Catholyte 20 mM NaOH; anolyte 20 mM H_3PO_4; ampholyte: 2% v/v Biolyte pH 3–10 Focusing: 8 kV; anodic mobilization 25 µm × 12 cm coated 280 nm	Hemoglobin		75
CIEF Catholyte 20 mM NaoH; anolyte 10 mM H_3PO_4; ampholyte: Biolyte pH 3–10 + 0.5% TEMED Focusing: 8 kV; cathodic mobilization 25 µm × 20 cm polyacrylamide coated 280 nm	Cytochrome c, hemoglobin A, phycocyanin		76

CIEF
Catholyte 20 mM NaOH; anolyte 10 mM H_3PO_4;
  ampholyte: 5% Pharmalyte 3–10, 0.05%
  methylcellulose
Focusing: 20 kV, 5 min
50 µm × 60 cm uncoated
280 nm
	Myoglobin, carbonic
	anhydrase
			77

CIEF Catholyte 20 mM NaOH; anolyte 10–100 mM H_3PO_4 ampholyte: 5% Pharmalyte pH 3–10 0.1% methylcellulose, 0.5–1.6% TEMED Focusing: 24 kV, 5 min 75 µm × 60 cm uncoated 280 nm	Cytochrome c, myoglobin, chymotrypsinogen A, β-lactoglobulin		78

CIEF
Catholyte 20 mM NaOH, 0.06–0.3% HPMC;
  anolyte 10 mM H_3PO_4;
  ampholyte: 2.5 or 5% Ampholine
Focusing: 20 kV
75 µm × 90 cm uncoated
Multiwavelength UV 196–320 nm
	Carbonic anhydrase
	cytochrome c
			79

CIEF
Catholyte 20 mM NaOH, 0.4% methylcellulose;
  ampholyte: 0.5% Servalyte 3–10, 0.4%
  methylcellulose
Focusing: 30 kV, 6 min
50 µm × 72 cm DB-1-coated
280 nm
	RNAse A, β-lactoglobulin A,
	carbonic anhydrase
			80

CIEF
Catholyte 20 mM NaOH;
  anolyte 25 mM H_3PO_4, 0.4% methylcellulose;
  ampholyte: 0.5–1% w/v Servalyte 3–10
Focusing: 13.5 kV, 2 min
mobilization: 13.5 kV, 0.5 psi rinse
280 nm
	RNAse, myoglobin, trypsin
	inhibitor, β-lactoglobulin,
	carbonic anhydrase,
	CCK flanking peptide
			81

Table 14.1 *Continued*

Separation mode Separation medium/buffer Voltage/current Capillary Detection	Analytes	Matrix	Ref.

CIEF
Catholyte 20 mM NaOH; anolyte 10 mM H_3PO_4;
  ampholyte: 2% Pharmalyte pH 3–10
Focusing: 3.5 kV, 2–4 min
100 μm × 4 cm square glass capillary
CCD absorption imaging system,
  light source: Ar^+ laser 496.5 and 514.5 nm
	Myoglobin, cytochrome c,
	hemoglobin, transferrin
			82

CIEF
Catholyte: 20 mmol l^{-1} NaOH, 0.015% w/v
  methylcellulose;
  ampholyte: 4.5% Ampholine 3.5–10,
  Pharmalyte 6.7–7.7; anolyte: 10 mM H_3PO_4
20 kV
75 μm × 60 cm uncoated
415 nm absorbance detection
	Hemoglobin variants		83

CIEF
Catholyte: 20 mM NaOH;
  anolyte: 10 mM H_3PO_4;
  ampholyte: 2% Pharmalyte pH 3–10
Focusing: 3 kV, 5 min
50 μm × 130 mm methylcellulose coated
280 nm
	Red cell glucose transporter		84

AMPD = 2-amino-2-methyl-1,3-propanediol, Brij 35 = polyethylene glycol dodecyl ether, BSA = bovine serum albumin, CACO = cacodylic acid, CCD = charge-coupled device, CGE = capillary gel electrophoresis, CHES = 2-(cyclohexylamino)ethanesulfonic acid, CHO = Chinese hamster ovary, CIEF = capillary isoelectric focusing, CZE = capillary zone electrophoresis, DOC = deoxycholate, EOF = electroosmotic flow, EPO = erythropoietin, FITC = fluorescein isothiocyanate, HDL = high-density lipoprotein, HPMC = (hydroxypropyl)methyl cellulose, IEF = isoelectric focusing, MECC = micellar electrokinetic chromatography, MES = 2-morpholinoethanesulfonic acid, MHEC = methylhydroxyethylcellulose, PBS = phosphate buffered saline, PEG = polyethylene glycol, PEO = polyethylene oxide, PFB = pentafluorobenzoyl-, PVA = poly(vinyl alcohol), SDS = sodium dodecyl sulfate, TEA = triethylamine, TEMED = N,N,N',N'-tetramethylethylenediamine, TFA = trifluoroacetic acid, TMAB = tetramethyl-1,3-diaminobutane, T = total acrylamide.

methylcellulose also acted as a viscosity-enhancing agent to minimize distortion of the focused zones during the mobilization step. Excellent linearity of the migration time as a function of the protein pI values was obtained ($r = 0.9988$) with standard deviations of less than 0.04 pH units for six measurements for all proteins.

CIEF has been successfully applied to the analysis of human serum transferrin [74], hemoglobin variants [75, 81–83], γ-globulins [73], hemoglobin A_{Ic} [73] and human red cell glucose transporter [84]. The CIEF electropherogram of hemoglobin variant mixture is presented in Figure 14.4 [81].

Table 14.1 oulines the methods published for protein analysis by CGE/SDS, MECC and CIEF in recent years. In many of these reports, the effects of several operational variables such as buffer concentration, pH, polymer concentration and temperature were investigated but due to space limitations, only the conditions which produced the best separations are listed.

In conclusion, the published reports reviewed herein on SDS/CGE and CIEF amply demonstrate that CE-based methods of protein analysis provide a significant improvement over their corresponding slab gel counterparts in terms of speed, resolution, automation and quantitation.

References

1. Jorgenson, J.W.; Lukacs, K.D. (1983) Capillary zone electrophoresis. *Science* **222**, 266–272.
2. Tsuda, T. (1994) Control of electroosmotic flow in capillary electrophoresis. In *Handbook of Capillary Electrophoresis*, Landers, J. (ed.), CRC Press, pp. 563–590.
3. Hayes, M.A.; Ewing, A.G. (1992) Electroosmotic flow control and monitor with an applied radial voltage for in capillary zone electrophoresis. *Anal. Chem.* **64**, 512–516.
4. Towns, J.K.; Regnier, F.E. (1991) Capillary zone electrophoretic separation of proteins using non-ionic surfactant coatings. *Anal. Chem.* **63**, 1126–1132.
5. Lauer, H.H.; McManigill, D. (1986) Capillary zone electrophoresis in untreated fused-silica tubing. *Anal. Chem.* **58**, 166–169.
6. Green, J.S.; Jorgenson, J.W. (1989) Minimizing adsorption of proteins on fused silica in capillary zone electrophoresis by addition of alkali metal salts to the buffer. *J. Chromatogr.* **478**, 63–70.
7. McCormick, R.M. (1988) Capillary zone electrophoretic separation of peptides and proteins using low pH buffers in modified silica capillaries. *Anal. Chem.* **60**, 2322–2328.
8. Hjertén, S. (1985) High performance electrophoresis: elimination of electroendoosmotic flow and solute adsorption. *J. Chromatogr.* **347**, 191–198.
9. Cobb, K.A.; Dolink, V.; Novotny, M. (1990) Electrophoretic separation of proteins with hydrolytically stable surface structure. *Anal. Chem.* **62**, 2478–2483.
10. Hjertén, S.; Kubo, K. (1993) A new type of pH- and detergent-stable coating for eliminating of electroosmosis and adsorption in capillary electrophoresis. *Electrophoresis* **14**, 390–395.
11. Huang, M.; Plocek, J.; Novotny, M.V. (1995) Hydrolytically stable cellulose-derivative coatings for capillary electrophoresis of peptides, proteins and glycoconjugates. *Electrophoresis* **16**, 396–401.
12. Tran, A.D.; Park, S.; Lisi, P.J.; Huynh, C.T.; Ryall, R.R.; Lane, P.A. (1991) Separation of carbohydrate-mediated microheterogeneity of recombinant human erythropoietin by free solution capillary electrophoresis. Effects of pH, buffer type and organic additives. *J. Chromatogr.* **542**, 459–471.
13. Chen, F.-T.A.; Kelly, L.; Palmieri, R.; Biehler, R.; Schwartz, H. (1992) Use of high ionic strength buffers for separation of proteins and peptides with capillary electrophoresis. *J. Liq. Chromatogr.* **15**, 1143–1149.
14. Rush, R.S.; Cohen, A.S.; Karger, B.L. (1991) Influence of column temperature on the electrophoretic behavior of myoglobin and a-lactalbumin in high-performance capillary electrophoresis. *Anal. Chem.* **63**, 1346–1350.
15. Chen, F.-T.A. unpublished results.
16. Gordon, M.G.; Lee, K-J.; Arias, A.A.; Zare, R.N. (1991) Protocol for resolving protein mixtures in capillary zone electrophoresis. *Anal. Chem.* **63**, 69–72.
17. Chen, F.-T.A.; Liu, C.M.; Hsieh, Y.Z.; Sternberg, J.C. (1991) Capillary electrophoresis – A new clinical tool. *Clin. Chem.* **37**, 14–19.
18. Chen, F.-T.A. (1991) Rapid protein analysis by capillary electrophoresis. *J. Chromatogr.* **516**, 69–78.
19. Cohn, E.J.; Edsall, J.T. (1942) *Proteins, Amino Acids, and Peptides as Ions and Dipolar Ions*, Reinhold, New York.

20 Bushey, M.M.; Jorgenson, J.W. (1989) Capillary electrophoresis of protein in buffers containing high concentration of zwitteric salts. *J. Chromatogr.* **480**, 301–305.
21 Fang, X.-H.; Zhu, T.; Sun, V.-H. (1994) Use of zwitterionic buffer additives to improve separation of proteins in capillary zone electrophoresis. *J. High Resolut. Chromatogr.* **17**, 749–752.
22 Emmer, A.; Jansson, M.; Roeraade. (1991) Improved capillary zone electrophoresis separation of basic proteins using a fluorosurfactant buffer additive. *J. Chromatogr.* **547**, 544–550.
23 Emmer, A.; Roeraade. (1994) Performance of zwitterionic and cationic fluorosurfactants as buffer additives for capillary electrophoresis of proteins. *J. Liq. Chromatogr.* **17**, 3831–3846.
24 Hines, H.B.; Brueggemann. (1994) Factors affecting capillary electrophoresis of ricin, a toxic glycoprotein. *J. Chromatogr.* **670**, 199–208.
25 Corradini, D.; Cannarsa, G. (1995) N,N,N',N'-tetramethyl-1,3-butadiamine as effective electrolyte additive for efficient electrophoretic separation of basic proteins in bare fused-silica capillaries. *Electrophoresis* **16**, 630–635.
26 Pedrosa, M.M.; Legaz, M.E. (1995) Separation of arginase isoforms by capillary zone electrophoresis and isoelectric focusing in density gradient column. *Electrophoresis* **16**, 659–669.
27 Oda, R.P.; Madden, J.C.; Spelsberg, T.C.; Landers, J.P. (1994) A,O-Bis-quaternary ammonium alkanes as effective buffer additives for enhanced capillary electrophoretic separation of glycoproteins. *J. Chromatogr.* **680**, 85–92.
28 Morbeck, D.E.; Madden, B.J.; McCormick, D.J. (1994) Analysis of microheterogeneity of glycoprotein chorionic gonadotropin with high-performance capillary electrophoresis. *J. Chromatogr.* **680**, 217–224.
29 Watson, E.; Yao, F. (1993) Capillary electrophoretic separation of human recombinant erythropoietin (r-HuEPO) glycoforms. *Anal. Biochem.* **210**, 389–393.
30 Okao, G.N.; Birrell, H.C.; Greenaway, M.; Haran, M.; Cameleri, P. (1994) The effect of phytic acid on the resolution of peptides and proteins in capillary electrophoresis. *Anal. Biochem.* **219**, 201–206.
31 Lindner, H.; Helinger, W.; Sarg, B.; Meraner, C. (1995) Effect of buffer composition on the migration order of histone H1 subtype. *Electrophoresis* **16**, 604–610.
32 Banks, P.R.; Paquette, D.M. (1995) Monitoring of a conjugation reaction between fluorescein isothiocyanate and myoglobin by capillary zone electrophoresis. *J. Chromatogr.* **693**, 145–154.
33 Paterson, G.R.; Hill, J.P.; Otter, D.E. (1995) Separation of b-lactoglobulin A, B and C variants of bovine whey using capillary zone electrophoresis. *J. Chromatogr.* **700**, 105–110.
34 Kalman, F.; Ma, S.; Fox, R.O.; Horvath, C. (1995) Capillary electrophoresis of S. nuclease mutants. *J. Chromatogr.* **705**, 135–154.
35 Kalman, F.; Ma, S.; Hodel, A.; Fox, R.O.; Horvath, C. (1995) Capillary electrophoresis of S. nuclease mutants. *Electrophoresis* **16**, 595–603.
36 Yao, Y.J.; Loh, K.C.; Chung, M.C.M.; Li, S.F.Y. (1995) Analysis of recombinant human tumor necrosis factor beta by capillary electrophoresis. *Electrophoresis* **16**, 647–653.
37 Mechref, Y.; El Rassi, Z. (1995) Fused-silica capillaries with surface-bound dextran layer crosslinked with diepoxypolyethyleneglycol for capillary electrophoresis of biological substances at reduced electroosmotic flow. *Electrophoresis* **16**, 617–624.
38 Swedberg, S.A. (1990) Characterization of protein behavior in high performance capillary electrophoresis using a novel capillary system. *Anal. Biochem.* **185**, 51–56.
39 Chen, F.-T.A.; Sternberg, J.C. (1994) Characterization of proteins by capillary electrophoresis in fused-silica columns: Review on serum protein analysis and application to immunoassays. *Electrophoresis* **15**, 13–21.
40 Zhao, Z.; Malik, A.; Lee, M.L. (1993) Solute adsorption on polymer-coated fused-silica capillary electrophoresis columns using selected protein and peptide standards. *Anal. Chem.* **65**, 2747–2752.
41 Piccoli, G.; Fiorani, M.; Biagiarelli, B.; Palma, F.; Vallorani, L.; De Bellis, R.; Stocchi, V. (1995) High performance capillary electrophoretic separation of proteins and peptides using a bonded hydrophilic phase capillary. *Electrophoresis* **16**, 625–629.

42 Cifuente, A.; Santos, J.M.; de Frutos, M.; Diez-Masa, J.C. (1993) Separation of basic proteins by capillary electrophoresis using cross-linked polyacrylamide-coated capillaries and cationic buffer additives. *J. Chromatogr.* **655**, 61–72.
43 Busch, M.H.A.; Kraak, J.C.; Poppe, H. (1995) Cellulose acetate-coated fused-silica capillaries for the separation of proteins by capillary zone electrophoresis. *J. Chromatogr.* **699**, 287–296.
44 de Jong, N.; Visser, S.; Olieman, C. (1993) Determination of milk proteins by capillary electrophoresis. *J. Chromatogr.* **652**, 207–213.
45 Cohen, A.S.; Karger, B.L. (1987) High-performance sodium dodecyl sulfate polyacrylamide capillary electrophoresis of peptides and proteins. *J. Chromatogr.* **397**, 409–417.
46 Tsuji, K. (1991) High-performance capillary electrophoresis of proteins. Sodium dodecyl sulfate–polyacrylamide gel-filled capillary column for the determination of recombinant biotechnology-derived proteins. *J. Chromatogr.* **550**, 823–830.
47 Ganzler, K.; Greve, K.S.; Cohen, A.S.; Karger, B.L.; Guttman, A.; Cooke, N.C. (1992) High-performance capillary electrophoresis of SDS-protein complexes using UV-transparent polymer networks. *Anal. Chem.* **64**, 2665–2671.
48 Lausch, R.; Scheper, T.; Reif, O.-W.; Schlosser, J.; Fleischer, J.; Freitag, R. (1993) Rapid capillary gel electrophoresis of proteins. *J. Chromatogr.* **654**, 190–195.
49 Guttman, A.; Horvath, J.; Cooke, N. (1993) Influence of temperature on the sieving effect of different polymer matrices in capillary SDS gel electrophoresis of proteins. *Anal. Chem.* **65**, 199–203.
50 Guttman, A.; Nolan, J.A.; Cooke, N. (1993) Capillary sodium dodecyl sulfate electrophoresis of proteins. *J. Chromatogr.* **632**, 171–175.
51 Hjertén, S.; Srichaiyo, T.; Palm, A. (1994) UV-transparent, replaceable agarose gels for molecular-sieve (capillary) electrophoresis for proteins and nucleic acids. *Biomed. Chrom.* **8** 73–76.
52 Nakatani, M.; Shibukawa, A.; Nakagawa, T. (1994) High-performance capillary electrophoresis of SDS–proteins using pullulan solution as separation matrix. *J. Chromatogr.* **672**, 213–218.
53 Benedek, K.; Thiede, S. (1994) High-performance capillary electrophoresis of proteins using sodium dodecyl sulfate–poly(ethylene oxide). *J. Chromatogr.* **676**, 209–217.
54 Gilges, M.; Kleemiss, M.H.; Schromburg, G. (1994) Capillary zone electrophoresis separations of basic and acidic proteins using poly(vinyl alcohol) coatings in fused silica capillaries. *Anal. Chem.* **66**, 2038–2046.
55 Karim, M.R.; Janson, J.-C.; Takagi, T. (1994) Size-dependent separation of proteins in the presence of sodium dodecyl sulfate and dextran in capillary electrophoresis: Effect of molecular weight of dextran. *Electrophoresis* **15**, 1531–1534.
56 Tsuji, K. (1994) Sodium dodecyl sulfate polyacrylamide gel- and replaceable polymer-filled capillary electrophoresis for molecular mass determination of proteins of pharmaceutical interest. *J. Chromatogr.* **662**, 291–299.
57 Goux, A.; Athias, A.; Persegol, L.; Lagrost, L.; Gambert, P.; Lallemant, C. (1994) Capillary gel electrophoresis analysis of apolipoproteins A-I and A-II in human high-density lipoproteins. *Anal. Biochem.* **218**, 320–324.
58 Simo-Alfonso, E.; Conti, M.; Gelfi, C.; Righetti, P.G. (1995) Sodium dodecyl sulfate capillary electrophoresis of proteins in entangled solutions of poly(vinyl alcohol). *J. Chromatogr.* **689**, 85–96.
59 Shieh, P.C.H.; Hoang, D.; Guttman, A.; Cooke, N. (1994) Capillary sodium dodecyl sulfate gel electrophoresis of proteins. I. Reproducibility and stability. *J. Chromatogr.* **676**, 219–226.
60 Guttman, A.; Shieh, P.; Lindahl, J.; Cooke, N. (1994) Capillary sodium dodecyl sulfate gel electrophoresis of proteins II. On the Ferguson method in polyethylene oxide gels. *J. Chromatogr.* **676**, 227–231.
61 Guttman, A.; Shieh, P.; Hoang, D.; Horvath, J.; Cooke, N. (1994) Effect of operational variables on the separation of proteins by capillary sodium dodecyl sulfate-gel electrophoresis. *Electrophoresis* **15**, 221–224.
62 Guttman, A. (1994) Comparison of the separation of proteins by sodium dodecyl sulfate-slab gel electrophoresis and capillary sodium dodecyl sulfate-gel electrophoresis. *Anal. Biochem.* **221**, 285–289.

63. Guttman, A. (1995) On the separation mechanism of capillary sodium dodecyl sulfate–gel electrophoresis of proteins. *Electrophoresis* **16**, 611–616.
64. Kinghorn, N.M.; Norris, C.S.; Paterson, G.R.; Otter, D.E. (1995) Comparison of capillary electrophoresis with traditional methods to analyse bovine whey proteins. *J. Chromatogr.* **700**, 111–123.
65. Tadey, T.; Purdy, W.C. (1993) Effect of detergents on the electrophoretic behaviour of plasma apolipoproteins in capillary electrophoresis. *J. Chromatogr.* **652**, 131–138.
66. Wu, N.; Sweedler, J.V.; Lin, M. (1994) Enhanced separation and detection of serum bilirubin species by capillary electrophoresis using a mixed anionic surfactant–protein buffer system with laser-induced fluorescence detection. *J. Chromatogr.* **654**, 185–191.
67. Strege, M.A.; Lagu, A.L. (1995) Capillary electrophoretic separations of biotechnology-derived proteins in *E. coli* fermentation broth. *Electrophoresis* **16**, 642–646.
68. Pedrosa, M.M.; Reyes, A.; Vicente, C.; Legaz, M.E. (1995) Analysis of the quaternary structure of catalase by capillary zone electrophoresis. *J. Chromatogr.* **697**, 571–578.
69. Pedersen, J.; Pedersen, M.; Soeberg, H.; Biedermann, K. (1993) Separation of isoforms of *Serratia marcescens* nuclease by capillary electrophoresis. *J. Chromatogr.* **645**, 353–361.
70. Kats, M.; Richberg, P.C.; Hughes, D.E. (1995) Conformational diversity and conformational transitions of a monoclonal antibody monitored by circular dichroism and capillary electrophoresis. *Anal. Chem.* **67**, 2943–2948.
71. Beattie, J.H.; Richards, M.P. (1995) Analysis of metallothionen isoforms by capillary electrophoresis: optimisation of protein separation conditions using micellar electrokinetic capillary chromatography. *J. Chromatogr.* **700**, 95–103.
72. Hjertén, S.; Zhu, M.-D. (1985) Adaptation of the equipment for high-performance electrophoresis to isoelectric focusing. *J. Chromatogr.* **346**, 265–270.
73. Hjertén, S.; Elenbring, K.; Kilar, F.; Liao, J.-L.; Chen, A.J.C.; Siebert, C.J.; Zhu, M.-D. (1987) Carrier-free zone electrophoresis, displacement electrophoresis and isoelectric focusing in a high-performance electrophoresis apparatus. *J. Chromatogr.* **403**, 47–61.
74. Kilar, F.; Hjertén, S. (1989) Separation of the human transferrin isoforms by carrier-free high-performance zone electrophoresis and isoelectric focusing. *J. Chromatogr.* **480**, 351–357.
75. Zhu, M.; Hansen, D.L.; Burd, S.; Gannon, F. (1989) Factors affecting free zone electrophopresis and isoelectric focusing in capillary electrophoresis. *J. Chromatogr.* **480**, 311–319.
76. Zhu, M.; Rodriguez, R.; Weir, T. (1991) Optimizing separation parameters in capillary electrophoresis. *J. Chromatogr.* **559**, 479–488.
77. Mazzeo, J.R.; Krull, I.S. (1991) Capillary isoelectric focusing of proteins in uncoated fused-silica capillaries using polymeric additives. *Anal. Chem.* **63**, 2852–2857.
78. Mazzeo, J.R.; Krull, I.S. (1992) Improvements in the method developed for performing isoelectric focusing in uncoated capillaries. *J. Chromatogr.* **606**, 291–296.
79. Thormann, W.; Caslavska, J.; Molteni, S.; Chmelik, J. (1992) Capillary isoelectric focusing with electroosmotic zone displacement and on-column multichannel detection. *J. Chromatogr.* **589**, 321–327.
80. Chen, S.-M.; Wiktorowicz, J.E. (1992) Isoelectric focusing by free solution capillary electrophoresis. *Anal. Biochem.* **206**, 84–90.
81. Huang, T.-L.; Shieh, P.C.H.; Cooke, N. (1994) Isoelectric focusing of proteins in capillary electrophoresis with pressure-driven mobilization. *Chromatographia* **39**, 543–548.
82. Wu, J.; Pawliszyn, J. (1994) Application of capillary isoelectric focusing with absorption imaging detection to the analysis of proteins. *J. Chromatogr.* **657**, 327–332.
83. Molteni, S.; Frischknecht, H.; Thormann, W. (1994) Application of dynamic capillary isoelectric focusing to the analysis of human hemoglobin variants. *Electrophoresis* **15**, 22–30.
84. Englund, A.-K.; Lundahl, P.; Elenbring, K.; Ericson, C.; Hjertén, S. (1995) Capillary and rotating-tube isoelectric focusing of a transmembrane protein, the human red cell glucose transporter. *J. Chromatogr.* **711**, 217–222.

15 Analysis of erythropoietin by capillary electrophoresis
F.-T.A. CHEN

15.1 Introduction

Human erythropoietin (EPO) is a glycoprotein hormone responsible for regulating the growth and maturation of erythroid progenitor cells [1]. The site of biosynthesis is in the liver in the fetus and switches to the kidney in adults [2–4]. It is maintained in circulation at about 10 pM under normal physiological conditions, and may be excreted in urine [5]. EPO was first purified in small quantities from the urine of aplatic anemia patients [5]. The first functional recombinant human erythropoietin was cloned, expressed [6] and made available as one of the most celebrated therapeutic drugs in recent years. Recombinant EPO (r-EPO) and natural EPO purified from the urine showed no significant difference in their functional activity *in vivo* or their carbohydrate structures [7, 8].

The molecular mass of EPO based on cDNA is 18 399 Da while the glycosylated r-EPO derived from Chinese hamster ovary (CHO) cells has an apparent molecular mass of about 34 000 Da [6]. Deglycosylation of the r-EPO from CHO cells with endo-β-N-acetylglucosaminidase F (N-glycanase) results in the formation of protein species with a molecular mass of 19 000 Da [6]. Thus, a significant proportion (40%) of the molecular mass in r-EPO is contributed by carbohydrates. The structural characterization of r-EPO is centered mostly in the microheterogeneity due to posttranslational sugar modification. The r-EPO contains 166 amino acids [6] with three N-linked sugar chains (N-24, N-38 and N-83) and one O-linked oligosaccharide chain (O-126), similar to that of the EPO derived from human urine. The oligosaccharides in r-EPO produced in CHO cells contain variable numbers of sialic acids that result in the formation of species with rather broad isoelectric points (pIs) of between 4.2 to 4.6. Desialylation of the r-EPO causes complete loss of its biological activity *in vivo* [9, 10], and the r-EPO species with tetraantennary oligosaccharide exhibits a substantially higher biological activity than that of the species with biantennary oligosaccharide structure [11]. Thus, the structure of the oligosaccharide chain in r-EPO has a profound effect on its biological activity. It is important to define the critical requirement of the glycoform heterogeneity of the recombinant proteins such as that in r-EPO in relation to the efficacy of its therapeutic utility, and to develop an analytical test to ensure the consistency of the heterogeneity between each production lot.

The structural characterization of oligosaccharides in the r-EPO derived from CHO [7–8, 11–13], that of a baby hamster kidney (BHK) [14, 15] and that of HeLa cells [16] indicates a general similarity of both the N- and O-linked sugars. The N-linked oligosaccharides profile from different lots of r-EPO of the same cell line (CHO) appeared to be similar [12]. The majority of di-, tri- and tetraantenarry structures in r-EPO are fully sialylated and linked almost exclusively from the non-reducing end of the sialic acid at α-[2–3] to galactosyl residue [13]. The characterization of r-EPO and its associated oligosaccharides were reported extensively by using a combination of many different analytical tools [6–16]. More recently, capillary electrophoresis (CE) methods have been shown as a powerful tool for the analysis of biomolecules, such as proteins, peptides, carbohydrates and DNA. The technique provides extremely high resolution power, rapid separation and minute sample requirement, and is well-suited for automation with real-time data analysis. This review focuses on the analysis of r-EPO related to capillary electrophoretic techniques, summarized in Table 15.1.

15.2 Native protein

CE separation of the r-EPO produced in CHO cells was first described using a phosphate–acetate buffer [17]. A fairly well-resolved set of isoforms of about five different species from the native protein was evident presumably due to the carbohydrate-mediated microheterogeneity in the r-EPO [17]. The use of a phosphate–acetate buffer clearly made significant improvements in resolution over the acetate-based buffer alone, so it was speculated that the long equilibration time may be reduced to a much shorter preequilibration of 30 min using phosphate buffer at pH 4.0 alone, indicating phosphate buffer is more favorable than acetate buffer. The speculated reason is that the prolonged contact of phosphate with the silica surface results in the formation of a phosphate–silicate surface. This dynamic interaction between the bulk phosphate buffer and fused-silica surface prevents protein adsorption to the capillary wall [18, 19].

Watson and Yao [20] reported the separation of the r-EPO by free zone capillary electrophoresis with nearly baseline resolution and comparable to the results obtained by the conventional gel-based isoelectric focusing (IEF) technique. The method involves the use of tricine-based buffer at pH 6.2 with 2.5 mM of 1,4-diaminobutane and urea. The use of 1,4-diaminobutane reduces the surface negative charge of the fused-silica column which results in a substantial reduction in the electroendoosmotic flow. Urea addition to the buffer system will prevent aggregation of r-EPO as well as its adsorption to the fused-silica capillary wall. The electrophoretic mobility of the isoforms is in the order of increasing number of sialic acids. Treatment of the r-EPO with neuraminidase results in the release of sialic acids from the non-reducing end of the oligosaccharides in r-EPO consistent with the relative migration order of each of the isoforms.

The r-EPO derived from *E. coli* is a non-glycosylated species consisting of 166 amino acids with pI of 8.8 and molecular mass estimated to be 19329 Da by CE [21] in an untreated fused-silica capillary using a SDS–polyethylene oxide replaceable polymer solution. The r-EPO derived from CHO cells is glycosylated with an apparent molecular mass of about 34000 Da [6]. The microheterogeneity of the r-EPO produced in CHO cells is primarily due to the variability in sialic acid content between species. The presence of sialic acids significantly reduces the pI between 4.2 to 4.6 [22]. Using sodium dodecyl sulfate (SDS)–polyethylene oxide (PEO) entangled polymer-based CE separation system, Benedek and Thiede [21] estimated the mass of glycosylated r-EPO as 44014 Da.

15.3 Peptide mapping and glycopeptides analysis

The utility of CE for the tryptic peptide mapping and evaulation of glycopeptide microheterogeneity of r-EPO was elegantly demonstrated by Rush *et al.* [23]. CE analysis of exhaustive trypsin digestion of the non-reduced r-EPO (derived from CHO cells) produced the electropherogram shown in Figure 15.1 [23]. A combination of heptanesulfonate and phosphate at pH 2.5 were used as the electrophoresis buffer. At buffer pH 2.5, the silica surface exhibited significant negative charge to cause peptide–silica wall interactions. Heptanesulfonate was added as an ion-pairing agent to the peptide which significantly enhanced the resolution of peptide in CE separation. Two major zones corresponding to non-glycosylated and glycosylated peptides were observed in the electropherogram. The non-glycopeptides migrate earlier, before 32 min, and the glycopeptides exhibit a substantially longer migration time, between 40 and 80 min [23]. The additional mass of the oligosaccharides in glycopeptides coupled with the partial ionization of the sialic acid residues of the sugars produced a considerable reduction in the charge-to-mass ratio that resulted in reduced electrophoretic mobility.

Eighteen discrete peptides are identified within the first 32 min from a possible 21 peptides predicted from complete tryptic digest of EPO (Figure 15.1). The three remaining peptides are sialic acid containing glycopeptides of the *N*- and *O*-linked glycosylation sites. The N-24 and N-38 asparagines are part of the same tryptic peptide fragment while N-83 asparagine and the serine-126 of the *O*-linked site are on the separate tryptic fragments. Clearly peptide mapping by the CE method provides extremely high sepa-

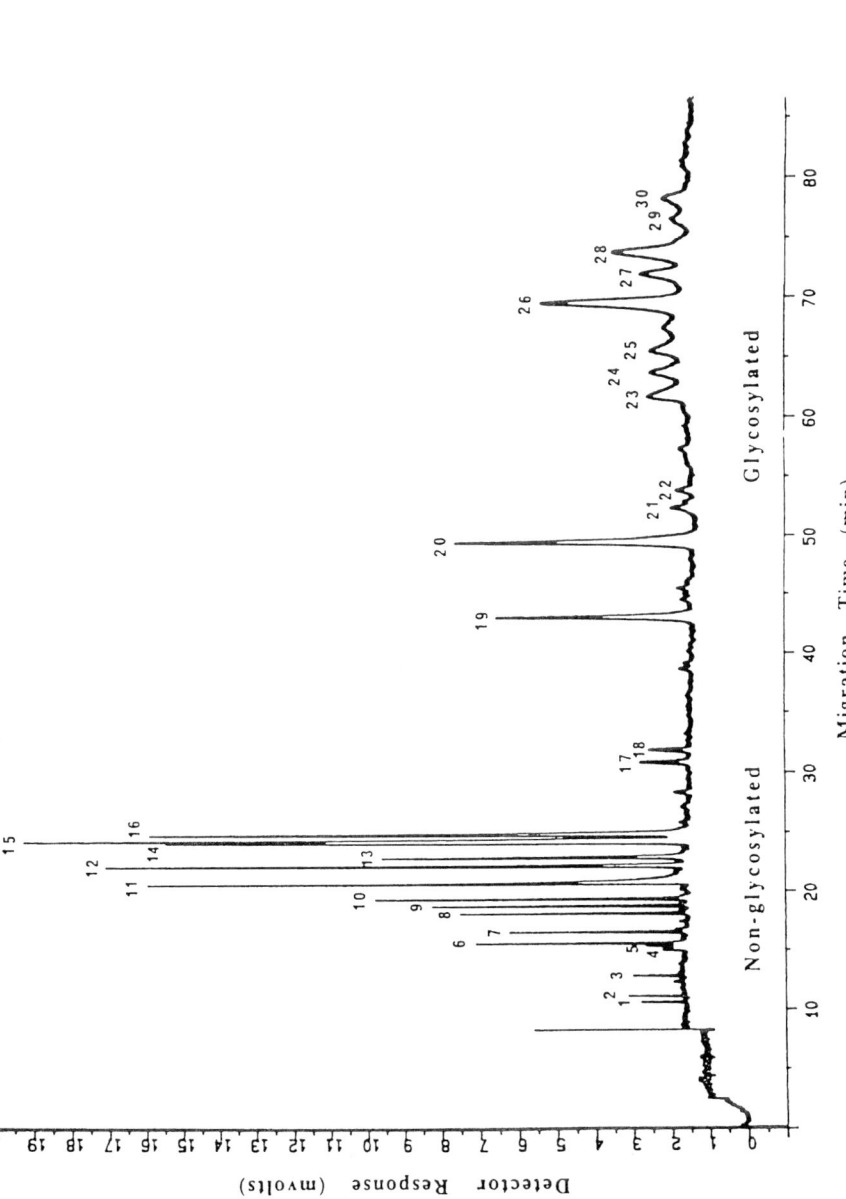

Figure 15.1 Electropherogram of mixture of tryptic digest of r-EPO in untreated fused-silica capillary. 50 μm × 75 cm (50 cm to detector); UV: 200 nm; 30°C; buffer: 40 mM phosphate, 0.1 M heptanesulfonic acid, pH 2.5; 16 kV/110 μA. Reprinted with permission from Ref. [23], copyright (1993) American Chemical Society.

ration efficiency which cannot be obtained readily by other analytical separation methods.

Both N- and O-linked glycopeptides were identified by comparing the electropherograms of the tryptic digest of r-EPO before and after pretreatment with N-glycanase [23]. The O-linked glycopeptides, peak 19 and 20 (Figure 15.1), presumably contain disaccharide of gal-β (1–3) gal NAc with mono- and disialyl residues [13, 24], respectively. The N-linked glycopeptides are quite heterogeneous, containing carbohydrate structures with bi-, tri- and tetraantennaries. More than 12 N-linked glycopeptide forms were evident from the electropherogram (peak 21 to 30) in Figure 15.1 [23]. Tryptic mapping by CE of the N-glycanase treated r-EPO resulted in two new distinct peptides at the expense of all the N-linked glycopeptides, which was consistent with the predicted tryptic mapping.

Inspection of the electropherogram of the tryptic digest of r-EPO showed great diversity in the glycopeptides region with the presence of many minor peaks, presumably due to further modifications of the sugar moieties. Attempts to characterize further the glycopeptides after CE separation were made [25]. Since the peptide mapping technique by CE method is reproducible, sequential collection of the peptides by CE using a 75 µm i.d. capillary could be achieved in less than four repeated runs, and a collective yield of 30 to 40 pmol pure fraction was obtained, enough for Edman sequence analysis and for matrix assisted laser desorption ionization (MALDI) mass spectrometry.

15.4 Carbohydrate analysis

Microheterogeneity due to the complex carbohydrates in recombinant proteins has been evaluated by many different analytical methods [7, 8, 11–16, 23–25]. Structural analysis of carbohydrates is extremely tedious and costly, and requires great expertise. Regardless of the analytical approaches, the sugar moiety on the protein has to be cleaved enzymatically or chemically into smaller manageable species in the form of glycopeptides or oligosaccharides. Direct analysis of oligosaccharides released from r-EPO by the CE method was explored by CE with UV detection at 190–200 nm [26]. The r-EPO derived from BHK cells was treated with peptide N-glycosidase (PNGase) to yield oligosaccharide mixtures. Direct CE analysis of the resulting mixture is shown and seven distinct species were evident from the electropherogram given by Hermentin et al. [26]. From the database of the calibrated migration times of the structurally well-characterized N-linked oligosaccharides, the carbohydrate structure of N-linked oligosaccharide species in r-EPO could be tentatively assigned.

Evangelista et al. [27] explored carbohydrate analysis using CE with laser-induced fluorescence (LIF) detection. Reductive amination of sugars with

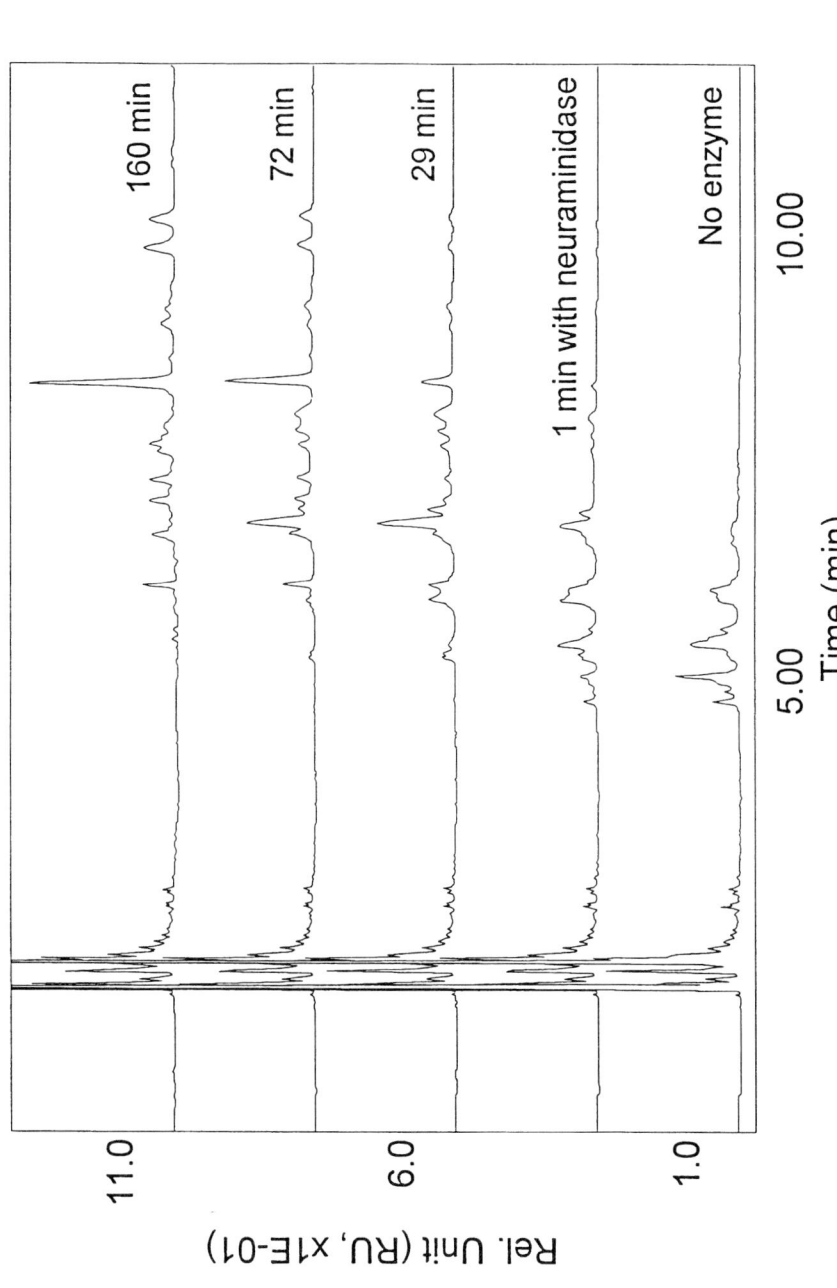

Figure 15.2 Electropherograms of APTS derivatized *N*-linked oligosaccharides of r-EPO and its desialylation catalyzed by neuraminidase. Column: untreated fused-silica of 19 μm × 25 cm (18 cm to detector); LIF with ex: 488 nm/em: 520 nm. Buffer: 0.15 M sodium phosphate, pH 2.5; 15 kV/25 μA.

Table 15.1 Erythropoietin analysis by capillary electrophoresis

Methods/detection Separation conditions	Analytes/matrix	Buffers/additives	Ref.
CZE / UV detection: 200 nm 75 μM × 27 cm (20 cm), 20 kV/18 μA; 25°C	Human r-EPO (CHO)	100 mM acetate–phosphate pH 4.0, equilibrate for >2 h	17
CZE / UV detection: 214 nm 75 μm × 57 cm (50 cm), 10 kV	Human r-EPO (CHO)	10 mM Tricine, 10 mM NaCl, 2.5 mM 1,4-diaminobutane pH 6.2	20
CZE / UV detection: 214 nm 100 μm × 27 cm (20 cm), 8.1 kV; 20°C	Human r-EPO (CHO), human r-EPO (*E. coli*)	100 mM Tris–CHES, 0.1% SDS, 3% PEO 100 kDa pH 8.5	21
CZE / UV detection: 214 nm 50 μm × 75 cm (50 cm), 16 kV/110 μA; 30°C	Tryptic digest of human r-EPO (CHO)	40 mM phosphate, 0.1 M heptanesulfonic acid, pH 2.5	23
CZE / UV detection: 194 nm 50 μm × 107 cm (100 cm), 30 kV; 30°C	Human r-EPO (CHO), N-linked oligosaccahrides	80 mM ammonium sulfate 20 mM sodium phosphate, 2.0 mM 1,5-diaminopentane, pH 7.0	26
CZE / UV detection: 214 nm 75 μm × 75 cm (50 cm), 9.4 kV/147 μA; 30°C	Tryptic digest of human r-EPO (CHO)	40 mM phosphate, 0.1 M heptanesulfonic acid, pH 2.5	25
CZE / LIF: ex:488 nm/em:520 nm 20 μm × 27 cm (20 cm), 15 kV/31 μA; 22°C	APTS derivatized sugars from PNGase F treated human r-EPO (CHO)	150 mM phosphate, pH 2.5	28

8-aminopyrene-1,4,6-trisulfonate (APTS) results in the derivatized product mixture that can be analyzed by CE/LIF. A derivatization and detection sensitivity of 2 pmol maltoheptaose with APTS was demonstrated. The N-linked oligosaccharides released from 50 μg (1.0 nmol) of the r-EPO (CHO) were reductively aminated with APTS to yield APTS–oligosaccharide mixtures. CE/LIF analysis of this mixture is shown in Figure 15.2 [28]. Neuraminidase treatment of this mixture resulted in gradual migration in the electropherograms, indicating a slower release of sialic acid from the N-linked oligosaccharides of the r-EPO. The front major peaks are derived from excess reagent background. A major species migrating at 8 min is presumably a single desialylated oligosaccharide that originally derived from fairly heterogeneous sialylation during the biosynthetic process.

15.5 Future trends in CE analysis of r-EPO

Recombinant therapeutic proteins such as human tissue plamin activator (tPA), Factor VIII and erythropoietin were approved with limited structural information about the carbohydrate. Study of the structure and function relationship of carbohydrates in r-EPO clearly indicated that the structure of oligosaccharides in r-EPO from different cell lines may vary. In a cell-line B8-300, the r-EPO produces a drastically reduced *in vivo* activity with the oligosaccharides rich in biantennary sugar chains while the highly active r-EPO contains mostly tetraantennary oligosaccharides [10]. CE analysis of glycopeptides has proved to be an extremely reliable method for monitoring the distribution of carbohydrate in r-EPO. The analysis of oligosaccharides in r-EPO by CE should provide a simple method to access such structural information.

References

1 Goldwasser, E.; Kung, C.K.H. (1968) Chemistry and purification of erythropoietin: progress in the purification of erythropoietin. *Ann. N.Y. Acad. Sci.* **149**, 49–53.
2 Jacobson, L.O.; Goldwasser, E.; Fried, W.; Plzak, L. (1957) Role of kidney in erythropoiesis. *Nature* **179**, 633–634.
3 Fried, W. (1972) The liver as a source of extrarenal erythropoietin production. *Blood* **40**, 671–677.
4 Zanjani, E.D.; Ascensao, J.L.; McGlave, P.B.; Banisadre, M.; Ash, R.C. (1981) Studies on the liver to kidney switch of erythropoietin production. *J. Clin. Invest.* **67**, 1183–1188.
5 Miyake, T.; Kung, C.K.H.; Goldwasser, E. (1977) Purification of human erythropoietin. *J. Biol. Chem.* **252**, 5558–5564.
6 Lin, F.K.; Suggs, J.; Lin, C.H.; Browne, J.; Smalling, R.; Egrie, J.C.; Chen, K.K.; Fox, G.M.; Martin, F.; Stabinsky, Z.; Badrawi, S.M.; Lai, P.S.; Goldwasser, E. (1985) Cloning and expression of human erythropoietin gene. *Proc. Natl Acad. Sci. USA* **82**, 7580–7584.
7 Sasaki, H.; Brothner, B.; Dell, A.; Fukuda, M. (1987) Carbohydrate structure of erythropoietin expressed in Chinese hamster ovary cells by a human erythropoietin cDNA. *J. Biol. Chem.* **262**, 12059–12076.
8 Takeuchi, M.; Takasaki, S.; Miyaki, H.; Kato, T.; Hoshi, S.; Kochibe, N.; Kobata, A. (1988) Comparative study of the asparagine-linked sugar chains of human erythropoietin purified from urine and culture medium of recombinant Chinese hamster ovary cells. *J. Biol. Chem.* **263**, 3657–3663.
9 Goldwasser, E.; Kung, C.K.H.; Ellason, J. (1974) On the mechanism of erythropoietin-induced differentiation. *J. Biol. Chem.* **249**, 4202–4206.
10 Kobata, A. (1992) Structures and functions of the sugar chains of glycoproteins. *Eur. J. Biochem.* **209**, 485–501.
11 Takeuchi, M.; Inoue, N.; Strickland, T.W.; Kubota, M.; Wada, M.; Shimizu, R.; Hoshi, S.; Kozutsumi, H.; Takasaki, S.; Kobata, A. (1989) Relationship between sugar chain structure and biological activity of recombinant human erythropoietin produced in Chinese hamster ovary cells. *Proc. Natl Acad. Sci. USA* **86**, 7819–7822.
12 Rice, K.G.; Takahashi, N.; Namiki, Y.; Tran, A.D.; Lisi, P.J.; Lee, Y.C. (1992) Quantitative mapping of the N-linked sialyloligosaccharides of recombinant erythropoietin: combination of direct high-performance anion-exchange chromatography and 2-aminopyridine derivatization. *Anal. Biochem.* **206**, 278–287.
13 Watson, E.; Bhide, A.; Van Halbeck, H. (1994) Structural determination of intact major sialylated oligosaccharide chains of recombinant human erythropoietin expressed in Chinese hamster ovary cells. *Glycobiology* **4**, 227–237.
14 Tsuda, E.; Goto, M.; Murakami, A.; Akai, K.; Ueda, M.; Kawanishi, G.; Takahashi, N.; Sasaki, R.; Chiba, H.; Ishihara, H.; Mori, M.; Tejima, S.; Endo, S.; Arata, Y. (1988) Comparative structural study of the N-linked oligosaccharides of urinary and recombinant erythropoietins. *Biochemistry* **27**, 5646–5654.
15 Nimtz, M.; Martin, W.; Wray, V.; Kloppel, K.D.; Augustin, J.; Conradt, H.S. (1993) Structures of sialylated oligosaccharides of human erythropoietin expressed in recombinant BHK-21 cells. *Eur. J. Biochem.* **213**, 39–56.
16 Ohashi, H.; Miyata, M.; Ishi, Y.; Takeuchi, M.; Takasago, A.; Suzuki, T.; Sudo, T. (1989) Purification and characterization of human recombinant erythropoietin expressed in human cervix carcinoma HeLa cells. *Trends Anim. Cell Culture Technol.* **1**, 115–120.

17. Tran, A.D.; Huynh, C.T.; Park, S.; Ryall, R.R.; Lisi, P.J.; Lane, P.A. (1991) Separation of carbohydrate-mediated microheterogeneity of recombinant human erythropoietin by free solution capillary electrophoresis. *J. Chromatogr.* **542**, 459–471.
18. McCormick, R.M. (1988) Capillary zone electrophoretic separation of peptides and proteins using low pH buffers in modified silica capillaries. *Anal. Chem.* **60**, 2322–2328.
19. Chen, F.-T.A.; Kelly, L.; Palmieri, R.; Biehler, R.; Schwartz, H. (1992) Use of high ionic strength buffers for the separation of proteins and peptides with capillary electrophoresis. *J. Liq. Chromatogr.* **15**, 114–119.
20. Watson, E.; Yao, F. (1993) Capillary electrophoretic separation of human recombinant erythropoietin (r-HuEPO) glycoforms. *Anal. Biochem.* **210**, 389–393.
21. Benedek, K.; Thiede, S.J. (1994) High-performance capillary electrophoresis of proteins using sodium dodecylsulfate–poly(ethylene oxide). *J. Chromatogr.* **676**, 209–217.
22. Narhi, L.O.; Arakawa, T.; Aoki, K.H.; Elmore, R.; Rohde, M.F.; Boone, T.; Strickland, T.W. (1991) The effect of carbohydrate on the structure and stability of erythropoietin. *J. Biol. Chem.* **266**, 23022–23026.
23. Rush, R.S.; Derby, P.L.; Strickland, T.W.; Rohde, M.F. (1993) Peptide mapping and evaluation of glycopeptide microheterogeneity derived from endoproteinase dogestion of erythropoietin by affinity high-performance capillary electrophoresis. *Anal. Chem.* **65**, 1834–1842.
24. Sasaki, H.; Ochi, N.; Dell, A.; Fukuda, M. (1988) Site-specific glycosylation of human recombinant erythropoietin expressed in Chinese hamster ovary cells. *Biochemistry* **27**, 8618–8626.
25. Boss, H.J.; Rohde, M.F.; Rush, R.S. (1995) Multiple sequenctial fraction collection of peptides and glycoproteins by high-performance capillary electrophoresis. *Anal. Biochem.* **230**, 123–129.
26. Hermentin, P.; Doenges, R.; Witzel, R.; Hokke, C.H.; Vliegenthart, J.P.G.; Kamerling, J.P.; Conradt, H.S.; Nimtz, M.; Brazel, D. (1994) A strategy for mapping *N*-glycans by high-performance capillary electrophoresis. *Anal. Biochem.* **221**, 29–41.
27. Evangelista, R.A.; Liu, M.-S.; Chen, F.-T.A. (1995) Characterization of 9-aminopyrene-1,4,6-trisulfonate derivatized sugars by capillary electrophoresis with laser-induced fluorescence detection. *Anal. Chem.* **67**, 2239–2245.
28. Evangelista, R.A.; Chen, F.-T.A.: unpublished results. We thank Dr Fu-Kuen Lin of Amgen for providing the r-EPO sample.

16 Assay of enzymes by capillary electrophoresis
F.-T.A. CHEN AND R.A. EVANGELISTA

16.1 Introduction

Enzymes are mostly proteins with specific catalytic activity corresponding to their substrates. They are essential components in living cells and a substantial part of modern biochemistry and molecular biology is devoted to their study. Nearly all biochemical reactions occur in complex mixtures with great speed through the mediation of enzymes with a high degree of specificity and great efficiency. In contrast to an antibody which binds specifically to the ground state of a molecule, it is generally recognized that enzymes bind to a transition state of a chemical reaction, which lowers the activation energy of the reaction pathway and consequently significantly increases the reaction rate. The catalytic efficiency of most enzymes is extremely high such that on a molar basis, they can catalyze the transformation of 10^3 to 10^6 moles of the substrate per minute. Enzymes provide some of the best chemical amplification processes in nature which have been used in many practical applications, including enzyme assay, enzyme-linked immunoassay (ELISA) and DNA probe assays.

The utility of enzymatic analysis is based on the unique structural binding specificity of the enzyme to the substrate and its high turnover rate. There are numerous chemical reactions that may be catalyzed by enzymes, and such reactions are widely used in biochemical and clinical analysis of substrates or for specific enzyme assays. The application of enzymatic analysis involves measurement of the change in substrate and product concentrations. Such changes, either in substrate or product, may be monitored by physical or chemical means. Enzyme activity may be measured from the rate of concentration changes in the substrate or product [1]. Capillary electrophoresis (CE) is proving to be a very powerful tool for separation of analytes based on the charge-to-mass ratio, and it is ideal for monitoring both reactant and product simultaneously for an enzyme catalyzed reaction.

Jorgenson and Lukacs first reported the use of CE with fluorescence detection to resolve fluorescamine-labeled peptides obtained from a tryptic digest of reduced and carboxymethylated lysozyme [2]. The short analysis time and extremely high resolution of the CE technique laid the foundations for CE as one of the most powerful tools in analytical chemistry. Methods of enzyme activity measurement and substrate identification and quantitation in complex biological mixtures by various CE techniques have been developed recently. This review describes the application of enzyme-based analysis by CE, and is categorized into off-line, on-line immobilized, postcolumn and on-line enzymatic reactions, summarized in Table 16.1.

16.2 Off-line analysis of enzyme catalysis and applications

CE can easily be adapted for assay of enzyme activity. The kinetics of reactant and product changes are monitored simultaneously by CE separation from an off-line enzymatic reaction mixture [3–21]. Pascual et al. [3] reported the assay of glutathione peroxidase activity based on the separation and quantitation of reduced and oxidized glutathione species simultaneously by the CE separation technique. Landers et al. [4] explored the potential use of CE for monitoring chloramphenicol acetyltransferase activity. Off-line enzymatic reaction mixtures were analyzed at defined time intervals by CE. The CE method yielded quantitative results similar to those obtained by radioisotope methods at a substantially reduced assay time and reagent consumption. Mulholland et al. [5] described the use of a synthetic peptide to monitor the tripeptidase activity specifically using CE with UV detection where the conventional protease assay method could not discriminate di- and tripeptidase activities. Vinther et al. [6] demonstrated an elegant CE procedure to monitor the process of a transpeptidation reaction of a recombinant peptide catalyzed by carboxypeptidase Y. The enzyme-catalyzed replacement of the C-terminal amino acid from an alanine to arginyl amide produced a substantial change in electrophoretic mobility. Lowther and Dunn [7] studied the kinetics of pepsin-catalyzed oligopeptide cleavage. The resulting peptide fragments with different charge-to-mass ratios can be readily separated from the original peptide substrate by the CE method. Dawson et al. [8] studied a novel CE procedure for detection and assay of the activities of protein kinases and phosphatases in rabbit skeletal muscle tissue extracts using well-defined peptide substrates. The enzyme-catalyzed multiple phosphorylation sites on the peptide substrate were well-separated and characterized by CE. The sensitivity of the CE-based protein kinase assay was about ten-fold less than that of the radioisotope assay. Similarly, a CE assay of adenosine diphosphate–glucose pyrophosphorylase activity was demonstrated by Roberts et al. [9]. Here enzyme catalysis involved transfer of the adenosine diphosphate moiety from adenosine-5′-triphosphate (ATP) to glucose-1-phosphate (G-1-P) to produce adenosine diphosphate-glucose (ADP-G) and pyrophosphate (PP). Using CE, the kinetics of changes in ADP-glucose and ATP could easily be obtained, and the method is simpler than the traditional method of using radioactive ATP-γ^{32}P substrate.

Table 16.1 Enzyme analysis by capillary electrophoresis

Methods Separation conditions Voltage applied, current Capillary length with i.d. (effective length) Detection	Analytes	Matrix	Ref.
MECC 0.1 M Sodium tetraborate, 0.1 M SDS, pH 8.2 16 kV/130 μA; 8 kV/45 μA; 30°C 75 μm × 27.5, 57.5 cm (20, 50 cm) 200 nm	Glutathione peroxidase in rat liver extract or calf hemolysate, glutathione as the substrate	Substrates in 50 mM phosphate, 0.5 mM EDTA, pH 7.2	3
CZE 0.1 M Na borate/tetraborate, pH 8.3 25 kV/32 μA; 25°C 50 μ × 57 cm (50 cm) 200 nm	Chloramphenicol acetyl transferase (CAT), purified enzyme from *E. coli*	Substrate in 20 mM Tris–HCl, pH 7.5	4
CZE 20 mM Na citrate, pH 2.5 30 kV; 45°C 50 μm × 72 cm (47 cm) 200 nm	Tripeptidase, partially purified from *Lactococcus* synthetic peptides as substrates	Substrate in final 6.6 mM Na phosphate, pH 7.5	5
CZE 0.1 M CAPS, pH 11.0 25 kV; 30°C 50 μm × 75 cm (50 cm) 200 nm	Carboxypeptidase Y transpeptidase activity recombinant peptide as substrate	r-Peptide in 1 M Arginine–NH$_2$ as buffer at pH 6.63–8.49	6
CZE 0.5 M Sodium phosphate, pH 2.5 8 kV; 15°C 25 μM × 24 cm, coated 200 nm	Pepsin synthetic oligopeptides as substrates	Oligopeptides in 0.1 M Na formate pH 3.5	7
CZE 0.15 sodium phosphate, pH 2.5 and 5.0 20 kV; 25°C 50 μm × 57 cm (50 cm) 200 nm	Protein kinases and phosphatase both with synthetic peptides	Synthetic peptide in 56 mM Tris–HCl, 25 mM Mg acetate, 0.56 mM ATP pH 7.5	8
CZE 0.4 M sodium borate, 1.0 M triethylammonium-propane sulfonate, 2 mM EDTA, pH 8.5 15 kV/100 μA; ambient temp. 75 μm × 50 cm 254 nm	Adenosine diphosphate-glucose pyrophosphorylase ATP and gluc-1-P are substrates	Substrates in 80 mM glycylglycine 5 mM MgCl$_2$, 5 mM DTT, 0.5 mg BSA pH 7.5	9
CZE 0.2 M sodium borate, pH 10.2 20 kV/30 μA; 22°C 20 μm × 27 cm (20 cm), LIF: ex: 543 nm/em: 590 nm	Proteases Cy3-labeled angiotensin II	Labeled substrate in 50 mM phosphate pH 6.7	10
CZE 0.1 M sodium borate, pH 10.2 20 kV/19 μA; 22°C 20 μm × 27 cm (20 cm), LIF: ex: 488 nm/em: 520 nm	Neuraminidase APTS-oligosaccharides	Labeled substrate in 0.2 M Na acetate pH 5.0	11
CZE 10 mM phosphate, 10 mM borate, 10 mM SDS, pH 9.3 29 kV 10 μm × 70 cm LIF: ex: 543 nm/em: 590 nm	Fucosyltransferase and fucosidase, oligosaccharide-rhodamine	Labeled substrate in 20 mM HEPES, 20 mM MnCl$_2$, 0.2% BSA, pH 7.0	12
CZE 20 mM tricine, pH 8.1 400 V cm^{-1} 10 μm × 45 cm LIF: ex: 488 nm/em: 520 nm	β-Galactosidase fluorescein di-β-galactopyranose	Labeled substrate in 20 mM Tricine pH 8.1	13

Table 16.1 *Continued*

Methods Separation conditions Voltage applied, current Capillary length with i.d. (effective length) Detection	Analytes	Matrix	Ref.
CGE 89 mM Tris/89 mM boric acid, 2 mM EDTA, 0.5% hydroxypropylmethyl cellulose, pH 8.3 175 V cm^{-1}; 25°C 100 μm × 47 cm (40 cm), coated LIF: ex: 488 nm/em: 520 nm	*Taq* DNA polymerase, PCR product	TBE buffer	14
CGE 89 mM Tris/89 mM boric acid, 2 mM EDTA, 3% polyacrylamide, pH 8.5 625 V cm^{-1}; 22°C 75 μm × 47 cm (40 cm), coated LIF: ex: 543 nm/em: 590 nm	*Taq* DNA polymerase, PCR product	TBE buffer	15
CGE 89 mM Tris/89 mM boric acid, 2 mM EDTA, 3% polyacrylamide, pH 8.5 625 V cm^{-1}; 22°C 75 μm × 47 cm (40 cm), coated LIF: ex: 633 nm/em: 665 nm	*Taq* DNA polymerase, PCR product, Cy5-labeled primers	TBE buffer	16
CZE 20 mM phosphate, pH 9.0 5 kV; 22°C 75 μm × 57 cm (50 cm) 200 nm	Glycosidase, ovine lentivirus	Glycoproteins in 0.2 M phosphate– citrate buffer pH 5.5	17
CZE 10 mM tricine, 10 mM NaCl, 6 M urea, 2.5 mM 1,4-diaminobutane, pH 6.2 10 kV; 25°C 75 μm × 57 cm (50 cm) 200 nm	Neuraminidase, human r-EPO	Glycoproteins in acetate buffer pH 5.0	18
CZE 20 mM phosphate, 27 mM dimethyl- β-cyclodextrin, pH 2.3 15 kV; ambient temp. 75 μm × 57 cm (50 cm) 214 nm	Cytidine deaminase, 2′-deoxy-3′-thiacytidine	Immobilized enzyme automated adjust pH with 20% acetic acid to constant pH 7.0	19
CZE 20 mM borate buffer, pH 9.4 25 kV; ambient temp. 50 μm × 88 cm (50 cm) 200 nm	γ-glutamyl transpeptidase off-line γ-GTP activity on PVDF membranes	Purified enzyme substrate mixture impregnated on PVDF membrane	28
CZE 10 mM phosphate, pH 7.0 250 V cm^{-1}; ambient temp. 75 μm × 60 cm (50 cm) Time-resolved LIF	Leucine aminopeptidase, leucine 4-methoxy- β-naphthylamine	Purified enzyme	30
CZE 192 mM glycine, 25 mM Tris, 0.2 mM G-6P 30 kV/10 μA; 25°C 50 μm × 75 cm (50 cm) 260 nm	G-6P-dehydrogenase, 0.2 mM NAD in buffer	Purified enzyme	33
CZE 38 mM AMPD; 23 mM aspartic acid, 25 mM hydroxyethylglycine, pH 8.5 30 kV/10 μA; 25°C 50 μM × 75 cm (50 cm) 260 nm	Lactate dehydrogenase, isozymes with 4 mM NAD and 200 mM L-lactate	Mixture of partially purified enzymes	34

ADP = adenosine-5′-diphosphate, AMPD = 2-amino-2-methyl-1,3-propanediol, APTS = 9-aminopyrene-1,3,6-trisulfonic acid, ATP = adenosine-5′-triphosphate, CAPS = 3-(cyclohexylamino)propane sulfonic acid, CAT = chloramphenical acetyl transferase, CZE = capillary zone electrophoresis, Cy3 and Cy5 = cyanine dyes that fluoresce at 590 and 670 nm, respectively, dsDNA = double-stranded DNA, DTT = dithiothreitol, EOF = electroosmotic flow, EPO = erythropoietin, HEPES = 4-(2-hydroxyethyl)piperazine-1-ethane sulfonic acid, LIF = laser-induced fluoresence, MECC = micellar electrokinetic chromatography, PCR = polymerase chain reaction, PVDF = poly(vinylidene difluoride), SDS = sodium dodecyl sulfate, TBE = 89 mM Tris, 89 mM boric acid, 2 mM EDTA, pH 8.3, TO = thiazole orange.

The above methods rely on the use of a well-defined substrate in a clean medium with very little potential analytical interference from substances in the reaction medium. It is essential to consider the sample matrix effect. To avoid interference from the sample matrix, Chen [10] designated a cyanine (Cy3) labeled angiotensin at the α-N-terminal as a stable substrate for probing catalysis of endo- and exopeptidases. The protease-catalyzed hydrolytic process of Cy3-labeled angiotensin was monitored by CE with laser-induced fluorescence (LIF) detection. Using a combination of specificity of endopeptidases and carboxypeptidases, the remnants of the original labeled species were characterized by CE/LIF. The method provides a general tool for studying the mechanism of protease-catalyzed hydrolysis of peptide. Similarly, a flurorescent-labeled sialyllactose was used for probing the specificity of sialidase activity by CE/LIF as shown in Figure 16.1 [11]. Zhao et al. [12] described the use of tetramethylrhodamine-labeled synthetic disaccharide as the substrate to monitor fucosyltransferase activity. Detection of 100 molecules of the prod-

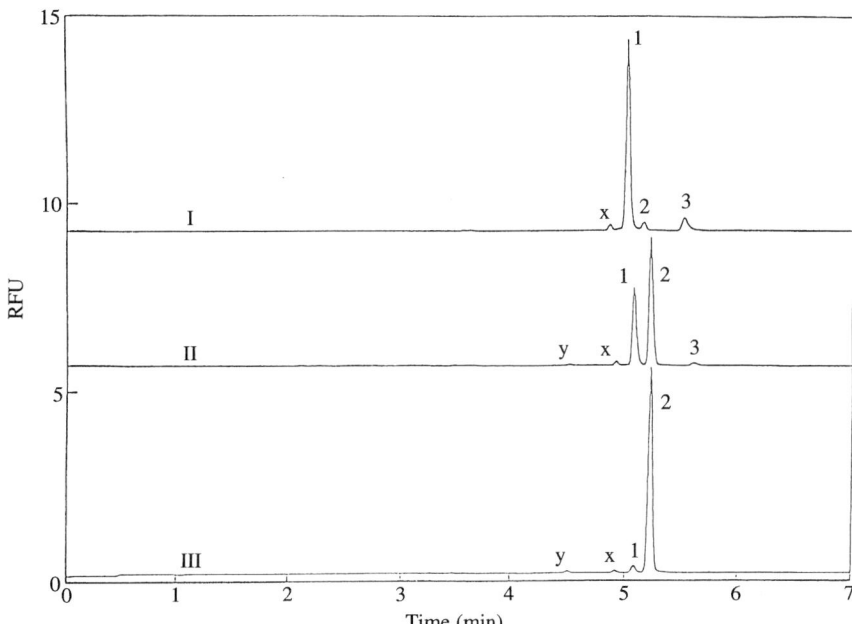

Figure 16.1 Electropherograms of APTS derivatized sialyllactose. Buffer: 140 mM MOPS, pH 7.0. Outlet: cathode. Applied potential, 25 kV/31 μA. (I) 4 μM APTS–sialyllactose (200 μl). (II) CE/LIF at 2 min after the addition of 2 μl neuraminidase (4 mU). (III) CE/LIF at 102 min after the addition of 2 μl neuraminidase (4 mU). Peak assignments: (1) Neu-5-Ac(2-3)-β-D-gal-(1-4)-D-glc-APTS; (2) L-lactose-APTS; (3) Neu-5-Ac(2-6)-β-D-gal-(1-4)-D-glc-APTS; (x) Neu-5-Ac(2-6)-β-D-gal-(1-4)-glcNAc-APTS. (y) N-acetyllactosamine. Reprinted with permission of Academic Press from Ref. [11].

uct, a tetramethylrhodamine-labeled trisaccharide, by CE with LIF detection was reported. Similarly, a few molecules of β-galactosidase were detected with fluorescein-di-β-galactopyranoside as the substrate and an appropriate off-line incubation time of the substrate with the enzyme [13]. Both substrate and product were analyzed by CE/LIF.

Analysis of the double-stranded DNA (dsDNA) product of polymerase chain reaction (PCR) by CE with direct UV detection at 254 nm showed that the dsDNA product is usually dwarfed by the excess primers. Schwartz and Ulfelder developed a highly sensitive detection method in CE analysis of dsDNA using thiazole orange (TO) as the intercalator dye [14]. Thiazole orange intercalates specifically with dsDNA and fluoresces intensely at 530 nm when excited by a 488 nm argon ion laser, while the free TO fluoresces poorly with emission at 580 nm. Using CE with LIF detection, the dsDNA PCR products and DNA restriction fragments were characterized and detected at nanogram per ml level [15]. Liu et al. [16] described the use of Cy5-labeled primers for PCR amplification of a Y chromosome-specific sequence using CE/LIF detection. A million-fold dilution of male DNA sample in female DNA could be PCR-amplified and detected.

The structural property of proteins can be probed with an enzyme that selectively catalyzes the removal of a specific structural moiety in combination with the CE separation technique [17, 18]. Glycoproteins often exist as heterogeneous species due to the microheterogeneity of oligosaccharides linked to the protein. Schmerr and Goodwin [17] reported that the antigenic property of the ovine lentivirus was associated with the surface sugar moiety of the glycoprotein. Using CE, the change in the electrophoretic mobility of the glycoprotein before and after treatment with glycosidic enzymes suggest that carbohydrate moieties contribute significantly to the surface charge of the lentiviruses. Watson and Yao [18] reported the separation of the recombinant erythropoietin (r-EPO) glycoforms by free-zone CE with baseline resolution. Treatment of the r-EPO with neuraminidase resulted in the release of sialic acids from the non-reducing end of the oligosaccharides in r-EPO which produced a decrease in electrophoretic mobility. The isoforms of r-EPO are due to variation in the number of sialic acids in both N-linked and O-linked oligosaccharides, and the CE migration is consistent with the degree of sialylation of the isoforms.

The use of CE to monitor the enzymatic transformation of a racemic mixture of an antiviral agent, 2′-deoxy-3′-thiacytidine (BCH 189) by cytidine deaminase using dimethyl-β-cyclodextrin (CD) as a chiral electrolyte additive in the running buffer system was demonstrated [19]. The enantiomers were separated with baseline resolution, and the cytidine deamination-catalyzed reaction resulted in the specific conversion of the (+) isomer to the uridine analog only, and a desirable product was retained.

Electrophoretically, the uridine analog migrated at a substantially longer time than the (−) isomer.

16.3 On-line immobilized enzymes and their applications

A capillary containing an enzyme immobilized on the inlet inner wall was used as a microreactor for the enzyme-catalysed reaction [22–24]. Nashabeh and El Rassi [22] reported the use of CE to monitor on-line catalytic hydrolysis of transfer RNA by the chemically immobilized ribonuclease T1. Immobilized protease was explored for on-line peptide mapping of β-casein [23], oxidized insulin B-chain and $α_1$-acid glycoprotein [24]. Enzymes are biotinylated and immobilized on the avidin anchored on the silica surface via silanyl moiety terminated with biotin. Reproducible peptide mapping by on-line CE analysis was obtained using immobilized carboxypeptidase Y for digestion of the oxidized insulin B-chain.

Yoshimoto et al. [25] demonstrated the application of CE techniques to study the kinetics of immobilized β-galactosidase. The enzyme was chemically immobilized by attachment the interior of the entire length of the capillary. The Michaelis–Menten constant (K_m) of the immobilized enzyme could be obtained by on-line introduction of a slug of substrate (o- or p-nitrophenyl β-galactoside) at varying concentrations, followed by CE separation of the product from the reactant.

Miller and Lytle [26] reported a technique for introducing a spherically porous particle into a restricted capillary at the inlet side of a CE system. A cell was trapped on the particle and the enzymatic activity of the cell was characterized by introducing a slug of substrate and incubated for a short period. The product was electrophoresed and the enzyme profile could be catagorized using different substrates.

16.4 Postcolumn enzyme assay

A postcolumn microreactor for on-line enzyme assay was attempted by Emmer and Roeraade [27] for CE analysis of glucose-6-phosphate dehydrogenase (G-6-PDH) and 6-phosphogluconin acid dehydrogenase (6-PGDH). Two on-line UV detectors were employed for the CE assay of enzymes. The first detector in an on-column UV detector to monitor the CE separation, while the second detector is mounted downstream after mixing the the enzyme with the substrate through a tee (T) of three-way joint tube to monitor the reaction products. Significant band broadening in postcolumn mixing at the T joint of the substrate with the enzyme was expected, and the time for the enzyme to act on the substrate was limited before the product reached the second detector owing to the very short period allowed for the reaction of enzyme and substrate. The T joint section is most necessary for postcolumn reaction. Much experience, know-how and skillful techniques are required to attain a successful result. A membrane pressure injection method was reported as an alternative method to the T joint. This will give easier control for postcolumn injection. A minimum detectable G6-PDH concentration was shown to be 0.2 mM, about 20 IU ml^{-1}.

Konse et al. [28] designed a membrane collection system in the outlet of the CE system. The membrane was used as a conveyer and the eluent from the CE separation was collected and moved to another spot on the membrane every 1 min. After completion of the CE separation of glutamyltranspeptidase (g-GTP), the membrane was incubated with the substrate to reveal the enzyme activity and quantitated off-line.

16.5 On-line analysis of enzyme activity and applications

A capillary may be used as a 'microreactor' where a plug of the enzyme is introduced in the capillary while the substrate or reactant may be injected into the capillary as a separate plug or be included in the electrophoresis buffer [29–36]. Bao and Regnier [29] reported a G-6-PDH enzyme assay system where the substrate G-6P and the coenzyme NAD (nicotinamide adenine dinucleotide) were included in the running buffer. During the electrophoretic separation process, substrates and coenzymes were in bulk solution and in contact with the migrating enzyme. The product produced by enzymatic reation continued moving at a different pace from the enzyme, leaving a constant trail of the product. When the applied potential was interrupted, the enzyme stopped in a discrete zone and a product was generated by the catalytic conversion of substrate and coenzyme by G-6-PDH and accumulated with time. One of the products, NADH, could then be electrophoresed and quantitated. Normally NADH is detected by UV at 340 nm. When sensitive detection is required, LIF detection is used as an alternative. The method is referred to as electrophoretically mediated microassay (EMMA). Leucine aminopeptidase activity in serum and urine samples was assayed by a method using L-leucine 4-methoxy-β-naphthylamide as the substrate, and the reaction product was monitored by a time-resolved LIF detection [30]. Similarly, alkaline phosphatase activity was assayed by EMMA using electrochemical and spectrophotometric detection [31]. Wu and Regnier [32] improved the assay by using capillary gel electrophoresis (CGE) for the separation of alkaline phosphatase (ALP) to reduce diffusion in the product zone during the incubation period. Detection of ALP at 7.6×10^{-12} M in gel-filled EMMA was reported.

Avila and Whitesides [33] demonstrated an enzyme assay by sequential introduction of plugs of glucose-6-phosphate dehydrogenase (G-6-PDH) and NAD as the coenzyme. In between the sample plug injections, G-6-PDH was electrophoresed for a short period to provide a small volume of buffer. Electrophoresis was initiated in a buffer containing glucose-6-phosphate (G-6P). NAD migrated through the zone of G-6-PDH, which resulted in the formation of NADH during the limited time of contact. The series of electropherograms in Figure 16.1 showed that the extent of conversion of NAD to NADH was a function of the amount of G-6-PDH in the enzyme plug [11]. The relatively sharp peaks of NAD and NADH indicate that dilution and diffusion in the contact zone where enzymatic reaction occured were insignificant.

Analysis of isozyme using a CE technique similar to that of the method described above was disclosed by Hsieh *et al.* [34]. Samples containing isozymes were electrophoresed in a buffer with a substrate for a predetermined period. When the isozymes could physically be separated into discrete zones, parking of the resulting separated system provided the enzymatic reaction in each zone. The products generated by the catalytic reaction in each zone could be separated and detected. Such isozyme assay systems using the CE technique may potentially be applied to the clinically important isoenzyme analysis such as the creatine kinase (CK), lactate dehydrogenase (LD) and alkaline phosphatase (ALP) isozyme assays. An example of LD isozyme assay is shown in Figure 16.2, where one of the LD reaction products, NADH, was monitored at 340 nm [34].

Xue and Yeung [35] applied a strategy similar to that of Hsieh *et al.* [34] to assay isozymes of lactate dehydrogenase (LD) in single human erythrocytes. The analysis of the LD-catalyzed reaction product, NADH, was detected by LIF using a 4 mW of combined 350–360 nm radiation from an argon ion laser, and fluorescence emission at 480 nm. In a later report by Tan and Yeung [36] on the assay of G-6-PDH enzyme activity, CE/LIF was used to monitor NADPH formation in the sample plug.

16.6 Future trends in CE analysis of enzymes

CE-based enzyme assays have been shown to be an extremely promising tool in method development for enzyme assays. With an automated system, CE-based enzyme assays are very simple to perform, highly reproducible and substantially more rapid with higher sensitivity than the conventional enzyme assay methods. The sequential nature of the CE instruments however, limits its throughput and future CE instruments with multicapillary arrays and microfabricated channels system will undoubtedly improve the analytical output.

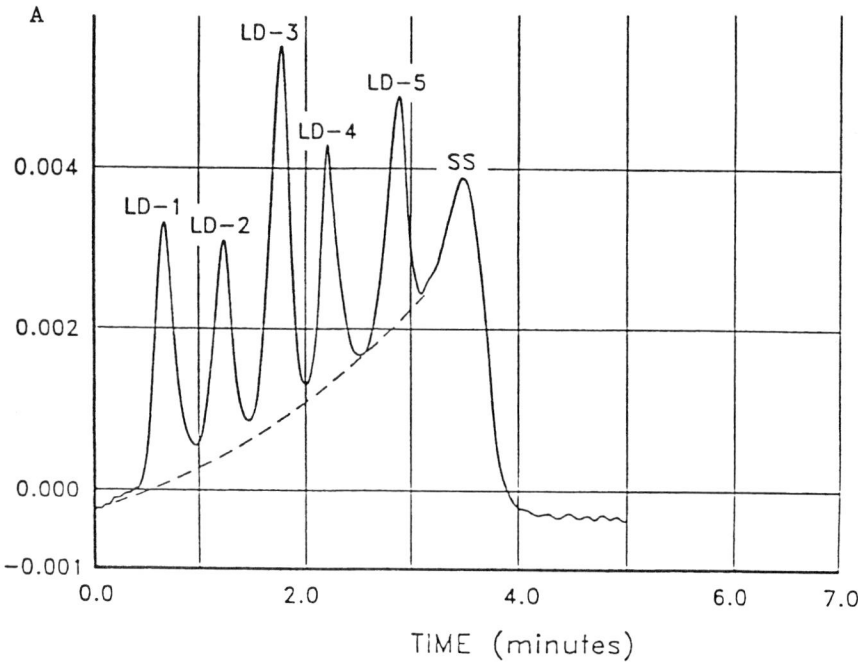

Figure 16.2 CE/UV of on-line enzyme analysis of human lactate dehydrogenase isozymes. SS, sample shock peak. Reprinted with permission of Beckman Instruments from Ref. [34].

References

1. Bergmeyer, H.U. (1974) *Methods of Enzyme Analysis*, Vol. 1, 2nd edn Engl., Verlag Chemie International, Deefield Beach, FL.
2. Jorgenson, J.W.; Lukacs, K.D. (1981) Zone electrophoresis in open-tubular glass capillaries: preliminary data on performance. *J. High Resolut. Chromatogr. Commun.* **4**, 230–231.
3. Pascual, P.; Martinez-Lara, E.; Barcena, J.A.; Lopez-Barea, J.; Toribio, F.J. (1992) Direct assay of glutathione peroxidase activity using high-performance capillary electrophoresis. *J. Chromatogr.* **581**, 49–56.
4. Landers, J.; Schuchard, M.D.; Subramaniam, M.; Simelich, T.P.; Spelberg, T.C. (1992) High-performance capillary electrophoresis analysis of chloramphenicol acetyl transferase activity. *J. Chromatogr.* **603**, 247–257.
5. Mulholland, F.; Movahedi, S.; Hague, G.R.; Kasumi, T. (1993) Monitoring tripeptidase activity using capillary electrophoresis. *J. Chromatogr.* **636**, 63–68.
6. Vinther, A.; Adelhorst, K.; Kirk, O. (1993) Using capillary electrophoresis in the optimization of carboxypeptidase Y catalyzed transpeptidation reaction. *Electrophoresis.* **14**, 486–491.
7. Lowther, W.T.; Dunn, B.N. (1994) Kinetics of enzyme-catalyzed oligopeptide cleavage monitored by capillary zone electrophoresis: Comparison to spectrophotometric and HPLC methods. *Lett. Peptide Synth.* **1**, 89–94.
8. Dawson, J.F.; Marion, P.B.; Holmes, F.B. (1994) A capillary electrophoresis-based assay for protein kinases and protein phosphatases using peptide substrates. *Anal. Biochem.* **220**, 340–345.

9   Roberts, M.W.; Preiss, J.; Okita, T.W. (1995) A capillary zone electrophoresis assay for the nucleoside transfer enzyme adenosine diphosphate-glucose pyrophosphorylase. *Anal. Biochem.* **225**, 121–126.
10 Chen, F.-T.A. (1995) Characterization of protease-catalyzed hydrolysis of cyanine-labeled angiotensin using capillary electrophoresis with laser-induced fluorescence detection. *Anal. Biochem.* **225**, 341–345.
11 Chen, F.-T.A.; Evangelista, R.A. (1995) Analysis of mono- and oligosaccharide isomers derivatized with 9-aminopyrene-1,4,6-trisulfonate by capillary electrophoresis with laser-induced fluorescence. *Anal. Biochem.* **230**, 273–280.
12 Zhao, J.Y.; Dovichi, N.J.; Hindsgaul, O.; Gosselin, S.; Palcic, M.M. (1994) Detection of 100 molecules of product formed in a fucosyltransferase reaction. *Glycobiology* **4**, 239–242.
13 Craig, D.; Arriaga, E.A.; Banks, P.; Zhang, Y.; Renborg, A.; Palcic, M.M.; Dovichi, N.J. (1995) Fluorescein-based enzymatic assay by capillary electrophoresis laser-induced fluorescence detection for the determination of a few β-galactosidase molecules. *Anal. Biochem.* **22**, 147–153.
14 Schwartz, H.E.; Ulfelder, K.J. (1992) Capillary electrophoresis with laser-induced fluorescence detection of PCR fragments using thiazole orange. *Anal. Chem.* **64**, 1737–1740.
15 Liu, M.S.; Zang, J.; Evangelista, R.A.; Rampal, S.; Chen, F.-T.A. (1995) Double-stranded DNA analysis by capillary electrophoresis with laser-induced fluorescence using ethidium bromide as an intercalator. *Biotechniques* **18**, 316–323.
16 Liu, M.S.; Rampal, S.; Evangelista, R.A.; Gregory, Lee, C.Y.; Chen, F.-T.A. (1995) Detection of amplified Y chromosome-specific sequence by capillary electrophoresis. *Fertility and Sterility* **64**, 447–451.
17 Schmerr, M.J.; Goodwin, K.R. (1993) Characterization by capillary electrophoresis of the surface glycoproteins of ovine lentiviruses before and after treatment with glycosidic enzymes. *J. Chromatogr.* **652**, 199–205.
18 Watson, E.; Yao, F. (1993) Capillary electrophoretic separation of human recombinant erythropoietin (r-HuEPO) glycoforms. *Anal. Biochem.* **210**, 389–393.
19 Rogan, M.M.; Drake, C.; Goodall, D.M.; Altria, K.D. (1993) Enantioselective enzymatic biotransformation of 2′-deoxy-3′-thiacytidine (BCH 189) monitored by capillary electrophoresis. *Anal. Biochem.* **208**, 343–347.
20 Thorsteinsdottir, M.; Beijersten, I.; Westerlund, D. (1995) Capillary electroseparations of enkephalin-related peptides and protein kinase A peptide substrates. *Electrophoresis* **16**, 563–573.
21 Lenz, V.J.; Gattner, H.J.; Leithauser, M.; Brandenburg, D.; Wollmer, A.; Hocker, H. (1994) Proteolyses of a fluorogenic insulin derivative and native insulin in reversed micelles monitored by fluorescence emission, reversed-phase high-performance liquid chromatography, and capillary zone electrophoresis. *Anal. Biochem.* **221**, 85–93.
22 Nashabeh, W.; Elrassi, Z. (1992) Enzymophoresis of nucleic acids by tandem capillary enzyme reactor–capillary zone electrophoresis. *J. Chromatogr.* **596**, 252–264.
23 Amankawa, L.N.; Kuhr, W.G. (1992) Trypsin-modified fused-silica capillary microreactor for peptide mapping by capillary zone electrophoresis. *Anal. Chem.* **64**, 1610–1613.
24 Licklider, L.; Kuhr, W.G. (1994) Optimization of on-line peptide mapping by capillary zone electrophoresis. *Anal. Chem.* **66**, 4400–4407.
25 Yoshimoto, Y.; Shibukawa, A.; Sasagawa, H.; Nitta, S.; Nakagawa, T. (1995) Michaelis–Menten analysis of immobilized enzyme by affinity capillary electrophoresis. *J. Pharm. Biomed. Anal.* **13**, 483–488.
26 Miller, K.J.; Lytle, F.E. (1994) Enzyme profiling of immobilized cells using CZE. *Anal. Chem.* **66**, 2420–2423.
27 Emmer, A.; Roeraade, J. (1994) Capillary electrophoresis, combined with an on-line micro post-column enzyme assay. *J. Chromatogr.* **666**, 375–381.
28 Konse, T.; Takahashi, T.; Nagashima, H.; Iwaoka, T. (1993) Blotting membrane micropreparation in capillary electrophoresis to evaluate enzyme purity and activity. *Anal. Biochem.* **214**, 179–191.
29 Bao, J.; Regnier, F.E. (1992) Ultramicro enzyme assays in a capillary electrophoretic system. *J. Chromatogr.* **608**, 217–224.

30 Miller, K.J.; Leesong, I.; Bao, J.; Regnier, F.E.; Lytle, F.E. (1993) Electrophoretically mediated microanalysis of leucine aminopeptidase in complex matrices using time-resolved laser-induced fluorescence detection. *Anal. Chem.* **65**, 3267–3270.
31 Wu, D.; Regnier, F.E.; Linhares, M.C. (1994) Electrophoretically mediated micro-assay of alkaline phosphatase using electrochemical and spectrophotometric detection in capillary electrophoresis. *J. Chromatogr.* **657**, 357–363.
32 Wu, D.; Regnier, F.E. (1994) Native protein separations and enzyme microassays by capillary zone and gel electrophoresis. *Anal. Chem.* **65**, 2029–2035.
33 Avila, L.Z.; Whitesides, G.M. (1993) Catalytic activity of native enzymes during capillary electrophoresis – An enzymatic microreactor. *J. Org. Chem.* **58**, 5508–5512.
34 Hsieh, Y.Z.; Chen, F.-T.A.; Sternberg, J.C.; Klein, G.; Liu, C.-M. (1993) Capillary zone electrophoretic analysis of isoenzymes. *U.S. Patent* 5,264,095, Nov. 23.
35 Xue, Q.; Yeung, E.S. (1994) Variability of intracellular lactate dehydrogenase isoenzymes in single human erythrocytes. *Anal. Chem.* **66**, 1175–1178.
36 Tan, W.H.; Yeung, E.S. (1995) Simultaneous determination of enzyme activity and enzyme quantity in single human erythrocytes. *Anal. Biochem.* **226**, 74–79.

17 Capillary electrophoresis-based immunoassays
R.A. EVANGELISTA AND F.-T.A. CHEN

17.1 Introduction

Immunoassays are extensively used in clinical and biochemical applications for quantitative detection of compounds of diagnostic significance [1–3]. The analytical utility of immunoassays is based on the extraordinary specificity and remarkably high affinity of antibody–antigen reactions. This unique characteristic of antibody–antigen binding allows the selective quantitation of an analyte at trace amounts in the presence of contaminants at much higher concentrations which would almost certainly interfere using other analytical methods. Such assays have been applied to various areas of clinical diagnosis such as endocrinology, oncology, autoimmune diseases, therapeutic drug monitoring, infectious diseases and forensic drug testing. More recently, immunoassays have been developed for environmental, food and agricultural applications such as detection of pesticide residues, food toxins and environmental carcinogens [4].

Among the wide variety of immunossay formats, those which are classified as homogeneous immunoassays are based on formation of a measurable signal which occurs as a result of antibody–antigen binding in solution without requiring physical separation. Examples of such signals are fluorescence polarization, enhancement or inhibition of enzyme activity or change in light scattering. Heterogeneous immunoassays are those that require physical separation of antibody-bound antigen from the free antigen. Such separations are accomplished by binding to surfaces of plastic microwell plates, latex partices or magnetizable beads followed by washing to remove unbound non-immunoreactive compounds. The separation of the free from the immunologically bound antigen can also be achieved by chromatographic or electrophoretic methods.

Capillary electrophoresis (CE) has been shown to be an exceptional technique for the study of specific binding of biological molecules. CE has been applied to the investigation of several non-covalent biospecific interactions such as antibody–antigen [5–9], protein–protein (human IgG–protein G) [10], protein–carbohydrate (lectin–sugar) [11, 12], receptor–ligand [13], DNA hybridization [14], protein–drug (BSA (bovine serum albumin)–warfarin) [15], protein–dye (conalbumin–arylaminonaphthalene sulfonates) [16] and DNA–intercalator binding [17–19].

CE, mainly because of its high resolution power, speed of separation, small sample requirement, low reagent consumption and potential for fully automated operation, provides a distinct advantage over other methods as a tool for the separation of free and antibody-bound antigens. In addition, the favorable kinetics in solution compared to that on solid surfaces facilitate efficient binding without long incubation times and use of excess antibody. In comparison, for example, to double-antibody sandwich immunoassays wherein a high concentration of first antibody is necessary for coating a plastic surface in the first step and excess second antibody solution is again added in the final step before extensive washing, competitive immunoassays by CE require only a nearly stoichiometric amount of a single antibody.

17.2 Immunoassays by CE with UV detection

The first reported example of antibody–antigen (Ab–Ag) reaction analyzed by CE, was the binding of human growth hormone to an antibody [5, 20]. Using free-solution CE combined with UV detection, separated peaks were observed for the free antigen and antibody. When the two species were mixed in solution, the peaks due to the complexes were formed with migration times between those of the free antigen and antibody. In this example, the molecular weight of the antigen is large enough to cause substantial addition to the size of the antibody to produce the observed shift in migration times. Although this method works well for identification of an isolated purified protein, UV absorbance detection is not practical for immunoassay of clinically important analytes at trace levels in biological samples such as serum, urine or cerebrospinal fluid because of the presence at higher concentrations of a large number of non-immunoreactive UV-absorbing species which can mask and interfere with both the antibody and antigen peaks. Thus, the sample matrix effect must be considered seriously.

In another example of immunoaffinity CE with UV detection, Phillips and Chmielinska [21] developed a method of analysis of the cyclic peptide cyclosporin A in tear fluid. This method differs from the others described in this review in that it involves solid-phase immunoseparation prior to electrophoresis. The first step involves a 10 min incubation of the sample with the first one-third of the capillary which is covalently coated will a Fab fragment of anticyclosporin A monoclonal antibody with captures the analyte and its metabolites from the sample solution. After two washes with neutral phosphate buffer to remove unbound materials, electroelution-separation was performed using phosphate buffer pH 1.5 which disrupts the Ab–Ag binding and acts as a separation buffer for electrophoresis of the peptide analyte and its metabolites. Another main difference between this method and others covered in this review is the fact that only the free unlabeled antigens are detected while, in the other reports, the electropherograms show the formation of the antibody–antigen complexes. The results of the immunoaffinity CE analysis correlated well with those from solid phase extraction (SPE) followed by reversed phase HPLC (high performance liquid chromatography). The method was able to measure cyclosporin A down to 6.2 ng ml^{-1} (5 nM) in human tear fluid with CVs (coefficient of variation) in the range 2.67% to 7.66%. Although the Fab is covalently bound on the capillary wall in the portion of the capillary used for immunoaffinity capture of the analyte, significant loss of the immobilized MAb (monoclonal antibody) was observed after 20–30 sample injections. Since the method involves UV detection, only analytes with high extinction coefficients can be analyzed with high sensitivity. In this case, the peptide analytes have high absorptivity at 214 nm due to the amide bonds and the aromatic amino acids.

A technique which can be considered related to the immunoconcentration described above is 'capillary electrophoretic immunosubtraction' in which a component of the biological sample is removed by capture with an antibody specific to that component [22]. This can be accomplished by passage of the sample through a column containing the antibody bound on solid support or by insolubilizing the conjugate. Comparison of CE analyses before and after the immunosubtraction step can then reveal the component that was immunologically removed. This method was demonstrated for the detection of IgG, IgA and lambda light chain monoclonal gammopathies.

17.3 Immunoassays by CE with laser-induced fluorescence (LIF) detection

Among the few available detection methods for CE analysis, LIF is best suited for quantitative detection of compounds at trace amounts. The detection limits achieved by LIF using common fluorescent labels are typically in the subnanomolar range, more than three orders of magnitude better than those obtained with UV absorbance detection. The amount of fluorescent analyte detected in the nanoliter injection volumes can be as low as a few thousand molecules [23, 24]. One great advantage provided by LIF detection using purified labeled species is that all of the non-labeled components of the mixture are not detected and thus clean electropherograms are obtained which contribute to improved detection limits. In the case of a single analyte determination using a fluor-labeled antigen (Ag*), the electropherograms would show only two peaks, that of Ag* and the antibody-labeled antigen complex (Ab–Ag*).

Schultz and Kennedy [6] reported on the immunoassay of insulin by CE/LIF using fluorescein-labeled insulin and anti-insulin antibody. The electropherograms, which were all completed within 3 min, showed more than one peak for the labeled antigen which were attributed to variation in the position of labeling in the protein. Each of the two major antigen peaks formed a different antibody–antigen complex peak after addition of the monoclonal antibody Fab fragment. Using an uncoated fused-silica capillary and the 442 nm helium–cadmium laser for excitation, the competitive immunoassay for insulin produced a detection limit of 3 nM with increasing peak area of the labeled competing species up to 200 nM insulin in the sample. Later they used polyacrylamide-coated capillaries and the 488 nm argon ion laser as excitation source for the CE/LIF immunochemical determination of insulin secreted from islets of Langerhans and obtained good reproducibility (average RSD 3.4%) [25].

Shimura and Karger [7] developed a CE/LIF-based immunoanalytical method for detection of methionyl recombinant human growth hormone using a Fab' antibody fragment singly labeled with tetramethylrhodamine. Of the three thiol groups resulting from reduction of the disulfide bonds in the hinge region of the antibody, two were oxidized under controlled conditions, leaving one SH for specific reaction with an iodoacetamide labeling reagent. Using a 488 nm argon ion laser for excitation and capillary isoelectric focusing (CIEF) in a polyacrylamide-coated capillary for simultaneous concentration and separation, the purified labeled antigen produced a sharp peak in the electropherogram. The antibody–antigen complex produced a narrow peak with shorter migration time than the labeled antigen. The method, termed affinity probe capillary electrophoresis, can detect down to 5 pmol (0.1 ng ml^{-1}) concentration of met-rhGH (methionyl recombinant human growth hormone) with less than or equal to 5% RSD in the time-normalized peak areas of the complex. The low detection limit was attributed to the concentration of the analytes during the focusing after filling almost the entire length of the capillary with the sample. It was estimated that the use of CIEF resulted in an improvement of over 100-fold in detection limits compared to normal capillary zone electrophoresis (CZE).

Chen [26] reported a CE/LIF-based non-competitive immunoassay for IgA using a fluorescein-labeled anti-IgA F(ab')$_2$ fragment which was charge-modified by succinylation of the amino groups to produce a species with longer migration times such that successful separation from the faster migrating antibody–antigen complex was achieved. The separation buffer was an alkaline borate buffer which has been successfully used for protein analysis by CE with UV detection [27–29]. The use of a buffer with pH greater than the pI (isoelectric point) of the proteins substantially reduced protein–silica interactions. The method was shown to be useful for detection of IgA in serum down to 0.1 mg ml^{-1} (0.3 nM). As in the case of immunochemical CE analysis of human growth hormone with UV [5] and LIF detection [7], such a non-competitive method based on the shift in migration time of the labeled antibody fragment upon binding is only applicable to antigens whose molecular weight is large enough to cause a significant change in the labeled antibody migration.

Chen and Sternberg [30] reported a CE/LIF-based competitive immunoassay for the cardiotonic drug digoxin using a tetramethylrhodamine-labeled antigen with a spacer arm composed of a short oligonucleotide. The anionic oligonucleotide linker produced a controlled increase in the migration time of the labeled antigen, similar to the effect of the protein succinylation mentioned above, to facilitate separation of the antibody–antigen complex from the free antigen. Using the green He–Ne laser (543 nm) as excitation source, the method was able to detect as low as 0.42 ng ml^{-1} (0.54 nM) digoxin in diluted serum.

Chen and Pentoney [31] presented another green He–Ne laser-based CE/LIF competitive immunoassay method for digoxin using a different label, B-phycoerythrin. The labeled antigen in this technique is unusual in the sense that the fluor label is a phycobiliprotein with a molecular weight 600 times greater than the analyte. In spite of the heterogeneity due to several digoxins attached to B-phycoerythrin at different sites, the labeled antigen exhibited a rather narrow CE/LIF peak. Addition of the antidigoxigenin Fab produced a very broad peak due to heterogeneity of the complex arising from binding of the antibody fragment at different sites of the large protein label. Due to the low pI of the protein label, the labeled antigen migrated with sufficiently longer migration time to allow resolution of the free labeled antigen from the complex. The electropherograms showed increasing Ag* signal with increasing serum-based digoxin calibrator concentration in the range 0.42 to 10.42 ng ml^{-1}.

Fuchs and co-workers [32] reported a CE/LIF competitive immunoassay for the hormone cortisol in undiluted serum using a Fab fragment of a monoclonal antibody and fluorescein-labeled cortisol oxime derivative as competing species. The use of a coated capillary allowed the assay to be performed in undiluted serum without the undesirable effects of adhesion of proteins on the capillary wall. The 488 nm argon ion laser was used as excitation source. The assay had a working range of 10^{-6} to 10^{-8} M, the clinically significant range for the hormone. Later improvement in the assay results after a modification in the assay system was reported, namely the use of polyclonal antibody with higher affinity [33]. In spite of the heterogeneity of the Ab–Ag* complexes derived from the polyclonal antibody, the measurement of the well-isolated sharp Ag* peak produced good precision.

Evangelista and Chen [34] demonstrated the use of CE/LIF for the analysis of structural specificity of the antibody–antigen reaction using morphine as an example. The selected fluor label, Cy5 cyanine dye (Cy5 = disulfonated cyanine dye with λ_{max} em = 667 nm), has several advantages for sensitive CE/LIF detection, namely high extinction coefficient (>200 000 M^{-1} cm^{-1}) in the far red visible region (λ_{max} = 650 nm) where there is no interference by endogenous fluorescent compounds in biological fluids, high quantum yield (0.28 for protein conjugates), good photostability and water solubility and negatively charged sulfonate groups producing electrophoretic migration times ideal for separation of the free labeled antigens from the antibody complex. The reactive dye has a flexible six-atom linker between the cyanine moeity and the amine-reactive group which becomes a spacer arm between the antigen and the label. Labeling of amine-containing compounds with the reactive Cy5 N-hydroxysuccinimide ester is a fast and simple procedure. Cy5 reactive dye has been used for the preparation of highly fluorescent proteins [35, 36] and oligonucleotides [37]. Furthermore, the 633 nm helium–neon laser or the 635 nm diode laser, both of which have the advantage of low cost, high stability and long lifetime, can be used as excitation source. The detection limit for the Cy5 label using a 20 µm × 25 cm fused-silica capillary using the red He–Ne laser for excitation is 5×10^{-11} M.

CE/LIF was shown to be useful for rapid screening of morphine antibodies to determine which one produced the best binding to the labeled antigen. The antibodies deemed unsuitable for assay had problems associated with low affinity or heterogeneity which were revealed by CE/LIF. In another application, the labeled antigen was used to determine crossreactivity of various opiates towards the morphine antibody by competitive binding. It was found that out of nine opiates tested, only hydromorphine and normorphine showed significant crossreactivity towards the antibody but at a much lower extent than morphine. This is presumably due to the fact that the fluorescent label is attached to the amine nitrogen and hence the antibody binds to the opposite side of the molecule which contains the alcohol, ether and the phenolic hydroxyl groups due to the stereochemistry of chemical structure of compounds. Thus, it was not surprising that the opiates in which any of these three groups were modified exhibited drastically lower affinity for the antibody. The same system was demonstrated for the competitive immunoassay for morphine which showed an increasing signal in the range 10^{-7} M to 10^{-5} M which includes the established cut-off concentration for a positive test result of 10^{-6} M morphine.

In another application of CE/LIF-based immunoassay using Cy5-labeled competing species, Pritchett et al. [38] reported an assay for the bioactive octapeptide angiotensin II in ten-fold diluted serum with a detection limit of 2×10^{-10} M and an upper assay limit of 2×10^{-8} M using essentially the same instrumentation and separation system as the above-mentioned morphine assay. Although the detection limit of the assay is not low enough to measure physiological concentrations of the peptide in serum (about 10^{-11} M), it serves as a model for the analysis of a synthetic or recombinant therapeutic peptide in complex biological fluid. The electropherograms and the dose–response plot are shown in Figure 17.1 [38].

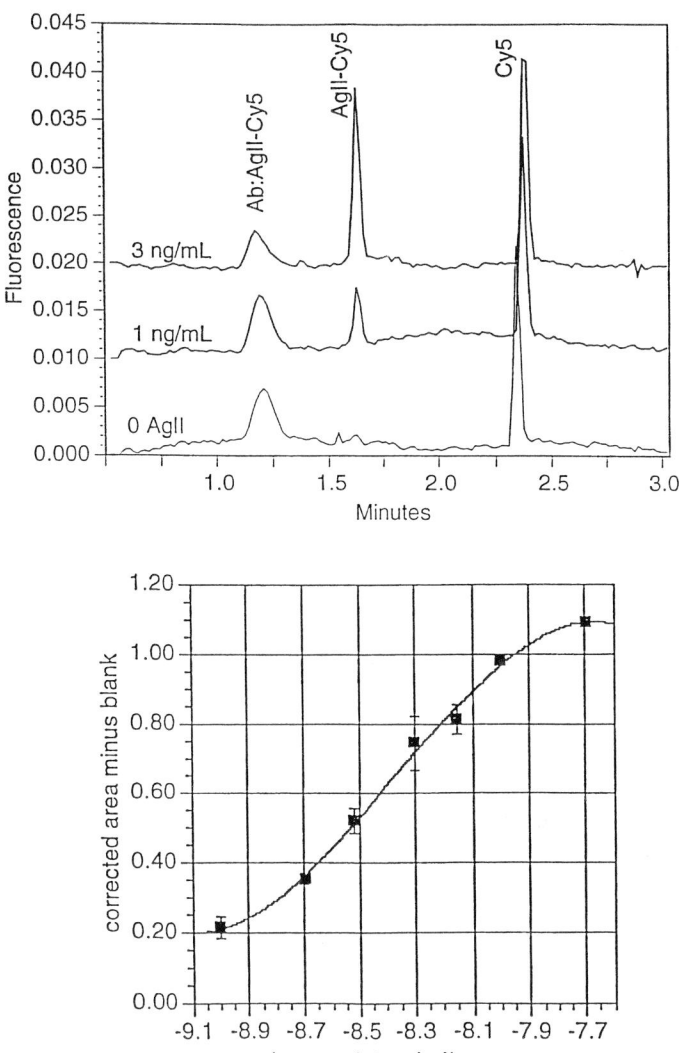

Figure 17.1 Competitive immunoassay of angiotensin II in diluted serum using rabbit polyclonal antibody and N-terminal Cy5-labeled angiotensin II as competing species. CE/LIF conditions: 80 mM borate at pH 10.25; 20 kV/24 µA; 20 µm × 25 cm fused silica; red He–Ne laser (633 nm) excitation; 665 nm fluorescence detection. Reprinted with permission from Ref. [38].

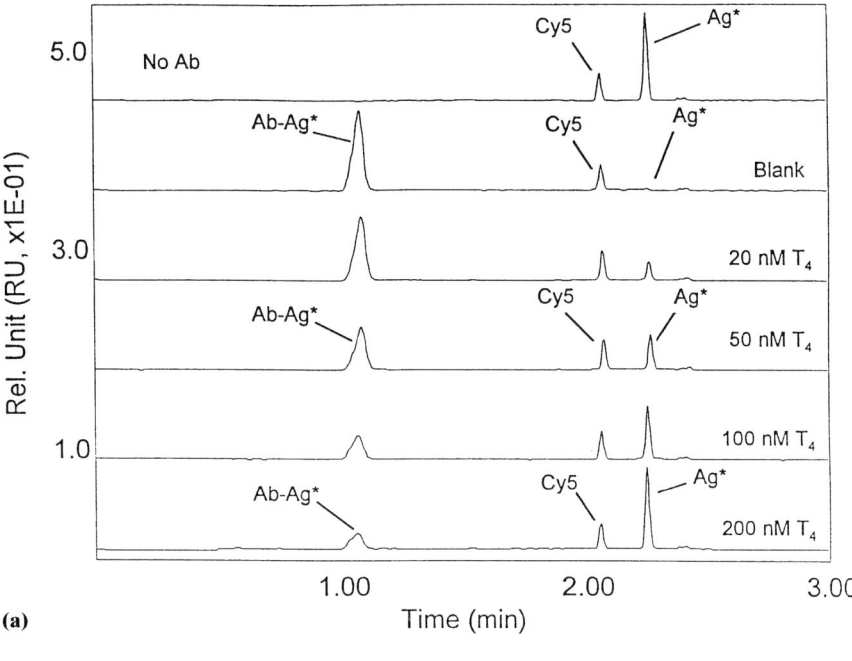

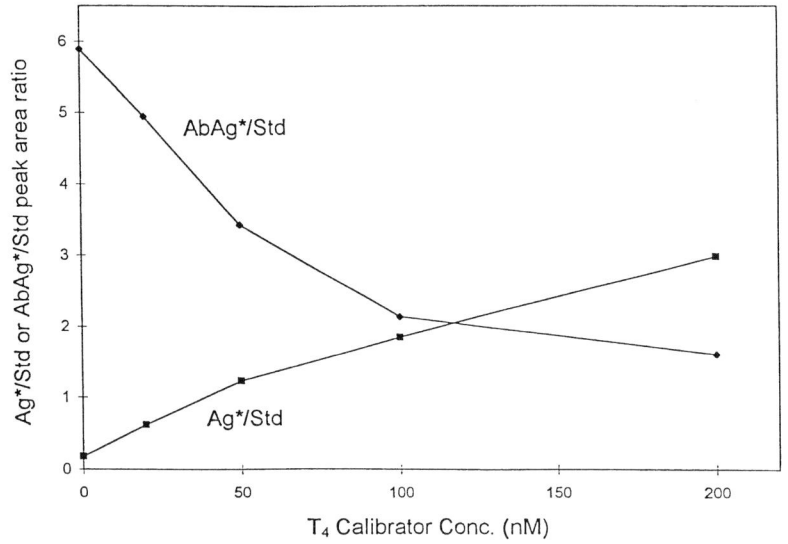

Figure 17.2 Competitive immunoassay of thyroxine using anti-T_4 monoclonal antibody and Cy5-labeled T_4 as competing species. (a) CE/LIF electropherograms: separation buffer, 300 mM borate at pH 8.5; 20 kV/19 µA; 20 µm × 25 cm fused-silica capillary; excitation 633 nm (red He–Ne laser); emission 665 nm. Total T_4–Cy5 (Ag* + Ab–Ag*) concentration in assay mixture = 25 nM. Cy5 diacid internal standard concentration (Std) = 1 nM. (b) Dose–response assay curve from average of triplicate runs: Ag*/Cy5 standard peak ratio vs. T_4 concentration and (Ab–Ag*)/Cy5 standard peak ratio vs. T_4 concentration.

In yet another example of competitive immunoassay by CE/LIF using Cy5-labeled antigen, the previously unpublished results of a model assay of thyroxine (T_4) in protein diluent are shown in Figure 17.2. Electropherograms were obtained using 300 mM borate buffer at pH 8.5 for separation at 20 kV/19 μA using 20 μm × 25 cm fused-silica capillary. The use of buffers with higher pHs produced unsatisfactory results presumably due to dissociation of the antibody–antigen complex arising from ionization of the phenolic group of the labeled T_4. A solution containing a 25:1 molar mixture of labeled antigen (T_4 attached to Cy5 at the amino nitrogen) and Cy5 diacid internal standard (Std) produced two peaks at only 3:1 ratio. The observed partial quenching of the fluor upon conjugation is presumably due to the effect of the four iodines in the molecule. Addition of a nearly stoichiometric amount of anti-T_4 MAb produced conversion of the sharp Ag* peak to the faster migrating larger MAb–Ag* complex peak. It is apparent that the signal that was partially lost in the covalent attachment of the label to the thyroxine was recovered upon binding of the antigen to the MAb. In the presence of increasing concentrations of T_4 calibrator, the Ag* peak increased gradually with concomitant decrease in the MAb–Ag* peak due to competition of the unlabeled antigen (the analyte) with the labeled antigen for the antibody binding site. Although the intensity of the fluor in the Ag* is different from that in the MAb–Ag* complex, both peaks exhibited change such that quantitation of the analyte can be made using either the increase in the Ag* peak or the decrease in the MAb–Ag* peak. The plot of Ag*/Std peak ratio vs. calibrator concentration was nearly linear in the range 20 nM to 200 nM of T_4 which includes the normal adult physiological range for T_4 (65–155 nM). The reproducibility of triplicate runs for both the Ag*/Std ratio and the AbAg*/Std ratio for each calibrator concentration was within 6%.

The results of reproducibility studies of the CE/LIF-based immunoassay using Cy5-labeled antigen as competing species are presented in Figure 17.3 with digoxin as a model analyte. The CE/LIF system consisted of a 20 μm × 27 cm fused-silica capillary with 80 mM borate as separation buffer and 633 nm (red He–Ne) excitation and 665 nm fluorescence detection. The labeled antigen was prepared by the reaction between 3-amino-3-deoxydigoxigenin and Cy5 reactive dye and was purified to homogeneity by reversed-phase HPLC. The antibody used was a polyclonal anti-digoxigenin Fab fragment. Formation of the Ab–Ag* complex produced a broad peak with shorter migration time than the unbound labeled antigen. The competition between the unlabeled digoxin in the sample and the labeled antigen for the limited amount of antibody resulted in the expected pattern of increasing Ag* peak and decreasing Ab–Ag* complex peak with increasing analyte concentration in the physiologically relevant range of 0.5 to 5 ng ml^{-1} (6.4×10^{-10} to 6.4×10^{-9} M). The reproducibility in the measurement of the Ag*/Std peak ratio for nine replicates was better for the higher

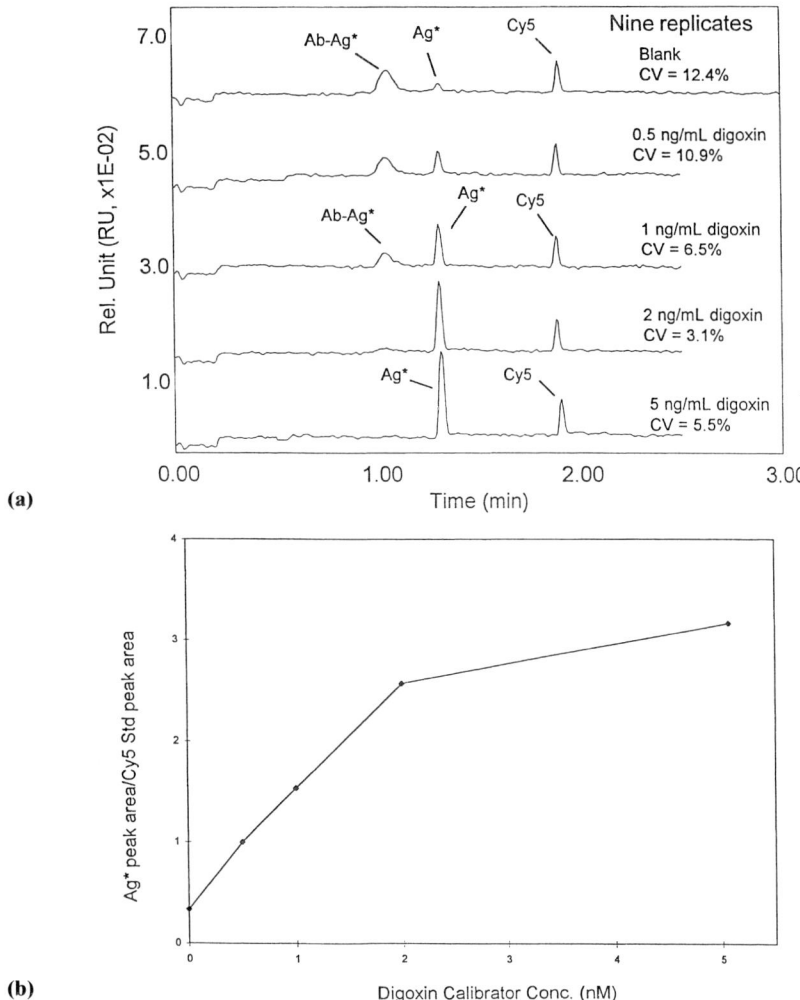

Figure 17.3 Competitive immunoassay of digoxin using Cy5-labeled 3-amino-3-digoxigenin as competing species. (a) CE/LIF electropherograms: separation buffer, 80 mM borate at pH 10.2; 20 kV/21 µA; 20 µm × 25 cm fused-silica capillary; excitation 633 nm (red He–Ne laser); emission 665 nm. Total Dig–Cy5 (Ag* + Ab–Ag*) concentration in assay mixture = 0.98 nM. Cy5 diacid internal standard concentration (Std) = 0.2 nM. (b) Dose–response assay curve from average of nine replicates: Ag*/Cy5 standard peak ratio vs. digoxin calibrator concentration.

concentration digoxin calibrators (3 to 6.5%) than for the low-end calibrator and the blank (10 to 12%), as expected from the size of the Ag* peak. The measurement of the Ag* peak area produces better precision because the broadness of the Ab–Ag* peak precludes exact peak area integration. The intra-assay reproducibility in migration times of Ag* and the internal standard was within 0.6%.

17.4 Multianalyte immunoassays

The speed and high resolution of CE separation not only allows separation of one free antigen from its antibody complex, but also permits analysis of several labeled species in a single run. Therefore, it is possible to perform simultaneous immunoassay of more than one analyte in a single sample in a single electrophoretic run. In principle, the number of analytes that can be analyzed in one sample in a single determination is limited only by the separation of the labeled antigens from each other and from the antibody–antigen complexes. Previously, simultaneous multianalyte immunoassays have been achieved by discrete placement of antibodies on distinct zones [39–43] on solid support which is placed in contact with the sample for binding of the different antigens to their respective antibodies or by using different fluorescent labels with no overlap of emission bands [44, 45].

A system to demonstrate the feasibility of simultaneous immunoassay of more than one analyte is presented in Figure 17.4 using digoxin and theophylline as model antigens. Each of the purified Cy5-labeled drugs produced a sharp CE/LIF peak with shorter migration times than the Cy5 Std. Addition of anti-theophylline MAb resulted in formation of the labeled theophylline–antibody complex with a migration time shorter than any of two labeled antigens without any effect on the labeled digoxin. Addition of a polyclonal anti-digoxigenin Fab fragment to the original mixture of two labeled drugs with internal standard resulted in capture of the labeled digoxin to form a broad peak due to the antibody–digoxin–Cy5 complex. Addition of the two antibodies to the solution containing the two labeled drugs resulted in capture of both labeled antigens to form two distinct Ab–Ag* complexes. It should be noted that the Ab–Ag* complex derived from the monoclonal antibody produces a narrower peak than that formed from the polyclonal antibody due to better selectivity.

Evangelista and Chen [46] demonstrated the application of CE/LIF for the simultaneous immunoassay of morphine and phencyclidine (PCP) in urine. The competing species used for the assay were morphine–Cy5 and PCP–Cy5.5 (Cy5.5 is tetrasulfonated cyanine dye with λ_{max} em = 695 nm). The two cyanine labels had different emission maxima (667 nm for Cy5 and 695 nm for Cy5.5) and thus the collection of fluorescence signal was made through a 690 nm filter where the emission of the two were about the same.

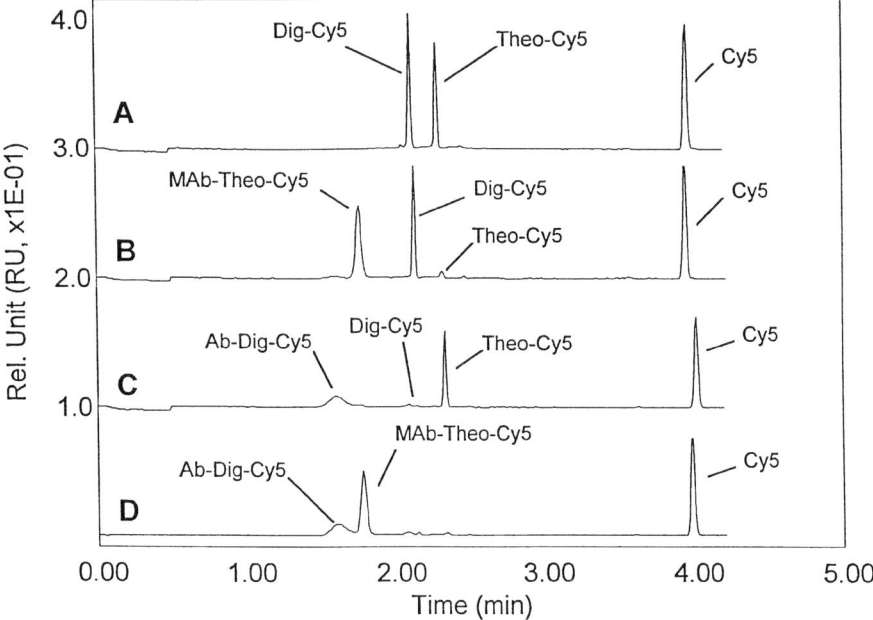

Figure 17.4 Simultaneous multianalyte immunoassay model. (A) Digoxigenin–Cy5 (10 nM), theophylline–Cy5 (10 nM) and Cy5 diacid internal standard (Std, 5 nM). CE/LIF: 200 mM borate at pH 10; 20 kV/28 µA; 20 µm × 25 cm fused silica; excitation 633 nm (red He–Ne laser); emission 665 nm. (B) Monoclonal anti-theophylline antibody added. (C) Polyclonal anti-digoxigenin Fab added to mixture in (A). (D) Monoclonal anti-theophylline antibody and polyclonal anti-digoxigenin Fab added to mixture in (A).

The two cyanine dye labels also differ in the number of sulfonate groups (two for Cy5 and four for Cy5.5) and this produced a difference in migration times which allowed individual monitoring of the binding of the labeled drugs to their respective antibodies in the same CE run. Since the two polyclonal antibodies are both IgG molecules which produced broad CE peaks, the two Ab–Ag* complexes could not be separated from each other but migrated much faster than the free labeled antigens due to their higher molecular weights. Consequently, only the peak areas of the free labeled antigens relative to the Cy5 Std could be used as quantitation parameters. The binding of each antigen to its respective antibody was independent of the other and thus each of the Ag* peak areas was proportional to the concentration of the corresponding analyte in the urine sample in the range 23 to 285 ng ml^{-1} for morphine (80 to 1000 nM) and 1.9 to 24 ng ml^{-1} (7.8 to 98 nM) for PCP.

The multianalyte CE/LIF competitive immunoassay system was extended to four drugs of abuse, morphine, PCP, THC–COOH (11-*nor*-Δ^8-tetrahydrocannabinol-9-carboxylic acid) (marijuana metabolite) and benzoylecgonine (BECG) (cocaine metabolite) [47]. As in the simultaneous dual immunoassay system, the four Ab–Ag* complexes comigrated to form one broad peak because the antibodies were all IgG molecules with molecular weight of 150 kDa. The quantifiable assay signals were the peak areas of the four labeled competing species, morphine–Cy5, PCP–Cy5.5, THC–Cy5 and benzoylecgonine–Cy5.5 normalized with that of the Cy5 Std. The tetrasulfonated label Cy5.5 was selected for the more basic drugs PCP and benzoylecgonine in order to offset their basicity to produce migration times which were longer than those of the Ab–Ag* complexes. The use of two different dyes produced good separation of the labeled competing species from each other, which was the main requirement for simultaneous multianalyte immunoassay in addition to separation of the free labeled antigens from the Ab–Ag* complexes. The four-drug assay system was shown to be applicable to the measurement of the amount of one drug in the absence of the others. The individual assay of morphine and THC–COOH shoed increasing peak areas of their respective labeled antigens with increasing drug concentrations without significant change in the other Ag* peaks. The commercially available antibodies used in the assay have passed crossreactivity tests to be useful for single-drug forensic testing, so this phenomen is not surprising since the four drugs are chemically and structurally unrelated to each other. The four-drug assay system produced increasing Ag* signals with increasing concentrations of all four drugs in urine (Figure 17.5) [47]. The detection limits were 50 nM for morphine, 5 nM for PCP, 3.6 nM for THC–COOH and 43 nM for benzoylecgonine which are one-twentieth of the established cut-off concentrations for positive test results for the four drugs of abuse.

For simultaneous immunoassay of multiple protein analytes such as those in a fertility panel (lutropin, follitropin, choriogonadotropin and prolactin) or immunoprotein panel (IgG, IgA, IgM, kappa and lambda chains), the use of labeled protein antigens is expected to produce unsatisfactory results due to the broadness of the labeled antigen peaks. The approach that can be envisaged for this application is the use of labeled peptides which have the same sequence as the epitope of the protein antigens. A CE/LIF system for the competitive immunoassay of a peptide has recently been developed [38]. Since purified labeled peptides produce narrow CE peaks, it is possible to prepare several labeled peptides chemically designed to produce different migration times and to develop multianalyte immunoassays based on the immunochemical binding of the peptides to their respective antibodies in competition with the protein antigens.

17.5 Labeling considerations

CE-based immunoassays with LIF detection in the visible wavelength region require preparation and purification of fluor-labeled antigens or anti-

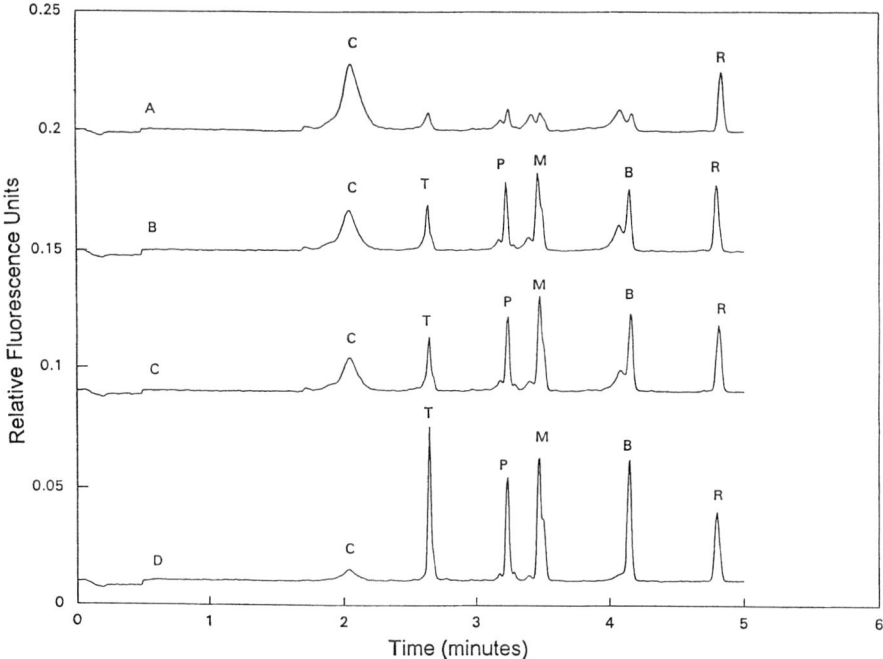

Figure 17.5 Simultaneous competitive immunoassay of morphine, PCP, THC–COOH and benzoylecgonine (BECG) in urine. (A) Drug-free urine. (B) 3.6 nM THC–COOH, 5 nM PCP, 50 nM morphine, 43 nM BECG (1/20th of cut-off conc). (C) 7.3 nM THC–COOH, 10 nM PCP, 100 nM morphine, 86 nM BECG (1/10th of cut-off conc). (D) 73 nM THC–COOH, 100 nM PCP, 1 mM morphine, 860 nM BECG (cut-off conc). Peaks: T, THC–Cy5; P, PCP–Cy5.5; M, MORPH–Cy5; B, BECG–Cy5.5; C, mixture of four antibody-labeled antigen complexes; R, Cy5 diacid internal reference. CE/LIF conditions: 200 mM borate at pH 10.2; 15 kV/32 µA; 20 µm × 25 cm fused-silica capillary; red He–Ne laser (633 nM) excitation; 665 nm fluorescence detection. Reprinted from *American Clinical Laboratory*, **14**(2), 27–29 (1995). Copyright 1995 by International Scientific Communications, Inc. [47].

bodies. Due to the availability of a wide variety of labeling reagents which react specifically with functional groups such as amino, thiol, carboxyl and carbonyl groups, labeling has become a simplified procedure. Several chromatographic methods such as gel filtration, ion exchange and HPLC are available for purification of the labeled species. Most of the CE/LIF-based immunoassays showing single-species peaks in the electropherogram of this review employ HPLC-purified labeled compounds.

Several considerations have to be dealt with in developing a CE/LIF immunoassay system. For competitive immunoassays of a low molecular weight analyte, HPLC purification is almost certainly a requirement in order to avoid formation of fluorescent impurity peaks which could lead to poorer assay sensitivity. The other important factor in small molecule immunoassays has to do with maintaining the avidity of the drug after conjugation with the fluor label. Thus, it is important to provide a spacer arm of sufficient distance between the label and the epitope to ensure immunochemical binding. The position of conjugation of the label with the drug is also of critical importance in the preparation of labeled antigens as competing species. For those assays employing commercially available antibodies, information about the structural regioselectivity of the antibody binding must be obtained from the supplier and then the chemistry of conjugation must be tailored such that the spacer arm and the label will not affect the binding site. Usually, this involves chemical synthesis of an amine analog of the drug which could then be reacted with any of the amine-reactive labeling reagents. For those who might want to produce their own antibodies, the immunogen used in raising the antibodies in animals must be composed of the drug conjugated with the carrier protein (bovine serum albumin or keyhole limpet hemocyanin) at exactly the same site as the attachment of the fluor in the labeled competing species. Another important factor to be considered is that electrophoretic migration of the resulting conjugate must be sufficiently different from that of the complex. For low molecular weight analytes, the binding of the labeled antigen does not significantly alter the migration of the antibody (mol. wt 150 kDa). In free solution CE, the introduction of negative charges on the labeled antigen produces migration times which are substantially longer than those of the complexes.

For labeling protein antigens, reaction with an amine-reactive reagent such as N-hydroxysuccinimide ester usually produces heterogeneity of labeling arising from a large number of amino groups at different sites in the protein. Thus, the labeled species usually produces CE peaks which are much broader than those obtained with labeled small molecules due to gross labeling heterogeneity. The labeling conditions have to be controlled such that the fluor-to-protein ratio is kept at a low level so as to maintain avidity and to avoid gross labeling heterogeneity which produces extremely broad peaks. Furthermore, there should be sufficient separation between the free and bound labeled species in spite of the broadness of the peaks. An example of modulation of electrophoretic migration times by succinylation of proteins to achieve this goal was reported by Chen for fluorescein-labeled IgA [26].

17.6 Assay considerations

A generalized scheme for competitive immunoassay is shown in the following equation

$$Ag + Ag^* + Ab \rightleftharpoons Ab\text{—}Ag + Ab\text{—}Ag^* + Ag^*$$

where the unlabeled antigen Ag (the analyte) in the biological sample competes with the labeled antigen (Ag^*) for binding with a limited amount of antibody (Ab). For a single-analyte assay system, only two peaks appear in the electropherogram, those of Ag* and Ab–Ag*. Although the analyte is captured by the antibody to produce the undetectable Ab–Ag complex, quantitation is performed by measurement of the Ag* and Ab–Ag* peak areas. In most of the competitive immunoassays reviewed here, the Ag* peak is a better quantitation parameter because the Ag* is a highly purified species producing a clean well-isolated peak in the electropherogram. Measurement of the Ag* peak results in a direct-read assay, meaning the Ag* peak increases with increasing analyte concentration in the working range of the assay.

It should be noted that the linear dynamic range of a competitive immunoassay is inherently narrow (one to two orders of magnitude). The response of the signal with increasing dose depends on the relative concentrations of the analyte and Ag*, the antibody concentration and the dissociation constants of the Ab–Ag* and Ab–Ag complexes. The lower detection limit is more important since a sample can always be diluted if the analyte concentration is higher than the upper linear limit of the assay. The detection limit to some degree is adjustable depending on the concentrations of Ag* and Ab. However, the concentration of Ag* can be lowered only to such an extent that its peak area can be accurately measured above the baseline noise. Furthermore, lowering both the Ag* and Ab concentrations can lead to dissociation of the Ab–Ag* complex if the affinity is not high enough. The upper limit of the working range of the assay can be raised by increasing both the Ag* and the equivalent amount of Ab. Thus, it is possible to perform immunoassays with more than one set of Ab–Ag* reagents, one with low amounts of Ab and Ag* to detect very low analyte concentrations and the other with higher Ab and Ag* concentrations for analyte amounts which are higher than the upper limit of the working range of the first set of reagents.

It has been mentioned that one advantage of CE/LIF-based immunoassay is the favorable kinetics of the antibody–antigen binding in solution compared to that where one of the reactants is immobilized on a solid support. Grossman *et al.* [5] reported that equilibrium binding was reached within 30s after mixing of reactants. Schultz and Kennedy [6] showed that only 100s was required for formation of the Ab–Ag* before CE separation. Similarly, Shimura and Karger [7] reported that the time required for association between a labeled Fab' and its met-recombinant human growth hormone antigen is less than 2 min. However, the order of addition of reactants can make a difference in competitive immunoassays. Since the rate of dissociation of the complex is much less than its formation, the displacement of an antigen that is already bound to an antibody by a competing species can require a long duration before equilibrium is reached. Thus, in all the CE/LIF-based competitive immunoassays reviewed herein summarized in Table 17.1, the immunoreaction was initiated

ASSAY CONSIDERATIONS

Table 17.1 Capillary electrophoresis-based immunoassays

Methods Separation conditions Detection	Compounds	Matrix	Ref.
CZE 100 mmol l^{-1} tricine pH 8.0 316 V cm^{-1}, 25 µA 50 µm × 95 cm (80 cm to detector) UV, 200 nm	Human growth hormone	Separation buffer	5, 20
Immunocapture/CZE 100 mmol l^{-1} phosphate pH 1.5 100 µA constant current 75 µm × 75 cm (60 cm to detector) UV, 214 nm	Cyclosporin A and metabolites	Human tears	21
Immunosubtraction/CZE 400 mmol l^{-1} borate pH 10.0 370 V cm^{-1} 27 µm × 20 cm UV, 214 nm	IgG, IgA lambda light chain	Diluted serum	22
CZE 50 mmol l^{-1} phosphate pH 7.5 with 25 mmol l^{-1} K$_2$SO$_4$ 25 µm × 25 to 30 cm LIF, 442 nm He–Cd ex/ 520 nm em	Human insulin	Phosphate buffer	6
CZE 20 mmol l^{-1} tricine pH 8.0 29 kV 50 µm × 30 cm, polyacrylamide-coated LIF, 488 nm Ar$^+$ ion ex/ 540 nm em	Human insulin, rat islet insulin	Ringer solution rat islets	25
IEF 200 V cm^{-1} (1 min), 400 V cm^{-1} (4 min) 10 mmol l^{-1} H$_3$PO$_4$ (anolyte) 20 mmol l^{-1} NaOH (catholyte) 75 µm × 15 cm polyacrylamide-coated LIF, 488 nm Ar$^+$ ion ex/ 580 nm em	Human growth hormone	Tris–Tween–BSA	7
CZE 80 mmol l^{-1} borate pH 10.0 20 kV/13.5 µA 20 µm × 25 cm LIF, 488 nm Ar$^+$ ion ex/ 520 nm em	Human IgA	Serum	26
CZE Beckman protein analysis buffer 7 kV/80 µA 75 µm × 25 cm LIF, green He–Ne 543 nm ex/590 nm em	Digoxin	Serum	30
CZE 80 mmol l^{-1} borate pH 10.0 200 V cm^{-1}/70 µA 75 µm × 37 cm LIF, green He–Ne 543 ex/580 nm em	Digoxin	Serum	31
CZE 20 mmol l^{-1} TAPS/AMPD 30 kV/17 µA 50 µm × 27 cm polyacrylamide coated LIF, Ar$^+$ ion 488 nm ex/520 nm em	Cortisol	Serum	32, 33

Table 17.1 *Continued*

Methods Separation conditions Detection	Compounds	Matrix	Ref.
CZE 90% 100 mmol l^{-1} borate pH 10.2 10% ethylene glycol 20 kV/17 μA 20 μm × 25 cm LIF, red He–Ne 633 nm ex/665 nm em	Morphine	PBS–protein diluent	34
CZE 300 mmol l^{-1} borate pH 8.5 20 kV/19 μA 20 μm × 25 cm LIF, red He–Ne 633 nm ex/665 nm em	Thyroxine	PBS–protein diluent	Previously unpublished
CZE 80 mmol l^{-1} borate pH 10.25 20 kV 20 μm × 25 cm LIF, red He–Ne 633 nm ex/665 nm em	Angiotensin II	Diluted serum	38
CZE 80 mmol l^{-1} borate pH 10.2 20 kV/21 μA 20 μm × 27 cm LIF, red He–Ne 633 nm ex/665 nm em	Digoxin	PBS–protein diluent	Previously unpublished
CZE 200 mmol l^{-1} borate pH 10.2 15 kV/32 μA 20 μm × 27 cm LIF, red He–Ne 633 nm ex/690 nm em	Morphine, PCP	Urine	46
CZE 200 mmol l^{-1} borate pH 10.2 15 kV/32 μA 20 μm × 27 cm LIF, red He–Ne 633 nm ex/690 nm em	Morphine, PCP, THC–COOH, benzoylecyonine	Urine	47

AMPD = 2-amino-2-methyl-1,3-propanediol, CZE = capillary zone electrophoresis, IEF = isoelectric focusing, LIF = laser-induced fluorescence, PBS = phosphate buffered saline, PCP = phencyclidine, TAPS = *N*-tris(hydroxymethyl)methyl-3-aminopropanesulfonic acid, THC–COOH = 11-*nor*-Δ^8-tetrahydrocannabinol-9-carboxylic acid, Theo = theophylline.

by addition of the antibody or antibody fragment to the mixture of the labeled antigen and the sample containing the analyte.

The slow rate of dissociation of Ab–Ag* complexes, which is even slower for high affinity antibodies, is the underlying reason for the stability of the complex during CE analysis even in the alkaline buffers used as separation media in the above examples. In the examples of CE/LIF-based competitive immunoassays using cyanine-labeled antigens described in this review, fast separation times were achieved by a combination of high voltage (15–20 kV), short columns (25–27 cm), narrow capillaries (20 μm) and strong electroendoosmotic flow at pH > 8, which sweeps all the species towards the cathode. The stability of the migrating antibody–antigen complexes during these short separation periods is a key factor in the successful application of

CE/LIF-based immunoassays for the analysis of biologically important compounds at low concentrations.

17.7 Summary

The articles reviewed here are testimonial to the potential of CE/LIF as a viable alternative to current methods of immunoassay. The short analysis times, the benefits of solution-phase immunochemical kinetics and the advantages of automated operation offer significant improvement in immunoassay technology. The use of Cy5 and related fluorescent dyes open the opportunity for the use of relatively inexpensive and very reliable semiconductor lasers in the far-red and near-infrared wavelengths. Work currently in progress on the development of parallel multicapillary arrays using diode lasers should lead to substantial increase in assay throughput to make CE/LIF-based even more attractive for routine analysis of clinically relevant compounds.

References

1. Price, C.P.; Newman, D.J. (eds) (1991) *Principles and Practice of Immunoassay*, Stockton Press, N.Y.
2. Weir, D.M. (ed.) (1986) *Handbook of Experimental Immunology. Volume 1: Immunochemistry*, 4th edn, Blackwell Scientific Publications, Oxford.
3. Wild, D. (ed.) (1994) *The Immunoassay Handbook*, Stockton Press, New York.
4. Vanderlaan, M.; Stanker, L.H.; Watkins, B.E.; Roberts, D.W. (eds) (1991) *Immunoassays for Trace Chemical Analysis, Monitoring Toxic Chemicals in Humans, Food, and the Environment*, American Chemical Society, Washington, D.C.
5. Grossman, P.D.; Colburn, J.C.; Lauer, H.H.; Nielsen, R.G.; Riggin, R.M.; Sittampalam, G.S.; Rickard, E.C. (1989) Application of free-solution capillary electrophoresis to the analytical scale separation of proteins and peptides. *Anal. Chem.* **61**, 1186–1194.
6. Schultz, N.M.; Kennedy, R.T. (1993) Rapid immunoassays using capillary electrophoresis with fluorescence detection. *Anal. Chem.* **65**, 3161–3163.
7. Shimura, K.; Karger, B. (1994) Affinity probe capillary electrophoresis: analysis of recombinant human growth hormone with a fluorescent labeled antibody fragment. *Anal. Chem.* **66**, 13–19.
8. Heegaard, N.H.H. (1994) Determination of antigen–antibody affinity by immuno-capillary electrophoresis. *J. Chromatogr.* **680**, 405–412.
9. Martin, L.M.; Rotondi, K.S.; Merrifield, R.B. (1994) Antibody binding constants by capillary electrophoresis. In *Peptides: Chemistry, Structure and Biology*: Proceedings of the Thirteenth American Peptide Symposium June 20–25, 1993, Hodges, R.S.; Smith J.A. (eds), ESCOM Science Publishers.
10. Reif, O.-W.; Lausch, R.; Scheper, T.; Freitag, R. (1994) Fluorescein isothiocyanate-labeled protein G as an affinity ligand in affinity/immunocapillary electrophoresis with fluoresence detection. *Anal. Chem.* **66**, 4027–4033.
11. Honda, S.; Taga, A.; Suzuki, K.; Suzuki, S.; Kakehi, K. (1992) Determination of the association constant of monovalent mode protein–sugar interaction by capillary zone electrophoresis. *J. Chromatogr.* **597**, 377–382.
12. Shimura, K.; Kasai, K. (1995) Determination of the affinity constants of concanavalin A for monosaccharides by fluorescence affinity probe capillary electrophoresis. *Anal. Biochem.* **227**, 186–194.

13 Fishman, H.A.; Orwar, O.; Scheller, R.H.; Zare, R.N. (1995) Identification of receptor ligands and receptor subtypes using antagonists in a capillary electrophoresis single-cell biosensor separation system. *Proc. Natl Acad. Sci. USA* **92**, 7877–7881.
14 Chen, J.W.; Cohen, A.S.; Karger, B.L. (1991) Identification of DNA molecules by precolumn hybridization using capillary electrophoresis. *J. Chromatogr.* **559**, 295–305.
15 Kraak, J.C.; Busch, S.; Poppe, H. (1992) Study of protein–drug binding using capillary zone electrophoresis. *J. Chromatogr.* **608**, 257–264.
16 Swaile, D.F.; Sepaniak, M.J. (1991) Laser-based fluorometric detection schemes for the analysis of proteins by capillary zone electrophoresis. *J. Liq. Chromatogr.* **14**, 869–893.
17 Schwartz, H.E.; Ulfelder, K.J. (1992) Capillary electrophoresis with laser-induced fluorescence detection of PCR fragments using thiazole orange. *Anal. Chem.* **64**, 1737–1740.
18 Demana, T.; Lanan, M.; Morris, M.D. (1991) Improved separation of nucleic acids with analyte velocity modulation capillary electrophoresis. *Anal. Chem.* **63**, 2795–2797.
19 Kim, Y.; Morris, M.D. (1994) Separation of nucleic acids by capillary electrophoresis in cellulose solutions with mono- and bis intercalating dyes. *Anal. Chem.* **66**, 1168–1174.
20 Nielsen, R.G.; Rickard, E.C.; Santa, P.F.; Sharknas, D.A.; Sittampalan, G.S. (1991) Separation of antibody–antigen complexes by capillary zone electrophoresis, isoelectric focusing and high-performance size-exclusion chromatography. *J. Chromatogr.* **539**, 177–185.
21 Phillips, T.M.; Chmielinska, J.J. (1994) Immunoaffinity capillary electrophoretic analysis of cyclosporin in tears. *Biomed. Chromatogr.* **8**, 242–246.
22 Liu, C.-M.; Wang, H.-P.; Chen, F.-T.A.; Sternberg, J.C.; Klein, G.L. (1993) Analysis of samples by capillary electrophoretic immunosubtraction. *U.S. Patent* 5,228,960, Jul. 20.
23 Cheng, Y.-F.; Dovichi, N.J. (1988) Subattomole amino acid analysis by capillary zone electrophoresis and laser-induced fluorescence. *Science* **242**, 562–564.
24 Wu, S.; Dovichi, N.J. (1989) High-sensitivity fluorescence detector for fluorescein isothiocyanate derivatives of amino acids separated by capillary electrophoresis. *J. Chromatogr.* **480**, 141–155.
25 Schultz, N.M.; Huang, L.; Kennedy, R.T. (1995) Capillary electrophoresis-based immunoassay to determine insulin content and insulin secretion from single islets of Langerhans. *Anal. Chem.* **67**, 924–929.
26 Chen, F.-T.A. (1994) Characterization of charge-modified and fluorescein-labeled antibody by capillary electrophoresis using laser-induced fluorescence. Application to immunoassay of low level immunoglobulin A. *J. Chromatogr.* **680**, 419–423.
27 Lee, K.-J.; Heo, G.S. (1991) Free solution capillary electrophoresis of proteins using untreated fused-silica capillaries. *J. Chromatogr.* **559**, 317–324.
28 Chen, F.-T.A. (1991) Rapid protein analysis by capillary electrophoresis. *J. Chromatogr.* **559**, 445–453.
29 Chen, F.-T.A.; Liu, C.-H.; Hsieh, Y.-Z.; Sternberg, J.C. (1991) Capillary electrophoresis – a new clinical tool. *Clin. Chem.* **37**, 14–19.
30 Chen, F.-T.A.; Sternberg, J.C. (1994) Characterization of proteins by capillary electrophoresis in fused-silica columns: Review on serum protein analysis and applications to immunoassays. *Electrophoresis* **15**, 13–21.
31 Chen, F.-T.A.; Pentoney, Jr, S.L. (1994) Characterization of digoxigenin-labeled B-phycoerythrin by capillary electrophoresis with laser-induced fluorescence. Application to homogeneous digoxin immunoassay. *J. Chromatogr.* **680**, 425–430.
32 Schmalzing, D.; Nashabeh, W.; Yao, X.-W.; Mhatre, R.; Regnier, F.E.; Afeyan, N.B.; Fuchs, M. (1995) Capillary electrophoresis-based immunoassay for cortisol in serum. *Anal. Chem.* **67**, 606–612.
33 Schmalzing, D.; Nashabeh, W.; Fuchs, M. (1995) Solution-phase immunoassay for determination of cortisol in serum by capillary electrophoresis. *Clin. Chem.* **41**, 1403–1406.
34 Evangelista, R.A.; Chen, F.-T.A. (1994) Analysis of structural specificity in antibody–antigen reactions by capillary electrophoresis with laser-induced fluorescence detection. *J. Chromatogr.* **680**, 587–591.
35 Mujumdar, R.B.; Ernst, L.A.; Mujumdar, S.R.; Waggoner, A.S. (1989) Cyanine dye labeling reagents containing isothiocyanate groups. *Cytometry* **10**, 11–19.

36 Mujumdar, R.B.; Ernst, L.A.; Mujumdar, S.R.; Lewis, C.J.; Waggoner, A.S. (1993) Cyanine dye labeling reagents: Sulfoindocyanine succinimidyl esters. *Bioconjugate Chem.* **4**, 105–111.
37 Chen, F.-T.A.; Tusak, A.; Pentoney, Jr, S.; Konrad, K.; Lew, C.; Koh, E.; Sternberg, J. (1993) Semiconductor laser-induced fluorescence detection in capillary electrophoresis using a cyanine dye. *J. Chromatogr.* **652**, 355–360.
38 Pritchett, T.J.; Evangelista, R.A.; Chen, F.-T.A. (1995) Peptide immunoassay using capillary electrophoresis with laser-induced fluorescence detection. *J. Capillary Electrophoresis* **2**, 145–149.
39 Parsons, R.G.; Kowal, R.; LeBlond, D.; Yue, V.T.; Neargarder, L.; Bond, L.; Garcia, D.; Slater, D.; Rogers, P. (1993) Multianalyte assay system developed for drugs of abuse. *Clin. Chem.* **39**, 1899–1903.
40 Buechler, K.F.; Moi, S.; Noar, B.; McGrath, D.; Villela, J.; Clancy, M.; Shenhav, A.; Colleymore, A.; Valkirs, G.; Lee, T.; Bruni, J.F.; Walsh, M.; Hoffman, R.; Ahmuty, F.; Nowakowski, M.; Buechler, J.; Mitchell, M.; Boyd, D.; Stiso, N.; Anderson, R. (1992) Simultaneous detection of seven drugs of abuse by the Triage™ panel for drugs of abuse. *Clin. Chem.* **38**, 1678–1684.
41 Kakabakos, S.E.; Christopoulos, T.K.; Diamandis, E.P. (1992) Multianalyte immunoassay based on spatially distinct fluorescent areas quantified by laser-excited solid-phase time-resolved fluorometry. *Clin. Chem.* **38**, 338–342.
42 Donohue, J.; Bailey, M.; Gray, R.; Holen, J.; Huang, T.M.; Keevan, J.; Mattimiro, C.; Putterman, C.; Stalder, A.; Defreese, J. (1989) Enzyme immunoassay system for panel testing. *Clin. Chem.* **35**, 1874–1877.
43 Ekins, R.; Chu, F.; Biggart, E. (1990) Multianalyte immunoassay: the immunological 'compact disc' of the future. *J. Clin. Immun.* **13**, 169–181.
44 Hemmila, I.; Holttinen, S.; Petterson, K.; Lovgren, T. (1987) Double-label time-resolved immunofluorometry of lutropin and follitropin in serum. *Clin. Chem.* **33**, 2281–2283.
45 Siitari, H. (1990) Dual label time-resolved fluoroimmunoassay for the simultaneous detection of adenovirus and rotavirus in faeces. *J. Virol. Meth.* **28**, 179–188.
46 Evangelista, R.A.; Chen, F.-T.A. (1994) Feasibility studies for simultaneous immunochemical multianalyte drug assay by capillary electrophoresis with laser-induced fluorescence. *Clin. Chem.* **40**, 1819–1822.
47 Evangelista, R.A.; Michael, J.M.; Chen, F.-T.A. (1995) Simultaneous multianalyte immunoassay of drugs of abuse by CE. *Am. Clin. Lab.* **14**, 27–29.

18 Analysis of antibodies by capillary electrophoresis
T. PRITCHETT

18.1 Introduction

High resolution analysis of antibodies can be one of the more difficult challenges that the analytical biochemist faces. The high molecular weight of intact antibodies makes the use of reverse-phase HPLC (high performance liquid chromatography), one of the highest resolution tools in the analyst's repertoire, extremely difficult if not impossible. In addition, since many antibodies differ only slightly in amino acid content and molecular weight, size-based separations are often of limited use. To complicate matters further, antibodies are glycosylated and otherwise modified post-translationally, meaning that actual samples will exist as a population of closely related species all of which must be separated for a thorough analysis.

Capillary electrophoresis (CE) is proving itself to be valuable for the analysis of antibodies. The primary reason for this is that CE is not a single technique, but rather a collection of related techniques each of which has a unique mechanism of separation. The modes of CE commonly used for protein analysis [1] include capillary zone electrophoresis (CZE), capillary isoelectric focusing (CIEF), micellar electrokinetic capillary chromatography (MECC), capillary gel electrophoresis in the presence of sodium dodecyl sulfate (SDS–CGE) and capillary isotachophoresis (CITP). With the exception of CITP, all the aforementioned modes of CE have been used for antibody analysis and will be discussed below (summarized in Table 18.1).

18.2 Analysis of antibodies using capillary zone electrophoresis

CZE, in which the separation mechanism is charge to mass ratio, has been widely used for the analysis of antibodies and antibody fragments. Additionally, CZE is the most widely used mode for separation of bound from free species during immunoaffinity CE. Following are examples of CZE separations of antibodies.

In a thorough study, Compton derived a semi-empirical model for the mobility of proteins in CZE and applied it to IgG heterogeneity analysis.

Table 18.1 Analysis of antibodies by capillary electrophoresis

Separation mode Separation medium/buffer Voltage/current Capillary Detection	Analytes	Ref.
CZE Phosphate pH 2.5, 5.6 and 6.4 12 kV 25 μm × 20 cm (17 cm) 200 nm	Chimeric antibodies	2
CZE 0.1 M Borax, 0.5 mM SDS, pH 10.0 5 kV/15°C 75 μm × 27 cm (20 cm) 280 nm	Monoclonal antibodies and conjugates with alkaline phosphatase	7
CZE, affinity CE Isotonic borate, pH 7.4 to 9.1 15–25 kV/18°C 50 μm × 57 cm (50 cm) 200 nm	Antibody–antigen affinity	10

CIEF
Catholyte: 20 mM NaOH,
  anolyte: 10 mM H_3PO_4,
  ampholyte: 0.5–1% w/v Pharmalyte 3–10
  TEMED
Focusing: 250 V cm^{-1}, 23°C,
  mobilization: 13.5 kV, 0.5 psi rinse
50 μm × 27 cm, neutral coated capillary
280 nm
	Monoclonal anti-CEA IgG antibody	12, 39

CZE 50 mM Borax, 25 mM LiCl, pH 8.3 5 kV/25°C 75 μm × 57 cm (50 cm) 214 nm	Chimeric monoclonal antibody	14

CIEF
Catholyte 20 mM NaOH,
  anolyte: 25 mM H_3PO_4, 0.4% methylcellulose,
  ampholyte: 0.5–1% w/v Servalyte 3–10
Focusing: 13.5 kV, 2 min
Mobilization: 13.5 kV, 0.5 psi rinse
50 μm × 27 cm, neutral coated capillary
280 nm
	Monoclonal anti-CEA IgA antibody	16

SDS–CGE 3% PEG 100 000, 0.1 M Tris–CHES, 0.1% SDS 300 V cm^{-1} 100 μm × 40 cm polyacrylamide coated UV 214 cm	IgG, heavy and light chain	18

Table 18.1 *Continued*

Separation mode Separation medium/buffer Voltage/current Capillary Detection	Analytes	Ref.
SDS–CGE 10% w/v dextran, 0.1% SDS, 100 mM Tris–CHES pH 8.6 370 V cm^{-1} 100 μm × 27 cm polyacrylamide coated 200 nm	Human IgG, heavy and light chain	19
SDS–CGE ABI/Perkin Elmer Prosort SDS–protein kit Linear polyacrylamide, TES 0.2% SDS, pH 7.0 385 V cm^{-1}, 30°C 55 μm × 42 cm (22 cm) 214 nm	Bovine g-globulin, non-reduced and reduced fragments	20
CGE/SDS Beckman eCAP SDS 14–200 kit (replaceable PEG/PEO) 300 V cm^{-1} 100 μm × 27 cm polyacrylamide coated 214 nm	IgG, heavy and light chain	22
CGE/SDS Beckman eCAP SDS 14–200 kit (replaceable PEG/PEO) 300 V cm^{-1} 100 μm × 27 cm polyacrylamide coated 214 nm	Monoclonal IgG antibodies, heavy and light chain	24
MECC 25 mM SDS, 12 mM borax, pH 9.4 30 kV, 25°C 214, 280 and 500 nm	Chimeric monoclonal IgG and its conjugate with doxirubicin	28
MECC 50 mM Taurine, 35 mM cholate, 0.1 M phosphate, 2% 1-propanol, pH 6.4 to 8.7 20 kV, 30°C	Monoclonal antibody to trypsin inhibitor	29
CZE 80 mM borate, pH 10 20 kV/13.5 μA, 23°C LIF: ex: 488 nm/em: 520 nm	Assay IgA in human serum with an affinity pure, fluorescein-labeled IgG antibody to human IgA	37
CZE 20 mM TAPS buffer, pH 8.8 1.1 kV cm^{-1}, 23°C 50 μm × 30 cm UV 214 nm	Antibody Fab fragments	3
CIEF Anode buffer: 10 mM phosphoric acid, cathode buffer: 20 mM NaOH 0.25 kV cm^{-1}, 23°C 50 μm × 37 cm UV 280 nm	Anti-CEA monoclonal antibody	17

Table 18.1 *Continued*

Separation mode Separation medium/buffer Voltage/current Capillary Detection	Analytes	Ref.
SDS–CGE eCAP SDS 14–200 gel buffer (Beckman) 0.25 kV cm^{-1}, 20°C 100 μm × 27 cm UV 214 nm	OKTcdr4a monoclonal antibody	24
MECC 12 mM borate buffer, pH 9.4, 25 mM SDS 30 kV, 25°C 75 μm × 50 cm UV 214 nm	Chimeric BR96 monoclonal antibody	27
Protein A affinity CE 40 mM borate buffer, pH 10.5 15 kV, 20°C 50 μm × 20 cm	Murine monoclonal antibody in cultivation medium, FITC–protein A	36

CEA: carcinoembryonic antigen, CIEF: capillary isoelectric focusing, CZE: capillary zone electrophoresis, CHES-2-(cyclohexylamino)ethanesulfonic acid, FITC: floorescein isothiocyanate, MECC: micellar electrokinetic capillary chromatography, SDS–CGE: sodium dodecyl sulfate–capillary gel electrophoresis, TEMED: tetramethyl ethylene diamine, TES:2{[2-hydroxy-1,1-bis-(hydroxymethyl)ethyl]amino}ethanesulfonic acid, TAPS: *N*-Tris [hydroxymethyl] methyl-3-aminopropane sulfonic acid.

It was found that mobility was influenced by protein valence, size and shape, and by the ionic strength, pH, temperature and viscosity of the buffer [2].

CZE is frequently used for analysis of antibody fragments. Mhatre *et al.* purified antibody Fab fragments using cation exchange chromatography and used CZE to monitor fraction purity (Figure 18.1) [3]. Vincentelli and Bihoreau used CZE to characterize five isoforms of a purified F(ab')$_2$ fragment of an IgG [4]. They found detectable heterogeneity even with only 80 amol of antibody. The hyphenated technique CZE–mass spectrometry is proving a powerful tool for biomolecule analysis [5] and has been applied to antibody fragment analysis by Kostiainen *et al.* [6]. These researchers analyzed a 37 residue peptide comprising the antigen binding site of a monoclonal antibody using CZE–electrospray mass spectrometry. Five impurities were also separated and identified using this method [6].

Analysis of antibody conjugates is another area where CZE has proven itself useful. CZE was used for the in-process control of enzyme-labeled monoclonal antibody conjugates by Harrington *et al.* [7]. Using methyl cellulose as an additive, these workers were able to separate IgG, the conjugate and the enzyme labeled alkaline phosphatase [7]. CZE was found to be an ideal technique by Althaus *et al.* [8] for determining polyclonal and

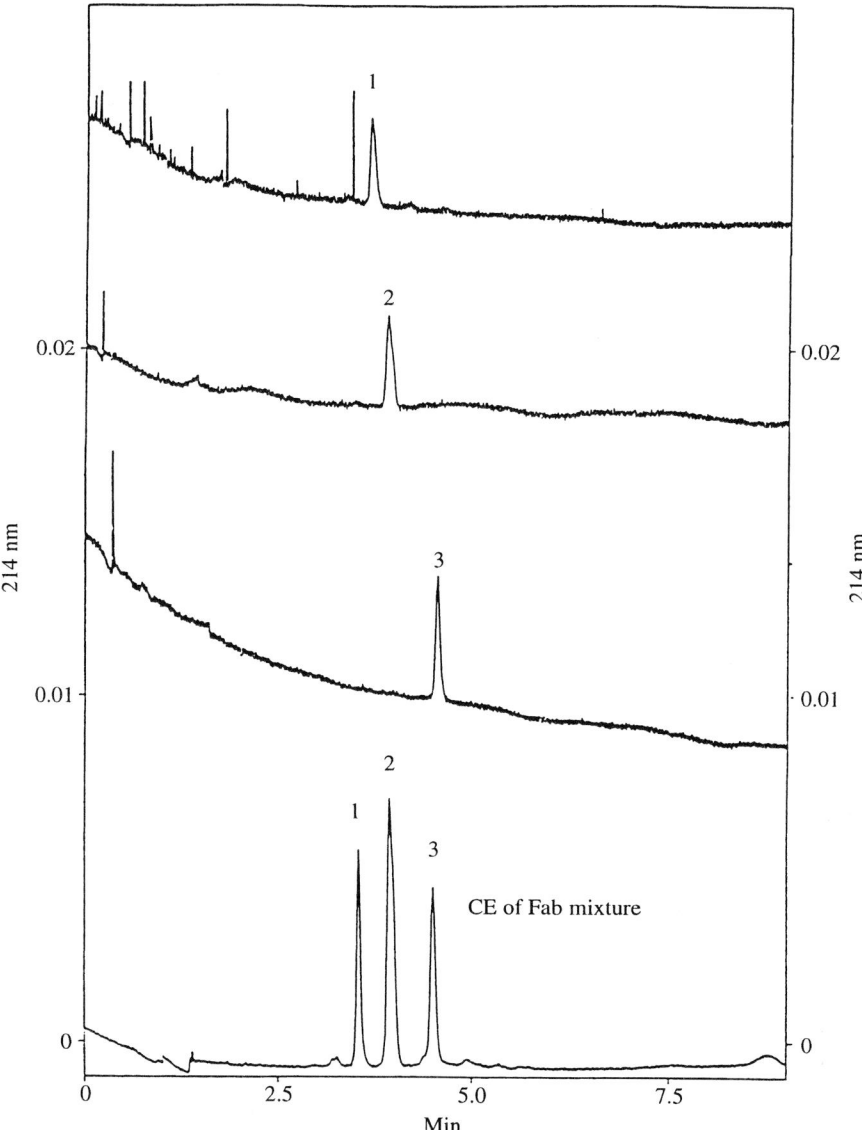

Figure 18.1 CZE analysis of fractions collected from cation-exchange chromatographic purification of antibody Fab fragments. Conditions: 50 μm i.d. × 30 cm coated silica capillary; 20 mM TAPS buffer, pH 8.8; 1100 V cm^{-1} field strength; UV detection at 214 nm. From Ref. [3] with permission.

monoclonal antibody transformation by peroxynitrite, which is suspected of being a biologically relevant endogenous neurotoxin [8]. Since most antibodies are variably glycosylated, analysis of carbohydrate groups is an important issue in antibody characterization. Hughes has reported the CZE analysis of underivatized *O*-linked and *N*-linked oligosaccharides released by hydrazinolysis from IgGs from several species [9].

CZE is also useful for assessing antibody activity. Heegard used CZE to determine the affinity of the interaction between anti-phsophotyrosine monoclonal antibodies and phosphotyrosine [10]. Evangelista and Chen used CZE for antibody screening and analysis of the structural specificity of an anti-morphine polyclonal antibody [11].

18.3 Analysis of antibodies using capillary isoelectric focusing

In CIEF, as in its traditional counterpart gel IEF, molecules are separated according to their unique isoelectric points. CIEF methods can be differentiated according to the means by which the focused zones are mobilized past the detector window. For reviews of CIEF refer to Pritchett and co-workers [12, 13].

Costello *et al.* [14] used CZE and CIEF with chemical mobilization to analyze humanized anti-Tac monoclonal antibody. Results were compared to those obtained using the traditional techniques of slab gel SDS–PAGE (polyacrylamide gel electrophoresis) and IEF. Because of simplicity, resolving power, efficiency and low sample volume requirements, these authors predicted in 1992 that CE had the potential to become the method of choice for analysis of monoclonal antibodies [14].

Silverman *et al.* used CIEF with chemical mobilization to resolve the isoforms of a murine monoclonal IgG and compared the results to slab gel IEF and cation exchange chromatography [15]. They found that the three techniques gave similar results, with the IEF methods giving the best resolution. The major difficulty they noted with CIEF was instability of the capillary coating, which is still a problem today.

Anti-CEA (carcinoembryonic antigen) monoclonal antibody was analyzed by CIEF with pressure/voltage mobilization by Huang *et al.* [16], while Pritchett used a one-step CIEF method employing limited electroosmotic flow for analysis of this antibody [12, 17]. This method, along with internal standards for pI determination, was carried out using a neutral coated capillary (Figure 18.2) [17]. The regression between migration time and pI was a first-order equation. The pIs of the anti-CEA peaks were determined by interpolation from this plot and ranged from pI 6.7 to pI 7.5. Adjacent peaks differed in pI by about 0.2 units.

18.4 Analysis of antibodies using SDS–capillary gel electrophoresis

SDS–CGE is the CE analog of traditional SDS–PAGE. As is the case for its slab gel counterpart, the mechanism of separation in SDS–CGE is relative

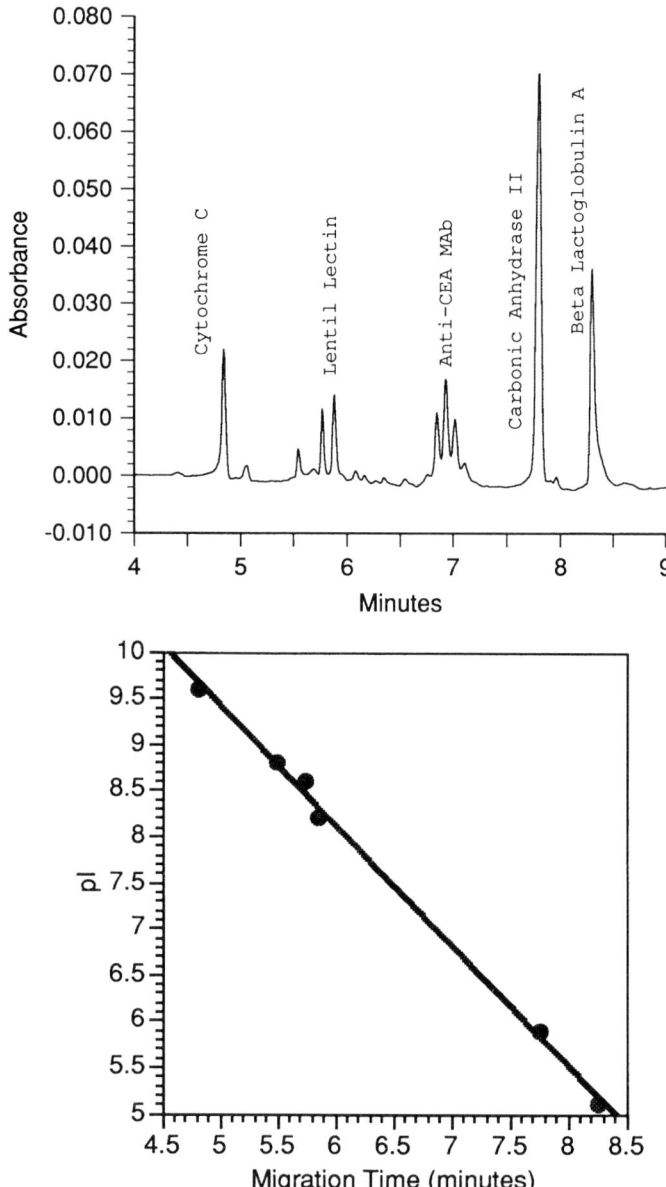

Figure 18.2 One-step CIEF analysis of anti-CEA monoclonal antibody with internal standards for pI determination. A calibration curve is shown below. Conditions: capillary, 50μm i.d. × 37cm neutral coated silica; anode buffer, 10mM phosphoric acid; cathode buffer, 20mM sodium hydroxide; run temperature, 23°C; polarity, cathode at the inlet; field strength, 250 V cm^{-1}; detection, 280nm; MAb concentration, 100μg ml^{-1}. The ampholyte solution contained 0.97ml purified water, 0.8ml hydroxypropylmethyl cellulose (4000cps), 0.2ml 3–10 ampholytes and 0.03ml N,N,N',N'-tetramethylethylenediamine per 2ml. From Ref. [17] with permission.

molecular mass. In one of the first reports of the application of SDS–CGE to antibody analysis, Ganzler *et al.* showed separation of human IgG heavy and light chains [18]. Similar separations were subsequently reported by Lausch *et al.* [19] and Werner *et al.* [20].

Determination of relative molecular masses (M_r) has always been a primary use of SDS gel electrophoresis. Guttman and Nolan reported comparisons of M_r determined by slab gel SDS–PAGE and SDS–CGE for 65 proteins, including five antibodies [21]. Discrepancies which were noted were ascribed to the non-ideal binding of SDS by post-translationally modified proteins. The authors recommended that Ferguson analysis (as described below) be used for these types of proteins.

Guttman *et al.* demonstrated the usefulness of SDS–CGE for performing Ferguson analysis of several glycoproteins, including an IgG [22]. Ferguson analysis, which involves performing a series of electrophoretic runs at varying gel concentrations, is a useful way of correcting for the non-ideal SDS binding of post-translationally modified proteins [23].

More recently, Kroon *et al.* used SDS–CGE to analyze intact and reduced monoclonal antibodies being developed as potential therapeutics [24]. The method proved useful for detecting protein contaminants and also as a means of detecting incompletely assembled antibody, a common occurrence during production of certain antibody types (Figure 18.3). These authors concluded that SDS–CGE was easy to perform and reproducible, making it an attractive technique for characterizing antibody quality.

18.5 Analysis of antibodies using micellar electrokinetic capillary chromatography

MECC, first reported by Terabe and co-workers [25], has recently been applied to the separation of antibodies. For charged species such as proteins, a number of separatory mechanisms come into play during MECC including electrophoretic and relative hydrophobicity considerations [26].

Alexander *et al.* used MECC and matrix-assisted laser desorption MS to analyze 16 different antibodies including polyclonal antibodies, monoclonal antibodies and a chimeric (human/mouse) monoclonal antibody [27]. The thermal stability of the chimeric monoclonal was studied in detail (Figure 18.4) [27]. Degradations observed included those deriving from disulfide bond disruption and from peptide bond scission. Disulfide-related instabilities included the loss of one light chain, the loss of one Fab arm, and formation of separated heavy and light chain species.

MECC has also been used to characterize a chimeric monoclonal antibody–cytoxin conjugate. The method was capable of separating the conjugated antibody from antibody heavy and light chains, antibody fragments and unconjugated cytotoxin. Conjugated antibody could be differen-

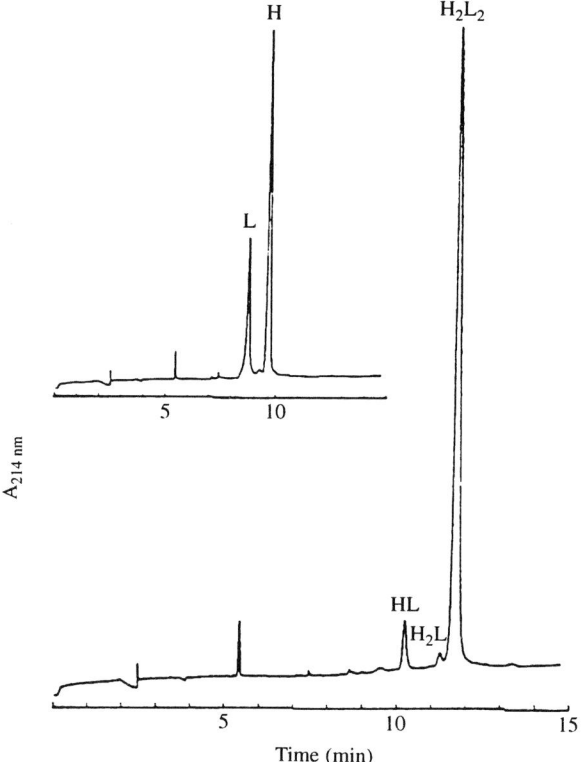

Figure 18.3 SDS–CGE analysis of OKTcdr4a monoclonal antibody unreduced and reduced with 2-mercaptoethanol. Conditions: 100 μm i.d. × 27 cm coated silica capillary; eCAP SDS 14–200 gel buffer (Beckman Instruments, Inc.); field strength of 250 V cm^{-1}; run temperature of 20°C; UV detection at 214 nm. H stands for antibody heavy chain and L for light chain. Incompletely assembled antibody is labeled HL and H_2L based on migration times relative to protein standards. From Ref. [24] with permission.

tiated from unconjugated antibody in the procedure due to the UV absorbance of the former at 500 nm [28].

Arentoft *et al.* developed the MECC method for analysis of monoclonal antibodies raised against protease inhibitors [29]. Buffers for this method contained cholate and zwitterionic compounds. The authors reported very efficient separations with the possibility of 245 000 plates m^{-1}, and also studied the effects of temperature, voltage, pH and buffer types [29]. The method also proved useful for analysing inhibitor–antibody and inhibitor–protease association complexes.

18.6 Quantitative analysis of antibodies

The development and validation of quantitative assays for antibodies is an important challenge faced by the analyst. Quantitative assays are crucial at all stages of antibody production, from unprocessed bulk lots of antibody in animal serum, hybridoma medium or ascites fluid, to purified antibody drug product. It is also likely that quantitative assays for antibodies being used as drugs, especially those assays used to evaluate dosage forms, will be closely scrutinized by regulatory agencies [30].

Traditionally, a variety of methods, including HPLC, ELISA (enzyme-linked immunoassay), UV–vis spectrophotometry, nephelometry and

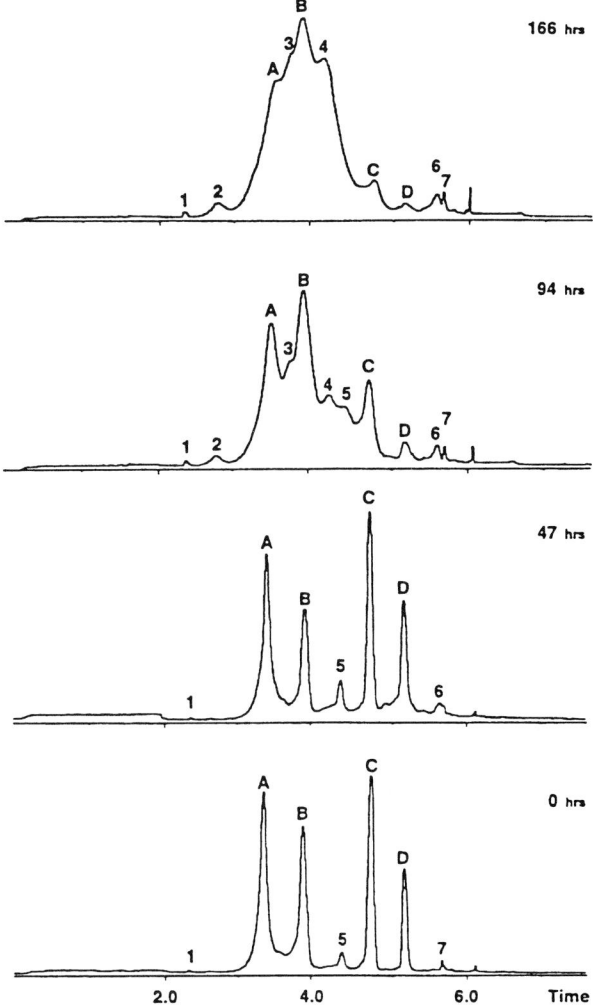

Figure 18.4 MECC analysis of heat degraded (66°C) chimeric BR96 monoclonal antibody. Conditions: 75 μm i.d. × 50 cm (length to the detector) uncoated fused-silica capillary; 12 mM borate, pH 9.4 running buffer containing 25 mM SDS; separations were performed at 30 kV and 25°C; UV detection at 214 nm. From Ref. [27] with permission.

even the semiquantitative SDS–PAGE/densitometry, have been used for antibody quantitation. This approach, however, has several disadvantages. First, many different instruments are required, increasing capital costs and consuming laboratory space. Also, new analysts must learn to operate each instrument, increasing training time. Finally, when matrix complexity and limited development time require the use of semiquantitative methods such as SDS–PAGE/densitometry, accuracy, precision and linearity often suffer.

CE offers several advantages for solving quantitative analysis problems including the following. (1) The ability to operate in numerous analytical modes with different mechanisms of separation (as detailed above). Thus, only one instrument need be purchased to achieve the multimatrix adaptability, resulting in considerable savings in money, analyst training time and laboratory space. (2) CE offers on-line detection and therefore full quantitative capability, even when performing gel techniques such as SDS–CGE. Thus, one can have the resolving power, rapid methods development and ability to tolerate complex samples typical of SDS–PAGE combined with the accuracy, precision, linearity and speed previously associated solely with HPLC. (3) Many manufacturers offer methods development kits. Many of these kits include coated capillaries designed to eliminate sample adsorption onto the capillary wall. Using these kits, analytical scientists are finding enhanced resolution, improved recovery, better resolution and greatly reduced method development time. (4) Well-developed CE methods have accuracy, precision and linearity comparable with those of HPLC methods [7, 31–35]. Detailed information for separation efficiency comparison between HPLC and CE is given in the Equipment section in this book (Part One).

In an innovative application of quantitative CE, Lausch et al. developed an affinity assay for the determination of IgG in cultivation media [36]. The affinity ligand was a recombinant fragment of protein A conjugated with a fluorescent dye, and separation of bound from free ligand was accomplished using CZE. The authors reported that IgG concentrations over a range of two orders of magnitude could be determined. Figure 18.5 shows electropherograms from assays of cultivation media containing from $5\,\mu g\,ml^{-1}$ to $60\,\mu g\,ml^{-1}$ IgG.

Low levels of immunoglobulin A (down to $6.6 \times 10^{-10}\,M$) in human serum were measured using a non-competitive CE immunoassay with laser-induced fluorescence (LIF) detection [37]. An IgA specific $F(ab)_2$ fragment labeled with a fluorescent dye was used as the affinity probe [38] in this assay. The electrophoretic mobility of the affinity probe was modified by succinylation to avoid comigration of the detection complex with the unbound probe.

Quantitative methods for monoclonal antibodies in a variety of analytical matrices were developed using three modes of capillary electrophoresis,

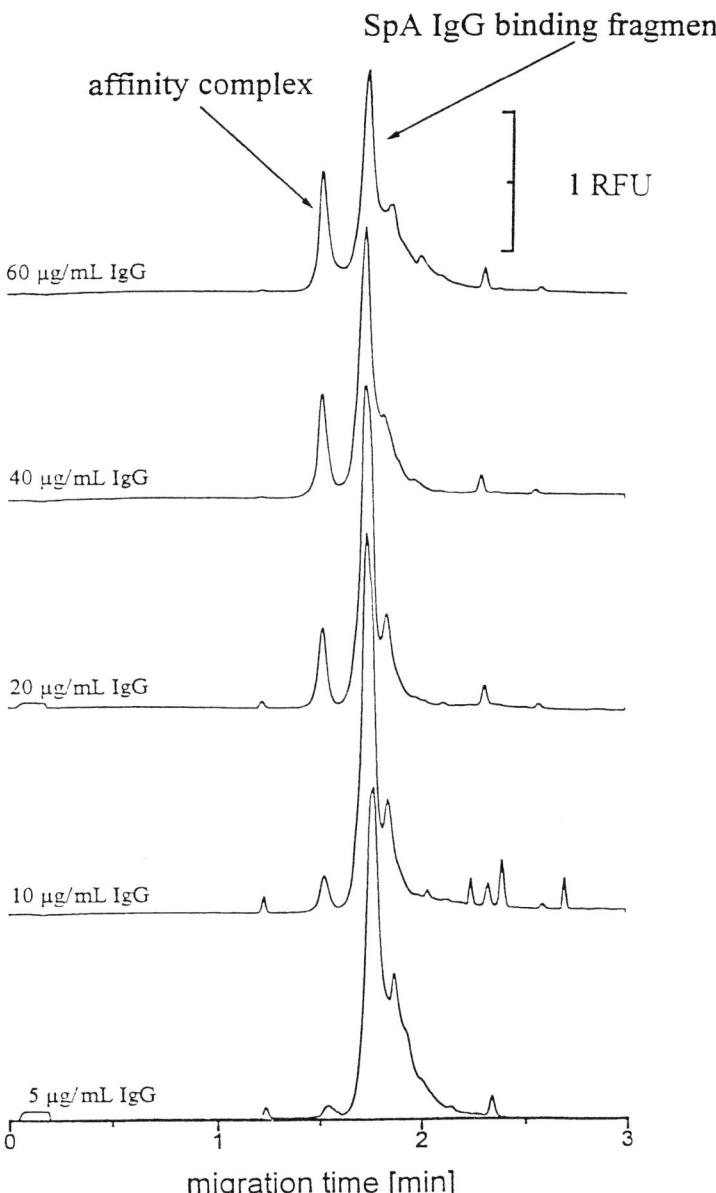

Figure 18.5 Quantitative analysis of a murine monoclonal antibody in cultivation medium using protein A affinity CE. Conditions: FITC–protein A concentration, 9.6×10^{-7} M; 50 μm i.d. × 20 cm (length to the detector) uncoated fused-silica capillary; 40 mM borate, pH 10.5 running buffer; separations were performed at 15 kV and 20°C. From Ref. [36] with permission.

CIEF, CZE and SDS–CGE [39]. Antitumor necrosis factor antibody in a model pharmaceutical dosage form including human serum albumin was quantitated using one-step CIEF with electroosmotic flow mobilization. Anti-CEA monoclonal antibody in serum-free hybridoma culture medium was quantitated using CZE, and the same antibody in serum-containing medium was quantitated using SDS–CGE.

Finally, the single capillary format of most CE instruments limits the throughput of the technique. Two approaches are being explored to overcome this limitation. The first approach involves the performance of CE in microfabricated systems with multiple microfluidic circuits [40]. The second approach involves the use of instruments employing multicapillary arrays. Such an instrument was recently introduced for the performance of clinical analyses by our company.

References

1. Landers, J.P.; Oda, R.P.; Spelsberg, T.C.; Nolan, J.A.; Ulfelder, K. (1993) Capillary electrophoresis: A powerful microanalytical technique for biological active molecules. *Biotechniques* **14**, 98–111.
2. Compton, B.J. (1991) Electrophoretic mobility modeling of proteins in free zone electrophoresis and its application to monoclonal antibody heterogeneity analysis. *J. Chromatogr.* **559**, 357–366.
3. Mhatre, R.; Nashabeh, W.; Schmalzing, D.; Yao, X.; Fuchs, M.; Whitney, D.; Regnier, F. (1995) Purification of antibody Fab fragments by cation-exchange chromatography and pH gradient elution. *J. Chromatogr.* **707**, 225–231.
4. Vincentelli, R.; Bihoreau, N. (1993) Characterization of each isoform of a F(ab')$_2$ by capillary electrophoresis. *J. Chromatogr.* **641**, 383–390.
5. Hofstadler, S.A.; Swanek, F.D.; Gale, D.C.; Ewing, A.G.; Smith, R.D. (1995) Capillary electrophoresis–electrospray ionization fourier transform ion cyclotron resonance mass spectrometry for direct analysis of cellular proteins. *Anal. Chem.* **67**, 1477–1480.
6. Kostiainen, R.; Lasonder, E.; Bloemhoff, W.; van Veelen, P.A.; Welling, G.W.; Bruins, A.P. (1994) *Biol. Mass Spectrom.* **23**, 346–352.
7. Harrington, S.J.; Varro, R.; Li, T.M. (1991) High-performance capillary electrophoresis as a fast in-process control method for enzyme-labelled monoclonal antibody conjugates. *J. Chromatogr.* **559**, 385–390.
8. Althaus, J.S.; Fici, G.C.; Von Voigtlander, P.F. (1995) *Res. Commun. Mol. Pathol. Pharmacol.* **87**, 359–366.
9. Hughes, D.E. (1994) Capillary electrophoretic examination of underivatized *O*-linked and *N*-linked oligosaccharide mixtures and immunoglobulin G antibody-released oligosaccharide libraries. *J. Chromatogr. B* **657**, 315–326.
10. Heegard, N.H.H. (1994) Determination of antigen–antibody affinity by immuno-capillary electrophoresis. *J. Chromatogr. A* **680**, 405–412.
11. Evangelista, R.A.; Chen, F.-T.A. (1994) Analysis of structural specificity of antibody–antigen reactions by capillary electrophoresis with laser-induced fluorescence detection. *J. Chromatogr. A* **680**, 587–591.
12. Pritchett, T. (1995) Isoelectric focusing of proteins by capillary electrophoresis. In *Molecular Biology: Current Innovations and Future Trends*, Griffin, A.M.; Griffin, H.G. (eds), Horizon Scientific Press, Wymondham, UK.
13. Schwartz, H.; Pritchett, T. (1994) New approaches to capillary isoelectric focusing of proteins. *Bio. Technol.* **12**, 408–409.
14. Costello, M.A.; Woititz, C.; De Feo, J.; Stremlo, D.; Wen, L.-F.; Palling, D.; Iqbal, K.; Guzman, N.A. (1992) Characterization of humanized anti-TAC monoclonal antibody by

traditional separation techniques and capillary electrophoresis. *J. Liq. Chromatogr.* **15**, 1081–1097.
15 Silverman, C.; Komar, M.; Shields, K.; Diegnan, G.; Adamovics, J. (1992) Separation of the isoforms of a monoclonal antibody by gel isoelectric focusing, high performance liquid chromatography and capillary isoelectric focusing, *J. Liquid Chromatogr.* **15**, 207–219.
16 Huang, T.-L.; Shieh, P.C.H.; Cooke, N. (1994) Isoelectric focusing of proteins in capillary electrophoresis with pressure-driven mobilization. *Chromatographia* **39**, 543–548.
17 Pritchett, T. (1994) Qualitative and quantitative analysis of monoclonal antibodies by one-step capillary isoelectric focusing. *Beckman Application Information Bulletin A-*1769.
18 Ganzler, K.; Greve, K.S.; Cohen, A.S.; Karger, B.L.; Guttman, A.; Cooke, N.C. (1992) High-performance capillary electrophoresis of SDS–protein complexes using UV-transparent polymer networks. *Anal. Chem.* **64**, 2665–2671.
19 Lausch, R.; Scheper, T.; Reif, O.-W.; Schlosser, J.; Fleischer, J.; Freitag, R. (1993) Rapid capillary gel electrophoresis of proteins. *J. Chromatogr.* **654**, 190–194.
20 Werner, W.; Demorest, D.M.; Stevens, J.; Wiktorowicz, J.E. (1993) Size-dependent separation of proteins denatured in SDS by capillary electrophoresis using a replaceable sieving matrix. *Anal. Biochem.* **212**, 253–258.
21 Guttman, A.; Nolan, J. (1994) Comparison of the separation of proteins by sodium dodecyl sulfate–slab gel electrophoresis and capillary sodium dodecyl sulfate–gel electrophoresis, *Anal. Biochem.* **221**, 285–289.
22 Guttman, A.; Shieh, P.; Lindahl, J.; Cooke, N. (1994) Capillary sodium dodecyl sulfate gel electrophoresis of proteins II. On the Ferguson method in polyethylene oxide gels. *J. Chromatogr.* **676**, 227–231.
23 Ferguson, K.A. (1964) Starch-gel electrophoresis: application to the classification of pituitary proteins and peptides. *Metab. Clin. Exp.* **13**, 985–1002.
24 Kroon, D.J.; Goltra, S.; Sharma, B. (1995) Analysis of monoclonal antibodies by sodium dodecyl sulfate–capillary gel electrophoresis with special reference to quantitation of half-antibody. *J. Capillary Electrophoresis* **2**, 34–39.
25 Terabe, S.; Otsuka, K.; Ichikawa, K.; Tsuchiya, A.; Ando, T. (1984) Electrokinetic separations with micellar solutions and open-tubular capillaries. *Anal. Chem.* **56**, 111–113.
26 Nielsen, K.R.; Foley, J.P. (1993) In *Capillary Electrophoresis: Theory and Practice*, Camilleri, P. (ed.), CRC Press, Boca Raton, FL.
27 Alexander, A.J.; Hughes, D.E. (1995) Monitoring of IgG antibody thermal stability by micellar electrokinetic capillary chromatography and matrix-assisted laser desorption/ionization mass spectrometry. *Anal. Chem.* **67**, 3626–3632.
28 Hughes, D.E.; Richberg, P. (1993) Capillary micellar electrokinetic, sequential multiwavelength chromatographic characterization of a chimeric monoclonal antibody–cytotoxin conjugate. *J. Chromatogr.* **635**, 313–318.
29 Arentoft, A.M.; Frokiaer, H.; Michaelsen, S.; Sorensen, H.; Sorensen, S. (1993) High-performance capillary electrophoresis for the determination of trypsin and chymotrypsin inhibitors and their association with trypsin, chymotrypsin and monoclonal antibodies. *J. Chromatogr.* **652**, 189–198.
30 Center for Drug Evaluation and Research (1987) *Guideline for Submitting Documentation for the Manufacture and Control of Drug Products*, Food and Drug Administration, Rockville, MD.
31 Bullock, J. (1993) Capillary zone electrophoresis and packed capillary column liquid chromatographic analysis of recombinant human interleukin-4. *J. Chromatogr.* **633**, 235–244.
32 Arcelloni, C.; Fermo, I.; Banfi, G.; Pontiroli, A.E.; Paroni, R. (1993) Capillary electrophoresis for protein analysis: Separation of human growth hormone and human insulin molecular forms. *Anal. Biochem.* **212**, 160–167.
33 Guzman, N.A.; Moschera, J.; Iqbal, K.; Malick, A.W. (1992) Effect of buffer constituents on the determination of therapeutic proteins by capillary electrophoresis. *J. Chromatogr.* **608**, 197–204.
34 Pande, P.G.; Nellore, R.V.; Bhagat, H.R. (1992) Optimization and validation of analytical conditions for bovine serum albumin using capillary electrophoreses. *Anal. Biochem.* **204**, 103–106.

35 Strege, M.A.; Lagu, A.L. (1993) Studies of migration time reproducibility of capillary electrophoretic protein separations. *J. Liquid Chromatogr.* **16**, 51–68.
36 Lausch, R.; Reif, O-W.; Riechel, P.; Scheper, T. (1995) Analysis of immunoglobulin G using a capillary electrophoretic affinity assay with protein A and laser-induced fluorescence detection. *Electrophoresis* **16**, 636–641.
37 Chen, F.-T.A. (1994) Characterization of charge-modified and fluorescein-labeled antibody by capillary electrophoresis using laser-induced fluorescence: Application to immunoassay of low level immunoglobulin A. *J. Chromatogr.* **680**, 419–423.
38 Shimura, K.; Karger, B.L. (1994) Affinity probe capillary electrophoresis: analysis of recombinant human growth hormone with a fluorescent labeled antibody fragment. *Anal. Chem.* **66**, 13–19.
39 Pritchett, T. (1995) Quantitative analysis of monoclonal antibodies using three modes of capillary electrophoresis. *Bio. Pharm.* **8**, 38–45.
40 Koutny, L.B.; Schmalzing, D.; Nashabeh, W.; Taylor, T.A.; Fuchs, M. (1995) Poster p-320 at High Performance Capillary Electrophoresis 95, Wurzburg, Germany.

19 Analysis of peptide hormones and model peptides by capillary electrophoresis

J.C. OSBORNE

19.1 Introduction

Polypeptide and protein hormones have been used extensively as model compounds in capillary electrophoresis. This is due primarily to their availability or ease of synthesis and their unique response to changes in environment or minor covalent modification. These compounds are water soluble and have half-lives from 10 to 30 min in plasma. Their conformation is a sensitive function of composition and biological activity can vary widely with structure. Many studies also include derivatives or analogs of polypeptide hormones to test conditions for optimal resolution of similar peptides.

The CE literature on polypeptide and protein hormones covers primarily their use as model peptides or proteins rather than their unique properties as hormones. This review is segmented by endocrine gland and hormone. Table 19.1 summarizes the separation conditions used for peptide hormones and model peptides. Recent reviews by Monning and Kennedy [1], Schöneich et al. [2] and that on 'Capillary zone electrophoresis of peptides' [3] cover the correlation of electrophoretic behavior with physicochemical properties of peptides and proteins in general. The analysis of model peptides are included in the references and Table 19.1.

19.2 Peptides from the pituitary gland

The pituitary gland, located at the base of the skull and weighing about 0.5 g, is known to secrete 11 protein hormones. Nine of these hormones (thyroid stimulating hormone (TSH), follicle stimulating hormone (FSH), luteinizing hormone (LH), growth hormone (GH), adrenocorticotropin (ACTH), α-melanocyte stimulating hormone (α-MSH), prolactin, β-endorphin and enkephalin) are secreted by the anterior pituitary lobe, and the remaining two (vasopressin and oxytocin) are secreted by the posterior pituitary lobe of the gland. TSH stimulates the thyroid gland, FSH and LH are involved in reproduction, GH stimulates growth of bone and muscle, and ACTH stimulates the adrenal cortex. Prolactin stimulates secretion of milk in the mammary gland and α-MSH is involved in pigmentation of the

Table 19.1 Capillary electrophoresis analysis of peptide hormones and other model peptides

Methods Separation conditions Voltage applied, current Capillary length with i.d. (effective length) Detection	Analytes	Ref.
CZE 80 mM Phosphate pH 2.2–2.6 25 kV, 20°C 75 μm × 50 cm, coated 214 nm	Met-enkephalin YGGFM (Lys-Gly-Gly-Phe-Met) Leu-enkephalin YGGEL (Lys-Gly-Gly-Phe-Leu)	4
CZE 30 mM Phosphate pH 2.5 25 kV 75 μm × 44 cm 200 nm	α-MSH, LHRH, TRH, bombesin, Leu-enkephalin, Met-enkephalin, oxytocin	5
MEKC 50 mM CHES, 7 mM SDS, 10% CH_3CN pH 9.0 20 kV 50 μm × 70 cm LIF He/CD, ex:325 nm/em: 440 nm	Tryptic digest of cytochrome c	6
CZE 20 mM Ammonium formate, pH 2.5 27 kV/13.5 μA; 30°C 100 μm × 98 cm 200 nm	Substance P Leucine–enkephalin-Lys	7
CE/TOF/MS 50 mM Acetic acid in 50:50, MeOH/water pH 4 30 kV 50 μm × 110 cm Detection by MS	Crystal violet gramicidin s, bag cell peptides (5–9 AA), *Aplesia californica*, egg laying hormone (36 AA)	8
CZE 20 mM Ammonium formate 27 kV, 85 μA: 30°C 100 μm × 98 cm (60 cm) 200 nm	Ekephalin analogs	10
CZE 20 mM Ammonium formate 27 kV/85 μA, 30°C 50 μm × 98 cm (68 cm) 200 nm	Neuropeptides, 14 opioid enkephalin analogs, dynorphin analogs, substance P, 5–17 AAs	10
CZE 20 mM Ammonium formate, pH 2.5 50 μM × 98 cm (68 cm) 30°C 200 nm	Substance P	11
CZE, 10 to 30 mM Phosphate, pH 3.0 7–30 kV; 25°C 50 μm × 48.5 cm UV detection	Enkephalin related peptides (3–6 AAs)	12

Table 19.1 *Continued*

Methods Separation conditions Voltage applied, current Capillary length with i.d. (effective length) Detection	Analytes	Ref.

CZE
20 mM Formic acid pH 3.8,
  25 mM phosphoric acid pH 2.2,
  50 mM Tris pH 7.5
25 kV, 25°C
74 μm × 57 cm
214 nm
	ACTH,
	  endorphins,
	  cholecystokinin
		13

CZE 20 mM Potassium phosphate; pH 2.6 12–25 kV, 20–25°C 50 μm × 50 cm 200 nm	Protein peptide mixtures	14
MEKC 25 mM Phosphate, 200 M borate, pH 7.0, 10 mM CTAB, 5–20% acetonitrile 15–25 kV 50 μm × 65 cm 215 nm	(Leu 13) Motilin, (Met 13) Motilin	15
CZE Good's buffer with NaCl, pH 8.5, Hepes, Tricine, Tris pH 8.5 12–20 kV, 23°C 50–75 μm × 57 cm 200 nm	Human growth hormone, human insulin, proinsulin	16
CZE 100 mM Glycine, pH 2.35 30 kV; 30°C 50 μm × 100 cm (80 cm) UV detection	hGH, bST, bHL (51–199 AAs)	17
IEF 10 mM Tris/HCl, 6% Pharmalyte 3–10 200 V cm^{-1} (1 min); 400 V cm^{-1} (4 min) 75 μm × 15 am, PA coated LIF with argon ion laser, ex: 488 nm, em: 580 nm	FAB fragment of anti-human growth hormone	18
MEKC 10 mM Sodium phosphate, 60 mM CHAPS pH 2.5 50 μm × 52 cm 15 kV; 30°C 200 nm	Oxytocin, vasopressin and its analogs	19
CZE 100 mM Phosphate pH 3.3 263 V cm^{-1} 22°C 75 μm × 57 cm 214 nm	Tryptic digest of progrowth hormone releasing hormone	20
CZE 500 mM Acetic acid, pH 4.8 16.5–17 kV/8 μA 75 μm × 120 cm 206 nm	Tetrapeptide c-terminal of growth hormone releasing peptide NH$_2$-Ala-Trp-D-Phe-Lys-NH$_2$	21

Table 19.1 *Continued*

Methods Separation conditions Voltage applied, current Capillary length with i.d. (effective length) Detection	Analytes	Ref.
CZE Various buffers pH 2.5–11.0 30 kV, 23°C 50 μm × 69 cm 200 nm	Luteinizing hormone releasing hormone amide and free acid	22
CZE 25 mM Triethylammonium phosphate with 0–30% organic modifiers: acetonitrile, methanol, ethanol, isopropanol 30 kV 50 μm × 60 cm 215 nm	Somatostatin analog peptides (6) (growth hormone release inhibiting hormone) 7–9 AAs	23
MEKC Borate–SDS buffer, pH 9.4 20 kV, 400 V cm^{-1} 30 μm × 50 cm, LIF, ex: 488 em: 630 nm	3-(2-furoyl)quinoline-carboxaldehyde, insulin B chain	24
CZE 50 mM Ammonium formate, pH 3.2 320 V cm^{-1}, 50 μm × 62 cm 200 nm	Insulin B peptidase digest	25
MEKC 10 mM Phosphate, 20–120 mM Tween-20, pH 1.5 10 kV 50 μm × 35 cm 214 nm	Leu-motilin, Met-motilin	26
CZE 25 mM Borate, 5 mM diaminopropane, pH 8.8 25 kV, 28°C 50 μm × 100 cm 200 nm	Human chorionic gonadotropin	28
CZE 0.1 M Sodium borate, 1 mM hexamethonium bromide, pH 8.4 25 kV 50 μm × 87 cm 200 nm	hCG glycoforms	29
CZE 20 mM Sodium citrate pH 2.5 30 kV; 30°C 200 nm	Human recombinant insulin-like growth factor rhIGF-I	30
CZE 100 mM Phosphoric acid pH 1.95 10 kV 75 μm × 60 cm 185 nm	Angiotension analogs	31

Table 19.1 *Continued*

Methods Separation conditions Voltage applied, current Capillary length with i.d. (effective length) Detection	Analytes	Ref.
CZE 30 mM Phosphate pH 2.4 20 kV 75 μm × 60 cm 214 nm	Bee venom, Melittin (26 AA)	32
CZE 300 mM Borate, 15 mM phytic acid, pH 9.2 10 kV, 30°C 50 μm × 27 cm 200 nm	Amino acids, peptides, Bradykinin analogs, β-lactoglobulin B, myoglobin	33
CZE 150 mM Phosphoric acid pH 2.0 20 kV, 25°C 50 μm × 50 cm 200 nm	Kemptide (LRRASLG), pp-1G peptide, (SPQPSRRGSESSEE)	34
CZE IEF pH 3–10 20 kV 100 μm × 47 cm Refractive index He–Ne CCD	Bovine, chicken cytochrome c peptide mapping	35
CZE 50 mM Phosphate pH 2.5 20 kV, 25°C 50 μm × 27 cm 214 nm	Amino acids	36
CZE 100 M Phosphate pH 2.5 20 kV 100 μm × 50 cm 214 nm	Superoxide dismutase peptide mapping	37
CZE 10 mM Citrate pH 2.5 30 kV, 30°C 100 μm × 122 cm 200 nm	rhIG-I peptide mapping	38
CZE 20 mM Ammonium acetate pH 5.5 30 kV 75 μm × 100 cm 214 nm	Synthetic heregulin-B fragment 14 AA	39
20 mM ε-amino-*n*-caproic acid pH 4.6 75 μm × 60 cm 214 nm	Synthetic heregulin fragment 63 AA	

ITP–CE
Leading: 10 mM Ammonium acetate pH 3.6,
  trailing: 50 mM acetic acid pH 3.1,
  background: 20 mM 6-aminocaproic acid pH 3.6
20–30 kV, 30°C
100 μm × 72.5 cm
280 nm
	Test peptides	40

Table 19.1 *Continued*

Methods Separation conditions Voltage applied, current Capillary length with i.d. (effective length) Detection	Analytes	Ref.
CZE 10 to 50 mM NaOH 20 kV 25 µm × 80 cm Electrochem detector Cu	Amino acids ASP-Phe methyl ester (aspartame) peptides, Gly-glyamide triglyine, hexaglycine, diglycine, tetraglycine, pentaglycine	41
20 mM Citrate pH 3.0, 7.5 kV, 30°C, 50 µm × 27 cm	Oligoglycines, oligoalanines	42
50 mM Phosphate pH 6.0, 10 kV, 25°C, 75 µm × 57 cm 200, 214 and 230 nm	Vancomycin binding di, tri, ala peptides	
CZE 20 mM Citrate pH 2.5, 20 kV, 30°C 122 cm length 200 nm	Histone rich peptides (21–38 AAs) synthetic and in saliva	43
CZE 20 mM Sodium tetraborate pH 9.2 30 kV, 30°C 50 µm × 72 cm 200 nm	Casein phosphopeptides	44
CZE 50 mM Phosphate, 25 mM SDS pH 6.1 25 kV 50 µm × 80 cm 220 nm	Phosphorylated (mono, di-) peptide isomers of insulin receptor fragment peptide (12 AA) 1142–1153	45
CZE Various buffers at pH 2–10, 5–100 mM 10–25 kV 75 µm × 63 cm, 210 nm	Peptides (9–10 AAs)	46
CZE 5 mM Borate, pH 9.2 13 kV 50 µm × 44 cm LIF with argon ion laser, ex: 488 nm, em: 520 nm	FITC-labeled peptides	47
CZE 20 mM Na citrate, pH 2.5 20 kV, 30°C 50 µm × 72 cm 200 nm	Human serum amyloid P heparin	48
CIEF 20 mM Phosphoric acid, 20 mM NaOH, 5% pharmalyte 400 V cm^{-1}, 3–10, 0.1% methyl cellulose, 16% TEMED 400 V cm^{-1} 50–75 µm × 60 cm 280 nm	Tryptic peptides of bovine/chicken cytochrome c peptide mapping, 8 peptides each	49

Table 19.1 *Continued*

Methods Separation conditions Voltage applied, current Capillary length with i.d. (effective length) Detection	Analytes	Ref.
CZE 100 mM Sodium phosphate pH 7.0 10 kV, 35°C 75 µm × 57 cm 200 nm	Synthetic amyloid b-A4 related peptides (1–40) (1–43)	50
CZE 20 mM Sodium citrate pH 2.5 30 kV, 45°C 200 nm	Ser-Pro-Arg-Gly Ser-Leu-Leu-Ser-Pro-Arg-Gly Ser-Arg-Leu-Leu-Ser-Pro-Arg-Gly	51
CZE 50 mM Phosphate pH 1.7 25 kV, 25°C 57 µm × 70 cm 206 nm	Multiple antigenic peptides, branching oligolysine core	52
CZE 80 mM Sodium phosphate pH 2.50 25 kV, 20°C 75 µm × 57.5 cm 214 nm	Peptide mapping tryptic digest of myoglobin	53
CZE 18 mM Boric acid, 5 mM sodium tetraborate, 5 mM sodium chloride pH 7.3 30 kV, 30°C 50 µm × 72 cm 200 nm	DNA binding (6 mer) to serum amyloid peptide (12 AAs) monomer and dimer	54
CZE 25 mM Tris/HCl, 2% v/v 1-propanol pH 3.0 8 kV 75 µm × 55 cm 215 nm, 410 nm	Tryptic digest of cytochrome c	55
CZE 20 mM Citrate pH 3.0 7.5 kV, 25°C 50 µm × 27 cm 200 nm	Oligoglycine (2–6), oligoalanine (2–6)	56
CZE 50 mM Phosphate pH 2.5 25 kV, 20–50°C 50 µm × 75 cm 214 nm	Polyglycine	57
CZE 20 mM Citrate pH 2.5 25 kV, 28°C, 50 µm × 57 cm 200 nm	Synthetic peptides to study disulfide dimerization	58
CZE 80 mM Sodium phosphate pH 2.5 25 kV 50 µm × 120 cm Electrochemical detection	Synthetic peptides	59

Table 19.1 *Continued*

Methods Separation conditions Voltage applied, current Capillary length with i.d. (effective length) Detection	Analytes	Ref.
CZE 50 mM Acetic acid 9.2–9.5 kV, 23°C 50 μm × 31 cm 206 nm	Enkephalin analog, diglycine, triglycine	60
CZE 20 mM Sodium citrate, pH 2.5–5.0 25 kV, 30°C, 50 μm × 72 cm, 200 nm	Vascoactive neuropeptide intestinal	61
10 mM sodium phosphate, 20 mM SDS pH 7.0 25 kV, 35°C, 50 μm × 72 cm, 200 nm	Peptide in rat brain (28 AAs)	
CZE 20 mM Phosphate pH 7.4 30 kV 50 μm × 70 cm 264 nm	Peptide binding to vancomycin	62
CZE 40 mM Sodium phosphate, 100 mM heptanesulfonic acid pH 2.5 16 kV, 30°C 50 μm × 75 cm 200 nm	Tryptic digest of recombinant human erythropoietin and sialidase treated peptides	63

AA: amino acid, ACTH: adrenocorticotropin, bHI: biosynthetic humaninsulin, bST: bovine somatostatin, CCD: charge-coupled device, CD: cyclodextrin, CHES: 2-(N'-cyclohexyl amino)ethanesulfonic acid, CRH: corticotropin releasing hormone, FITC: fluorescein isothiocyanate, hGH: human growth hormone, IEF: isoelectric focusing, ITP: isotachophoresis, MEKC: micellular electrokinetic chromatography SDS: sodium dodecyl sulfate, TOF/MS: time of flight/mass spectrometry, TEMED: tetramethylethylene diamine.

skin. β-Endorphin and enkephalin are endogenous opiates that raise pain thresholds. Vasopressin inhibits the flow of urine and at high levels increases blood pressure, and oxytocin acts on smooth muscles during contraction and milk secretion.

The enkephalins (5 AA (amino acid)) and corresponding analogs (3–6 AA) have been extensively studied using CE technology. The properties of these small peptides are quite sensitive to composition and analogs are prepared easily by peptide synthesis. Resolution in capillary zone electrophoresis (CZE) is improved with decreasing pH and most studies have been carried out at low ionic strength in acidic buffers at pH 2–4 [4–12]. Detection is routinely based on absorption in the far UV at 190 to 214 nm. Fang *et al.* [8] have reported use of on-line time of flight mass spectroscopy and Smith and co-workers [9] have used electron spray ionization in line with mass spectroscopy for detection.

The 13 residue α-MSH [5], the 39 residue ACTH [13] and the 31 residue endorphins [13] and their analogs (8–31 residues) are also well resolved at low ionic strength in acidic buffers. Piccoli *et al.* [14] have also demonstrated good resolution of ACTH from model peptides and proteins under similar conditions. The endorphins have been well resolved at neutral pH in the presence of cetyltrimethylammonium bromide (CTAB) [15] based on selective partitioning between micellular and bulk phases during electrophoresis.

The largest hormone secreted by the anterior pituitary, GH (191 AA), was well separated from corresponding molecular analogs by Arcelloni *et al.* [16] in alkaline buffers. GH was also included in the model proteins used in the electrophoretic mobility studies by Rickard *et al.* [17]. Shimura and Karger [18] have used isoelectric focusing with laser-induced fluorescent (LIF) detection to separate free and GH bound fluorescently labeled Fab fragments of anti-human GH IgG. This approach, which gave picomolar sensitivity, may prove useful for quantitation of analytes by combining CE with immunoassay formats.

Perrett *et al.* [5] demonstrated baseline separation of oxytocin from model peptides at low ionic strength in acidic buffer. Sutcliffe and Corran [19] evaluated cationic, anionic, zwitterionic and neutral micelles in micellar electrokinetic chromatography (MEKC) of oxytocin and vasopressin analogs. Good separation was obtained at low pH in the presence of the zwitterionic detergent cholamidopropyl dimethylamino-1-propane sulfonate (CHAPS).

19.3 Peptides from the hypothalamus

The hypothalamus, technically a part of the brain, is known to secrete six protein hormones. Each of these hormones acts directly on the anterior pituitary gland causing activation or inhibition of secondary hormone release. Thyrotropin releasing hormone (TRH) stimulates release of TSH and prolactin, luteinizing hormone releasing hormone (LHRH) causes release of LH and FSH, and corticotropin releasing hormone (CRH) causes release of ACTH. Growth hormone releasing hormone (GHRH) stimulates release of GH and somatostatin suppresses release of GH and TSH.

Studies of intact GHRH have not been reported. Chang *et al.* [20] obtained good resolution of the 11 tryptic digest fragments of GHRH in moderate ionic strength acidic buffers and Kasicka *et al.* [21] obtained sharp profiles of a synthetic c-terminal tetrapeptide at high ionic strength (0.5 M) in acetic acid. Heit *et al.* [22] have investigated the mobility of LHRH (10 AA) and its corresponding free acid as a function of pH and Perrett *et al.* [5] have shown good resolution of LHRH and TRH (3 AA) from model peptides at low ionic strength in acidic buffer. Somatostatin (14 AA) was also included in model peptides and proteins to demonstrate correlation of electrophoretic mobility with physiochemical properties [17]. Idei *et al.* [23] demonstrated rapid baseline separation of somatostatin analogs (7–9 AA) in acidic buffer in the presence of organic modifiers.

19.4 Peptides from the pancreas

The pancreas, a large organ weighing about 250 g responsible for secreting enzymes and chemicals involved in digestion, is known to produce three protein hormones. Insulin is secreted into the bloodstream shortly after a meal and increases the flow of glucose and amino acids into cells and stimulates storage of excess metabolites as fat and glycogen. Glucagon, also secreted shortly after meals, stimulates the conversion of amino acids into glucose (gluconeogenesis) and acts in glycogenolysis. The remaining protein hormone, pancreatic polypeptide, acts on the gastrointestinal tract.

Insulin was included in the model peptide studies of Richard *et al.* [17]. Arcelloni *et al.* [16] demonstrated separation of insulin from proinsulin in alkaline buffer. Yashima *et al.* [15] obtained baseline separation of bovine, porcine, sheep and equine insulins using MEKC in alkaline buffer using CTAB and organic modifiers. Dovichi and co-workers [24] used a solid phase fluorescent labeling technique on the β-chain of insulin and obtained sharp profiles of the fluorescent derivatives in borate buffers containing SDS. Licklider and Kuhr [25] have used insulin as a model protein in studies of on-line protease digestion used for peptide mapping.

19.5 Peptides from the gastrointestinal tract

The gastrointestinal tract is known to produce nine protein hormones. Six of these hormones, motilin, vasoactive intestinal peptide, gastric inhibitory polypeptide bombesin, neurotensin and substance P act on the gastrointestinal tract directly to stimulate relaxation and contraction of smooth muscle and gastric secretion. Gastrin acts on the stomach to increase secretion of gastric acid, secretin causes secretion of enzymes in the pancreas and cholecystokinin causes contraction of the gall-bladder and secretion of digestive enzymes by the pancreas. Substance P (11 AA) and bombesin (14 AA) have been well resolved from similar peptides at low ionic strength in acidic buffers [5, 7, 10, 11]. The Leu- and Met-Motilin (22 AA) analogs have been separated using MEKC [15, 26]. In the presence of Tween-20, human Leu^{13} and Met^{13} Motilin were baseline resolved in acidic buffer [26]. The use of an organic modifier was requried for similar resolution of porcine Leu^{13} and Met^{13} in the presence of CTAB at neutral pH [15]. Janssen and co-workers [13] have demonstrated separation of sulfated c-terminal fragments (1–8 AA) of cholecystokinin in neutral buffer.

19.6 Polypeptides from the placenta

During pregnancy, the placenta secretes one known protein hormone, human chorionic gonadotropin (hCG). This hormone stimulates secretion of progesterone in the ovary and androgens in the testes, as well as longevity of the corpus luteum. The concentration of hCG increases early in pregnancy and is the basis for most pregnancy tests. This hormone, a non-covalent dimer of non-identical α (92 AA) and β (145 AA) subunits is approximately 35% by weight in carbohydrate and exists as several isoforms due to the variations in sialic acid contents in oligosaccharides. These isoforms have different biological activities [27] and may be diagnostic of endocrine disorders. McCormick and co-workers [28] have demonstrated separation of the isoforms of the isolated and combined α and β subunits of hCG in the presence of 1,3-diaminopropane in alkaline borate buffer. This separation was quite sensitive to buffer composition and the concentation of diaminopropane. Landers and co-workers [29] in studies of α,ω-bisquaternary ammonium alkanes, demonstrated good resolution of hCG using hexamethonium bromide in alkaline borate buffer. The effect of these compounds on glycoform resolution is believed to involve interaction with the capillary wall to decrease electroosmotic flow and non-specific analyte absorption, as well as specific binding to analyte in the bulk phase.

19.7 Polypeptides from the kidney

One of the most important hormone secreted in the kidney is erythropoietin (EPO). It is a glycoprotein hormone responsible for regulating the growth and maturation of erythroid progenitor cells. The site of biosynthesis is in the liver in the fetus and switches to the kidney in adults. The analysis of EPO by CE is described in Chapter 15.

References

1. Monning, C.A.; Kennedy, R.T. (1994) Capillary electrophoresis. *Anal. Chem.* **66**, 280R–314R.
2. Schöneich, C.; Kwok, S.K.; Wilson, G.S.; Rabel, S.R.; Stobaugh, J.F.; Williams, T.D.; Vander Velde, D.G. (1993) Separation and analysis of peptides and proteins. *Anal. Chem.* **65**, 67R–84R.
3. McCormick, R.M. (1994) Capillary zone electrophoresis of peptides. In *Handbook of Capillary Electrophoresis*, Landers, J.L. (ed.); CRC Press, Boca Raton, FL. Chap. 12, pp. 287–323.

4. Castagnola, M.; Cassiano, L.; Messana, I.; Nocca, G.; Rabino, R.; Rossetti, D.V.; Giardina, B. (1994) Capillary zone electrophoresis of peptides: Prediction of the electrophoretic mobility and resolution. *J. Chromatogr.* **656**, 87–97.
5. Perrett, D.; Birch, A.; Ross, G. (1994) Capillary electrophoresis for peptides, including neuropeptides. *Biochem. Soc. Trans.* **22**, 127–131.
6. Cobb, K.A.; Novotny, M.V. (1992) Selective determination of arginine-containing and tyrosine-containing peptides using capillary electrophoresis and laser-induced fluorescence detection. *Anal. Biochem.* **200**, 149–155.
7. Lee, H.G.; Desiderio, D.M. (1994) Preparative capiallry zone electrophoresis of synthetic peptides – conversion of an autosampler into a fraction collector. *J. Chromatogr.* **686**, 309–317.
8. Fang, L.L.; Zhang, R.; Williams, E.R.; Zare, R.N. (1994) On-line time-of-flight mass spectrometric analysis of peptides separated by capillary electrophoresis. *Anal. Chem.* **66**, 3696–3701.
9. Wahl, J.H.; Goodlett, D.R.; Udseth, H.R.; Smith, R.D. (1993) Use of small-diameter capillaries for increasing peptide and protein detection sensitivity in capillary electrophoresis–mass spectrometry. *Electrophoresis* **14**, 448–457.
10. Lee, H.G.; Desiderio, D.M. (1994) Optimization of the loading limit for capillary zone electrophoresis of synthetic opioid and tachykinin peptides: A study of the interactions among the amount of peptide, resolution, saturation, injection volume and capillary diameter. *J. Chromatogr.* **662**, 35–45.
11. Lee, H.G.; Desiderio, D.M. (1994) Capillary zone electrophoresis of synthetic opioid and tachykinin peptides. *J. Chromatogr.* **667**, 271–283.
12. Thorsteinsdottir, M.; Isaksson, R.; Westerlund, D. (1995) Performance of aminosilylated fused-silica capillaries for the separation of enkephalin-related peptides by capillary zone electrophoresis and micellar electrokinetic chromatography. *Electrophoresis* **16**, 557–563.
13. Langenhuizen, M.H.J.M.; Janssen, P.S.L. (1993) Capiallry zone electrophoresis of pharmaceutical peptides. *J. Chromatogr.* **638**, 311–318.
14. Piccoli, G.; Fiorani, M.; Biagiarelli, B.; Palma, F.; Vallorani, L.; Debellis, R.; Stocchi, V. (1995) High-performance capillary electrophoretic separation of proteins and peptides using a bonded hydrophilic phase capillary. *Electrophoresis* **16**, 625–629.
15. Yashima, T.; Tsuchiya, A.; Morita, O.; Terabe, S. (1992) Separation of closely related large peptides by micellar electrokinetic chromatography with organic modifiers. *Anal. Chem.* **64**, 2981–2984.
16. Arcelloni, C.; Fermo, I.; Banfi, G.; Pontiroli, A.E.; Paroni, R. (1993) Capillary electrophoresis for protein analysis – separation of human growth hormone and human insulin molecular forms. *Anal. Biochem.* **212**, 160–167.
17. Rickard, E.C.; Strohl, M.M.; Nielsen, R.G. (1991) Correlation of electrophoretic mobilities from capillary electrophoresis with physicochemical properties of proteins and peptides. *Anal. Biochem.* **197**, 197–207.
18. Shimura, K.; Karger, B. (1994) Affinity probe capillary electrophoresis: Analysis of recombinant human growth hormone with a fluorescent labeled antibody fragment. *Anal. Chem.* **66**, 9–15.
19. Sutcliffe, N.; Corran, P.H. (1993) Comparison of selectivities of reversed-phase high performance liquid chromatography, capillary zone electrophoresis and micellar capillary electrophoresis in the separation of neurohypophyseal peptides and analogues. *J. Chromtogr.* **636**, 95–103.
20. Chang, J.P.; Smiley, D.L.; Coleman, M.R. (1994) Free-solution capillary electrophoresis of tryptic digest fragments of a recombinant porcine pro-growth hormone releasing hormone (2–76) OH. *J. Liq. Chromatogr.* **17**, 1899–1916.
21. Kasicka, V.; Prusik, Z.; Smekal, O.; Hlavacek, J.; Barth, T.; Weber, G.; Wagner, H. (1994) Application of capillary and free-flow zone electrophoresis and isotachophoresis to the analysis and preparation of the synthetic tetrapeptide fragment of growth hormone-releasing peptide. *J. Chromatogr.* **656**, 99–106.
22. Heit, M.; McFarland, A.; Bock, R.; Riviere, J. (1994) Isoelectric focusing and capillary zone electrophoretic studies using luteinizing hormone releasing hormone and its analog. *J. Pharm. Sci.* **83**, 654–656.

23 Idei, M.; Mezo, I.; Vandasz, Z.; Horvath, A.; Teplan, I.; Keri, G. (1992) Capillary electrophoretic analysis of somatostatin analog peptides. Effect of organic solvents as buffer modifiers. *J. Liq. Chromatogr.* **15**, 3181–3192.
24 Pinto, D.M.; Arriaga, E.A.; Sia, S.; Li, Z.; Dovichi, N.J. (1995) Solid-phase fluorescent labeling reaction of picomole amounts of insulin in very dilute solutions and their analysis by capillary electrophoresis. *Electrophoresis* **16**, 534–540.
25 Licklider, L.; Kuhr, W.G. (1994) Optimization of on-line peptide mapping by capillary zone electrophoresis. *Anal. Chem.* **66**, 4400–4407.
26 Yashima, T.; Tsuchiya, A.; Morita, O. (1992) Separation of closely related large peptides by micellar electrokinetic chromatography with organic modifiers. *Anal. Chem.* **64**, 2981–2984.
27 Grotjan, Jr, H.E.; Cole, L.A.; Keel, B.A.; Grotjan, Jr, H.E. (eds) (1989) *Microheterogeneity of Glycoprotein Hormones*, CRC Press, Boca Raton, FL, Chapter 11, p. 219.
28 Morbeck, D.E.; Madden, B.J.; McCormick, D.J. (1994) Analysis of the microheterogeneity of the glycoprotein chorionic gonadotropin with high performance capillary electrophoresis. *J. Chromatogr.* **680**, 217–224.
29 Oda, R.P.; Madden, B.J.; Spellsbert, T.C.; Landers, J.P. (1994) α,ω-Bis-quaternary ammonium alkanes as effective buffer additives for enhanced capillary electrophoretic separation of glycoproteins. *J. Chromatogr.* **680**, 85–92.
30 Hilser, V.J.; Worosila, S.E.; Rudnick, S.E. (1993) Protein and peptide mobility in capillary zone electrophoresis – a comparison of existing models and further analysis. *J. Chromatogr.* **630**, 329–336.
31 Lim, B.C.; Sim, M.K. (1994) Determination of angiotensins by capillary electrophoresis. *J. Chromatogr.* **655**, 127–131.
32 Qi, S.; Zhu, T.; Zhao, T.; Fang, X.H.; Sun, Y.L. (1994) Purity control of different bee venom melittin preparations by capillary zone electrophoresis. *J. Chromatogr.* **658**, 397–403.
33 Okafo, G.N.; Birrell, H.C.; Greenaway, M.; Haran, M.; Camilleri, P. (1994) The effect of phytic acid on the resolution of peptides and proteins in capillary electrophoresis. *Anal. Biochem.* **219**, 201–206.
34 Dawson, J.F.; Boland, M.P.; Holmes, C.F.B. (1994) A capillary electrophoresis-based assay for protein kinases and protein phosphatases using peptide substrates. *Anal. Biochem.* **220**, 340–345.
35 Vonguyen, L.; Wu, J.Q.; Pawliszyn, J. (1994) Peptide mapping of bovine and chicken cytochrome C by capillary isoelectric focusing with universal concentration gradient imaging. *J. Chromatogr.* **657**, 333–338.
36 Sziele, D.; Bruggemann, O.; Doring, M.; Freitag, R.; Schugerl, K. (1994) Adaptation of a microdrop injector to sampling in capillary electrophoresis. *J. Chromatogr.* **669**, 254–258.
37 Stromqvist, M. (1994) Peptide mapping using combinations of size-exclusion chromatography, reversed-phase chromatography and capillary electrophoresis. *J. Chromatogr.* **667**, 304–310.
38 Rudnick, S.E.; Hilser, V.J.; Worosila, G.D. (1994) Comparison of the utility of capillary zone electrophoresis and high-performance liquid chromatography in peptide mapping and separation. *J. Chromatogr.* **672**, 219–229.
39 Castagnola, M.; Cassiano, L.; Messana, I.; Nocca, G.; Rabino, R.; Rossetti, D.V.; Giardina, B. (1994) Capillary zone electrophoresis of peptides: Prediction of the electrophoretic mobility and resolution. *J. Chromatogr.* **656**, 87–97.
40 Witte, D.T.; Agard, S.; Larsson, M. (1994) Improved sensitivity by on-line isotachophoretic preconcentration in the capillary zone electrophoretic determination of peptide-like solutes. *J. Chromatogr.* **687**, 155–166.
41 Ye, J.N.; Baldwin, R.P. (1994) Determination of amino acids and peptides by capillary electrophoresis and electrochemical detection at a copper electrode. *Anal. Chem.* **66**, 2669–2674.
42 Goodall, D.M. (1993) Studies of binding equilibria for peptides and pharmaceuticals using capillary electrophoresis. *Biochem. Soc. Trans.* **21**, 125–129.
43 Lal, K.; Xu, L.; Colburn, J.; Hong, A.L.; Pollock, J.J. (1992) The use of capillary

electrophoresis to identify cationic proteins in human parotid saliva. *Arch. Oral Biol.* **37**, 7–13.
44 Adamson, N.J.; Reynolds, E.C. (1995) High performance capillary electrophoresis of casein phosphopeptides containing 2–5 phosphoseryl residues: relationship between absolute electrophoretic mobility and peptide charge and size. *Electrophoresis* **16**, 525–528.
45 Tadey, T.; Purdy, W.C. (1995) Capillary electrophoretic resolution of phosphorylated peptide isomers using micellar solutions and coated capillaries. *Electrophoresis* **16**, 574–579.
46 Cifuentes, A.; Poppe, H. (1995) Effect of pH and ionic strength of running buffer on peptide behavior in capillary electrophoresis: theoretical calculation and experimental evaluation. *Electrophoresis* **16**, 516–524.
47 Zhao, J.Y.; Waldron, K.C.; Miller, J.; Zhang, J.C.; Harke, H.; Dovichi, N.J. (1992) Attachment of a single fluorescent label to peptides for determination by capillary zone electrophoresis. *J. Chromatogr.* **608**, 239–242.
48 Heegaard, N.H.H.; Robey, F.A. (1992) Use of capillary zone electrophoresis to evaluate the binding of anionic carbohydrates to synthetic peptides derived from human serum amyloid-p component. *Anal. Chem.* **64**, 2470–2482.
49 Mazzeo, J.R.; Martineau, J.A.; Krull, I.S. (1993) Peptide mapping using EOF-driven capillary isoelectric focusing. *Anal. Biochem.* **208**, 323–329.
50 Sweeney, P.J.; Darker, J.G.; Neville, W.A.; Humphries, J.; Camilleri, P. (1993) Electrophoretic techniques for the analysis of synthetic amyloid beta-A4-related peptides. *Anal. Biochem.* **212**, 179–184.
51 Mulholland, F.; Hague, G.R. (1992) Monitoring 4-methoxy-2,3,6-trimethylbenzenesulphonyl deprotection of arginine-containing synthetic peptides using capillary zone electrophoresis. *J. Chromatogr.* **589**, 380–384.
52 Lambros, T.; Lema, P.; Liu, W.; Bongers, J.; Felix, A.M.; Heimer, E.P. (1993) Capillary electrophoresis so multiple antigenic-peptide (MAPS) of the malaria circumsporozoite protein epitopes. *J. Liq. Chromatogr.* **16**, 2039–2048.
53 Castagnola, M.; Cassiano, L.; Rabino, R.; Rossetti, D.; Assi, F. (1991) Peptide mapping through the coupling of capillary electrophoresis and high-performance liquid chromatography – map prediction of the tryptic digest of myoglobin. *J. Chromatogr.* **572**, 51–58.
54 Heegaard, N.H.H.; Robey, F.A. (1993) Use of capillary zone electrophoresis for the analysis of DNA – binding to a peptide derived from amyloid P component. *J. Liq. Chromatogr.* **16**, 1923–1939.
55 Huang, M.X.; Plocek, J.; Novotny, M.V. (1995) Hydrolytically stable cellulose-derivative coatings for capillary electrophoresis of peptides, proteins and glycoconjugates. *Electrophoresis* **16**, 396–401.
56 Survay, M.A.; Goodall, D.M.; Wren, S.A.C.; Rowe, R.C. (1993) Oligoglycines and oligoalanines as tests for modelling mobility of peptides in capillary electrophoresis. *J. Chromatogr.* **636**, 81–86.
57 Chen, N.; Wang, L.; Zhang, Y.K. (1993) Influence of column temperature and physicochemical properties on the electrophoretic behaviour of polyglycine peptides in free-solution capillary electrophoresis. *J. Chromatogr.* **644**, 175–182.
58 Landers, J.P.; Oda, R.P.; Liebenow, J.A.; Spelsberg, T.C. (1993) Utility of high resolution capillary electrophoresis for monitoring peptide homo- and hetero-dimer formation. *J. Chromatogr.* **652**, 109–117.
59 Deacon, M.; Oshea, T.J.; Lunte, S.M.; Smyth, M.R. (1993) Determination of peptides by capillary electrophoresis electrochemical detection using on-column Cu (II) complexation. *J. Chromatogr.* **652**, 377–383.
60 Kasicka, V.; Prusik, Z.; Pospisek, J. (1992) Conversion of capillary zone electrophoresis to free-flow zone electrophoresis using a simple model of their correlation – application to synthetic enkephalin-type peptide analysis and preparation. *J. Chromatogr.* **608**, 13–22.
61 Soucheleau, J.; Denoroy, L. (1992) Determination of vasoactive intestinal peptide in rat brain by high-performance capillary electrophoresis. *J. Chromatogr.* **608**, 181–188.
62 Chu, Y.H.; Avila, L.Z.; Biebuyck, H.A.; Whitesides, G.M. (1993) Using affinity capillary

electrophoresis to identify the peptide in a peptide in a peptide library that binds most tightly to vancomycin. *J. Org. Chem.* **58**, 648–652.

63 Rush, R.S.; Derby, P.L.; Strickland, T.W.; Rohde, M.F. (1993) Peptide mapping and evaluation of glycopeptide microheterogeneity derived from endoproteinase digestion of erythropoietin by affinity high performance capillary electrophoresis. *Anal. Chem.* **65**, 1834–1842.

20 Affinity analysis by capillary electrophoresis
J.E. WIKTOROWICZ

20.1 Introduction

The ability of biomolecules to recognize other molecules (molecular recognition) is fundamental to biochemistry, and serves as the basis of molecular medicine and therapeutics. Receptor–ligand, enzyme–substrate (inhibitor), antibody–antigen are the general classes of such interactions that have found important application in the treatment of human disease. The measurement of the thermodynamic parameters that govern molecular recognition is an important step in characterizing any such molecular system and is generally termed affinity analysis. Affinity analysis is therefore broadly defined as any technique that exploits the chemical affinity of one molecule for another in order to measure the thermodynamic parameters that govern the interaction.

To succeed, the affinity complex (receptor–ligand, enzyme–substrate, antibody–antigen, etc.) must behave differently from the individual ligands. For example, the change in the size of the complex with respect to the free ligands might result in measurable differences in diffusion rates (as in immunodiffusion or equilibrium dialysis), measurable differences in molecular weight (as in size exclusion techniques) [1], and/or differences in sedimentation (as in ultracentrifugation) [2]. Alternatively, spectrophotometric properties of the complex might be modified by the interaction [3] (measured by fluorescence quenching or spectral shifts), or electrophoretic migration might be altered (as in migration inhibition assays, or chiral electrophoresis). In all forms, however, the fundamental concept remains that the formation of complexes is dependent upon the binding constant, kinetic rates of binding and the relative concentration of ligands and receptors, as well as other physical parameters, such as temperature, pH, etc. As long as the concentration of bound (complexed) ligands can be measured and compared to the concentrations of free ligands in an equilibrium reaction, the binding parameters and kinetic constants can be measured.

These relationships have been developed long ago and the applications are well known. The most widely used approach involves separation based upon exploitation of the size difference between the complex and one of the ligands, such as a drug–receptor complex and the free drug, respectively. Three approaches have been developed to characterize the relative concentrations of these equilibrium species: the Hummel–Dreyer method [1], the vacancy peak method (indirect detection method) [4] and the frontal analysis method [5].

The application of capillary electrophoresis (CE) to affinity analyses is novel and brings the characteristic advantages of CE, i.e. speed, flexibility, absolute quantification, minute sample volumes and automation to affinity analyses, summarized in Table 20.1. The power of CE in this application and others, lies in its ability to exploit a free solution format. In this configuration, electrophoresis is performed without the need for rigid gels as anticonvective media. In fact, it may be performed in a simple aqueous buffer or a complex entangled polymer system. The beauty of CE is that the electrophoretic environment can be absolutely controlled by the user. All that is needed is an understanding of the analytical technique to be exploited and the molecule(s) to be analyzed.

20.2 Size/mobility-based affinity CE

The technique of CE can be exquisitely applied for the analysis of affinity-based interactions. All three size-based strategies mentioned above have been applied to CE in an important paper that investigated the efficacy of each size-based strategy in the CE format [6]. The application of these strategies is summarized in Figure 20.1 and below.

All of these methods require that the free drug migrate differently from the protein–drug complex. The Hummel–Dreyer method (Figure 20.1(a)) requires, in addition, that the drug exhibit absorbance at the analysis wavelength. This imparts a high background when the drug is incorporated into the separation buffer. Addition of the drug to the sample ensures that the sample slug initially contains the same absorbance as the separation buffer and is in equilibrium with the protein. Electrophoresis causes the protein–drug complex to migrate out of the sample slug, leaving a region of low absorbance (negative peak) that migrates with the free drug. Clearly the magnitude of the negative peak represents the amount of drug in equilibrium with the protein. Inclusion of the drug in the separation buffer at an equal concentration to that originally in the sample ensures that the equilibrium condition is met.

The vacancy peak method (indirect detection method) (Figure 20.1(b)) permits the momentary violation of the equilibrium stipulation. The capillary is filled with protein and drug, imparting a high background absorbance. Buffer containing neither is injected and electrophoresis performed. At the front edge of the buffer, the slower migrating of the two (protein or drug) is 'captured' in the buffer plug and exists in the free form, as the other rapidly migrates away when released from the complex. Similarly, at the back edge of the buffer plug, the more rapidly migrating component is free.

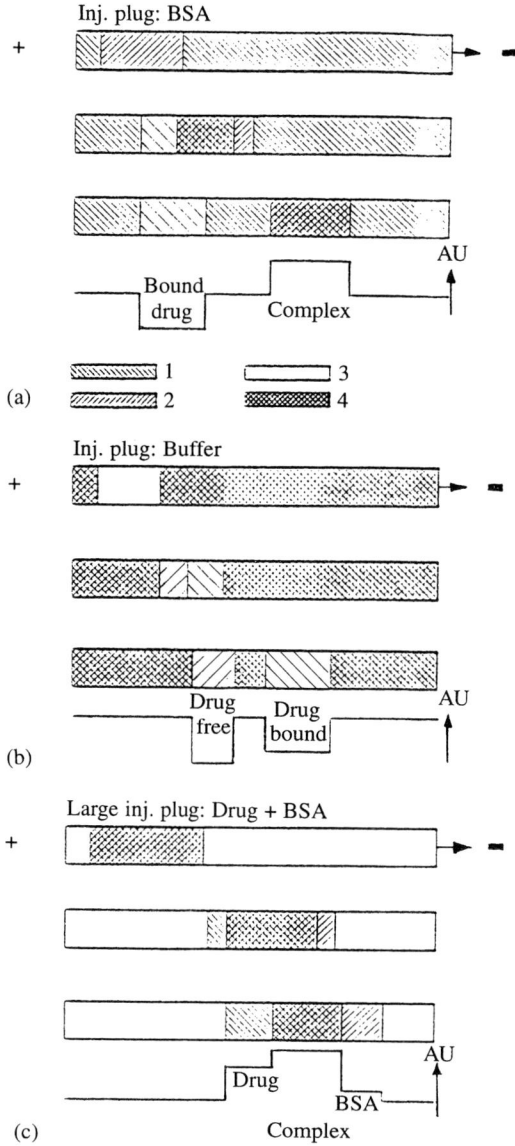

Figure 20.1 (a) Schematic representation of the Hummel–Dreyer capillary zone electrophoresis (CZE) method [1]. 1 = Drug; 2 = BSA; 3 = buffer; 4 = BSA–drug complex. (b) Schematic representation of the vacancy peak CZE method. (c) Schematic representation of the frontal analysis CZE method. Reproduced from Ref. [6] with the kind permission of Elsevier Science – NL, Sara Burgerhartstraat 25, 1055 KV Amsterdam, Netherlands.

This continues until both fronts meet and interact. This process continues until at some point in time, the equilibrium is re-established as the formation of the complex will alter the mobility of the front and prevent either free component from migrating through the zone of interaction. When equilibrium is achieved the complex and free drug will migrate as distinct, negative peaks each reflecting their respective equilibrium concentrations.

The frontal analysis approach (Figure 20.1(c)) directly measures the concentration of drug in the free and bound forms by selecting the detection wavelength and filling the capillary with only buffer. In this scheme, a large sample plug containing drug, protein and buffer is injected. Once again since the mobilities of the drug and protein differ, the sample plug contains a gradient of free protein, complex and free drug (depending on their relative mobilities). Selection of the wavelength to correspond to the maximum of the drug permits detection and quantification of only the drug and its complex. Using all three methods, Kraak *et al.* [6] examined the behavior and affinity parameters of a bovine serum albumin–warfarin affinity system (Figure 20.2).

20.3 General applications

Significant power can be realized from the free solution CE approach if the mobility of the receptor changes upon complexation with a ligand. This condition is most likely to occur if the ligand imparts significant charge density or frictional coefficient to the complex. Under these circumstances it may even be possible to separate and quantify intermediate [7] or enantiomeric [8] forms of the receptor–ligand complex.

In general the approach taken to study the binding constants (K_b), stoichiometry and dissociation rates (k_{on} and k_{off}) of receptor–ligand complexes involves holding the concentration of one component constant (more often but not always the larger of the two, such as a receptor or enzyme) while changing the concentration of the other component in successive runs. The measurement of free ligand is required to process the data and obtain an estimate of the constant. In addition, stoichiometric studies need to discriminate between weak (fast off-rate) and tight (slow off-rate) binding systems.

20.3.1 Measurement of the binding constant

If the electrophoretic mobility of the protein is changed upon complexation with the ligand, then the binding constant, K_b, and the kinetic constants, k_{on} and k_{off}, may be determined by Scatchard analysis, and by analyzing the peak shapes, respectively. The magnitude of the change in mobility as a

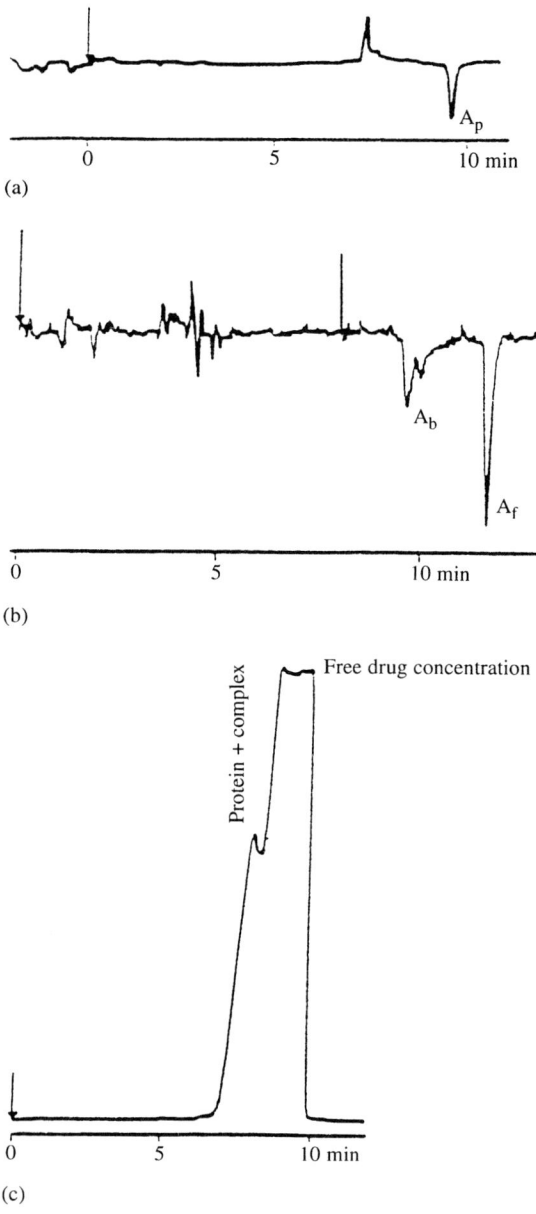

Figure 20.2 Typical electropherograms obtained with (a) the Hummel–Dreyer method, (b) the vacancy peak method and (c) frontal analysis method. A_p = sample peak, A_b = bound (drug) peak, A_f = free (drug) peak. Reproduced from Ref. [6] with the kind permission of Elsevier Science – NL.

function of the variable concentration of ligand in the separation buffer yields K_b by conventional Scatchard analysis as follows [6, 9]:

$$\Delta t_L / [L] = K_b \Delta t^{max} - K_b \Delta t_L \qquad (20.1)$$

where Δt_L is the migration time difference of a protein at a concentration of ligand [L] and the migration time at [L] = 0, and Δt^{max} is the value of Δt at saturating concentrations of L. When electroosmotic flow (EOF) is negligible, Δt_L is proportional to $\Delta \mu_L$, the mobility difference. This method has been used to determine the binding constants of multiple FMOC-peptides (fluorenylmethoxycarbonyl-labeled peptides) to vancomycin [8].

In the presence of significant EOF, equation 20.1 is modified to reflect its contribution to migration time. That is, corrected mobility is used as follows [10]:

$$\Delta \mu_{P,L} / [L] = K_b \Delta \mu^{max}_{P,L} - K_b \Delta \mu_{P,L} \qquad (20.2)$$

where $\Delta \mu_{P,L}$ is the change in EOF-corrected mobility of the protein as a function of the concentration of added ligand. This method was used to demonstrate the utility of affinity analysis by CE in estimating the binding constants between carbonic anhydrase B and charged benzenesulfonamides [10].

The underlying assumptions in this type of analysis include (1) that the interaction between the species is monovalent, (2) that the ligand is present in a vast excess over the protein and (3) that equilibrium is achieved over the course of the run.

20.3.2 Measurement of the kinetic constants

As stated above, the determination of binding constants by CE is greatly facilitated when the mobility of the complex is different from the protein alone. This suggests that during the course of the electrophoretic experiment, the rate of dissociation (k_{off}) will cause the peak to display a heterogeneous behavior (band broadening) representative of the magnitude of the off rate. The faster the rate, the more time the protein will migrate according to its 'free' mobility and the peak will exhibit either tailing or fronting depending upon whether the free protein migrates slower or faster than the complex. In order to quantify this behavior, a simulation model [11, 12] was created that changed the peak widths as free protein was distributed for fractional time increments with complexed ligand. The magnitude of the protein:ligand contribution was measured in a separate experiment with excess ligand. The simulation was completed when the total time for the simulation equaled the experimentally determined migration time for the uncomplexed protein. Using the experimentally estimated binding constants, the authors were able to approximate the experimental

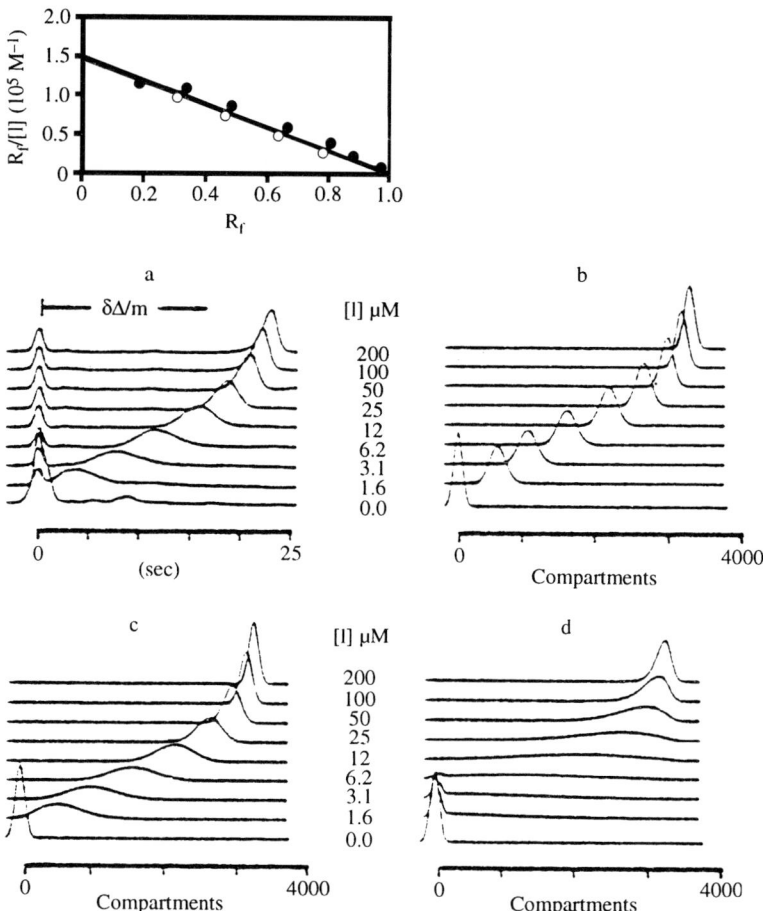

Figure 20.3 Migration time of carbonic anhydrase changes with increasing concentration of charged ligand in the electrophoresis buffer. The non-mobile peak is due to horse heart myoglobin. Stacked electropherograms (a) were obtained experimentally. Electropherograms (b–d) were generated by simulation. The graph is a Scatchard plot of the experimental (●) data from (a) and simulation (○) data (c). Reproduced from Ref. [11] with the kind permission of the American Chemical Society (copyright 1993).

electropherograms by using different combinations of k_{off} and k_{on} (Figure 20.3).

20.3.3 Measurement of the binding stoichiometry

The determination of binding stoichiometry by CE can be accomplished in tight-binding (slow-off) systems by simple inclusion into the separation

GENERAL APPLICATIONS

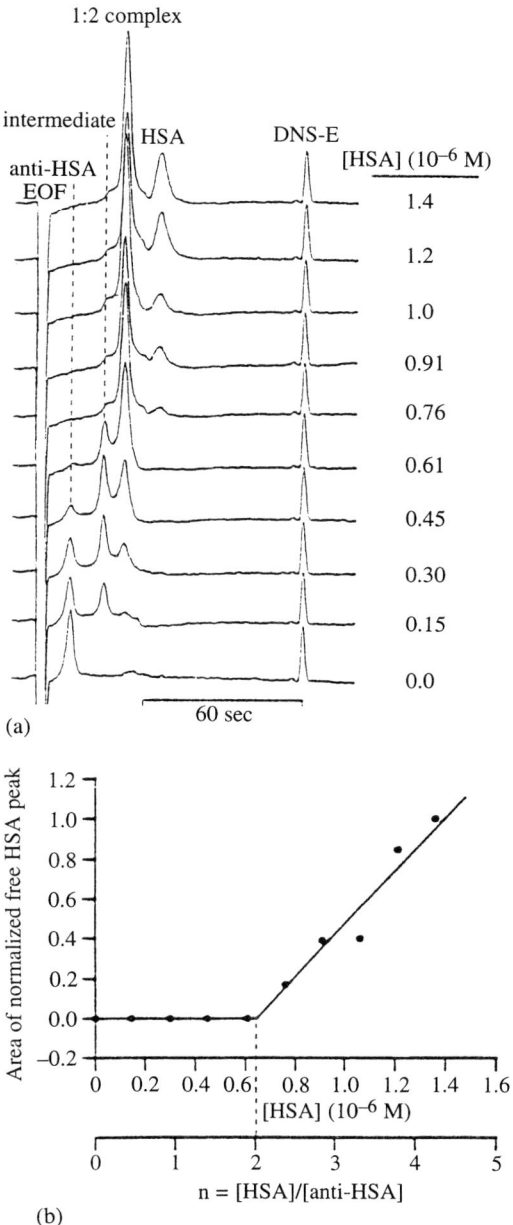

Figure 20.4 Determination of the binding stoichiometry of human serum albumin (HSA) to its mouse monoclonal IgG antibody (anti-HSA) (a) The total electrophoresis time in each experiment was 4.0 min at 10 kV using 100 min as the detection wavelength. DNS-E was dansylglutamic acid used as an internal standard. The intermediate species was tentatively assigned to the 1:1 complex. (b) A plot of the concentration of free ligand vs. the ratio of [HSA]/[anti-HSA] gives a sharp break at the stoichiometric point. Reproduced from Ref. [7] with the kind permission of the American Chemical Society (copyright 1994).

buffer of a known concentration of receptor protein. Successive, various concentrations of ligand are then sequentially injected into the capillary. When an internal standard is included, errors due to injection can be eliminated and the integral of the free ligand peaks can be normalized. A plot of the normalized free ligand area vs. the concentration of ligand of each injection yields a sharp inflection point representing the point of equivalence (Figure 20.4).

Measurement of the stoichiometry of weakly interacting systems is slightly more complex. Under these circumstances, the receptor must be saturated ($[\text{ligand}]_{\text{sample}}$ and $[\text{ligand}]_{\text{buffer}} \gg K_a$) before and after introduction into the capillary [7]. Electrophoresis in the presence of several concentrations of ligand in the sample and buffer yields a series of electropherograms with variable free ligand peak areas, negative and positive (Figure 20.5). Plotting the normalized peak areas of free ligand vs. the concentrations of the ligands permits interpolation to zero free ligand peak area. Subtraction of this value from the concentration of ligand added to the buffer and division of the remainder by the concentration of receptor yields the binding stoichiometry of the system. In this way Chu *et al.* [7] calculated the binding stoichiometry of carbonic anhydrase B with an arylsulfonamide affinity ligand of 1:1. A non-interacting internal standard was added in order to correct for injection errors.

20.4 Immobilized affinity analysis

A subset of the free solution CE approaches to affinity analyses involves the immobilization of one of the components of an affinity system. Several approaches lend themselves to this strategy. Certainly the capillary itself may provide an adequate adsorption surface for the immobilization of either ligand, protein or inhibitor, as many a frustrated CE user can testify. This approach has been recently utilized to perform classical Michaelis–Menten kinetics of β-galactosidase and *o*- and *p*-nitrophenols [14]. In this work, the enzyme was covalently immobilized on the capillary surface through a bridging group, and the substrate injected into the capillary. Electrophoresis was performed and the reaction products migrated towards the detector. This permitted direct estimation of the concentration of products and the calculation of kinetic constants by Lineweaver–Burk analysis.

The first instance of base-specific detection of DNA molecules was reported using poly(9-vinyladenine) immobilized on polyacrylamide-filled capillaries. This report demonstrated the utility of this strategy for the separation and quantification of DNAs of specific sequence [16]. Injection of variable lengths of poly-dT was followed by electrophoresis in the presence of urea. The separation of the various poly-dT length isomers was

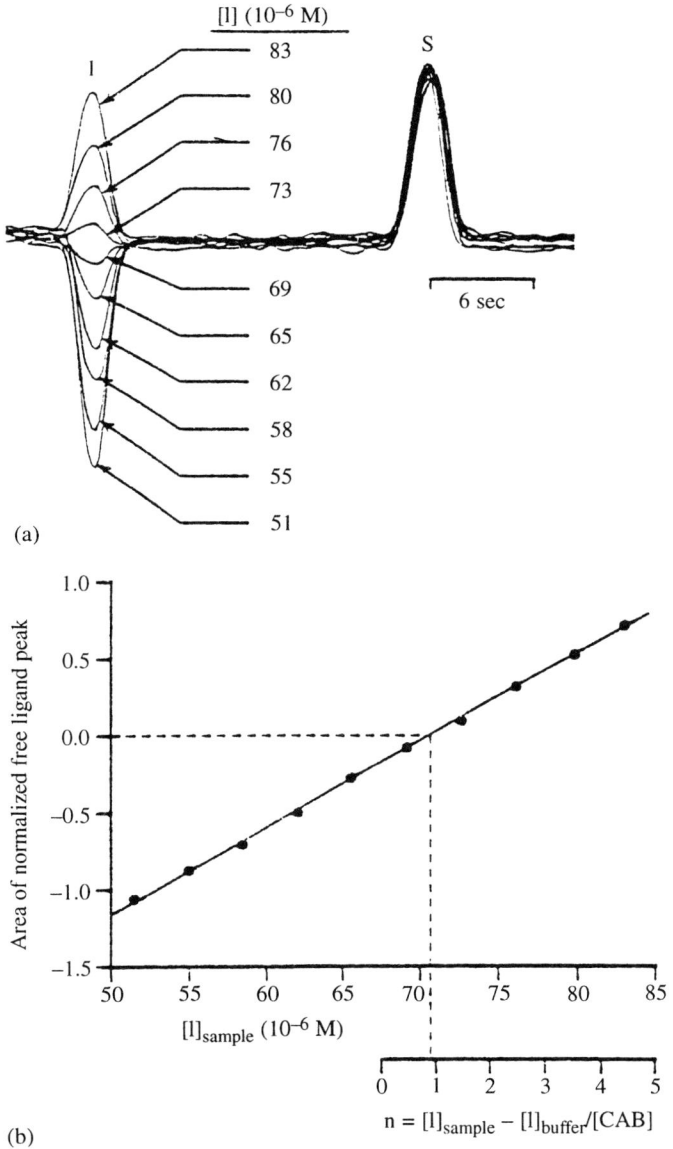

Figure 20.5 Determination of binding stoichiometry of a low affinity system: the carbonic anhydrase–arylsulfonamide interaction. (a) Affinity capillary electrophoresis of samples having a constant concentration of bovine carbonic anhydrase B and various concentrations of affinity ligand. 2-Iodobenzoic acid (S) was used as the internal standard for integration calibration. (b) Experimental data according to panel (a). Reproduced from Ref. [7] with the kind permission of the American Chemical Society (copyright 1994).

Table 20.1 Conditions for capillary electrophoresis affinity analysis

Methods	Compounds	Matrix	Ref.
67 mM Phosphate, pH 7.4, 200 V cm^{-1}, 60 cm × 50 µm, 340 nm	Warfarin–BSA	Buffer	6
192 mM Glycine–25 mM Tris, pH 8.3, 444 V/cm, 27 cm × 50 µm, 200 nm	Carbonic anhydrase, B-arylsulfonamides	Buffer	7
192 mM Glycine–74 mM Tris, pH 8.7, 370 V cm^{-1}, 27 cm × 50 µm, 200 nm	HSA–anti-HSA (mouse monoclonal)	Buffer	7
192 mM Glycine–25 mM Tris, pH 8.3, 370 V cm^{-1}, 27 cm × 50 µm, 200 nm	Streptavidin–fluorescein	Buffer	7
10 mM Phosphate, pH 7.1, ~666 V cm^{-1}, 45 cm × 50 µm, 200 nm	Vancomycin–synthetic peptides	Buffer	8
192 mM Glycine–25 mM Tris, pH 8.3, 444 V cm^{-1}, 27 cm × 50 µm, 200 nm	Carbonic anhydrase, B-arylsulfonamides	Buffer	11
10 mM Phosphate, pH 7.0, 300 V cm^{-1}, 32 cm × 75 µm, 200 nm	HSA–, BSA–ibuprofen, flurbiprofen, ketoprofen	Buffer	13
20 mM MOPS/NaOH, pH 7.1, 238 V cm^{-1}, 42 cm × 75 µm, 410 nm	Immobilized β-galactosidase–o-, p-nitrophenols	Buffer	14
50 mM Tris/HCl/1.5 mM NaCl, pH 7.4, 278 V cm^{-1}, 72 cm × 50 µm, 200 nm	Amyloid P peptides–anionic carbohydrates	Buffer	15
0.1 M Tris–borate/variable urea, pH 8.3, 200 V cm^{-1}, 50 cm × 100 µm, 260 nm	Poly(9-vinyladenine) polyacrylamide–T_nA_n oligonucleotides	Buffer	16–19

BSA = bovine serum albumin, HSA = human serum albumin, MOPS = 4-morpholinopropane sulfonic acid.

facilitated by the slow ramping of temperature from 30°C to 60°C. It was also shown using this gel formulation that sequence isomers of the general formula, T_5A, could be separated using varying concentrations of urea [17, 18]. In addition, the association constants for the three forms T_6, A_3T_3 and A_6 (obviously zero) were calculated [19]. One can easily contemplate the application of immobilized gene probe analogs serving as sequence-specific diagnostic markers.

20.5 Conclusions

By virtue of the ability of exploit the advantages of a free solution electrophoretic separation, the limits to the capability of CE to replace more conventional analysis schemes are yet to be defined. The rapid analysis time, essentially non-destructive sample use and precise control of separation parameters, as well as physical and chemical environments, have permitted the replacement of many time- and sample-consuming analytical techniques.

The wide application of CE to new techniques is not without its hazards, however. Model systems, such as the ones demonstrating affinity analyses in this chapter, must take into consideration the unique disadvantage of CE, i.e. the potential for wall interactions. Clearly, severely retarded species would prevent accurate estimation of dissociation rates at the very minimum. To this end many strategies have been created to minimize this problem, and sufficient flexibility is available to the investigator to tailor the capillary wall carefully to accommodate their system. While simple and straightforward to use, it is nevertheless incumbent upon the investigator to understand the properties of their analytes and to select carefully the analytical system to take advantage of CE fully. With this caveat in mind, we have only scratched the surface of the number of new ways CE can be utilized to address the needs of the biochemist.

References

1. Hummel, J.P.; Dreyer, W.J. (1962) Measurement of protein binding phenomena by gel filtration. *Biochim. Biophys. Acta* **63**, 530.
2. Boudier, C.; Bieth, J.G. (1992) The proteinase:mucus proteinase inhibitor binding stoichiometry. *J. Biol. Chem.* **267**, 4370–4375.
3. Chriswell, C.D.; Schilt, A.A. (1975) New and improved techniques for applying the mole ratio method to the identification of weak complexes in solution. *Anal. Chem.* **47**, 1623–1629.
4. Sebille, B.; Thuaud, N.; Tillement, J.P. (1979) Equilibrium saturation chromatographic method for studying the binding of ligands to human serum albumin by high-performance liquid chromatography. *J. Chromatogr.* **180**, 103–110.
5. Cooper, P.F.; Wood, G.C. (1968) Protein-binding of small molecules: New gel filtration method. *J. Pharm. Pharmacol.* **20**, 1503.
6. Kraak, J.C.; Busch, S.; Poppe, H. (1992) Study of protein–drug binding using capillary zone electrophoresis. *J. Chromatogr.* **608**, 257–264.
7. Chu, Y.-H.; Lees, W.J.; Stassinopoulos, A.; Walsh, C.T. (1994) Using affinity capillary electrophoresis to determine binding stoichiometries of protein–ligand interactions. *Biochemistry.* **33**, 10616–10621.
8. Chu, Y.-H.; Whitesides, G.M. (1992) Biocatalytic resolution of tertiary α-substituted carboxylic acid esters: Efficient preparation of a quaternary asymmetric carbon center. *J. Org. Chem.* **57**, 3524–3525.
9. Chu, Y.-H.; Avila, L.Z.; Biebuyck, H.A.; Whitesides, G.M. (1992) Use of affinity capillary electrophoresis to measure binding constants of ligands to proteins. *J. Med. Chem.* **35**, 2915–2917.
10. Gomez, F.A.; Avila, L.Z.; Chu, Y.-H.; Whitesides, G.M. (1994) Determination of binding constants of ligands to proteins by affinity capillary electrophoresis: Compensation for electroosmotic flow. *Anal. Chem.* **66**, 1785–1791.
11. Avila, L.Z.; Chu, Y.-H.; Blossey, E.C.; Whitesides, G.M. (1993) Use of affinity capillary electrophoresis to determine kinetic and equilibrium constants for binding of arylsulfonamides to bovine carbonic anhydrase. *J. Med. Chem.* **36**, 126–133.

12 Dose, E.V.; Guiochon, G.A. (1991) High resolution modeling of capillary zone electrophoresis and isotachophoresis. *Anal. Chem.* **63**, 1063–1072.
13 Sun, P.; Hoops, A.; Hartwick, R.A. (1994) Enchanced albumin protein separations and protein–drug binding constant measurements using anti-inflammatory drugs as run buffer additives in affinity capillary electrophoresis. *J. Chromatogr.* **661**, 335–340.
14 Yoshimoto, Y.; Shibukawa, A.; Sasagawa, H.; Nitta, S.; Nakagawa, T. (1995) Michaelis–Menten analysis of immobilized enzyme by affinity capillary electrophoresis. *J. Pharm. Biomed. Anal.* **13**, 483–488.
15 Heegaard, N.H.; Robey, F.A. (1992) Use of capillary zone electrophoresis to evaluate the binding of anionic carbohydrates to synthetic peptides derived from serum amyloid P component. *Anal. Chem.* **64**, 2479–2482.
16 Baba, Y.; Tsuhako, M.; Sawa, T.; Akashi, M.; Yashima, E. (1992) Specific base recognition of oligodeoxynucleotides by capillary affinity gel electrophoresis using polyacrylamide–poly(9-vinyladenine) conjugated gel. *Anal. Chem.* **64**, 1920–1925.
17 Baba, Y.; Tsuhako, M.; Sawa, T.; Akashi, M. (1993) Effect of urea concentration on the base-specific separation of oligodeoxynucleotides in capillary affinity gel electrophoresis. *J. Chromatogr.* **652**, 93–99.
18 Baba, Y.; Tomisaki, R.; Tsuhako, M.; Sawa, T.; Inami, Y.; Kishida, A.; Akashi, M. (1993) Detection of mismatch positions on the DNA–polyvinyladenine hybrids using capillary affinity gel electrophoresis. *Nucleic Acids Symp. Ser.* **29**, 81–82.
19 Baba, Y.; Inoue, H.; Tsuhako, M.; Sawa, T.; Kishida, A.; Akashi, M. (1994) Evaluation of the selective binding ability of oligodeoxynucleotides to poly(9-vinyladenine) using capillary affinity gel electrophoresis. *Anal. Sci.* **10**, 967–969.

21 Analysis of nucleic acids by capillary electrophoresis
K. KITAGISHI

21.1 Introduction

This chapter gives descriptions of the separation of bases, substituted bases, nucleosides, nucleotides, oligonucleotides, DNA and RNA. The separation modes for these compounds are included in the text as well as in Table 21.1.

21.2 Bases, nucleosides, nucleotides and their related compounds

Bases which are constituents of nucleic acids include purines and pyrimidines. Substituted purines, theophylline, caffeine used for drugs and therapeutic purposes, can be analyzed with micellar electrokinetic chromatography (MEKC). Theophylline and its analogs in serum can be quantitated after extraction with ethyl acetate [1]. The substituted purines in serum, saliva and urine samples can be determined even without any sample treatment, i.e. by the direct injection of the body fluids [2]. These techniques will be applicable to routine analysis of therapeutic drug monitoring in body fluids.

Since nucleosides are electrically neutral compounds at neutral pH, separation cannot be achieved by free zone capillary electrophoresis (FZE). For this reason, MEKC is employed to resolve these compounds. An anionic surfactant, SDS (sodium dodecyl sulfate), as micelles causes a satisfactory separation of five major dideoxynucleosides in 10 min [3]. Kaneta *et al.* reported that addition of glucose to the SDS micellar solution improved the resolution of nucleosides by both extending the elution range and changing the selectivity. Nine nucleosides can be completely separated upon the addition of 1 M glucose in MEKC [4].

Fluorescein–ethylenediamine conjugates of four major deoxynucleotides can be resolved with FZE. Control of pH is an important factor for resolution. The four conjugates could be completely separated at pH 10.4, maintaining the fluorescence intensity of the fluorophore [5]. Fluorescein–ethylenediamine–deoxyadenosine-5'-monophosphate was exposed to ^{60}Co and H_2O_2 and analyzed with FZE in laser-induced fluorescence (LIF) detection mode. The purpose of the separation is the detection of DNA adducts damaged by exposure to toxic chemicals and radiation. The ex-

Table 21.1 Analysis of nucleic acids

Methods	Compounds	Origin, matrix	Ref.
25 mM Phosphate buffer, pH 8.0, 80 mM SDS 21 kV, 26.5°C, 72 cm (50.5 cm to detector) × 50 μm (i.d.), 274 nm	Theophylline and its analogs	Serum, extracted with ethyl acetate and dissolved in the distilled water	1
6 mM $Na_2B_4O_7$, 10 mM Na_2HPO_4, pH 9, 75 mM SDS 20 kV (75 μA), 40°C, 90 cm (70 cm to detector) × 75 μm (i.d.), 195–320 nm 15 kV (32 μA), 40°C, 37 cm to detector × 50 μm (i.d.), 280 nm	Substituted purines	Serum, saliva, urine	2
50 mM Phosphate buffer, pH 6.5, 40 mM SDS 20 kV, 21°C, 80 cm (52 cm to detector) × 75 μm (i.d.) with a Z-shaped flowcell of a light path of 3 mm, 260 nm	Dideoxyribonucleosides	Commercial products from Sigma; synthetic samples, dissolved in buffer solution	3
50 mM Phosphate buffer, PH 7.0, 150 mM SDS, 1.0 M glucose 35 μA, 20°C, 70 cm (50 cm to detector) × 50 μm (i.d.), 210 nm	Nucleosides (adenosine, guanosine, cytidine, uridine, thymidine, 2′-deoxyadenosine, 2′-deoxyguanosine, 2′-deoxycytidine, deoxyuridine)	Commercial products from Tokyo Kasei Co., dissolved in 0.02 M HCl	4
10 mM Tris–borate buffer, pH 8.7 and pH 10.4 with NaOH, 10% acetonitrile 15 kV, 95 cm (50 cm to detector) × 75 μm (i.d.) LIF, ex. 488 nm (Ar laser)	Fluorescein–ethylenediamine conjugates of deoxyadenosine-, deoxyguanosine-, deoxycytidine-, thymidine-5′-monophosphates	Prepared by chemical reaction in laboratory, dissolved in Tris–borate buffer at pH 8.7	5
10 mM Tris–borate buffer, pH 8.7 and pH 10.4 with NaOH, 10% acetonitrile 15 kV or 30 kV, 95 cm × 75 μm (i.d.) LIF, ex. 488 nm (Ar laser)	Fluorescein–ethylenediamine-deoxyadenosine-5′-monophosphate and its exposed samples to ^{60}Co and H_2O_2	^{60}Co irradiation or H_2O_2 reaction of F-ED-dAMP, dissolved in Tris–borate buffer at pH 8.7 with 10% acetonitrile	6
30 mM Phosphate–50 mM Tris–HCl buffer pH 5.28 −20 kV, 65 cm × 50 μm (i.d.), Ucon-coated capillary, 254 nm	UTP, CTP, ATP, GTP, UDP, CDP, ADP, GDP, XMP, UMP, CMP, AMP, GMP	Human lymphoma cell, extracted with perchloric acid	7

Table 21.1 *Continued*

Methods	Compounds	Origin, Matrix	Ref.
20 mM Sodium carbonate/bicarbonate buffer, pH 9.6 (FZE) 16 mM Sodium carbonate/bicarbonate buffer, pH 9.6 with 20 mM SDS (MEKC) 12–18 kV, 62–76 cm × 75 µm (i.d.), 279 nm	2′-Deoxy nucleosides, 2′-deoxy-5′-monophosphate nucleotides and benzo[a]pyrene guanosine monophosphate	Modifications with diol-epoxide in laboratory, dissolved in 50 mM Tris at pH 8.1–8.7 or 50 mM citrate buffer at pH 5.0	8
30 mM Ammonium bicarbonate, pH 7.9, 10 mM SDS 30 kV, 100 cm (75 cm to detector) × 50 µm (i.d.) Fluorescence, ex. 246 nm, em. 370 nm	Benzo[a]pyrene-deoxyguanosine-5′-monophosphate, benzo[a]pyrene-tetrahydrotetrols	Modification of calf thymus DNA with diol-epoxide and hydrolyzed with DNAase I in laboratory, dissolved in aqueous solution	9
Tris–borate/urea purchased from Beckman 14.1 kV, 30°C, 47 cm (total) polyacrylamide gel-filled capillary, 254 nm or 265 nm	21-mer phosphorothioate oligonucleotides, ISIS 2922, and its deletion products	Synthesized products, dissolved in Tris–borate buffer with 7 M urea	11
100 mM Tris–borate, 2 mM EDTA, pH 8.5, in the absence or presence of 7 M urea 400 V cm^{-1}, 25°C, 47 cm (40 cm to detector) × 100 µm (i.d.), polyacrylamide gel-filled capillary, 254 nm	Homo- and heterooligonucleotides in the 10–20-mer range	Commercial products from Pharmacia, dissolved in water	12
Polyacrylamide gel-filled capillary, 7 M urea 11.1 kV, 30°C, 37 cm × 100 µm (i.d.), 254 nm	Heterogeneous oligo-DNAs	Supplied from Takara Co.; synthesized with Milligen DNA synthesizer	13
0.1 M Tris–borate, pH 8.6, 5 M or 7 M urea 8% T and 5% C polyacrylamide gel contained 0.05–0.4% poly(9-vinyladenine) filled in capillary 200–214 V cm^{-1}, ambient to 50°C, 42–50 cm (22–30 cm to detector) × 100 µm (i.d.), 260 nm	Oligodeoxy-nucleotides	Commercial products from Pharmacia; synthesized with DNA synthesizer (ABI), diluted with water	14, 15, 16
50 mM Tris–phosphate buffer, pH 7.0, 0–100 mM SDS, 0 M or 7 M urea 10–30 kV, 30°C, 72 cm (50 cm to detector) × 50 µm (i.d.), 260 nm	Diastereomers of phosphoramidate bridged oligonucleotides	Dissolved in aqueous solution	17
0.1 M Tris–borate buffer, pH 8.1, 2.5 mM EDTA, 0.1% SDS, 7 M urea	DNA restriction fragment mixture, DRIgest™ III,	Commercial products from Pharmacia;	18

Table 21.1 *Continued*

Methods	Compounds	Origin, Matrix	Ref.
15 kV, 27°C, 30–50 cm × 75 μm (i.d.), 260 nm	from 72 to 23 130 bps	dissolved in buffer solution	
0.5% Hydroxypropyl methyl cellulose, 89 mM Tris, 89 mM borate, 2 mM EDTA, pH 8.5, 0.1 mM EDTA 175–350 V cm^{-1}, 25°C, 57 cm (50 cm to detector) × 100 μm (i.d.) dimethylpolysiloxane-coated capillary, 254 nm	PCR-amplified rDNA internal transcribed spacer and intergenic spacer from ectomycorrhizal fungi	*Laccaria laccata* and *Laccaria bicolor*, dissolved in PCR amplified solution	19
0.2–0.5% hydroxyethyl cellulose, 89 mM Tris–89 mM borate, 5 mM EDTA 301.3 V cm^{-1}, 30°C, 50 cm (35 cm to detector) × 50 μm (i.d.), 260 nm	DNA restriction digest samples (øX174/Hae III)	Commercial products from Bethesda Research Labs, dissolved in aqueous. solution	20
75 mM Tris–phosphate, 1 mM EDTA, pH 7.5 50–400 V cm^{-1}, 25°C, 55–105 cm (25–75 cm to detector) × 50 μm (i.d.), polyacrylamide gel-filled capillary containing 8.3 M urea, LIF, ex. 488 nm (Ar laser), em. 4 bandpass filters (540 nm, 560 nm, 580 nm, 610 nm)	M13mp18 single-stranded DNA labeled with 4 fluorescent dyes	Synthesized products, dissolved in 5:1 formamide: 0.01 × running buffer (or 50 mM EDTA at pH 8.0)	21, 34, 35
0.5% Hydroxyethyl cellulose, 100 mM Tris–borate pH 8.7, 0.1 mM EDTA, 0.635–6.35 μM ethidium bromide 37.8 μA, 25°C, 70 cm × 100 μm (i.d.) DB-17 capillary, 260 nm	pBR 322 Hae III digested DNA; PCR amplified DNA	Commercial products from Sigma from human chromosome, dissoved in salt-free aqueous solution	22
0.2, 0.4, 0.6% Methyl cellulose (mol. wt = 86000), 100 mM Tris, 100 mM borate, 2 mM EDTA, pH 8.0 +20 kV and −30 kV, 50 cm (45 cm to detector) × 100 μm (i.d.) and 100 cm (80 cm to detector) × 75 μm (i.d.) bare capillary and polyacrylamide-coated capillary, 260 nm	1 kb DNA ladder; DNA molecular weight marker V (Boeringer Mannheim)	Commercial products from Bethesda Research Labs and Boeringer Mannheim, dissolved in aqueous solution	23
0.5% (Hydroxypropyl)methyl cellulose, 89 mM Tris, 89 mM borate, 2 mM EDTA, pH 8.5	PCR amplified RFLP samples of ERBB2 oncogene;	Oncogen locus on human chromosome;	24

Table 21.1 *Continued*

Methods	Compounds	Origin, Matrix	Ref.
175 V cm^{-1}, 25°C, 57 cm (50 cm to detector) × 100 μm (i.d.) DB-17 capillary, 260 nm	DNA restriction digest samples (øX174/Hae III)	commercial products from New England Biolabs, dissolved in desalting solution of PCR-RFLP samples and diluted øX174/Hae III marker with water	
0.5% (Hydroxypropyl)methyl cellulose, 90 mM Tris, 90 mM borate, 2 mM EDTA, pH 8.5, 0–7.5% glycerol 228 V cm^{-1}, 22°C, 57 cm (50 cm to detector) × 100 μm (i.d.) DB-17 capillary, 260 nm	DNA restriction digest samples (øX174/Hae III)	No descriptions, diluted with water	25
Commercial polyacrylamide gel-filled capillary; two kinds of commercial sieving buffer and capillary, 10 μM ethidium bromide 210 V cm^{-1}, 30°C, 40 or 50 cm to detector × 100 μm (i.d.), 260 nm	øX174/Hinf I; øX174/Hae III; DNA marker XI (Boeringer Mannheim); HUMTHO1 alleic ladder	Commercial products from Gibco BRI, Boeringer Mannheim and Promega Co. from human mitochondria, dissolved in desalting solution of PCR products	26
100 mM Tris, 100 mM borate, 0.1 mM EDTA, pH 8.7, 1.0% hydroxyethylcellulose, 1.27 μM ethidium bromide, 5 ng ml^{-1} YO-PRO-1 38 μA, 25°C, 60 cm to detector × 100 μm (i.d.) DB-17 capillary, LIF, ex. 488 nm (Ar laser)	HUMTHO1 alleles; vWA alleles; MBP alleles	Human chromosome; human myelin, diluted with YO-PRO-1 solution	27
89 mM Tris, 89 mM borate, 2 mM EDTA, pH 8.5, 1/15–2/15 μM TOTO, 0.5% methylcellulose, 40–50 cm × 50 or 100 μm (i.d.) DB-17 capillary 40 cm × 50 μm (i.d.) polyacrylamide gel-filled capillary (3% T, 3% C) 8–15 kV, 25°C, LIF, ex. 488 nm (Ar laser), em. 530 nm bandpass filter	DNA restriction digest samples (øX174/Hae III); PCR products (apolipoprotein B; VNTR locus D1S80; mitochondrial DNA)	Commercial products from BRL laboratories from human genes, prestained with TOTO solution	28

Table 21.1 Continued

Methods	Compounds	Origin, Matrix	Ref.
0.75% Hydroxyethyl cellulose (Mn = 105 K), 40 mM Tris–acetate buffer, 1 mM EDTA, pH 8.0, monomeric (ethidium bromide, propidium-2, propidium-3, thiazole orange, (N,N'-tetramethylpropanediamino)propyl thiazole orange, oxazole orange) or dimeric (9-aminoacridine hydrochloride) dye is added to tunning and column buffer 125 V cm^{-1}, 50 cm (25 cm to detector) × 100 μm (i.d.) poly(ethyleneglycol)-coated capillary LIF, ex. 488 nm (Ar laser), em. green filter (515–545 nm) or red filter (>590 nm)	DNA restriction digest samples (øX174/Hae III)	Commercial products from Pharmacia, diluted into 0.4 mM Tris–acetate, 0.1 mM Na$_2$EDTA, pH 7.0 and prestained with dye solution	29
1.0% Hydroxyethyl cellulose, 100 mM Tris–borate buffer, 1 mM EDTA, pH 8.7 or 8.9, 1.27 μM ethidium bromide, 50 ng ml^{-1} YO-PRO-1 15 kV, 20°C or 25°C, 57 cm (50 cm to detector) × 50 or 100 μm (i.d.) DB-17 coated capillary LIF, ex. 488 nm (Ar laser), em. 520 nm	PCR products of human mitochondrial DNA (HV1A; HV1B; HV2)	Human mitochondria, dissolved in desalting solution	30
6–10% Linear polyacrylamide, 89 mM Tris–borate buffer containing EDTA, 5% glycerol 200 V cm^{-1}, 5–50°C, 20 cm (14 cm to detector) × 100 μm (i.d.) polyacrylamide-coated capillary LIF, ex. 488 nm (Ar laser), em. two bandpass filters of 520 nm and 560 nm	IGF1–BP3 PCR products (primers are labeled with fluorescent dyes)	Human plasmid, diluted with water	31
0.8% Hydroxyethyl cellulose (Mn = 438 K), 45 mM Tris, 45 mM borate, 1 mM EDTA, pH 8.3 80 V cm^{-1}, 22°C, 50 cm × 75 μm (i.d.) polyacrylamide-coated capillary LIF, ex. 488 nm (Ar laser), em. bandpass filter of 520 nm and longpass filter of 590 nm	HUMTHO1 PCR products (primers are labeled with fluorescent dyes)	Human chromosome, desalted by dialysis and dissolved in aqueous solution	32

Table 21.1 *Continued*

Methods	Compounds	Origin, Matrix	Ref.
0.4% Hydroxypropylmethyl cellulose, 1 × TBE, 0.1 M NaCl 100 V cm^{-1}, 28.6 cm × 32 µm (i.d.) LIF, ex. 594 nm (He–Ne laser)	DNA resrtriction digest samples (øX174/Hae III) fluorescently labeled using deoxynucleotidyl transferase	Commercial products from Gibco BRL, dissolved in 1 × TBE	33
100 mM Tris–borate, 2.5 mM EDTA, pH 7.6, 7 M urea 300 V cm^{-1}, 75–94 cm (50–75 cm to detector) × 75 µm (i.d.), polyacrylamide gel-filled capillary (3% T/5% C) LIF, ex. 488 nm (Ar laser), em. 520 nm	JOE-labeled chain-termination sequencing reaction products	Synthesized with Cyclone DNA synthesizer, dissolved in 80% formamide and 8 mM EDTA	36
TBE (1.08% Tris, 0.55% borate, 0.07% EDTA) 150–200 V cm^{-1}, 31 cm × 50 µm (i.d.), polyacrylamide gel-filled capillary (6% T/5% C) containing 7 M urea, (20% formamide) LIF, ex. 534.5 nm (He–Ne laser), em. 590 nm bandpass filter	M13mp18 single-stranded DNA fragments labelled with tetramethylrho-damine isothiocyanate	Supplied and/or synthesized products, dissolved in formamide containing trace EDTA	37
100 mM Tris–borate, 7 M urea, pH 8.3 82–200 V cm^{-1}, 25°C, 20–100 cm to detector × 75 µm (i.d.), linear polyacrylamide-filled capillary (9% T) containing 7 M urea LIF, ex. 488 nm (Ar laser), em. 520 nm	M13mp18 single-stranded DNA fragments labeled with FITC	Supplied and/or synthesized products, mixed with formamide	38
Mixed solution of 1.5 PEO (Mn = 8000 K) and 1.4% PEO (Mn = 600 K), 89 mM Tris 89 mM borate, 2 mM EDTA, 3.5 M urea 12 kV, 45 cm (35 cm to detector) × 75 µm (i.d.) polyacrylamide-coated capillary or bare capillary LIF, ex. 543 nm (He–Ne laser), em. two RG610 long-pass filters	DNA fragments from standard Sanger reactions	Prepared by standard Sanger reactions with commercial DNA sequencing instrument (ABI), dissolved in 5:1 formamide–50 mM aqueous EDTA solution	39
0.5% Hydroxyethyl cellulose, 10 mM Tris–borate buffer, 0.1 mM EDTA, pH 8.7, 25 mM NaCl, 1.27 µM ethidium bromide 15 kV, 35°C, 72 cm (50 cm to detector) × 50 µm (i.d.), 260 nm	PCR-amplified DNA fragments of androgen receptor mRNA	Human fibroblast, dissolved in 10 mM Tris–HCl and 1 mM EDTA at pH 8.0	40

Table 21.1 Continued

Methods	Compounds	Origin, Matrix	Ref.
1% Hydroxyethyl cellulose, 89 mM Tris–89 mM borate buffer, 2 mM EDTA, 630 ng ml^{-1} YO-PRO-1 7 kV, 25°C, 20 cm to detector × 50 μm (i.d.) DB-17-coated capillary LIF, ex. 488 nm (Ar laser), em. 510 nm	RT-PCR- amplified DNA fragments of HCV RNA	Human serum of hepatitis C patient, dissolved in 10 mM Tris, 50 mM KCl and 1.5 mM MgCl$_2$ at pH 8.3	41
LIFluor dsDNA 1000 Kit (Beckman) 7.4 kV, 37 cm × 100 μm (i.d.) LIF, ex. 488 nm (Ar laser)	PCR-amplified and RT-PCR- amplified DNA fragments of HIV DNA and RNA	Human peripheral blood mononuclear cells, dissolved in aqueous solution	42

F-Ed-dAMP: fluorescein–ethylenediamine-deoxyadenosine-5′-monophosphate, UTP: uridine triphosphate, CTP: cytidine triphosphate, ATP: adenosine triphosphate, GTP: guanosine triphosphate, UDP: uridine diphosphate, CDP: cytidine diphosphate, ADP: adenosine diphosphate, GDP: guanosine triphosphate, XMP: xanthine monophosphate, UMP: uridine monophosphate, CMP: cytidine monophosphate, AMP: adenosine monophosphate, GMP: guanosine monophosphate, i.d.: inner diameter, ex.: excitation, em.: emission, DB-17: capillary column (available from J&W Scientific), MW: molecular weight, RFLP: restriction fragment length polymorphism, YO-PRO-1: 1-(4-[3-methyl-2,3-dihydro-(benzo-1,3-oxazole)-2-methyidene]-quinolinium)-3-trimethylammonium propane diiodide, vWA: Von Willenbrand Factor, MBP: myelin basic protein, TOTO: 1,1′-(4,4,7,7-tetramethyl-4,7-diazundecamethylene)-bis-4-[3-methyl-2,3-dihydro-(benzo-1,3-thiazole)-2-methylidene]-quinolinium tetraiodide, VNTR: variable number tandem repeat, Mn: number average molecular weight, HV: hypervariable area, IGF1-BP3: insulin-like growth factor 1-binding protein 1, TBE: Tris–borate–EDTA, HCV: hepatitis C virus, RT: reverse transcription, HIV: human immunodeficiency virus.

posed sample shows an electropherogram with more peaks than the non-exposed sample [6].

Nucleotide pools in human lymphoma cells were determined by FZE. Over 12 nucleotides were separated in 35 min and quantitated with a Ucon-coated column and UV detection [7]. Non-adducted nucleoside and nucleotide components were separated in FZE at pH 9.6. MEKC provided improved separations for these compounds. One of DNA adduct nucleotides, benzo[a]pyrene guanosine monophosphate, can be easily monitored by the capillary electrophoresis (CE) technique, if its concentration is high enough [8]. When the concentration is relatively low, a stacking technique for sample injection [8] or fluorescence detection [9] enhances the detection sensitivity. Further reading for MEKC separation of bases, nucleosides and nucleotides can be found in a recent review [10].

21.3 Oligonucleotides

Most studies of oligonucleotides in capillary electrophoresis are based on capillary gel electrophoresis (CGE), generally under conditions such that oligonucleotides are denatured in the presence of urea. The CE separation of oligonucleotides can be applied to RNA analysis, since both oligonucleotide and RNA are single stranded compared with the native double-stranded form of DNA. Srivatsa et al. [11] developed the quantitative assay of a 21-mer phosphorothioate oligonucleotide, ISIS 2922, a pharmaceutical product, in CGE. Routine drug analysis can be achieved for ISIS 2922 and its deletion products [11]. Guttman et al. [12] investigated the migration behavior of single-stranded oligonucleotides using denaturing and non-denaturing polyacrylamide capillary gel. The equations to predict the migration time of an oligonucleotide dependent upon its base composition derived from migration properties of homo- and heterooligonucleotides in the 10–20-mer range provide the migration model [12]. The effects of the secondary structure of oligonucleotides upon electrophoretic migration can be evaluated for heterogeneous oligo-DNA with a gel-filled capillary [13].

Baba and co-workers developed capillary affinity gel electrophoresis, incorporating an affinity ligand, poly(9-vinyladenine), in polyacrylamide gel matrix [14–16]. Because the poly(9-vinyladenine)–polyacrylamide conjugated gel has a strong affinity for thymidylic acids in nucleic acids, oligo(dT) migrates slower than oligo(dA) [14]. The affinity depends upon the number of sequential thymidylic acids for oligodeoxynucleotide isomers [15]. Manipulation of capillary temperature enables the optimization of base-specific separation of oligodeoxynucleotides, by using oligo(dA)$_{12-18}$ and oligo(dT)$_{12-18}$ [16].

An example of CE separations of oligonucleotides without a gel-filled capillary is MEKC to resolve diastereomers. Diastereomers of phosphoramidate-bridged unnatural oligonucleotides can be separated in MEKC with anionic micelles of SDS or cationic micelles of dodecyltrimethylammonium bromide. Electrophoretic resolution is controlled by factors such as micelle concentration, addition of urea and/or propan-2-ol and temperature [17].

21.4 Molecular sieving separation of DNA restriction fragments and polymerase chain reaction (PCR) products

One of the main purposes for developing the separation mode of molecular sieving in CE is DNA analysis including DNA sequencing and DNA fragment analysis. These analyses have conventionally been carried out by gel electrophoresis. CE with gel-filled capillaries can be substituted for conventional gel electrophoresis. Designs of various gel-filled capillaries and gel formation procedures are described in 'Gel-filled capillaries' (section 4.3) of Chapter 4. Polyacrylamide gel-filled capillaries are widely used in CE.

A DNA restriction fragment mixture could be roughly separated in FZE with a Tris–borate buffer containing 7M urea and 0.1% SDS, injecting a heat-treated sample [18]. To maintain the easy operation of FZE and to improve the separation efficiency, molecular sieving with non-gel polymer solution has been developed and this separation technique has recently played an important role in DNA analysis. Molecular sieving in low viscosity polymer solutions is preferable for rapid, stable and reproducible measurements, since sieving media can easily be formed and replaced, in combination with the resolving power of gel-based CE analysis [19]. A large number of references have been published for non-gel sieving media in CE, as mentioned in 'Electrolytes in polymer sieving' (section 5.4).

DNA molecules have generally the same electric density independent of the molecular size, due to the regular shape and the ionizable phosphate groups located at intervals. Since the electric mobility of a separand depends on its electric density, DNA molecules migrate at the same mobility in the absence of sieving medium. Double-stranded DNA migrates at a velocity corresponding to its size in sieving media independent of the base composition [20], essentially according to the Ogston model. Single-stranded DNA ranging from 46 to 760 bases similarly migrates in denaturing polyacrylamide gel. The migration behavior can be explained by the combined model of Ogston, reptation and biased-reptation regimes [21]. In contrast, for single-stranded oligonucleotides of less than 20-mer, the electrophoretic mobility depends upon base composition in polyacrylamide gel under both denaturing and non-denaturing conditions due to their secondary structure as mentioned in section 21.3 [13].

Although the separation resolution with polymer solutions was formerly considered to be relatively lower than that with gel-filled capillaries, DNA separations at comparable resolution to gel-filled capillaries have been reported [22]. The modification of capillary inner surface is not essential for DNA separation, because DNA molecules are not retained in the capillary inner wall due to their negative charges. However, the capillary surface is often coated in order to improve the separation efficiency induced by the suppression of electroosmotic flow (EOF) [23]. More information on coating methods can be obtained from 'Techniques for coating the capillary surface' (section 4.2). The addition of an intercalating dye, ethidium bromide, enhances the separation [22, 24]. When glycerol was added to the polymer sieving solution in Tris–borate buffer in the ratio up to 7.5% (v/v), the resolution increased considerably. Cheng and Mitchelson speculated that the formation of dimeric 1:2 borate:di-diol complexes compared with complexes of borate–glycerol–cellulose derivative reduced the effective pore size of the polymer [25]. The accuracy of size determination for PCR product analysis is based on the calibration plot of DNA markers. The proper choice of DNA calibration markers which are blunt-ended and non-phosphorylated guarantees an accurate estimation [26].

The sensitivity of DNA fragments in UV detection is at µg/ml level. LIF–CE generally shows better sensitivity by more than two orders of UV detection. For DNA samples at lower concentrations than the UV detection limit, LIF–CE can provide a much greater sensitivity. Intercalating dyes such as YO-PRO-1, TOTO, YOYO and ethidium bromide can be used for on-column staining. An argon ion laser is used to provide excitation at 488 nm for this technique [27–29]. The detection limits shown in the studies are less than 500 pg ml^{-1} DNA [27], 500 fg µl^{-1} DNA [28] and 40~200 fg µl^{-1} DNA [29].

Quantitation of PCR products by LIF–CE was examined by Butler *et al.* [30]. Using an internal standard, the precision of peak migration time was below 0.1% of relative standard deviation (RSD) with a peak area precision of 3.0% of RSD. In comparison to slab-gel electrophoresis, hybridization with chemiluminescence detection (i.e. slot blot) and fluorescence spectrophotometry, LIF–CE shows high sensitivity and good precision and resolution [30].

Instead of intercalating dyes, DNA covalently labeled with fluorescent dye, such as fluorescein and rhodamine, is also available for fluorescence detection. As for DNA sequencing, fluorescently labeled primers can be used in PCR generation of double-stranded DNA [21]. Two-dye labeling in LIF–CE has recently been developed. Hebenbrock *et al.* applied two-dye labeling which had already been used in slab-gel electrophoresis to CE, in which forward and reverse primers are labeled with two different dyes [31]. Wang *et al.* designed an energy-transfer dye-labeled primer having a donor fluorescein at the 5' end and an acceptor rhodamine at the seventh base. Addition of an intercalating dye, thiazole orange or 9-aminoacridine, to the sieving solution enhances the resolution in LIF–CE similarly to those cases in UV detection [32].

Fluorescence detection with primers has the disadvantage of being inapplicabile in analysis of restriction of endonuclease digests. Dovichi's group reported 3'-end labeling of DNA using terminal deoxynucleotidyl transferase in order to overcome this problem [33].

21.5 DNA sequencing

The analysis of DNA fragments produced in sequencing reactions requires high resolution which enables baseline separation for difference in length of a single nucleotide. Denaturing gel-filled capillaries with LIF detection is generally utilized for DNA sequencing in CE. Both the multiple fluorophore approach using four different fluorophore-conjugated primers

[34, 35] and the single fluorophore approach using single fluorophore labeling primers having JOE (2′,7′-dimethoxy-4′,5′-dichloro-6-carboxyfluorescein) [36], tetramethylrhodamine [37] or fluorescein isothiocyanate (FITC) [38] have been applied.

Luckey et al. [35] evaluated the factors which determine resolution as a function of fragment length. The contribution of four factors, injection, diffusion, thermal gradients and detection volume, was determined and directed toward optimum separation [35]. Manabe et al. [38] investigated the relationship between resolution and analysis time, changing the linear polyacrylamide concentration filling the capillaries, capillary length and electric field strength for the separation of FITC–p(dT)$_{16-500}$. Single-base resolution of M13mp10 DNA fragments up to 520 nucleotides was shown using 9% T linear polyacrylamide and an electric field length of 100 V cm^{-1} [38].

Fung and Yeung [39] reported ultra-high resolution of DNA fragments from the Sanger reaction, when a replaceable solution of commercially available poly(ethylene oxide) (PEO) in two size ranges is used as the sieving matrix under denaturing conditions. The mixed PEO matrix yields identical separation performance for small DNA fragments and higher resolution for large ones, compared to a polyacrylamide gel-filled capillary. This means that replaceable polymer solutions can be an alternative to gel-filled capillaries in DNA sequencing in CE [39].

21.6 Possible application to RNA analysis

The CE separation of RNA has rarely been described, probably due to the greater difficulties in the handling of RNA samples than DNA. If we can purify a sufficient amount of RNA sample at a concentration higher than its detection limit, many separation methods for oligonucleotides and DNA are, in principle, applicable for RNA.

The chemical structure of RNA is very close to that of DNA. Major differences are: (1) one of the four constituent bases is different, uracil for RNA instead of thymine for DNA, (2) the sugar moiety is ribose in RNA and deoxyribose in DNA and (3) RNA molecules are generally single stranded while DNAs are complementarily double stranded. Accordingly, analytical methods for single-stranded DNA could be applied to RNA analysis, if the concentration of the sample prepared is greater than the detection limit [12, 13, 21].

Another possibility is RNA reverse transcription to DNA followed by PCR amplification in order to solve the problems of sample preparation [40–42]. The intergenic spacer and the internal transcribed spacer of ribosomal RNA gene were amplified as PCR-amplified rDNA fragments, and sized by the molecular sieving technique in CE [19].

21.7 Conclusion

Analysis of bases, nucleosides, nucleotides, oligonucleotides or DNA in several matrices and DNA sequences have been successfully attained by CE combined with LIF detection or PCR method in recent development. However, a problem still remains for RNA analysis mostly due to the tiny amount in the matrix, making PCR the most appropriate method for RNA detection, if applicable, as the amount of RNA is less than the detection limit.

RNA analysis by CE will be the subject of further improvement and innovation in the near future as the new technique is keenly required in several fields.

References

1. Lee, K.-J.; Heo, G.S.; Kim, N.J.; Moon, D.-C. (1992) Separation of theophylline and its analogues by micellar electrokinetic chromatography: application to the determination of theophylline in human plasma. *J. Chromatogr.* **577**, 135–141.
2. Thormann, W.; Minger, A.; Molteni, S.; Caslavska, J.; Gebauer, P. (1992) Determination of substituted purines in body fluids by micellar electrokinetic capillary chromatography with direct sample injection. *J. Chromatogr.* **593**, 275–288.
3. Singhal, R.P.; Hughbanks, D.; Xian, J. (1992) Separation of dideoxyribonucleosides in trace amounts by automated liquid chromatography and capillary electrophoresis. *J. Chromatogr.* **609**, 147–161.
4. Kaneta, T.; Tanaka, S.; Taga, M.; Yoshida, H. (1992) Effect of addition of glucose on micellar electrokinetic capillary chromatography with sodium dodecyl sulphate. *J. Chromatogr.* **609**, 369–374.
5. Li, W.; Moussa, A.; Giese, R.W. (1992) Capillary electrophoresis of fluorescein–ethylenediamine-5′-deoxynucleotides. *J. Chromatogr.* **608**, 171–174.
6. Li, W.; Moussa, A.; Giese, R.W. (1993) Capillary electrophoresis with laser fluorescence detection for profiling damage to fluorescein-labeled deoxyadenylic acid by background, ionizing radiation and hydrogen peroxide. *J. Chromatogr.* **633**, 315–319.
7. Shao, X.; O'Neill, K.; Zhao, Z.; Anderson, S.; Malik, A.; Lee, M. (1994) Analysis of nucleotide pools in human lymphoma cells by capillary electrophoresis. *J. Chromatogr.* **680**, 463–468.
8. Norwood, C.B.; Jackim, E.; Cheer, S. (1993) DNA adduct research with capillary electrophoresis. *Anal. Biochem.* **213**, 194–199.
9. Harvey, S.D.; Bean, R.M.; Udseth, H.R. (1992) High-resolution separation and detection of DNA adducts of benzo[a]pyrene. *J. Microcol. Sep.* **4**, 191–198.
10. Terabe, S.; Chen, N.; Otsuka, K. (1994) Micellar electrokinetic chromatography. In *Advances in Electrophoresis*, Chrambach, A.; Dunn, M.J.; Radola, B.J. (eds), VCH, Weinheim, Germany, Vol. 7, pp. 87–153.
11. Srivatsa, G.S.; Batt, M.; Schuette, J.; Carlson, R.H.; Fitchett, J.; Lee, C.; Cole, D.L. (1994) Quantitative capillary gel electrophoresis assay of phosphorothioate oligonucleotides in pharmaceutical formulations. *J. Chromatogr.* **680**, 469–477.
12. Guttman, A.; Nelson, R.J.; Cooke, N. (1992) Prediction of migration behavior of oligonucleotides in capillary gel electrophoresis. *J. Chromatogr.* **593**, 297–303.
13. Satow, T.; Akiyama, T.; Machida, A.; Utagawa, Y.; Kobayashi, H. (1993) Simultaneous determination of the migration coefficient of each base in heterogeneous oligo-DNA by gel filled capillary electrophoresis. *J. Chromatogr.* **652**, 23–30.
14. Baba, Y.; Tsuhako, M.; Sawa, T.; Akashi, M.; Yashima, E. (1992) Specific base recognition of oligodeoxynucleotides by capillary affinity gel electrophoresis using polyacrylamide–poly(9-vinyladenine) conjugated gel. *Anal. Chem.* **64**, 1920–1925.

15 Akashi, M.; Sawa, T.; Baba, Y.; Tsuhako, M. (1992) Specific separation of oligodeoxynucleotides by capillary affinity gel electrophoresis (CAGE) using poly(9-vinyladenine)–polyacrylamide conjugated gel. *J. High Resolut. Chromatogr.* **15**, 625–626.

16 Baba, Y.; Tsuhako, M.; Sawa, T.; Akashi, M. (1993) Temperature-programmed capillary affinity gel electrophoresis for the sensitive base-specific separation of oligodeoxynucleotides. *J. Chromatogr.* **632**, 137–142.

17 Bevan, C.D.; Mutton, I.M.; Pipe, A.J. (1993) Resolution of diastereomeric phosphoramidate bridged unnatural oligonucleotides by micellar electrokinetic chromatography. *J. Chromatogr.* **636**, 113–123.

18 Cohen, A.S.; Najarian, D.; Smith, J.A.; Karger, B.L. (1988) Rapid separation of DNA restriction fragments using capillary electrophoresis. *J. Chromatogr.* **458**, 323–333.

19 Martin, F.; Vairelles, D.; Henrion, B. (1993) Automated ribosomal DNA fingerprinting by capillary electrophoresis of PCR products. *Anal. Biochem.* **214**, 182–189.

20 Grossman, P.D.; Soane, D.S. (1991) Capillary electrophoresis of DNA in entangled polymer solutions. *J. Chromatogr.* **559**, 257–266.

21 Luckey, J.A.; Smith, L.M. (1993) A model for the mobility of single-stranded DNA in capillary gel electrophoresis. *Electrophoresis* **14**, 492–501.

22 McCord, B.R.; Jung, J.M.; Holleran, E.A. (1993) High resolution capillary electrophoresis of forensic DNA using a non-gel sieving buffer. *J. Liq. Chromatogr.* **16**, 1963–1981.

23 MacCrehan, W.A.; Rasmussen, H.T.; Northrop, D.M. (1992) Size-selective capillary electrophoresis (SSCE) separation of DNA fragments. *J. Liq. Chromatogr.* **15**, 1063–1080.

24 Ulfelder, K.J.; Schwartz, H.E.; Hall, J.M.; Sunzeri, F.J. (1992) Restriction fragment length polymorphism analysis ERBB2 oncogene by capillary electrophoresis. *Anal. Biochem.* **200**, 260–267.

25 Cheng, J.; Mitchelson, K.R. (1994) Glycerol-enhanced separation of DNA fragments in entangled solution capillary electrophoresis. *Anal. Chem.* **66**, 4210–4214.

26 Williams, P.E.; Marino, M.A.; Del Rio, S.A.; Turni, L.A.; Devaney, J.M. (1994) Analysis of DNA restriction fragments and polymerase chain reaction products by capillary electrophoresis. *J. Chromatogr.* **680**, 525–540.

27 McCord, B.R.; McClure, D.L.; Jung, J.M. (1993) Capillary electrophoresis of polymerase chain reaction-amplified DNA using fluorescence detection with an intercalating dye. *J. Chromatogr.* **652**, 75–82.

28 Srinivasan, K.; Girard, J.E.; Williams, P.; Roby, R.K.; Weedn, V.W.; Morris, S.C.; Kline, M.C.; Reeder, D.J. (1993) Electrophoretic separations of polymerase chain reaction-amplified DNA fragments in DNA typing using a capillary electrophoresis–laser induced fluorescence system. *J. Chromatogr.* **652**, 83–91.

29 Zhu, H.; Clark, S.M.; Benson, S.C.; Rye, H.S.; Glazer, A.N.; Mathies, R.A. (1994) High-sensitivity capillary electrophoresis of double-stranded DNA fragments using monomeric and dimeric fluorescent intercalating dyes. *Anal. Chem.* **66**, 1941–1948.

30 Butler, J.M.; McCord, B.R.; Jung, J.M.; Wilson, M.R.; Budowle, B.; Allen, R.O. (1994) Quantitation of polymerase chain reaction products by capillary electrophoresis using laser fluorescence. *J. Chromatogr.* **658**, 271–280.

31 Hebenbrock, K.; Williams, P.M.; Karger, B.L. (1995) Single strand conformational polymorphism using capillary electrophoresis with two-dye laser-induced fluorescence detection. *Electrophoresis* **16**, 1429–1436.

32 Wang, Y.; Ju, J.; Carpenter, B.A.; Atherton, J.M.; Sensabaugh, G.F.; Mathies, R.A. (1995) Rapid sizing of short tandem repeat alleles using capillary array electrophoresis and energy-transfer fluorescent primers. *Anal. Chem.* **67**, 1197–1203.

33 Figeys, D.; Renborg, A.; Dovichi, N.J. (1994) Labeling of double-stranded DNA by ROX-dideoxycytosine triphosphate using terminal deoxynucleotidyl transferase and separation by capillary electrophoresis. *Anal. Chem.* **66**, 4382–4383.

34 Luckey, J.A.; Drossman, H.; Kostichka, A.J.; Mead, D.A.; D'Cunha, J.; Norris, T.B.; Smith, L.M. (1990) High speed DNA sequencing by capillary electrophoresis. *Nucleic Acids Res.* **18**, 4417–4421.

35 Luckey, J.A.; Norris, T.B.; Smith, J.M. (1993) Analysis of resolution in DNA sequencing by capillary gel electrophoresis. *J. Phys. Chem.* **97**, 3067–3075.

REFERENCES

36 Cohen, A.S.; Najarian, D.R.; Karger, B.L. (1990) Separation and analysis of DNA sequence reaction products by capillary gel electrophoresis. *J. Chromatogr.* **516**, 49–60.
37 Chen, D.Y.; Swerdlow, H.P.; Harke, H.R.; Zhang, J.Z.; Dovichi, N.J. (1991) Low-cost, high-sensitivity laser-induced fluorescence detection for DNA sequencing by capillary gel electrophoresis. *J. Chromatogr.* **559**, 237–246.
38 Manabe, T.; Chen, N.; Terabe, S.; Yohda, M.; Endo, I. (1994) Effects of linear polyacrylamide concentrations and applied voltages on the separation of oligonucleotides and DNA sequencing fragments by capillary electrophoresis. *Anal. Chem.* **66**, 4243–4252.
39 Fung, E.N.; Yeung, E.S. (1995) High-speed DNA sequencing by using mixed poly(ethylene oxide) solutions in uncoated capillary columns. *Anal. Chem.* **67**, 1913–1919.
40 Nathakarnkitkool, S.; Oefner, P.J.; Bartsch, G.; Chin, M.A.; Bonn, G.K. (1992) High-resolution capillary electrophoretic analysis of DNA in free solution. *Electrophoresis* **13**, 18–31.
41 Felmlee, T.A.; Mitchell, P.S.; Ulfelder, K.J.; Persing, D.H.; Landers, J.P. (1995) Capillary electrophoresis for the post-amplification detection of a hepatitis C virus-specific DNA product in human serum. *J. Capillary Electrophoresis* **2**, 125–130.
42 Lu, W.; Han, D.-S.; Yuan, J.; Andrieu, J.-M. (1994) Multi-target PCR analysis by capillary electrophoresis and laser-induced fluorescence. *Nature* **368**, 269–271.

Part Three

Pharmaceutical Science Applications

22 Analysis of pharmaceuticals by capillary electrophoresis
T.J. O'SHEA

22.1 Introduction

In the pursuit of safe and efficacious drugs, pharmaceutical companies must subject possible candidates to biological screening processes that involve toxicity and pharmacological evaluation. For those that successfully proceed through this phase, the manufacture is scaled up and the drug is subjected to further scrutiny. This involves more extensive biological testing and product development. Specifications are established for acceptability of all future production batches. This involves establishing limits of byproducts in production resulting from impurities in the starting materials, isomerization, enantiomeric impurities, residual solvents and degradates. Dosage form and formulation must also be developed and suitable methods implemented for their analysis. In addition stability characteristics of the bulk drug substance and the formulation must be established.

As the reader can appreciate, the diverse analytical needs require numerous assays to be developed in the drug development process. Capillary electrophoresis with its inherent separation capabilities offers an alternative to the more established separation techniques for many applications in pharmaceutical analysis. The application of CE across the full spectrum of pharmaceutical analysis continues to expand and is the subject of several review articles [1–3]. Practical guidelines for strategies to method development [4], and for controlling resolution and quantitation [5] for separations of pharmaceutical related compounds by CE and micellar electrokinetic chromatography (MEKC) have been published. The aim of this section is to provide the reader with representative applications of CE relating to the pharmaceutical industry focusing primarily on publications over the period 1992–1995 (Table 22.1). Therapeutic drug monitoring and chiral analysis of pharmaceuticals will be covered in other chapters of this book.

22.2 Method validation

The ultimate objective of the method validation process is to generate the highest standard of analytical results possible. To obtain such results, all the

Table 22.1 Summary of applications of capillary electrophoresis to pharmaceutical analysis

Method Separation conditions Applied voltage (kV) Column dimensions: L (cm) × ID (μm) Detection	Compound	Reference
Antiasthmatics MEKC 0.02 M Borate (pH 9.2), 0.05 M SDS 12 35 × 50 UV 210 nm	Theophylline	16
CZE 0.01 M Tris (pH 5.0) 12 50 × 75 UV 214 nm	Salbutamol	17
Antibiotics CZE 0.1 M Borate (pH 7.06) 15 50 × 50 PDA	Antibiotics	18
CZE 0.02 M Phosphate (pH 7.3) 1 kV min^{-1} to 15 min 24 × 50 UV 200 nm	β-Lactam antibiotics	19
MEKC 0.02 M Borate, 0.2 M SDS 30 72 × 50 UV 210 nm	β-Lactam antibiotics	20
ITP 5 mM Tris–HCl (pH 7.0), 0.01 M barium hydroxide (pH 10.0) 10 23 × 500 UV 254 nm	β-Lactam antibiotics	21
CZE 0.02 M Phoshate–borate (pH 7.0) 10 57 × 75 UV 254 nm	Sulfonamides	24

MEKC
0.025 M Phosphate–borate (pH 8.5),
  0.1 M SDS,
  0.01 M TAB
12
50 × 50
UV 240 nm
	Sulfonamides	26

Table 22.1 *Continued*

Method Separation conditions Applied voltage (kV) Column dimensions: L (cm) × ID (μm) Detection	Compound	Reference
CZE 0.02 M Acetate (pH 6.8), 20% methanol 26 100 × 75 API–MS	Sulfonamides	28
CZE 4.3 mM Phosphate (pH 7.5) 20 μA 111.9 × 75 UV 260 nm	Tetracyclines	31

Anticancer
MEKC
0.025 M Tris–phosphate (pH 8.5),
  0.05 M SDS,
  25% methanol
25
87 × 50
UV 200 nm
	Taxol	34

CZE 0.075 M Borate (pH 9.2) 25 97 × 50 LIF He–Cd 325 nm	6-Mercaptopurine	35
CZE 0.01 M MES (pH 5.5) 30 50 × 50 ECD +150 mV	6-Mercaptopurine	23
CZE 0.025 M Acetate, 50:50 acetonitrile:methanol 30 55 × 50 UV 254 nm	Tamoxifen	36
Antiulcer CZE 0.02 M Phosphate (pH 7.0) 25 80 × 50 UV 220 nm	Cimetidine	38

MEKC
3.3 mM Trizma base,
  9.4 mM phosphate (pH 6.4)
  9.8 mM HTAB
−20
60 × 50
UV 228 nm
	Cimetidine	39

Table 22.1 *Continued*

Method Separation conditions Applied voltage (kV) Column dimensions: L (cm) × ID (μm) Detection	Compound	Reference
CZE 0.05 M Borax, 2 mM HP-β-CD 20 57 × 50 UV 214 nm	Ranitidine	10
Antidepressants CZE 0.025 M Acetate in acetonitrile 25 60 × 75 UV 214 nm	Imipramine	43
ITP 0.01 M Acetate (pH 4.6)– β-alanine 15 16 × 300 Conductivity detection	Imipramine maprotiline	44
Analgesics MECC 0.015 M Phosphate (pH 11.0), 0.025 M SDS 30 50 × 60 UV 214 nm	Acetaminophene salicylic acid	47
MECC 0.02 M Borate– 0.02 M phosphate (pH 9.0), 0.025 M sodium cholate, 0.05 M sodium deoxycholate 20 57 × 75 UV 214 nm	Salicylic acid	48
CZE 0.025 M Citrate (pH 4.2) 20 60 × 50 UV 214 nm	Codeine	49
Cardiovascular drugs MECC 0.08 M Phosphate (pH 6.8), 0.015 M HTAB, 2.5% 2-propanol −20 50 × 50 UV 214 nm	β-Blockers	52

Table 22.1 *Continued*

Method Separation conditions Applied voltage (kV) Column dimensions: L (cm) × ID (μm) Detection	Compound	Reference

MECC
0.02 M Borate (pH 8.5),
  2% Brij 35,
  0.02 M SDS
15
70 × 100
UV 200 nm
	Enalapril	55

MECC 0.02 M Borate (pH 9.5), 0.03 M SDS 20 67 × 100 UV 214 nm	Hydrochlorothiazide, chlorothiazide	56
CZE 0.06 M CAPS (pH 10.6) 25 20–70 × 50 UV 220 nm	13 Diuretics	57

MEKC
0.1 M Borate (pH 8.1),
  0.05 M SDS,
  15% acetone
25
57 × 75
UV 200 nm
	Verapamil, diltiazem,
	atenolol, nicardipine,
	nifedipine
		59

CZE 0.01 M MES (pH 5.5) 30 50 × 50 ECD +150 mV	Captopril	23
Chinese herbal drugs CZE 0.02 M Phosphate (pH 7.5) 25 70 × 75 UV 254 nm	Glycyrrhizin, glycyrrhetinic	60

CZE
0.01 M Phosphate–
  12.5 mM borate,
  0.02 M SDS
30
100 × 75
UV 275 nm
	Baicalein, baicalin, wogonin
	wogonin-7-*O*-glucuronide
		62

CZE Barium hydroxide– 0.02 M isoleucine (pH 10.0) 28 60 × 75 UV 185 nm		

Table 22.1 *Continued*

Method Separation conditions Applied voltage (kV) Column dimensions: L (cm) × ID (μm) Detection	Compound	Reference

CZE
0.03 M TMAC–
  0.07 M HP-β-CD–
  0.01 M SDS (pH 2.0)
28
90 × 50
UV 210 nm
	Ephedrine alkaloids	64

Drugs of abuse
CZE
9 mM Phosphate–
  9 mM borate (pH 9.0),
  0.045 M SDS,
  6.9 mM β-CD SBE IV
30
78 × 50
LIF Kr-fluoride 248 nm
	Heroin	70

CZE
0.01 M borate (pH 1.5),
  0.025 M HTAB,
  1% ethanolamine,
  11% DMSO
−15
72 × 75
UV 254 nm
	Amphetamines	72

CZE 0.05 Borate (pH 9.2) 20 40 × 50 UV 200 nm	Cocaine, morphine	75
Miotic agents MEKC 0.02 M Borate (pH 9.3), 5% w/v SDS 20 57 × 75 UV 214 nm	Pilocarpine	76
Protein/peptide drugs CZE – coated capillary 0.1 M Phosphate (pH 2.5) 12 20 × 25 UV 206 nm	Leucinostatins	82
CZE 0.1 M Phosphate (pH 6.0) 20 92 × 50 UV 195 nm	Human growth hormone	84

Table 22.1 *Continued*

Method Separation conditions Applied voltage (kV) Column dimensions: L (cm) × ID (μm) Detection	Compound	Reference
CZE 0.05 M Borate (pH 8.3), 0.025 M LiCl 10 57 × 75 UV 200 nm	Recombinant cytokines	85
CZE SDS non-acrylamide polymer gel −14.1 40 × 100 UV 280 nm	Recombinant bovine somatotropin	87
MEKC 0.2 M Phosphate (pH 2.5), 0.03 M PAPS 15 60 × 75 UV 214 nm	Polymyxins	88
Miscellaneous CZE 1.6 mM HNO_3– 2 mM $CuSO_4$ 25 64 × 75 UV 24 nm	Alendronate	92
CZE 0.023 M Phosphate (pH 7.0) 30 64.5 × 50 UV 260 nm	Penciclovir	93

MEKC
0.04 M Phosphate (pH 8.0),
  0.104 M SDS,
  3% methanol
20
58.7 × 50
UV 254 nm
	15 Anti-inflammatory drugs	94

API: atmospheric pressure ionization, CAPS: 3-(cyclohexylamino)-1-propanesulfonic acid, DMSO: dimethyl sulfoxide, HP: hydroxypropyl, HTAB: hexadecyltrimethylammonium bromide, MES: 2-(*N*-morpholino)ethanesulfonic acid, PAPS: 3-(*N,N*-dimethylhexadecylammonium)propane sulfonate, SDS: sodium dodecyl sulfate, SBE: sulfobutyl ether, TAB: tetrabutylammonium bromide, TMAC: tetramethylammonium chloride.

variables of the assay should be considered, including sampling procedure, sample preparation, separation parameters, storage conditions, recovery, specificity, etc. Validated methods for the determination of active drug, drug impurities and degradates are requirements of the worldwide regulatory agencies to support new drug applications. The criterion for what constitutes a valid assay has received considerable attention in the literature and from regulatory agencies.

Numerous CE applications have been developed in this industry that prove the technique can provide the high quality data necessary for validation. This is evident from the number of validated assays referenced in this paper. Notably a recent intercompany cross-validation exercise demonstrated the successful transfer of CE procedures between independent pharmaceutical laboratories [6].

22.3 Impurity analysis

The safety of pharmaceuticals is critically related to their purity, therefore exhaustive efforts are made to ensure the purity of bulk drugs. Analytical methodology is required to assess the presence of impurities in pharmaceuticals. CE has been demonstrated to provide the requisite features required for purity determination. It is now recognized as a viable option for this task within the pharmaceutical industry. A review of the quantitative aspects of the application of CE to the analysis of drug-related impurities has been presented [7]. Ng *et al.* monitored the presence of impurities in a basic drug substance related to ondansetron [8]. Significant improvement in the peak shape over the established HPLC analysis was reported. The application of a fractional factorial design approach to the development of purity assay was investigated [9].

Experiments were designed to evaluate the relative effects of several CE operating parameters upon resolution and migration time. The use of high–low injection volumes for CE was shown to provide enhancement in a study of drug-related impurities [10]. Specific improvements included increases in both the total number and total level of impurities detected, improved detection limits and precision. High voltages coupled to short capillaries were demonstrated to render analysis times of under 2 min for the impurity analysis of salbutamol and fluparoxon [11].

Altria reported on the capability of CE in the determination of two dimeric impurities in salbutamol sulfate drug substance [12]. The results obtained agreed with those found by high performance liquid chromatography (HPLC) and thin layer chromatography (TLC). In another study, CE and HPLC were employed independently to quantify salbutamol-related impurity levels. Satisfactory agreement was obtained between both techniques. Furthermore confirmation of peak homogeneity was determined for fractions collected from both assays by reanalysis with the alternative technique illustrating the powerful capability of combining the techniques [13].

Anionic impurities presenting in pravastatin sodium, taxol and iopiperidol were determined using capillary ion electrophoresis [14]. Resolution of several anions including bromide, chloride, fluoride and phosphate was achieved using a cationic surfactant. Detection was accomplished by indirect UV employing Waters Anion-BT as the visualization agent. An exercise was conducted to exemplify the need for normalization of peak areas with migration times for impurity content analysis by CE [15]. Unless this normalization is adopted, it makes the cross correlation of percentage area/area impurity results between CE and other separation techniques such as LC unfeasible to relate accurately.

22.4 Antiasthmatics

The determination of seven active ingredients in theophylline tablets by MEKC has been described [16]. Complete electrophoretic analysis was accomplished in 8 min. Two internal standards were employed, the second served to improve the precision of a slow migrating peak. The improved quantitative precision provided by peak height rather than peak area measurements is worthy of note.

CE, isotachophoresis (ITP) and HPLC were compared for the determination of the antiasthmatic drugs salbutamol and turbutaline sulfate [17]. In the comparison of the measured peak areas/zone lengths for the different techniques, regression lines were obtained. Slopes of approximately one and intercepts of zero validated the three techniques. Application of the techniques in the analysis of several pharmaceutical dosage forms gave comparable results covering the labeled values. Analysis of drug-related impurities in several antiasthmatic drugs using CE has been extensively investigated by Altria [7, 11, 12].

22.5 Antibiotics

Antibiotics are a class of drugs widely prescribed as chemotherapeutic agents in the treatment of infections. The separation of six antibiotics including penicillin-G, ampicillin and amoxicillan by CE system equipped with a photodiode array (PDA) detector has been described [18]. The utility of the PDA detector was demonstrated for confirmation of peak identity and purity. CE was employed to study the mechanism of hydrolysis of a β-lactamase inhibitor catalysed by Tris (tris(hydroxymethyl)aminomethane) buffer and Tem-2 β-lactamase [19]. In both instances CE proved to be ideal in the identification of the products of reaction, the interpretation of the mechanism of hydrolysis and the determination of kinetic parameters and stoichiometry. The anionic surfactant, SDS, in combination with the ion-pairing reagent pentanesulfonic acid was employed to separate several cephalosporins by CE [20]. ITP was also employed for the analysis of a series of β-lactam antibiotics including penicillin G and cephalosporin C in pharmaceutical preparations [21]. Another β-lactamic antibiotic investigated by CE was cephradine [22]. The developed assay was statistically compared with an established HPLC procedure. Analysis of variance confirmed that there was no significant difference between the results obtained by both methods. Electrochemical detection (ECD) using a gold/mercury amalgam electrode was demonstrated for the sensitive detection of D-penicillamine [23].

Effective mobilities for 16 sulfonamides were measured as a function of pH in two different electrolyte systems. The objective was to calculate their respective pK_a values and mobility at infinite dilution [24]. The optimized separation was applied for the determination of these antibiotics in meat extracts. Ricci and Cross studied for separation of 22 sulphonamides and three dihydrofolate reductase inhibitors in a single electrophoretic run [25]. However, due to the diverse pK_a values, separation of all compounds was unsuccessful. MEKC has also been implemented for the analysis of sulfonamides. Dang et al. reported the CE separation and determination of several sulfonamides employing SDS as a micellar phase and tetrabutylammonium bromide as a buffer modifier [26]. The application of the assay to the quantitative analysis of active ingredients in commercial tablets was described. In another report β-cyclodextrin (CD) was employed as a buffer additive to enable the separation of seven sulfonamides [27]. The assay was used for the analysis of these compounds in their tablet form in the concentration range of 0.1 to 1.0 µg ml^{-1}.

CE with mass spectrometry (MS) distinguishes analytes by both their differences in electrophoretic mobilities and structural information. A number of groups have reported the use of ion-spray–atmospheric pressure ionization MS as a detector in CE for the analysis of sulfonamides. In this interface mode, MS data indicates mostly parent ion and less fragmentation, resulting in insufficient information for chemical structure identification, especially for neutral compounds. Therefore, CE–MS–MS is more appropriate and recommendable for the purpose of identification (editor's data). Henion and co-workers used this approach for the structural determination and confirmation of sulfametazine in a synthetic mixture comprising of six sulfonamides [28]. In another report, CE–MS was used for the analysis of a variety of sulfonamides and their potentiators in shellfish extracts at the low ppm level [29]. Tandem MS was used to provide spectral information on these analytes.

Besides penicillins and sulfonamides, CE has been investigated for the analysis of a range of other antibiotics. Ciprofloxacin, a quinolone antibiotic, presented formidable challenges for separation due to its limited pH solubility [30]. In this study CE was preferred to LC because of its capability operating at pH 1.5. Utilizing pH mediated sample stacking a detection limit of 1×10^{-6} M was achieved.

The electrophoretic behavior of several tetracyclines was characterized by CE in phosphate buffers in the pH range 4–11 [31]. A complete set of acid–base equilibrium constants and electrophoretic mobilities was calculated and subsequently used to assess the optimum separation parameters. The analysis of tetracyclines in pharmaceutical formulations was satisfactorily achieved under the predicted optimum conditions. The addition of EDTA to the separation buffer was illustrated by others, to improve the assay selectivity for tetracycline and its degradation products [32]. Aminoglycoside antibiotics were determined by MEKC with indirect UV detection in the anionic mode with a reversed EOF [33]. Employing the cationic surfactant CTAB, both neutral and charged antibiotics were resolved in combined antibiotic formulations. As an application example, neomycin and hydrocortisone were determined in otosporin eardrops.

22.6 Anticancer drugs

Taxol, a diterpene amide isolated from the bark of the Pacific yew has generated considerable interest due to its antitumor and antileukemic activities. Due to environmental concerns, the extraction of taxol from the needles rather than the bark has been explored. A study using MEKC demonstrated the feasibility of the technique for the analysis of taxol in crude needle and bark extracts [34]. In comparison to HPLC, MEKC offered baseline resolution of taxol from other taxines in the extract, less solvent waste, a smaller sample requirement and the simultaneous detection of taxol, cephalomannine and baccatia III in a relatively simple electrophoretic run. Although UV–vis detection is favored by most researchers performing pharmaceutical analysis, electrochemical [23] and laser-induced fluorescence [35] detection were used for the determination of 6-mercaptopurine. These detection modes addressed the sensitivity limitation of UV detection.

A non-aqueous separation medium was utilized for the determination of tamoxifen and its metabolites to overcome its aqueous solubility limitation [36]. The effects of various organic salts and different compositions of organic mixtures on the separation were investigated. A study by Corran and Sutcliffe compared MEKC with reversed phase HPLC for the analysis of gonadorelin and five therapeutically significant analogs [37]. MEKC with cetrimide micelles provided a complete separation of the six components and compared favorably with a gradient LC assay.

22.7 Antiulcer drugs

Cimetidine, an H_2 receptor antagonist, was analyzed in several commercially available formulations [38]. Samples were dissolved, diluted and extracted with petroleum ether prior to electrophoretic analysis. Analysis of over 60 samples from different preparations yielded RSD values between 1.9 and 6.4%. An electrochromatographic solid phase extraction (SPE) enabled a 10–15-fold increase in sensitivity due to condensation [39]. A two component polymer mixture of polyethylene oxide–polydextran as a separation medium for CE was demonstrated to be effective in the determination of famotidine in complex sample matrices [40]. An exercise was conducted to illustrate the necessity for signal normalization for the reproducible quantitation of ranitidine by CE [41].

22.8 Antidepressants

Seven tricyclic antidepressants including imipramine, nortriptyline and nordoxepin were separated by CE [42]. A systematic method development approach was presented to obtain separation of the structurally related compounds. Inclusion of methanol in the separation buffer was found necessary to accomplish resolution of all the components. Non-aqueous CE was demonstrated for the analysis of imipramine and several other cationic drug substances [43]. High separation efficiency was reported without the use of surfactants or complexing agents in the electrophoresis medium. The use of different organic solvents enabled significant selectivity changes to occur for the analytes examined. ITP was employed by others [44] for the analysis of imipramine, domipramine and maprotiline. Limits of detection reported were approximately $200\,ng\,ml^{-1}$ for injection volumes of $50\,\mu l$.

22.9 Antimigraine drugs

Quantitative determination of sumatripan levels in subcutaneous injection solutions by CE has been described [45]. Satisfactory cross-correlation was demonstrated between sumatripan content results by CE and HPLC. In a further study, extended electrokinetic injection sampling times were demonstrated to result in a limit of detection for sumatripan of $2\,ng\,ml^{-1}$ [46].

22.10 Analgesics

Analgesics are drugs widely used for the relief of pain that are commonly available without prescription. Swartz outlined an approach for the electrophoretic method development of the analysis of analgesics [47]. Detailed in the study was the systematic evaluation of the effects of buffer pH, ionic strength, separation voltage and concentration of SDS on migration time and selectivity of a group of common analgesics. The separation and simultaneous determination of the active ingredients of an analgesic tablet formulation of caffeine, paracetamol, dextropropoxyphene, acetylsalicylic acid and chlorpheniramine were reported using MEKC [48]. As SDS did not resolve the two basic components, the separation was completed by using a mixture of the bile salts sodium cholate and sodium deoxycholate.

An MEKC assay was developed for the determination of paracetamol and caffeine in a capsule dosage form [6]. An injection standard was found to improve the precision associated with hydrodynamic injection. The assay was successfully implemented in seven different companies that all demonstrated acceptable assay performance.

CE was used as a quality control tool for the determination of codeine and its byproducts [49]. The assay was applied to tablet and liquid dosage formulations. Correction of area for migration times was found to improve reproducibility in those cases where environmental conditions lead to increased migration time variability [50]. Normalization of peak area was demonstrated for the quantitative determination of codeine, codeine derivatives and emetine in five antitussive drugs.

22.11 Cardiovascular drugs

There is a diverse range of drugs developed for the treatment of cardiovascular disease including β-blockers, antihypertensive agents, vasodilators and diuretics. β-Adrenergic blocking agents are used in the treatment of hypertension and ischemic heart disease. Sun *et al.* utilized the polymer chitosan immobilized on the surface of a fused-silica capillary for the analysis of pindolol [51]. A study of the effect of organic modifiers in MEKC on the elution and separation of 11 common β-blockers was reported [52]. Except for acetonitrile, the modifiers lengthened the migration time and in appropriate volumes they improved resolution values.

These effects were attributed to the increase in viscosity of the buffer caused by the addition of the organic phase. The effect of buffer pH on the elution order of these β-blockers was the subject of another investigation by the same group [53]. The relationship between the structure of the analytes described by molecular and molecular connectivity indices and the elution order and separation of the β-blockers was explored. Enapapril, an inhibitor of angiotensin converting enzyme (ACE) commonly prescribed for the treatment of hypertension, was analyzed by MEKC with SDS [54]. The procedure enabled the separation of the *cis* and *trans* rotamers of enapapril from its degradates. Similar methodology was described employing both SDS and the non-ionic detergent Brij 35 for the determination of enalapril and its degradates diketopiperazine (a cyclization product) and enalaprilat (a hydrolysis product) [55]. The assay was validated and used to support quality control analysis. CE with ECD was used for the determination of captopril, another ACE inhibitor [23]. MEKC procedure was developed and validated for the quality control analysis of the diuretic drug substances, hydrochlorothiazide and chlorothiazide [56]. Some less-reported subtle factors that yield the precision and accuracy necessary in quantitative assay were discussed. The control of inadvertent hydrodynamic flow and buffer replenishment were found to be essential for precise quantitation.

An effective procedure for the optimization of resolution in CE was presented [57]. The separation of 13 diuretics was achieved through independent control of the capillary length and analysis time, while all other parameters were kept constant. This was accomplished by using $[Cu(NH_3)_4]^{2+}$ as a modifier in the electrolyte solution which has a pronounced effect on electroosmosis but negligible effect on selectivity. Quadratic regression models were employed to predict the effect of these parameters on the resolution of each analyte pair. The correlation of resolution with frictional coefficients and pK_a values in CE of four diuretics was studied to ascertain the most ideal separation conditions [58]. Complete separation of the six cardiovascular drugs atenolol, verapamil, nifedipine, nicardipine, diltiazem and amlodipine proved difficult using MEKC with either of the commonly used organic modifiers acetonitrile or methanol [59]. However, the use of acetone proved successful, demonstrating the utility of this solvent as an organic modifier in MEKC.

22.12 Chinese herbal drugs

Chinese herbal preparations are extensively used in eastern Asia and suitable assay procedures are required for analysis. A CE assay was developed by Chen and Sheu [60] for *Glycyrrhizae Radix*, a herbal drug used as an expectorant and detoxicant. The same group reported the simultaneous determination of 12 bioactive components in several herbal drug preparations [61]. Compared with a developed LC analysis, the CE method was more attractive owing to its shorter analysis time. A methodology for the determination of the flavonoids, baicalein, baicalin, wogonin and wogonin-7-*O*-glucuronide in traditional Chinese medicinal formulations was described [62]. A simple and rapid method for the determination of six ephedrine alkaloids in *Ephedra Herba* by CE was described [63]. Another assay employing hydroxypropyl-β-CD was used in the electrophoretic separation, identification and quantitation of several ephedrine compounds present in *Ephedra Herba* [64]. Analysis of ephedrine and pseudoephedrine in 19 Chinese herbal preparation by CE was reported [65]. Consecutive injections could be accomplished in 5 min inclusive of a column conditioning process. Quaternary alkaloid constituents of the herbal preparations *Phellodendri Cortex* [66] and *Coptidis Rhizoma* [67] were determined by CE. Simultaneous analysis of six flavonoids and four alkaloids in *Scute Coptis* was described using MEKC [68].

22.13 Drugs of abuse

Assays for the determination of illicit drug substances are important requirements in forensic science. Several applications have demonstrated that CE is a viable and in some instances a superior technique in this field. In a study by Weinberger and Lurie [69] MEKC was found to provide significantly greater efficiency, selectivity, peak symmetry and faster analysis time compared to HPLC for a range of illicit drug substances. For a complex mixture consisting of acid and neutral impurities presenting in a heroin sample, MEKC resolved twice the number of peaks as HPLC. Illicit cocaine and its basic impurities were analyzed by MEKC without the peak-tailing exhibited in reversed-phase HPLC using bonded phase columns.

Lurie *et al*. [70] described an improved MEKC separation of the impurities in illicit heroin for intelligence purposes. Using the charged β-CD sulfobutyl ether IV in combination with SDS, a significant improvement in resolution of the impurities was observed. Enhancement of detection sensitivity was achieved using LIF detection. The phenanthrene-like heroin impurities exhibit high native fluorescence using 248 nm laser excitation.

The limit of detection of one of the solutes, acetylthebaol, was 1.8 ng ml^{-1}, 500-fold more sensitive than UV detection. Another CE assay exploiting the detection power of LIF was developed for the determination of morphine [71]. The method is based on competitive immunochemical binding of a fluorescent morphine-Cy5 dye conjugate that permitted levels down to 5×10^{-8} M to be detected.

Quantitation of complex seizures of amphetamines and related substances was analyzed using MEKC with cetyltrimethylammonium bromide (CTAB) micelles [72]. The addition of dimethylsulfoxide and ethanolamine to the borate buffer was required to obtain resolution of the eight amphetamines. In comparison with an established GC procedure, the CE method had the same order of reproducibility, however CE was faster and permitted simultaneous resolution of several components. Wernly and Thormann demonstrated MEKC application of common drugs of abuse and their metabolites in urine [73]. They additionally demonstrated the utility of a multiwavelength UV detection (UV photodiode array) in MEKC for the confirmation testing of a range of these substances [74]. The determination of cocaine and morphine in the hair of heroin and cocaine users has been described [75]. Using CE at 214 nm detection, sensitivity was sufficient to detect 0.15 ng mg^{-1} of morphine in hair.

22.14 Miotic agents

The chemical stability of pilocarpine, a miotic agent used in the treatment of glaucoma was studied using MEKC [76]. Employing SDS as the micelle, the assay achieved baseline resolution of pilocarpine and three potential degradation products in a single analysis. β-CD was used for pilocarpine analysis by CE [77]. The separation of pilocarpine from its *trans* epimer enabled quantitative analysis of a commercial ophthalmic preparation. L-Epinephrine also employed for the control of simple glaucoma was analyzed in ophthalmic solutions using heptakis-β-CD as a buffer additive [78]. Quantitation of l-epinephrine and the determination of the D-/L-epinephrine ratio was performed using an internal standard to improve the reproducibility.

22.15 Protein/peptide drugs

Many techniques have been implemented for the assay and quality control of proteins and peptides in pharmaceutical formulations including enzyme-linked immunosorbent assay (ELISA), radioimmunoassay (RIA), polyacrylamide gel electrophoresis (PAGE) and HPLC. Many of these assays, however, have certain limitations including complex and time-consuming method development, poor detection limits and imprecision and are subject to interferences.

The capability of CE for the analysis of proteins and peptides particularly in complex sample mixtures is well documented. Most recently the technique has experienced an increase in its application to protein/peptide analysis in drug formulations.

A general strategy was described for the development of capillary zone electrophoresis (CZE) procedure in routine pharmaceutical peptide analysis [79]. For basic and neutral peptides, low pH buffers should be tested first, and for acidic peptides the optimization should initiate with buffers of neutral pH. Further refinement in resolution should be examined by pH fine tuning and/or higher buffer concentration.

The analysis of recombinant proteins presents a unique challenge due to non-specific adsorption of the proteins on the silica capillary wall. Numerous efforts have been made to decrease these unwanted interactions. Trimethylammonium propylsulfonate addition to the migration buffer was found to be effective preventing the binding of the humanized monoclonal antibody anti-TAC to the fused-silica capillary [80]. Another approach adapted by Tsuji and Little involved coating the capillary wall with an amphiphatic polymer [81]. The modified capillary was used for the characterization of a recombinant chimeric glycoprotein developed as a vaccine for respiratory syncytial virus. The migration time and efficiency were investigated as a function of buffer composition, ionic strength and pH. Polymer-coated CE capillaries were also used for the analysis of several leucinostatins [82]. The method permitted the separation of the peptides in a crude sample matrix extracted from culture broth. Insulin and its deamidation products were successfully resolved by CE using incorporation of zwitterions and acetonitrile in the migration buffer [83]. The reproducibility was improved by washing the analytical capillary between sample runs. A simple free solution CE assay was found to be adequate for the simultaneous quantitation of biosynthetic human growth hormone and related impurities in a pharmaceutical formulation [84].

Formulation mixtures of the recombinant cytokins leukocyte-A interferon and interleuken-1 α were separated by CE and monitored at 200 nm [85]. Buffer composition was observed to play an important role in the selectivity of the separation. Improved resolution was obtained on addition of lithium chloride to the borate buffer. Precise temperature control was another factor found to be essential to provide acceptable assay performance as the separation profile and stability of the proteins were influenced by as little as 5°C variation in temperature.

CE as a complimentary tool to micro-HPLC was applied to the analysis of the recombinant human protein interleukin-4 [86]. Separations for both the parent protein and its enzymatic digest were developed for the purpose of characterizing protein purity and identity.

Other modes of CE have also been utilized for protein drug analysis. The size-based separation of recombinant bovine somatotropin and its aggregates was examined using SDS non-acrylamide polymer gel CE [87]. The migration time for over 140 injections remained constant, although a gradual loss in efficiency was noted. High efficiency separations of basic polypeptides from the polymyxin group with similar or equal mass-to-charge ratios were achieved by MEKC [88]. The inclusion of 3-(N,N-dimethylhexadecylammonium)propanesulfonate, in addition to forming the micellar pseudophase, coated the silica wall dynamically preventing adsorption.

As many protein and peptide drugs exhibit strong biological potency, dosage forms usually contain a few micrograms of the active drug in the presence of milligram amounts of excipients. Furthermore, detection of proteins is difficult owing to the lack of usable chromophores in easily accessible spectral regions. To address this issue, Guzman et al. developed a procedure for increasing the detection sensitivity in the CE analysis of protein drug substances [89]. The procedure involved derivatization of amino functional groups with fluorescamine and then employing the derivative as a UV chromophore (monitored at 280 nm) to enhance detection sensitivity. Significant improvements in the detectability of several protein drug substances were reported.

MS was also employed to overcome sensitivity limitations. Electrospray ionization (EI)–MS was interfaced to CE (CE–MS in EI interface) and applied to the analysis of recombinant bovine and porcine somatotropin mixtures [90]. Detection limits in the femtomole range were reported. Naproxen was conjugated with lysozyme which served as a carrier to target the anti-inflammatory drug to the kidney. The reaction products of the conjugation were analysed by CE with ionspray MS detection [91]. Native lysozyme and its conjugates with one, two and three naproxen molecules were separated and their identities confirmed by MS.

22.16 Miscellaneous

Alendronate, used in the treatment of a variety of bone diseases, was analyzed in pharmaceutical dosage forms by CE [92]. The assay was based on on-line complex formation between alendronate and copper sulfate present in the electrolyte solution. This enabled direct UV detection to be used via the formation of an alendronate–Cu^{2+} chromophore. Validation of the assay proved its suitability for the analysis of tablet and i.v. dosage forms.

Biodegradation kinetic studies of the antiviral compound penciclovir were performed using CE [93]. Hydrodynamic injection utilizing ionic strength mediated stacking improved the UV detection sensitivity to a limit of detection of 0.5 µg ml^{-1} comparable to that obtained with HPLC. MEKC with SDS was used simultaneously to separate 15 non-steroidal anti-inflammatory drugs including diclofinac, fenoprofen and naproxen [94]. Reported detection limits employing UV detection at 254 nm ranged from 0.1 to 10 µg ml^{-1} for the drugs investigated.

A study was carried out using MEKC to detect trace components in bulk pharmaceuticals with an emphasis on the identification of differences among manufacturers that could be used for source verification in suspect/counterfeit cases [95]. Chemical profiling of β-lactam antibiotics was the focus of the study.

References

1. Sadeka, J.; Polonsky, J.; Shintani, H. (1994) Current advancement of pharmaceutical analysis by capillary zone electrophoresis, micellar electrokinetic chromatography and isotachophoresis. *Pharmazie* **49**, 631–641.
2. Smith, N.W.; Evans, M.B. (1994) Capillary zone electrophoresis in pharmaceutical and biomedical analysis. *J. Pharm. Biomed. Anal.* **12**, 579–611.
3. Li, S.F.Y.; Ng, C.L.; Ong, C.P. (1995) Pharmaceutical analysis by capillary electrophoresis. In *Advances in Chromatography*, Brown, P.R. and Grushka, E. (eds), Marcel Dekker, pp. 199–257.
4. Watzig, H.; Dette, C. (1994) Capillary electrophoresis – Review. *Pharmazie* **49**, 83–96.
5. McLaughlin, G.M.; Nolan, J.A.; Lindahl, J.L.; Palmeiri, R.H.; Anderson, K.W.; Morris, S.C.; Morrison, J.A.; Bronzert, T.J. (1992) Pharmaceutical drug separations by HPCE: Practical guidelines. *J. Liq. Chromatogr.* **15**, 961–1012.
6. Altria, K.D.; Clayton, N.G.; Hart, M.; Harden, R.C.; Hevizi, J.; Makwana, J.V.; Portsmouth, M.J. (1994) An inter-company cross-validation exercise on capillary electrophoresis testing of dose uniformity of paracetomol in formulations. *Chromatographia* **39**, 180–184.
7. Altria, K.D. (1993) Quantitative aspects of the application of capillary electrophoresis to the analysis of pharmaceuticals and drug related impurities. *J. Chromatogr.* **646**, 245–257.
8. Ng, C.L.; Ong, C.P.; Lee, H.K.; Li, S.F.Y. (1994) Determination of pharmaceuticals and related impurities by capillary electrophoresis. *J. Chromatogr.* **680**, 579–586.
9. Altria, K.D.; Filbey, S.D. (1994) The application of experimental design to the robustness testing of a method for the determination of drug-related impurities by capillary electrophoresis. *Chromatographia* **39**, 306–310.
10. Altria, K.D. (1993) High and low injection volumes in CE for improved quantitative determination of drug related impurities. *Chromatographia* **35**, 493–496.
11. Altria, K.D. (1993) High speed determination of drug related impurities by capillary electrophoresis employing commercial instrumentation. *J. Chromatogr.* **636**, 125–132.
12. Altria, K.D. (1993) Determination of salbutamol-related impurities by capillary electrophoresis. *J. Chromatogr.* **634**, 323–328.
13. Altria, K.D. (1993) Peak homogeneity determination and micro-preparative fraction collection by capillary electrophoresis for pharmaceuticals. *J. Chromatogr.* **633**, 221–225.
14. Nair, J.B.; Izzo, C.G. (1993) Anion screening for drugs and intermediates by capillary ion electrophoresis. *J. Chromatogr.* **640**, 445–461.
15. Altria, K.D. (1993) Essential peak area normalization for quantitative impurity content determination by capillary electrophoresis. *Chromatographia* **35**, 177–182.
16. Dang, Q.; Yan, L.; Sun, Z.; Ling, D. (1993) Separation and simultaneous determination of the active ingredients in theophylline tablets by micellar electrokinetic capillary chromatography. *J. Chromatogr.* **630**, 363–369.
17. Ackermans, M.T.; Beckers, J.L.; Everaerts, F.M.; Seleen, I.G.J.A. (1992) Comparison of isotachophoresis, capillary electrophoresis and high performance liquid chromatography for the determination of salbutamol, terbutaline sulphate and fenoterol hydrobromide in pharmaceutical dosage forms. *J. Chromatogr.* **590**, 341–353.
18. Yeo, S.K.; Lee, H.K.; Li, S.F.Y. (1991) Separation of antibiotics by high performance capillary electrophoresis with photodiode-array detection. *J. Chromatogr.* **585**, 133–137.
19. Okafo, G.N.; Cutler, P.; Knowles, D.J.; Camilleri, P. (1995) Capillary electrophoresis study of the hydrolysis of a β-lactamase inhibitor. *Anal. Chem.* **67**, 3697–3701.
20. Sciacchitano, C.J.; Mopper, B.; Specchio, J.J. (1994) Identification and separation of five cephalosporins by micellar electrokinetic capillary chromatography. *J. Chromatogr.* **657**, 395–399.
21. Tsikas, D.; Hofrichter, A.; Brunner, G. (1990) Capillary isotachophoretic analysis of β-lactam antibiotics and their precursors. *Chromatographia* **30**, 657–662.
22. Emaldi, P.; Fapanni, S.; Baldini, A. (1995) Validation of a capillary electrophoresis method for the determination of cephradine and its related impurities. *J. Chromatogr.* **711**, 339–346.
23. O'Shea, T.J.; Lunte, S.M. (1993) Selective detection of free thiols by capillary

electrophoresis–electrochemistry using a gold/mercury amalgam microelectrode. *Anal. Chem.* **65**, 247–250.

24 Ackermans, M.T.; Beckers, J.L.; Everaerts, F.M.; Hoogland, H.; Tomassen, M.J.H. (1992) Determination of sulphonamides in pork meat extracts by capillary zone electrophoresis. *J. Chromatogr.* **596**, 101–109.

25 Ricci, M.C.; Cross, R.F. (1993) Capillary electrophoresis separation of sulphonamides and dihydrofolate reductase inhibitors. *J. Microcol. Sep.* **5**, 207–215.

26 Dang, Q.; Sun, Z.; Ling, D. (1992) Separation of sulphonamides and determination of the active ingredients in tablets by micellar electrokinetic capillary chromatography. *J. Chromatogr.* **603**, 259–266.

27 Ng, C.L.; Lee, H.K.; Li, S.F.Y. (1993) Determination of sulphonamides in pharmaceuticals by capillary electrophoresis. *J. Chromatogr.* **632**, 165–170.

28 Johansson, I.M.; Pavelka, R.; Henion, J.D. (1991) Determination of small drug molecules by capillary electrophoresis–atmospheric pressure ionization mass spectrometry. *J. Chromatogr.* **559**, 515–528.

29 Pleasance, S.; Thibault, P.; Kelly, J. (1992) Comparison of liquid-junction and coaxial interfaces for capillary electrophoresis–mass spectrometry with application to compounds of concern to the aquaculture industry. *J. Chromatogr.* **591**, 325–339.

30 Altria, K.D.; Chanter, Y.L. (1993) Validation of a capillary electrophoresis method for the determination of a quinolone antibiotic and its related impurities. *J. Chromatogr.* **652**, 459–463.

31 Tavares, M.F.M.; McGuffin, V.L. (1994) Separation and characterization of tetracycline antibiotics by capillary electrophoresis. *J. Chromatogr.* **686**, 129–142.

32 Zhang, C.; Sun, Z.; Ling, D.; Zhang, Y. (1992) Separation of tetracycline and its degradation products by capillary zone electrophoresis. *J. Chromatogr.* **627**, 281–286.

33 Ackermans, M.T.; Everaerts, F.M.; Beckers, J.L. (1992) Determination of aminoglycoside antibiotics in pharmaceuticals by capillary zone electrophoresis with indirect UV detection coupled with micellar electrokinetic capillary chromatography. *J. Chromatogr.* **606**, 229–235.

34 Chan, K.C.; Alvarado, A.B.; McGuire, M.T.; Muschik, G.M.; Issaq, H.J.; Snader, K.M. (1994) High-performance liquid chromatography and micellar electrokinetic chromatography of taxol and related taxanes from bark and needle extracts of *Taxus* species. *J. Chromatogr.* **657**, 301–306.

35 Rabel, S.R.; Trueworthy, R.; Stobaugh, J.F. (1993) Recent developments utilizing capillary electrophoresis with laser-induced fluorescence for the determination of 6-mercaptopurine metabolites. *J. High Resolut. Chromatogr.*, **16**, 326–327.

36 Hg, C.L.; Lee, H.K.; Li, S.F.Y. (1994) Separation of tamoxifen and metabolites by capillary electrophoresis with non-aqueous buffer system. *J. Liq. Chromatogr.* **17**, 3847–3857.

37 Corran, P.H.; Sutcliffe, N. (1993) Identification of gonadorelin (LHRH) derivatives: comparison of reversed-phase high-performance liquid chromatography and micellar electrokinetic chromatography. *J. Chromatogr.* **636**, 87–94.

38 Arrowood, S.; Hoyt, A.M. (1991) Determination of cimetidine in pharmaceutical preparations by capillary zone electrophoresis. *J. Chromatogr.* **586**, 177–180.

39 Soini, H.; Tsuda, T.; Novotny, M.V. (1991) Electrochromatographic solid phase extraction for determination of cimetidine in serum by micellar electrokinetic capillary chromatography. *J. Chromatogr.* **559**, 547–558.

40 Soini, H.; Riekkola, M.L.; Novotony, M.V. (1994) Mixed polymer networks in the direct analysis of pharmaceuticals by capillary electrophoresis. *J. Chromatogr.* **680**, 623–634.

41 Altria, K.D. (1993) Essential peak area normalization for quantitative impurity content determination by capillary electrophoresis. *Chromatographia* **35**, 177–182.

42 Salomon, K.; Burgi, D.S.; Helmer, J.C. (1991) Separation of seven tricyclic antidepressants using electrophoresis. *J. Chromatogr.* **549**, 375–385.

43 Bjornsdottir, I.; Hansen, S.H. (1995) Comparison of separation selectivity in aqueous and non-aqueous capillary electrophoresis. *J. Chromatogr.* **711**, 313–322.

44 Buzinkaiova, T.; Sadecka, J.; Polonsky, J.; Vlaisicova, E.; Korinkova, V. (1993) Isotachophoresis analysis of some antidepressants. *J. Chromatogr.* **638**, 231–234.

45 Altria, K.D.; Filbey, S.D. (1993) Quantitative pharmaceutical analysis by capillary electrophoresis. *J. Liq. Chromatogr.* **16**, 2281–2292.

46 Altria, K.D. (1993) Optimization and improvement of sensitivity in capillary electrophoresis for quantitation of selected pharmaceuticals. *LC–GC* **11**, 438–442.
47 Swartz, M.E. (1991) Method development and selectivity control for small molecule pharmaceutical separations by capillary electrophoresis. *J. Liq. Chromatogr.* **14**, 923–938.
48 Broonkerd, S.; Lauwers, M.; Detaevernier, M.R.; Michotte, Y. (1995) Separation and simultaneous determination of the components in an analgesic tablet formulation by micellar electrokinetic chromatography. *J. Chromatogr.* **695**, 97–102.
49 Korman, M.; Vindevogel, J.; Sandra, P. (1993) Separation of codeine and its byproducts by capillary zone electrophoresis as a quality control tool in the pharmaceutical industry. *J. Chromatogr.* **645**, 366–370.
50 Korman, M.; Vindevogel, J.; Sandra, P. (1993) Application of capillary electrophoresis to the quality control of pharmaceutical formulations: effect of area correction on quantitation of antitussives. *J. Microcol. Sep.* **5**, 525–530.
51 Sun, P.; Landman, A.; Hartwick, R.A. (1994) Chitosan coated capillary with reversed electroosmotic flow in capillary electrophoresis for the separation of basic drugs and proteins. *J. Microcol. Sep.* **6**, 403–407.
52 Lukkari, P.; Vuorela, H.; Riekkola, M. (1993) Effects of organic mobile phase modifiers on elution and separation of β-blockers in micellar electrokinetic capillary chromatography. *J. Chromatogr.* **655**, 317–324.
53 Lukkari, P.; Vuorela, H.; Riekkola, M. (1993) Effect of buffer solution pH on the elution and separation of β-blockers by micellar electrokinetic capillary chromatography. *J. Chromatogr.* **652**, 451–457.
54 Qin, X.; Ip, Dominic.P.; Tsai, E.W. (1992) Determination and rotamer separation of enalapril maleate by capillary electrophoresis. *J. Chromatogr.* **626**, 251–258.
55 Thomas, B.R.; Ghodbane, S. (1993) Evaluation a mixed micellar electrokinetic capillary electrophoresis method for validated pharmaceutical quality control. *J. Liq. Chromatogr.* **16**, 1983–2006.
56 Thomas, B.R.; Fang, X.G.; Chen, X.; Tyrrell, R.J.; Ghodbane, S. (1994) Validated micellar electrokinetic capillary chromatography method for quality control of the drug substance hydrochlorothiazide and chlorothiazide. *J. Chromatogr.* **657**, 383–394.
57 Jumppanen, J.H.; Riekkola, M.; Haario, H. (1994) Optimization of resolution of 13 diuretics in CZE by controlling capillary length and electroosmotic flow velocity. *J. Microcol. Sep.* **6**, 595–604.
58 Jumppanen, J.H.; Siren, H.; Riekkola, M.L.; Soderman, O. (1993) Correlation of resolution with frictional coefficients and pK_a values in capillary electrophoresis of four diuretics. *J. Microcol. Sep.* **5**, 451–457.
59 Bretnall, A.E.; Clarke, G.S. (1995) Investigation and optimization of the use of micellar electrokinetic chromatography for the analysis of six cardiovascular drugs. *J. Chromatogr.* **700**, 173–178.
60 Chen, H.; Sheu, S. (1993) Determination of glycyrrhizin and glycyrrhetinic acid in traditional Chinese medicinal preparations by capillary electrophoresis. *J. Chromatogr.* **653**, 184–188.
61 Sheu, S.; Chen, H. (1995) Simultaneous determination of twelve constituents of I-tzu-tang, a Chinese herbal preparation, by high performance liquid chromatography and capillary electrophoresis. *J. Chromatogr.* **704**, 141–148.
62 Liu, Y.; Sheu, S. (1994) Capillary electrophoresis determination of baicalein, baicalin, wogonin and wogonin-7-*o*-glucuronide in traditional Chinese medicinal preparations. *J. High Resolut. Chromatogr.* **17**, 559–560.
63 Liu, Y.; Sheu, S. (1992) Determination of ephedrine alkaloids by capillary electrophoresis. *J. Chromatogr.* **600**, 370–372.
64 Flurer, C.L.; Lin, L.A.; Satzger, R.D.; Wolnik, K.A. (1995) Determination of ephedrine compounds in nutritional supplements by cyclodextrin-modified capillary electrophoresis. *J. Chromatogr.* **669**, 133–139.
65 Liu, Y.; Sheu, S. (1993) Determination of ephedrine and pseudoephedrine in Chinese herbal preparations by capillary electrophoresis. *J. Chromatogr.* **637**, 219–223.
66 Liu, Y.; Sheu, S. (1993) Determination of quaternary alkaloids for *Phellodendri Cortex* by capillary electrophoresis. *J. Chromatogr.* **634**, 329–333.
67 Liu, Y.; Sheu, S. (1992) Determination of quaternary alkaloids for *Coptidid Rhizoma* by capillary electrophoresis. *J. Chromatogr.* **623**, 196–199.

68 Li, K.L.; Sheu, S.J. (1995) Determination of flavonoids and alkaloids in the *Scute Coptis* herb couple by capillary electrophoresis. *Anal. Chim. Acta* **313**, 113–120.
69 Weinberger, R.; Lurie, I.S. (1991) Micellar electrokinetic capillary chromatography of illicit drug substances. *Anal. Chem.* **63**, 823–827.
70 Lurie, I.S.; Chan, K.C.; Spratley, T.K.; Casale, J.F.; Issaq, H.J. (1995) Separation and detection of acidic and neutral impurities in illicit heroin via capillary electrophoresis. *J. Chromatogr.* **669**, 3–13.
71 Evangelista, R.A.; Chen, F.A. (1994) Analysis of structural specificity in antibody–antigen reactions by capillary electrophoresis with laser induced fluorescence detection. *J. Chromatogr.* **680**, 587–591.
72 Trenerry, V.C.; Robertson, J.; Wells, R.J. (1995) Analysis of illicit amphetamine seizures by capillary electrophoresis. *J. Chromatogr.* **708**, 169–176.
73 Wernly, P.; Thormann, W. (1991) Analysis of illicit drugs in human urine by micellar electrokinetic capillary chromatography with on-column fast scanning polychrome absorption detection. *Anal. Chem.* **63**, 2878–2882.
74 Wernly, P.; Thormann, W. (1992) Drug of abuse in human urine using stepwise solid-phase extraction and micellar electrokinetic capillary chromatography. *Anal. Chem.* **64**, 2155–2159.
75 Tagliaro, F.; Poiesi, C.; Aiello, R.; Dorizzi, R.; Ghielmi, S.; Marigo, M. (1993) Capillary electrophoresis for the investigation of illicit drugs in hair: determination of cocaine and morphine. *J. Chromatogr.* **638**, 303–309.
76 Charmen, W.N.; Humbersone, A.J.; Charman, W.N. (1992) Analysis of pilocarpine and its degradation products by micellar electrokinetic capillary chromatography. *Pharm. Res.* **9**, 1219–1223.
77 Baeyens, W.; Weiss, G.; Van Der Weken, G.; Van Den Bossche, W. (1993) Analysis of pilocarpine and its *trans* epimer isopilocarpine, by capillary electrophoresis. *J. Chromatogr.* **638**, 319–326.
78 Peterson, T.E.; Trowbridge, D. (1992) Quantitation of l-epinephrine and determination of the d-/l-epinephrine enantiomer ratio in a pharmaceutical formulation by capillary electrophoresis. *J. Chromatogr.* **603**, 298–301.
79 Langenhuizen, M.H.J.M.; Janssen, P.S.L. (1993) Capillary zone electrophoresis of pharmaceutical peptides. *J. Chromatogr.* **638**, 311–318.
80 Guzman, N.A.; Moschera, J.; Iqbal, K.; Malick, A.W. (1992) Effect of buffer constituents on the determination of therapeutic proteins by capillary electrophoresis. *J. Chromatogr.* **608**, 197–204.
81 Tsuji, K.; Little, R.J. (1992) Charge-reversed, polymer-coated capillary column for the analysis of a recombinant chimeric glycoprotein. *J. Chromatogr.* **594**, 317–324.
82 Quaglia, M.G.; Fanali, S.; Nardi, A.; Rossi, C.; Ricci, M. (1992) Separation of leucinostatins by capillary zone electrophoresis. *J. Chromatogr.* **593**, 259–263.
83 Mandrup, G. (1992) Rugged method for the determination of deamidation products in insulin solutions by free zone capillary electrophoresis using an untreated fused-silica capillary. *J. Chromatogr.* **604**, 267–281.
84 Dupin, P.; Galinou, F.; Bayol, A. (1995) Analysis of recombinant human growth hormone and its related impurities by capillary electrophoresis. *J. Chromatogr.* **707**, 396–400.
85 Guzman, N.A.; Ali, H.; Moschera, J.; Iqbal, K.; Malick, A.W. (1991) Assessment of capillary electrophoresis in pharmaceutical applications. *J. Chromatogr.* **559**, 307–315.
86 Bullock, J. (1993) Capillary zone electrophoresis and packed capillary liquid chromatographic analysis of recombinant human interleukin-4. *J. Chromatogr.* **633**, 235–244.
87 Tsuji, K. (1993) Evaluation of sodium dodecyl sulfate non-acrylamide, polymer gel-filled capillary electrophoresis for molecular size separation of recombinant bovine somatotropin. *J. Chromatogr.* **652**, 139–147.
88 Kristensen, H.; Hansen, S. (1993) Separation of polymixins by micellar electrokinetic capillary chromatography. *J. Chromatogr.* **628**, 309–315.
89 Guzman, N.A.; Moschera, J.; Bailey, C.A.; Iqbal, K.; Malick, A.W. (1992) Assay of protein drug substances present in solution mixtures by fluorescamine derivatization and capillary electrophoresis. *J. Chromatogr.* **598**, 123–131.
90 Tsuji, K.; Baczynskyj, L.; Bronson, G.E. (1992) Capillary electrophoresis–electrospray

mass spectrometry for the analysis of recombinant bovine and porcine somatotropins. *Anal. Chem.* **64**, 1864–1870.
91 Kostiainen, R.; Franssen, E.J.F.; Bruins, A.P. (1993) Capillary zone electrophoresis–ionspray mass spectrometry of a synthetic drug–protein conjugate. *J. Chromatogr.* **647**, 361–365.
92 Tsai, E.W.; Singh, M.M.; Lu, H.H.; Brook, M.A. (1992) Application of capillary electrophoresis to pharmaceutical analysis. *J. Chromatogr.* **626**, 245–250.
93 Hsu, L.C.; Constable, D.J.C.; Orvos, D.R.; Hannah, R.E. (1995) Comparison of high performance liquid chromatography and capillary electrophoresis in penciclovir biodegradation kinetic studies. *J. Chromatogr.* **669**, 85–92.
94 Maboundou, C.W.; Paintaud, G.; Berard, M.; Bechtel, P.R. (1994) Separation of fifteen non-steroidal anti-inflammatory drugs using micellar electrokinetic capillary chromatography. *J. Chromatogr.* **657**, 173–183.
95 Flurer, C.L.; Wolnik, K.A. (1994) Chemical profiling of pharmaceuticals by capillary electrophoresis in the determination of drug origin. *J. Chromatogr.* **674**, 153–163.

23 Analysis of medicinal plants: comparison of capillary electrophoresis with high performance liquid chromatography
P.G. PIETTA

23.1 Introduction

Capillary zone electrophoresis (CZE) is a technique which was originally applied to the analysis of biological macromolecules, mainly protein and DNA [1].

In CZE the analytes elute from the capillary (cathode) in order of decreasing positive charge, and the separation relies on differences in the charge-to-mass ratio. Thus, smaller positive charged analytes have shorter migration times than their negative counterparts. CZE allows the separation of neutral species only in the presence of complexing agents such as borate and carbohydrates, which possess a charge which makes the electrophoretic separation possible. However, separation of neutral and hydrophobic compounds has been achieved based on the presence of micelles in the electrolyte (micellar electrokinetic capillary chromatography, MEKC) [2]. Sodium dodecyl sulphate (SDS) is normally added to the buffer to form negatively charged micelles, which migrate to the anode, that is in the opposite direction to the driving force, the electroosmotic flow (EOF). The analytes (mainly neutral and hydrophobic species) interact with SDS micelles, and their migration to the cathode (dictated by EOF) is retarded. As a result, species differing in hydrophobicity may be separated, and for this reason MEKC is properly considered a hybrid technique that combines both electrophoretic and reversed-phase chromatographic separation principles.

In the last few years, many studies on CZE and MEKC of nutrients [3], drugs [4] and environmental analytes [5] have been published. The first paper dealing with flavonoid drugs (*Ginkgo bioba*) using MEKC appeared in 1991 [6], disclosing this approach to several other studies. In a recent review [7], the influence of different parameters such as voltage, pH, micelle concentration, complexing agents and organic solvents on the separation of a variety of plant secondary metabolites has been described.

The aim of this chapter is to draw up a survey of medicinal plants analyzed by CE mainly focusing on the description of methods developed since 1993.

23.2 Analysis of medicinal plants

In Table 23.1, a list of medicinal plants and related CE is presented.

The content of five quaternary alkaloids (berberine, palmatine, jatrorrhizine, phellodendrine and magnoflorine) in *Phellodendri cortex* was determined by CZE using a background electrolyte of 0.5 M sodium acetate (pH 4.6)–acetonitrile in a ratio of 1:1 [8]. By this method, a systematic comparative study of 31 commercial samples of various *Phellodendron* species was performed [9]. Berberine was the predominant component in almost all the examined samples, accounting for 40% of the total alkaloids in *Phellodendron amurense Rupr.* and *Phellodendron chinese*, respectively.

A quantitative determination of berberine, palmatine, phellodendrine and magnoflorine from the bark of *Phellodendron wilsonii* was described using combined CE–ion trap mass spectrometry (CE–ion trap MS) [10]. The study represents the first example of the use of this combined technique for a sensitive assay of isoquinoline alkaloids.

Sample analysis was performed on a modified CE apparatus using a capillary having a total length of 90 cm with the first 25 cm housed in a cartridge and the remainder extended to the ion spray CE–MS interface. The buffer was 60 mM ammonium acetate, pH 4.5, 40% methanol and the separation was performed with 30 kV at the cathode (ion spray interface). The quantities of berberine and palmatine in the extracts were obtained from standard curves (in the range from 0.6 to 16 pg), and the quantitative result was $13.9\,\mu g\,mg^{-1}$ and $0.23\,\mu g\,mg^{-1}$ in dry bark, respectively.

Various *Coptidis rhizoma* containing Chinese medicinal preparations were analyzed by CZE [11]. The samples were extracted with 70% methanol and the clear filtrate injected directly into CE. Berberine, coptisine and palmatine were separated within 8 min using a buffer of 0.2 M sodium acetate–acetonitrile at a ratio of 1:1 without interference from other components.

Compared with HPLC for the analysis of alkaloids in complex Chinese herbal formulations, the preferable benefit of CZE was further proved in case of *Ephedrae Herba* (MaHuang) [12]. An electrolyte containing 0.01 M valine adjusted to pH 10.0 with ammonia allowed a rapid and reproducible determination of ephedrine and pseudoephedrine in 19 different *Ephedrae Herba* preparations.

The imidazole alkaloids from *Pilocarpus jaborandi*, i.e. the optical isomers pilocarpin and isopilocarpine were separated with CZE by adding the phosphate buffer (0.1 M, pH 6.9) and β-cyclodextrin (β-CD, 0.1 M) [13]. Gaus *et al.* [14] separated cardiac glycosides from *Digitalis lanata* by MEKC modified by CDs, urea and sodium cholate. To attain satisfactory separation of most cardenolides which have similar chromatographic behavior, a step or linear gradient system of HPLC is required, which is quite reproduc-

Table 23.1 Summary of applications to the analysis of medicinal plants

Methods Separation conditions Detection	Medicinal plant, compounds	References
CZE 0.5 M Sodium acetate, pH 4.6, 50% acetonitrile UV 280 nm	*Phellodendron* species isoquinoline alkaloids	8–10
CZE–MS 60 mM Sodium acetate, pH 4.5, 40% methanol Ion trap mass spectrometry		10
CZE 0.2 M Sodium acetate, pH 4.6, 50% acetonitrile UV 254 nm	*Coptidis rhizoma* isoquinoline alkaloids	11
CZE 10 mM Valine (pH 10 with ammonia) UV 185 nm	*Ephedrae herba* protoalkaloids	12
CZE 0.1 M Phosphate, pH 6.9, 0.1 M β-cyclodextrin UV 217 nm	*Pilocarpus jaborandi* imidazole alkaloids	13
MEKC 30 mM Borate, pH 9.3, 50 mM SDS, 20 mM δ-cyclodextrin or 25 mM sodium cholate UV 225 nm	*Digitalis lanata* cardiac glycosides	14
MEKC 5 mM Phosphate, 50 mM SDS, 2% methanol UV 340 nm (for piperine), DAD Electrochemical at 0.9 V vs. Ag/AgCl (for capsaicin and dihydrocapsaicin)	*Capsicum/Piper* capsacinoids/piperine	15
MEKC 30 mM Borate, pH 8.5, 50 mM SDS UV 270 nm, DAD	*Tilia* flavonol-glycosides	16
MEKC 10 mM Phosphate, 12 mM sodium borate, pH 9.75, 20 mM SDS UV 275 nm	*Scutellaria baicalensis* flavones	21
CZE 0.2 M Borate, pH 8.6 UV 214 nm	*Pueraria lobata* isoflavones	22
MEKC 10 mM Lauryltrimethylammonium chloride–5 mM borate–10 mM phosphate–10% acetonitrile UV 280 nm	*Angelica radix* ferulic acid	23

Table 23.1 *Continued*

Methods Separation conditions Detection	Medicinal plant, compounds	References
CZE 25 mM Borate, pH 10.1, 0.7 mM β-cyclodextrin UV 215 nm, DAD	*Propolis* phenolic acids/quercetin	24
MEKC 25 mM Tris–phosphate, pH 8.5, 50 mM SDS, 25% acetonitrile UV 230 nm, DAD	*Taxus brevifolia* diterpenes	25
MEKC 10 mM Borate, 10 mM phosphate, pH 9, 25 mM SDS, 5% acetonitrile UV 214 nm	Opium opium alkaloids	26,27
CZE 0.5 M 6-Aminocaproic acid, pH 4.0 30 mM heptakis-(2,6-di-*O*-methyl)-β-cyclodextrin UV 214 nm, DAD		28
MEKC 25 mM borate, pH 8.3, 50 mM SDS UV 270 nm, DAD	*Arnica* species *Heteroteca inuloides* flavonol- and flavon-glycosides	29
MEKC 25 mM Borate, PH 8.2, 50 mM SDS UV 210 nm, DAD	*Eleutheroccoccus senticosus* eleutherosides B, E	30
CZE 66 mM Phosphate, pH 2.4 40% methanol UV 280 nm, DAD	*Chelidonium majus* isoquinoline alkaloids	31
MEKC 18 mM Phosphate, pH 7–30 mM SDS– 5% acetonitrile UV 220 nm, DAD	*Arnica species* sesquiterpene lactones	32
MEKC 20 mM Borate, pH 8.3–30 mM SDS UV 280, DAD	*Echinacea species* echinacoside, cynarine, phenolic acids	33
MEKC 20 mM Borate, pH 8.3–30 mM SDS UV 200 nm	*Stevia rebaudiana*	34

DAD = diode array detection.

ible. These compounds are also very hydrophobic and in MEKC they tend to migrate very close to the micellar front. For these reasons different CDs were investigated as 'hosts' for the separation. Optimal conditions were 30 mM borate–50 mM SDS–20 mM δ-CD at pH 9.3 for the separation of a critical pair differing only in the anomeric form of the sugar moiety (glucodigifucoside and glucodigiglucomethyloside).

Khaled et al. [15] determined the pungent compounds of *Capsium* fruits, namely capsaicin and dihydrocapsaicin, and of *Piper*, i.e. piperine, by MEKC. The separation was possible in the presence of SDS as surfactant and with the addition of methanol as organic modifier (5 mM phosphate, 50 mM SDS, 2% methanol). Simultaneous dual detection was required for selective UV detection of piperine at 340 nm and selective electrochemical detection of capsaicins at 0.9 V vs. Ag/AgCl because piperine comigrated with capsaicin under the conditions used.

MEKC was applied for the determination of nine different flavonol glycosides from *Tilia* [16] using a buffer of 30 mM borate (pH 8.5) and 50 mM SDS. The analytes were monitored by 'on-line' UV diode array detection to identify peaks as quercetin or kaempferol derivatives. As expected, the migration order of the flavonol glycosides was different from the elution order in HPLC as well as the spectral behavior which will be mentioned later. The influence of the organic modifier, buffer, surfactant concentration and pH was studied for electrophoretic mobilities and separation selectivity of selected flavonoids from medicinal plants [17, 18]. The organic solvent modifies the interaction between micelles and analytes, thus reducing migration times and resolution. SDS improves the separation at pH 8.3, whereas it has less or no effect at higher pH than 8.3. At pH 10.5, the separation is mainly regulated by ionization of the hydroxy groups and borate complexation of the carbohydrate moieties. Ring B substitution and the presence of a free hydroxyl group at C7 of ring A play an important role, whereas *O*-methoxylation is less significant. The extent and the kind of glycosylation also contribute in determining different degrees of complexation with borate, thus affecting the mobilities. Concerning the 'on-line' spectra, wavelengths were found to be shifted (about 10 nm) in comparison with those measured during HPLC. This difference is ascribed to the different solvents, i.e. borate–buffer–SDS at pH 8.3 in MEKC and 2-propanol–tetrahydrofuran–water in HPLC. Nevertheless, the UV spectrum in ring C region absorbing at 300–400 nm is indicative of the flavonoid class, whereas the shape of the maximum or its first derivative in ring B region absorbing at 240–280 nm enables distinction of the specific aglycone. Combining chromatographic, electrophoretic and 'on-line' spectral data, identification of the most common flavonoids from medicinal plants was achieved [19].

A buffer of 20 mM SDS, 10 mM sodium dihydrogenphosphate and 12.5 mM sodium borate (pH 9.75) was found to be the most suitable for MEKC analysis for six major flavonoids in crude *Scutellaria baicalensis* roots [20, 21]. Oroxylin A 7-*O*-glucuronide, baicalein, wogonin 7-*O*-glucuronide, baicalin, wogonin and oroxylin A migrated well and separated in this order within 25 min, and the samples of *Scutellariae radix* were analyzed without any pretreatment.

Baicalin was the predominant component (around 122 mg g^{-1} in dry root) followed by baicalein and wogonin 7-*O*-glucuronide (around 30 mg g^{-1}), whereas the other flavones were present in smaller amounts.

A simple CE for the analysis of isoflavones in soy beans and in Kudzu roots (*Pueraria lobata*) was developed by Shihabi *et al.* [22]. Powdered samples were extracted with 66% acetonitrile to dissolve the slightly soluble isoflavones and after centrifugation aliquots were applied to the instrument by pressure injection for 10 s, using 0.2 M borate buffer at pH 8.6. Daidzin, genistin, daidzein and genistein were detected at 214 nm, and the major isoflavones were daidzin and daidzein for soy bean seeds and Kudzu roots, respectively.

Chen and Shen determined the amounts of ferulic acid, nicotinic acid and phthalic acid in *Angelicae radix* and related medicinal preparations by reversed electroosmotic flow CE [23]. The electrolyte consisted of 10 mM lauryltrimethylammonium chloride–5 mM borate–10 mM phosphate and acetonitrile (9:1), and the aromatic acids were separated within 3 min. *Propolis* samples were analyzed without any precleaning by CZE, using 25 mM borate adjusted to pH 10.1 with 1 M sodium hydroxide in the presence of 0.7 mM β-CD [24]. Due to the presence of contaminants in the complex *Propolis* matrix, after each analysis the capillary needed to be washed out with a cleansing solution (methanol–concentrated hydrochloric acid, 5:1 v/v). Thus, the advantage presented by the direct injection of the untreated sample is partly lost by this cleaning procedure. Addition of SDS to the electrolyte did not improve the separations, while β-CD resulted in a satisfactory separation of some typical phenolic components such as 3,4-dimethoxycinnamic acid, 3-hydroxy-4-methoxycinnamic acid (isoferulic acid), 4-hydroxy-3-methoxycinnamic acid (ferulic acid), 3,4-dihydroxycinnamic acid (caffeic acid) and quercetin. The detection wavelength of these components was selected with a UV photodiode array detector, and 215 nm was selected for the determination of *Propolis* samples.

Chan *et al.* compared HPLC and MEKC in the separation of taxol, cephalomannine and baccatin III from bark and needle extracts of *Taxus* species [25]. The electrophoresis buffer consisted of 25 mM Tris–phosphate (pH 8.5), 50 mM SDS and 25% acetonitrile, while HPLC was performed in the isocratic mode with methanol–acetonitrile–50 mM ammonium acetate (5:8:12, pH 4.4) as the mobile phase. In both cases, detection was at 230 nm. HPLC separation of these compounds from impurities indicated difficulty, especially those from the needle extracts. In contrast, MEKC attained significant separation of the main component taxol from cephalomannine and baccatin III and from other interfering compounds in the needle and bark samples.

Opium analysis remains an important task, due to the difficulty of separation in a short period as components possess contrasting properties. Indeed, a major problem associated with HPLC methods is their inability to separate opium alkaloids, as some of them are very similar and tend to coelute in isocratic elution. Gradient HPLC elution may improve separation and reproducibility is satisfactory. MEKC may also overcome this kind of difficulty.

Krogh et al. [26] separated some opium alkaloids using a buffer of 10 mM borate, 10 mM sodium dihydrogenphosphate, 25 mM SDS and 5% acetonitrile at pH 9.0. The pH and the amount of acetonitrile were found to be critical factors affecting the resolution of the mixtures. Illicit drugs were dissolved in the buffer and new samples were injected every 13 min. A similar approach was developed by Trennery et al. [27], who compared the MEKC method with a fully validated HPLC procedure. Results indicated that the superior resolving power of CE permitted identification and quantification of additives in a range of complex samples, which will be impossible in HPLC. A further improvement in the separation of opium alkaloids was obtained by adding 30 mM of heptakis(2,6-di-O-methyl)-β-CD to the buffer (0.5 M 6-aminocaproic acid, pH 4.0) [28]. These conditions allowed separation of normorphine, morphine, codeine and thebaine which are not separated in the presence of surfactants such as SDS or cetyltrimethylammonium bromide. Falsification in the flowers of *Arnica montana* and *Arnica chamissonis* ssp. *foliosa* by blending them with those from *Heterotheca inuloides* was rapidly separated by MEKC [29]. *Arnica montana* flowers contain the flavonoids patuletin-3-O-glucoside, quercetin-3-O-glucoside, 6-methoxykaempferol-3-O-glucoside, kaempferol-3-O-glucoside together with 1,5-dicaffeoylquinic acid and chlorogenic acid, whereas *Arnica chamissonis* ssp. *foliosa* flowers are additionally characterized by the presence of two flavone glycosides. The specific components of *Heterotheca inuloides* are rutin, quercetin-3-O-galactoside and quercetin-3-O-arabinoside. Using 25 mM borate, pH 8.3 and 50 mM SDS, it was possible to obtain rapidly the 'finger-printing' of each drug and to detect possible falsification from *Heterotheca inuloides* in *Arnica* flowers. The LOD (limit of detection) of impurities was 0.1 µg ml^{-1} and it allowed detection of adulteration levels even lower than those achievable by LC–MS.

The main components, namely eleutheroside B and E and chlorogenic acid, of the roots of *Eleutherococcus senticosus* were separated by MEKC [30]. These compounds were most appropriately separated using 25 mM borate buffer at pH 8.2 containing 50 mM SDS. The content of eleutheroside B and E in *Eleutherococcus senticosus* was indifferent. Extraction and formulation were carried out after precleaning of the samples through a C18 SPE column to remove mostly chlorogenic acid.

Six different isoquinoline alkaloids from *Chelidonium majus* (leaves and roots) were separated by CZE using 66 mM phosphate buffer at pH 2.4 containing 40% methanol [31]. This large percentage of the organic modifier was found to be essential in order to obtain satisfactory resolution of the complex mixtures present in leaf and root extracts. Actually major attention was devoted to leaf extracts because of their predominant phytotherapeutic interest.

The electrophoretic run lasted less than 14 min and among the peaks sanguinarine, cheleryihryne, berberine and chelidonine were easily identified by comparison with standards. Alkaloids associated with two other peaks needed to be isolated by semipreparative HPLC and identified as coptisine and D,L-stylopine by UV and MS.

Sesquiterpenelactones of the *helenalide* type from *Arnica montana* L. and *Arnica chamissonis less* ssp. *foliosa* were separated by MEKC [32]. The electrolyte was a solution of 30 mM SDS in 18 mM phosphate, pH 7.0 and acetonitrile (17:3). This paper was the first application of MEKC to sesquiterpenelactones analysis, and the results are comparable with those achievable by HPLC.

The characteristic components of *Echinacea anqustifolia*, *Echinacea purpurea* and *Echinacea pallida* (roots and aerial parts) were easily separated by MEKC using 20 mM borate, pH 8.3 and 30 mM SDS. It was possible to obtain 'finger-prints' of each drug and to discriminate among them in mixtures [33].

A buffer of 20 mM borate, pH 8.3 and 30 mM SDS yielded the separation of diterpene glycosides from *Stevia rebaudiana* leaves and extracts. Stevioside, rebaudioside A and steviolbioside were significantly separated from other cinnamic acid and flavonol derivatives [34].

23.3 Conclusions

CZE and MEKC represent valuable techniques for the study of medicinal plants. They can be considered complementary to HPLC, and are particularly suitable in the analysis of complex natural matrices, where more significant separation than HPLC is achieved. The possibility of injecting small volumes enables CE to be applicable when the amount of sample is limited, as with HPLC fractions obtained in preliminary analysis or semipreparative collection of herbal drugs.

The combination of CE with MS broadens the potential of this technique, and it is likely that its range of application in the field of medicinal plants will be further developed and enlarged.

However, CE–MS and LC–MS have a restriction which must be solved when applying neutral compounds, since MS has no detection especially under atomic pressure chemical ionization (APCI), so when neutral compounds applicable to MEKC are involved this inferiority of LC–MS or CE–MS in APCI mode must be considered seriously.

References

1. Deyl, Z.; Struzinsky, R. (1991) Capillary zone electrophoresis: its applicability and potential in biochemical analysis. *J. Chromatogr.* **569**, 63–122.
2. Terabe, S.; Otsuka, K.; Ichikawa, K.; Tsuchiya, A.; Ando, T. (1984) Electrokinetic separations with micellar solutions and open-tubular capillaries. *Anal. Chem.* **56**, 111–113.
3. Zeece, M. (1992) Capillary electrophoresis: a new analytical tool for food science. *Trends Food Sci. Technol.* **3**, 6–10.
4. Sadecka, J.; Polonsky, J.; Shintani, H. (1994) Current advancement of pharmaceutical analysis by capillary zone electrophoresis, micellar electrokinetic chromatography and isotacophoresis. *Pharmazie* **49**, 631–641.
5. Blumley, W.C.; Brownrigg, C.M.; Grange, A.H. (1994) Capillary liquid chromatography–mass spectrometry and micellar electrokinetic chromatography techniques in environmental analysis, *J. Chromatogr.* **680**, 635–644.
6. Pietta, P.G.; Mauri, P.L.; Rava, A.; Sabbatini, G. (1991) Application of micellar electrokinetic capillary chromatography to the determination of flavonoid drugs. *J. Chromatogr.* **549**, 367–373.
7. Tomas-Barberan, F. (1995) Capillary electrophoresis: a new technique in the analysis of plant secondary metabolites. *Phytochem. Anal.* **6**, 177.
8. Liu, Y.M.; Shen, S.J. (1993) Determination of quaternary alkaloids from *Phellodendri cortex* by capillary electrophoresis. *J. Chromatogr.* **634**, 329–333.
9. Liu, Y.M.; Shen, S.J.; Chiou, S.; Chong, H.; Chen, Y. (1993) A comparative study on commercial samples of *Phellodendri cortex*. *Planta Med.* **59**, 376–378.
10. Henion, J.D.; Mordehai, A.V.; Cai, J. (1994) Quantitative capillary electrophoresis–ion spray mass spectrometry on a benchtop ion trap for the determination of isoquinoline alkaloids. *Anal. Chem.* **66**, 2103–2109.
11. Liu, Y.M.; Shen, S.J. (1993) Determination of coptisine, berberine and palmatine in traditional Chinese medical preparations by capillary electrophoresis, *J. Chromatogr.* **639**, 323–328.
12. Liu, Y.M.; Shen, S.J. (1993) Determination of ephedrine and pseudoephedrine in Chinese herbal preparations by capillary electrophoresis. *J. Chromatogr.* **637**, 219–223.
13. Bayens, W.; Weiss, G.; Van der Wexen, G.; Van den Bossche, W.; Dewaele, C. (1993) Analysis of pilocarpine and its *trans* epimer, isopilocarpine, by capillary electrophoresis. *J. Chromatogr.* **638**, 319–326.
14. Gaus, H.J.; Treumann, A.; Kreis, W.; Bayer, E. (1993) Separation of cardiac glycosides by micellar electrokinetic capillary electrophoresis. *J. Chromatogr.* **635**, 319–327.
15. Khaled, M.Y.; Anderson, M.R.; McNair, H.M. (1993) Micellar electrokinetic capillary chromatography of pungent compounds using simultaneous on-line ultraviolet and electrochemical detection. *J. Chromatogr. Sci.* **31**, 259–264.
16. Pietta, P.G.; Mauri, P.L.; Bruno, A.; Zini, L. (1993) High-performance liquid chromatography and micellar electrokinetic chromatography of flavonol glycosides from *Tilia*. *J. Chromatogr.* **638**, 357–361.
17. Pietta, P.G.; Mauri, P.L.; Bruno, A.; Gardana, C. (1994) Influence of the structure on the behavior of flavonoids in capillary electrophoresis. *Electrophoresis* **15**, 1326.
18. Pietta, P.G.; Mauri, P.L.; Zini, L.; Gardana, C. (1994) Optimization of separation selectivity in capillary electrophoresis of flavonoids. *J. Chromatogr.* **680**, 175–179.
19. Pietta, P.G.; Mauri, P.L.; Gardana, C. (1994) Identification of the main flavonoids by use of chromatographic, electrophoretic and 'on-line' spectral data. *J. High Resolut. Chromatogr.* **17**, 616–618.
20. Liu, Y.M.; Shen, S.J. (1994) Determination of six major flavonoids in *Scutellariae radix* by micellar electrokinetic capillary electrophoresis. *Anal. Chim. Acta* **288**, 221–226.
21. Liu, Y.M.; Shen, S.J. (1994) Capillary electrophoresis determination of baicalein, baicalin,

wogonin and wogonin-7-*O*-glucuronide in traditional Chinese medicinal preparations. *J. High Resolut. Chromatogr.* **17**, 559–560.
22 Shihabi, Z.K.; Kute, T.; Garcia, L.L.; Hinsdale, M. (1994) Analysis of isoflavones by capillary electrophoresis. *J. Chromatogr.* **680**, 181–185.
23 Chen, H.R.; Sheu, S.J. (1994) Determination of the aromatic acids in traditional Chinese medicinal preparations by reversed electroosmotic flow capillary electrophoresis. *Chim. Pharm. J. (Taipei)* **46**, 145–153.
24 Chi, H.; Hsiek, A.K.; Ng, C.L.; Lee, H.K.; Li, S.F.Y. (1994) Determination of components in propolis by capillary electrophoresis and photodiode array detection. *J. Chromatogr.* **680**, 593–597.
25 Chan, K.C.; Alvarado, A.B.; McGuire, M.T.; Muschik, G.M.; Issaq, H.J.; Snader, K.M. (1994) High-performance liquid chromatography and micellar electrokinetic chromatography of taxol and related taxones from bark and needle extracts of *Taxus* species. *J. Chromatogr.* **657**, 301–306.
26 Krogh, M.; Brekke, S.; Tonnesen, F.; Rasmussen, K.E. (1994) Analysis of drug seizures of heroin and amphetamine by capillary electrophoresis. *J. Chromatogr.* **674**, 235–240.
27 Trennery, V.C.; Wells, R.J.; Robertson, J. (1994) The analysis of illicit heroin seizures by capillary zone electrophoresis. *J. Chromatogr. Sci.* **23**, 1–6.
28 Biornsdottir, I.; Hansen, S.H. (1995) Determination of opium alkaloids in opium by capillary electrophoresis. *J. Pharm. Biomed. Anal.* **13**, 687–693.
29 Pietta, P.G.; Mauri, P.L.; Bruno, A.; Merfort, I. (1994) MEKC as an improved method to detect falsifications in the flowers of *Arnica montana* and *A. chamissonis*. *Planta Med.* **60**, 369–372.
30 Pietta, P.G.; Mauri, P.L.; Gardana, C.; Zini, L. (1994) Micellar electrokinetic chromatographic/ultraviolet diode array analysis of *Eleuterococcus senticosus*. *Phytochem. Anal.* **5**, 305–308.
31 Pietta, P.G.; Mauri, P.L.; Gardana, C.; Colombo, M.L.; Tomè, F. (1995) Capillary electrophoresis of isoquinoline alkaloids from *Chelidonium majus* L. *Phytochem. Anal.* **6**, 196–202.
32 Mauri, P.L.; Pietta, P.G.; Merfort, I.; Willuhn, G. Separation of sesquiterpenelactones from *Arnicae flos* DAB 10 by MEKC. *Planta Med.*, in press.
33 Mauri, P.L.; Pietta, P.G.; Catalano, G.; Bauer, R. Micellar Electrokinetic capillary chromatography of *Echinaceae angustifolia*, *E. purpurea* and *E. pallida Planta Med.*, in press.
34 Mauri, P.L.; Catalano, G.; Gardana, C.; Pietta, P.G. Analysis of *Stevia rebaudiana* glycosides by capillary electrophoresis. *Electrophoresis*, in press.

24 Quantitative applications of the resolution of enantiomers by capillary electrophoresis
K.D. ALTRIA

24.1 Introduction

The separation and quantitation of enantiomers are important applications of capillary electrophoresis. Other chapters cover the background to the separation selectivities that can be obtained in free solution capillary electrophoresis (FSCE) and micellar electrokinetic capillary chromatography (MECC) using a variety of chiral selectors such as cyclodextrins (CDs), bile salts and crown ethers. This chapter covers the validation and quantitative application of chiral CE methods. Table 24.1 covers the majority of quantitative applications with selected details of selectivity modes, performance and application. The validation criteria applied to chiral CE methods are similar to those employed when assessing HPLC (high performance liquid chromatography) methods. Standard criteria assessed include linearity, repeatability, reproducibility, precision and accuracy. HPLC and CE may be employed in tandem to confirm accuracy. The majority of quantitative reports have focused on the analysis of chiral pharmaceuticals and agrochemicals. Application areas include enantiomeric purity testing of single enantiomer compounds, reaction rate monitoring, stability testing and enantiomeric separation of drugs in clinical samples.

24.2 Method validation

Prior to consideration of CE as a viable analytical technique for the separation and quantitation of enantiomers, it is important to consider the performance levels that may be obtained for key criteria. These criteria are similar to those employed in the validation of HPLC methods and include limits of detection and quantitation for the undesired enantiomer, linearity of detector response, recovery, precision, freedom from interference and method robustness.

24.2.1 Detection limits

The detection limits that are possible for determination of trace undesired enantiomers are generally the most important performance measurement

Table 24.1 Range of drugs chirally resolved by CE

Compounds	Separation conditions	Comments	Ref.
Low pH, CDs			
10 Basic drugs including ephedrine (racemates)	CE, 70 cm × 50 μm i.d., 200 nm 30°C, buffer: 30 or 60 mM Tris(hydroxymethyl) aminomethane-base to pH 2.4–3.3 with H_3PO_4. CD dissolved in Tris–phosphate buffer	<0.5% Trace enantiomer	23
BCH 189	CE, 57 cm × 75 μm i.d., 214 nm ambient temperature, 15 kV, buffer: NaH_2PO_4 to pH 2.3 with H_3PO_4 + 27 mM dimethyl-β-CD	Quantitative monitoring of enzymatic biotransformation, LOD of 0.5%	26
BCH 189 and others	CE, 57 cm × 75 μm, 32 mM dimethyl-β-CD in 100 mM NaH_2PO_4 to pH 4 with citric acid, 210 nm, 13 kV	LOD of 0.1% for BCH 189, cross-correlation with HPLC data	24
Bupivacaine (and other basic drugs)	CE, 50 cm × 75 μm i.d., 220 nm 18–20 kV Buffer: 10 mM heptakis (2,3,6-tri-*o*-methyl)-β-CD or heptakis (2,6-di-*o*-methyl)-β-CD, Trizma base (18 mM, pH 2.8–3.0 with H_3PO_4 and methylhydroxyethylcellulose 1000 or 4000 (0.1%, w/w). Several modifiers such as 0.05 mM hexadecyl-trimethylammonium bromide were added in buffer systems	Assay in plasma, internal standard, good linearities	17
Clenbuterol	CE, 57 cm × 50 μm i.d., 214 nm 30 kV, ambient Buffer: 30 mM hydroxypropyl-β-CD in 50 mM $Na_2B_4O_7 \cdot 10H_2O$, pH adjusted to 2.2 with H_3PO_4	Method transfer exercise between 7 drug companies	6
Denopamine, trimetoquinol, and timepidium	CE, 50 cm × 75 μm i.d., 214 nm 14 kV, 23°C Buffer: 25 mM phosphate buffer of pH 2.7 containing 5% β-CD	0.1% LOD, Spiked-recovery experiments – recovery rates 95–104%	14
Dihydropyridine and racemic amines	MEKC, 60 cm × 75 μm i.d. 214 nm, 20 kV, 25°C Buffer: 20 mM citric acid pH 2.5 containing 2% (w/v) of carboxymethyl-β-CD	Various neutral and ionizable CDs used for separations, syrup samples analyzed	28
Epinephrine	CE, 50 cm × 75 μm i.d., 206 nm 15 kV (30–40 μA) Buffer: 10 mM Tris/18 mM methyl-β-CD to pH 2.4 with H_3PO_4	Pseudoephedrine internal standard, formulation analysis	3
Epinephrine (adrenaline)	Low pH, CD	Commercial adrenaline solution analyzed	22

Table 24.1 *Continued*

Compounds	Separation conditions	Comments	Ref.
Fluparoxan	CE, 57 cm × 50 μm i.d., 214 nm 16 kV, 25°C Buffer: (10 mM borax, 10 mM Tris, 150 mM β-CD, 6 M urea)–isopropanol (4:1 v/v), adjusted pH 2.5 with H_3PO_4	Full method validation, 1% LOD	9
Lilly drug LY248686	CE, 50 cm × 50 μm i.d., 214 nm 23 kV, 30°C Buffer: 50 mM Tris–phosphate buffer at pH 2.35, 4 mM hydroxypropyl-β-CD	Full validation, 0.1% LOD	10
Naproxen	CE, 20 cm × 25 μm i.d., 230 nm 14 kV (3 μA), 20°C Buffer: 200 mM 2-(N-morpholino)-ethanesulfonic acid/tetrabutyl ammonium hydroxide, 10 mM hydroxypropyl-β-CD, 0.4% polymeric additives, pH 5.0	Validation including 0.1% LOD, repeatability	7
Picumeterol	Low pH CD	Study on the effects of peak are normalization	21
Phenoxy acid herbicides	CE, 63.1 cm × 50 μm i.d., 200 nm 30 kV, 30°C Buffer: 30 or 50 mM lithium acetate and adjusted to pH 4.8 with acetic acid. CDs dissolved in lithium acetate buffer	<1% LOD, good linearities, applied to production batches	13a
5 Phenoxy acid herbicides	MECC Buffer: 0.1% (w/w) methylhydroxy ethycellulose 30000 and SDS to a 0.1 M in 0.58 M acetic acid and adjusting pH to 3.7 with 1 M tri(hydroxymethyl) aminomethane (Tris)	Good migration time precisions and resolutions	13b
Quinagolide, Sandoz ENX 792q	Low pH, CD	0.2% inactive ENX 792 detected	25
Verapamil and norverapamil	CE, 18 cm × 75 μm i.d., 200 nm 12 kV (90–100 μA), 15°C Buffer: 60 mM phosphate buffer pH 2.5 with 60 mM trimethyl-β-CD	Assay in plasma, acceptable validation data	4
High pH, CDs Phenylalanine (dansylated)	pH 9 with CD	25 μm capillary, Analysis in nutritional formulation 70s	27
Sandoz drug SDZEAA 494s (and amino acids)	CE, 60 cm × 75 μm i.d., 214 nm 10 kV Buffer: 0.1 M borate buffer of pH 9.5 and 20 mM γ-CD	Compound dansylated precolumn, LOD of 0.1%	30a

Table 24.1 Continued

Compounds	Separation conditions	Comments	Ref.
Warfarin	CE, 50 cm × 50 µm i.d., 310 nm 20 kV (70 µA), 25°C Buffer: 0.1 M sodium phosphate buffer (pH 8.35): 8 mM methyl-β-CD–methanol (49:1, v/v)	Enantioselective metabolism studies conducted, internal standard used for improved precision	5
SDS, MECC, CDs			
Bristol–Myers Squibb drug BMS-180431-09	SDS, MEECC with CD	Validation including LOD of 0.06%, recoveries, robustness	11
Nalbuphine and other diastereoismers	SDS, MECC, with THF or ACN	LOD or 0.05% 0.005% with z-cell	2
9 NSAIDs and barbiturates (racemates)	MECC, SDS, CD	Low pH with CD also used, analysis in plasma	18
Organic acids	CE, 45.4 cm × 50 µm i.d., 206 nm 20 kV, 20°C Buffer: (2-hydroxy)propyl-β-CD (0–80 mM) or 0–15 mM β-CD was dissolved in 50 mM phosphate buffer pH 7 or 8		19
Valine (OPA derivative)	CE, 50 cm × 75 µm i.d., 230 nm 20 kV, 30°C Buffer: acetate 50 mM, pH 4.45: 50 mM glacial acetic acid–50 mM sodium acetate (1:1, v:v). alpha, β or methylated CDs in buffer	0.13% LOD, automated derivatization	30b
Carbohydrates			
Oxamnique	CE, 50 cm × 50 µm i.d., 246 nm, 20 kV, 30°C Buffer: 50 mM sodium dihydrogen phosphate (pH 3) +3 mM heparin	0.23% m/m LOD, RSD for 10 injections <1%	8
Bile salt or Taurocholate with MECC			
Serotonin agonist and derivatives	Bile salt MECC + CD	GITC derivatives separated, <1% LOD	16
Various including amphetamine	MECC, 57 cm × 50 µm i.d., 210 nm Buffer: 10 mM phosphate–borate buffer (pH 9.0) containing 50 mM sodium taurocholate and this solution was adjusted to pH 11.7 with 1 M NaOH	Samples derivatized with GITC preseparation	15

ACN = acetonitrile, GITC = 2,3,4,6-tetra-*o*-acetyl-β-D-glucopyranosyl isothiocyanate, NSAID = non-steroidal anti-inflammatory drug, OPA = *ortho*-phthalaldehyde, SDS = sodium dodecyl sulfate, THF = tetrahydrofuran.

as single enantiomer chiral compounds may often be produced to high enantiomeric purities, with <1% of the undesired enantiomers present as an enantioimpurity. Detection levels of <1% trace enantiomer by CE have been demonstrated by several workers (Table 24.1). Altria *et al.* [1] indicate an expanded section of the analysis of a batch of the antiviral compound BCH 189 (2'-deoxy-3'-thiacytidine), containing 0.3% of the undesired enantiomer, using a low pH electrolyte with dimethyl-β-CD as the chiral selector. Clearly the signal-to-noise ratio indicates that a limit of detection of <0.1% is achievable. Use of modified capillary detection windows can lead to further improvements in sensitivity [2].

24.2.2 Precision

Injection precision in CE is generally poorer than that in HPLC. This imprecision is largely attributable to the difficulties involved in reproducibly injecting nanoliter volumes of samples into the capillary. Many quantitative reports have utilized internal standards to alleviate this problem and acceptable precision data have been reported in those cases [3–5]. Each enantiomer essentially acts as an internal standard for the other and good precision for peak area ratios for enantiomeric mixtures are therefore obtained. For example, in an intercompany cross-validation exercise concerning the separation of the racemic drug clenbuterol, all seven participating companies obtained relative standard deviation (RSD) values of <1% ($n = 10$) for peak area ratios [6]. Migration time reproducibility is also required to confirm peak identity for each enantiomer and is generally acceptable with RSD values of less than 2% obtainable. Relative migration time (RMT) of one enantiomer compared to the other should be considerably better with RSD values of <1% being obtained [7, 8]. For example in the intercompany exercise, all companies reported RMT precisions of <0.05% RSD [6].

24.2.3 Linearity

A method may be employed to analyze a wide range of enantiomeric mixture compositions and it is therefore essential to demonstrate suitable linearity of detector response (peak area) for both sample concentration and trace enantiomeric impurity levels. To demonstrate this a single enantiomeric compound is spiked with levels of the undesired enantiomer covering the likely concentration range that may be encountered in routine testing. Adequate and acceptable detector linearity for both main peak concentration and for varying levels of the undesired enantiomer have been reported with typical linearity data of greater than 0.998 described [7, 9–11].

24.2.4 Selectivity

Comigration of related impurities with either of the enantiomer peaks is possible. Therefore, it is necessary to establish the migration position of all available related substances. This procedure has been performed in a number of studies including the chiral separations of Lilly drug [10], a cholesterol lowering agent [11], salbutamol [12], phenoxy acid herbicides [13] and amino acids [30]. Rogan et al. indicate a simultaneous significant separation of salbutamol and a range of chiral and achiral impurities [12].

24.2.5 Recovery

To demonstrate the accuracy of recovery of the method, experiments have been conducted in which single enantiomeric compounds have been spiked with known levels of their stereoisomers [7, 9–11, 14, 15]. The peak area ratio obtained by CE quantitatively confirmed the spiking level. These experiments also serve to demonstrate the accuracy of the methods under assessment. Noroski et al. indicate satisfactory recovery data obtained during validation of a method for enantiopurity determination of a cholesterol-lowering drug [11].

24.2.6 Cross validation

Combinations of HPLC and CE are being increasingly applied in method validation studies. Agreement between these two techniques strengthens the validity of the data generated by the final method of choice. Altria et al. present an acceptable agreement between enantiomeric purity results generated by both CE and HPLC methods [1]. However, the HPLC method required a base extraction step and was therefore less attractive than the CE counterpart. Good agreement between HPLC and CE has also been reported in the enantiomeric purity testing of phenoxy acid herbicides [13] and batches of a serotonin agonist [16].

24.2.7 Freedom from interference

Freedom from interference is especially important in the analysis of formulations and clinical samples where endogenous peaks have a possibility of masking the peaks of interest, but depending on compounds of interest a direct injection of clinical samples can be attainable. This has been demonstrated by analyzing appropriate samples of dissolving solvents [9] or biosamples [4, 5, 17, 18]. However, for example, uric acid in blood when injected directly indicated interference on analysis with admixtures. Therefore, pretreatment such as solid phase extraction (SPE) or dialysis prior to CE application will be required preferably to attain freedom from interference when handling complicated matrix compounds such as biosamples.

24.2.8 Robustness evaluation

The effect of small deliberate changes in operating parameters is assessed as part of method validation. Factors investigated include electrolyte concentration, pH, CD concentration, voltage, temperature and sample concentration [7, 9–11]. Many chemically derivatized CDs have been used (Table 24.1) to achieve separations and variability in the extent of the derivatization can lead to changes in the degree of selectivity [10, 19]. Therefore, it is essential to demonstrate that the method can be repeated using CD from different manufacturers and lots from the same supplier. Given the number of factors to be examined it may be appropriate to employ an experimental design such as Plackett–Burman to identify the critical parameters affecting the resolution [20].

24.2.9 Method transfer

A successful method transfer took place between seven independent laboratories of a method for the resolution of clenbuterol enantiomers. Each company independently obtained chiral resolution using the specified conditions which involved use of a low pH electrolyte containing hydroxypropyl-β-CD [13].

24.3 Quantitation

In CE the composition of an enantiomeric mixture can be reported in several ways [1]. The most common approach is to describe percent area/area (or %m/m [8]) levels for the undesired enantiomer. This is calculated as:

$$\% \text{ area}/\text{area} = \frac{\text{area}_1/MT_1 \times 100}{\text{area}_2/MT_2}$$

where 1 = desired enantiomer (or first migrating enantiomer of a racemic mixture), 2 = undesired enantiomer (or second migrating enantiomer of a racemic mixture), MT = migration time.

The need to divide the area of a peak by its migration time is essential to compensate for different migration speeds through the detector [21]. Without this compensation the later migration enantiomer would be overestimated [21]. Alternatively, enantiomeric purity may be reported as enantiomeric excess (e/e) which is calculated as:

$$\%\text{e}/\text{e} = (E_1 - E_2)/(E_1 + E_2) \times 100$$

where E_1 = amount of desired enantiomer, E_2 = amount of undesired enantiomer. It is important to check response factors as the two enantiomers can have different UV absorbance activities when complexed into chelating agents such as CDs [22].

24.4 Quantitative application areas

The quantitative applications of chirally selective CE methods have focused on enantiomeric purity testing of products, analysis of formulations, reaction rate monitoring and both forensic and clinical applications. The use of CE for chiral separations currently constitutes 22% of the routine use of CE in pharmaceutical laboratories [29].

24.4.1 Enantiomeric purity testing

Several applications involving low pH electrolytes containing CDs have been reported for the enantiomeric purity testing of drugs and herbicides [7, 9–19, 13, 14, 23–25]. For example, Nielen demonstrated chiral separation of both norephedrine and ephedrine within 14 min [23]. The method was capable of detecting <1% of the undesired enantiomers; detector linearities of >0.999 and precisions of <2% RSD for peak areas were also reported. Application of a chirally selective MECC methods has been described for the optical purity testing of batches of drug substances [2, 11, 16].

24.4.2 Reaction rate monitoring

The robust nature of chiral CE separations makes the monitoring of chirally selective processes possible. To date only one report has appeared concerning this area [26]. A chirally selective CE method was employed to monitor the enzymatic transformation of a racemate to a single enantiomer. The reaction was monitored over a 51 h period. The final product was shown to contain only 0.5% of the undesired stereoisomer.

24.4.3 Formulation stability testing

Interconversion of enantiomers can occur with time, therefore it is often necessary to incorporate a chirally specific assay in stability studies when analyzing chiral compounds. The requirement of this assay is both speed and simplicity as many samples may be analyzed. Peterson and Trowbridge monitored the enantiomeric purity of L-epinephrine in a pharmaceutical formulation employing L-pseudoepinephrine as an internal standard [3].

Peak area ratio precisions of 1.8% RSD with 99% recoveries and detector linearities of >0.998 were reported. Stored samples were tested and enantiomeric purity results were within specification. Baseline resolution of the phenylalanine enantiomers in a pharmaceutical dose form was achieved within 70 s using 35 kV applied across a 50 cm × 25 µm capillary [27]. Chiral separation of ephedrine and doxylamine contained in cough syrups was obtained using carboxylated CD at low pH [28].

24.4.4 Clinical and forensic analysis

Applications of CE in therapeutic drug level monitoring in biofluids is covered in Chapter 27 of this book. Often extensive sample workup is required in HPLC due to the presence of matrix components. These components may mask the peaks of interest or impair chromatographic performance. However in CE it has been shown that biosamples may often be directly analysed without sample pretreatment. A number of chiral bioassay applications have been reported [4, 5, 17, 18].

Gareil *et al.* have performed studies concerning the enantioselective metabolism of warfarin in patients undergoing warfarin therapy [5]. CE analysis of plasma samples confirmed the preferential metabolism of the (−) enantiomer. A detection limit of 0.2 mg l^{-1} for each warfarin enantiomer was reported. Verapamil and norverapamil levels in plasma were determined at 10 ng ml^{-1} concentrations [3].

24.5 Conclusions

CE has been shown to be a useful addition to the techniques available for the resolution and quantitation of enantiomers. Methods have been validated and can give similar performance levels to those obtained by HPLC. Successful method transfer has also been demonstrated. Particular features of chirally selective CE methods may include simplicity, ruggedness and low cost. Application areas of quantitative CE analysis reported to date include enantiomeric purity testing of drugs and herbicides, reaction rate monitoring, stability testing and the analysis of clinical and forensic samples. Undoubtedly the number of quantitative applications areas will greatly expand within the near future as CE becomes more established across a variety of industries.

References

1 Altria, K.D.; Goodall, D.M.; Rogan, M.M. (1994) Quantitative applications and validation of the resolution of enantiomers by capillary electrophoresis. *Electrophoresis* **15**, 824–827.

2. Williams, R.C.; Edwards, J.F.; Ainsworth, C.R. (1994) Analysis of diastereoisomeric impurities in chiral pharmaceutical compounds by capillary electrophoresis. *Chromatographia* **38**, 441–445.
3. Peterson, T.E.; Trowbridge, D. (1992) Quantitation of *l*-epinephrine and determination of the *d*-/*l*-epinephrine enantiomer ratio in a pharmaceutical formulaion by capillary electrophoresis. *J. Chromatogr.* **603**, 298–301.
4. Dethy, J.M.; De Broux, S.; Lesne, M.; Longstreth, J.; Gilbert, P. (1994) Stereoselective determination of verapmil and norverapmil by capillary electrophoresis. *J. Chromatogr.* **654**, 121–127.
5. Gareil, P.; Gramond, J.P.; Guyon, F. (1993) Separation and determination of warfarin enantiomers in human plasma samples by capillary zone electrophoresis using a methylated β-cyclodextrin-containing electrolyte. *J. Chromatogr.* **615**, 317–325.
6. Altria, K.D.; Harden, R.C.; Hart, M.; Hevizi, J.; Hailey, P.A.; Makwana, J.; Portsmouth, M.J. (1993) An inter-company cross-validation exercise on capillary electrophoresis. 1. Chiral analysis of clenbuterol. *J. Chromatogr.* **641**, 147–153.
7. Guttman, A.; Cooke, N. (1994) Practical aspects in chiral separation of pharmaceuticals by capillary electrophoresis II. Quantitative separation of naproxen enantiomers. *J. Chromatogr.* **685**, 155–159.
8. Abushoffa, A.M.; Clark, B.J. (1995) Resolution of the enantiomers of oxamniquine by capillary electrophoresis and high-performance liquid chromatography with cyclodextrins and heparin as chiral selectors. *J. Chromatogr.* **700**, 51–58.
9. Altria, K.D.; Walsh, A.R.; Smith, N.W. (1993) Validation of a capillary electrophoresis method for the enantiomeric purity testing of fluparoxan. *J. Chromatogr.* **645**, 193–196.
10. Rickard, E.C.; Bopp, R.J. (1994) Optimization of a capillary electrophoresis method to determine chiral purity of a drug. *J. Chromatogr.* **680**, 609–621.
11. Noroski, J.E.; Mayo, D.J.; Moran, M. (1995) Determination of the enantiomer of a cholesterol-lowering drug by cyclodextrin-modified micellar electrokinetic chromatography. *J. Pharm. Biomed. Anal.* **13**, 52–54.
12. Rogan, M.M.; Goodall, D.M.; Altria, K.D. (1994) Enantiomeric separation of salbutamol and related impurities using capillary electrophoresis. *Electrophoresis* **15**, 808–817.
13a. Nielen, M.W.F. (1993) (Enantio-) separation of phenoxy acid herbicides using capillary zone electrophoresis. *J. Chromatogr.* **637**, 81–90.
13b. Garrison, A.W.; Schmitt, P.; Kettrup, A. (1994) Separation of phenoxy acid herbicides and their enantiomers by high-performance capillary electrophoresis. *J. Chromatogr.* **688**, 317–327.
14. Nishi, H.; Nakumura, K.; Nakai, H.; Sato, T.; Terabe, S. (1995) Enantiomeric separation of trimetroquinol, denopamine, and timepidium by capillary electrophoresis and HPLC and the application of capillary electrophoresis to the optical purity testing of the drugs. *Chromatographia* **40**, 638–644.
15. Lurie, I.S. (1992) Micellar electokinetic capillary chromatography of the enantiomers of amphetamine, methamphetamine and their hydroxyphenethylamine precursors. *J. Chromatogr.* **605**, 269–275.
16. Okafo, G.N.; Rana, K.K.; Camilleri, P. (1994) Improved separation of diastereoisomers in capillary electrophoresis using a mixture of β-cyclodextrin and sodium taurocholate. *Chromatographia* **39**, 627–630.
17. Soini, H.; Riekkola, M-L.; Novotony, M.V. (1992) Chiral separations of basic drugs and quantitation of bupivacaine enantiomers in serum by capillary electrophoresis with modified cyclodextrin buffers. *J. Chomatogr.* **608**, 265–274.
18. Francotte, E.; Cherkaoul, S.; Faupel, M. (1993) Separation of the enantiomers of some racemic nonsteroidal aromatase inhibitors and barbiturates by capillary electrophoresis. *Chirality* **5**, 516–526.
19. Valko, I.E.; Billeiet, H.A.H.; Frank, J.; Luyben, K.Ch.A.M. (1994) Effect of degree of substitution of (2-hydroxy) propyl-β-cyclodextrin on the enantioseparation of organic acids by capillary electrophoresis. *J. Chromatogr.* **678**, 139–144.
20. Rogan, M.M.; Altria, K.D.; Goodall, D.M. (1994) Plackett–Burman experimental design in chiral capillary electrophoresis. *Chromatographia* **38**, 723–729.
21. Altria, K.D. (1993) Essential peak area normalisation in capillary electrophoresis. *Chromatographia* **35**, 177–182.

22 Fanali, S.; Bocek, P. (1990) Enantiomeric resolution by using capillary zone electrophoresis: resolution of racemic trytophan and determinaiton of the enantiomer composition of commercial pharmaceutical epinephrine. *Electrophoresis* **11**, 757–760.
23 Nielen, M.W.F. (1993) Chiral separation of basic drugs using cyclodextrin-modified capillary zone electrophoresis. *Anal. Chem.* **65**, 885–893.
24 Altria, K.D.; Goodall, D.M.; Rogan, M.M. (1994) Quantitative applications and validation of the resolution of enantiomers by capillary electrophoresis. *Electrophoresis* **15**, 824–827.
25 Kuhn, R.; Stoecklin, F.; Erni, F. (1992) Chiral separations by host–guest complexation with cyclodextrins and crown ethers by capillary zone electrophoresis. *Chromatographia* **33**, 32–36.
26 Rogan, M.M.; Drake, C.; Goodall, D.M.; Altria, K.D. (1993) Enantioselective enzymatic biotransformation of 2′-deoxy-3′-thiacytidine (BCH 189) monitored by capillary electrophoresis. *Anal. Biochem.* **208**, 343–347.
27 Sepaniak, M.J.; Cole, R.O.; Clark, B.K. (1992) Use of native and chemically modified cyclodextrins for the capillary electrophoretic separation of racemates. *J. Liq. Chromatogr.* **15**, 1023–1040.
28 Schmitt, T.; Engelhardt, H. (1995) Optimization of enantiomeric separations in capillary electrophoresis by reversal of the migration order and using different derivatized cyclodextrins. *J. Chromatogr.* **697**, 561–570.
29 Altria, K.D.; Kersey, M. (1995) Capillary electrophoresis and pharmaceutical analysis: a survey of the industrial application and their status of in the United States and United Kingdom. *LC–GC* **13**, 40–46.
30a Werner, A.; Nassauer, T.; Keichle, P.; Erni, F. (1994) Chiral separation by capillary zone elctrophoresis of an optically active drug and amino acids by host–guest complexation with cyclodextrins. *J. Chromatogr.* **666**, 375–379.
30b Houben, R.J.H.; Gielen, H.; Van der Wal, S. (1993) Automated preseparation derivatization on a capillary electrophoresis instrument. *J. Chromatogr.* **634**, 317–322.
31 Quang, C.; Khaledi, M.G. (1994) Chiral separations of acidic compounds by dextrin-mediated capillary zone electrophoresis. *J. High Resolut. Chromatogr. Comm.* **17**, 609–612.

25 Applications of chiral capillary electrophoresis (cyclodextrin–capillary zone electrophoresis; CD–CZE)

A. AUMATELL

25.1 Introduction

Method development can be time consuming for new and advanced capillary electrophoresis users. In order to help accelerate CE method development, chiral CE applications have been summarized into ten groups (amino acids, sympathomimetics, stimulants, anesthetics/sedatives, anticoagulants, antidepressants, non-steroidal anti-inflammatory, polycyclic aromatic compounds, cationic drugs/solutes and anionic solutes). Also, for each group Tables 25.1–25.12 list the solute separated and buffer conditions employed.

25.2 Amino acids

Non-aromatic amino acids Gly, Ala, Val, Leu, ILe, Met, Pro, Ser, Thr, Cys, Asn, Gln, Asp, Glu, Lys and Arg exhibit low UV absorbance. Nevertheless, amino acids have been separated by CE and detected directly, after derivatization, or by indirect detection, as given in Table 25.1. Derivatized amino acids have been resolved optically by CE employing native cyclodextrins CD [1–7], modified CDs [6–11], charged CDs [8, 9] and crown ethers [12–15] as chiral buffer additives. Furthermore, β-CD and dextran buffer coadditives have been applied in the optical resolution of Dns-amino acids [4]. Also, these amino acid derivatives have been resolved optically by incorporating SBE-β-CD into a buffered solution and Cb-β-CD into a liquid gel, respectively [8, 9]. Lindner *et al.* [6] separated optically FMOC, AQC-, DNP- and DNB-amino acid derivatives employing γ-CD and modified CDs (Me-β-CD and HP-β-CD) as chiral additives. However, the majority of the derivatized amino acids were resolved employing HP-β-CD. Yamashoji *et al.* [7] showed the CE optical elution order of Ala β-naphthylamide, employing β-CD and DAc-β-CD chiral selectors, corresponded to an NMR study of guest–host complex. The guest penetrates into the β-CD cavity from the secondary hydroxyl side, but into the DAc-β-CD cavity from the primary hydroxyl side. These changes of selectivity were ascribed to differences in the shape and size of the CD cavity and/or in their

Table 25.1 Separation of amino acids and derivatives

Analyte	Chiral selector	CE buffer	Ref.
Dns-Asp, -Glu, -Thr, -Val	(10 mM) γ-CD	(10 mM) Phosphate–(6 mM) borate pH 9.3	1
Dns-Asp, -Gln	(20 mM) γ-CD	(100 mM) Borate pH 9.5	2
Dns-Glu, -His, -Lys	(20 mM) γ-CD	(100 mM) Phosphate pH 2.6	2
Dns-Phe	(5 mM) γ-CD	(25 mM) Phosphate pH 8.0	3
Dns-Leu, -Asp, -Glu, -NorVal, -Ser, -Thr, -Trp, -Val, -Met, -NorLeu,	(5 mM) β-CD	(10 mM) Phosphate pH 7.2– (0–10%) dextran	4
DNP-Glu, -Met, -NorLeu	β-CD	(10 mM) TEAA pH 4.7, (10–25%) MeOH	5
Dns-Phe	(10 mM) Sulfonated-β-CD	(20 mM) Borate–phosphate pH 7.0	8
Dns-Asp, -Glu, -Leu, -Met, -NorLeu, -Phe, -Ser, -Thr, -Val	(20 mM) Cb-β-CD	(100 mM) Tris–(250 mM) boric acid pH 8.3–(2%) acrylamide	9
FMOC-Ala, -Phe, -Val, -Leu	(30 mM) Me-β-CD; (30 mM) γ-CD	(20 mM) Borate pH 9.2, (0–20%) MeOH	6
AQC-Ala, -Asp, -His, -Leu, -Lys, Phe, -Ser, -Tyr, -Val, -Met, -Pro, -Thr; DNP-Ala, -Gln, -Phe, Pro, -Ser, -Thr, -Trp, -Val, -Leu, -His, -Ile; -Leu, -Met; Dns-GABA, -Leu, -Met, -Val; DNB-Phe; CBZ-Ala	(30 mM) HP-β-CD	(70 mM) Phosphate pH 7.0– (0–20%) MeOH-(0–1%) *t*-butylmethylether	6
Ala β-naphthylamide	(10 mM) β-CD; (10 mM) DAc-β-CD; γ-CD	(50 mM) Phosphate pH 2.0	7
Dns-Phe	(25 mM) HP-β-CD	(6 mM) Borate–(10 mM) phosphate pH 4.0	10
Dns-Phe	(1%) HP-β-CD	(100 mM) Tris–(100 mM) boric acid pH 8.3–(5 mM) spermidine	11
DAT-Phe-Gly, -Trp; DBT-Gln, -Leu, -Met, -Phe, -Ser, -Thr, -Trp, -Val	–	(25 mM) Phosphate pH 5.8– (3%) PVP	17
OPA-*N*-acetylCys	–	(155 mM) Borate pH 9.0– (5%) PEG	18
HomoPhe, Phenylglycinol, Trp	(10 mM) 18C6H4	(10 mM) Tris–citrate pH 2.5	12
Gly-Phe, Gly-Trp, Val	(10 mM) 18C6H4	(10 mM) Tris pH 2.2– (6 mM) BTAC	13
His	(10 mM) 18C6H4	(20 mM) Tris pH 3.5	13
Dopa	(10 mM) 18C6H4	(10 mM) Tris–citrate pH 2.2	14
Ala, GlyGlyLeu, Ile, Leu, NorVal, Thr, Val	(15 mM) 18C6H4	(5 mM) Tris–citrate pH 2.2– (6 mM) BTAC	14
GlyPhe, His, Phe, Trp, Tyr	(10 mM) 18C6H4	(10 mM) Tris–citrate pH 2.2	14
Phe, Trp, Tyr	(30 mM) 18C6H4; (50 mM) α-CD	(50 mM) Phosphate pH 2.2–2.5	15

CBZ = carboxybenzyl derivative; OPA = *o*-phthaldialdehyde derivative; GABA = γ-aminobutyric acid.

hydrogen bonding abilities before and after acetylation. Stereoisomers of Dns-Phe were resolved using HP-β-CD as chiral additive and electroosmotic flow (EOF) was reversed using spermidine as buffer coadditive [10, 11]. DBT and DAT amino acid derivatives were optically resolved employing a network of linear PVP [16]. The optical elution order for derivatized non-aromatic amino acids were opposite to that of derivatized aromatic amino acids. This was explained in terms of π–π interactions between the aromatic moieties in the analyte and the amido-group of PVP. Alternately, acetylcystein was resolved chirally by forming a distereoisomer with o-phthaldialdehyde [17]. Kuhn and co-workers [12–15] showed that amino acid stereoisomers were directly resolved employing 18C6H4 as chiral additive. Furthermore, amino acids were detected indirectly using BTAC in CE buffer.

25.3 Sympathomimetics

Most sympathomimetic drug substances (β-agonists, β-antagonists and agenergic agents) have a hydroxyphenylamine structure. There have been reports of the use of native CDs [2, 15, 18–27], derivatized CDs [3, 11, 18, 20–22, 25, 26, 28–43], charged CDs [11, 28, 29, 37], proteins [44, 45], antibiotics [46] and crown ethers [12–14] as buffer chiral additives for optical separation of drug enantiomer by capillary zone electrophoresis (CZE). CDs appear to be the most popular chiral additive for optical resolution of sympathomimetic drug substances [3, 11, 15, 18–43]. Furthermore, most of these CD–CE methods use low pH buffers with varying concentrations of native CD (ranging from 12 to 200 mM) [15, 18–27] and modified CDs (ranging from 1 to 120 mM) [3, 11, 18, 20–22, 25–38, 40–43]. However, among the modified β-CDs used, DM-β-CD and HP-β-CD have resolved optically the largest number of chiral compounds [3, 11, 18, 20, 22, 25–28, 30–38, 40–43]. Tables 25.2–25.4 summarize CE buffer conditions for some sympathomimetic drug substances. Also, sections 25.3.1–25.3.3 discuss CE chiral applications for β-agonists, β-antagonists and agenergic agents, respectively.

25.3.1 β-Agonists

Quang and Khaledi [24] optically resolved hydroxyphenylamines employing β-CD as chiral selector in combination with a short chain tetra-alkyl ammonium salt for EOF reversal (Table 25.2). Wren [36] described a chiral CD–CZE separation theory, where chiral resolution increases proportionally with CD up to a point where further CD concentration decreases chiral resolution. This was demonstrated using propranolol, practolol, atenolol and ephedrine as model compounds employing DM-β-CD and β-CD as chiral selectors. Furthermore, Wren and Rowe [39]

Table 25.2 Separation of β-agonists

Analyte	Chiral selector	CE buffer	Ref.
Propranolol	(20 mM) β-CD	(50 mM) Phosphate pH 2.5–(100 mM) tetrabutylammonium phosphate	24
Propranolol	(14 mM) β-CD	(50 mM) Phosphate pH 3.0	26
Propranolol	(5.5 mM) DM-β-CD	(50 mM) Phosphate pH 3.0	26
Propranolol	(3.7 mM) DM-β-CD	(40 mM) Phosphate pH 3.0–(5%) MeOH	39
Propranolol	(3.7 mM) DM-β-CD	(40 mM) Phosphate pH 3.0–(2%) acetonitrile	39
Atenolol, oxprenolol, metoprolol	(14–37 mM) DM-β-CD	(40 mM) Phosphate pH 3.0	40
Bambuterol	(15 mM) DM-β-CD	(50 mM) Phosphate pH 2.5	22
Practolol, atenolol	(30–40 mM) DM-β-CD	(50 mM) Phosphate pH 2.5	36
Alprenolol, pindolol, sotalol	(15 mM) DM-β-CD	(100 mM) Phosphate/triethanolamine pH 3.0–(30%) MeOH	31
Atenolol	(30 mM) DM-β-CD	(100 mM) Phosphate/triethanolamine pH 3.0	31
Propanolol	(15 mM) TM-β-CD	(100 mM) Phosphate/triethanolamine pH 3.0	31
Pindolol	(30 mM) DM-β-CD	(20 mM) Phosphate/HCl pH 2.2	11
Pindolol, propranolol	(2%) CM-β-CD	(20 mM) Citrate pH 2.5	11
Propranolol	(10 mM) HP-β-CD	(200 mM) TAPSO/TBAH pH 7.0–(0.4%) hydrophilic polymeric additive	41
Propranolol	(10 mM) HP-β-CD	(25 mM) Phosphate pH 2.5	3
Atenolol	(28 mM) DM-β-CD	(20 mM) Tris/phosphate pH 2.4	33
Propranolol	(28 mM) HE-β-CD	(20 mM) Tris/phosphate pH 2.4	33
Propranolol	(28 mM) HP-β-CD	(20 mM) Tris/phosphate pH 2.4	33
Sotalol	(30 mM) HP-β-CD	(50 mM) Phosphate pH 3.3	25
Alprenolol, atenolol, cis-nadolol, labetalolol, oxprenolol, pindolol, propranolol, transnadolol	(120 mM) HP-β-CD	(100 mM) Citrate/(19.27 mM) phosphate pH 2.5	37
Atenolol	(2–4.6 mM) SBE-β-CD	(20 mM) Citrate/(3.85 mM) phosphate pH 2.5	37
Alprenolol, atenolol, metoprolol, oxprenolol	(25 mM) Rifamycin B	(100 mM) Phosphate pH 7.0–(40%) propan-2-ol	46
Alprenolol, metoprolol, pindolol, propranolol, pseudolabetolol	(40 mg ml)$^{-1}$ Cellobiohydrolase hydrolase I	(400 mM) Phosphate pH 5.1–(20%) propan-2-ol	45
Alprenolol, oxprenolol	Glycoprotein α-1-acid	(4 mM) Phosphate pH 6.8–(4%) propan-2-ol	44
Metoprolol	Glycoprotein α-1-acid	(2 mM) Phosphate pH 6.8–(2%) propan-2-ol	44

HE-β-CD, hydroxyethyl-β-cyclodextrin; CM-β-CD, carboxymethyl-β-cyclodextrin; TM-β-CD, trimethyl-β-cyclodextrin; TAPSO, 3-(N-trishydroxymethyl)-methyl-amino)-2-hydroxypropane sulfonic acid; TBAH, tetrabutylammonium hydroxide.

Table 25.3 Separation of β-antagonists

Analyte	Chiral selector	CE buffer	Ref.
Clenbuterol, picumeterol	(16 mM) β-CD	(100 mM) Citrate/(200 mM) phosphate	27
SDZ EAA 494 s	(20 mM) γ-CD	(100 mM) Borate pH 11.0	2
Terbutaline	(15 mM) β-CD	(50 mM) Phosphate pH 2.5	22
Terbutaline	(25 mM) HP-β-CD	(50 mM) Phosphate	22
Terbutaline	(5 mM) DM-β-CD	(50 mM) Phosphate	22
Terbutaline sulfate	(15 mM) β-CD	(50 mM) Phosphate pH 10.6	22
Terbutaline sulfate	(20 mM) DM-β-CD	(50 mM) Phosphate pH 11.6	22
Terbutaline sulfate	(20 mM) HP-β-CD	(50 mM) Phosphate pH 11.6	22
Salbutamol	(12 mM) DAc-β-CD	(100 mM) Phosphate pH 3.0	21
Salbutamol, salbutamoldeoxy, salbutamolaldehyde, salbutamoldimer	(112 mM) DM-β-CD	(50 mM) Phosphate/citrate pH 2.5	30
Terbutaline	(15 mM) HP-β-CD	(100 mM) Phosphate/triethanolamine pH 3.0	31
Salbutamol	(30 mM) DM-β-CD	(50 mM) Phosphate pH 3.3	25
Clenbuterol	(30 mM) HP-β-CD	(50 mM) Phosphate pH 3.3	25
Clenbuterol	(30 mM) HP-β-CD	(50 mM) Borate/phosphate	38
Cimaterol, clenbuterol, pirbuterole, salbutamol, terbutaline	(120 mM) HP-β-CD	(100 mM) Citrate/(19.27 mM) phosphate pH 2.5	37
Cimaterol, clenbuterol, terbutaline	(2–4.6 mM) SBE-β-CD	(20 mM) Citrate/(3.85 mM) phosphate pH 2.5	37
Clenbuterol	(2 mM) SBE-β-CD	(20 mM) Citrate/(3.854 mM) phosphate pH 2.5–(5%) propan-l-ol	37
Clenbuterol	(1 mM) SBE-β-CD	(50 mM) Phosphate pH 3.1	29
Salbutamol, terbutaline	(25 mM) Rifamycin B	(100 mM) Phosphate pH 7.0–(40%) propan-2-ol	46

proposed a model for the role of organic solvents in CD–CZE, where the affinity between solute and CD cavity was reduced by the addition of an organic modifier. This change in equilibrium constant of the host–guest complex can lead to an increase or decrease in the apparent mobility differences between the optical isomers and hence separation. Nevertheless, chiral separation of isomers will depend on whether the concentration of CD is above or below the optimum value for the original system. However, Aumatell et al. [37] employed a high concentration of chiral additive (120 mM HP-β-CD, which is a complex mixture of related CDs) to resolve β-agonists optically. Optical separation of the drug racemates appeared due to the high CD concentration enabling one of the CDs to form guest–host complexes with a certain racemate that approach optimum separation by CE, whilst another might be more suitable in CE separation of another racemate. Alternatively, Armstrong et al. [46] employed rifamycin B, a macrocyclic antibiotic, as a chiral additive resolving hydroxyphenylamine enantiomers optically. Rifamycin B absorbs strongly in both the UV and

Table 25.4 Separation of agenergic agents

Analyte	Chiral selector	CE buffer	Ref.
Ephedrine isomers	–	(10 mM) Val/ammonia pH 10	47
Ephedrine, epinephrine, isoproterenol, pseudoephedrine	(20 mM) β-CD	(50 mM) Phosphate pH 2.5–(150 mM) tetrabutylammonium phosphate	29
Epinephrine	γ-CD	(20 mM) Borate/phosphate pH 7.0–(10%) propan-2-ol	19
Ephedrine	(200 mM) γ-CD	(40 mM) Phosphate pH 3.0–(0.05%) HPMC	20
Ephedrine	(20 mM) β-CD; (50 mM) DM-β-CD; (50 mM) γ-CD; (50 mM) Me-β-CD	(40 mM) Phosphate pH 3.0–(0.05%) PVA	20
Lofexidine	(20 mM) HP-β-CD; (50 mM) γ-CD	(50 mM) Phosphate pH 3.0–(0 to 20%) MeOH	18
Ephedrine	(1.55%) Succinyl-β-CD; (2%) CM-β-CD	(20 mM) Citrate pH 2.2–2.8	11
Ephedrine	(30 mM) DM-β-CD	(20 mM) Phosphate/HCl pH 2.2	11
Ephedrine	(1%) Succinyl-β-CD; (1%) CM-β-CD	(20 mM) Tris pH 5.7–6.2	11
Ephedrine	(2%) DM-β-CD	(20 mM) Phosphate pH 5.7	11
Ephedrine	(15 mM) DM-β-CD	(50 mM) Phosphate pH 2.5	22
Isoprenaline	(15 mM) DM-β-CD	(100 mM) Phosphate/triethanolamine pH 3.0	31
Ephedrine	(30 mM) DM-β-CD	(100 mM) Phosphate/triethanolamine pH 3.0–(30%) MeOH	31
Norverapanmil	(25 mM) TM-β-CD	(100 mM) Phosphate/triethanolamine pH 3.0	31
Ephedrine, isoprenaline	(30 mM) HP-β-CD	(100 mM) Phosphate/triethanolamine pH 3.0	31
Epinephrine	(9 mM) DM-β-CD	(20 mM) Tris/phosphate pH 2.4	33
Epinephrine	(18 mM) DM-β-CD	(10 mM) Tris/phosphate pH 2.4	32
Ephedrine	(18 mM) DM-β-CD	(20 mM) Phosphate pH 2.5–(0.1%) hydroxypropyl cellulose	28
Methylephedrine, methylpseudoephedrine, norephedrine	(18 mM) DM-β-CD	(20 mM) Phosphate pH 2.5–(10 mM) tetrabutylammonium bromide	28
Ephedrine, methylephedrine, methylpseudoephedrine, norephedrine	(40 mM) SBE-β-CD	(20 mM) Borate pH 10.0	28
Epinephrine, isoproterenol, norepinephrine	(20 mM) DM-β-CD	(50 mM) Phosphate pH 2.5	34
Isoproterenol, metaproterenol	(30 mM) DM-β-CD	(25 mM) Phosphate pH 2.5	43
Ephedrine, norephedrine	(40 g l^{-1}) DM-β-CD	(30 mM) Tris/phosphate pH 2.5	35
Ephedrine	(50 mM) DM-β-CD	(50 mM) Phosphate pH 2.5	36
Ephedrine	(16.3 mM) β-CD	(50 mM) Phosphate pH 3.3	25
Methylephedrine, norephedrine, norfenefrine	(30 mM) DM-β-CD	(50 mM) Phosphate pH 3.3	25

Table 25.4 *Continued*

Analyte	Chiral selector	CE buffer	Ref.
Etilefrine, isoprenaline, lofexidine, pholedrine	(30 mM) HP-β-CD	(50 mM) Phosphate pH 3.3	25
Ephedrine, epinephrine, pseudoephedrine	(120 mM) HP-β-CD	(100 mM) Citrate/(19.27 mM) phosphate pH 2.5	37
Ephedrine	(12 mM) DAc-β-CD	(100 mM) Phosphate pH 3.0	21
Etilefrine, metaproterenol (oriciprenaline), oxedrine (oxyephedrine)	(12 mM) β-CD (12 mM) DAc-β-CD	(100 mM) Phosphate pH 3.0	21
Etilefrine, isoprenaline, lofexidine	(1 mM) SBE-β-CD	(50 mM) Phosphate pH 3.1	29
Norephedrine, normetanephrine	(10 mM) 18C6H4	(10 mM) Tris/citrate pH 2.2	14
ENX 792, Quinagolide	(20 mM) β-CD	(50 mM) Phosphate pH 2.5	15
Norephedrine	(10 mM) 18C6H4	(10 mM) Tris/citrate pH 2.5	12
Norepinephrine	(20 mM) HP-β-CD/ (5 mM) 18C6H4	(30 mM) Tris/citrate pH 2.5	12
Norephedrine, norphenylephrine	(30 mM) 18C6H4	(100 mM) Tris pH 2.5	13
Quinagolide	(20 mM) 18C6H4/ (20 mM) α-CD	(10 mM) Tris/citrate	13
Eperisone	(50 μM) Ovomucoid	(50 mM) Phosphate pH 5.0–(500 mM) o-phosphorylethanolamine	48
Ephrine, isoproterenol, metaproterenol, epinephrine, norepinephrine, norphenylephrine, pseudoephrine	(25 mM) Rifamycin B	(100 mM) Phosphate pH 7.0–(40%) propan-2-ol	46

visible spectral regions, thus analytes were easily visualized by indirect detection. Also, hydroxyphenylamines were separated optically by injecting a plug of cellobiohydrolase I (chiral selector)/buffer and electrophoresing the analyte through the plug [45]. The electrophoresis was conducted such that the UV absorbing cellobiohydrolase migrated in the opposite direction to the analyte enantiomers, thus not hindering analyte detection. Nevertheless, alprenolol, oxprenolol and metoprolol enantiomers were separated by capillary electrochromatography (CEC) using a capillary packed with a chiral immobilized α_1-acid glycoprotein stationary phase [44].

25.3.2 β-Antagonists

pH and CD type were shown to be important parameters for optimization of chiral separation [22, 31, 37]. However, the degree of CD substitution and position of substitution can affect chiral separation of optical isomers. Branch et al. [21] optically resolved β-antagonists and agenergic agents using β-CD and DAc-β-CD, respectively, as chiral additives (see Tables 25.3 and 25.4). Furthermore, NMR and CE results showed β-CD formed stronger complexes with drug analytes than with DAc-β-CD. Also, buffer electrolyte composition was shown to affect the selectivity for chiral separation of pharmaceutical drug compounds [31]. This was explained in terms of EOF reduction. Aumatell et al. [37] showed the versatility of HP-β-CD and sodium sulfobutylether-β-CD as chiral additives in the direct chiral resolution of β-antagonists. Also, steric factors are relevant for chiral recognition, at least one polar group with the proper hydrogen bonding ability must be in proximity to the stereogenic center, such as hydroxyl groups and/or secondary amines.

25.3.3 Agenergic agents

Liu and Sheu [47] resolved isomers of ephedrine in Chinese herbal preparations by CZE employing a Val buffer at pH 10 (CE buffer conditions given in Table 25.4). This work showed ephedrine resolution decreased with buffer pH and at pH 10 isomers of ephedrine were satisfactorily resolved within 3 min. Furthermore, Belder and Schomburg [20] showed ephedrine enantiomers optically resolved by CD–CZE using a capillary coated with polybren (cationic coating) and polyvinyl alcohol (thermally immobilized non-ionic coating), respectively. Also, Aumatell and Guttman [43] optically resolved enantiomers of metaproterenol and isoproterenol in <40s, achieving >1 million theoretical plates, employing a non-ionic coated capillary with an optimized DM-β-CD buffer system at pH 2.5. Yoshida and Sato [48] showed eperisone and chlorpheniramine enantiomers resolved optically using ovomucoid, a protein from egg white, as chiral selector (Tables 25.4 and 25.11). Furthermore, the CE conditions were easily optimized by adjusting ovomucoid and organic modifier concentration.

25.4 Stimulants

Stimulants such as R-amphetamine, a central nervous stimulant, are prescribed for treatment of hyperkinetic behavior disorders in children [49]. However, amphetamines are illicit drug substances and chiral CE separation and identification of drug optical isomers of forensic significance can provide information to source. Table 25.5 lists analyte and buffer conditions. Enantiomers of amphetamine, methydimethoxyamphetamine, methylamphetamine, methyldimethoxyethylamphetamine and

Table 25.5 Separation of stimulants

Analyte	Chiral selector	CE buffer	Ref.
Octopamine	(20 mM) DM-β-CD	(50 mM) Phosphate pH 2.5	34
Octopamine	(30 mM) HP-β-CD	(50 mM) Phosphate pH 3.3	25
Amphetamine, methydimethoxyamphetamine, methylamphetamine, methyldimethoxy-ethylamphetamine, methyldimethoxymethyl-amphetamine,	(120 mM) HP-β-CD	(100 mM) Citrate/(19.27 mM) phosphate pH 2.5	37
Amphetamine, methylamphetamine	(4.6 mM) SBE-β-CD	(20 mM) Citrate/(3.85 mM) phosphate pH 2.5	37
Methyldimethoxyamphetamine	(2 mM) SBE-β-CD	(20 mM) Citrate/(3.854 mM) phosphate pH 2.5–(5%) propan-1-ol	37
Octopamine	(25 mM) Rifamycin B	(100 mM) Phosphate pH 7.0–(40%) propan-2-ol	46

methyldimethoxymethylamphetamine have been directly resolved optically employing HP-β-CD and SBE-β-CD, respectively, as chiral additives [37]. Optical differentiation of 'R' and 'S' amphetamine was attributed to steric hindrance between the hydroxypropyl groups in HP-β-CD and the optically active methyl substituent in amphetamine. The enantiomers of octopamine were separated optically employing DM-β-CD, HP-β-CD and rifamycin B, respectively, as chiral additive [25, 34, 46].

25.5 Anesthetics and sedatives

Anesthetics (ketamine) and sedatives (metomidate, hexobarbital, zopiclone, oxomemazine) have been resolved optically by CE employing chiral additives [5, 8, 11, 18, 25, 29, 34, 31, 44, 50]. Table 25.6 gives the various buffer conditions used. Blaschke and Hempel [50] resolved isomers of zolpidem and enantiomers of zopiclone in urine by CD–CZE employing a helium–cadmium laser detector. Alternatively, hexobarbital was separated electrochromatographically with β-CD packed into a 50 μm (inside diameter) silica capillary column [5]. However, hexobarbital was also resolved by CZE employing charged CDs [8], native CDs [11], modified CDs [8, 11] and glycoprotein α-1-acid [44], respectively, as chiral additive. Furthermore, variations in CD and electrolyte concentration, CD type, organic modifier and temperature were shown to be important parameters for chiral separation of ketamine, metomidate, oxomemazine and zopiclone [18, 25, 29, 31, 34].

Table 25.6 Separation of anesthetics and sedatives

Analyte	Chiral selector	CE buffer	Ref.
N-Demethylzopiclone, zopiclone, zopiclone-N-oxide	(16.3 mM) β-CD	(50 mM) Phosphate pH 2.8	50
Hexobarbital	β-CD	(10 mM) TEAA pH 4.7–(25%) MeOH	5
Metomidate	(20 mM) β-CD	(50 mM) Phosphate pH 3.0–(0 to 20%) MeOH	18
Hexobarbital	(1 mM) Sulfonated-β-CD	(20 mM) Borate/phosphate pH 7.0	8
Hexobarbital	OM-β-CD	(20 mM) Tris/hydrochloric acid pH 7.0	8
Ketamine	(20 mM) DM-β-CD	(50 mM) Phosphate pH 2.5	34
Hexobarbital	(1.56%) β-CD	(100 mM) Tris–boric acid–EDTA pH 8.3	11
Hexobarbital	(1.56%) β-CD	(100 mM) Tris–boric acid–EDTA pH 8.3–(5 mM) spermidine	11
Hexobarbital	(0.7%) HP-β-CD	(100 mM) Tris/(100 mM) Boric acid pH 8.3–(5 mM) spermidine	11
Hexobarbital	(0.7%) HP-β-CD	(100 mM) Tris/(100 mM) boric acid pH 8.3	11
Hexobarbital	(1%) HP-β-CD	(100 mM) Tris/boric acid pH 8.4–(0.5–1 mM) CTAB	11
Oxomemazine	(15 mM) HP-β-CD	(100 mM) Phosphate/triethanolamine pH 3.0	31
Metomidate, zopiclone	(30 mM) HP-β-CD	(50 mM) Phosphate pH 3.3	25
Ketamine	(16.3 mM) β-CD	(50 mM) Phosphate pH 3.3	25
Metomidate	(1 mM) SBE-β-CD	(50 mM) Phosphate pH 3.1	29
Hexobarbital	Glycoprotein α-1-acid	(2 mM) Phosphate pH 5.5–(2%) propan-2-ol	44

EDTA, ethylenediaminetetra-acetic acid; CTAB, cetyltrimethylammonium bromide; OM-β-CD, monokis-6-o-octamethylene-permethyl-β-CD.

Table 25.7 Separation of anticoagulants

Analyte	Chiral selector	CE buffer	Ref.
Carvedilol	(16.3 mM) β-CD	(50 mM) Phosphate pH 3.3	25
5-Chlorowarfin, warfarin	(8 mM) Me-β-CD	(100 mM) Phosphate pH 8.4–(2%) MeOH	51
Verapamil	(25 mM) TM-β-CD	(100 mM) Phosphate/triethanolamine pH 3.0	31
Gallopamil	(20 mM) HP-β-CD	(100 mM) Phosphate/triethanolamine pH 3.0	31
Phenoprocoumon, warfarin	(2.5%) Glucidex 2 (maltodextrin)	(10 mM) Tris/phosphate pH 7.0	52
p-Chlorophenprocoumon, phenoprocoumon, warfarin	(2.5%) Glucidex 2 (maltodextrin)	(10 mM) Phosphate pH 7.1	53
p-Chlorowarfarin	(3%) Glucidex 2 (maltodextrin)	(10 mM) Tris/phosphate pH 7.0	54
Warfarin	(25 µM) Avidin	(50 mM) Phosphate pH 6.0–(10%) propan-2-ol	55
Bamethan	(25 mM) Rifamycin B	(100 mM) Phosphate pH 7.0–(40%) propan-2-ol	46

25.6 Anticoagulants

Anticoagulant drug enantiomers have been directly resolved optically by CE employing chiral additives, as given in Table 25.7. Chiral separation of carvedilol and verapamil enantiomers was demonstrated employing β-CD and TM-β-CD as chiral additive, respectively [25, 31]. However, enantiomers of warfarin and 5-chlorowarfarin were optically resolved in human plasma in <15 min by employing Me-β-CD as chiral additive [51]. Furthermore, D'Hulst and Verbeke [52–54] resolved enantiomers of non-steroidal anti-inflammatory agents, anticoagulants and cephalosporin antibiotics using maltodextrin (an oligosaccharide) as chiral selector (Tables 25.7, 25.9 and 25.11). In addition, optical selectivity of linear oligosaccharide appeared to emulate amylose (α-(1-4)-linked glucose oligomers) and polymers, due to the balanced hydrophilic–hydrophobic surface resulting from the helical conformation. Also, chiral separation increased with maltodextrin concentration and decreasing dextrose equivalent (defined as the percentage of reducing sugars calculated as glucose on dry substance basis). Avidin, an egg white protein, was shown to be a suitable chiral additive for optical resolution of warfarin enantiomers [55]. However, the choice of pH and organic modifier was important for chiral resolution. Alternatively, bamethan enantiomers were resolved optically employing rifamycin B as chiral modifier [46].

25.7 Antidepressants

Antidepressants, such as 4,5-dihydrodiazepam, fluoxetine, ALO3152, ALO3363, norfluoxetine, imafen, mianserin, nefopam and nomifensine have been directly resolved chirally by CE employing CDs as chiral additives [8, 25, 29, 33, 56–59]. The chiral additive concentrations ranged from 9 mM–100 mM, 50 μm–100 mM, 9 mM–30 mM for β-CD [8, 33, 56–59], charged β-CDs [8, 29, 58, 59] and modified β-CDs [25, 33, 58], respectively (Table 25.8 lists the analyte and buffer conditions). However, *cis*- and *trans*-dothiepin were resolved using β-CD and propan-1-ol (an organic modifier) in CE buffer [56]. Similarly, GR50360A enantiomers were chirally resolved employing a CE buffer modified with β-CD, propan-2-ol and urea [57].

25.8 Non-steroidal anti-inflammatory drugs

Rawjee and co-workers [60, 61] described a CD-based chiral CE model. This model showed that chiral selectivity can vary depending on whether only the non-dissociated (type I), only the dissociated (type II), or both

Table 25.8 Separation of antidepressants

Analyte	Chiral selector	CE buffer	Ref.
trans-Dothiepin/cis-dothiepin	(10 mM) β-CD	(40 mM) Phosphate pH 4.7– (10%) propan-1-ol pH 4.7	56
GR50360 A	(100 mM) β-CD	(10 mM) Tris/borate (adjusted to pH 2.7 with dilute phosphoric acid)–(20%) propan-2-ol/(5 M) urea	57
ALO3363	(9 mM) β-CD	(20 mM) Tris/(10 mM) phosphate pH 11.0	33
ALO3152	(9 mM) DM-β-CD	(20 mM) Tris/(10 mM) phosphate pH 11.0	33
4,5-Dihydrodiazepam	(20 mM) β-CD;(5 mM) sulfonated-β-CD	(20 mM) Citrate pH 2.3	8
Fluoxetine	(1 mM) β-CD; HP-β-CD; Me-β-CD; SBE-β-CD	(1%) Triethylamine/acetic acid pH 5.5–(10%) acetonitrile	64
Norfluoxetine	(7.5 mM) SBE-β-CD	(1%) Triethylamine/acetic acid pH 5.5	64
Mianserine, imafen	(50 μM–1 mM) SBE-β-CD	(50 mM) Phosphate pH 3.1	29
Imafen, mianserin, nefopam, nomifensine	(30 mM) HP-β-CD	(50 mM) Phosphate pH 3.3	25
Mianserine and analogs	(100 mM) SBE-β-CD; (20 mM) β-CD; (5 mM) carboxymethyl-β-CD	(50 mM) Phosphate pH 3.0	59

forms of enantiomer (type III) complex selectively with the resolving agent. The model was tested with fenoprofen, ibuprofen and naproxen employing β-CD and HP-β-CD as chiral selector, respectively. Furthermore, pH was a major factor influencing the separation of enantiomers by CE. Also, chiral selector type and concentration can vary the guest–host complex, thus directly affecting the chiral separation of enantiomers. However, four non-steroidal anti-inflammatory (NSAF) drugs were optically resolved employing monokis-6-o-octamethylene-permethyl-β-CD immobilized in a silica capillary [8]. Nevertheless, NSAF drugs have been resolved optically employing proteins (avidin) and oligosaccharides (maltodextrins) as chiral additives [52, 53, 55] and some CE buffer conditions for the optically resolved NSAF drugs are given in Table 25.9.

25.9 Polycyclic aromatic compounds

Generally, polycyclic aromatic compounds (PAC) display low water solubility. However, PACs have been resolved optically by CZE using buffers with organic modifiers (alcohols) and chiral selectors (CDs), as given in Table 25.10. Furthermore, benzoin, a neutral PAC, was optically resolved using β-CD and glycoprotein α-1-acid, respectively, as chiral additive modified with an organic solvent (MeOH, propan-2-ol) [5, 44]. Charged and neutral PAC enantiomers have been successfully resolved under acid [8, 10, 13, 14, 42] and alkaline [8] conditions by modifying the buffer with chiral selector, such as native CDs [5, 8], charged CDs (sulfonated-β-CD) [8], modified CDs [8, 10, 42], glycoproteins [44] and crown ethers [13, 14].

25.10 Cationic drugs and solutes

Generally, cationic drug and organic compounds can be candidates for CZE separation if the analyte is charged. Table 25.11 shows tabulated CE buffer conditions for optical separation of cationic drugs and solutes. McKillop *et al.* [62] separated the Z and E isomers of 2-(3-pentenyl) pyridine with a pH 2.5 phosphate buffer by CZE. These isomers appeared to migrate according to molecular shape. Furthermore, by considering the shape of the molecule, migration order was predictable and subsequently confirmed by NMR. Armstrong *et al.* [63] resolved mephobarbitol (antispasmodic), optically by employing a CD immobilized capillary column. However, Gu and Fu [64] separated *p*-nitrophenyl-2-amino-3-hydroxypropanone enantiomers with β-CD (chiral additive) and HPMC

Table 25.9 Separation of non-steroidal anti-inflammatory drugs

Analyte	Chiral selector	CE buffer	Ref.
Fenoprofen, ibuprofen	(15 mM) β-CD	(200 mM) MES/NaOH pH 4.5– (0.2–2%) HEC	60
Etodolac,	(20 mM) γ-CD	(20 mM) Borate/phosphate pH 7.0	8
Carprofen, cicloprofen, flurbiprofen	OM-β-CD	(20 mM) Tris/HCl pH 7.0	8
Ibuprofen	OM-β-CD	(20 mM) Borate/phosphate pH 7.0	8
Fenoprofen, naproxen	(5–20 mM) HP-β-CD	(200 mM) MES pH 4.4–(0.2%) HEC	61
Flurbiprofen	(25 μM) Avidin	(50 mM) Phosphate pH 6.0–(10%) EtOH	55
Ibuprofen, ketoprofen	(25 μM) Avidin	(50 mM) Phosphate pH 6.0–(10%) propan-2-ol	55
Suprofen, carprofen, flurbiprofen, ibuprofen, indoprofen	(2.5%) Glucidex 2 (maltodextrin)	(10 mM) Ammonium phosphate pH 7.0	52
Carprofen, fenoprofen	(2.5%) Glucidex 2 (maltodextrin)	(10 mM) Phosphate pH 7.0	52
Fenoprofen, flurbiprofen, ibuprofen, indoprofen, suprofen	(2.5%) Glucidex 2 (maltodextrin)	(10 mM) Tris/phosphate pH 7.0	52
Flurbiprofen, ibuprofen	(10%) Maltrin 150; (10%) mylose STD; (2%) glucidex 2	(10 mM) Phosphate pH 7.1	53

HEC, hydroxyethyl cellulose; MES, morpholinoethanesulfonic acid; TBAH, tetrabutylammonium hydroxide; OM-β-CD, monokis-6-*o*-octamethylene-permethyl-β-CD.

Table 25.10 Separation of polycyclic aromatic compounds

Analyte	Chiral selector	CE buffer	Ref.
Benzoin	β-CD	(10 mM) TEAA pH 4.7–(15%) MeOH	5
1,1'-Binaphthyl-2,2'-diamine, Tröger's base	(10 mM) β-CD	(20 mM) Citrate pH 2.3	8
1,1-Binaphthyl-2,2-diylhydrogenphosphate, hexahelicene-7,10-dicarboxylic acid	(10 mM) β-CD	(20 mM) Borate/phosphate pH 7.0	8
1,1'-Binaphthyl-2,2'-diylhydrogenphosphate	Chirasil-β-Dex	(20 mM) Borate/phosphate pH 7.0	8
1,1'-Binaphthyl-2,2'-diol, 1,1'-Binaphthyl-2,2'-diylhydrogenphosphate	(10 mM) Sulfonated-β-CD	(20 mM) Borate/phosphate pH 7.0	8
1,1'-Binaphthyl-2,2'-dicarboxilic acid, 1,1'-binaphthyl-2,2'-diol, 1,1'-binaphthyl-2,2'-diylhydrogenphosphate	(30 mM) γ-CD	(20 mM) Borate/phosphate pH 9.0	8
1,1'-Binaphthyl-2,2'-diylhydrogenphosphate, hexahelicene-7,10-dicarboxylic acid	(10 mM) Permethylated-β-CD	(20 mM) Borate/phosphate pH 7.0	8
1,1'-Binaphthyl-2,2'-dicarboxilic acid	(10 mM) Glycosylated-α-CD	(6 mM) Borate/(10 mM) phosphate pH 4.0	10
1,1-Binaphthyl diylhydrogenphosphate	(10 mM) α-CD-(10 mM) β-CD	(6 mM) Borate/(10 mM) phosphate pH 4.0	10
3,5-DNB-HPhe	(9.7 mM) 2-O-(S)-NEC-β-CD	(50 mM) Phosphate pH 6.5	42
3,5-DNB-HPhe	(9.7 mM) 6-O-(R)-NEC-β-CD	(50 mM) Phosphate pH 6.5	42
3,5-DNB-Phe	(6.52 mM) β-CD; (9.7 mM) 2-O-(S)-NEC-β-CD; (9.7 mM) 6-O-(R)-NEC-β-CD	(50 mM) Phosphate pH 6.5	42
3,5-DNB-PG	(9.7 mM) 2-O-(S)-NEC-β-CD; (9.7 mM) 6-O-(R)-NEC-β-CD	(50 mM) Phosphate pH 6.5	42
Naphthylethylamine	(30 mM) 18C6H4	(100 mM) Tris pH 2.5	13
Naphthylethylamine	(10 mM) 18C6H4	(10 mM) Tris/citrate pH 2.2	14
Benzoin	Glycoprotein α-1-acid	(5 mM) Phosphate pH 6.5–(5%) propan-2-ol	44

NEC, 1-(1-naphthyl)ethylcarbamoylated; 3,5-DNB-PG, N-(3, 5-dinitrobenzoyl)phenylglycine; 3,5-DNB-Phe, N-(3, 5-dinitrobenzoyl)-Phe and 3,5-DNB-HPhe, 3,5-DNB-homophenylalanine.

(0.1%) organic modifier. Interestingly, enantiomers of GR57732A were separated employing β-CD (100 mM) modified CE buffer containing 5 M urea and propan-2-ol (30%) [57]. Alternatively, Heuermann and Blaschke [25] optically resolved dimethindenes (antihistaminic), ambucetamide

Table 25.11 Separation of cationic drugs and solutes

Analyte	Chiral selector	CE buffer	Ref.
2-(3-Pentenyl)-pyridine *E* and *Z* isomer	–	(40 mM) Phosphate pH 2.5	62
Mephobarbitol	β-CD	(50 mM) Phosphate pH 7.8	63
p-Nitrophenyl-2-amino-3-hydroxypropanone	(30 mM) β-CD	(20 mM) Tris/citric acid pH 4.5–(0.1%) HPMC	64
GR57732 A	(100 mM) β-CD	(10 mM) Tris/borate (adjusted to pH with dilute phosphoric acid) pH 2.7–(20%) propan-2-ol/(5 M) urea	57
Isopilocarpine, pilocarpine	(9 mM) β-CD; (9 mM) DM-β-CD	(40 mM) Tris/phosphate pH 2.4	33
Organon analogs	(20 mM) β-CD; (40 g l⁻¹) DM-β-CD; (60 mM) α-CD	(30 mM) Tris/phosphate pH 2.5	35
Doxylamine	(30 mM)) DM-β-CD	(20 mM) Phosphate/HCl pH 2.2	11
Doxylamine	(2%) CM-β-CD	(20 mM) Citrate pH 2.5	11
Tocainide analogs	(200 mM) γ-CD; (20 mM) β-CD; (50 mM) DM-β-CD; (50 mM) γ-CD; (50 mM) Me-β-CD	(40 mM) Phosphate pH 3.0–(0.05%) PVA	20
Synephrine	(30 mM) DM-β-CD	(50 mM) Phosphate pH 3.3	25
Chlorpheniramine, dimethindene	(30 mM) HP-β-CD	(100 mM) Phosphate/triethanolamine pH 3.0	31
6-Methoxy-dimethindene, *N*-demethyl-dimethindene, *N*-dimethyl-6-methoxy-dimethindene, dimethindene, dimethindene-*N*-oxide, ambucetamide, mefloquine	(30 mM) HP-β-CD	(50 mM) Phosphate pH 3.3	25
6-Methoxy-dimethindene, *N*-demethyl-dimethindene	(30 mM) HP-β-CD	(50 mM) Phosphate pH 3.3	65
Tioconazole	(10 mM) HP-β-CD	(30 mM) Phosphate/citrate pH 4.3–(25%) MeOH	66
Econazole, enilconazole, ketocanazole, miconazole	(0.1–1 mM) SBE-β-CD	(50 mM) Phosphate pH 9.0–(10%) MeOH	18
Mefloquine, dimethindene	(50 μM) SBE-β-CD	(50 mM) phosphate pH 3.1	29
cis/trans Decaline derivatives	(30 mM) 18C6H4	(6 mM) Benzyltrimethylammonium chloride/NaOH pH 3.7	13
Phenylglycinol	(30 mM) 18C6H4	(100 mM) Tris pH 2.5	13
p-Aminophenylalamine, phenylalaninol	(10 mM) 18C6H4	(10 mM) Tris/citrate pH 2.2	14
Chlorpheniramine	(250 μM) ovomucoid	(10 mM) Phosphate pH 5.0–(9%) propan-2-ol	48
Metamephrine, synephrine	(25 mM) Rifamycin B	(100 mM) Phosphate pH 7.0–(40%) propan-2-ol	46
Pentobarbital	Glycoprotein α-1-acid	(2 mM) Phosphate pH 5.5–(2%) propan-2-ol	44

(antispasmodic) and mefloquine (antimalarial) stereoisomers with a pH 3.3 CE buffer with HP-β-CD chiral additive. Furthermore, this buffer system was also investigated for resolving dimethindene and metabolites in human urine [65]. Nevertheless, enantiomers of econazole, enilconazole, ketocanazole and miconazole (antifungal agents), mefloquine (antimalarial agent) and dimethindenes (antihistaminic) were chirally separated employing SBE-β-CD, a charged CD additive [18, 29].

25.11 Anionic solutes

Chiral selectors, such as native CDs, modified CDs, avidin, crown ethers and copper complexes, incorporated in CE buffer were successfully applied for resolving acidic compounds chirally [8, 16, 20, 55, 67]. Table 25.12 lists analyte and CE buffer conditions. Phenyl compounds substituted with lin-

Table 25.12 Separation of anionic solutes

Analyte	Chiral selector	CE buffer	Ref.
DBT-atrolatic acid, DBT-mandelic acid	–	(25 mM) Phosphate pH 5.8–(2%) PVP	16
2,4-Dichlorophenoxypropionic acid	(200 mM) γ-CD; (50 mM) HE-β-CD; (50 mM) HP-β-CD; (50 mM) HP-γ-CD	(40 mM) Phosphate pH 3.0–(0.05%) HPMC	20
3-Phenylbutyric acid	(200 mM) γ-CD; (50 mM) HP-α-CD; (50 mM) HP-β-CD; (50 mM) HP-γ-CD	(40 mM) Phosphate pH 3.0–(0.05%) HPMC	20
3-Phenyllactic acid	(50 mM) α-CD; (200 mM) γ-CD; (50 mM) HP-α-CD (50 mM) HP-β-CD	(40 mM) Phosphate pH 3.0–(0.05%) HPMC	20
Madelic acid	(50 mM) γ-CD; (50 mM) HE-β-CD; (50 mM) HP-β-CD; (50 mM) HP-γ-CD	(40 mM) Phosphate pH 3.0–(0.05%) HPMC	20
Phenylpropionic acid	(50 mM) HP-α-CD; (50 mM) HP-β-CD	(40 mM) Phosphate pH 3.0–(0.05%) HPMC	20
1-Phenylethanol	Chirasil-B-Dex	(20 mM) Borate/phosphate pH 7.0	8
Vanilmandelic acid	(25 µM) Avidin	(50 mM) Phosphate pH 4.0	55
2-Phenyllactic acid, 3,4-dihydroxymandelic acid, 3-phenyllactic acid, m-hydroxymandelic acid, mandelic acid, p-hydroxymandelic acid,	(8 mM) Cu(II)/(16 mM) aspartame	(20 mM) Acetate pH 4.4	67
2-Amino-3-phenyl-butyric acid, 3-amino-3 phenyl-propionic acid, pseudo-2-amino-3-phenyl butyric acid	(10 mM) 18C6H4	(10 mM) Tris/citrate pH 2.2	14

HPMC, hydroxypropylmethyl cellulose.

ear acids have been optically resolved using native CDs and modified CDs, respectively, employing a pH 3 CE buffer and a coated (polybren) capillary [20]. However, enantiomers of 1-phenylethanol were separated chirally at pH 7 using a 80 cm (effective length) capillary column coated with a 0.20 μm film of Chirasil-B-Dex [8]. Also, stereoisomers of α-hydroxy acids were optically resolved using a copper(II)-S-amino acid and copper(II) aspartame complex [67].

25.12 Chiral method development

Guttman et al. [3] described a CD array chiral analysis (CACA) by CE in an easy step-by-step process for developing quickly optimized chiral separation methods. The pH, CD type and concentration, field strength and temperature are key parameters affecting optical resolution of enantiomers in CACA by CE. CACA analysis can start by determining the analyte pK_a (to predetermine buffer pH) and maximum wavelength (for low detection limits). In the next step, a CD array consisting of two CD concentrations (β-CD {3 mM and 15 mM}, γ-CD {10 and 50 mM}, DM-β-CD {10 mM and 50 mM} and HP-β-CD {10 mM and 50 mM}) is used for CE separation. As a result of the CD array, if chiral separation is found, CD concentration, applied electric field and separation temperature can be optimized. Conversely, if no chiral separation is obtained at pH < pK_a or pH > pK_a, then a pH = pK_a buffer system can be employed with the CD array and/or with other chiral selectors to achieve chiral separation.

Abbreviations

Ala, alanine; AQC, 6-aminoquinolyl-N-hydroxysuccinimidyl carbamoyl; Arg, arginine; Asn, asparagine; Asp, aspartic acid; BTAC, benzyltrimethylammonium chloride; Cb-β-CD, carbamoylated-β-cyclodextrin; CD, cyclodextrin; CE, capillary electrophoresis; 18C6H4, (18) crown-6-tetracarboxylic acid; Cys, cysteine; CEC, capillary electrochromatography; CZE, capillary zone electrophoresis; DAc-β-CD, (2,6-di-o-acetyl)-β-CD; DAT, O,O'-diacetyl-s-tartrate; DBT, O,O'-dibenzoyl-s-tartrate; DM-β-CD, dimethyl-β-cyclodextrin; DNB-, dinitrobenzoyl derivative; DNP-, dinitrophenyl derivative; Dns-, 5-dimethylaminonaphthylsulfonyl derivative; EOF, electroosmotic flow; FMOC-, 9-fluroenylmethoxycarbonyl derivative; Gln, glutamine; Glu, glutamic acid; Gly, glycine; His, histidine; HP-β-CD, hydroxypropyl-β-cyclodextrin; HPMC, hydroxypropylmethyl cellulose; Leu, leucine; Ile, isoleucine; Lys, lysine; Me-β-CD, methyl-β-cyclodextrin; MeOH, methanol; Met, methionine; NorLeu, norleucine; NorVal, Norvaline; PEG, polyethylene glycol; Phe, phenylalanine; Pro, proline; PVP, polyvinylpyrrolidone; SBE-β-CD, sulfobutylether-β-cyclodextrin; Ser, serine; TEAA, triethylammonium acetate, Thr, threonine; TM-β-CD, trimethyl-β-cyclodextrin; Tris, tris(hydroxy)aminomethane; Trp, tryptophan; Tyr, tyrosine; UV, ultraviolet; Val, valine

References

1 Copper, C.L.; Davis, J.B.; Cole, R.O.; Sepaniak, M.J. (1994) Separations of derivatized amino acid enantiomers by cyclodextrin-modified capillary electrophoresis: Mechanistic and molecular modeling studies. *Electrophoresis* **15**, 785–792.
2 Werner, A.; Nassauer, T.; Kiechle, P.; Erni, F. (1994) Chiral separation by capillary zone electrophoresis of an optically active drug and amino acids by host–guest complexation with cyclodextrins. *J. Chromatogr.* **666**, 375–379.
3 Guttman, A.; Aumatell, A.; Brunet, S.; Nelson, C. (1995) Cyclodextrin array chiral analysis (CACA). *Amer. Lab.* December, 18–22.
4 Sun, P.; Barker, G.E.; Mariano, G.J.; Hartwick, R.A. (1994) Enhanced chiral separation of dansylated amino acids with cyclodextrin–dextran polymer network by capillary electrophoresis. *Electrophoresis* **15**, 793–798.
5 Li, S.; Lloyd, D.K. (1994) Packed capillary electrochromatographic separation of the enantiomers of neutral and anionic compounds using β-cyclodextrin as chiral selector: Effects of operating parameters and comparison with free-solution capillary electrophoresis. *J. Chromatogr.* **666**, 321–335.
6 Lindner, W.; Böhs, B.; Seidel, V. (1995) Enantioselective capillary electrophoresis of amino acid derivatives on cyclodextrin. Evaluation of structure–resolution relationships. *J. Chromatogr.* **697**, 549–560.
7 Yamashoji, Y.; Ariga, T.; Asano, S.; Tanaka, M. (1992) Chiral recognition and enantiomeric separation of alanine β-naphthylamide by cyclodextrins. *Anal. Chim. Acta* **268**, 39–47.
8 Mayer, S.; Schurig, V. (1994) Enantiomer separation using mobile and immobile cyclodextrin derivatives with electromigration. *Electrophoresis* **15**, 835–841.
9 Cruzado, I.D.; Vigh, G. (1992) Chiral separations by capillary electrophoresis using cyclodextrins-containing gels. *J. Chromatogr.* **608**, 421–425.
10 Sepaniak, M.J.; Cole, R.O.; Clark, B.K. (1992) Use of native and chemically modified cyclodextrins for the capillary electrophoretic separation of enantiomers. *J. Liq. Chromatogr.* **15**, 1023–1040.
11 Schmitt, T.; Engelhardt, H. (1995) Optimization of enantiomeric separation in capillary electrophoresis by reversal of the migration order and using different derivatized cyclodextrins. *J. Chromatogr.* **697**, 561–570.
12 Kuhn, R.; Steinmetz, C.; Bereuter, T.; Haas, P.; Erni, F. (1994) Enantiomeric separation in capillary zone electrophoresis using a chiral crown ether. *J. Chromatogr.* **666**, 367–373.
13 Kuhn, R.; Wagner, J.; Walbroehl, Y.; Bereuter, T. (1994) Potential and limitations of an optically active crown ether for chiral separation in capillary zone electrophoresis. *Electrophoresis* **15**, 828–834.
14 Kuhn, R.; Erni, F.; Bereuter, T.; Häusler, J. (1992) Chiral recognition and enantiomeric resolution based on host–guest complexation with crown ethers in capillary zone electrophoresis. *Anal. Chem.* **64**, 2815–2820.
15 Kuhn, R.; Stoecklin, F.; Erni, F. (1992) Chiral separations by host–guest complexation with cyclodextrin and crown ether in capillary zone electrophoresis. *Chromatographia* **33**, 32–36.
16 Schützner, W.; Caponecchi, G.; Fanali, S.; Rizzi, A.; Kenndler, E. (1994) Improved separation of diastereomeric derivatives of enantiomers by a physical network of linear polyvinylpyrrolidone applied as pseudophase in capillary zone electrophoresis. *Electrophoresis* **15**, 769–773.

17 Dette, C.; Wätzig, H. (1994) Separation of enantiomers of *N*-acetylcysteine by capillary electrophoresis after derivatization by *O*-phthaldialdehyde. *Electrophoresis* **15**, 763–768.
18 Chankretadze, B.; Endresz, G.; Blaschke, G. (1995) Enantiomeric resolution of chiral imidazole derivatives using capillary electrophoresis with cyclodextrin-type buffer modifiers. *J. Chromatogr.* **700**, 43–49.
19 Szemán, J.; Ganzler, K. (1994) Use of cyclodextrins and cyclodextrin derivatives in high-performance liquid chromatography and capillary electrophoresis. *J. Chromatogr.* **668**, 509–517.
20 Belder, D.; Schomburg, G. (1994) Chiral separations of basic and acidic compounds in modified capillaries using cyclodextrin-modified capillary zone electrophoresis. *J. Chromatogr.* **666**, 351–365.
21 Branch, S.K.; Holzgrabe, U.; Jefferies, T.M.; Mallwitz, H.; Matchett, M.W. (1994) Chiral discrimination of phenethylamines with β-cyclodextrin and heptakis (2,3-di-*o*-acetyl) β-cyclodextrin by capillary electrophoresis and NMR spectroscopy. *J. Pharm. Biomed. Anal.* **12**, 1507–1517.
22 Pálmarsdóttir, S.; Edholm, L.E. (1994) Capillary zone electrophoresis for separation of drug enantiomers using cyclodextrins as chiral selectors: Influence of experimental parameters on separation. *J. Chromatogr.* **666**, 337–350.
23 Sirén, H.; Jumppanen, J.H.; Manninen, K.; Riekkola, M.-L. (1994) Introduction of migration indices for identification: Chiral separation of some β-blockers by using cyclodextrins in micellar electrokinetic capillary chromatography. *Electrophoresis* **15**, 779–784.
24 Quang, C.; Khaledi, M.G. (1993) Improved chiral separation of basic compounds in capillary electrophoresis using β-cyclodextrin and tetraalkylammonium reagents. *Anal. Chem.* **65**, 3354–3358.
25 Heuermann, M.; Blaschke, G. (1993) Chiral separation of basic drugs using cyclodextrins as chiral pseudo-stationary phases in capillary electrophoresis. *J. Chromatogr.* **648**, 267–274.
26 Wren, S.A.C.; Rowe, R.C. (1992) Theoretical aspects of chiral separation in capillary electrophoresis: Initial evaluation of a model. *J. Chromatogr.* **603**, 235–241.
27 Altria, K.D.; Goodall, D.M.; Rogan, M.M. (1992) Chiral separation of β-amino alcohols by capillary electrophoresis using cyclodextrins as buffer additives. I. Effect of varying operating parameters. *Chromatographia* **34**, 19–24.
28 Dette, C.; Ebel, S.; Terabe, S. (1994) Neutral and anionic cyclodextrins in capillary zone electrophoresis: Enantiomeric separation of ephedrine and related compounds. *Electrophoresis* **15**, 799–803.
29 Chankvetadze, B.; Endresz, G.; Bladchke, G. (1994) About some aspects of the use of charged cyclodextrins for capillary electrophoresis enantioseparation. *Electrophoresis* **15**, 804–807.
30 Rogan, M.M.; Altria, K.D.; Goodall, D.M. (1994) Enantiomeric separation of salbutamol and related impurities using capillary electrophoresis. *Electrophoresis* **15**, 808–817.
31 Bechet, I.; Paques, P.; Fillet, M.; Hubert, P.; Crommen, J. (1994) Chiral separation of basic drugs by capillary zone electrophoresis with cyclodextrin additives. *Electrophoresis* **15**, 818–823.
32 Peterson, T.E.; Trowbridge, D. (1992) Quantitation of *l*-epinephrine and determination of the *d*-/*l*-epinephrine enantiomer ratio in a pharmaceutical formulation by capillary electrophoresis. *J. Chromatogr.* **603**, 298–301.
33 Peterson, T.E. (1993) Separation of drug stereoisomers by capillary electrophoresis with cyclodextrins. *J. Chromatogr.* **630**, 353–361.
34 Schutzner, W.; Fanali, S. (1992) Enantiomers resolution in capillary zone electrophoresis by using cyclodextrins. *Electrophoresis* **13**, 687–690.
35 Nielen, M.W.F. (1993) Chiral separation of basic drugs using cyclodextrin-modified capillary zone electrophoresis. *Anal. Chem.* **65**, 885–893.
36 Wren, S.A.C. (1993) Theory of chiral separation in capillary electrophoresis. *J. Chromatogr.* **636**, 57–62.
37 Aumatell, A.; Wells, R.J.; Wong, D.K. (1994) Enantiomeric differentiation of a wide range of pharmacologically active substances by capillary electrophoresis using modified β-cyclodextrins. *J. Chromatogr.* **686**, 293–307.
38 Altria, K.D.; Harden, R.C.; Hart, M.; Hevizi, J.; Hailey, P.A. (1993) Inter-company cross-

validation exercise on capillary electrophoresis: I. Chiral analysis of clenbuterol. *J. Chromatogr.* **641**, 147–153.
39 Wren, S.A.C.; Rowe, R.C. (1992) Theoretical aspects of chiral separation in capillary electrophoresis: II. The role of organic solvent. *J. Chromatogr.* **609**, 363–367.
40 Wren, S.A.C.; Rowe, R.C. (1993) Theoretical aspects of chiral separation in capillary electrophoresis: III. Application to β-blocker. *J. Chromatogr.* **635**, 113–118.
41 Guttman, A.; Cooke, N. (1994) Practical aspects of chiral separations of pharmaceuticals by capillary electrophoresis: I. Separation optimization. *J. Chromatogr.* **680**, 157–163.
42 Gahm, K.-H.; Stalcup, A.M. (1995) Capillary zone electrophoresis study of naphthylethylcarbamoylated β-cyclodextrins. *Anal. Chem.* **67**, 19–25.
43 Aumatell, A.; Guttman, A. (1995) Ultra fast chiral separation of basic drugs by capillary electrophoresis, *J. Chromatogr.* A **717**, 229–234.
44 Li, S.; Lloyd, D.K. (1993) Direct chiral separation by capillary electrophoresis using capillaries packed with an α_1 acid glycoprotein chiral stationary phase. *Anal. Chem.* **65**, 3684–3690.
45 Valtcheva, L.; Mohammad, J.; Pettersson, G.; Hjertén, S. (1993) Chiral separation of β-blockers by high-performance capillary electrophoresis based on non-immobilized cellulase as enantioselective protein. *J. Chromatogr.* **638**, 263–267.
46 Armstrong, D.W.; Rundlett, K.; Reid, G.L. (1994) Use of a macrocyclic antibiotic, rafamycin B, and indirect detection for the resolution of racemic amino alcohols by capillary electrophoresis. *Anal. Chem.* **66**, 1690–1695.
47 Liu, Y.-M.; Sheu, S.-J. (1993) Determination of ephedrine and pseudoephedrine in Chinese herbal preparations by capillary electrophoresis. *J. Chromatogr.* **637**, 219–223.
48 Yoshida, Y.; Sato, T. (1994) Optical resolution by electrokinetic chromatography using ovomucoid as a pseudo-stationary phase. *J. Chromatogr.* **666**, 193–201.
49 MIMS Australia (1994) Promail Printing Group, Sydney, (April/May) p. 69.
50 Blaschke, G.; Hempel, G. (1995) Metabolism of the hypnotic drug zopiclone and zolpidem investigated by capillary electrophoresis. Presented at the *7th International Symposium on High Performance Capillary Electrophoresis*, Wurtzburg, Germany, January, Paper 216.
51 Gaveil, P.; Gramond, J.P.; Guyon, F. (1993) Separation and determination of warfarin enantiomers in human plasma samples by capillary zone electrophoresis using a methylated β-cyclodextrin-containing electrolyte. *J. Chromatogr.* **615**, 317–325.
52 D'Hulst, A.; Verbeke, N. (1994) Quantitation in chiral capillary electrophoresis: Theoretical and practical considerations. *Electrophoresis* **15**, 854–863.
53 D'Hulst, A.; Verbeke, N. (1992) Chiral separation by capillary electrophoresis with oligosaccharides. *J. Chromatogr.* **608**, 275–287.
54 D'Hulst, A.; Verbeke, N. (1994) Separation of the enantiomers of coumarinic anticoagulant drugs by capillary electrophoresis using maltodextrins as chiral modifiers. *Chirality* **6**, 225–229.
55 Tanaka, Y.; Matsubara, N.; Terabe, S. (1994) Separation of enantiomers by affinity electrokinetic chromatography using avidin. *Electrophoresis* **15**, 848–853.
56 Clark, B.J.; Barker, P.; Large, T. (1992) The determination of the geometric isomers and related impurities of dothiepin in a pharmaceultial preparation by capillary electrophoresis. *J. Pharm. Biomed. Anal.* **10**, 723–726.
57 Smith, N.W. (1993) Separation of positional isomers and enantiomers using capillary zone electrophoresis with neutral and charged cyclodextrins. *J. Chromatogr.* **652**, 259–262.
58 Piperaki, S.; Penn, S.G.; Goodall, D.M. (1995) Systematic approach to treatment of enantiomeric separations in capillary electrophoresis and liquid chromatography II. A study of the enantiomeric separation of fluoxetine and norfluoxetine. *J. Chromatogr.* **700**, 59–67.
59 Endresz, G.; Chankvetadze, B.; Blaschke, G. (1995) About the chiral resolution mechanism in HPCE using charged cyclodextrin selectors. Presented at the *7th International Symposium on High Performance Capillary Electrophoresis*, Wurtzburg, Germany, January, Paper 211.
60 Rawjee, Y.Y.; Staerk, D.U.; Vigh, G. (1993) Capillary electrophoretic chiral separations with cyclodextrin additives I. Acids: Chiral selectivity as a function of pH and the concentration of β-cyclodextrin for fenoprofen and ibuprofen. *J. Chromatogr.* **635**, 291–306.

61 Rawjee, Y.Y.; Vigh, G. (1994) A peak resolution model for the capillary electrophoretic separation of the enantiomers of weak acids with hydroxypropyl-β-cyclodextrine-containing background electrolytes. *Anal. Chem.* **66**, 619–627.

62 Mckillop, A.G.; Smith, R.M.; Rowe, R.C.; Wren, S.A.C. (1995) Separation and identification of the Z and E isomers of 2-(3-pentenyl) pyridine by capillary electrophoresis and nuclear magnetic resonance spectroscopy. *J. Chromatogr.* **700**, 69–72.

63 Armstrong, D.W.; Tang, Y.; Ward, T.; Nichols, M. (1993) Derivatized cyclodextrins immobilized on fused-silica capillaries for enantiomeric separations via capillary electrophoresis, gas chromatography, or supercritical fluid chromatography. *Anal. Chem.* **65**, 1114–1117.

64 Gu, J.; Fu, R. (1994) Separation of chiral isomers of *p*-nitrophenyl-2-amino-3-hydroxypropanone by capillary zone electrophoresis using cyclodextrins as chiral selector. *J. Chromatogr*, **667**, 367–370.

65 Heuermann, M.; Blaschke, G. (1994) Simultaneous enantioselective determination and quantification of dimethindene and its metabolite *N*-dimethyl-dimethindene in human urine using cyclodextrins as chiral additives in capillary electrophoresis. *J. Pharm. Biomed. Anal.* **12**, 753–760.

66 Penn, S.G.; Goodall, D.M.; Loran, J.S. (1993) Differential binding of tioconazole enantiomers to hydroxypropyl-β-cyclodextrin studied by capillary electrophoresis. *J. Chromatogr*, **636**, 149–152.

67 Desiderio, C.; Aturki, Z.; Fanali, S. (1994) Separation of alpha-hydroxy acid enantiomers by high performance capillary electrophoresis using copper (II)-L-amino acid and copper(II) aspartame. *Electrophoresis* **15**, 864–869.

26 Separation of enantiomeric compounds by micelle-mediated capillary electrokinetic chromatography

M.E. SWARTZ, J.R. MAZZEO AND E.R. GROVER

The separation of enantiomers by the mode of capillary electrophoresis termed micelle electrokinetic chromatography is reviewed. Topics include indirect separations accomplished by derivatization with chiral reagents and direct separations using cyclodextrin-modified, natural product and synthetic chiral micelles.

26.1 Introduction

From its beginning in the early 1980s [1–3], the use of capillary electrophoresis (CE) for the separation of a wide variety of compounds has increased dramatically. The use of high currents and a narrow-bore capillary for effective heat dissipation has led to fast, highly efficient separations. Due to this inherent high efficiency, CE has been investigated almost since its inception for the analysis of enantiomeric mixtures. The format and scale allow for the use of reagents that would prove either too expensive or problematic (e.g. a high detector background that limits sensitivity) for use in other techniques. While the predominant chromatographic method used for the separation of enantiomeric mixtures is high performance liquid chromatography (HPLC) utilizing chiral stationary phases [4], there are at present in excess of 50 different commercial chiral stationary phases available. The proliferation of chiral HPLC stationary phases is a result of the fact that the separations are characterized by compound specific methods and low efficiency. Often, changes in the mobile phase result in unpredictable changes in enantioselectivity, making method development a tedious and time consuming process. The limitation in efficiency dictates the goal of developing HPLC phases with higher alpha values. However in view of the large number of chiral phases available, it is difficult to find a single chiral phase or selectand which shows a sufficient selectivity or alpha for a broad range of compounds. What is needed is a technique that offers increased efficiency to take advantage of the low alphas often encountered. High efficiencies decrease the number of selectands needed and a broader applicability is realized. It is for this reason (high efficiencies) that CE has significantly increased in popularity. Many different chiral selectands have been successfully employed for enantioselective separations in the free solution mode of CE, as described elsewhere in this book. However, much as the success realized by CE would not have been possible without a significant breakthrough that occurred in the mid-1980s. This was when Terabe *et al.* first reported the use of a mode of CE that has come to be known as micellar electrokinetic chromatography (MEKC) [5]. By the addition of a surfactant (sodium dodecyl sulfate, SDS) to the buffer at solution concentrations above its critical micelle concentration (CMC), micelles are formed that allow the separation of both ionic and neutral compounds simultaneously.

By logical extension, the MEKC mode has been successfully applied to the separation of a number of enantiomeric mixtures using both indirect and direct methods. Indirect methods employ derivatization of the solute by an enantiomerically pure reagent followed by the separation of the resulting diastereomers utilizing typical MEKC conditions. Direct methods utilize the solute in its native form and provide a chiral environment to induce the separation of enantiomers. In the MEKC mode, direct separations can be accomplished in one of two ways, either by addition of a separate and distinct chiral selectand along with a surfactant, or by the use of a chiral surfactants in the place of or in combination with achiral surfactants in the buffer. Chiral surfactants can be further subdivided into two classes, natural products and synthetic.

If properly designed and implemented, micelle-mediated CE separations of enantiomeric mixtures have several distinct advantages when compared to free solution enantiomeric separations, or in comparison with other analytical techniques in general (e.g. HPLC). First and foremost is the ability to separate complex mixtures of both chiral and achiral components. This separation is possible due to the solutes general hydrophobic interaction with the micelle in addition to the enantiomeric recognition made possible by the chiral selectand. This advantage greatly extends the applicability of the technique to a broad range of compound classes. Second, in some instances, it is possible to reverse exactly the migration order of the enantiomers. This reversal is important for the quantitation of trace levels of enantiomeric impurities, where it is desirable to have the trace enantiomer peak elute first before the larger, often tailed peak of the major enantiomer. All of the inherent advantages of CE in general can also be maintained, such as high efficiencies, low volumes of predominately aqueous solvents, low sample volume requirements, and a fast simple and straightforward method development process.

This paper reviews the use of micelle-mediated MEKC for the separation of enantiomeric compounds. It is organized according to the type of analysis (indirect versus direct) and is further divided into sections according to the type of micelle additive or micelle employed. Conditions, analytes and references are summarized in Tables 26.1–26.4. For additional background information on the general theory of MEKC several good reviews of the principles of MEKC and their application are available and can be consulted for more information [6–10].

26.2 Indirect separation of enantiomers by MEKC

Indirect separations are common in HPLC, where conventional reverse phase packing materials can be used to separate diastereomers [11]. Prior to indirect separations, enantiomers are converted to diastereomers by a chemical reaction with an enantiomerically pure reagent. Since MEKC is a chromatographic technique and is arguably at least a distant cousin of reverse phase HPLC, it is no surprise that these HPLC separations also work in the MEKC mode using standard surfactants like SDS. In order to be successful, however, there are certain requirements on the part of the derivatization reagent, the derivatization reaction and the enantiomers of interest. First and foremost, the enantiomer must have an appropriate functional group that lends itself to the derivatization reaction. The reagent must be of high enantiomeric purity and have suitable MEKC properties. The reaction should be selective and proceed with rapid kinetics, but without racemization of either reactants or products. In addition, both enantiomers should react at the same rate, and the reaction should proceed to completion, or at least be reproducible. Some of the more popular reagents that satisfy these requirements include 2,3,4,6-tetra-*o*-acetyl-β-D-glucopyranosyl isothiocyanate (often referred to as GITC), 1-fluoro-2,4-dinitrophenyl-5-L- or D-alanine amide (L- or D-Marfey's reagent), and (+) or (−)-1-(9-fluorenyl)-ethyl chloroformate (FLEC).

For the separation of several chiral amino acids, Kang and Buck used OPA (*ortho*-phthalaldehyde) and chiral mercaptan derivatives [12]. The separation was shown to be influenced by both the concentration of surfactant and organic modifier used. Nishi *et al.* achieved the simultaneous resolution of ten GITC-derivatized DL-amino acids [13]. The effects of surfactant concentration and buffer pH were also reported. Tran *et al.* also investigated the separation of DL-amino acids and peptide isomers, but as derivatives of L- or D-Marfey's reagent [14]. Both free solution CE and MEKC results were reported. Lurie utilized GITC for the analysis of the enantiomers of amphetamine, methamphetamine and their hydroxyphenethylamine precursors [15]. The effects of organic modifier type and concentration, voltage, temperature and SDS concentration on resolution were described. The application of the methodology to a forensic sample was also presented and compared with an HPLC separation. Houben *et al.* established separation conditions for the determination of D-valine in excess L-valine by preseparation automated derivatization with OPA and *N*-acetylcysteine [16]. UV detector sensitivity limitations relative to HPLC fluorescence detection were discussed. Leroy *et al.* used both MEKC and cyclodextrin-modified MEKC to separate the OPA and NDA (naphthalene-2,3-dicarboxaldehyde) derivatives of several amine and thiol drugs [17]. Wan *et al.* employed FLEC as a derivatizing reagent for the analysis of the enantiomers of several racemic amino acids [18]. The concentration of SDS and pH were shown to be critical parameters. The indirect method developed was found to be superior to a direct method using cyclodextrin-modified MEKC (see below). Chan *et al.* also reported the use of FLEC for the analysis of amino acid enantiomers [19], utilizing laser-induced fluorescence detection for improved sensitivity. Okafo *et al.* used β-cyclodextrin-modified taurodeoxycholate micelles to separate the diastereomers obtained following derivatization of a seratonin (5-HT) agonist and related *N*-alkyl analogs with GITC [20]. The results of the CE method were compared with results obtained by reverse phase HPLC and showed good agreement for the analysis of samples with varying degrees of enantiomeric impurity.

In practice, indirect approaches are usually used only as a last resort. The added complexity of the sample preparation, derivatization and the very real likelihood that racemization can occur (affecting quantitative results) make it less attractive than other techniques. In the case of amino acid analysis, however, sample preparation and derivatization may be much less of a concern since derivatization is often performed for detectability. However, the potential for racemization still must be addressed. In the rare instances where this is the technique of choice, simplified chromatographic conditions can be used and often increased sensitivity is obtained. In addition, although it has not been reported, migration order reversal of the enantiomeric pairs as an aid in quantitation should be possible due to the availability of both isomers of some derivatizing reagents. A summary of the use of indirect methods for the separation of enantiomers in MEKC is presented in Table 26.1.

26.3 Direct separation of enantiomers by cyclodextrin-modified MEKC

Cyclodextrins (CDs) are toroidal cyclic oligomers consisting of D-glucose units. The most common of these are α-, β- and γ-cyclodextrin consisting of six, seven and eight D-glucose units, respectively. Cyclodextrins can also be derivatized at the 2, 3 or 6-position rim hydroxyl groups with methyl or hydroxypropyl groups commonly employed. Cyclodextrins can be employed in combination with micelles in the MEKC mode of CE to provide a third phase imparting chirality to the buffer. When enantiomers form inclusion complexes with CDs, the stability of the complex is different for each enantiomer, resulting in differential solute migration and chiral recognition.

Table 26.1 Indirect separation of enantiomers by chiral micelle electrokinetic chromatography

Separation conditions Detection	Racemic compounds	Reference
7.5 mM pH 9.6 Borate, 150 mM SDS, 5.5% methanol 75 μm i.d. by 70 cm length capillary UV 340 nm	OPA-derivatized amino acids; Tyr, Trp, Leu, Val, Arg, Lys	12
20 mM pH 9.0 Phosphate–borate, 200–250 mM SDS 50 μm i.d. by 60 cm length capillary UV 210 nm	19 GITC-derivatized amino acids	13
100 mM pH 8.5 Borate, 200 mM SDS, 5% ACN 75 μm i.d. by 57 cm length capillary UV 340 nm	Ala, Leu, Val, Phe, Trp, Asp, Glu, derivatized with Marfey's reagent	14
10 mM pH 9.0 Phosphate–borate, 100 mH SDS, 20% methanol 50 μm i.d. by 48 cm length capillary UV 210 nm	GITC-derivatized amphetamine, methamphetamine, ephedrine, pseudoephedrine, norephedrine, norpseudoephedrine	15
580 mM Acetic acid adj. to pH 3.7 with Tris, 100 mM SDS, 0.1% MEC 50 μm i.d. by 70 cm length capillary UV 340 nm	OPA-derivatized D,L-valine	16
100 mM pH 9.5 Borate, 10 mM SDS 75 μm i.d. by 70 cm length capillary UV 335 nm	OPA-derivatized N-acetyl-cysteine, norephedrine, haptaminol, amphetamine	17
20 mM Borate, 15 mM phosphate, pH 9.2, 20 mM SDS 25 μm i.d. by 62.5 cm length capillary UV 254 nm	(−)-FLEC-derivatized Ser, Ala, Thr, Val, Nval, Met, Ile, Leu, Nle, Phe	18
10 mM Phosphate pH 6.8 or 5 mM borate pH 9.2, 25 mM SDS 50 μm i.d. by 67 cm length capillary UV 200 nm, 10 kV, 20°C	(+)-FLEC-derivatized Ser, Ala, Val, Met, Leu, Phe, Trp	19
30 mM Sodium dihydrogen phosphate, 30 mM boric acid, pH 7.02, 50 mM sodium taurodeoxycholate, 20 mM β-CD 50 μm i.d. by 57 cm length capillary UV 254 nm, 25 kV, 25°C	GITC-derivatized serotonin (5-HT) agonist [11]3-aminoalkyl-6-carboxamido-1,2,3-tetrahydrocarbazole and related N-alkyl analogs	20

i.d. = internal diameter; MEC = methylhydroxyethylcellulose.

The first reference to the use of CD-modified MEKC was reported by Nishi and Terabe in 1990 using γ-CD in combination with SDS for the separation of dansylated amino acids [21]. Nishi *et al.* also applied this approach to the separation of various barbiturates and determined that the additives such as D-camphor-10-sulfonate or L-menthoxyacetic acid improved the selectivity [22]. In all, five different CDs were evaluated and γ-

CD was found to provide the best results. Francotte et al. reported the effects of pH, CD type and concentration, as well as the effects of both chiral and achiral buffer additives on the separation of barbiturates and barbiturate and non-steroidal aromatase inhibitor enantiomers [23]. The resolution of R- and S-chlorpheniramine obtained both with and without SDS using β-CD in combination with urea has been reported by Otsuka and Terabe [24]. Aumatell and Wells used β-CD in combination with SDS and 1-propanol for the separation of racemethorphan and racemorphan enantiomers, in addition to several other related compounds [25]. Detection limits of 20 ppb in urine were obtained. Siren et al. found that a combination of β- and α-CD in SDS provided the best separation of propranolol, alprenolol and atenolol enantiomers [26]. The use of a migration index system incorporating two marker compounds helped to distinguish between the different enantiomers. Desiderio et al. used β-CD in combination with SDS to separate the enantiomers of some anticonvulsant drugs and their metabolites [27]. The study concluded that CD-modified MEKC was a simple and attractive method for the determination of the oxidative metabolism of mephenytoin and phenytoin by direct injection of deglucuronidated urine.

The use of γ-CD in association with SDS for the separation of the enantiomers of diniconazole, uniconazole and structurally related compounds [28, 29] has been reported by Furata and Doi. The type of CD, its concentration, and the type and concentration of an organic modifier vital to the separation were studied [28]. In addition, NMR was used to probe the separation mechanism and results were compared with those obtained with HPLC on a β-CD stationary phase [29]. Ueda et al. have reported the use of both β- and γ-CD and SDS for the separation of CBI (1-cyano-2-substituted-benz[f]isoindole)-derivatized amino acids [30]. Under the conditions employed, γ-CD was found to provide more enantiomeric selectivity. Attomole detection limits using laser-induced detection were obtained. The advantageous use of γ-CD for the separation of dansylated D,L-amino acid enantiomers was confirmed by Terabe et al. [31]. A combination of both β- and γ-CD and the addition of an organic modifier improved the separation.

A method for the resolution of the enantiomers of a new cholesterol lowering drug utilizing hydroxypropyl-β-CD in SDS was reported by Noroski et al. [32]. Amongst the various parameters studied, resolution of the enantiomers was shown to be affected most by the borate buffer concentration and the hydroxypropyl-β-CD and SDS concentrations. The method was judged to be preferable to HPLC due to the decreased toxicity and smaller quantities of the solvents used, lower operational costs, simplicity, superior enantiomeric resolution and decreased analysis time. Data supporting the validation of the method were provided and revealed the method to be accurate and rugged. A 0.06% w/w detection limit of enantiomeric excess was obtained. Flurer et al. also used hydroxypropyl-β-CD and SDS for the separation, identification and quantification of the enantiomers of ten stereoisomers in the ephedrine family [33]. Calibration plots of the ephedrines were linear over the range 4–100 µg ml^{-1}, and the method was used for the analysis of nutritional supplements. Furata and Doi compared HPLC using a β-CD stationary phase and MEKC with both β- and γ-CD for the enantiomeric separation of a thiazole derivative [34]. The enantiomers were successfully separated by both analytical techniques, while a dependence on the type of CD utilized was shown.

MEKC and free solution CD were compared for the separation of aminoglutethimide enantiomers by Anigbogu et al. [35]. SDS in combination with α-, β- and γ-CD were all evaluated in addition to various buffer parameters. Prunonosa et al. used γ-CD and SDS to quantitate cicletanine enantiomers in plasma following a single therapeutic oral dose [36]. Linearity from 10 to 500 ng ml^{-1} and a LOD (limit of detection) of 10 ng ml^{-1} for each enantiomer were reported. Precision data at the LOQ (limit of quantitation) (25 ng ml^{-1}) and target concentration (500 ng ml^{-1}) were also presented. Desiderio et al. also employed γ-CD and SDS for the enantioselective separation of racemic trans-1,2-dihydro-1,2-dihydroxychrysene [37]. The resolution of enantiomers was found to be strongly influenced by the surfactant concentration and the addition of an organic modifier. The method allowed a fast and interference free analysis of an in vitro metabolic sample extract without pretreatment. The separation of the enantiomers of trimetoquinol hydrochloride, related substances and some additional drugs using a β-CD polymer in combination with SDS was reported by Nishi et al. [38]. The best separations were obtained using 5–7% β-CD polymer in acidic solutions, with resolution increasing at higher concentrations. The separations utilizing the β-CD polymer/SDS combination compared favorably with those obtained from five other types of CDs in the free solution CE mode.

Bile salt surfactants (see below) have also been utilized in combination with various CDs. This approach has been most commonly employed for the separation of dansylated D- and L-amino acids [39–41]; however, other applications for compounds such as mephenytoin and fenoldopam also exist [41]. Both β- and γ-CD in combination with sodium taurodeoxycholate, sodium deoxycholate or taurodeoxycholic acid have been utilized. Aumatell and Wells reported the enantiomeric differentiation of a wide range of pharmacologically active substances by the addition of various β-CDs to both sodium taurodeoxycholate and SDS [42].

The use of CDs in combination with either SDS or bile salts is advantageous for the separation of complex mixtures of both achiral and chiral compounds. The separation of complex mixtures is accomplished by the CD's ability to superimpose a chiral environment on the existing achiral hydrophobic micelle. The resulting buffers are often quite complex, however, and their composition and development are not readily intuitive due to the lack of understanding of the mechanisms involved. A limitation that has more seriously impacted the widespread use of this technique is the lack of availability of both CD enantiomers. This prevents migration order reversal which is shown to be important in quantitation, particularly in instances where resolution is compromised. In addition, lot-to-lot variations and the purity of many of the CD reagents deviates significantly, which is certain to raise method validation issues. A summary of the direct separation of enantiomers by CD-modified MEKC is presented in Table 26.2.

26.4 Direct separation of enantiomers by MEKC with natural product chiral micelles formed from bile salt surfactants

Bile salts are natural product surfactants that have a hydroxy-substituted steroid backbone. In much the same way that SDS has a hydrophilic 'head' and a hydrophobic 'tail', bile salts have both hydrophilic and hydrophobic 'faces' that also promote the formation of micelles [43]. The result is a helical structure with the hydrophilic region in the interior of the micelle [44]. Their polar nature (relative to SDS) generally causes reductions in k' and expands the applicability of MEKC to very hydrophobic compounds, since bile salt micelles have low aggregation numbers [45]. In addition, since the bile salt monomers are chiral, the separation of enantiomeric mixtures is also possible. This was first reported in 1989 by Terabe et al., who used a bile salt (taurodeoxycholate) to separate dansylated amino acid enantiomers [46, 47]. Although the resolution in this early electropherogram is not complete, significant improvements have been made since.

The use of taurodeoxycholate for the separation of the enantiomers of diltiazem, trimetoquinol and related compounds has been reported by Nishi et al. [48–50]. One study included the enantiomeric purity testing of S-trimetoquinol to the 1% impurity level (R-form impurity) of five different batches of drug produced [49]. The effects of various buffer parameters such as pH, type and concentration of bile salt and temperature were shown to be important for successful method development. Cole and co-workers used sodium deoxycholate in combination with methanol to separate the enantiomers of a binaphthyl compound [51, 52]. See et al. also utilized sodium deoxycholate for the separation of the enantiomers of triaza aromatic ligand compounds of iron(II) [53]. The influence on resolution of 13 different organic buffer additives was investigated, with the addition of acetone providing superior resolution.

Smith and co-workers utilized another natural product chiral micelle, N-D-gluco-N-methylalkanamide (MEGA) to separate enantiomers [54, 55]. When used in combination with borate buffers, MEGA yields micelles with adjustable surface charge densities allowing the MEKC elution window to

Table 26.2 Direct separation of enantiomers by cyclodextrin-modified micelle electrokinetic chromatography

Separation conditions Detection	Racemic compounds	Reference
20 mM pH 9.0 Phosphate, 30 mM γ-CD, 50 mM SDS 50 μm i.d. by 70 cm length capillary UV 210 nm, 20 kV, ambient	2,2'-dihydroxy-1,1'-dinaphthyl and 2,2,2-trifluoro-1-(9-anthryl)ethanol	21
20 mM pH 9.0 Phosphate–borate, 50 mM SDS, one of five different CDs (15–40 mM), one of three different additives; methanol, D-camphor-10-sulfonate or L-menthoxyacetic acid 50 μm i.d. by 65 cm length capillary UV 220 nm, 20 kV, ambient	Thiopental, pentobarbital, phenobarbital, barbital, 2,2'-dihydroxy-1,1'-dinaphthyl and 2,2,2-trifluoro-1-(9-anthryl)ethanol, 1,1'-binaphthyl-2,2'-diylhydrogenphosphate	22
20 mM pH 7.0 Phosphate–borate, 100 mM SDS, 30 mM of one of four different CDs, 15% methanol 75 μm i.d. by 57 cm length capillary UV 214 nm, 12.5 kV, 25°C	Butabarbital, secobarbital, hexobarbital, mephobarbital	23
50 mM pH 3.0 Phosphate, 50 mM SDS, 5 M urea, 100 mM β-CD 50 μm i.d. by 26 cm length capillary UV 214 nm, 10 kV, 25°C	Chlorpheniramine	24
50 mM pH 9.05 Borate, 50 mM SDS, 60 mM β-CD, 20% (v/v) 1-propanol 50 μm i.d. by 100 cm length capillary UV 200 nm, 30 kV, 30°C	Racemethorphan and racemorphan	25
40 mM pH 9.3 Borate, 32 mM SDS, 12 mM β-CD, 6 mM α-CD 50 μm i.d. by 67 cm length capillary UV 200 nm, 18 kV, 22°C	Propranolol, alprenolol, atenolol	26
8.4 mM Phosphate, 5.6 mM tetraborate, pH 9.1, 95 mM SDS, 40 mM β-CD, 8% (v/v) 2-propanol 75 μm i.d. by 91 cm length capillary UV 192 nm, 13–20 kV	Mephenytoin, 4-hydroxymephenytoin, 4-hydroxyphenytoin	27
100 mM pH 9.0 Borate, 100 mM SDS, 2 M urea, 50 mM γ-CD, 5% (v/v) 2-methyl-2-propanol 75 μm i.d. by 57 cm length capillary UV 254 nm, 15 kV, 25°C	Diniconazole, uniconizole	28
100 mM pH 9.0 Borate, 100 mM SDS, 2 M urea, 50 mM γ-CD, 5% (v/v) 2-methyl-2-propanol 75 μm i.d. by 57 cm length capillary UV 254 nm, 15 kV, 25°C	Diniconazole, uniconizole	29
100 mM pH 9.0 Borate, 50 mM SDS, 10 mM γ-CD 50 μm i.d. by 70 cm length capillary LIF (He–Cd) 442/490 nm, 15 kV	Naphthalene-2,3-dicarboxaldehyde derivatized Thr, Asp, Ser, Tyr, Phe, Ile, Leu	30

Table 26.2 *Continued*

Separation conditions Detection	Racemic compounds	Reference
100 mM pH 8.6 Borate, 100 mM SDS, 50 mM β-CD, with and without 10% methanol (v/v) 50 μm i.d. by 70 cm length capillary UV 210 nm, 20 kV	Dansyl-derivatized Thr, Ser, Aba, Val, Nva, Met, Leu, Glu, Asp, Trp, Phe, Nle	31
100 mM pH 9.3 Borate, 30 mM SDS, 10 mM hydroxypropyl-β-CD, with and without 10% methanol (v/v) 50 μm i.d. by 50 cm length capillary UV 200 nm, 20 kV, 35°C	Cholesterol-lowering drug	32
30 mM pH 2.0 Tetramethylammonium chloride, 10 mM SDS, 70 mM hydroxypropyl-β-CD 50 μm i.d. by 90 cm length capillary UV 210 nm, 28 kV, 31°C	Ephedrine and related compounds	33
50 mM Phosphate, 100 mM borate pH 9.0, 100 mM SDS, 2 M urea, 50 mM β-CD 75 μm i.d. by 57 cm length capillary UV 254 nm, 15 kV, 25°C	4-[1-(2-Fluoro-4-biphenyl)ethyl]-2-methylaminothiazole	34
10 mM Phosphate, 6 mM borate, pH 9.00, 50 mM SDS, 17.5 mM β-CD, with and without 5% methanol 50 μm i.d. by 50 cm length capillary UV 205 nm, 15 kV	Aminoglutethimide	35
100 mM Sodium borate, pH 8.6, 110 mM SDS, 25 mM γ-CD, 10% ACN 75 μm i.d. by 57 cm length capillary UV 214 nm, 15 kV, 35°C	Cicletanine in plasma	36
25 mM Phosphate, pH 7.8, 50 mM SDS, 20 mM γ-CD, 7.4% 2-propanol (v/v) 50 μm i.d. by 50 cm length capillary UV 220 nm, 15 kV, 25°C	Chry-*trans*-1,2-diOH (a *trans*-1,2-dihydrodiol metabolite of chrysene)	37
25 mM Phosphate, pH 2.7, 10 mM SDS, 5% β-CD polymer, 2 M urea 50 μm i.d. by 40 cm length capillary UV 220 nm, 15 kV, ambient	Trimetoquinol, laudanosoline, norlaudanosoline	38
10 mM pH 7.8 Phosphate–borate, 80 mM taurodeoxycholate, 40 mM β-CD 75 μm i.d. by 100 cm length capillary UV 230 nm, 20 kV	Dansyl-derivatized Thr, Leu, Trp, Nle, Nvl	39
30 mM Phosphate, 10 mM boric acid, pH 7.02, 40 mM taurodeoxycholic acid, 20 mM β-CD 50 μm i.d. by 57 cm length capillary UV 254 nm, 9 or 20 kV	CBI derivatives of baclofen and 3 amino-phosphoric acid analogs; CBI and dansyl-derivatized His, Glu, Asp, Phe, Tyr, Arg, Ser, Gly, Val, Ile, Met, Trp, Ala, Pro (multiple runs)	40

Table 26.2 *Continued*

Separation conditions Detection	Racemic compounds	Reference
30 mM Phosphate, 10 mM borate pH 7.2, 50 mM taurodeoxycholate, 20 mM β-CD 50 μm i.d. by 57 cm length capillary UV 254 nm, 9 kV	Dansyl-derivatized Thr, Asp, Trp, Val, Nvl, Glu	41
50 mM Sodium borate, pH 9.5, 50–120 mM sodium taurodeoxycholate, 30–60 mM hydroxypropyl-β-CD, 5–20% propan-1-ol(v/v) 75 μm i.d. by 57 cm length capillary UV 200 nm, 30 kV, 23°C	Nadolol, diclofensine, pindolol, metoprolol, timolol, clenbuterol, cimaterol, methylamphetamine, methyldimethoxyethylamphetamine, and methyldimethoxymethylamphetamine	42

i.d. = internal diameter.

be tailored to suit the separation needs. When modified with CDs, MEGA micelles have been employed for the separation of dansyl-derivatized amino acids [54], and medicarpin and vesitone (55). In addition, Ishihama and Terabe have reported the use of the natural product chiral micelles glycyrrhizic acid (GRA) and β-escin in combination with SDS for the separation of dansyl- and PTH (phenylthiohydantoin)-derivatized D,L-amino acids [56].

Separations using bile salts or other natural product additives are often characterized by long analysis times as a consequence of working at low pH where electroosmotic flow is decreased or non-existent. Long analysis times are not a serious drawback, however, if the separation cannot be achieved in any other manner. While chiral natural product micelles have enjoyed some success in this regard, the factor responsible for their success (polar nature resulting in reduction in k') also limits their enantioselectivity. A more serious limitation, however, common to many natural product chiral selectands, is the unavailability of both enantiomers for migration order reversal. Migration order reversal is particularly important in this instance, when baseline resolution is often not obtained. A summary of the use of natural product chiral micelles for the separation of enantiomers in MEKC is listed in Table 26.3.

26.5 Direct separation of enantiomers by MEKC with synthetic chiral surfactants

Another way to impart chirality in a MEKC electrolyte is to utilize synthetic chiral surfactants. The use of synthetic surfactants, if properly designed, is potentially the most powerful enantioselective MEKC technique, for a

Table 26.3 Direct separation of enantiomers by natural product chiral micelle electrokinetic chromatography

Separation conditions Detection	Racemic compounds	Reference
50 mM pH 3.0 Phosphate, 50 mM taurodeoxycholate, 40 degrees 50 μm i.d. by 70 cm length capillary UV 220 nm, 17 kV, 40°C	Dansyl-derivatized Trp, Nle, Leu, Phe, Met	46
50 mM pH 3.0 Phosphate, 50 mM taurodeoxycholate, 40 degrees 50 μm i.d. by 70 cm length capillary UV 220 nm, 19 kV, 35–40°C	Dansyl-derivatized Trp, Nle, Leu, Phe, Met	46
20 mM pH 7.0 Phosphate–borate, 50 mM sodium taurodeoxycholate 50 μm i.d. by 50 cm length capillary UV 220 nm, 20 kV, ambient	Laudanosoline, norlaudanosoline, trimetoquinol hydrochloride, laudanosine	48
20 mM pH 7.0 Phosphate–borate, 50 mM sodium taurocholate 50 μm i.d. by 65 cm length capillary UV 210 nm, 20 kV, ambient	Diltiazem, trimetoquinol and related compounds	49
20 mM pH 7.0 or 9.0 Phosphate–borate, 50 mM sodium taurodeoxycholate 50 μm i.d. by 65 cm length capillary UV 210 nm, 20 kV, ambient	Diltiazem, trimetoquinol and related compounds; carboline derivatives and 2,2-dihydroxy-1,1-diaphythyl	50
10 mM pH 9.0 Phosphate, 50 mM sodium deoxycholate, 12% methanol 50 μm i.d. by 75 cm length capillary UV detection, 15 kV	1,1-bi-2-Naphthol, 1,1-binaphthyl diyl hydrogen phosphate 1,1-binaphthyldicarboxylic acid	51
pH 9.0 10 mM Phosphate, 6 mM borate, 50 mM sodium cholate, 15% methanol 50 μm i.d. by 65 cm length capillary UV 229 nm, 20 kV	Bi-2-napthol	52
10 mM pH 9.0 Phosphate–borate, 40 mM deoxycholic acid, 15% acetone 50 μm i.d. by 63 cm length capillary UV 232 nm, 20 kV	Triaza aromatic ligand compounds of iron (II)	53
150 mM pH 10.0 Borate, 75 mM decanoyl-N-methylglucamide, 15 mM γ-cyclodextrin 50 μm i.d. by 80 cm length capillary UV 254 nm, 15 kV	Dansyl-derivatized Val, Asp, Leu, Phe, Met	54
50 mM decyl-β-D-glucopyranoside, 150 mM 1-butaneboronic acid, pH 11.0, 15 mM heptakis-(2,6-O-methyl)-β-CD (or γ-CD) 50 μm i.d. by 80 cm length capillary UV 254 nm, 15 kV	Medicarpin, vestitone, dansyl-derivatized Val, Glu, Leu, Phe, Thr, Met	55
10 mM phosphate/borate, pH 7.0, 10 mM SDS, 30 mM glycyrrhizic acid, 50 mM octyl-β-D-glucoside 52 μm i.d. by 80 cm length capillary Fl 330 nm ex, 550 nm em, 25 kV	Dansyl-derivatized Ser, Thr, Val, Leu, Phe	56

i.d. = internal diameter; Fl = fluorescence; ex = excitation; em = emission.

number of reasons. First, the molecule synthesized can still posses all of the inherent advantages of MEKC using 'traditional' micelles, including high efficiencies, tolerance of complex sample matrices, use of small volumes of predominately aqueous solvents and simple straightforward methods development according to accepted theory. Second, both achiral and chiral separations can be performed. It should also be possible, through rational synthetic design and systematic evaluation of structural perturbations, to tailor the enantioselectivity of the surfactant. Finally, it is possible to synthesize both enantiomers in high purity so that exact migration order reversal can be obtained. The purity of the reagents is also an important consideration for method validation.

Dobashi et al. first described the use of synthetic chiral surfactants in MEKC in 1989 [57, 58]. Using (S)-N-dodecanoylvaline, neutral amino acid derivative enantiomers were separated. Otsuka and co-workers subsequently performed enantiomeric MEKC separations with the same surfactant, again utilizing various amino acid derivative enantiomers [59–61]. By adding methanol and/or urea to the buffer, or by using a mixed micelle, improved peak shape was obtained. Otsuka and co-workers also reported the use of (S)-N-dodecanoyl-L-glutamate, digitonin–sodium taurodeoxycholate (bile salt) mixed micelles, and (S)-N-dodecanoyl-L-serine for the separation of amino acid enantiomers, as well as a racemic mixture of the neutral compound benzoin [62, 63]. The separation of the enantiomers of a wide range of racemic compounds employing dodecyl-β-D-glucopyranoside was investigated by Tickle et al. [64] and Jones et al. [65]. High solubility in aqueous media, low CMC and low UV detector background at wavelengths above 200 nm were the reported advantages.

Mazzeo et al. substituted a carbamate group in (S)-N-dodecoxycarbonylvaline for the amide group in (S)-N-dodecanoylvaline and obtained a lower detector background and increased enantioselectivities for 10 out of the 12 pharmaceutical amine compounds studied [66]. Surfactant concentration, pH and organic modifiers were utilized to optimize partitioning of the solutes into the micelle. Higher concentrations of the surfactants (S)- and (R)-N-dodecoxycarbonylvaline were necessary to accomplish the separation of the acidic AQC-derivatized amino acid enantiomers reported by Swartz et al. [67]. It was necessary to increase the surfactant concentration until the hydrophobicity of the buffer overcame charge repulsion between the negatively charged micelles and solutes. The pH does not play a role in the separation since the amino acids studied were negatively charged (acidic) over the entire pH range employed (pH 7–10). Swartz et al. extended the use of (S)- and (R)-N-dodecoxycarbonylvaline to include several piperidine–dione type compounds [68]. The use of an additional synthetic chiral surfactant, (R)-N-dodecoxycarbonylproline, for the separation of glutethimide enantiomers was also reported.

A resolution equation for MEKC was developed by Mazzeo et al. utilizing the synthetic chiral surfactants (S)-N-dodecoxycarbonylvaline and (S)-2-[(1-oxododecyl)amino]-(3S)-methyl-1-sulfooxypentane for the separation of benzoin enantiomers [69]. The influence of the migration window and solute–micelle partitioning was experimentally determined and agreed with predictions based on the equation. In addition, the ability to obtain very high resolution values by migration window manipulation was demonstrated for the separation of N-methylpseudoephedrine enantiomers. The validation of the separation of ephedrine enantiomers using (S)- and (R)-N-dodecoxycarbonylvaline was reported by Swartz et al. [70]. Linearity, precision, LOD, LOQ and the utility of migration order reversal in quantitation and identification were highlighted. (R)-N-dodecoxycarbonylvaline has also been utilized by Swartz and Brown, in combination with photodiode array spectral analysis and spectral contrast techniques, to distinguish between the diastereomers and enantiomers of ephedrine and pseudoephedrine [71].

Two novel cysteine-based anionic chiral surfactants were successfully used as chiral selectands in MEKC for the separation of cromakalin, fenoldopam and related compounds by de Biasi et al. [72]. Wang and Warner reported the use of a chiral micelle polymer, poly(sodium N-undecylenyl-L-valinate) for the separation of the enantiomers of 1,1'-bi-2-naphthol and laudanosine [73]. A comparison study of the micelle polymer versus the corresponding non-polymerized surfactant showed a number of advantages of the polymerized micelle; first, the elimination of the dynamic equilibrium between monomer and micelle resulted in better resolution, second, a wider concentration range as a result of the lack of a CMC for the polymerized micelles made their use more practical in MEKC and third, an improved mass transfer rate due to the rigidity of the polymer reduced peak broadening. An approach applying microemulsions of (2R, 3R)-di-n-butyl tartrate in combination with SDS has been presented by Aiken and Huie for the separation of ephedrine enantiomers [74]. Cohen et al. employed the technique of metal–ligand exchange chromatography using copper(II) and didecyl-L-alanine in mixed micelles with SDS for the separation of six dansylated amino acid enantiomers [75]. Table 26.4 summarizes the use of synthetic chiral surfactants in MEKC for the separation of enantiomers.

26.6 Conclusion

MEKC using chiral surfactants or additives has shown great promise as a solution to many chiral compound separation problems. While capillary electrophoresis has long been thought of as complementary to other analytical techniques, the separation of enantiomers is one area where CE can be seen to have significant advantages. High efficiencies allow the separation of a wider range of compounds given identical alpha, and lead to the

Table 26.4 Direct separation of enantiomers by synthetic chiral surfactant micelle electrokinetic chromatography

Separation conditions Detection	Racemic compounds	Reference
50 mM Phosphate 25 mM borate, pH 7.0, 12.5 mM sodium dodocylvaline, 12.5 mM SDS, 10% (v/v) methanol 50 μm i.d. by 64 cm length capillary UV 254 nm, 10.5 kV, ambient	N-(3,5-dinitrobenzoyl)-O-isopropyl ester derivatives of Ala, Val, Leu, Phe	57
50 mM Phosphate 25 mM borate, pH 7.0, 12.5 mM sodium dodocylvaline, 12.5 mM SDS, 10% (v/v) methanol 50 μm i.d. by 64 cm length capillary UV 254 nm, 10.5 kV, ambient	N-(3,5-dinitrobenzoyl)-O-isopropyl ester derivatives of Ala, Val, Leu, Phe	58
50 mM Phosphate, pH 3.0, 50 mM SDS, 25 mM digitonin 50 μm i.d. by 63 cm length capillary UV 260 nm, 20 kV, ambient	PTH-derivatives of Ala, Val, Trp, Nle, Aba	59
50 mM Phosphate, pH 7.0, 25 mM SDVal, 10% methanol, 5 M urea 50 μm i.d. by 63 cm length capillary UV 260 nm, 15 kV, ambient	PTH-derivatives of Nva, Met, Trp, Nle	60
50 mM Phosphate, pH 7.0, 50 mM SDVal, 30 mM SDS, 10% methanol, 0.5 M urea 50 μm i.d. by 65 cm length capillary UV 260 nm, 20 kV, ambient	PTH-derivatives of Ser, Nva, Val, Trp, Nle, Aba	61
50 mM Phosphate, pH 9.0, 50 mM SDS, 75 mM sodium N-dodecanoyl-L-glutamate, 1 M urea, 10% v/v methanol 50 μm i.d. by 55 cm length capillary UV 260 nm, 32 μA, ambient	PTH-derivatives of Nva, Trp, Nle, Aba, Met	62
50 mM Borate, pH 11.0, 75 mM N-dodecanoyl-L-serine, 75 mM SDS, 20% (v/v) methanol or isopropanol, 1 M urea 50 μm i.d. by 65 cm length capillary UV 260 nm, 20 kV, ambient	PTH-derivatives of Met, Nva, Val, Trp, Nle, Aba	63
30 mM Sodium dihydrogen phosphate, 10 mM boric acid, pH 8.0, 45 mM dodecyl-β-D-glucopyranoside monophosphate or monosulfate, up to 10% (v/v) methanol 50 μm i.d. by 57 cm length capillary UV 214 nm, 20–28 kV, 30°C	Dansyl-derivatized amino acids, cromakalin, 1,1'-dinaphthyl-2,2'-diyl hydrogenphosphate, mephenytoin, hydroxymephenytoin, 3,4-dimethyl-5,7-dioxo-2-phenyperhydro-1,4-oxazepine, metoprolol, ephedrine, Troger's base, hexobarbital, phenobarbital, fenoldopam	64
30 mM Phosphate, pH 7.0, 10 mM dodecyl-β-D-glucopyranoside 50 μm i.d. by 27 cm length capillary UV 214 nm, 10 kV, 25°C	Dansyl-derivatized Val, Phe, Trp	65

Table 26.4 Continued

Separation conditions Detection	Racemic compounds	Reference
25 mM Phosphate/borate, varying basic pHs, 10–100 mM (R)- or (S)-N-dodecoxycarbonylvaline, 0–30% ACN 50 μm i.d. by 60 cm length capillary UV 214 nm, 12 kV, ambient	Pseudoephedrine, ephedrine, N-methylpseudoephedrine, nor-ephedrine, norphenylephrine, octopamine, bupivacaine, homotropine, ketamine, terbutaline, and four β-blockers	66
25 mM Phosphate/borate, pH 9.25, 100 mM (R-N-dodecoxycarbonylvaline, 50 μm i.d. by 60 cm length capillary UV 254 nm, 15 kV, 30°C	12 AQC-derivatized amino acids	67
25 mM Phosphate/borate, pH 9.25, 80 mM (S)-N-dodecoxycarbonylvaline, or (S)-N-dodecoxycarbonylproline 50 μm i.d. by 60 cm length capillary UV 214 nm, 15 kV, 30°C	Glutethimide, aminoglutethimide, cyclohexylaminoglutethimide, pyridoglutethimide, phenglutarimide	68
25 mM Phosphate/acetate, various pHs, various conc. of (S)-N-dodecoxycarbonylvaline, or (S)-2-[(1-oxododecyl)amino]-(3S)-methyl-1-sulfooxypentane 50 μm i.d. by 60 or 45 cm length capillary, coated and uncoated UV 214 nm, 15 kV, 30°C	Benzoin, N-methylpseudoephedrine	69
25 mM Phosphate/borate, pH 8, 50 mM (R)-N-dodecoxycarbonylvaline, 50 μm i.d. by 60 cm length capillary UV 214 nm, 15 kV, 30°C	Ephedrine validation	70
25 mM phosphate/borate, pH 8.0, 50 mM (R)-N-dodecoxycarbonylvaline, 50 μm i.d. by 100 cm length capillary PDA UV 214 nm, 30 kV, ambient	Ephedrine, pseudoephedrine	71
25 mM Phosphate 50 mM borate, pH 7.0, 25 mM (2R,4R)-dodecylcysteine 50 μm i.d. by 67 cm length capillary UV 254 nm, 12.5 kV, 20°C	Cromalkin and fenoldopam	72
25 mM Borate, pH 9.0, 0.5% w/v poly(sodium N-undecylenyl-L-valinate) 50 μm i.d. by 67 cm length capillary UV 290 nm, 12 kV	1,1′-bi-2-Naphthyl, laudenosine	73
15 mM tris-Hydroxyaminomethane, pH 8.1, 0.6% w/w SDS, 1.2% w/w 1-butanol, 0.5% w/w (2R,3R)-di-n-butyl tartrate 75 μm i.d. by 45 cm length capillary UV 215 nm, 25 kV	Ephedrine	74
10 mM Sodium acetate, 20 mM SDS, 5 mM N,N didecyl L-alanine, 25 mM copper(II), 10% v/v glycerol 75 μm i.d. by 80 cm length capillary UV 260 nm, 25 kV, 25°C	Dansyl-derivatized Thr, Met, Leu	75

i.d. = internal diameter.

likelihood that a smaller number of chiral selectands need to be evaluated during method development. With the novel synthetic surfactants currently in development, the increased resolution and the ability to reverse migration order aid in the quantitation and identification of enantiomeric impurities. Improvements in enantioselectivity and decreased detector background resulting in increased sensitivity and linearity/dynamic range have also been made. It is only a matter of time before these improvements ultimately result in the evolution of CE techniques from a research mode to the transfer of validated methods in routine use.

References

1. Jorgenson, J.W.; Lukacs, K.D. (1981) Zone electrophoresis in open tubular glass capillaries. *Anal. Chem.* **53**, 1298.
2. Jorgenson, J.W.; Lukacs, K.D. (1981) High resolution separations based on electrophoresis ond electroosmosis. *J. Chromatogr.* **218**, 209.
3. Jorgenson, J.W.; Lukacs, K.D. (1983) Capillary zone electrophoresis. *Science* **222**, 266.
4. Zief, M.; Crane, L.J. (eds) (1988) *Chromatographic Chiral Separations*, Marcel Dekker, New York.
5. Terabe, S.; Otsuka, K.; Ando, T. (1985) Electrokinetic chromatography with micellar solution and open-tubular capillary. *Anal. Chem.* **57**, 834.
6. Weinberger, R. (1993) *Practical Capillary Electrophoresis*, Academic Press, Boston, MA.
7. Terabe, S.; Chen, N.; Otsuka, K. (1994) In Micellar electrokinetic capillary chromatography. In *Advances in Electrophoresis*, Volume 7, Chrambach, A.; Dunn, M.J.; Radola, B.J. (eds), VCH Publishers, New York, Chapter 2, pp. 89–153.
8. Nielson, K.R.; Foley, J.P. (1993) In Micellar electrokinetic capillary chromatography. *Capillary Electrophoresis, Theory and Practice*, Camilleri, P. (ed.), CRC Press, Boca Raton, FL, Chapter 4, pp. 117–161.
9. Khaledi, M. (1994) Micellar electrokinetic capillary chromatography. In *Handbook of Capillary Electrophoresis*, Landers, J.P. (ed.), CRC Press, Boca Raton, FL, Chapter 3, pp. 43–93.
10. Swartz, M.; Brown, P. (1996) Micelle mediated capillary electrophoretic separation of enantiomeric compounds. In *The Impact of Stereochemistry on Drug Development and Use*, Wainer, I.; Aboul-Enein, H. (eds), John Wiley and Sons, New York, in press.
11. Gal, J. (1987) Resolution of optical isomers by derivatization with 2,3,4,6-tetra-O-acetyl-beta-D-glucopyranosyl isothiocyanate (GITC). *LC–GC* **5**, 106.
12. Kang, L.; Buck, R.H. (1992) Separation and enantiomer determination of OPA-derivatized amino acids by using capillary zone electrophoresis. *Amino Acids* **2**, 103.
13. Nishi, H.; Fukuyama, T.; Matsuo, M. (1990) Resolution of optical isomers of 2,3,4,6-tetra-O-acetyl-beta-D-glucopyranosyl isothiocyanate (GITC) derivatized D,L-amino acids by micellar electrokinetic chromatography. *J. Microcol. Sep.* **2**, 234.
14. Tran, A.D.; Blanc, T.; Leopold, E.J. (1990) Free solution capillary electrophoresis and micellar electrokinetic resolution of amino acid enantiomers and peptide isomers with L- and D-Marfey's reagent. *J. Chromatogr.* **516**, 241.
15. Lurie, I.S. (1992) Micellar electrokinetic capillary chromatography of the enantiomers of amphetamine, methamphetamine, and their hydroxyphenethylamine precursors. *J. Chromatogr.* **605**, 269.
16. Houben, R.J.H.; Gielen, H.; van der Wal, Sj. (1993) Automated preseparation derivatization on a CE instrument. *J. Chromatogr.* **634**, 317.
17. Leroy, P.; Bellucci, L.; Nicolas, A. (1995) Chiral derivatization for the separation of racemic amino acids and thiol drugs by LC and CE. *Chirality* **7**, 235.
18. Wan, H.; Andersson, P.E.; Engstrom, A.; Blomberg, L.G. (1995) Direct and indirect chiral

separations of amino acids by CE. *J. Chromatogr.* **704**, 179.
19. Chan, K.C.; Muschik, G.M.; Issaq, H.J. (1995) Enajntiomeric separation of amino acids using micellar electrokinetic chromatography after pre-column derivatization with the chiral reagent 1-(9-fluorenyl)-ethylchloroformate. *Electrophoresis* **16**, 504.
20. Okafo, G.N.; Rana, K.K.; Camilleri, P. (1994) Improved separation of diastereomers in CE using a mixture of β-cyclodextrin and sodium taurodeoxycholate. *Chromatographia* **39**, 627.
21. Nishi, H.; Terabe, S. (1990) Applications of micellar electrokinetic chromatography to pharmaceutical analysis. *Electrophoresis* **11**, 691.
22. Nishi, H.; Fukuyama, T.; Terabe, S. (1991) Chiral separation by cyclodextrin-modified micellar electrokinetic chromatography. *J. Chromatogr.* **553**, 503.
23. Francotte, E.; Cherkaoui, S.; Faupel, M. (1993) Separation of the enantiomers of some racemic nonsteroidal aromatase inhibitors and barbiturates by capillary electrophoresis. *Chirality* **5**, 516.
24. Otsuka, K.; Terabe, S. (1993) Optical resolution of chlorpheniramine by cyclodextrin added capillary zone electrophoresis and cyclodextrin modified micellar electrokinetic chromatography. *J. Liq. Chromatogr.* **16**, 945.
25. Aumatell, A.; Wells, R.J. (1993) Chiral differentiation of the optical isomers of racemethorphan and racemorphan in urine by capillary zone electrophoresis. *J. Chromatogr. Sci.* **31**, 502.
26. Siren, H.; Jumppanen, J.; Manninen, K.; Riekkola, M. (1994) Introduction of migration indices for identification; Chiral separation of some beta-blockers by using cyclodextrins in micellar electrokinetic capillary chromatography. *Electrophoresis* **15**, 779.
27. Desiderio, C.; Fanali, S.; Kupfer, A.; Thormann, W. (1994) Analysis of mephenytoin, 4-hydroxymephenytoin and 4-hydroxyphenytoin enantiomers in human urine by cyclodextrin micellar electrokinetic capillary chromatography: simple determination of a hydroxylation polymorphism in man. *Electrophoresis* **15**, 87.
28. Furuta, R.; Doi, T. (1994) Chiral separation of diniconazole, uniconizole, and structurally related compounds by cyclodextrin-modified micellar electrokinetic chromatography. *Electrophoresis* **15**, 1322.
29. Furuta, R.; Doi, T. (1994) Enantiomeric separation of diconazole and uniconizole by cyclodextrin modified micellar electrokinetic chromatography. *J. Chromatogr.* **676**, 431.
30. Ueda, T.; Kitamura, F.; Mitchell, R.; Metcalf, T.; Kuwana, T.; Nakamoto, A. (1991) Chiral separation of naphthalene-2,3-dicarboxaldehyde labeled amino acid enantiomers by cyclodextrin-modified micellar electrokinetic chromatography with laser induced fluorescence detection. *Anal. Chem.* **63**, 2979.
31. Terabe, S.; Miyashita, Y.; Ishihama, Y.; Shibata, O. (1993) Cyclodextrin modified micellar electrokinetic chromatography: separation of hydrophobic and enantiomeric compounds. *J. Chromatogr.* **636**, 47.
32. Noroski, J.E.; Mayo, D.J.; Moran, M. (1995) Determination of the enantiomers of a cholesterol lowering drug by cyclodextrin-modified micellar electrokinetic chromatography. *J. Pharm. Biomed. Anal.* **13**, 45.
33. Flurer, C.L.; Lin, L.A.; Satzger, R.D.; Wolnik, K.A. (1995) Determination of ephedrine compounds in nutritional supplements by cyclodextrin-modified CE. *J. Chromatogr.* **669**, 133.
34. Furuta, R.; Doi, T. (1995) Enantiomeric separation of a thiazole derivative by HPLC and MEKC. *J. Chromatogr.* **708**, 245.
35. Anigbogu, V.C.; Copper, C.L.; Sepaniak, M.J. (1995) Separation of stereoisomers of aminoglutethimide using three CE techniques. *J. Chromatogr.* **705**, 343.
36. Prunonosa, J.; Obach, R.; Diez-Cascon, A.; Gouesclou, L. (1992) Determination of cicletanine enantiomers in plasma by HPCE. *J. Chromatogr.* **574**, 127.
37. Desiderio, C.; Fanali, S.; Sinibaldi, M.; Polcaro, C. (1995) Cyclodextrin modified MECC for the chiral direct resolution of (+),(−)-*trans*-1,2-dihydrodiol metabolite of chrysene *in vitro* activated by rat liver microsome S9 fraction. *Electrophoresis* **16**, 784.
38. Nishi, H.; Nakamura, K.; Nakai, H.; Sato, T. (1994) Chiral separation of drugs by CE using β-cyclodextrin polymer. *J. Chromatogr.* **678**, 333.
39. Lin, M.; Wu, N.; Barker, G.E.; Sun, P.; Huie, C.W.; Hartwick, R.A. (1993) Enantiomeric separation by cyclodextrin modified micellar electrokinetic chromatography using bile

salt. *J. Liq. Chromatogr.* **16**, 3667.
40 Okafo, G.N.; Camilleri, P. (1993) Direct chiral resolution of amino acid derivatives by CE. *J. Microcol. Sep.* **5**, 149.
41 Okafo, G.N.; Bintz, C.; Clarke, S.E.; Camilleri, P. (1992) Micellar electrokinetic capillary chromatography in a mixture of taurodeoxycholic acid and beta-cyclodextrin. *J. Chem. Soc. Chem. Commun.* 1189.
42 Aumatell, A.; Wells, R.J. (1994) Enantiomeric differentiation of a wide range of pharmacologically active substances by cyclodextrin modified MECC using a bile salt. *J. Chromatogr.* **688**, 329.
43 Small, D.M.; Nair, P.P.; Kritchevsky (eds) (1971) In *The Bile Acids*, Volume 1, Marcel Dekker, New York, p. 119.
44 Pavel, N.V.; Giglio, E.; Eposito, G.; Zanabi, A. (1987) Bile salt micelle structure. *J. Phys. Chem.* **91**, 356.
45 Attwood, D.; Florence, A.T. (1983) In *Surfactant Systems*, Chapman & Hall, London, pp. 185–188.
46 Terabe, S. (1989) Electrokinetic chromatography: an interface between electrophoresis and chromatography. *Trends Anal. Chem.* **8**, 129.
47 Terabe, S.; Shibata, M.; Miyashita, Y. (1989) Chiral separation by electrokinetic chromatography with bile salt micelles. *J. Chromatogr.* **480**, 403.
48 Nishi, H.; Fukuyama, T.; Matsuo, M.; Terabe, S. (1990) Chiral separation of trimetoquinol hydrochloride and related compounds by micellar electrokinetic chromatography using sodium taurodeoxycholate solutions and application to optical purity determinations. *Anal. Chem. Acta* **236**, 281.
49 Nishi, H.; Fukuyama, T.; Matsuo, M.; Terabe, S. (1990) Chiral separations of diltiazem, trimetoquinol, and related compounds by micellar electrokinetic chromatography with bile salts. *J. Chromatogr.* **515**, 233.
50 Nishi, H.; Fukuyama, T.; Matsuo, M.; Terabe, S. (1989) Chiral separation of optical isomeric drugs using micellar electrokinetic chromatography and bile salts. *J. Microcol. Sep.* **1**, 234.
51 Cole, R.O.; Sepaniak, M.J.; Hinze, W.L. (1990) Optimization of binaphthyl enantiomer separations by capillary zone electrophoresis using mobile phases containing bile salts and organic solvents. *J. High Resolut. Chromatogr. Commun.* **13**, 579.
52 Cole, R.O.; Sepaniak, M.J. (1991) The use of bile salt surfactants in micellar electrokinetic capillary chromatography. *LC–GC* **10**, 380.
53 See, M.M.; Elshihabi, S.; Burke, Jr, J.A.; Bushey, M.M. (1995) Resolution effects of organic additives on the MEKC enantiomeric separations of tri aza aromatic ligand compounds of iron (II). *J. Microcol. Sep.* **7**, 199.
54 Smith, J.T.; Nashbeth, W.; El Rassi, Z. (1994) MECC with *in situ* charged micelles. *Anal. Chem.* **66**, 1119.
55 Smith, J.T.; El Rassi, Z. (1994) MECC with *in situ* charged micelles. *Electrophoresis* **15**, 1248.
56 Ishihama, Y.; Terabe, S. (1993) Enantiomeric separation by micellar electrokinetic chromatography using saponins. *J. Liq. Chromatogr.* **16**, 933.
57 Dobashi, A.; Ono, T.; Hara, S.; Yamaguchi, J. (1989) Optical resolution of enantiomers with chiral mixed micelles by electrokinetic chromatography. *Anal. Chem.* **61**, 1984.
58 Dobashi, A.; Ono, T.; Hara, S.; Yamaguchi, J. (1989) Enantioselective hydrophobic entanglement of enantiomeric solutes with chiral functionalized micelles by electrokinetic chromatography. *J. Chromatogr.* **480**, 413.
59 Otsuka, K.; Terabe, S. (1990) Effects of methanol and urea on optical resolution of phenylthiohydantoin-D,L-amino acids by micellar electrokinetic chromatography with sodium *N*-dodecanoyl-L-valine. *Electrophoresis* **11**, 982.
60 Otsuka, K.; Terabe, S. (1990) Enantiomeric resolution by micellar electrokinetic chromatography with chiral surfactants. *J. Chromatogr.* **515**, 221.
61 Otsuka, K.; Kawahara, J.; Tatekawa, K.; Terabe, S. (1991) Chiral separation by micellar electrokinetic chromatography with sodium *N*-dodecyl-L-valinate. *J. Chromatogr.* **559**, 209.
62 Otsuka, K.; Kashihara, M.; Kawaguchi, Y.; Koike, R.; Terabe, S. (1993) Optical resolution by high performance capillary electrophoresis: micellar electrokinetic chromatography with sodium *N*-dodecanoyl L-glutamate and digitonin. *J. Chromatogr.* **652**, 253.

63. Otsuka, K.; Karuhaka, K.; Higashimori, M.; Terabe, S. (1994) Optical resolution of amino acid derivatives by micellar electrokinetic chromatography with N-dodecanoyl-L-serine. *J. Chromatogr.* **680**, 317.
64. Tickle, D.C.; Okafo, G.N.; Camilleri, P.; Jones, R.F.D.; Kirby, A.J. (1994) Glucopyranoside based surfactants as psuedostationary phases for chiral separations in capillary electrophoresis. *Anal. Chem.* **66**, 4121.
65. Jones, R.F.D.; Camilleri, P.; Kirby, A.J.; Okafo, G.N. (1994) The synthesis and micellar properties of a novel anionic surfacant. *J. Chem. Soc. Chem. Commun.* **11**, 1311.
66. Mazzeo, J.R.; Grover, E.R.; Swartz, M.E.; Petersen, J.S. (1994) Novel chiral surfactant for the separation of enantiomers by micellar electrokinetic capillary chromatography. *J. Chromatogr.* **680**, 125.
67. Swartz, M.E.; Mazzeo, J.R.; Grover, E.R.; Brown, P.R. (1995) Separation of amino acid enantiomers by micellar electrokinetic capillary chromatography using synthetic chiral surfactants. *Anal. Biochem.* **231**, 65.
68. Swartz, M.E.; Mazzeo, J.R.; Grover, E.R.; Brown, P.R.; Aboul-Enein, H.Y. (1996) Separation of piperidine-2,6-dione drug enantiomers by micellar electrokinetic capillary chromatography using synthetic chiral surfactants. *J. Chromatogr. A* **724**, 307–316.
69. Mazzeo, J.R.; Swartz, M.E.; Grover, E.R. (1995) A resolution equation for EKC based on electrophoretic mobilities. *Anal. Chem.* **67**, 2966.
70. Swartz, M.E.; Mazzeo, J.R.; Grover, E.R.; Brown, P.R. (1996) Validation of enantiomeric spearations by micellar electrokinetic chromatography using synthetic chiral surfacants. *J. Chromatogr.*, in press.
71. Swartz, M.E.; Brown, P.R. (1966) Use of mathematically enhanced spectral analysis and spectral contrast techniques for the liquid chromatographic and capillary electrophoretic detection and identification of pharmaceutical compounds. *Chirality* **8**, 67–76.
72. de Biasi, V.; Senior, J.; Zukowski, J.A.; Haltiwanger, R.C.; Eggleston, D.S.; Camilleri, P. (1994) Chiral discrimination in capillary electrophoresis using novel anionic surfactants related to cysteine. *J. Chem. Soc. Chem. Commun.* **11**, 1311.
73. Wang, J.; Warner, I.M. (1994) Chiral separations using micellar electrokinetic capillary chromatography and a polymerized chiral micelle. *Anal. Chem.* **66**, 3773.
74. Aiken, J.H.; Huie, C.W. (1993) Use of a microemulsion system to incorporate a lipophilic chiral selector in electrokinetic capillary chromatography. *Chromatographia* **35**, 448.
75. Cohen, A.S.; Paulus, A.; Karger, B.L. (1987) High performance capillary electrophoresis using open tubes and gels. *Chromatographia* **24**, 15.

27 Therapeutic drug monitoring by capillary electrophoresis

Z.K. SHIHABI

27.1 Introduction

Therapeutic drug monitoring (TDM) is an area which has been expanding rapidly in the last three decades in clinical laboratories. Because of the continuous introduction of new drugs this area is expected to grow even further. TDM can be performed by several techniques, including capillary electrophoresis (CE). CE offers several unique advantages in comparison to other methods which will be discussed in this chapter. In contrast to its advantages, CE has several limitations which require thoughtful or clever manipulation to overcome.

The general emphasis here is on the practical aspects. The aim of this review is not to be a complete review of the literature but to:

1. summarize the recent literature on developments in this area;
2. illustrate the advantages and disadvantages of CE;
3. show where this technique is useful and what assays are most suitable for this technique;
4. compare CE to other common methods used for TDM;
5. discuss the special manipulation required for successful use of this technique in routine work;
6. discuss the analysis of selected drugs where they have been analyzed successfully by CE;
7. speculate on the future of CE.

27.1.1 History

In the 1960s, TDM was in its infancy and limited to a few drugs, including phenobarbital, phenytoin and occasional salicylates, compared to about 30 routine different tests being offered now. The most common instrument for detection was the UV spectrophotometer. It soon became clear that many drugs and endogenous substances interfered with this method. Gas chromatography (GC) became popular in the 1970s for drug analysis because of its greater specificity which avoided interfering substances and offered a larger menu of drugs which could be analyzed such as theophylline, carbamazepine and primidone. Although GC could monitor a larger number of drugs, it required tedious sample extraction and an experienced technician. Basic drugs were difficult to analyze due to their binding onto the silica column packing. Derivatization procedures were often required for analysis. The danger from the use of a flame detector and the need for a skilled operator were also problems for the routine clinical chemistry laboratory to address, especially in the face of emergency work requirements. Heat-unstable compounds are not applicable to GC analysis. Degradation of compounds of interest in a heated column is also problematic, resulting in artifact formation and poor reproducibility of quantitation. Presently GC is no longer used routinely for TDM, but in conjunction with mass spectrometry (MS), it is used for identification of the chemical structure of drugs of abuse as well as their metabolites.

At the same time, several researchers demonstrated that clinical improvements among patients were not related to dosage but more related to the serum level of the drug. For example, the number of seizures decreased when the level of antiepileptic drug was brought within a therapeutic window. Other researchers demonstrated that aminoglycoside monitoring decreased both morbidity and length of stay in hospital. This spurred more interest in TDM. At the present time some drugs like cyclosporine and tacrolimus (FK506) cannot be prescribed for patient treatment without regular monitoring of their serum level.

Late in the 1970s, HPLC (high performance liquid chromatography) became the technique of choice for TDM and further increased the menu of analyzed drugs. Several books and reviews have been published about TDM by HPLC. The main reasons for its popularity were the simplicity of sample extraction (mainly protein precipitation with acetonitrile in many cases in addition to solid phase extraction) and the relative ease of instrument operation compared to GC.

The popularity of HPLC continued simultaneously with the appearance of the immunoassay method appeared in the 1980s. The immunoassay method is a highly expensive technique. The HPLC method has been innovative since its infancy by supplying several types of column, such as fast HPLC columns, 2–3 cm packed with smaller even particles of around 2 μm diameter with pure silica completely free from heavy metal impurities and completely end capped, or capillary HPLC, so that HPLC is still the primary method utilized for TDM of new drugs and for special drugs that have several metabolites with a similar structure.

CE started with home-made instruments in the 1980s. The commercial instruments were introduced in about 1989. The use of CE for TDM is still in its infancy. Its potential has not been realized by clinical or analytical laboratories. Thus, the aim here is to bring attention to the potential of this technique in this field and summarize the work which has been done so far.

27.1.2 Role of CE in therapeutic drug monitoring

Several factors have kept clinical and analytical chemists from tapping into this technique:

1. the belief that CE is not sensitive enough for drug assays;
2. the belief that CE requires elaborate sample preparation;
3. under-estimating the potential of CE in chiral separation;
4. the expense of the instrumentation;
5. the new government regulations for diagnostic tests.

CE offers several general advantages for TDM:

1. CE has a relatively lower reagent cost when compared with GC, HPLC and immunoassays.
2. Considering sample preparation and gradients, CE is easier than GC and HPLC. In many instances the sample can be injected directly with little or no preparation. However, it is more difficult and more time consuming than the automated immunoassays. CE often requires sample pretreatment prior to application to avoid interference of admixtures from the matrix.
3. In many instances CE is faster than HPLC and GC depending on the type of HPLC or GC.
4. Considering theoretical plate numbers, CE has a much higher number than HPLC, but this also differs depending on eluent or buffer or size of support material.
5. CE is more suited to analysis of peptides and proteins than HPLC, GC or agarose electrophoresis, especially larger molecular weight proteins.
6. CE as well as HPLC or GC can resolve stereoisomers. For practical reasons, this field is untapped in TDM. In TDM, chiral separation is essential because different chemical and biological characteristics depend on the different chemical stereoisomers. It is certain that analysis of racemates is old-fashioned in current analytical chemistry.

The disadvantages of CE are:

1. Without sample stacking (concentration in the capillary), low sensitivity limits the detection because of the short light pathlength. However, this problem can be remedied in most cases by sample stacking especially by inclusion of acetonitrile in the sample. Detection limits of $<1 \mu g\, ml^{-1}$ can be achieved easily with CE without any need for sample concentration. Since the majority of the drugs analyzed in serum are between 2 and $40 \mu g\, ml^{-1}$, the detection limit is not a problem for such drugs.
2. Lower reproducibility: HPLC offers much better reproducibility than CE. However, since the CE technique is still in its infancy, precision is

Table 27.1 Therapeutic drug monitoring: conditions for analysis of some model compounds in serum by capillary electrophoresis

Buffer	Conditions	Capillary	Compounds	Ref.
MECC				
• Phosphate 25 mM, pH 8.0, 50 mM SDS	210 nm, 30 kV	72 cm × 50 μm	Barbiturates	17
• Borate 9 mM; phosphate 15, pH 7.8, SDS 50 mM	Scanning, 30 kV	44 cm × 50 μm	Barbiturates	5
• Borate 6 mM; phosphate 10 mM, pH ~9.0, 75 mM SDS	Scanning, 20 kV	35–45 cm × 50 μm	Theophylline	4
• Borate 6 mM; phosphate 10 mM, pH ~9.0, 75 mM SDS	Scanning, 20 kV	35–45 cm × 50 μm	Caffeine	4
• Borate 100 mM, pH 8.4; SDS 55 mM	214 nm, 16.5 kV	50 cm × 50 μm	Felbamate	6
• Borate 100 mM, pH 8.4; SDS 55 mM	214 nm, 16.5 kV	50 cm × 50 μm	Phenobarbital	6
• SDS 20 mM; 10 mM Phosphate, 10 mM borate, pH 7.4, 40% acetonitrile	200 nm, 10 kV	57 cm × 75 μm	Cyclosporine	10
CZE				
• Borate 300 mM, pH 8.5	254 nm, 13 kV	25 cm × 50 μm	Theophylline	7
• Borate 220 mM, pH 8.8	254 nm, 12 kV	25 cm × 50 μm	Iohexol	28
• Capso buffer 63 mM, pH 9.7	254 nm, 15 kV	25 cm × 50 μm	Suramin	31
• Borate 10 mM, pH 9.36, or phosphate, 10 mM, pH 2.3	254 nm, 25 kV	57 cm × 50 μm	Nicotinic acid	43
• Borate 300 mM, pH 8.5	254 nm, 11 kV	25 cm × 50 μm	Pentobarbital	18
• Acetate buffer, 130 mM, pH 4.8	214 nm, 10 kV	42 cm × 50 μm	Lamotrigine	20

expected to improve. We will discuss how the precision of the CE technique may be improved below.
3. Although the CE technique is simple, it requires a certain amount of knowledge for successful analysis, which is identical to HPLC or GC.

CE competes with two well established analytical procedures, immunoassay, a popular and convenient method for routine tests, and HPLC or GC. It is unlikely that CE will replace immunoassays in the near future. CE can compete with HPLC for drugs for which immunoassays are not available. Of course, this will include all the new drugs and a few of the old ones. With the emergence of fast separations in terms of time (seconds) and instrument miniaturization (etched glass chips), TDM by CE may become the trend of the future. Table 27.1 lists some of the tests which have been analysed by CE.

27.2 Practical aspects

27.2.1 Sample preparation for CE

It is very easy to run pure standards in CE but running serum samples is much more difficult. The proteins and salts in the serum greatly affect separation, quantitation and reproducibility. The problem of matrix effects is not encountered in HPLC. In CE, it requires understanding for a successful analysis [1].

Three general techniques which we found to be suitable for analysis of drugs in serum are described below.

27.2.1.1 Direct serum injection. In micellar electrokinetic capillary chromatography (MECC) a surfactant such as sodium dodecyl sulfate (SDS) is added to the buffer at a concentration above the critical micellar concentration [2, 3]. Neutral or hydrophobic compounds separate on the basis of their partition between the aqueous phase of the buffer and the micellar pseudophase. The separation is analogous to that of the reversed phase in HPLC. This technique is well suited for the analysis of neutral, non-polar and some ionized compounds. The surfactants form micelles and also solubilize serum proteins. Thus, serum can be injected directly into the capillary, provided the drugs can partition between the two phases and migrate before the serum proteins.

Thormann and co-workers [4, 5] have successfully applied this technique to the analysis of several drugs such as theophylline, caffeine and barbiturates. We have applied this technique to the analysis of the new antiepileptic drug felbamate [6]. The simplicity and the low sample volume necessary for the assay render this method suitable for monitoring the levels of these drugs in pediatric patients [4–6].

If the concentration of the analyte is high, simple dilution of serum (or urine) can also be used for analysis by capillary zone electrophoresis (CZE) [1].

27.2.1.2 Acetonitrile deproteinization/stacking. Deproteinization with acetonitrile is used often in HPLC. However, in CZE removing protein with acetonitrile causes sample concentration in the capillary, termed 'stacking' [1, 7, 8]. The presence of acetonitrile and high concentration of salts (around 50 mM of NaCl) in the final mixture allows the injection of up to 50% of the capillary volume with sample which causes a 5–30-fold concentration [8]. This allows many compounds to be determined at concentrations of 1 µg ml^{-1} or less [18]. Deproteinization can eliminate the need for washing the capillary with NaOH between sample runs, thus promoting analytical speed. This method can be used in CZE but not in MECC [9].

27.2.1.3 Extraction. Solvent extraction followed by solvent evaporation, used traditionally for HPLC and GC, can be used to remove the excess of proteins and salts present in serum; however, this technique is inappropriate for routine analysis as it causes artifact formation by reacting the solvent with the compound of interest. Artifact formation during solvent extraction is described in detail in Chapter 33.

In the last few years, more emphasis has been placed on using solid phase extraction (SPE). The advantage of this technique is that it permits better sample clean-up than can be obtained by solvent extraction, especially since several kinds of support are commercially available. An automated on-line SPE which can be connected to HPLC or CE with an auto-injector is available commercially and is fully capable of automatic drug analysis.

Drug extraction from urine with minimal contamination by endogenous compounds remains a challenging problem. Sample extraction was the major problem for HPLC application to routine drug analysis. Automated SPE treatment prior to HPLC is one candidate (see Chapter 33).

27.2.2 General guidelines for improving separation in CE

1. Use of high ionic strength electrophoresis buffers slows the migration but improves the separation and allows for better sample concentration by stacking.
2. Borate buffers are the best as starting buffers for analysis of anionic compounds by CZE, while borate and SDS are good for analysis by MECC.
3. Lower voltage slows the migration time, but in many cases improves the separation.
4. In the experience of the author, a capillary of around 50 µm in diameter

and around 50 cm in length is a good compromise for speed and sensitivity allowing in most cases levels close to 1 mg l^{-1} to be detected.
5. When a new method is introduced, the following will be important to check: (i) the plot of injection time vs. peak height or peak area, (ii) linearity vs. concentration, (iii) reproducibility of both quantitation and migration time, (iv) the importance of calculation based on an internal standard and (v) the extent of capillary washing needed for good reproducibility.

27.2.3 Basic principles of TDM

The lack of a correlation between drug dose, serum level and patient response is the primary argument for serum monitoring. Some drugs have a narrow therapeutic range, e.g. phenytoin or theophylline, which necessitates closely monitoring the serum levels of the drug. Other drugs such as acetaminophen have a wide range obviating the need for monitoring except in special situations where overdose is suspected. It is important to monitor serum levels of drugs in the steady state which usually is achieved after 5–6 half-lives of the drug. Most orally administred drugs are monitored at trough level which occurs following the longest daily interval without drug. Drug interaction is another major problem which leads to frequent monitoring and dose adjustment.

27.3 Drug analysis by CE

27.3.1 Common drugs

Cyclosporine: Cyclosporine is an immunosuppressant. Since its metabolites crossreact with antibodies in the parent compound, its determination by HPLC presents an advantage over immunoassays for patient monitoring. Many problems are encountered in the analysis of this drug: low levels of the drug in serum, absence of a characteristic wavelength for its detection and low solubility in aqueous buffers. Most workers who have attempted this assay by HPLC found that selection of the appropriate analytical column as well as the detection method is critical. The values for cyclosporine determination by HPLC are about half those obtained by polyclonal immunoassays and about 70% those by monoclonal antibodies due to the specificity of the HPLC method, indicating that immunoassay is insufficient for selectivity when metabolites or compounds of similar chemical structure are present (see Chapter 33 of this book on toxins).

Cyclosporine has been determined by MECC after solvent extraction with ether, at 10 kV in a capillary 57 cm × 75 μm (i.d.) with detection at

200 nm [10]. The running buffer was composed of 10 mM sodium phosphate, 10 mM sodium borate, pH 7.4 containing 20 mM SDS and 40% acetonitrile. The results were 30% lower than monoclonal immunoassay, similar to HPLC results.

Tricyclics: The tricyclics are used as antidepressants. They are very difficult to measure regardless of the method. Most of the antibodies available for commercial immunoassays have different affinities for the different tricyclics (e.g. imipramine, desipramine, amitriptyline) and their metabolites. Several immunoassays have resorted to SPE to measure individual tricyclics specifically. Because these drugs are chemically basic and present in the serum at a low level, a sufficient end-capped column with a high theoretical plate number is important for their separation and determination by HPLC. The absorbance of these compounds is much greater at 200 nm than at 254 nm. Salomon *et al.* [11] have separated seven antidepressants from aqueous standards by CE using a CAPSO (3-(cyclohexylamino)-2-hydroxy-1-propanesulfonic acid) buffer at pH 9.5. The addition of methanol was important for the separation.

Benzodiazepines: The measurement of benzodiazepines is also difficult because of the low level of these compounds. Several benzodiazepines including the urinary metabolite of flurazepam were separated by CZE (ammonium acetate buffer pH 2.5) and detected by MS [12]. Evenson and Wiktorowicz [13] separated several benzodiazepines by MECC in about 15 min. More research is required to measure these compounds in serum by CE.

Antiasthmatic: Theophylline is often used for the treatment of chronic obstructive pulmonary diseases. Theophylline and caffeine both are used frequently to treat apnea in the newborn. These two compounds, in addition to a few other purines, were measured by MECC with direct injection of serum, urine or saliva [4]. A borate buffer, pH 9.0 in the presence of 75 mM SDS, was used for the separation which was accomplished in about 15 min [4]. We have found that theophylline can be measured by CZE in borate buffer, 300 mM, pH 8.5 with a good correlation coefficient ($r = 0.98$) to an immunoassay method [7]. Johansson *et al.* [14] also used CZE to measure theophylline in phosphate–borate buffers. Caffeine, dyphylline, theobromine and theophylline were separated by MECC in a 30 mM borate buffer, pH 9.3 containing 30 mM SDS and 30% acetonitrile [13].

As we see, both CZE and MECC have been used for analysis of the antiasthmatic drugs; CZE allows a larger sample volume to the introduced into the capillary resulting in a higher sensitivity while MECC allows direct sample injection without treatment. However, pretreatment is often required to avoid interference of admixtures from the matrix (also described in Chapter 33 of this book on toxins).

Antibiotics: Nishi [15] separated several penicillins such as ampicillin, amoxicillin and benzylpenicillin by MECC using different concentrations of SDS and phosphate–borate buffers. Cefixime, a cephalosporin antibiotic, was determined in urine by direct injection into the capillary in 50 mM phosphate buffer, pH 6.8 [16]. The separation of the different metabolites required more additives to the buffer. Several sulfonamides were separated by CZE using ammonium acetate buffer after solvent extraction with MS detection [12].

Antiepileptic: Several antiepileptic drugs (ethosuximide, phenytoin, primidone, valproic acid, phenobarbital and carbamazepine) have been analyzed by MECC after ethyl acetate extraction using 50 mM SDS in phosphate buffer, 25 mM, pH 8.0 [17]. The separation was completed in 14 min. Thormann *et al.* [5] measured several barbiturates such as phenobarbital, pentobarbital, amobarbital and butalbital in serum and urine by MECC with direct serum injection. However, urine required extraction. The buffer was composed of 9 mM $Na_2B_4O_7$, 15 mM NaH_2PO_4 and SDS 50 mM at pH 7.8. Evenson and Wiktorowicz [13] separated several antiepileptic drugs by MECC in 30 mM borate buffer, pH 9.3 containing 30 mM SDS and 30% acetonitrile. The separation was completed in about 15 min. Serum samples were injected after SPE treatment. Pentobarbital was analyzed by CZE after acetonitrile deproteinization using an electrophoresis buffer of 300 mM borate at pH 8.5 [18]. The assay can be completed in about 5 min (Figure 27.1).

In the last year several new antiepileptic drugs have been approved by the US FDA (Food and Drug Administration). These include gabapentin, felbamate and lamotrigine. None of these has a commercial immunoassay at the present time.

Gabapentin (neurotin) structurally is close to the neurotransmitter GABA (γ-aminobutyric acid) and to other amino acids in general. It is not chromogenic. Hence it had to be derivitized by fluorescamine which was incorporated in the deproteinization reagent acetonitrile. The derivatized compound can be detected by UV–vis or fluorescence. The pH 7.4 of the serum was a good buffer for the derivatization, reaction. Thus, several steps were combined in one, deproteinization, derivitazation and buffering. Analysis by CE with UV at 200 nm was achieved within 12 min with a sensitivity of 1 mg l^{-1} [19].

Felbamate is a neutral compound. MECC was used to analyze this compound by injecting serum directly into the capillary after adding an internal standard [6]. The assay was rapid, about 5 min with sensitivity of 5 mg l^{-1} and no interference. One of the main problems in such an analysis is to produce migration of serum proteins far from the drugs being analyzed, thus requiring complete washing before next run. The correct amount of SDS and ionic strength of the buffer are manipulated to achieve this effect [6]. Phenobarbital was also analyzed by the same method [6] (Figure 27.2).

Lamotrigine is another antiepileptic drug with a basic character. Basic

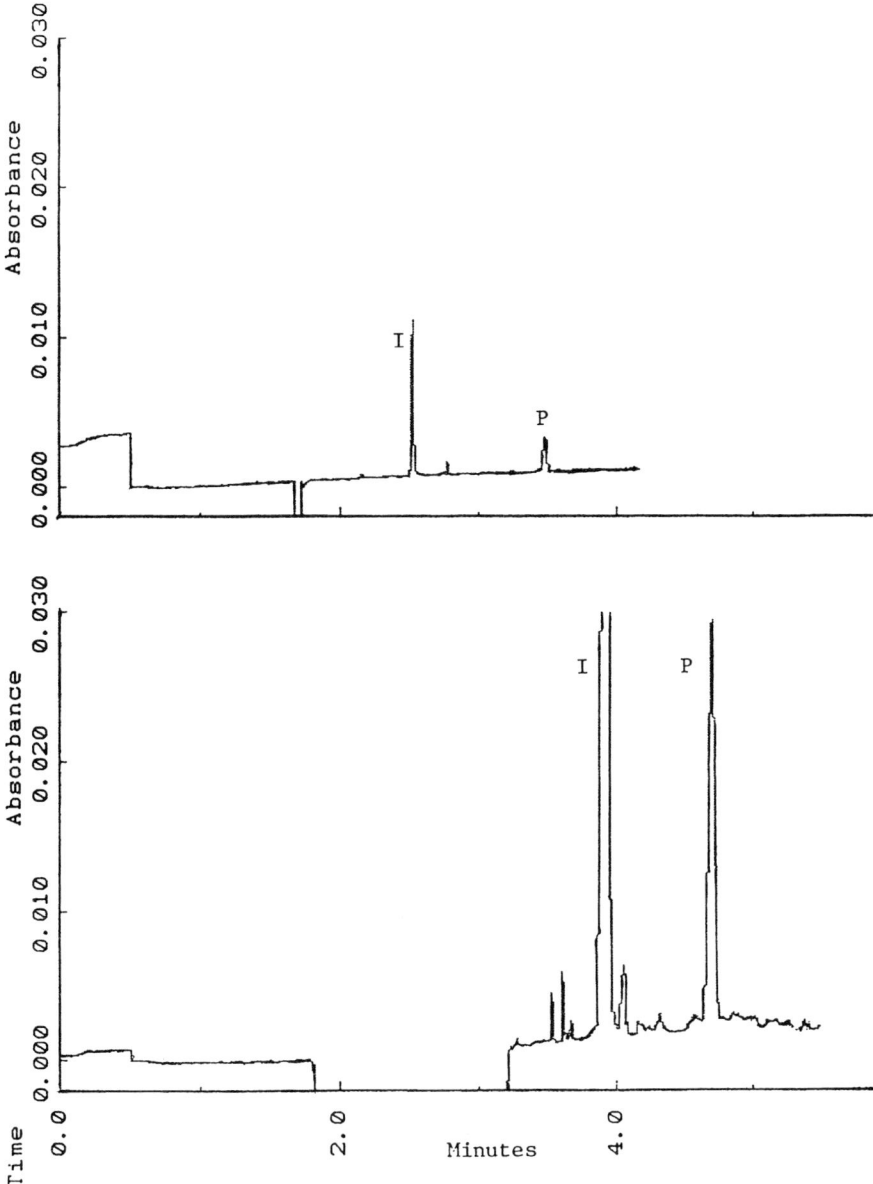

Figure 27.1 Electropherogram of a patient on pentobarbital, P (12 mg l^{-1}). Serum, 100 µl, was deproteinized with 200 µl acetonitrile containing an internal standard, isobutyl methyl xanthine 80 mg l^{-1}, I. The tubes were centrifuged for 30 s at 15 000× g. The supernatant was injected into the instrument. CE conditions (Beckman C): capillary 30 cm × 50 µm, voltage 12 kV, wavelength 214 nm, buffer 300 mM, pH 8.5. (Above) 2% of the capillary is filled with sample, and (below) 40% of the capillary is filled with sample. Repinted with permission of Marcel Dekker, Inc., from (Ref. [18]).

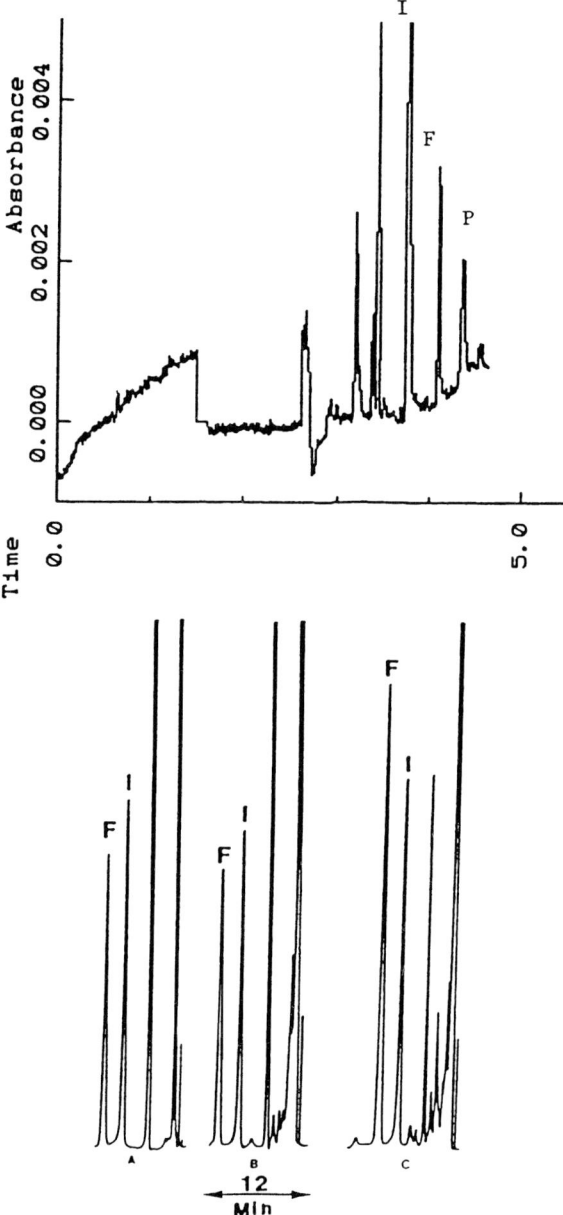

Figure 27.2 Comparsion of analysis of felbamate by CE and HPLC. (Above) CE analysis; a patient receiving felbamate (53 mg l^{-1}); F and phenobarbital (18 mg l^{-1}; P) and (below) HPLC analysis: A, an aqueous standard (50 mg l^{-1}); B, control; and C, and patient on felbamate (78 mg l^{-1}) from Ref. [6]. With permission from the American Association for Clinical Chemistry. For CE analysis (Beckman CE): equal volumes of serum and the internal standard (acetacetanilide 200 mg l^{-1}; I) were mixed and injected into the instrument. Capillary 50 cm × 50 μm, wavelength 214 nm, fixed current at 38 μA, pressure injection 5 s. Buffer: boric acid 100 mM, pH 8.4 containing 55.4 mM SDS.

compounds in general are difficult to analyze by GC or HPLC and the HPLC columns used for such analysis have to be deactivated or end capped completely. In CE, basic drugs migrate rapidly and can be analyzed easily in aqueous standards. However, their analysis in serum is more difficult than that of acidic compounds. Acetonitrile deproteinization yields a supernatant with a pH of 7.4. A buffering step is required to lower the pH below the pK_a of the analytes and below the pH of the running buffer to depress ionization of acidic compounds. Unfortunately this complicates the analysis and introduces extra ions which interfere with the assay.

Double solvent extraction can be helpful but is not very suitable for routine work. We found that acetonitrile deproteinization followed by addition of acetic acid to lower the pH below the pK_a of the compound is important to depress ionization [20] (Figure 27.3).

We see again that both CZE and MECC have been used for analysis of antiepileptic drugs.

Antiarrhythmic: The majority of anitarrhythmic drugs are basic compounds. Some of these compounds also act as antihypertensive agents. Their analysis by CZE presents some advantages as well as disadvantages. The compounds migrate rapidly in the first part of the electropherogram; however they are not as easy to separate and not as reproducible as the acidic compounds. Lukkari *et al.* [21] have separated ten β-adrenergic blockers such as propramolol, oxyprenalol and nadolol and determined them in urine after extraction. Propramolol was determined by direct serum injection after sample concentration and pretreatment on the capillary where the drug interacts with a small bed of protein-coated C18 column [22]. The sample was eluted and the capillary switched for separation. The detection was improved by two orders of magnitude. Amiodarone, a highly hydrophobic compound, can be separated in buffers of organic solvents [23]. Evenson and Wiktorowicz [13] separated procainamide, *N*-acetylprocainamide, disopyramide and chlorodisopyramide by MECC.

Analgesics: Many analgesics are available over the counter including ibuprofen, salicylates and acetaminophen. Because of their wide therapeutic windows, routine monitoring of these drugs is unnecessary. Acetaminophen overdose is encountered occasionally. In such cases careful monitoring of the half-life is important. High doses of ibuprofen have been advocated recently for the treatment of cystic fibrosis [24], hence monitoring serum level is important. Many methods have been described for the analysis of some of these drugs in tablets or pure form [25–27]. We measured ibuprofen in serum by CZE after acetonitrile deproteinization in borate 200 mM, pH 8.5 at 17 kV and 214 nm using a capillary 60 cm × 50 μm [72].

Renal function and contrast agents: Glomerular filtration is one of the main functions of the kidney. Traditionally this function has been estimated by creatinine clearance. However, the current research results indicate that creatinine clearance is not always the best monitor to evaluate renal func-

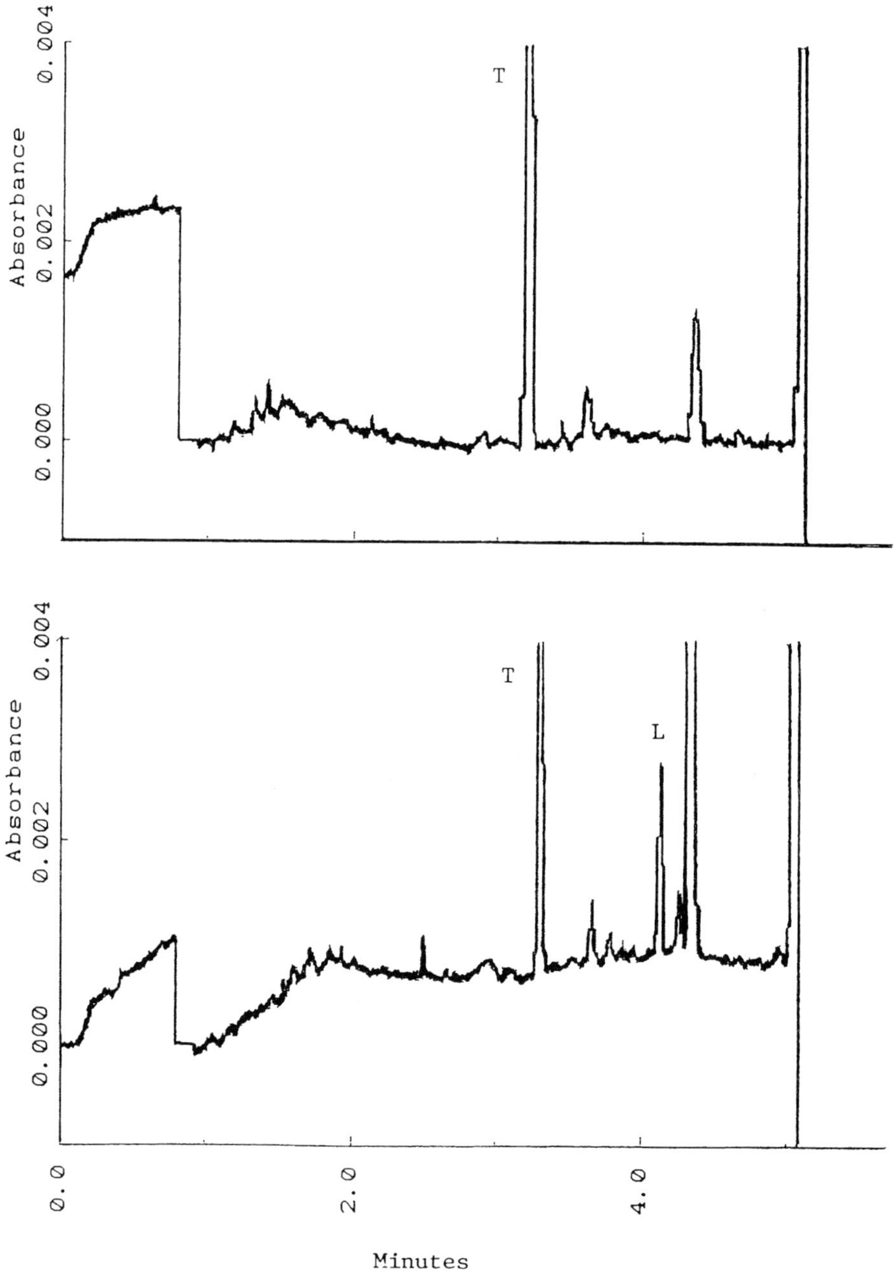

Figure 27.3 Analysis of lamotrigine by CE. (Above) patient free from the drug and (below) patient receiving lamotrigine (3.8 mg l^{-1}). (L = lamotrigine, T = tyramine, internal standard). Serum (50 μl) was mixed with 100 μl acetonitrile containing 40 mg l^{-1} tyramine as an internal standard. The mixture was centrifuged for 30 s at 14 000 × g and the supernatant was mixed with 100 μl acetic acid 0.9 M and injected into the capillary (42 cm × 50 μm, i.d.). The electrophoresis buffer was sodium acetate 130 mM, pH 4.8. The CE was set at 210 nm and 10 kV (Beckman CE).

tion. Several iodinated compounds such as iothalamic acid and iohexol are more appropriate to provide a better measurement of renal clearance. Iohexol [28] was assayed by CZE after serum deproteinization by acetonitrile. The analysis was rapid (<5 min). Values as low as 5 mg l^{-1} can be measured by this technique. Isovue is another candidate compound which was also measured by CZE after acetonitrile deproteinization [29]. Conditions for analysis of iopamidol and a few other related compounds (iopamidol, iohexol, iothalamic acid, p-aminohippuric acid) in 1% NaCl treated with acetonitrile are described in Figure 27.4.

Antitumor drugs: Methotrexate is used for treatment of leukemia. Levels as low as 0.1 nM have been determined after SPE and concentration using CZE with a buffer, 5 mM MES, 5 mM Tris and 1 mM NaCl, pH 6.7 with laser-induced fluorescence [30]. The method correlated well with an enzymatic immunoassay method. Suramin is a polysulfonated naphthylurea used for the treatment of African sleeping sickness and also for prostate tumors. Serum levels of this drug were measured by CZE in CAPSO buffer, 63 mM, pH 9.7 [31]. The test can be completed in less than 5 min. Cytosine-β-arabinoside [32] in serum was determined after SPE using a running buffer of citrate 40 mM, pH 2.5. The separation was rapid about 5 min and detection limit was about 0.5 μM.

27.3.2 *Miscellaneous non-common drugs*

Many new and experimental but less common drugs have been analyzed by CE. Often the drugs themselves and their metabolites can be measured by CE. Although TDM is mainly performed on serum, many drugs have been detected in urine by CE. Here we discuss some of these drugs and the main features of these methods.

Fosfomycin, an antibacterial agent, was measured in serum and other biological fluids after acetonitrile precipitation and separation using borate buffer with indirect detection at 254 nm with a sensitivity of about 10 μg ml^{-1} [33]. 7-OH coumarin, the predominant metabolite of coumarin, was determined in urine and serum after solvent extraction with an electrophoresis buffer of 0.25 M phosphate, pH 7.5 [34]. The assay was very rapid (<2 min) by CE. Hsu *et al.* [35] studied the degradation of the antiviral compound penciclovir by both HPLC and CE.

The antiulcer drug, famotidine, was analyzed by direct urine injection using an electrophoresis buffer of aminocaproic acid, pH 4.5, containing dextran, polyethylene oxide and methanol [36]. Methylflavone carboxylic acid, the metabolite of the flavoxate, a smooth muscle antispasmodic, was detected in urine without treatment by MECC in an electrophoresis buffer, 20 mM phosphate and 10 mM tetraalkylammonium bromide with a detection limit of 0.2 mg l^{-1} [37]. Haloperidol, a neuroleptic drug, and its metabolites were determined by CZE using an ammonium acetate buffer,

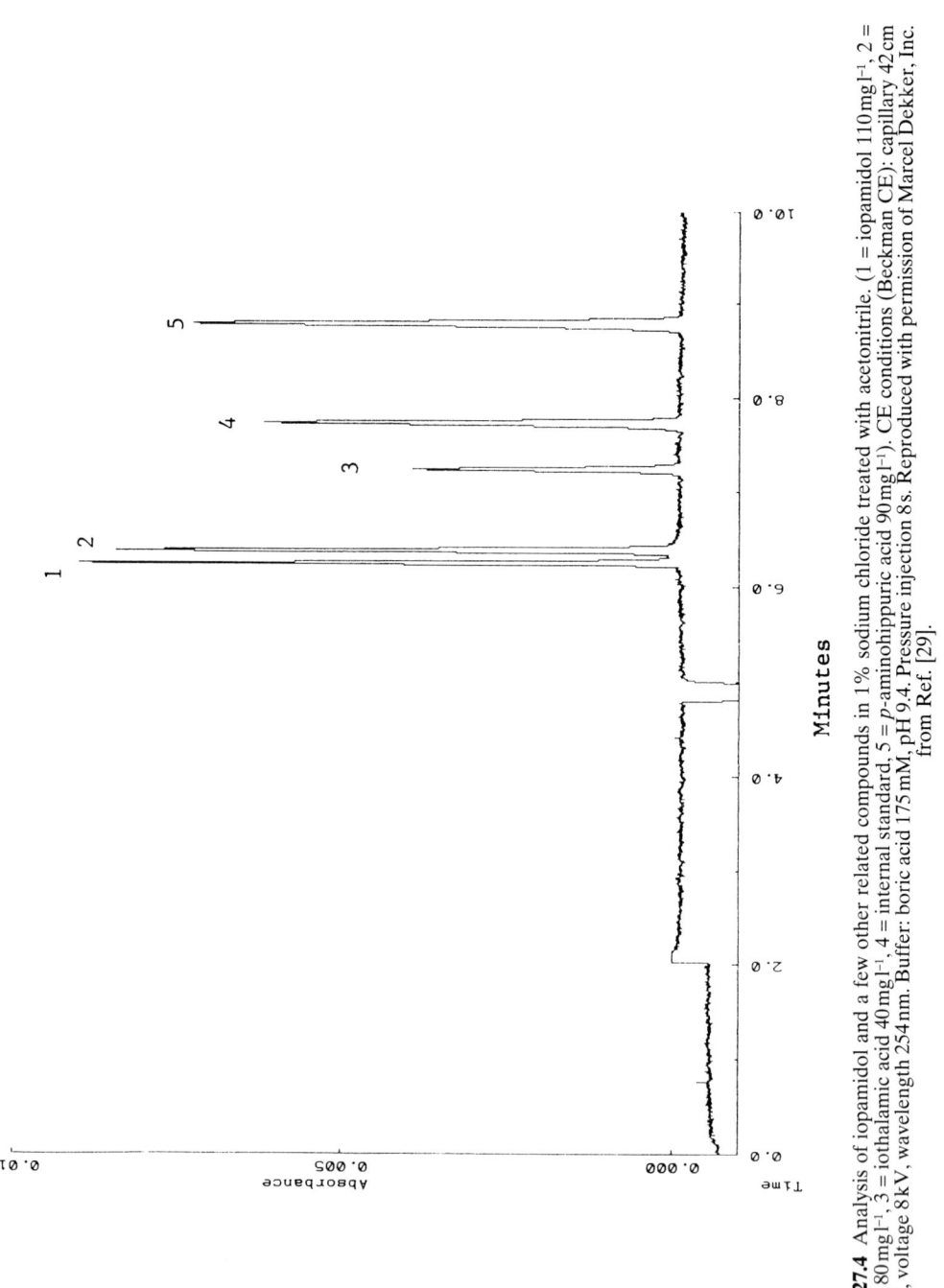

Figure 27.4 Analysis of iopamidol and a few other related compounds in 1% sodium chloride treated with acetonitrile. (1 = iopamidol 110 mg l^{-1}, 2 = iohexol 80 mg l^{-1}, 3 = iothalamic acid 40 mg l^{-1}, 4 = internal standard, 5 = p-aminohippuric acid 90 mg l^{-1}). CE conditions (Beckman CE): capillary 42 cm × 50 μm, voltage 8 kV, wavelength 254 nm. Buffer: boric acid 175 mM, pH 9.4. Pressure injection 8 s. Reproduced with permission of Marcel Dekker, Inc. from Ref. [29].

pH 4.1 [38]. Prunonsa *et al.* [39] described an assay based on solvent extraction with diethyl ether and MECC for serum cicletanine, an antihypertensive drug. Several diuretics were separated by MECC in less than 10 min [13]. Naproxen, an anti-inflammatory drug, has been analyzed in serum by CE after solvent extraction [40]. UV detection was sensitive for concentrations of 0.5–25 µg ml^{-1}, while laser-induced fluorescence was necessary for levels of 0.01–0.5 µg ml^{-1}. Antipyrine, a marker compound confirming a liver oxidative function [41, 42], was analyzed in saliva directly by MECC using a borate buffer at pH 9.6, containing SDS. The analysis was rapid at 1–4 min, without any sample treatment. Nicotinic acid which is used for treatment of hyperlipidemia was determined in serum by CZE after simple solvent extraction with detection at 254 nm [43]. The RSD (relative standard deviation) was about 7%, which is relatively greater than that of HPLC analysis of pharmaceutical compounds. The advantages of CE for monitoring drug metabolism have been reviewed by Naylor *et al.* [44].

27.3.3 Therapeutic drug monitoring by combination of CE and immunoassay

Some drugs such as digoxin are below the detection limits of spectrophotometric detectors requiring the use of immunoassays. Chen and Evangelista [45] have described a system which has potential for simultaneous detection of several drugs based on a combination of immunoassay and CE. In this system, prepared drug–fluorescent conjugates were mixed with antibody and unknown urine samples. After the competition reaction, the free labeled drug was separated from the bound by CE. Since this method requires many complicated steps to prepare the tagged drugs, it is more suited for commercial companies.

27.3.4 Comparison of CE with HPLC

It is accepted that CE and HPLC are complementary techniques especially for complicated separations. However, CE is faster than conventional HPLC but not as fast as fast HPLC and capillary HPLC. CE may produce faster separation with sufficient resolution for purity checking and drug identification; however, HPLC in general tends to give better precision for quantitation and for the capacity factor.

27.4 Chiral separation

Although isomers have almost identical chemical structures, in many instances they have different biological effects. Chiral separations have generated great interest in CE. Detailed information can be found in Chapter 25. The principles of this technique depend on the selective interaction of one isomer with the cavity of special compounds such as cyclodextrins (CD), heparin, Chaps (3[(cholamidopropyl) dimethylammoniol]-1-propane sulfonate) or Big Chaps during electrophoresis.

Soini et al. [46] have shown that the anesthetic drug bupivacaine can be separated in serum at the therapeutic level after solvent extraction into R and S isomers with satisfactory precision and sensitivity within the range of 0.95–1.9 µg ml^{-1} based on the addition of β-CD. Dextromethorphan is used as antitussive agent while the L-isomer is a narcotic analgesic. The optical isomers of the racemethorphan and racemorphan were resolved by the use of propanol-modified β-CD after SPE of urine with detection limits to 20 ppm [47]. Warfarin enantiomers (S/R) in serum were separated by methylated β-CD with a detection limit of 0.2 mg l^{-1} [48]. The S and R forms of cicletinine, an antihypertensive furopyridine drug, were separated and measured by MECC/CDs after solvent extraction with ether [49]. The detection was as low as 10 µg l^{-1} with an RSD of 4–10%.

27.5 Quantitation and reproducibility

Peak area (or height) is important for quantitation while migration time is important for drug identification. Unfortunately, the reproducibility of these parameters in CE is less than that in HPLC. The capacity factor in HPLC rarely changes within a run. However in CE it changes more often, thus resulting in a greater deviation with the CE method. The migration time is more predictable in CE compared to peak height or area and can thus be corrected by adding a reference or an internal standard [50]. Dose and Guichon [51] reported an RSD of 1% for the migration time and peak area using two internal standards (Ids). Siren et al. [52] have shown that multiple standards of close migration bracketing the analyzed compound greatly improve the reproducibility. Adding two or more Ids into the matrix is troublesome in routine analysis especially when the sample matrix is complex. Under multi-Id addition conditions, the RSD is below 1%. An RSD (of mobility) of 0.01–0.03% can be obtained for compounds with a pK_a far from the pH of the electrolyte solution [53]. Based on our work, as well as others, the majority of the RSD for migration time can be below 2% and under ideal conditions below 0.5%. The RSD of peak height or area is 3–7% and under ideal conditions can be below 2% [54]. In general, but not always, peak area values show less variation than peak height [50, 54, 55].

Imprecision in peak height/area is related to a great extent to two factors: injection volume (or time) and capillary wall effects. Very short and very long injection times both lead to increased RSD. The RSD of peak height (area) is inversely related to sample concentration [56]. The precision improves with higher sample concentration especially when peak area instead of peak height is used. In the absence of the sample extraction, Ids slightly improve the precision of peak height or area. In practical work, a visual inspection of Id peak height is adequate. If the peak height of Id is quite different from the other samples, then the injection is repeated. Correction for peak area based on Id is very important if the sample is extracted.

A thorough washing [57] with either NaOH (0.1–1 N), phosphoric acid (0.1 M) or acetonitrile to decrease wall effects is important to attain satisfactory reproducibility of the migration time and peak height (area). The precision of peak height vs. area depends on the software of the integrator and the sharpness of the peak. In CE, peak area depends on the velocity of the analyte. For accuracy a corrected peak area based on the migration time is important and gives in many instances better accuracy than peak height especially for wide peaks such as those encountered in protein analysis. Peak area also gives better linearity than peak height.

It is important when implementing a new analysis to check both linearity and reproducibility based on peak height and area with or without correction for Id to determine which gives better precision.

27.6 Drug screening for drugs of abuse and forensic applications

Forensic analysis is a very specialized area of drug analysis. Although it is not the aim of this chapter, it relates to TDM. Many of the drugs mentioned used for therapeutic purposes can be abused or administered accidentally in large doses. Thus, the progress of CE for screening urine and other body fluids for drugs of abuse or in toxicological studies is discussed briefly here.

The utility of capillary electrophoresis in forensic science has been reviewed by Northrop [58]. Northrop *et al.* [59] have described the analysis of 34 compounds (mostly drugs of abuse) by MECC using a short capillary of 25 cm × 50 µm and a mixed buffer of borate and phosphate. The analysis is complete in 40 min. Acidic, basic and neutral compounds were all separated satisfactorily at the same time. Since an isomer can be a controlled substance while its enantiomer may not be, CE is very useful as a separation method. Chee and Wan [60] described screening for 17 basic drugs such as amphetamines, codeine, diazepam and methaqualone in urine and serum by CE. They used a running buffer of phosphate 50 mM, pH 2.35 and detection at 214 nm. Thormann and co-workers [61–65] have described in several publications the screening of drugs of abuse in serum and urine. Urine samples were extracted with chloroform–isopropanol (9:1). Several benzodiazepines (e.g. diazepam, lorazepam, oxazepam) were separated from human urine by MECC after hydrolysis and solvent extraction [61]. The methods were shown to be superior to immunoassay in terms of sensitivity and specificity. They compared three variations of CE (MECC, CZE and capillary isotachophoresis) for screening serum and urine in patients with overdoses of drugs such as barbiturates, acetaminophen and salicylates. Isotachophoresis required careful attention to the leading and terminating buffers. MECC was useful in resolving acidic, basic and neutral drugs in the same run. Serum or urine can be injected directly without extraction provided the drug concentration is high with the analysis complete within 30 min [62].

Wernly and Thormann [63] described analysis of several drugs of abuse in urine such as benzoylecogonine, amphetamines and methaqualone. After SPE treatment of 5 ml of urine, the drugs were analyzed by MECC using borate–phosphate buffer containing 75 mM SDS (pH 9.1). Levels down to 100 mg l^{-1} can be detected with a multiwavelength detector. The metabolite of cannabinoids, tetrahydrocannabinol carboxylic acid, can be determined in urine by MECC under similar conditions to the above down to 10 mg l^{-1} after SPE [64]. Thormann et al. [65] have reviewed the forensic application of CE and tabulated many drugs which have been analyzed by MECC and CZE.

Hair analysis is used occasionally in forensic drug analysis because it can indicate the chronic use of drugs. Tagliaro et al. [66] have shown that cocaine and morphine can be detected in hair by hydrolysis and extraction. The borate buffer of 50 mM at pH 9.2 was used combined with detection at 214 and 238 nm. Cocaine or morphine, as little as 0.15 ng mg^{-1}, can be detected using 100 mg of hair. Tagliaro et al. [67] showed the separation of 20 illicit drugs and pointed to the advantages of CE for hair analysis of illicit drugs as no derivatization is needed and both MECC and CZE can be used to confirm the drugs. Jelinkova et al. [68] have shown that alcohol can modify the low sulfur keratin of hair which can be detected by CE.

Occupational exposure to organic solvents has been monitored by measuring hippuric acid and methylhippuric acid in urine by MECC [69]. The analysis was performed using 20 mM phosphate buffer at pH 8.0 containing 100 mM dodecyltrimethylammonium bromide and 4 M urea. The detection limits were around 1.5–2.5 μg ml^{-1}.

27.7 Concluding remarks

Because of the continuous introduction of new drugs, TDM will keep growing. Usually, new drugs do not have a commercial immunoassay procedure (i.e. a commercially available immunoassay kit), thus the simplicity of the analysis by CE will be tempting for these drugs. The general notion that

CE is not sensitive enough is not true, but CE is less reproducible compared with HPLC. As we have demonstrated, simple methods of stacking allow levels down to 1 µg ml^{-1} to be analyzed without sample concentration. Furthermore, the use of specialized detectors such as laser-induced fluorescence and MS can greatly enhance the sensitivity. CE–MS–MS is more appropriate and desirable. Their detection methods can be useful in studying drug metabolism. Compounds which lack a chromophore can be analyzed by indirect detection or by UV–vis or fluorescence detection after derivatization.

In routine analysis, speed, simplicity and reproducibility are most important considerations as well as the purpose of the research. Reproducibility is especially necessary in quantitation analysis. Changing the instrument settings and capillaries from one test to another is a major burden for routine work where emergency tests are requested or where several different technicians have to operate the same instrument. Our approach for routine assays has been to compromise between several tests by selecting conditions which are suitable as a group but may not be optimum for a single test. Thus several tests can be performed using the same capillary buffer, voltage etc. with minimum change. For example, we select common conditions to perform analyses of iohexol, isovue, ibuprofen and serum proteins on the same capillary.

It is expected in the future that manufacturers of CE instruments will build special instruments designed solely for TDM where the capillary size and conditions are programmed and optimized for several groups of tests. Great emphasis will be placed on achieving better reproducibility. The use of organic solvents [23, 70] for electrophoresis instead of aqueous buffers have some advantages for analyzing compounds which are not water soluble. The feasibility of instrument miniaturization (silicon-etched) [71] and fast separation in terms of time (seconds) may change the field of TDM and shift it towards CE for many drugs.

References

1. Shihabi, Z.K.; Garcia, L.L. (1994) Effects of sample matrix on separation by capillary electrophoresis. In *Handbook of Electrophoresis*, Landers, J.P. (ed.), CRC Press, Boca Raton, pp. 537–548.
2. Terabe, S. (1994) Micellar electrokinetic capillary chromatography. *Adv. Electrophoresis* **7**, 87–153.
3. Khaledi, M.G. (1994) Micellar electrokinetic capillary chromatography. In *Handbook of Electrophoresis*, Landers, J.P., (ed.), CRC Press, Boca Raton, pp. 43–93.
4. Thormann, W.; Minger, A.; Molteni, S.; Caslavska, J.; Gebauer, P. (1992) Determination of substituted purines in body fluids by micellar electrokinetic capillary chromatography with direct sample injection. *J. Chromatogr.* **593**, 275–288.
5. Thormann, W.; Meier, P.; Marcolli, C.; Binder, F. (1991) Analysis of barbiturates in human serum and urine by high performance capillary electrophoresis–micellar

electrokinetic capillary chromatography with on-column multi-wavelength detection. *J. Chromatogr.* **545**, 445–460.
6. Shihabi, Z.K.; Oles, K.S. (1994) Felbamate measured in serum by two methods: HPLC and capillary electrophoresis method. *Clin. Chem.* **40**, 1904–1908.
7. Shihabi, Z.K. (1993) Sample matrix effects in capillary electrophoresis. II Acetonitrile deproteinization. *J. Chromatogr.* **652**, 471–475.
8. Shihabi, Z.K. (1995) Sample stacking by acetonitrile–salt mixtures. *J. Capillary Electrophoresis* **2**, 267–271.
9. Shihabi, Z.K.; Hinsdale, M.E. (1995) Sample matrix effects in micellar electrokinetic capillary electrophoresis. *J. Chromatogr.* **669**, 75–83.
10. Huie, R.; Wang, H.P.; Van Dreal, P.; Wong, S.S. (1995) Quantitation of cyclosporine using capillary electrophoresis. *Clin. Chem.* **41**, S116.
11. Salomon, K.; Burgi, D.S.; Helmer, J.C. (1991) Separation of seven tricyclic antidepressants using capillary electrophoresis. *J. Chromatogr.* **549**, 375–385.
12. Johansson, I.M.; Pavelka, R.; Henion, J.D. (1991) Determination of small drug molecules by capillary electrophoresis–atmospheric pressure ionization mass spectrometry. *J. Chromatogr.* **559**, 515–528.
13. Evenson, M.A.; Wiktorowicz, J.E. (1992) Automated capillary electrophoresis applied to therapeutic monitoring. *Clin. Chem.* **38**, 1847–1852.
14. Johansson, M.; Rydberg, M.; Schmekel, B. (1993) Determination of theophylline in plasma using different capillary electrophoretic systems. *J. Chromatogr.* **625**, 487–493.
15. Nishi, H. (1990) Separation and determination of apoxicillin in human plasma by micellar electrokinetic chromatography with direct sample injection. *J. Chromatogr.* **515**, 245–255.
16. Honda, S. (1992) Determination of cefixime and its metabolites by high-performance capillary electrophoresis. *J. Chromatogr.* **590**, 364–368.
17. Lee, K-J.; Heo, G.S.; Kim, N.J.; Moon, D.C. (1992) Analysis of antiepileptic drugs in human plasma using micellar electrokinetic capillary chromatography. *J. Chromatogr.* **608**, 243–250.
18. Shihabi, Z.K. (1993) Serum pentobarbital assay by capillary electrophoresis. *J. Liq. Chromatogr.* **16**, 2059–2068.
19. Garcia, L.L.; Shihabi, Z.K.; Oles, K.S. (1995) Determination of gabapentin in serum by capillary electrophoresis. *J. Chromatogr.* **669**, 157–162.
20. Shihabi, Z.K. (in press) Serum lamotrigine analysis by capillary electrophoresis. *J. Chromatogr.*
21. Lukkari, P.; Siren, H.; Pantsar, M.; Riekkola, M.-L. (1993) Determination of ten β-blockers in urine by micellar electrokinetic capillary chromatography. *J. Chromatogr.* **632**, 143–148.
22. Morita, I.; Sawada, J. (1993) Capillary electrophoresis with on-line sample pretreatment for the analysis of biological samples with direct injection. *J. Chromatogr.* **641**, 375–381.
23. Zhang, C.-X.; Heeren, F.V.; Thormann, W. (1995) Separation of hydrophobic, positively chargeable substances by capillary electrophoresis. *Anal. Chem.* **67**, 2070–2077
24. Konstin, M.W.; Byard, P.J.; Hoppel, C.L.; Davis, P.B. (1995) Effect of high-dose of ibuprofen in patients with cystic fibrosis. *New Engl. J. Med.* **332**, 848.
25. Nishi, H.; Terabe, S. (1990) Separation and determination of the ingredients of a cold medicine by micellar electrokinetic chromatography with bile salts. *J. Chromatogr.* **498**, 313–323.
26. Nishi, H.; Terabe, S. (1990) Effect of surfactant structures on the separation of cold medicine ingredients by micellar electrokinetic chromatography. *J. Pharm. Sci.* **79**, 519–523.
27. Schwartz, M.E. (1991) Method development and selectivity control for small molecule pharmaceutical separations by capillary electrophoresis. *J. Liq. Chromatogr.* **14**, 923–938.
28. Shihabi, Z.K.; Constantinescu, M.S. (1992) Iohexol in serum determined by capillary electrophoresis. *Clin. Chem.* **38**, 2117–2120.
29. Shihabi, Z.K.; Rocco, M.V.; Hinsdale, M.E. (1995) Analysis of the contrast agent iopamidol by capillary electrophoresis. *J. Liq. Chromatogr.* **18**, 3825–3832.
30. Roach, M.; Gozel, P.; Zare, R.N. (1988) Determination of methotrexate and its major metabolite, 7-hydroxymethotrexate using capillary zone electrophoresis and laser in-

duced fluorescence detection. *J. Chromatogr.* **426**, 129–140.
31. Garcia, L.L.; Shihabi, Z.K. (1993) Suramin determination by capillary electrophoresis. *J. Liq. Chromatogr.* **16**, 2049–2057.
32. Lloyd, D.K.; Cypess, A.M.; Wainer, I.W. (1991) Determination of cytosine–D-arabinoside in plasma using capillary electrophoresis. *J. Chromatogr.* **568**, 117–124.
33. Baillet, A.; Pianetti, G.A.; Taverna, M.; Mahuzier, G.; Baylocq-Ferrier, D. (1993) Fosfomycin determination in serum by capillary zone electrophoresis with indirect ultraviolet detection. *J. Chromatogr.* **616**, 311–318.
34. Bogan, D.P.; Deasy, B.; O'Kennedy, R.; Smyth, M.R.; Fuhr, U. (1995) Determination of free and total 7-hydroxycoumarin in urine and serum by capillary electrophoresis. *J. Chromatogr.* **663**, 371–378.
35. Hsu, L.C.; Constable, D.J.; Orvos, D.R.; Hannah, R.E. (1995) Comparison of high-performance liquid chromatography and capillary zone electrophoresis in penciclovir biodegradation kinetic studies. *J. Chromatogr.* **669**, 85–92.
36. Soini, H.; Riekkola, M.-L.; Novotny, M.V. (1994) Mixed polymer network in the direct analysis of pharmaceuticals in urine by capillary electrophoresis. *J. Chromatogr.* **680**, 623–634.
37. Zhang, C.-X.; Sun, Z.-P.; Ling, D.-K.; Zheng, J.-S.; Li, X.-Y. (1993) Determination of 3-methylflavone-8-carboxylic acid, the main metabolite of flavoxate, in human urine by capillary electrophoresis with direct injection. *J. Chromatogr.* **612**, 287–294.
38. Tomlinson, A.J.; Benson, L.M.; Landers, J.P.; Scanlan, G.F.; Fang, J.; Gorrod, J.W.; Naylor, S. (1993) Investigation of metabolism of the neuroleptic drug haloperidol by capillary electrophoresis. *J. Chromatogr.* **652**, 417–426.
39. Prunonsa, J.; Obach, R.; Diez-Cascon, A.; Gouesclou, L. (1992) Comparison of high-performance liquid chromatography and high-performance capillary electrophoresis for the determination of cicletanine in plasma. *J. Chromatogr.* **581**, 219–226.
40. Soini, H.; Novotny, M.V.; Reikkola, M.-L. (1992) Determintion of naproxen in serum by capillary electrophoresis with ultraviolet absorbance and laser-induced fluorescence detection. *J. Microcol. Sep.* **4**, 313–318.
41. Brunner, L.J.; DiPiro, J.T.; Feldman, S. (1993) Serum antipyrine concentrations determined by micellar electrokinetic capillary chromatography. *J. Chromatogr.* **622**, 98–102.
42. Perrett, D.; Ross, G.A. (1995) Rapid determination of drugs in biological biofluids by capillary electrophoresis. Measurement of antipyrine in saliva for pharmacokinetic studies. *J. Chromatogr.* **700**, 179–186.
43. Zarzycki, P.K.; Kowalski, P.; Nowakowska, J.; Lamparczyk, H. (1995) High-performance liquid chromatographic and capillary electrophoretic determination of free nicotinic acid in human plasma and separation of its metabolites by capillary electrophoresis. *J. Chromatogr.* **709**, 203–208.
44. Naylor, S.; Benson, L.M.; Tomlinson, A.J. (1994) Monitoring drug metabolism by capillary electrophoresis. In *Handbook of Electrophoresis*, Landers, J.P. (ed.), CRC Press, Boca Raton, pp. 459–491.
45. Chen, F.-T.A.; Evangelista, R.A. (1994) Feasibility studies for simultaneous immunochemical multianalyte drug assay by capillary electrophoresis with laser-induced fluorescence. *Clin. Chem.* **40**, 1819–22.
46. Soini, H.; Riekkola, M.-L.; Novotny, M.V. (1992) Chiral separations of basic drug and quantitation of bupivocaine enantiomers in serum by capillary electrophoresis with modified cyclodextrin buffers. *J. Chromatogr.* **608**, 265–274.
47. Aumatell, A.; Wells, R.J. (1993) Chiral differentiation of the optical isomers of racemethorphan and recemorphan in urine by capillary zone electrophoresis. *J. Chromatogr. Sci.* **31**, 502–508.
48. Gareil, P.; Gramond, J.P.; Guyon, F. (1993) Separation and determination of warfarin enantiomers in human plasma samples by capillary zone electrophoresis using a methylated β-cyclodextrin-containing electrolyte. *J. Chromatogr.* **615**, 317–325.
49. Prunonsa, J.; Obach, R.; Diez-Cascon, A.; Gouesclou, L. (1992) Determination of cicletanine enantiomers in plasma by high-performance capillary electrophoresis. *J. Chromatogr.* **574**, 127–133.
50. Shihabi, Z.K.; Hinsdale, M.E. (1995) Some variables affecting reproducibility in capillary electrophoresis. *Electrophoresis* **16**, 2159–2163.

51 Dose, E.V.; Guichon, G.A. (1991) Internal standardization technique for capillary zone electrophoresis. *Anal. Chem.* **63**, 1154–1158.
52 Siren, H.; Jumppanen, J.H.; Manninen, K.; Reikkola, M.-L. (1994) Introduction of migration indices for identification: chiral separation of some β-blockers by using cyclodextrins in micellar electrokinetic capillary chromatography. *Electrophoresis* **15**, 779–784.
53 Jumppanen, J.H.; Reikkola, M.-L. (1995) Marker techniques for high-accuracy identification in CZE. *Anal. Chem.* **67**, 1060–1066.
54 Hoyt, A.M. (1993) Quantitative analysis with capillary electrophoresis. In *Capillary Electrophoresis Technology*, Guzman, N.A. (ed.), Marcel Dekker, New York, pp. 705–714.
55 Watzig, H. (1995) Appropriate calibration functions for capillary electrophoresis. I. Precision and sensitivity using peak areas and heights. *J. Chromatogr.* **700**, 1–7.
56 Watzig, H.; Dette, C. (1993) Precise quantitative capillary electrophoresis. Methodological and instrumental aspects. *J. Chromatogr.* **636**, 31–38.
57 Smith, S.C.; Strasters, J.K.; Khaledi, M.G. (1991) Influence of operating parameters on reproducibility in capillary electrophoresis. *J. Chromatogr.* **559**, 57–68.
58 Northrop, D.M. (1993) The utility of capillary electrophoresis in forensic science. In *Capillary Electrophoresis Technology*, Guzman, N.A. (ed.), Marcel Dekker, New York, pp 673–691.
59 Northrop, D.M.; McCord, B.R.; Butler, J.M. (1994) Forensic applications of capillary electrophoresis. *J. Capillary Electrophoresis.* **1**, 158–188.
60 Chee, G.L.; Wan T.S. (1993) Reproducible and high-speed separation of basic drugs by capillary zone electrophoresis. *J. Chromatogr.* **612**, 172–177.
61 Schafroth, M.; Thormann, W.; Allemann, D. (1994) Micellar elektrokinetic capillary chromatography of benzodiazepines in human urine. *Electrophoresis* **15**, 72–78.
62 Caslavska, J.; Lienhard, S.; Thormann, W. (1993) Comparative use of three electrokinetic capillary methods for the determination of drugs in body fluids. Prospects for rapid determination of intoxications. *J. Chromatogr.* **638**, 335–342.
63 Wernly, P.; Thormann, W. (1991) Analysis of illicit drugs in human urine by micellar electrokinetic capillary chromatography with on-column fast scanning polychrome absorption detection. *Anal. Chem.* **63**, 2878–2882.
64 Wernly, P.; Thormann, W. (1992) Confirmation testing of 11-nor-delta 9-tetrahydrocannabinol-9-carboxylic acid in urine with micellar electrokinetic capillary chromatography. *J. Chromatogr.* **608**, 251–256.
65 Thormann, W.; Molteni, S.; Caslavska, J.; Schmutz, A. (1994) Clinical and forensic application of capillary electrophoresis. *Electrophoresis* **15**, 3–12.
66 Tagliaro, F.; Poiesi, C.; Aiello, R.; Dorizzi, R.; Ghielmi, S.; Marigo, M. (1993) Capillary electrophoresis for the investigation of illicit drugs in hair: determination of cocaine and morphine. *J. Chromatogr.* **638**, 303–309.
67 Tagliaro, F.; Smyth, W.F.; Turrina, S.; Deyl, Z.; Marigo, M. (1995) Capillary electrophoresis: a new tool in forensic toxicology. Application and prospects in hair analysis for illicit drugs. *Forensic Sci. Int.* **70**, 93–104.
68 Jelinkova, D.; Deyl, Z.; Miksik, I.; Tagliaro, F. (1995) Capillary electrophoresis of hair proteins modified by alcohol intake in laboratory rats. *J. Chromatogr.* **709**, 111-119.
69 Lee, K.-J.; Lee, J.J.; Moon, D.C. (1994) Application of micellar elektrokinetic capillary chromatography for monitoring hippuric and methylhippuric acid in human urine. *Electrophoresis* **15**, 98–102.
70 Leung, G.N.; Tang, H.P.O.; Twinnie, S.C.; Wan, T.S. (1995) Separation of basic drugs with non-aqueous capillary electrophoresis. *J. Chromatogr. A* **738**, 141–154.
71 Jacobson, S.C.; Ramsey, J.M. (1995) Microchip electrophoresis with sample stacking. *Electrophoresis* **16**, 481–486.
72 Shihabi, Z.; Hinsdale, M.E. (in press) Analysis of ibuprofen in serum by capillary electrophoresis. *J. Chromatogr. B*.

Part Four

Bioscience Applications

28 Biomedical applications and biological systems
P.L. WEBER

28.1 Introduction

The potential of capillary electrophoresis (CE) in the analysis of the many compounds of biological interest can been deduced from a quick scan of the variety of topics covered in other chapters of this book. These chapters have dealt with analysis of drugs and endogenous compounds present as either standard aqueous solutions or in biological fluids routinely used for clinical analysis, such as serum, plasma, urine or cerebrospinal fluid. In this chapter the review will focus on those reports that involve tissue analysis, particularly of the brain. Also reviewed will be the use of CE in the analysis of single cells in which the human erythrocyte seems to be the most popular cell studied. Table 28.1 summarizes the conditions used, compounds analyzed and gives a sample matrix for cited applications. Also reviewed are studies on the use of microdialysis and related methods for obtaining samples for CE analysis. The literature covered includes only those papers published in 1992 or later.

28.2 Tissue analysis

The analysis of tissue samples by CE appears attractive because of the small sample volume (nanoliters or less) requirement of CE coupled with its high efficiency. Samples from discrete locations within a particular tissue may be obtained. Likewise, CE is conducive to the low sample volumes (a few microliters) encountered in microdialysates of tissues such as the brain.

28.2.1 Brain tissue

The brain is a tissue particularly well suited to CE since many of the more interesting analytes in this tissue are small, charged molecules separable by capillary zone electrophoresis (CZE). If possible, the direct analysis of underivatized samples is preferred due to ease of analysis and problems inherent with any derivatization procedure. Employing electrochemical detection, Malone *et al.* [1] determined the concentrations of tryptophan and kynurenine in the extracellular fluid of the rat brain. Samples were obtained by microdialysis after tryptophan loading. Standard solutions of

BIOMEDICAL APPLICATIONS

Table 28.1 Summary of biomedical applications* of capillary electrophoresis

Separation conditions, voltage applied, capillary length and i.d., detection method	Compounds analyzed	Sample matrix	Ref.
20 mM Borate, pH 9 20 kV, 80 cm × 50 μm i.d., electrochemical detection	Kynurenine and tryptophan metabolites	Rat brain microdialysate	1
20 mM Carbonate, pH 9.7, 18 kV, 105 cm × 50 μm i.d., electrochemical (Cu) detection	Trp, Ala, Gly, Thr, Ser, Glu, Asp	Rat brain microdialysate	2
50 mM Borate–phosphate, pH 9.3 with 16.5 mM α-CD, 30 kV, 60 cm × 50 μm i.d., UV 185, 214 nm	Gangliosides	Rat brain homogenate, deer antler, apricot seed	3
20 mM Borate, pH 9.0, 30 kV, 115 cm × 50 μm i.d., visible 420 nm	Amino acids, including GABA and taurine	Rat brain homogenate	4
10 mM Borate, pH 9.25, 425 V cm^{-1}, electrochemical detection (ECD)	Glu, Asp	Rat brain homogenate	5
20 mM Borate, pH 9.0, 30 kV, 100 cm × 50 μm i.d., electrochemical detection	GABA, Ala, Glu, Asp	Rat brain microdialysate	6
100 mM Tris, pH 8.65, 25 kV, 27 cm (14 cm to detector) × 25 μm i.d., fluorescence	Asp, Glu	Rat brain microdialysate	7
110 mM Phosphate, pH 7.0, 15 kV, 72 cm (52.5 cm to detector) × 50 μm i.d., fluorescence	Noradrenaline, dopamine	Rat brain microdialysate	8
20 mM Carbonate, pH 9.5, 21 (or 30) kV, 30 cm (20 cm to detector) × 12 (or 15) μm i.d., fluorescence	Glu, Asp	Rat brain microdialysate	9, 10

Table 28.1 *Continued*

Separation conditions, voltage applied, capillary length and i.d., detection method	Compounds analyzed	Sample matrix	Ref.
†30 mM Carbonate with 20 mM SDS, pH 9.5, 30 kV, 47 cm × 50 μm i.d., fluorescence	Glu	Striatum of rat brain microdialysate	11
50 mM Borate, pH 8.3, with 25 mM LiCl, 25 kV, 125 cm (100 cm to detector) × 75 μm i.d., fluorescence	Luteinizing hormone-releasing hormone, β-endorphine, neuropeptide Y	Ewe median eminence microdialysate	12
20 mM Citrate, pH 2.5, 30°C, 25 kV, 72 cm × 50 μm i.d., UV 200 nm †10 mM Phosphate with 20 mM SDS, pH 7.0, 35°C, 25 kV, 72 cm × 50 μm i.d., UV 200 nm	Vasoactive intestinal peptide	Rat brain homogenate	13
100 mM Borate, pH 8.5, 15 kV, 57 cm (50 cm to detector) × 75 μm i.d., UV 214 nm	Immunocomplex of scrapie prion protein and antibody	Sheep brain stem homogenate	14
10 mM Sulfate with 2.5% electroosmotic modifier, 15 kV, 60 cm × 100 μm i.d., UV 214 nm	Nitrate, nitrite	Rat brain cortex homogenate	15
100 mM Phosphate, pH 7.0, 27 kV, 75 cm (60 cm to detector) × 75 μm i.d., column coated with polyethylene glycol polyether, UV 200 nm	Citokines, including interleukins-1,6,8; γ-interferon; TNF; MCP-1; RANTES	Renal glomerular and interstitial tissue extracts	16
20 mM Citrate, pH 2.5, 4.5 or 5.5, 25 kV, 72 cm × 50 μm i.d., UV 200 nm	Neurotensin	Rat duodenum or adrenal glands	17
100 mM Phosphoric acid, pH 1.95, 10 kV, 60 cm × 75 μm i.d., UV 185 nm	Angiotensin I, II and III	Rat lung homogenate	18

Table 28.1 *Continued*

Separation conditions, voltage applied, capillary length and i.d., detection method	Compounds analyzed	Sample matrix	Ref.
10 mM Tris–HCl, pH 9.1, 30 kV, 100 cm × 75 μm i.d., UV 185 or 214 nm	Metallothionein isoforms	Rat liver	19
20 mM Phosphate, pH 8.8, −20 kV, 100 cm (50 cm to detector) × 10 μm i.d., direct fluorescence (0.5 mM 6-aminoquinoline and 0.2 mM cetyltrimethyl-ammonium bromide, pH 3.8 used for indirect fluorescence)	Glutathione, Na, K	Single human erythrocyte	22
50 mM Borate, pH 9.1, −29 kV, 110 cm (64 cm to detector) × 20 μm i.d., fluorescence	Carbonic anhydrase, hemoglobin A_0, methemoglobin	Single human erythrocyte	23
1% Glucose, 0.1 mM fluorescein, 0.5 mM Tris, pH 8.5, 25 kV, 70 cm (55 cm to detector) × 14 μm i.d., indirect fluorescence	Lactate, pyruvate	Single human erythrocyte	24
5 mM Lactate, 5 mM NAD+, 30 mM phosphate, pH 7.3, 30 kV, 65 cm (50 cm to detector) × 20 μm i.d., fluorescence	Lactate dehydrogenase isoenzymes activity	Single human erythrocyte	25
50 mM Borate with 1.3×10^{12} antibody-coated particles l⁻¹, pH 9.2, 20 kV, 65 cm (40 cm to detector) × 20 μm i.d., light scattering (particle-counting immunoassay)	Glucose-6-phosphate dehydrogenase amount	Single human erythrocyte	26
0.67 mM G-6-P, 0.5 mM NADP+, 35 mM triethanolamine with 2.5×10^{12} particles l⁻¹, pH 8.4, 30 kv, 75 cm (50 cm to	Glucose-6-phosphate amount and activity	Single human erythrocyte	27

Table 28.1 *Continued*

Separation conditions, voltage applied, capillary length and i.d., detection method	Compounds analyzed	Sample matrix	Ref.
detector) × 20 μm i.d., light scattering and fluorescence			
20 mM Tricine, pH 8.0, 29 kV, 30 cm (15 cm to detector) × 50 μm i.d., column coated with linear polyacrylamide, fluorescence	Rat insulin	Single cell and secretion – islets of Langerhans	28
100 mM Citrate, pH 2.3, 30 kV, 65 cm (45 cm to detector) × 16 μm i.d., fluorescence	Epinephrine, norepinephrine	Single bovine adrenal medullary cell	29
20 mM Phosphate, pH 6.8, 25 kV, 80 cm (60 cm to detector) × 50 μm i.d., UV	Primary amines	Single bovine adrenal medullary cell	30
25 mM MES, pH 5.65, 25 kV, 75 cm × 25 μm i.d., electrochemical detection	Dopamine	Single nerve cell of *Planoris corneus*	32, 33
100 mM Borate, pH 9.5, 30 kV, (100 cm (85 cm to detector) × 17 μm i.d., fluorescence	Dopamine, Ala, Gly, Glu, Asp, taurine	Single rat pheochromocytoma cells	34
Borate, pH 9.2 with 20 mM SDS, 20 kV, 67 cm × 50 μm i.d., fluorescence	Egg laying hormone peptide	Single cell of *Aplysia californica*	35
50 mM Phosphate, pH 5.0, 20 kV, 100 cm × 75 μm i.d., radioactivity	^{35}S-Methionine labeled components	Single cell of *Aplysia californica*	36

MES, 2-morpholinoethanesulfonic acid.
All methods involve the use of CZE except those denoted with (†) which are MECC.

other important tryptophan metabolites, such as kynurenic acid, were separated. Zhou and Lunte [2] used an electrochemical detector with a copper microelectrode to quantitate the concentrations of seven amino acids, including the neurotransmitter glutamate, in a rat brain microdialyzate. Yoo et al. [3] used CE with a cyclodextrin (CD) additive to analyze an extract from a rat brain homogenate and successfully separated two gangliosides. Additionally, gangliosides from apricot seed and deer antler were identified by the method. Often, brain tissue samples are derivatized prior to CE analysis primarily to increase sensitivity. An additional benefit is an increase in selectivity since the derivatizing reagent selectively labels particular classes of compounds. One of the more popular reagents is naphthalene-2,3-dicarboxaldehyde (NDA) mixed with cyanide (CN). Among the many desirable features of the reagent are the UV, fluorescent and electrochemical properties of the derivatives. Weber et al. [4], utilizing an internal standard, quantitated the important neurochemicals GABA (γ-aminobutyric acid), taurine and aspartate as well as five other amino acids in an NDA-derivatized rat brain homogenate. Also, separation of 15 amino acids and amine standards including norepinephrine and dopamine was accomplished with UV detection. O'Shea et al. [5] obtained a dramatic increase in sensitivity for NDA-derivatized glutamate and aspartate in rat brain samples by using CE with electrochemical detection (ECD). O'Shea et al. [6] applied this method to the analysis of rat cortex microdialysis samples to monitor the levels of glutamate, aspartate and alanine before and after an influx of high concentrations of potassium ion. Noteworthy is the fact that peaks in the microdialyzate were positively identified by voltammetric characterization – the electrochemical equivalent of identifying a compound by its UV spectrum. Zhou et al. [7] performed continuous *in vivo* on-line monitoring of aspartate and glutamate in the rat hippocampus microdialyzate of a freely moving animal. In this system the sample is never manipulated off-line as the microdialyzate was derivatized on-line with NDA and directly analyzed by CE with fluorescence detection. The effect of a high potassium influx was studied by sampling the microdialysate at 2 min intervals. NDA derivatization of analytes in microdialysates of rat cortex samples was also performed by Robert et al. [8]. CE with fluorescence detection of basal samples and those obtained after desipramine and idazoxan treatment indicated changes in levels of noradrenaline and dopamine. Hernandez et al. [9] also analyzed microdialysate samples, each collected throughout the course of 20 min, from the corpus striatum of a freely moving rat. Samples derivatized with fluorescein isothiocyanate (FITC) off-line and analyzed by CE with fluorescence detection indicated a 50% decrease in glutamate levels after the administration of haloperidol. In subsequent work [10], they used a more sensitive fluorescence detector arrangement to enable separation and detection of NDA-derivatized glutamate and aspartate at levels an order of magnitude lower than in their previous work. Dawson et al. [11] used a similar method to analyze changes in NDA-derivatized glutamate in striatal microdialysates in order to help establish the method as a general procedure for the measurement of glutamate in brain samples. Infusion of a calcium-free perfusate resulted in a 60% decrease in glutamate while tetrodotoxin treatment gave a 90% reduction, thus establishing the neuronal basis for the observed changes in glutamate concentration.

Peptides and protein of the brain have also been analyzed by CE. Advis and Guzman [12] reported the simultaneous determination of three neuropeptides present in microdialyzates obtained from ewe median eminence. Samples collected at 10 min intervals were derivatized with fluorescamine and subjected to CE with fluorescence detection. Detection of the neuropeptides β-endorphine and neuropeptide Y, present in concentrations of $10\,\text{pg}\,\text{l}^{-1}$ or greater, was good while that of luteinizing hormone-releasing hormone, present in concentrations of $1.0\,\text{pg}\,\text{l}^{-1}$ or less, was marginal. The method represents a significant advantage over radioimmunoassay (RIA) methods which are restricted to the determination of one analyte at a time. Soucheleau and Denoroy [13] analyzed rat brain for vasoactive intestinal peptide using both CZE and micellar electrokinetic capillary chromatography (MECC) with UV detection. Homogenate from the cortex, hippocampus and cerebellum were subjected to solid phase extraction (SPE) and high performance liquid chromatography (HPLC) purification prior to CE analysis. Quantitation gave results similar to literature values obtained by RIA, but the CE method, though less sensitive due to the UV detection limit, indicated a higher molecular specificity. Schmerr et al. [14] studied methods for detecting scrapie prion protein, a marker found in the infected brains of sheep or goats suffering from scrapie. The protein in brain samples was solubilized and reacted with an antibody specific for this protein. This immunocomplex gave a detectable peak on CZE with UV detection. Further treatment of the complex with a second fluorescent antibody, specific to the first, afforded enhanced detection when subjected to CE with fluorescence detection. Finally, Meulemans and Delsenne [15] described a simple method using CE with direct detection at 214 nm for the determination of sub-$\mu\text{g}\,\text{ml}^{-1}$ levels of nitrite and nitrate in a homogenate of rat brain cortex.

28.2.2 *Other tissues*

Analysis of various tissues for enzymatic activity is another use of CE that is reviewed elsewhere in this book. Studies of tissues other than brain for non-enzymic substances are less extensive. Phillips and Kimmel [16] studied the pathobiology of renal disease in HIV-positive patients by CE analysis of tissue-bound citokines in frozen biopsy samples. Separation in polyethylene glycol (PEG)-coated capillaries with UV detection was followed by fraction collection of peaks and chemiluminescence-enhanced enzyme-linked immunoassay (CHEM–ELISA) of the fractions. Results from UV detection correlated well with that from CHEM–ELISA indicating the method is suitable for routine screening purposes. Denoroy [17] used CZE with UV detection to identify a peak corresponding to neurotensin in homogenates from rat duodenum and adrenal glands using pure synthetic neurotensin. Homogenates had been subjected to SPE and HPLC prior to electrophoresis. Positive identification was provided by employing different pH values for the running buffer which yielded the same electrophoretic mobility. Lim and Sim [18] separated angiotensins present in a rat lung homogenate following purification by dialysis using CE with UV detection. Metallothionein isoforms from rabbit liver were separated by CE by Beattie et al. [19]. A two-solvent extraction system was utilized. Subsequent studies by Richards [20] indicated that a polyamine-coated column might prove useful for these separations.

28.3 Single cell analysis

CE is particularly well suited to the analysis of single cells in various modes. In whole cell analysis, the inner diameter of the typical CE column (5 to 100 μm) is the same order as that of many cells such as the human erythrocyte with a diameter of 7 μm. This allows the entire single cell to be positioned in the end of capillary and lysed, the lysate reacted with derivatizing agents if necessary and the mixture electrophoresed. Likewise, CE utilizes small sample volumes which are compatible with the volumes of cells such as the erythrocyte with a volume of 90 fl. For extremely large cells such as snail neurons with a volume of approximately 1 nl, the cytoplasm of the cell can be sampled by using microinjectors capable of sampling volumes of about 100 pl. In order to show that CE analysis of single cells yields accurate results, in many studies the average of multiple single analyses is compared to results obtained on larger samples by more traditional methods.

28.3.1 Erythrocyte

An excellent review of the analysis of single human erythrocytes by CE has been written by the dominant researcher in the field, Ed Yeung [21]. Summations of the amounts of various chemical species per cell are given along with a discussion on the various forms of fluorescence detection method utilized. Hogan and Yeung [22] used CE with fluorescence detection to analyze single human erythrocytes for the reduced form of glutathione in cells treated with oxidizing or reducing agents or in untreated cells. The method used for chemical derivatization showed total glutathione levels comparable to macroscopic values. Also, an indirect detection method gave rough estimations of Li, Na and K levels. Lee and Yeung [23] studied 29 individual cells for carbonic anhydrase and hemoglobin A levels by monitoring the native fluorescence of peaks separated by CE. Even though all cells had similar volumes, they noted cell-to-cell variations over one order of magnitude for hemoglobin levels and a direct intracellular correlation between the two proteins. Xue and Yeung [24] determined pyruvate and lactate levels in 27 cells using indirect fluorescence detection in a buffer containing fluorescein. Large variations in individual levels were noted but averages for all cells were close to literature values. They also analyzed 36 cells for the enzymatic activity of lactate dehydrogenase isozymes [25]. In this method, the enzymes are separated into different zones in the CE column where they are allowed to catalyze the reaction of lactate and NAD^+ (nicotinamide adenine dinucleotide) in the buffer for a few minutes while the high voltage is turned off. The NADH formed appears as a peak by fluorescence detection when electrophoresis is resumed. Rosenzweig and Yeung [26] used CE to quantitate glucose-6-phosphate dehydrogenase (G-6-PDH) in an erythrocyte. The CE detection system monitored the light scattering of agglutinated particles containing enzyme–antibody complexes. The average amount for 25 cells is 34 zmol per cell. Tan and Yeung [27] used CE to determine simultaneously both the activity and the concentration of G-6-PDH in single erythrocytes. The immunoassay detection technique in [26] and the enzyme assay method in [25] were applied to a single CE separation. Analysis of 30 cells showed that, on the average, 35% of the enzyme was active, with individual cells ranging from 6% to 80%. In all of these studies conducted by Yeung's group [21–27], large intercellular variations in analyte levels are observed and are due not only to differences in individual cell volumes but also other factors such as cell age.

28.3.2 Other cells

Researchers have studied single cells other than erythrocytes. Schultz et al. [28] have used CE to analyze single islets of Langerhans for insulin content and insulin secretion. In this immunoassay procedure, FITC-labeled insulin binding to specific antibody fragments, is released in proportion to the level of insulin present and is quantitated by CE with laser-induced fluorescence detection. Insulin levels assayed were in close agreement with previously reported values. Chang and Yeung [29] determined the amounts of epinephrine (E) and norepinephrine (NE) in individual adrenal medullary cells by CE with native fluorescence detection. The ratio of E to NE was constant for each gland analyzed and agreed well with literature values. Single cells of the same type were also analysed by Hoyt *et al.* [30]. On-line preconcentration of the NDA-derivatized cellular components was followed by CE to give a profile of the primary amine contents.

Larger single cells, such as invertebrate neurons with volumes 3–4 orders larger than erythrocytes permit studies which sample only part of the cytoplasm using microinjectors. Ewing [31] describes this method as well as methods used for whole cell analysis in a review. Kristensen *et al.* [32] used a microinjection procedure followed by CE with ECD to sample the subcellular compartments of a single nerve cell of *Planorbis corneus*. Quantitation of dopamine in one compartment consisting of the cytoplasm easily released transmitter vesicles compared to another compartment consisting of non-functional storage vesicles indicated a ratio of three parts of the former to one of the latter. Sulzer *et al.* [33] used this system to study the effect of amphetamine on compartmental release and concluded that amphetamine acts at the synaptic vesicles rather than at the plasma membrane to promote reverse transport and dopamine release. Gilman and Ewing [34] determined the concentrations of dopamine and the amino acids alanine, taurine, glycine, glutamate and aspartate in single rat pheochromocytoma cells using CE with fluorescence detection. The use of on-column derivatization with NDA and a narrow-bore capillary with an etched tip aided in producing successful results. The average dopamine concentration of six cells was almost identical to previously reported work. Large cells from *Aplysia californica* were analyzed for trace level peptides labeled with fluorescamine by Shippy *et al.* [35]. A CE with a specially designed fluorescence detector enabled detection of an egg-laying hormone at low nM levels. Also, unique peak patterns of the electropherograms were used to characterize three different cell types (the metacerebral, B1 and bag cell). Jankowski *et al.* [36] analyzed these B1 cells for components labeled with ^{35}S-methionine present in the cell culture medium. The components were collected off-column by a peptide-binding membrane and detected with a bioimaging analyzer.

28.4 Sampling of tissues by microdialysis

The on-line *in vivo* microdialysis method was first introduced in 1988 for rat brain catecholamine analysis using HPLC combined with ECD. Subsequently, microdialyzate samples were analyzed by other separation techniques, including CE. A limited number of CE studies have appeared which focus on the use of microdialysis and related techniques to sample biological tissues such as the brain. Microdialysis is particularly compatable with

CE since it provides a clean sample ready for analysis. It should be noted that microdialysis operates at a typical perfusion rate of 1 µl min^{-1} and, thus, requires the analysis of small volumes for a high sampling rate. Again, this is well suited to CE with sample volumes of a few nanoliters or less, as opposed to a few microliters in HPLC. In most applications, as exemplified in Roussin et al. [37], microdialysate samples are collected off-line and then analyzed by CE. Tellez et al. [38] used CE with UV detection to monitor levels of injected phenobarbital in microdialysates obtained simultaneously from the jugular vein and brain of rats.

Due to the problems associated with the manipulation of small volume samples off-line, on-line interfaces between microdialysis and CE are being developed. Lada et al. [39] report the use of a microdialysis system interfaced to a CE with UV detection for the analysis of ascorbic acid in the caudate nucleus of rat brain. Very low perfusion rates of 79 and 155 nl min^{-1} were used to achieve the high recovery rates necessary for UV detection. Hogan et al. [40] demonstrated the use of on-line interface for microdialysis operating at a more typical perfusion rate of 1 µl min^{-1}. CE with sensitive fluorescence detection enabled separation in under 60 s of antineoplastic SR 4233 and its main metabolite SR 4317. Zhou et al. [7] extended this work by modifying the system so that chemical derivatization of the microdialysate could be accomplished on-line before CE analysis. Thus, endogenous compounds not natively fluorescent can be made so, providing a sensitive method for the analysis of a variety of compounds in the microdialyzate.

In a related sampling technique, ultrafiltration as described by Linhares and Kissinger [41] can also be used to assay tissue samples for CE analysis, provided there is sufficient fluid turnover. Though flow rates are more difficult to control and on-line implementation may be difficult, the high recovery rates (over 90% vs. 5–30% for microdialysis) make this an attractive sampling method for some applications. Linhares and Kissinger [42] used ultrafiltration to sample subcutaneous tissue for administered theophylline. MEKC of the samples exhibited good reproducibility and results comparable to HPLC analysis.

28.5 Conclusion

In summary, CE analysis of tissue samples, espcially of those from the brain, is proving to be an active area of research. The suitability of microdialysis to CE is especially attractive and, with the recent advances in on-line interfacing of the two, interest from the neuroscience community is likely to increase. Future reports are likely to extend the application of these methods to a much wider variety of analytes in the brain as well as in other tissues. Likewise, the reported advances in single cell analysis by CE allow new questions to be raised at the cellular level. Extended studies on cells other than the erythrocyte will probably be made and add to fundamental knowledge of individual cellular behavior.

References

1. Malone, M.M.; Zuo, H.; Lunte, S.M.; Smyth, M.R. (1995) Determination of tryptophan and kynurenine in brain microdialysis samples by capillary electrophoresis with electrochemical detection. *J. Chromatogr.* **700**, 73–80.
2. Zhou, J.; Lunte, S.M. (1995) Direct determination of amino acids by capillary electrophoresis/electrochemistry using a copper microelectrode and zwitterionic buffers. *Electrophoresis* **16**, 498–503.
3. Yoo, Y.S.; Kim, Y.S.; Jhon, G.; Park, J. (1993) Separation of gangliosides using cyclodextrin in capillary zone electrophoresis. *J. Chromatogr.* **652**, 431–439.
4. Weber, P.L.; O'Shea, T.J.; Lunte, S.M. (1994) Separation and determination of the amino acid neurotransmitters in rat brain by capillary electrophoresis. *J. Pharm. Biomed. Anal.* **12**, 319–324.
5. O'Shea, T.J.; Greenhagen, R.D.; Lunte, S.M.; Lunte, C.E.; Smith, M.R.; Radzik, D.M.; Watanabe, N.J. (1992) Capillary electrophoresis with electrochemical detection employing an on-column Nafion joint. *J. Chromatogr.* **593**, 305–312.
6. O'Shea, T.J.; Weber, P.L.; Brammel, B.P.; Lunte, C.E.; Lunte, S.M. (1992) Monitoring excitatory amino acid release *in vivo* by microdialysis with capillary electrophoresis–electrochemistry. *J. Chromatogr.* **608**, 189–195.
7. Zhou, S.Y.; Zuo, H.; Stobaugh, J.F.; Lunte, C.E.; Lunte, S.M. (1995) Continuous *in vivo* monitoring of amino acid neurotransmitters by microdialysis sampling with on-line derivatization and capillary electrophoresis separation. *Anal. Chem.* **34**, 594–599.
8. Robert, F.; Bert, L.; Denoroy, L.; Renaud, B. (1995) Capillary zone electrophoresis with laser-induced fluorescence detection for the determination of nanomolar concentrations of noradrenaline and dopamine: application to brain microdialysate analysis. *Anal. Chem.* **67**, 1838–1844.
9. Hernandez, L.; Tucci, S.; Guzman, N.; Paez, X. (1993) *In vivo* monitoring of glutamate in the brain by microdialysis and capillary electrophoresis with laser-induced fluorescence detection. *J. Chromatogr.* **652**, 393–398.
10. Hernandez, L.; Joshi, N.; Murzi, E.; Verdeguer, P.; Mifsud, J.C.; Guzman, N. (1993) Collinear laser-induced fluorescence detector for capillary electrophoresis: Analysis of glutamic acid in brain microdialysates. *J. Chromatogr.* **652**, 399–405.
11. Dawson, L.A.; Stow, J.M.; Douriash, C.T.; Routledge, C. (1995) Analysis of glutamate in striatal microdialysates by capillary electrophoresis and laser-induced fluorescence detection. *J. Chromatogr.* **700**, 81–87.
12. Advis, J.P.; Guzman, N.A. (1993) Capillary electrophoresis coupled to fluorescence detection for the determination of *in vivo* release of multiple neuropeptides from ewe median eminence. *J. Liq. Chromatogr.* **16**, 2129–2148.
13. Soucheleau, J.; Denoroy, L. (1992) Determination of vasoactive intestinal peptide in rat brain by high-performance capillary electrophoresis. *J. Chromatogr.* **608**, 181–188.
14. Schmerr, M.J.; Goodwin, K.R.; Cutlip, R.C. (1994) Capillary electrophoresis of the scrapie prion protein from sheep brain. *J. Chromatogr.* **680**, 447–453.
15. Meulemans, A.; Delsenne, F. (1994) Measurement of nitrite and nitrate levels in biological samples by capillary electrophoresis. *J. Chromatogr.* **660**, 401–404.
16. Phillips, T.M.; Kimmel, P.L. (1994) High-performance capillary electrophoretic analysis of inflammatory citokines in human biopsies. *J. Chromatogr.* **656**, 259–266.
17. Denoroy, L. (1994) Detection of neurotensin in tissues by capillary zone electrophoresis. *Electrophoresis* **15**, 46–50.

18. Lim, B.C.; Sim, M.K. (1994) Determination of angiotensins by capillary electrophresis. *J. Chromatogr.* **665**, 127–131.
19. Beattie, J.H.; Richards, M.P.; Self, R. (1993) Separation of metallothionein isoforms by capillary electrophoresis. *J. Chromatogr.* **632**, 127–135.
20. Richards, M.P. (1994) Application of a polyamine-coated capillary to the separation of metallothionein isoforms by capillary zone electrophoresis. *J. Chromatogr.* **657**, 345–355.
21. Yeung, E.S. (1994) Chemical analysis of single human erythrocytes. *Acc. Chem. Res.* **27**, 409–414.
22. Hogan, B.L.; Yeung, E.S. (1992) Determination of intracellular species at the level of a single erythrocyte via capillary electrophoresis with direct and indirect fluorescence detection. *Anal. Chem.* **64**, 2841–2845.
23. Lee, T.T.; Yeung, E.S. (1992) Qualitative determination of native proteins in individual erythrocytes by capillary zone electrophoresis with laser-induced fluorescence detection. *Anal. Chem.* **64**, 3045–3051.
24. Xue, Q.; Yeung, E.S. (1994) Indirect fluorescence determination of lactate and pyruvate in single erythrocytes by capillary electrophoresis. *J. Chromatogr.* **661**, 287–295.
25. Xue, Q.; Yeung, E.S. (1994) Variability in intracellular lactate dehydrogenase isozymes in single human erythrocytes. *Anal. Chem.* **66**, 1175–1178.
26. Rosenzweig, Z.; Yeung, E.S. (1994) Laser-based particle-counting microimmunoassay for the analysis of single human erythrocytes. *Anal. Chem.* **66**, 1771–1776.
27. Tan, W.; Yeung, E.S. (1995) Simultaneous determination of enzyme activity and enzyme quantity in single human erythrocytes. *Anal. Biochem.* **226**, 74–79.
28. Schultz, N.M.; Huang, L.; Kennedy, R.T. (1995) Capillary electrophoresis-based immunoassay to determine insulin content and insulin secretion from single islets of Langerhans. *Anal. Chem.* **67**, 924–929.
29. Chang, H.; Yeung, E.S. (1995) Determination of catecholamines in single adrenal medullary cells by capillary electrophoresis and laser-induced native fluorescence. *Anal. Chem.* **67**, 1079–1083.
30. Hoyt, A.M.; Beale, S.C.; Larmann, J.P.; Jorgenson, J.W. (1993) Preparation and evaluation of an on-line preconcentrator for capillary electrophoresis. *J. Microcol. Sep.* **5**, 325–330.
31. Ewing, A.G. (1993) Microcolumn separations of single nerve cell components. *J. Neurosci. Methods* **48**, 215–224.
32. Kristensen, H.K.; Lau, Y.Y.; Ewing, A.G. (1994) Capillary electrophoresis of single cells: observation of two compartments of neurotransmitter vesicles. *J. Neurosci. Methods* **51**, 183–188.
33. Sulzer, D.; Chen, T.; Lau, Y.Y.; Kristensen, H.; Rayport, S.; Ewing, A. (1995) Amphetamine redistributes dopamine from synaptic vesicles to the cytosol and promotes reverse transport. *J. Neurosci.* **15**, 4102–4108.
34. Gilman, S.D.; Ewing, A.G. (1995) Analysis of single cells by capillary electrophoresis with on-column derivatization and laser-induced fluorescence detection. *Anal. Chem.* **67**, 58–64.
35. Shippy, S.A.; Jankowski, J.A.; Sweedler, J.A. (1995) Analysis of trace level peptides using capillary electrophoresis with UV laser-induced fluorescence. *Anal. Chim. Acta* **307**, 163–171.
36. Jankowski, J.A.; Tracht, S.; Sweedler, J.V. (1995) Assaying single cells with capillary electrophoresis. *Trends Anal. Chem.* **14**, 170–176.
37. Roussin, A.; Verdeguer, P.; Hernandez, L. (1993) Combining microdialysis with capillary electrophoresis for in vivo analysis. *Analusis* **21**, m42–m45.
38. Tellez, S.; Forges, N.; Roussin, A.; Hernandez, L. (1992) Coupling of microdialysis with capillary electrophoresis: a new approach to the study of drug transfer between two compartments of the body in freely moving rats. *J. Chromatogr.* **581**, 257–266.
39. Lada, M.W.; Schaller, G.; Carriger, M.H.; Vickroy, T.W.; Kennedy, R.T. (1995) Online interface between microdialysis and capillary zone electrophoresis. *Anal. Chim. Acta* **307**, 217–225.
40. Hogan, B.L.; Lunte, S.M.; Stobaugh, J.F.; Lunte, C.E. (1994) On-line coupling of in vivo microdialysis sampling with capillary electrophoresis. *Anal. Chem.* **66**, 596–602.

41 Linhares, M.C.; Kissinger, P.T. (1993) Capillary ultrafiltration: in vivo sampling probes for small molecules. *Anal. Chem.* **64**, 2831–2835.
42 Linhares, M.C.; Kissinger, P.T. (1993) Pharmokinetic studies using micellar electrokinetic capillary chromatography with in vivo capillary ultrafiltration probes. *J. Chromatogr.* **615**, 327–333.

29 Analysis of selected common clinical tests by capillary electrophoresis

Z.K. SHIHABI

29.1 Introduction

Capillary electrophoresis (CE) is a new analytical technique in the clinical laboratory for separation and quantification of a wide variety of molecules based not only on charge, but also on size, hydrophobicity and stereospecificity. It offers certain advantages for routine clinical analysis; however, it has not found wide acceptance in the clinical community. One of the main reasons for this underutilization is the fact that clinical chemistry laboratories, particularly in the USA, rely to a great extent on commercial instruments specifically built and, furthermore, approved by the governmental agencies for specific tests. Such commercial instruments are on the horizon for CE. Capillary electrophoresis offers the following advantages in clinical analyses:

- higher theoretical plate number;
- no staining for proteins and peptide analysis;
- rapid analysis time;
- ability to analyze low concentrations of small molecules in the presence of large amounts of proteins;
- lower cost per test;
- full automation, not always restricted to CE assays, but also applicable to those of high performance liquid chromatography (HPLC).

At the same time, CE offers many challenges such as lower sensitivity of detection, problems with avoiding sample matrix interference, a need for better precision, selectivity, reproducibility and higher instrument cost. Later in this review we will discuss how to overcome some of these problems.

CE resembles both agarose electrophoresis (AE) as well as HPLC. Obviously, numerous clinical tests can be adapted to CE; however, some are better suited than others for analysis by this method. In this chapter we will focus on those tests better suited for CE, especially those tests which are more common (routine). In general, proteins, peptides and small molecules in the presence of large amounts of proteins can be analyzed more easily by CE compared to other techniques. We discuss progress in these assays with emphasis on the practical aspects of analysis. Table 29.1 lists some of the

Table 29.1 Analysis of selected common clinical tests: conditions for some model compounds analyzed by capillary electrophoresis

Buffer	Conditions	Capillary	Matrix	Compounds	Ref.
CZE					
• 0.7 g Boric acid, 0.7 sodium carbonate 0.5 g PEG; 100 ml H$_2$O	214 nm, 6–7 kV	32 cm × 50 μm	Serum	Proteins	*
• 0.7 g Boric acid, 0.7 sodium carbonate 0.5 g PEG; 100 ml H$_2$O	214 nm, 6–7 kV	32 cm × 50 μm	Urine	Proteins	*
• 0.7 g Boric acid, 0.7 sodium carbonate 0.5 g PEG; 100 ml H$_2$O	214 nm, 6–7 kV	32 cm × 50 μm	Serum	Cryoglobulins	*
• 0.7 g Boric acid, 0.7 sodium carbonate 0.5 g PEG; 100 ml H$_2$O	214 nm, 6–7 kV	32 cm × 50 μm	CSF	Proteins	*
• Borax buffer 50 mM, pH 11.0	214 nm, 15 kV	20 cm × 25 μm	CSF	Proteins	27
• Tricine, 1.1 mM, pH 8.2	214 nm, 7 kV	30 cm × 30 μm	Blood cells	Hemoglobins	*
• Borate 175 mM, pH 8.4	214 nm, 15 kV	42 cm × 50 μm	Tissue	Cathepsin D	*
• Boric acid 7 g, sodium carbonate 7 g, H$_2$O 1000 ml	280 nm, 11 kV	50 cm × 50 μm	Different	Purine/pyrimidine	63
• Borate, 25 mM, pH 10	214 nm, 20 kV	65 cm × 50 μm	Serum	Phenylalanine	67
CZE (indirect detection)					
• 10 mM Sodium chromate; 0.5 mM tetradecylammonium Br	254 nm, 25 μA	60 cm × 75 μm	Urine	Oxalate	61
MECC					
• 20 mM Borate, pH 9.4	200–300 nm, 20 kV	44 cm × 75 μm	Cord blood	Purines	64

*Unpublished data.

tests which have been analyzed by CE; however, many tests await development and represent a challenge for the clinical chemist as will be pointed out.

29.2 Serum proteins

29.2.1 Clinical significance

The test for serum proteins is performed routinely on a daily basis in most large hospitals. Changes in serum proteins, as evident by electrophoresis, can precede patient symptoms. Thus, in many community hospitals serum proteins are tested on admission as part of a screening profile. Serum proteins can detect several disorders such as low albumin due to renal failure, or increased gamma globulins due to infection; however, the real significance is the detection of monoclonal gammapathies which are difficult to diagnose by physical examination or by other techniques.

29.2.2 Present methods of analysis

Serum proteins are comprised of more than 100 proteins at varying concentrations. They are separated into 5–12 bands by agarose gel electrophoresis (AG), the most common method in clinical laboratories. AG is a time consuming method because of the need for staining and destaining steps. Serum proteins can be detected by several other techniques such as HPLC, polyacrylamide gel electrophoresis and isoelectric focusing; however, the clinical significance of the bands separated by AG has been tested and accepted through the years. A new separation by CE initially has to mimic these separations. Hopefully, in the future more elaborate separations can open new avenues for diagnosis.

29.2.3 CE separation

Protein separation is well suited for assay by CE since the peptide bond can be detected directly at 214 nm. Obviously this avoids the staining steps needed for AG and simplifies instrument automation. Serum protein analysis by CE also represents a challenge. Proteins, especially those with a high pI (isoelectric point), tend to adsorb strongly onto the capillary walls thus distorting the separation and affecting the precision of the migration time. Several approaches are used to decrease protein binding and to obtain a higher theoretical plate number in CE such as the use of high ionic strength buffers [1], addition of zwitterions [2], use of coated capillaries [1, 3–5] or operating at a pH far away from the isoelectric point of the protein [7, 8].

Several researchers [6, 7, 9] previously attempted to separate serum proteins by CE. Some of these separations did not resemble closely those obtained by AG, while others had a lengthy analysis time. Serum proteins can be separated using different buffers and at different pH such as borate [9–14], tricine [13], glycine [13] and Tris [6] with pH of 8–11. In general, the separation is better if the pH is above 8 to keep all the proteins including the immunoglobulins negatively charged. The observed separation changes slightly based on the selected conditions producing 5–12 zones, each representing a mixture of several proteins as in Figure 29.1.

The first successful separation of serum proteins by CE which resembled that of AG was described by Chen *et al.* [10, 12]. Using borate buffers they separated serum proteins into bands similar to those obtained by AG. Lehmann *et al.* [14] have described a similar separation of serum proteins by CZE using borate buffer 30 mM at pH 10. However, the separation was relatively long, about 20 min. On the other hand, Chen [12] has described a serum protein separation in narrower capillaries which was completed in about 90 s. The speed of the CE system can be contrasted to 1–2 h for AG.

Klein and Jolliff [15] described a prototype of a commercial instrument for performing CE on six samples simultaneously using borate buffers in six capillaries of narrow diameter (25 µm). The narrow capillaries have less detection sensitivity but much better resolution than the wider ones. The separation for the six samples was accomplished in about 5 min. Because of the high concentration of proteins in serum the samples were diluted in phosphate buffer, pH 7, which induced stacking. Serum electropherograms from patients with different disorders were described to illustrate the clinical significance of this system [15].

Dolnik [13] investigated the effect of several electrolytes and different co-ions on the separation of serum proteins with emphasis on the separation of alpha 1-globulin from albumin. Good separation was obtained by using a mixture of 100 mM of methylglucamine and 100 mM of ε-aminocaproic acid. An interesting separation was obtained by using 60 mmol of methylglucamine and 30 mM of lauric acid resulting in several extra unknown peaks relative to the other buffers which were ascribed to hydrophobic interaction.

Kim *et al.* [16] compared 38 serum samples from patients with different disorders by CE and AG. The correlation coeffecient was high (0.872, 0.93 and 0.96) for albumin, alpha 2- and γ-globulin, respectively, while it was low (0.535, 0.434) for alpha 1- and β-globulin fractions, respectively. The coefficients of variation (CVs) for peak area were 5–16% and those for migration times were 1–3%. These results are comparable to those obtained with AG. The lower correlation results of the alpha 1- and β-globulin can be explained by the heterogeneity of these two bands.

We used a mixture of carbonate and borate for serum protein separations

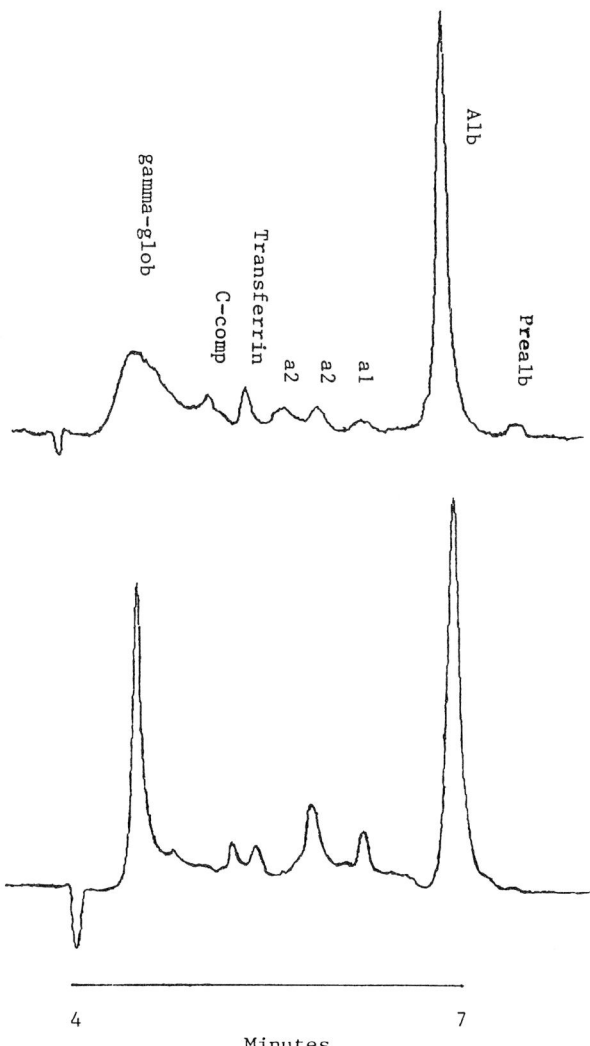

Figure 29.1 Serum protein analysis as described in section 29.9.1. Top: patient with slight polyclonal increase of gamma globulins; bottom: patient with monoclonal gammapathy.

(Figure 29.1). Polyethylene glycol (PEG) is added to the buffer to improve the separation based on size in addition to charge but more importantly to improve the reproducibility (for further details see section 29.9). Serum protein separation by CE has been reviewed [17, 18]. Based on our experience, serum protein analysis by CZE has definite advantages over AG and it has matured enough to be practical for routine analysis.

29.2.4 Capillary isoelectric focusing (CIEF)

Gel isolectric focusing is a powerful separating tool in many aspects of research. However, in routine clinical analysis of serum proteins it has not gained much acceptance. This fact is due in part to the difficulty of interpreting the gamma region by this technique. Isoelectric focusing can be performed in capillaries too, either with or without gel. It is also a powerful technique for separating proteins. It has the advantage of concentrating the sample as well as separating the proteins. In CE the technique is performed in two steps, first, focusing of the proteins and second, moving the proteins in front of the detector while the voltage is turned on. The first step is very easy while the second step is more difficult and can be accomplished by several techniques such as pressure, gravity, EOF, changing the electrolytes at the anode, etc. [19]. At the present time commercial CE instruments are not built for this purpose [17, 20, 21]. Isofocusing is better suited for detection by a pixel charge-coupled device (CCD) which eliminates the second step, i.e. moving the focused samples in frond of the detector [19–21]. In addition, several capillaries (samples) can be monitored simultaneously with a greater precision. The cost of a CCD in the UV range at the present time is very high. Few studies have been described for CIEF of serum proteins [17, 22]. However, this is an area which is in need of further improvement to make it more suitable for routine use in the laboratory. Isotachophoresis has also been tried for separating serum proteins [17], but difficulties in the analysis and in interpreting the data restrict this technique mainly to research purposes.

29.2.5 Specific serum proteins

For practical reasons, some of the bands in CE represent single proteins. For example, prealbumin is an important protein for nutritional assessment especially since it has a short half-life, reflecting recent changes in the nutritional status of the patients. It is the slowest migrating peak on serum electrophoresis. Because of low concentration, its peak height is small. Albumin separates well by CE and its level can be quantitated directly or by the internal standard method. Small molecules can work well as internal standards. An ideal internal standard will be a compound migrating close to the compound of interest and which has an almost identical chemical structure and peak height. The level of albumin reflects both the synthesis and degradation. It decreases in liver, renal and nutritional disorders. Alpha 1–antitrypsin and transferrin were separated by CE, but the bands, just as in AG, represent several proteins.

29.2.6 Immunofixation (electrophoresis)

Immunoglobulins are composed of heavy and light chains. The full chains are classified into IgG, IgA, IgM, IgD and IgE. The light chains are composed of κ and λ units. Classification is performed on the basis of the reaction with specific antibodies and is useful for patient treatment and follow-up. Klein and Jolliff [15] described a method suitable for CE for immunofixation based on reacting the serum proteins with specific antibodies bound to a solid matrix. The sample is assayed before and after binding. The difference between the two represents the specific type of the monoclonal spike. Under these conditions the sample has to be treated with the five common antibodies for proper classification. This technique is easier than performing immunofixation by AG where the sample has to be reacted with the same antibodies followed by staining and destaining.

29.2.7 Cryoglobulins

Cryoglobulins are immunoglobulins which reversibly precipitate from serum at cold temperatures. They can be monoclonal, mixed polyclonal–monoclonal or mixed polyclonal–polyclonal immunoglobulins [23]. These proteins can precipitate in the extremities and the skin of the patient where the temperature can drop below 37°C. They also can precipitate in the kidney. Mixed cryoglobulins are associated with chronic liver disorders, especially hepatitis C, rheumatoid arthritis and systemic lupus erythematosus, while the monoclonal cryoglobulins are associated with monoclonal spikes, especially Waldenstrom's macroglobulinemia [23].

Identification of cryoglobulins is important for patient diagnosis since certain types are associated with certain disorders, patient treatment and interpretation of abnormal laboratory results such as falsely elevated sedimentation rate. The cryoprecipitates can be rich in rheumatoid factors, or in hepatitis C virus.

Cryoglobulins are detected by precipitating a 500 μl aliquot of serum at 4°C, centrifuging and dissolving the precipitate in 100 μl borate buffer. This is followed by electrophoresis using the same conditions as those for serum proteins. The electropherogram of the cryopreciptate is compared to that of serum (Figure 29.2) [23].

The main advantages of CE over AG for cryoglobulin analysis are the higher sensitivity, speed and improved quantitation. Thus, sample volume can be about ten times smaller than that for AG. Values as low as $1\,\mathrm{mg\,l^{-1}}$ can be detected using only 0.5 ml of serum. Precipitating a larger volume of serum gives further sensitivity.

29.3 Urinary proteins

Urine contains several proteins of clinical interest such as myoglobin, hemoglobin, albumin, transferrin and light chains of the immuoglobulins as well as many peptides. In general, urinary proteins are more difficult to measure by CE because of their low concentrations in the presence of

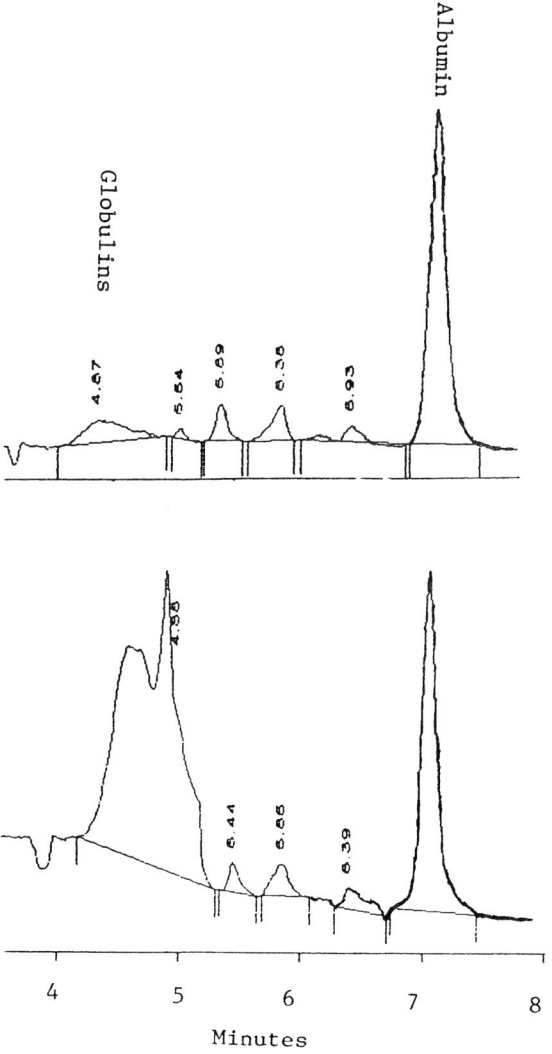

Figure 29.2 Electrograms of cryoglobulins. 500 µl serum were precipitated in the cold for 5 days, centrifuged and dissolved in 100 µl of diluted electrophoresis buffer (1:5). Top, patient serum; bottom, patient cryoprecipitate (type II, 8.48 g l^{-1}).

numerous UV-absorbing compounds which interfere with the measurement (Figure 29.3). Thus, concentration and clean-up steps are recognized to be important. We tried several techniques including dialysis and alcohol precipitation. These two methods have certain advantages and disadvantages. They can remove the low molecular weight compounds and also can concentrate the sample 2–10 times (Figure 29.3). Because of the small

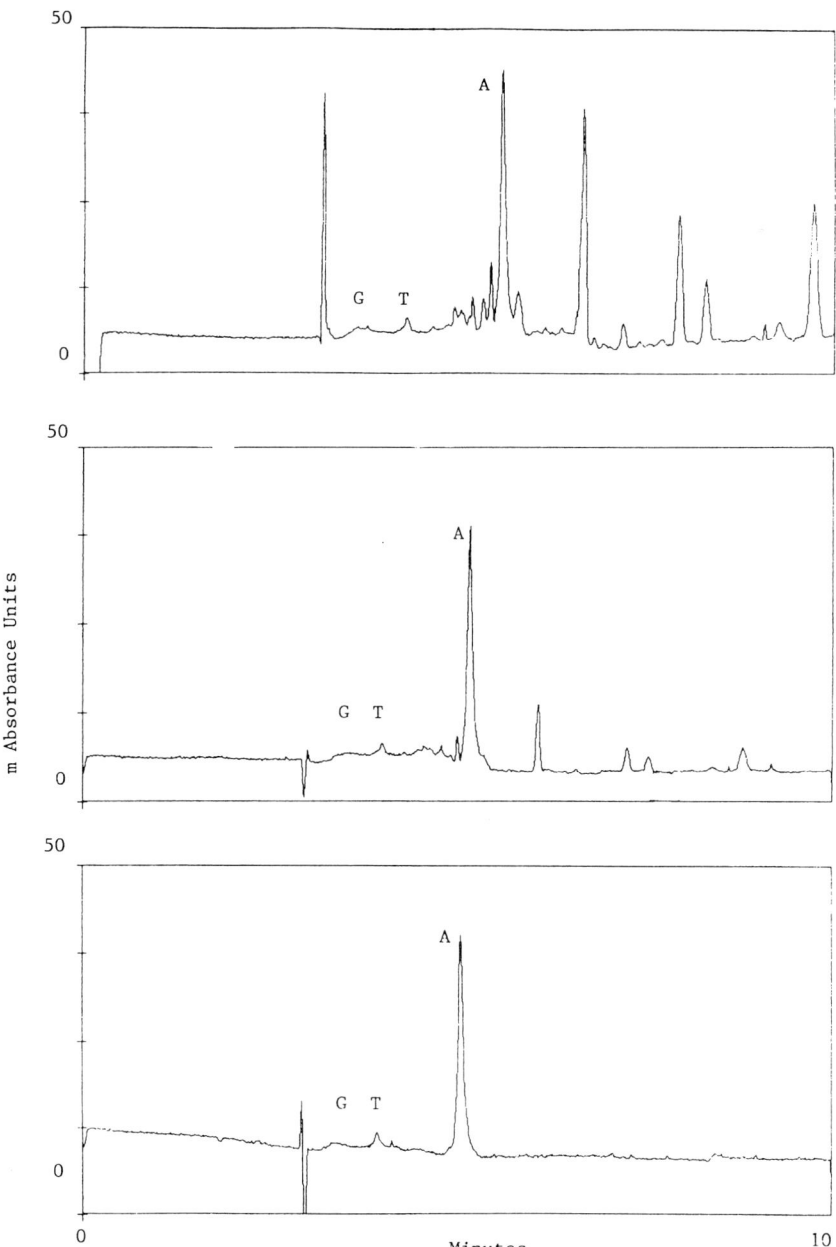

Figure 29.3 Effect of the clean-up by dialysis or alcohol precipitation on urinary proteins from a patient with non-selective proteinuria. Top: untreated urine; middle: dialyzed urine, 4 h, against water; bottom: precipitated with ethanol (200 μl urine precipitated with 1 ml ethanol at 4°C and finally dissolved in 200 μl in a diluted running buffer). Conditions: Beckman instrument equipped with a 50 μm × 40 cm capillary at 250 V cm^{-1}, 214 nm, 24°C, 4 s pressure injection and Beckman urine CE protein running buffer.

volume of samples needed for CE (5–100 µl), dialysis can be performed as rapidly as 1–3 h. The same buffers and conditions for serum proteins are basically used for analysis of urine. Alcohol precipitation is much faster than dialysis and tends to give cleaner electropherograms than dialysis (Figure 29.3).

An interesting analysis of urinary proteins was described by Lim *et al.* [24]. Diluted urine is reacted with lissamine 20, separated by MECC (micellar electrokinetic chromatography) and detected with laser-induced fluorescence. Unfortunately, the peaks have not been identified and do not resemble those by AG [24].

Hemoglobin and myoglobin can be present in urine in some pathological conditions, e.g. in trauma, burns, hemolysis and infection. Both can be detected in the urine directly without any treatment by their absorbency at 415 nm by CZE using borate buffer of 150 mM, pH 8.7 [25]. The separation between hemoglobin and myoglobin is improved by the addition of PEG. The minimum detection level is <15 mg l^{-1}.

29.4 Cerebrospinal fluid (CSF) proteins

Proteins in CSF occur at about 100 times lower concentrations than those in serum. In addition to that, very few UV-absorbing compounds occur in CSF. The main clinical significance of CSF protein electrophoresis is in the detection of the oligoclonal bands in multiple sclerosis. Direct analysis of CSF proteins without concentration can be performed as a rapid screening step [26]. However, because of the high resolution needed in the gamma region to resolve the oligoclonal bands, a 10–20 fold sample concentration is required [27].

The advantages of CE for analysis of CSF are speed, small sample volumes and avoidance of staining and destaining procedures. CSF protein separation can be accomplished in less than 10 min with CE against 2 h for AE. Oligoclonal banding which is helpful in the diagnosis of multiple sclerosis can be detected by this technique. Furthermore, CE can reveal several types of compounds which are not detected by AG such as peptides and nucleotides. These compounds might be clinically useful for patient diagnosis.

Interestingly, if the proteins are removed by an acetonitrile deproteinization, several compounds can be detected in the supernatant (Figure 29.4). The chemical characteristics and clinical significance of these compounds have not been explored (Figure 29.4). Note in Figure 29.4 that the sample fills around 20% of the capillary volume. This is possible because of the use of acetonitrile in the sample to induce sample concentration 'stacking' [28] (see Chapter 27 on Therapeutic Drug Monitoring).

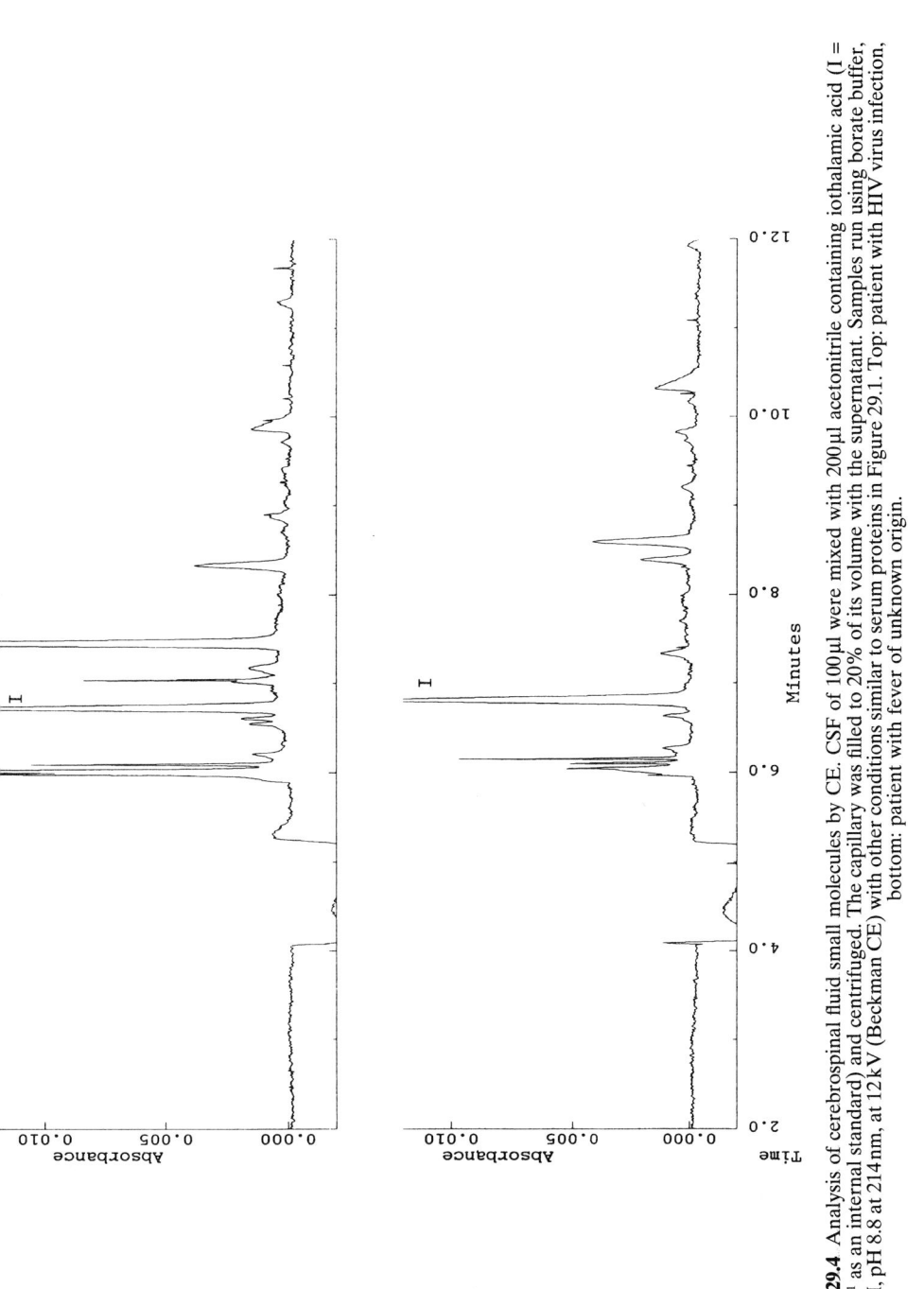

Figure 29.4 Analysis of cerebrospinal fluid small molecules by CE. CSF of 100μl were mixed with 200μl acetonitrile containing iothalamic acid (I = 10mg l^{-1} as an internal standard) and centrifuged. The capillary was filled to 20% of its volume with the supernatant. Samples run using borate buffer, 200mM, pH 8.8 at 214nm, at 12kV (Beckman CE) with other conditions similar to serum proteins in Figure 29.1. Top: patient with HIV virus infection, bottom: patient with fever of unknown origin.

29.5 Serum lipoproteins

The lipoproteins are of great interest because of their association with atherosclerosis and coronary heart disease. They are divided into five main groups: chylomicrons, VLDL, IDL, LDL and HDL (very low-, intermediate-, low- and high-density lipoproteins). The different fractions have different clinical significance. These fractions have been separated by different techniques including electrophoresis, chromatography, immunoassays and ultracentrifugation. Schmitz and Mollers [29] separated lipoproteins by isotachophoresis. The serum was incubated with Sudan B black, a lipophilic dye, and mixed with some spacers [29]. The sample was injected between a leading and a terminating buffer and the separation was monitored at 570 nm. Fourteen fractions can be detected by this method in less than 1 h. The separation was improved to mimic biologic function and attained a better separation (14 fractions) compared to other methods. The authors showed that different disorders could be detected by this method. Tady and Purdy [30] have described the separation of LDL and HDL after ultracentrifugation in borax buffer of 30 mM at pH 8.3 containing 0.1% SDS. They used both untreated and polyacrylamide-coated capillaries. Lehmann *et al.* [14] isolated HDL by ultracentrifugation and were able to separate apo I and apo II by CE. Goux *et al.* [31] separated apo I and II after sample ultracentrifugation on coated capillary by MECC. Apo I gave a single peak while apo II was heterogeneous. Cruzado *et al.* [32] found that both SDS and acetonitrile in the buffer are important for separation of HDL, LDL and lipoprotein (a).

29.6 Enzymes

29.6.1 General

Enzymes can be measured by mass as in the immunometric and direct light absorbancy methods or by catalytic activity. These two approaches have advantages as well as disadvantages. The enzyme as well as the substrates or the products can be measured by CE. CE is suited for the assay of enzymes where the substrate is expensive or extra sensitivity is needed. Analysis of enzymes by CE can be categorized into four types: direct determination, incubation in the capillary, on-line postcapillary reactor and incubation outside the capillary.

29.6.1.1 Direct determination. Enzymes which are present in high concentration can be measured as mass directly by their light absorbency. For example, the protease enzyme, savinase, which is used as an ingredient in washing powder was determined directly by its absorbance at 200 nm by using MECC for the separation [33]. Also carbonic anhydrase was analyzed using CIEF [21].

29.6.1.2 Incubation in the capillary. Most of the CE methods measure the amount of enzymes by their catalytic activity on a substrate. Bao and Regnier [34] and Avila and Whitesides [35] have described enzymatic analysis of glucose-6-phosphate dehydrogenase by CE in which the capillary is used as a microreactor. It is filled with the substrate, coenzyme and the running buffer. After injecting the enzyme and mixing it electrophoreticaly, the potential is turned to zero to allow for product accumulation. The potential is turned on again to separate the products from the substrate. The detection limit of this method appeared to be about three orders lower than conventional methods, around 1 pg of glucose-6-phosphate dehydrogenase. Wu and Regnier [36] described a similar analysis for alkaline phosphatase and β-galactosidase in which both of the enzymes are assayed simultaneously using a gel filled capillary to decrease the band broadening due to product diffusion. The lowest detection limit for alkaline phosphatase was 5.2×10^{-20} M. Miller *et al.* [37] assayed the enzyme leucine aminopeptidase based on the previous principles but using a non-fluorescent amide derivative which produced a fluorescent derivative with a lifetime of 26 ns after the reaction. About 400 enzyme molecules can be detected by this method. The enzyme, substrate (e.g. alcohol, glucose), product or inhibitors can be measured by such principles.

29.6.1.3 On-line postcapillary reactor. Emmer and Roeraade [38] separated glucose-6-phosphate and 6-phosphogluconic dehydrogenase by CE, then added the substrate after the separation using a postcapillary reactor.

29.6.1.4 Incubation outside the capillary. If a long incubation step is needed, then it is more convenient to perform it outside the instrument. Proteolytic enzymes with low activity are quite suited for analysis by CE by this method. Since an absorbency of 1×10^{-5} unit can be measured by the instrument, high sensitivity can be achieved.

Landers *et al.* [39] have measured the enzyme chloramphenicol acetyl transferase activity by CE. The enzyme and the substrate were incubated outside the capillary and the products acetyl chloramphenicol and CoA were separated by CE. Glutathione peroxidase was measured in cell-free preparations by incubating the enzyme and the substrate outside the capillary, then separating by CE the reduced and oxidized glutathione [40]. The method compared well with an HPLC and with a coupled assay [40]. The enzymes 2,5-oligoadenylate synthetase and 2,5-oligoadenylate diastrase which are part of the interferon-related 2-5A system were determined by capillary isotachophoresis [41].

29.6.2 Proteolytic enzymes

Because of their low concentration in tissues it is difficult to measure enzymes directly by mass. However, they can be measured more easily by their catalytic activity. Special enzymes such as peptidases are suitable for analysis by CE since the peptide bond can be detected at 215 nm without staining or the need to use labeled substrates. CE can separate the products of the reaction from the substrates and gives basic information on the structure of peptides such as ionization, charge to mass ratio, etc. Angiotension converting enzyme [42] and carboxypeptidase Y [43] were assayed by CE. A tripeptidase from *Lactococcus lactis* was measured rapidly (10 min) with CE without derivatization [44]. The enzyme was reacted outside the capillary with the substrate Gly-Gly-Phe and the products were separated using a citrate buffer for electrophoresis. CE was more useful than the traditional colorimetric assay in the detection of other contaminating enzymes [44].

Cathepsin D is a proteolytic lysosomal enzyme with an optimum pH of 3.5. It is secreted from some tumor cells, aiding the cells in metastasizing. Tissue enzyme levels have been found to be a good predictor of tumor malignancy in general and of breast carcinoma in particular [45].

Initially the enzyme was assayed by its catalytic activity on several proteins and more recently by immunoassays [45]. Both of these methods are time consuming requiring about a day to perform, leading to high costs. We measured this enzyme by its catalytic activity outside the capillary. After incubating the tissue homogenates with a buffered hemoglobin, acetonitrile was added to terminate the reaction and precipitate the hemoglobin. The tubes are centrifuged and the supernatant is injected into the capillary. A specific peptide is cleaved and separated by CZE in borate buffer and detected at 214 nm in less than 5 min (Figure 29.5). This peptide is not produced by the action of pepsin or trypsin. Human hemoglobin acted as a better substrate than bovine hemoglobin. The test compared well with a radioenzymatic immunoassay. The method demonstrates the advantages of CZE for assay of proteolytic enzymes in general [74].

29.6.3 Isoenzymes

For the analysis of isoenzymes the substrate is incorporated in the running buffer. The sample is introduced and the voltage is turned on for a short period to separate the isoenzymes. After that, the electric field is turned off and the isoenzymes are left in the capillary to act on the substrate. The products can be moved in front of the detector by pressure or voltage. The five isoenzymes of LDH have been separated by this technique [15, 46]. Determination of isoenzymes of other enzymes of clinical interest like

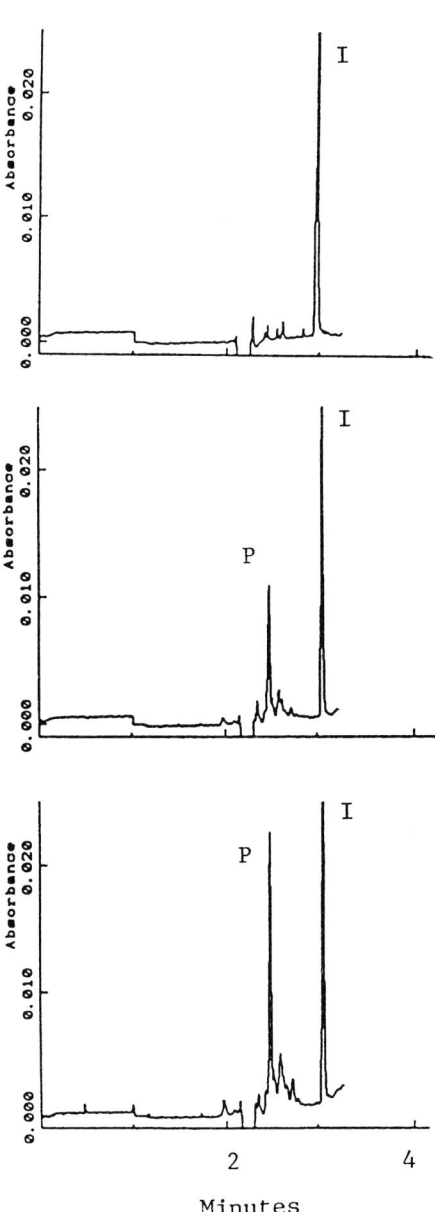

Figure 29.5 Electrograms of cathepsin D. The test is performed as described in section 29.9 (Beckman CE). Reaction: top, at zero time; middle, at 10 min; and bottom, at 20 min. P = peptide released from the action of cathepsin D, I = internal standard (see detail in section 29.9.3).

creatine kinase, alkaline phosphatase and serum aspartate aminotransferase have not been attempted by CE.

29.7 Hemoglobin variants, A_{1c} and globins

The common hemoglobin (Hb) variants are A, F, S, C. The common globin chains of the hemoglobin molecule are α, β, γ, δ, E. However, there are numerous other less common variants. Chen et al. [10] showed that the common variants could be separated in untreated capillaries by CZE. Sahin et al. [47] tried several buffers, e.g. borate, Tris and barbital for the separation of Hb variants in uncoated capillaries. The separation was best using Tris 1 M or barbital 50 mM (pH 8.0–9.0). HbF separated satisfactorily from HbA at pH 8.0. Because of the small charge difference (pI), the hemoglobin variants do not separate satisfactorily by CZE especially between HbA and HbF [47]. Based on our work, for satisfactorily separation of hemoglobin variants by CZE, a high buffer concentration, a narrower capillary >30 μm (i.d.), minimum volume of sample size and low voltages are required (Figure 29.6). Both Tris 0.8 M and Tricine 1.1 M buffers at pH 8–8.4 indicate a good separation. The separation between HbA and HbF is better at pH 7.9. The separation by CZE resembles very closely that of the alkaline separation by AG.

Although the CE instruments are not well designed for CIEF as pointed out earlier [20], several workers have successfully separated these variants much better by CIEF than by CZE. Zhu and co-workers [48, 49], in addition to the common variants, were able to separate HbA_{1c}, G Philadelphia and Bart's using coated capillaries. Molteni et al. [50] used methylcellulose in uncoated capillaries to suppress the EOF during electrophoresis. They were also able to separate A_{1c}, A, F, C, S, E and A_2. The separation compared well with HPLC and gel isoelectric focusing. Hempe and Craver [51] separated the major hemoglobin variants including A_{1c} on a coated capillary. They were able to separate S from D-Los Angeles, which is difficult to achieve by other methods. The CV for the HbA was <2%, while that for A_2, F and A_{1c} was 1–11%. Yao et al. [52] and Yao and Regnier [53] described coatings for capillaries which were suited for CIEF of the hemoglobin including HbA_{1c}. Conti et al. [54] described the problems involved in using coated capillaries in the CIEF of the hemoglobin F separation in cord blood samples.

The variants have been also analyzed by CIEF with special absorption imaging detectors. These types of detection devices do not need the peaks to be moved to the detector and can detect several capillaries simultaneously with better precision than the commercial instruments [21].

The globin chains which are useful for investigating the thalassemias have been determined by CZE in phosphate buffer either at pH 11.8 [55] or at

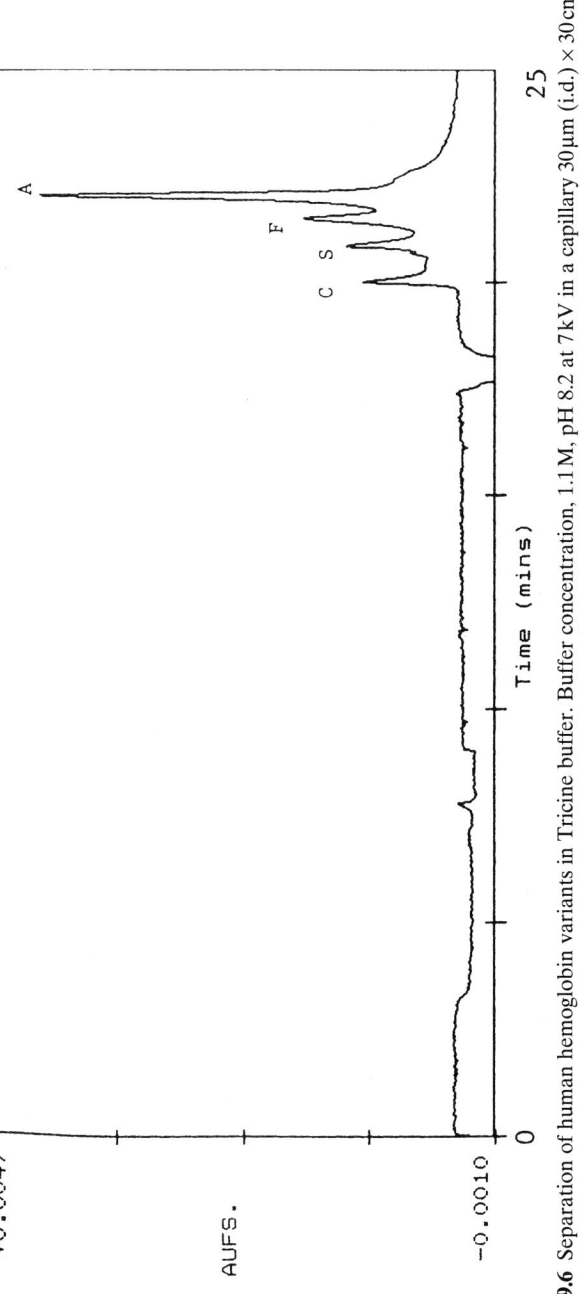

Figure 29.6 Separation of human hemoglobin variants in Tricine buffer. Buffer concentration, 1.1 M, pH 8.2 at 7 kV in a capillary 30 μm (i.d.) × 30 cm; wavelength 214 nm (see details in section 29.9).

pH 2.5–4.5 [49, 56] after acetone precipitation. Furthermore, the tryptic digests of the globin chains were analyzed by CZE using 50 mM phosphate buffer at pH 2.5 with detection at 200 nm for the study of the variants [50, 57]. The analytical result was found to be better than with HPLC.

One of the additional advantages of hemoglobin analysis by CE is the ability to quantify HbA_2. This fraction which is useful for the diagnosis of thalassmia constitutes a small percentage (2–5%) and cannot be quantitated accurately by agarose gel electrophoresis. No attempt has been made in CZE to mimic the acid (citrate) separation of hemoglobin by AG.

29.8 Miscellaneous tests

There are several tests which are used in very few specialized clinical laboratories to aid in patient diagnosis. Simple analytical methods for these tests may make them more widespread. Here we discuss briefly a few of these tests since some are described elsewhere in this book.

Many anions which are of clinical interest such as sulfate, phosphate, aspartate and gluconate have been separated rapidly within a few minutes by CE using indirect detection [58, 59]. Urinary oxalate and citrate which are important in stone formation have been determined by CE using indirect detection [60, 61]. Urinary porphyrins have been determined by MECC at pH 11 with fluorescence detection [62]. Xanthine and other purine and pyrimidine bases in biological fluids were determined, after sample deproteinization with acetonitrile, at levels of <1 mg l^{-1} by CZE using a borate–carbonate buffer [63]. Purine bases in human cord plasma were analyzed by CZE in borate buffer [64]. The samples were deproteinized with trichloroacetic acid and neutralized with potassium carbonate before injection into the capillary [64]. The sensitivity of the method was 0.5 µM. Petucci et al. [65] have determined several compounds such as hypoxanthine, pseudouridine, hippuric acid and uric acid in normal and uremic serum by CE in 150 mM borate buffer at pH 9.0 after removing the proteins by ultrafiltration. The limit of detection was 1–6 µM.

Homocysteine and other sulfur-containing amino acids, being important in the diagnosis of homocystinuria, were analyzed by CZE after derivatization with bromobimane and detection by fluorescence [66]. Tagliaro et al. [67] described the determination of phenylalanine by CZE after sample deproteinization with ethanol containing an internal standard (diprophylline) using borate buffer of 25 mM at pH 10.0.

Caslavska et al. [68] described the feasibility of analysis of the urinary metabolites of catecholamines (vanillylmandelic acid (VMA), homovanillic acid (HVA)) and urinary 5-hydroxyindole acetic acid by MECC using a modified detector which measures fluorescence with excitation at 220 nm and detection at 340 nm.

Methylmalonic acid is used clinically to diagnose the inherited metabolic disorder methyl malonic aciduria in the newborn and also more often to diagnose vitamin B_{12} deficiency. This compound has been determined in serum by CE after reaction with pyrenyldiazomethane and detection with laser-induced fluorescence. The assay was linear between 0.1–200 µM. The assay compared well with GC–MS [69]. Proline and hydroxyproline are important in collagen synthesis and degradation was determined in urine by CZE after reaction with UV detection [70].

Therapeutic drug analysis is a major testing area in the clinical laboratories. This subject is discussed in detail in Chapter 27.

29.9 Practical aspects

29.9.1 General method for serum protein analysis

Sample: serum (10 µl) is diluted with 250 µl water.

Electrophoresis buffer: 0.7 g sodium carbonate, 0.7 g boric acid and 0.5 g polyethylene glycol (PEG 8000) dissolved in 100 ml of water.

Capillary: 32 cm length, 50 µm i.d.

CE conditions: injection, 0.1% of the capillary volume (for cerebrospinal fluid (CSF) and hemoglobin the least amount of sample which can be injected).

Voltage: 6–7 kV, detection wavelength 214 nm.

Chromatogram: see Figure 29.1.

29.9.2 Method for urine clean-up

Mix 200 µl urine with 1 ml cold ethanol. After 5 min, centrifuge, wash the precipitate with 1 ml cold ethanol. Centrifuge again and dissolve the precipitate in 100 µl electrophoresis buffer diluted with water (1:5).

Urine can be dialysed against water. However, the dialysis is much slower (about 2–6 h) and leaves many UV absorbing materials which makes interpretation of electropherogram difficult compared to alcohol precipitation (Figure 29.3).

29.9.3 Method for analysis of cathepsin D

Reagents:

1. Cathepsin buffer: formic acid 0.3 M at pH 3.4 containing 12.5 g l^{-1} NaCl and 0.5 ml l^{-1} of Triton × 100.
2. Hemoglobin solution: lyophilized human hemoglobin, 24 g l^{-1} in water.

3. Substrate: equal volumes of hemoglobin solution and cathepsin buffer are mixed and kept on ice.
4. Deproteinization reagent: iothalamic acid at a concentration of 70 mg l^{-1} in acetonitrile.

Hemoglobin in formate buffer (150 µl) is mixed with 50 µl tissue homogenate and incubated for 20 min at 37°C. At the end of the incubation period, 500 µl deproteinizing reagent was added, mixed and centrifuged for 30 s. The supernatant was injected for 10 s into a capillary of 50 µm (i.d.) × 42 cm length at 360 V cm^{-1}, 241 nm. The electrophoresis buffer was borate, 175 mM at pH 8.4.

29.9.4 Recommended guidelines for CE in clinical analysis

The sample matrix is very important in CE. Sample overloading and sample stacking (concentration) can occur simply by properly manipulating the sample ionic strength, volume, pH, etc. In general, the sample should be dissolved in a buffer ten times less concentrated than the electrophoresis buffer. For small molecules including many small peptides the sample can be dissolved in a mixture made up of 1% NaCl and acetonitrile in a ratio of 1:2, v/v. Under these conditions a 5–20 fold increase in sensitivity can be obtained [28, 71], see also Chapter 27 on therapeutic drug monitoring, TDM.

- Borate buffers with concentration >150 mM are a good starting point for analysis of biological samples.
- High pH decreases the binding of the basic proteins to the capillary wall due to deionization of basic proteins.
- The use of lower voltages increases the theoretical plate numbers when stacking is employed and avoids heat denaturation of proteins [72].
- Capillaries of 50 µm (i.d.) × 24–50 cm give faster separations with satisfactory sensitivity.
- Addition of 0.1–0.5% PEG to the electrophoresis buffer when injecting on the column samples with higher concentrations of proteins improves the precision and the separation especially for compounds with significantly different molecular weights.
- Proteins are quantitated more accurately by peak area than by height provided the concentration is high enough. Also, the linearity is better when peak area using an integrator rather than peak height is used.

29.9.5 Quantitation and reproducibility

Proteins adsorb onto the capillary surface changing EOF. This causes deterioration in precision. For example, it has been pointed out that when as small a segment as 2% of the capillary length is fouled with protein, the efficiency deteriorated by about 50% [73].

A sufficient washing after each sample-run with 0.1–1 M NaOH is critical in CZE to attain a good reproducibility, especially for migration time. In general, the reproducibility of migration time is less than 1% while the reproducibility of the corrected peak area is less than 5%. For further details refer to Chapter 27 on TDM.

29.10 Concluding remarks

CE offers special advantages as an analytical tool in clinical laboratories. At the present time the separation of serum proteins and hemoglobin variants resembles closely that of agarose gel electrophoresis. The analysis of these compounds has matured enough to be practical for routine diagnosis. With better buffers, additives and coated capillaries, greater resolution may be achieved to open new avenues of clinical diagnosis. Enzymes that are difficult to analyze by conventional methods as pointed out earlier, may be more conveniently assayed by CE especially where the sample size is limited, e.g. fine needle aspirates, or where extra sensitivity is needed. Proteolytic enzymes are well suited for analysis by CE. Little attention has been paid to lipoprotein analysis. Application of CE for some specialized clinical tests has been discussed. It is expected that more applications will be explored in the near future.

The application of capillary electrophoresis in clinical analysis is expected to become more popular with the introduction of commercial instruments designed specifically for the analytical laboratory. The driving forces for a widespread use of CE at the present time are simplicity, speed and low costs. However, in the future, better separation methods for protein analysis, as well as the analysis of new and more difficult analytes, will be the major driving force. The use of CE instruments with multicapillaries (channels) is important for rapid analysis in the routine laboratory. Specialized CE instruments designed specifically for isofocusing will enhance protein separation, speed the analysis and improve the precision. Specialized detectors such as the electrochemical detector will help in many clinical tests such as for the catecholamines and their metabolites and many specific amino acids. CE can offer also basic physical and structural information for proteins such as charge (pI), size, isoform or interaction with other compounds.

References

1 Chen, F.A.; Kely, L.; Palmieri, R.; Biehler, R.; Schwart, H. (1992) Use of high ionic strength buffers for the separation of proteins and peptides with capillary electrophoresis. *J. Liq. Chromatogr.* **15**, 1143–1162.
2 Bushey, M.M.; Jorgenson, J.W. (1989) Capillary electrophoresis of proteins in buffers containing high concentrations of zwitterionic salts. *J. Chromatogr.* **480**, 301–310.

3. Cifuentes, A.; Santos, J.M.; Frutos, M.; Diez-Masa, J.C. (1993) High-efficiency capillary electrophoretic separation of basic proteins using coated capillaries and cationic buffer additives. *J. Chromatogr.* **652**, 161–170.
4. Smith, J.T.; Rassi, Z.E. (1993) Capillary zone electrophoresis of biological substances with fused silica capillaries having zero or constant electroosmotic flow. *Electrophoresis* **14**, 396–406.
5. Hjertén, S.; Kubo, K. (1993) A new type pH- and detergent-stable coating for elimination of electroosmosis and adsorption in (capillary) electrophoresis. *Electrophoresis* **14**, 390–395.
6. Jorgenson, J.W.; Lukacs, K.D. (1993) Capillary zone electrophoresis. *Science* **222**, 166–172.
7. Lee, K.J.; Heo, G.S. (1991) Free solution capillary electrophoresis of proteins using untreated fused-capillaries. *J. Chromatogr.* **559**, 317–324.
8. Zhu, M.; Rodriguez, D.; Hansen, D.; Wehr, T. (1990) Capillary electrophoresis of proteins under alkaline conditions. *J. Chromatogr.* **516**, 123–132.
9. Gordon, M.J.; Lee, K.J.; Arias, A.A.; Zare, R.N. (1991) Protocol for resolving protein mixtures in capillary zone electrophoresis. *Anal. Chem.* **63**, 69–72.
10. Chen, F.A.; Liu, C.M.; Hsieh, Y.; Sternberg, J. (1991) Capillary electrophoresis – a new clinical tool. *Clin. Chem.* **37**, 14–19.
11. Chen, F.A.; Sternberg, J. (1994) Characterization of proteins by capillary electrophoresis in fused-silica columns: Review on serum protein analysis and application to immunoassays. *Electrophoresis* **15**, 13–21.
12. Chen, F.A. (1991) Rapid protein analysis by capillary electrophoresis. *J. Chromatogr.* **559**, 445–453.
13. Dolnik, V. (1995) Capillary zone electrophoresis of serum proteins: study of separation variables. *J. Chromatogr.* **709**, 99–110.
14. Lehmann, R.; Liebich, H.; Grubler, G.; Voelter, W. (1995) Capillary electrophoresis of human serum proteins and apolipoproteins. *Electrophoresis* **16**, 998–1001.
15. Klein, J.L.; Jolliff, C.R. (1994) Capillary electrophoresis for the routine clinical laboratory. In: *Handbook of Electrophoresis*, Landers, J.P. (ed.), CRC Press, Boca Raton, pp. 419–418.
16. Kim, J.W.; Park, J.H.; Park, J.W.; Doh, H.J.; Heo, G.S.; Lee, K.J. (1993) Quantitative analysis of serum proteins separated by capillary electrophoresis. *Clin. Chem.* **39**, 689–692.
17. Reif, O.W.; Lausch, R.; Freitag, R. (1994) High-performance capillary electrophoresis of human serum and plasma proteins. *Adv. Chromatogr.* **34**, 1–56.
18. Landers, J.P. (1995) Clinical capillary electrophoresis. *Clin. Chem.* **41**, 495–509.
19. Kilar, F. (1994) Isoelectric focusing in capillaries. In: *Handbook of Electrophoresis*, Landers, J.P. (ed.), CRC Press, Boca Raton, pp. 96–110.
20. Wu, J.; Pawliszyn, J. (1994) Dual detection for capillary isoelectric focusing with refractive index and absorption imaging detector. *Anal. Chem.* **66**, 867–873.
21. Wu, J.; Pawliszyn, J. (1995) Application of capillary isoelectric focusing with absorption imaging detection to the quantitative determination of human hemoglobin variants. *Electrophoresis* **16**, 670–673.
22. Zhu, M., Rodriguez, R.; Wehr, T. (1991) Optimizing separation parameters in capillary isoelectric focusing. *J. Chromatogr.* **559**, 479–488.
23. Shihabi, Z.K. (in press) Analysis and general classification of serum cryoglobulins by capillary zone electrophoresis. *Electrophoresis*.
24. Lim, H.B.; Lee, J.J.; Lee, K.J. (1995) Simple and sensitive laser-induced fluorescence detection for capillary electrophoresis and its application to protein separation. *Electrophoresis* **16**, 674–678.
25. Shihabi, Z.K. (1995) Myoglobinuria detection by capillary electrophoresis. *J. Chromatogr.* **669**, 53–58.
26. Shihabi, Z.K. (1991) Capillary electrophoresis of cerebrospinal fluid proteins. *Ann. Clin. Lab. Sci.* **21**, 346.
27. Chaing, S.H.; Highsmith, E.; Silverman, L.H.; Chapman, J.F. (1993) Analysis of cerebrospinal fluid proteins by capillary electrophoresis. *Clin. Chem.* **39**, 1132.
28. Shihabi, Z.K. (1995) Sample stacking by acetonitrile–salt mixtures. *J. Capillary Electrophoresis* **2**, 267–271.

29 Schmitz, G.; Mollers, C. (1994) Analysis of lipoproteins with analytical capillary isotachophoresis. *Electrophoresis* **15**, 31–39.
30 Tady, T.; Purdy, W.C. (1992) Characterization of plasma apolipoproteins by capillary electrophoresis. *J. Chromatogr.* **583**, 111–115.
31 Goux, A.; Athias, A.; Persegol, L.; Lagrost, L.; Gambert, P. (1994) Capillary gel electrophoresis analysis of apoliporotein A-I and A-II in human high-density lipoproteins. *Anal. Biochem.* **218**, 320–324.
32 Cruzado, I.D.; Hu, Z.; McNeal, C.J.; McFarlane (1995) Characterization of lipoprotein a by capillary zone electrophoresis. *J. Chromatogr.* **717**, 33–39.
33 Vinther, A.; Petersen, J.; Soeberg, H. (1992) Capillary electrophoretic determination of the protease savinase in cultivation broth. *J. Chromatogr.* **608**, 205–210.
34 Bao, J.; Regnier, F.E. (1992) Ultramicro enzyme assay in capillary electrophoretic system. *J. Chromatogr.* **608**, 217–224.
35 Avila, L.Z.; Whitesides, G.M. (1993) Catalytic enzyme activity during capillary electrophoresis: an enzymatic 'microreactor'. *J. Org. Chem.* **58**, 5508–5512.
36 Wu, D.; Regnier, F.E. (1992) Native protein separations and enzyme microassays by capillary zone and gel electrophoresis. *Anal. Chem.* **65**, 2029–2035.
37 Miller, K.J.; Leesong, I.; Bao, J.; Regnier, F.E.; Lytle, F.E. (1993) Electrophoretically mediated microanalysis of leucine aminopeptidase in complex matrices using time-resolved laser-induced fluorescence detection. *Anal. Chem.* **65**, 3267–3270.
38 Emmer, A.; Roeraade, J. (1994) Capillary electrophoresis, combined with an on-line micro post-column reactor. *J. Chromatogr.* **662**, 375–381.
39 Landers, J.P.; Schuchard, M.D.; Subramaniam, M.; Sismelich, T.P.; Spelsberg, T.C. (1992) High-performance capillary electrophoretic analysis of chloramphenicol acetyl transferase activity. *J. Chromatogr.* **603**, 247–257.
40 Pascual, P.; Martinez-Lara, E.; Barcena, J.A.; Lopez-Berea, J.; Toribio, F. (1992) Direct assay of glutathione peroxidase activity using high-performance capillary electrophoresis. *J. Chromatogr.* **581**, 49–56.
41 Bruchet, G.; Budenbender, M.; Schmidt, K.-H.; Birk, A.; Treuner, J.; Niethammer, D. (1994) Determination of 2-5A synthetase 2-5A phosphodiesterase in neuroblastoma cells by analytical capillary isotachophoresis: Effect of cytokines and comparison with radioenzymatic methods. *Electrophoresis* **15**, 40–45.
42 Shihabi, Z.K. (1992) Clinical application of capillary electrophoresis. *Ann. Clin. Lab. Sci.* **22**, 398–405.
43 Vinther, A.; Adelhorst, K.; Kirk, O. (1993) Using capillary electrophoresis in the optimization of a carboxypeptidase Y catalyzed transpeptidation reaction. *Electrophoresis* **14**, 486–491.
44 Mulholland, F.; Movahedi, S.; Hague, G.R.; Kasumi, T. (1993) Monitoring tripeptidase activity using capillary electrophoresis. Comparison with the ninhydrin assay. *J. Chromatogr.* **636**, 63–68.
45 Kute, T.E.; Shao, Z.M.; Sugg, N.K.; Long, R.T.; Russell, G.B.; Case, L.D. (1992) Cathepsin D as a prognostic indicator for node-negative breast cancer patients using both immunoassays and enzymatic assays. *Cancer Res.* **52**, 5198–5203.
46 Uji, Y.; Okabe, H.; Karmen, A.; Wang, H.P. (1995) Lactate dehydrogenase isoenzyme analysis by capillary electrophoresis. *Clin. Chem.* **41**, S236.
47 Sahin, A.; Laleila, Y.R.; Ortancil, R. (1995) Haemoglobin analysis by capillary zone electrophoresis. *J. Chromatogr.* **709**, 121–126.
48 Zhu, M.; Rodriguez, R.; Wehr, T.; Siebert, C. (1992) Capillary electrophoresis of hemoglobin and globin chains. *J. Chromatogr.* **608**, 225–237.
49 Zhu, M.; Wehr, T.; Levi, V.; Rodriguez, R.; Shiffer, R.; Cao, Z.A. (1993) Capillary electrophoresis of abnormal hemoglobins associated with α-thallasemias. *J. Chromatogr.* **652**, 119–129.
50 Molteni, S.; Frischknecht, H.; Thormann, W. (1994) Application of dynamic capillary isoelectric focusing to the analysis of human hemoglobin variants. *Electrophoresis* **15**, 22–30.
51 Hempe, J.M.; Craver, R.D. (1994) Quantitation of hemoglobin variants by capillary electrophoresis. *Clin. Chem.* **40**, 2288–2295.
52 Yao, X.-W.; Wu, D.; Regnier, F.E. (1993) Manipulation of electroosmotic flow in capillary electrophoresis. *J. Chromatogr.* **636**, 21–29.

53 Yao, X.-W.; Regnier, F.E. (1993) Polymer- and surfactant-coated capillaries for isoelectric focusing. *J. Chromatogr.* **632**, 185–193.
54 Conti, M.; Gelfi, C.; Righetti, P.G. (1995) Screening of umbilical cord blood hemoglobins by isoelectric focusing in capillaries. *Electrophoresis* **16**, 1485–1491.
55 Ong, C.-N.; Liau, L.S.; Ong, H.Y. (1992) Separation of globins using free zone capillary electrophoresis. *J. Chromatogr.* **576**, 346–350.
56 Ferranti, P.; Malorni, A.; Pucci, P.; Fanali, S.; Nardi, A.; Ossicini, L. (1991) Capillary zone electrophoresis and mass spectrometry for the characterization of genetic variants of human hemoglobin. *Anal. Biochem.* **194**, 1–8.
57 Ross, G.A.; Lorkin, P.; Perrett, D. (1993) Separation and tryptic digest mapping of normal and variant hemoglobins by capillary electrophoresis. *J. Chromatogr.* **636**, 69–79.
58 Jones, W.R.; Jandik, P. (1992) Various approaches to analysis of difficult sample matrices of anions using capillary ion electrophoresis. *J. Chromatogr.* **608**, 385–393.
59 Weston, A.; Brown, P.R.; Jandik, P.; Heckenberg, A.L.; Jones, W.R. (1992) Optimization of detection sensitivity in the analysis of inorganic cations by capillary electrophoresis using indirect photometric detection. *J. Chromatogr.* **608**, 395–402.
60 Wildman, B.J.; Jackson, P.E.; Jones, W.R.; Alden, P.G. (1991) Analysis of anion constituents of urine by inorganic capillary electrophoresis. *J. Chromatogr.* **546**, 459–466.
61 Holmes, R.P. (1995) Measurement of oxalate and citrate by capillary electrophoresis and indirect ultraviolet absorbance. *Clin. Chem.* **41**, 1297–1301.
62 Weinberger, R.; Sapp, E.; Moring, S. (1990) Capillary electrophoresis of urinary porphryrins with absorbance and fluorescence detection. *J. Chromatogr.* **516**, 271–285.
63 Shihabi, Z.K.; Hinsdale, M.E.; Bleyer, A.J. (1995) Xanthine analysis in biological fluids by capillary electrophoresis. *J. Chromatogr.* **669**, 163–169.
64 Grune, T.; Ross, G.A.; Schmidt, H.; Siems, W.; Perrett, D. (1993) Optimized separation of purine bases and nucleosides in human cord plasma by capillary zone electrophoresis. *J. Chromatogr.* **636**, 105–111.
65 Petucci, C.J.; Kants, H.L.; Strein, T.G.; Veening, H. (1995) Capillary electrophoresis as a clinical tool determination of organic anions in normal and uremic serum using photodiode-array detection. *J. Chromatogr.* **668**, 241–251.
66 Jellum, E.; Thorsrud, A.K.; Time, E. (1991) Capillary electrophoresis for diagnosis and studies of human disease, particularly metabolic disorders. *J. Chromatogr.* **559**, 455–465.
67 Tagliaro, F.; Moretto, S.; Valentini, R.; Gambaro, G.; Antonioli, C.; Moffa, M.; Tato, L. (1994) Capillary zone electrophoresis determination of phenylalanine in serum: A rapid, inexpensive and simple method for the diagnosis of phenylketonuria. *Electrophoresis* **15**, 94–97.
68 Caslavska, J.; Gassmann, E.; Thormann, W. (1995) Modification of a tunable uv–visible capillary electrophoresis detector for simultaneous absorbance and fluorescence detection: profiling of body fluids for drugs and endogenous substances. *J. Chromatogr.* **709**, 147–156.
69 Schneede, J.; Ueland, P.M. (1995) Application of capillary electrophoresis with laser-induced fluorescence detection for routine determination of methylmalonic acid in human serum. *Anal. Chem.* **67**, 812–819.
70 Guzman, N.; Moschera, J.; Iqbal, K.; Malik, W. (1992) Quantitative assay for the determination of proline and hydroxyproline by capillary electrophoresis. *J. Liq. Chromatogr.* **15**, 1163–1178.
71 Shihabi, Z.K.; Garcia, L.L. (1994) Effects of sample matrix on separation by capillary electrophoresis. In *Handbook of Electrophoresis*, Landers, J.P. (ed.), CRC Press, Boca Raton, FL, pp. 537–548.
72 Vinther, A. (1991) Mathematical model describing dispersion in free solution capillary electrophoresis under stacking conditions. *J. Chromatogr.* **559**, 3–26.
73 Towns, J.K.; Regnier, F.E. (1992) Impact of polycation adsorption on effeciency and electroosmotically driven transport in capillary electrophoresis. *Anal. Chem.* **64**, 2473–2476.
74 Shihabi, Z.; Kute, T. (in press) Analysis of cathepsin D from breast tumors by CE. *J. Chromatogr.*

30 Gene analysis and nucleic acid sequencing
H.-M. WENZ AND J.E. WIKTOROWICZ

30.1 Introduction

Since the inception and popularization of capillary electrophoresis (CE) during the 1970s and 1980s, the application of the technique to the analysis of biomolecules has captured the imagination and energies of scientists in academia and industry. Central to this fascination is the ability to accomplish high resolution separations without the need to use rigid anti-convective media.

Since discrimination between two or more analytes in electrophoresis is related to the relative effects of charge and frictional coefficient (molecular size) on mobility, nucleic acids in particular cannot be separated in free solution alone as they exhibit the same charge-to-mass ratio [1]. Therefore, to effect electrophoretic separation of DNA fragments based on their size, a sieving medium is required.

Typically, two different separation matrix approaches have been utilized. DNA sequencing and some forms of fragment analysis have traditionally employed polyacrylamide in a slab format. Attempting to bring these applications to CE, early workers found that under the high fields typically used for DNA analysis, the gel was seen to extrude from the capillary. Although later polymerization schemes utilized capillaries covalently coated with acrylamide, thereby anchoring the polyacrylamide gel to the capillary wall, voids were found to arise from the tendency for acrylamide to shrink during polymerization. Thus when challenged with real sequencing reactions, most rigid, crosslinked polyacrylamide-filled capillaries acquire intractable bubbles that at best diminish current flow (and risk arcing) and at worst cause severe loss in resolution.

Because of this (and other difficulties inherent in covalent capillary coating technology), recent attention has turned to flowable media. This second approach consists of a non-crosslinked linear polymer network (usually of lower viscosity), with a more dynamic pore structure. Since this technology grants several advantages, including replenishment of separation polymers between runs, much of recent development work has focused on comparatively low viscosity polymer networks to accomplish molecular sieving (for recent reviews, see Dovichi [2] and Guttman [3]) for the fragment analysis of double-stranded DNA.

In general, these polymer networks have been based on soluble linear chains such as polyacrylamide [4–6], agarose [7], polyethylene oxide [8] and various cellulose derivatives [9, 10]. In addition to sieving, these materials should ideally possess the ability to coat the capillary to prevent wall interactions from destroying resolution [6]. By exploiting this strategy, the advantages of CE also include a high degree of operator efficiency (no gel polymerization or pouring).

Finally, a critical variable in obtaining maximum resolution in a CE system is the width of the sample after injection. While samples can be injected hydrostatically, this often results in inadequate signal strength and decreased resolution [11] and is not suitable for capillaries filled with highly viscous polymer matrices. Electrokinetic injection is generally available on all commercial CE equipment and imparts several advantages, particularly with samples in low concentrations. As anticipated, no sample bias was observed with electrokinetic injection of equimolar DNA fragments between 72 and 1353 bp, i.e. a monotonic increase in peak height was seen with increasing fragment length [12].

In summary, the use of CE for nucleic acid analysis is attractive for numerous reasons:

1. Flowable, liquid separation media circumvent cumbersome gel pouring and sample loading.
2. High mass sensitivity eliminates the need to label DNA with carcinogenic stains or radioactive DNA precursors (mass sensitivity is in the order of 10^{-15} mol in the UV mode and 10^{-18}–10^{-21} mol using fluorescence detection [13]).
3. Efficient Joule heat dissipation permits high field strengths resulting in high throughput (typical analysis times between 5 and 45 min) and high resolution (<1 bp).
4. On-line detection permits accurate quantification of fragments.
5. On-line detection also facilitates multicolor chemistries necessary for highly reproducible size information (by using an internal standard which compensates for run-to-run variations).
6. Visualization of structural and conformational features of DNA fragments even at elevated temperatures.

In this chapter, only the most recent work will be discussed as it relates to the analysis of DNA for fragment analysis and sequencing by CE. The reader is referred to any other books that review the historical development of DNA analysis (e.g. Ref. [10]).

30.2 Fragment and gene analysis

30.2.1 Detection of sequence-induced anomalous migration of dsDNA

Bent or curved DNA is nearly ubiquitous and has been correlated with a variety of biological processes [14]. DNA fragments with such features often exhibit a retarded mobility in non-denaturing polyacrylamide slab gels, especially at low temperatures. Wenz [12] and Berka *et al.* [15] have shown that CE and replaceable polymer solutions are useful for the detection of anomalous migration. The results indicate the increased migration time of a DNA fragment, containing the T antigen binding site, relative to a mutated fragment with identical length [12]. It is noticeable that even at elevated temperatures the shift in mobility is still evident.

30.2.2 Mutation detection

As more gene sequences and their associated mutations become known through the successes of the Human Genome Project, the demand for more rapid screening methodologies in order to stimulate further the information flow increase. Because the diseased state of a gene results from an altered DNA sequence, sequencing is the only unequivocal means for mutation detection. As long as conventional DNA sequencing (manual or automated) remains prohibitive because of constraints in throughput and cost, alternate screening methods will be in demand. Technologies that are capable of detecting differences between DNA fragments that result from altered secondary structures created by an altered nucleotide sequence (mutation) have recently come to the forefront of mutation research. This section will discuss the use of CE to discriminate between such DNA fragments.

One of the most popular techniques, especially for the detection of single point mutations, is the detection of single-strand conformational polymorphisms (SSCP) [16]. Fundamental to this method is the observation that the denatured strands of a wild-type gene frequently assume a different intrastrand conformation when subjected to electrophoresis under non-denaturing conditions than its mutated allele. These different conformations result in different mobilities and can be resolved by electrophoresis.

This method has deservedly received considerable attention from the scientific community. For the reasons enumerated above, CE is readily adaptable and tailor-made for this application. Several recent articles have demonstrated the use of capillary electrophoresis for rapid, ambient temperature-tolerant SSCP analyses [17–20]. In methodologies that rely on the detection of differences in mobilities caused by secondary structures, it is imperative to verify that mobility differences are truly correlated with sequence changes and not with variations in run conditions from injection to injection. The use of an internal size standard helps to compensate for these variations. The use of an internal size standard, labeled with a different fluorophor than the polymerase chain reaction (PCR)–SSCP sample resulted in excellent reproducibility of migration times (average of six runs) of the two SSCP strands with a coefficient of variation (CV) of 0.05% for the forward and 0.06% for the reverse strand [21]. The ability of linear polymers to separate DNA structural polymorphs of 200 to 300 bp in length opens the possibility that even larger fragments might be accessible to SSCP analysis.

CE has proven to be more useful than other methods for detecting DNA mutations based on differential fragment mobilities, such as heteroduplex analysis. In these cases, a known wild-type DNA fragment is mixed with a suspected mutated fragment. After denaturation, they are allowed to reanneal, resulting in the formation of homo- and heteroduplexes. Because heteroduplexes contain regions of base pair mismatch, these fragments will migrate more slowly than the homoduplexes. In this way, fragments containing mutations can be identified. The use of CE in this application, called heteroduplex mobility assay (HMA), under native and denaturing conditions has been recently described [22]. This method is well suited for the detection of multiple mutations in a DNA fragment, such as in the identification of different viral subtypes. Figure 30.1 shows the result of screening for sequence differences in a 700 bp fragment of two different HIV subtypes in the env gene. In this case only one strand of one subtype is labeled with a fluorescent dye. The resulting electropherogram using laser-based detection (LIF) shows increased sensitivity and selectivity, since only fluorescent analytes can be selectively detected. Because of this selectivity, a less complex pattern of peaks (only two peaks) is obtained compared to a UV-based system, where at least four peaks as a whole are detected [23]. This results from the hybridization of the labeled strand to its complementary strand, forming a homoduplex, and to the hybridization of the complementary strand of the second fragment, forming the heteroduplex. Cheng *et al.* [23] report the use of hydroxypropylmethylcellulose and non-denaturing conditions for the detection of a single-nucleotide mismatch in the 5.8S rRNA gene of *H. annosum* in a UV-based system.

Khrapko *et al.* [24] utilized a denaturing (denaturing agents and/or heat) milieu within the capillary leading to a partial melt of the heteroduplex molecule. This results in the reduction of the electrophoretic mobility relative to the unaffected homoduplex. This technique was shown to be suitable for the detection of low frequency mutations and for the detection of rare variants by genetic screening of pooled samples. The result shows that as low as 0.03% of a heteroduplex can be detected in a mixture [24].

Because the narrow dimensions of a capillary permit precise internal

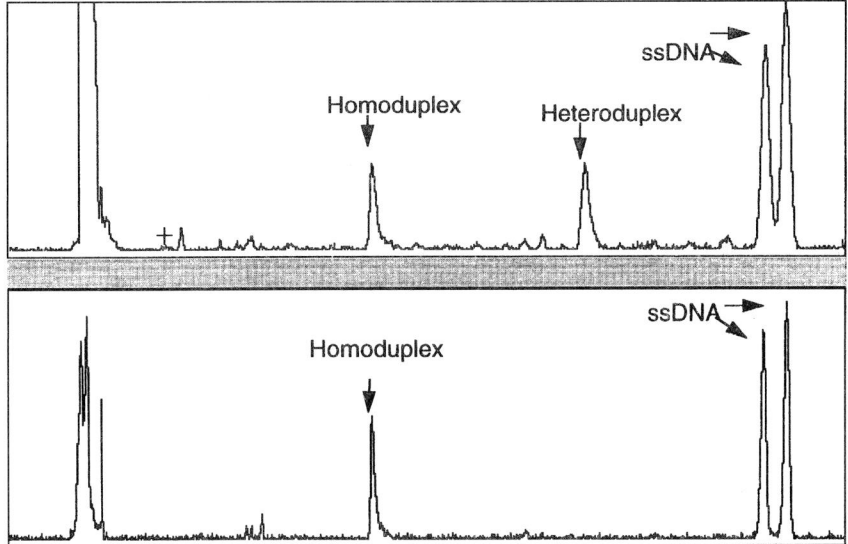

Figure 30.1 Detection of sequence differences of the env gene in HIV subtypes. Conditions: 3% GeneScan Polymer, uncoated capillary of 75 μm i.d., Ld (length to detector) 30 cm, Lt (total length) 41 cm, 268 V cm^{-1}, 30°C, run time approximately 13 min. (Above) mix of subtype A and B PCR HIV DNAs. (Below) subtype A alone. By permission of Perkin Elmer Corp. Analysis was performed on an Applied Biosystems Model 310 Genetic Analysis System.

temperature control, it is possible to alter the internal temperature by adjusting the power within the capillary. Exploiting this principle, Gelfi *et al.* [25] established a temporal thermal gradient within a capillary that is filled with a linear polyacrylamide polymer containing a constant amount of a denaturant. The temperature at any given moment is modeled by taking into account the exact capillary diameter and length, the buffer specific conductivity at 25°C, its thermal coefficient of conductivity, the voltage applied and the Biot number. The differences in the melting profiles between homo- and heteroduplexes, and the resulting differences in mobility are used to detect mutated fragments. Five different mutations in two gene systems could be detected.

30.2.3 Determination of length variations for mutation detection, mapping of heritable diseases, carrier detection and forensic identification

DNA polymorphisms, as revealed by differences in fragment lengths after restriction endonuclease digestion, have been linked to a variety of genetic diseases such as Duchenne muscular dystrophy or retinitis pigmentosa. CE has been shown to be a fast alternative to either Southern analysis or conventional slab gel electrophoresis [26–28]. Because this test requires the demonstration of either two fragments as a result of the creation of a restriction site, or of only a larger fragment, indicative of the loss of the restriction site, the requirements for resolution are relatively lower. Therefore, analysis conditions can be chosen which yield fast run times, typically less that 8 min.

The use of highly informative short tandem repeats (STRs or microsatellites) with repeat length of 1–5 bp [29] has increased in recent years for different applications. One of the most important is in deciphering the inheritance pattern of disease-causing genes, and ultimately their isolation and characterization [30], and in the field of forensic identification [31]. Because of the small repetition length of STRs, the demand on the separation medium is high [11, 32]. Figure 30.2 shows an example of the analysis of four different forensic STRs (HUMTHO1, HUMFES, HUMvWA and HUMF13A) by laser-induced fluorescence (LIF). The use of different fluorophors allows the combination of markers which overlap in

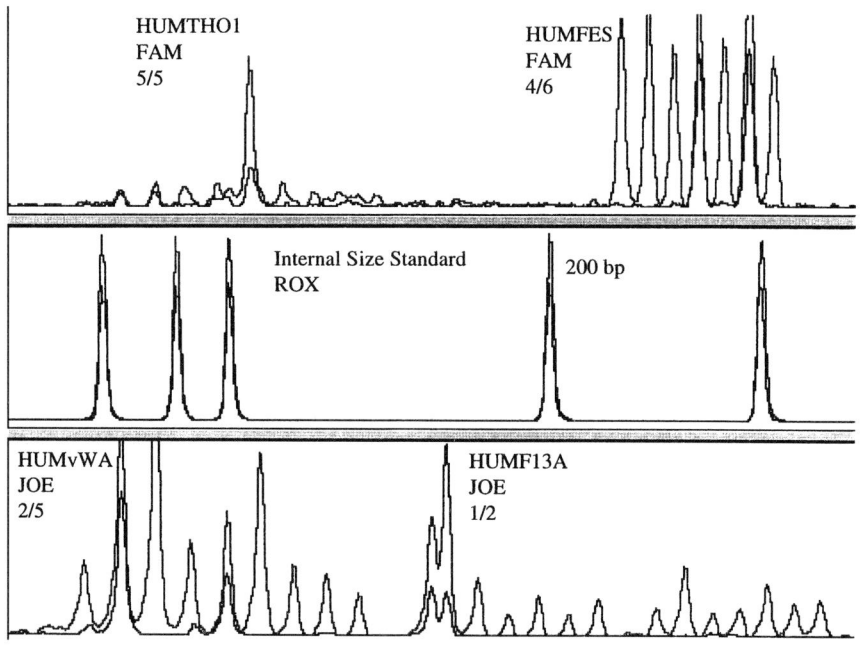

Figure 30.2 High resolution separation of a forensic quadruplex ladder. 3% GeneScan Polymer/6.6 M urea, uncoated capillary of 75 μm i.d., Ld (length to detector) 30 cm, Lt (total length) 41 cm, 317 V cm^{-1}, 30°C, run time approximately 15 min. The top and bottom panels show an overlay of an unknown sample with allelic size standards. Overlapping allele regions were labeled with different fluorophor primers (FAM, JOE) and electrophoresed in a single run. The genotype for each marker for the unknown sample is indicated. The match was done by overlaying an internal size standard, which was labeled with a third fluorophor, ROX (middle panel). Analysis was performed on an Applied Biosystems Model 310 Genetic Analysis System.

CUSTOMER P.O. NO.	ACCOUNT NO.	INVOICE DATE	INVOICE NO.
6100710	0000019	10/16/03	689912

QUAN. SHIP	ISBN EDITION	SHIP ADVICE	DESCRIPTION	SERIES	LIST PRICE
1	0751403598	SHINTANI	HBK CAPILLARY ELE		452.00
01406		6100710			

UPS

c/o Kluwer Academic Publishers
Return Address

PAGE 1 OF 1   0310089

THIS IS NOT AN INVOICE

size (multiplexing), thereby increasing the throughput. These samples can be discriminated by their different fluorescence properties.

In this example, the match between an unknown sample and individual peaks of the different allelic ladders ('genotyping') was aided by adding an internal size standard to each sample, correcting for run-to-run variations. Most alleles differ in size by 4 bp, with the exception of the first two alleles in the HUMF13A ladder, which are 2 bp apart. Because samples in CE are run sequentially rather than side-by-side as with a slab gel system, it is very important that the precision in sizing (reproducibility) is very high to ensure accurate genotyping. This is especially valid in cases where no allelic ladders are available, against which an unknown sample can be matched (see previous example). Butler *et al.* [33] use a dual internal standard approach to the STR typing of the HUMTHO1 in 97 population samples. They demonstrated that the CV for sizing seven alleles in the range of 179 to 203 bp is between 0.07% and 0.08%. The authors have shown for 88 consecutive injections that the CV for two alleles of the MFD 11 locus (114 and 118 bp) are 0.25% and 0.28%, respectively.

30.2.4 Quantitative assays

Quantitative assays, as for the determination of the abundance of a certain class of mRNAs by reverse transcription–PCR (RT–PCR), or for the multiplication or loss of certain regions of DNA, is possible by CE. By virtue of on-line detection, CE overcomes the poor quantification of slab gel electrophoresis due to the non-stoichiometric binding of the dye [34], and avoids the use of radiolabeled DNA precursors with all its associated problems. Several groups have used CE in combination with RT–PCR for the quantitative assessment of the expression of certain genes. Lu *et al.* [35] assessed the quantification of HIV-1, using LIF. Rossomando *et al.* [36] screened for the polio virus and Gelfi *et al.* [34] quantified the expression of basic fibroblast growth factor (bFGF) with competitive RT–PCR. All of the groups compared the quantitative data obtained by CE with quantitative data obtained after radiolabeling and calculating the PCR products from the resulting autoradiographs. Better sensitivity and equal or better reproducibility with CE was found. Lu *et al.* [35] report a gain in sensitivity by CE combined with LIF of four orders of magnitude over the classical HIV p24 antigen enzyme-linked immunosorbent assay (ELISA). Stalbom *et al.* [37] describe a competitive RT–PCR assay for the quantification of rat gastric H^+, K^+-ATPase mRNA. This shows an excellent precision with a CV of better than 4% on a UV-based system using an internal standard.

Kuypers *et al.* [38] describe a competitive PCR assay to quantify residual lymphoma cells carrying a translocation between chromosomes 14 and 18.

Samples were analyzed on a UV-based CE instrument and compared to autoradiographs. Both methods yield data within each other's error, with CE showing a sufficient reproducibility.

Arakawa *et al.* [28] describe the use of CE for the detection of single base mutations by an amplification refractory mutation system (ARMS). In this technique, a known mutation is discriminated from wild-type DNA by the use of a mutation-specific primer. Normal and patient samples are amplified in different tubes and analyzed in two consecutive CE runs. The result is displayed either as a 'yes' or 'no' signal. The use of mutation/wild-type specific primers, each labeled with a different color fluorophor should increase both throughput and accuracy of the test result.

Clearly the biomedical research community has recognized the utility of CE for mutation detection. Other fields that have identified DNA technology as important in their context will adapt the tools of the biomedical community. Thus, the future will see increased use of DNA analysis by CE including the field of agriculture and animal husbandry [39]. This chapter serves as an introduction to a field with enormous growth potential and a further demonstration that the limits of the utility of CE to the biological community have yet to be defined.

30.3 Sequence analysis

30.3.1 Instrumentation

It is almost a truism that in order for an automated technology successfully to supersede an entrenched manual technology, the automated approach must at least match (if not exceed) the performance of the manual technique, as well as provide greater throughput and increased operator efficiency. This particularly applies to DNA sequencing by CE. Historically, CE has demonstrated the former, but because of the sequential nature of single capillary commercial systems, sample electrophoretic throughput is considered to be a disadvantage. This calculation ignores, of course, pre- and postseparation sample, gel and data processing time, as well as the time the operator actually spends working on the electrophoretic system (making reagent solutions, cleaning glass plates, etc.).

Because of this perceived disadvantage, and because of the challenging nature of batchwise approaches, most DNA sequencing instrument development has focused on multiple capillary systems. All are based on LIF detection to permit high sensitivity and selection of signal from different dye labels. Notable in these efforts, in particular, are confocal sequencing detection schemes [40], axial beam excitation of capillary arrays [41] and sequencing on micro fabricated capillary chips [41a]. Other variations on the capillary array approach have appeared, each addressing a special feature of the capillary array [42]. It is not the intent of this paper to review the instrument issues extensively, however, it can be confidently stated that no one approach adequately satisfies all of the necessary components of a sequencing system, yet.

30.3.2 Separation chemistry

Until recently, most investigations of CE for DNA sequencing utilized rigid, crosslinked polyacrylamide, despite the difficulties described above. Theoretical arguments have been proposed that the high degree of resolution necessary for DNA sequencing cannot be expected for DNAs of molecular mass > 100 bp using entangled polymer networks that are neither concentrated nor extremely viscous [43]. These are based on the notion that in order to achieve the highest resolution, the relaxation time of the entangled polymers (the average length of time during which entanglement between polymer chains and, therefore, sieving occurs) should be orders of magnitude longer than the residence time of an analyte molecule (the average length of time during which the analyte frictionally interacts with the sieving network). According to these calculations the relaxation time and residence time are 5.9×10^{-4} s and about 1.1×10^{-5} s, respectively; clearly far below the resolution threshold established in their theoretical arguments.

Although the arguments are compelling, empirical evidence suggests that entangled polymer networks may yet find reasonable application in DNA sequencing by CE. It is clear that alternatives to crosslinked polyacrylamide are essential for DNA sequencing by CE to be successful. A widely recognized but poorly emphasized fact is that time after time, when duplication of the early reports using crosslinked polyacrylamide for DNA sequencing is attempted, regardless of the effort expended, the polymer fails after a few injections (see Ueno and Yeung [41], for the most recent example).

In this regard, two new reports have recently appeared that describe novel non-crosslinked polymer systems for DNA sequencing. One report [44], describes a binary separation mixture of poly(ethylene oxide) (PEO) of 600 000 and 8 000 000 M_n. The rationale for the binary mixture is to permit better resolution for a wider range of DNA fragments (up to 400 bp). To this end the authors report adequate resolution for base calling for one run of fragments up to 108 bp. Comparison of the performance of this polymer system beyond this size range to that observed in this laboratory for conventional crosslinked polyacrylamide in slab gels suggests a one-to-one correspondence [44]. The authors report that under these conditions, the resolution is insufficient for base calling beyond 420 bp.

A second report [45] describes a novel synthetic separation polymer consisting of polyethylene glycols end-capped with micelle-forming

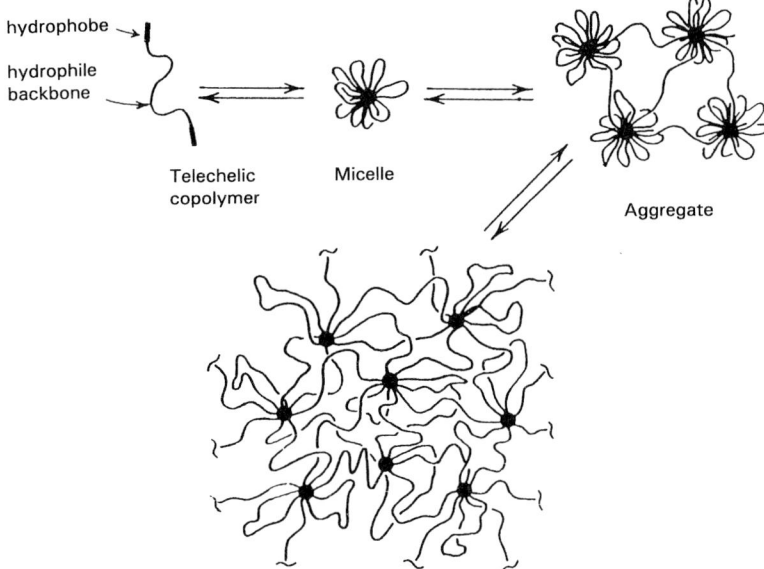

Figure 30.3 Transition of association structures of hydrophobically modified telechelic water-soluble polymer with increasing concentration. Above the critical micelle concentration the copolymer associates into micellar-like structures. As the concentration is increased these structures aggregate, forming aggregates that become larger as their concentration is further increased. They then form a continuous network displaying an average pore size related to the length of the hydrophile backbone. Reproduced from Ref. [45] with kind permission from VCH Verlagsgesellschaft mBH.

fluorocarbon tails (Figure 30.3). This material self assembles into networks of sufficient viscosity to satisfy the theoretical arguments discussed above for high resolution systems, but under shear stress are capable of flowing, thereby allowing capillary refilling under moderate (1500 psi) pressures (relative to the viscosity of the polymer mixture 'at rest'). This is because under sufficient shear force the gel, or aggregate phase, breaks down, creating a lubricating layer at the polymer–capillary interface.

The ability of the polymer to permit high resolution DNA separations is demonstrated in Figure 30.4 [45]. For the polymer mixture (1:1) of $(C_6F_{13})_2PEG(35K):(C_8F_{17})_2PEG(35K)$ ($M_n = 35\,000$), the point to maximum resolution is at approximately 460 bp (Figure 30.4c). The resolution is sufficient to permit base calling to 465 bp as shown in Figure 30.5 [45]. Further increases in resolution may be achievable by manipulating the length of the linker (hydrophile backbone in Figure 30.3 [45]).

30.4 Conclusions

Clearly the future of DNA sequencing and gene analysis by CE is dependent primarily on the creation of reproducible separation systems that

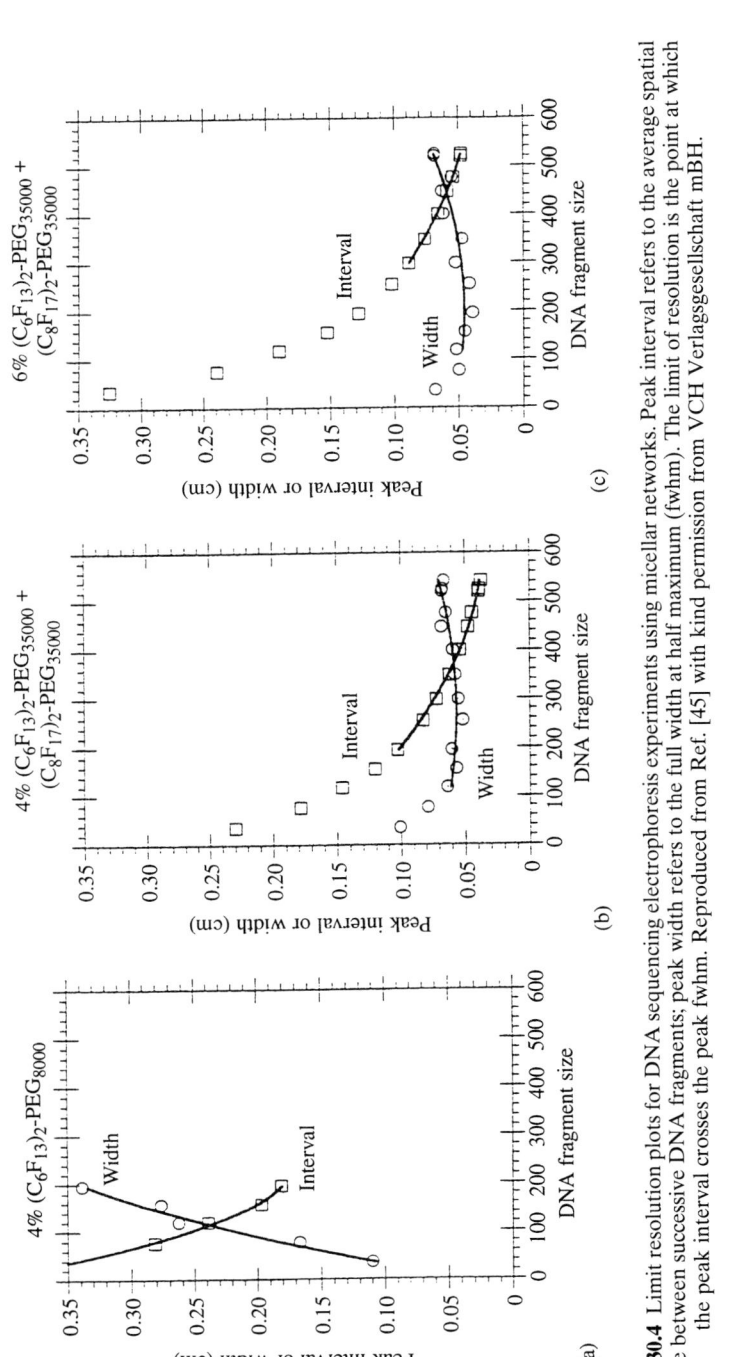

Figure 30.4 Limit resolution plots for DNA sequencing electrophoresis experiments using micellar networks. Peak interval refers to the average spatial distance between successive DNA fragments; peak width refers to the full width at half maximum (fwhm). The limit of resolution is the point at which the peak interval crosses the peak fwhm. Reproduced from Ref. [45] with kind permission from VCH Verlagsgesellschaft mBH.

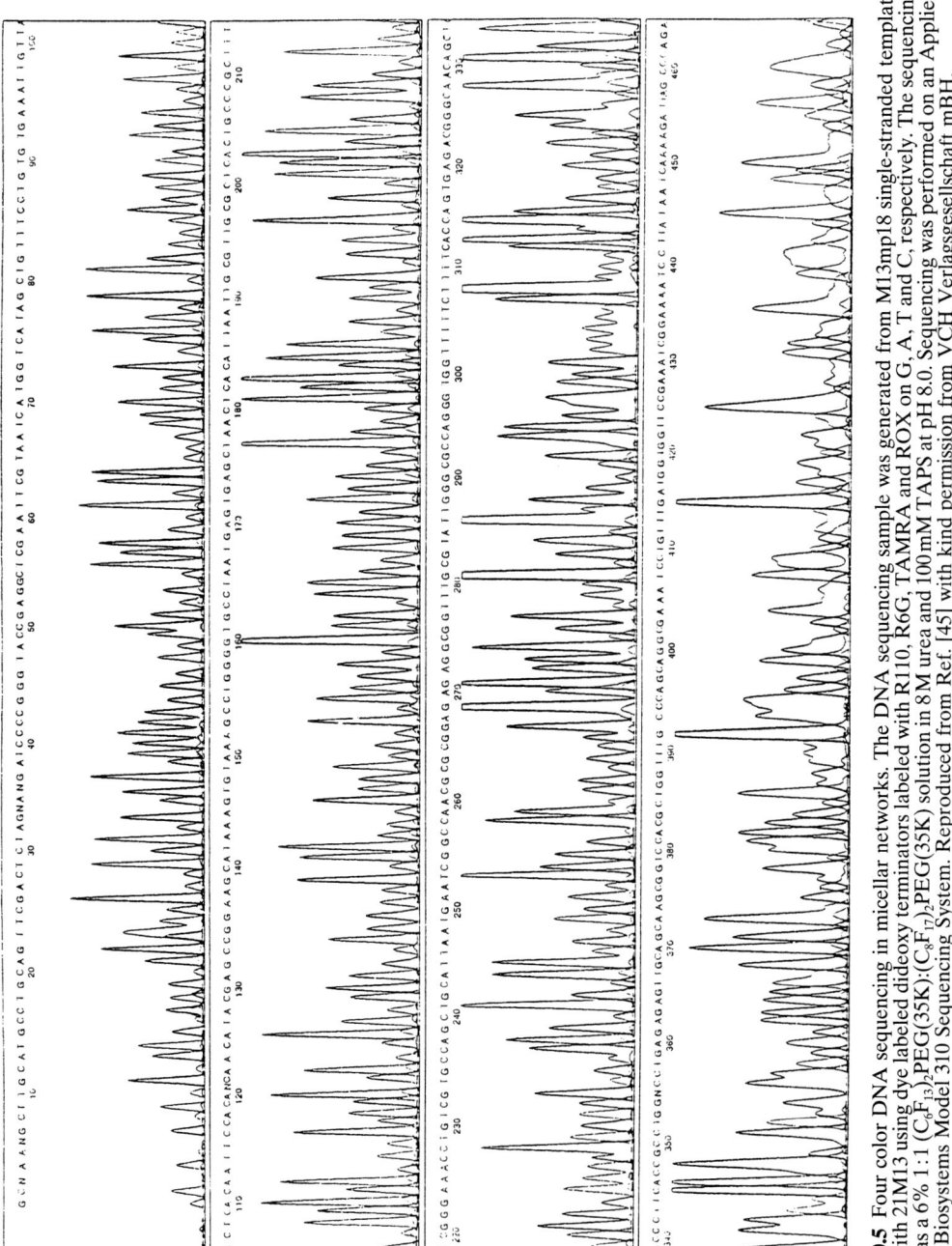

Figure 30.5 Four color DNA sequencing in micellar networks. The DNA sequencing sample was generated from M13mp18 single-stranded template primed with 21M13 using dye labeled dideoxy terminators labeled with R110, R6G, TAMRA and ROX on G, A, T and C, respectively. The sequencing matrix was a 6% 1:1 $(C_6F_{13})_2PEG(35K):(C_8F_{17})_2PEG(35K)$ solution in 8M urea and 100mM TAPS at pH 8.0. Sequencing was performed on an Applied Biosystems Model 310 Sequencing System. Reproduced from Ref. [45] with kind permission from VCH Verlagsgesellschaft mBH.

Table 30.1 Conditions for gene analysis

Methods	Compounds	Matrix	Reference
3% GeneScan polymer,[a] genetic analysis buffer, 240 V cm^{-1}, 50 cm × 75 µm, UV detection (ABD 270A-HT)	Bent DNA (anomalous migration)	Restriction digest	12
6% LPA,[b] 1 × TBE, 300 V cm^{-1}, 50 cm × 75 µm (coated), 260 nm	Anomalous migrating DNA	Restriction digest	15
3% GeneScan polymer, 1 × TBE/10% glycerol, 312 V cm^{-1}, 41 cm × 75 µm, laser excitation (ABD Prism 310)	SSCP (single-strand conformational polymorphism)	PCR buffer	21
3% GeneScan polymer, 0.5 × TBE, 268 V cm^{-1}, 41 cm × 75 µm, laser excitation (ABD Prism 310)	HMA (heteroduplex mobility assay)	PCR buffer	21
3% GeneScan polymer, genetic analysis buffer/6.6M urea, 317 V cm^{-1}, 41 cm × 75 µm, laser excitation (ABD Prism 310)	Forensic STR (single tandem repeat)	PCR buffer	21
0.5% HPMC,[c] 1 × TBE, 228 V cm^{-1}, 57 cm × 100 µm (DB-17 coated), 260 nm	HMA	Restriction digest	23
6% LPA, 1 × TBE/3.3M urea/20% formamide, variable V cm^{-1}, 75 µm (coated), laser excitation	Denaturation analysis for mutation detection	Water	24

Table 30.1 Continued

Methods	Compounds	Matrix	Reference
1% HEC,[d] 1 × TBE, 185 V cm^{-1}, 27 cm × 50 μm (DB-17 coated), laser excitation	PCR products	Water	33
0.5% HPMC, 1 × TBE, 176 V cm^{-1}, 100 μm (DB-17 coated), 260 nm	Quantitative competitive RNA/PCR	Dialyzed	37
4% LPA, 1 × TAE, 176 V cm^{-1}, 31 cm × 75 μm (polyacrylamide coated), 260 nm	Quantitative competitive DNA/PCR	Water	38
1.4% PEO[e] (M_n = 600K)/1.5% PEO (M_n = 8M)/3.5M urea, 267 V cm^{-1}, 45 cm × 75 μm (variably coated), laser excitation	DNA sequencing	Formamide/EDTA	44
6% (C$_6$F$_{13}$)$_2$PEG(35K): (C$_8$F$_{17}$)$_2$PEG (35K)[f]/8M urea/100 mM TAPS, pH 8, 200 V cm^{-1}, 47 cm × 75 μm (DB-210 coated), laser excitation	DNA sequencing	Cycle sequencing buffer/formamide	45

[a] Perkin-Elmer proprietary capillary electrophoresis separation polymer, [b] LPA: linear polyacrylamide, [c] HPMC: hydroxypropylmethylcellulose, [d] HEC: hydroxyethylcellulose, [e] PEO: polyethylene oxide, [f] PEG: polyethylene glycol.

exploit the advantages of CE versus conventional polyacrylamide gels. The attractiveness of CE certainly lies in the high degree of automation, flexibility in its application and the speed of separations. Table 30.1 summarizes conditions for gene analysis.

Improvements in instrumentation and in the chemistry of separation media will further stimulate its acceptance. While these have been challenging problems and most scientists have chosen to focus on the instrument issues (multiple capillary systems), recent successes are demonstrating that DNA analytical systems (hardware and chemistry) of great performance are sure to come in the near future.

Acknowledgments

The authors wish to express their thanks to Drs Barnett Rosenblum, James Robertson, Ben Johnson and Steve Menchen for their generous contribution of samples and/or unpublished work.

References

1. Olivera, B.M.; Baine, P.; Davidson, N. (1964) Electrophoresis of nucleic acids. *Biopolymers* **2**, 245–257.
2. Dovichi, N.J. (1994) In *Handbook of Capillary Electrophoresis*, Landers, L.P. (ed.), CRC Press, Boca Raton, pp. 369–387.
3. Guttman, A. (1994) In *Handbook of Capillary Electrophoresis*, Landers, J.P. (ed.), CRC Press, Boca Raton, pp. 129–143.
4. Sudor, J.; Foret, F.; Bocek, P. (1991) Pressure refilled polyacrylamide columns for the separation of oligonucleotides by capillary electrophoresis. *Electrophoresis* **12**, 1056–1058.
5. Heiger, D.N.; Cohen, A.S.; Karger, B.L. (1990) Separation of DNA restriction fragments by high performance capillary electrophoresis with low and zero crosslinked polyacrylamide using continuous and pulsed electric fields. *J. Chromatogr.* **516**, 33–48.
6. Werner, W.; Demorest, D.M.; Stevens, J.; Wiktorowicz, J.E. (1993) Size-dependent separation of proteins denatured in SDS by capillary electrophoresis using a replaceable sieving matrix. *Anal. Biochem.* **212**, 253–258.
7. Bocek, P.; Chrambach, A. (1991) Capillary electrophoresis of DNA in agarose solutions at 40°C. *Electrophoresis* **12**, 1059–1061.
8. Zhu, M.D.; Hansen, D.L.; Burd, S.; Gannon, F. (1989) Factors affecting free zone electrophoresis and isoelectric-focusing in capillary electrophoresis. *J. Chromatogr.* **480**, 311–319.
9. Chin, A.M.; Colburn, J.C. (1989) Counter-migration capillary electrophoresis (CMCE) in DNA restriction fragment analysis. *Am. Biotech Lab.* **7**, 16.
10. Grossman, P.D. (1992) In *Capillary Electrophoresis: Theory and Practice*, Grossman, P.D.; Colburn, J.C. (eds), Academic Press, San Diego, pp. 215–233.
11. Williams, P.E.; Marino, M.A.; Del Rio, S.A.; Turni, L.A.; Devaney, J.M. (1994) A practical approach to DNA restriction fragment and PCR product analysis by capillary electrophoresis. *J. Chromatogr.* **680**, 525–540.
12. Wenz, M.H. (1994) Capillary electrophoresis as a technique to analyze sequence-induced anomalously migrating DNA fragments. *Nucleic Acids Res.* **22**, 4002–4008.
13. Olefirowicz, T.M.; Ewing, A.G. (1990) In: *Capillary Electrophoresis: Theory and Practice*, Grossman, P.D.; Colburn, J.C. (eds), Academic Press, San Diego, pp. 45–85.
14. Hagerman, P.J. (1990) Sequence-directed curvature of DNA. *Ann. Rev. Biochem.* **59**, 755–781.

15 Berka, J.; Pariat, Y.F.; Müller, O.; Hebenbrock, K.; Heiger, D.N.; Foret, F.; Karger, B.L. (1995) Sequence dependent migration behavior of double-stranded DNA in capillary electrophoresis. *Electrophoresis* **16**, 377–388.
16 Orita, M.; Iwahana, H.; Kanazawa, H.; Hayashi, K.; Sekiya, T. (1989) Detection of polymorphisms of human DNA by gel electrophoresis as single-strand conformation polymorphisms. *Proc. Natl Acad. Sci. USA* **86**, 2766–2770.
17 Kuypers, A.W.H.M.; Willems, P.M.W.; van der Schans, M.J.; Linssen, P.C.M.; Wessels, H.M.C.; de Bruijn, C.H.M.M.; Everaerts, F.M.; Mensink, E.J.B.M. (1993) Detection of point mutations in DNA using capillary electrophoresis in a polymer network. *J. Chromatogr.* **621**, 149–156.
18 Cheng, C.; Kasuga, T.; Watson, N.D.; Mitchelson, K.R. (1995) Enhanced single-stranded DNA conformation polymorphism analysis by entangled solutions capillary electrophoresis. *J. Capillary Electrophoresis* **2**, 24–29.
19 Hebenbrock, K.; Williams, P.M.; Karger, B.L. (1995) Single strand conformational polymorphism using capillary electrophoresis with two-dye laser-induced fluorescence detection. *Electrophoresis* **16**, 1429–1436.
20 Perkin Elmer Corp. (1995) *The ABI Prism 310. GeneScan Chemistry Guide.* Foster City, CA.
21 Wenz, H.M. in preparation.
22 White, M.B.; Carvalho, M.; O'Brien, S.; Derse, D.; Dean, M. (1992) Detecting single base substitutions as heteroduplex polymorphisms. *Genomics* **12**, 301–306.
23 Cheng, J.; Kasuga, T.; Mitchelson, K.R.; Lightly, E.R.T.; Watson, N.D.; Martin, W.J.; Atkinson, D. (1994) Polymerase chain reaction heteroduplex polymorphism analysis by entangled solution capillary electrophoresis. *J. Chromatogr.* **677**, 169–177.
24 Khrapko, K.; Hanekamp, J.S.; Thilly, W.G.; Belenkii, A.; Foret, F.; Karger, B.L. (1994) Constant denaturant capillary electrophoresis (CDCE): a high resolution approach to mutational analysis. *Nucleic Acids Res.* **22**. 364–369.
25 Gelfi, C.; Righetti, P.G.; Cremonesi, L.; Ferrari, M. (1994) Detection of point mutations by capillary electrophoresis in liquid polymers in temporal thermal gradients. *Electrophoresis* **15**, 1506–1511.
26 Ulfelder, K.J.; Schwartz, H.E.; Hall, J.M.; Sunzeri, F.J. (1992) Restriction fragment length polymorphism analysis of ERBB2 oncogene by capillary electrophoresis. *Anal Biochem.* 260–267.
27 Del Principe, D.; Iampieri, M.P.; Germani, D.; Menichelli, A.; Novelli, G.; Dallapiccola, B. (1993) Detection by capillary electrophoresis of restriction fragment length polymorphism: Analysis of a polymerase chain reaction-amplified product of the DXS 164 locus in the dystrophy gene. *J. Chromatogr.* **638**, 277–281.
28 Arakawa, H.; Uetanaka, K.; Maeda, M.; Tsuji, A. (1994) Analysis of polymerase chain reaction product by capillary electrophoresis and its application to the detection of single base substitution in genes. *J. Chromatogr.* **664**, 89–98.
29 Weber, J.L. (1990) In *Genome Analysis Volume I: Genetic and Physical Mapping*, Cold Spring Harbor Laboratory Press, pp. 159–181.
30 Ziegle, J.S.; Su, Y.; Corcoran, K.P.; Nie, L.; Mayrand, P.E.; Hoff, L.B.; McBride, L.J.; Kronick, M.N.; Diehl, S.R. (1992) Application of automated DNA sizing technology for genotyping microsatellite loci. *Genomics* **14**, 1062–1031.
31 Caskey, C.T.; Edwards, A.; Hammond, H.A. (1989) DNA: The history and future use in forensic analysis. In *Proc. Int. Symp. on the Forensic Aspects of SNA Analysis*, U.S. Government Printing Office, Washington, D.C., pp. 3–9.
32 McCord, B.R.; Jung, J.M.; Holleran, E.A (1993) High resolution capillary electrophoresis of forensic DNA using a non-gel sieving buffer. *J. Liq. Chromatogr.* **16**, 1963–1981.
33 Butler, J.M.; McCord, B.R.; Jung, J.M.; Lee, J.A.; Budowle, B.; Allen, R.O. (1995) Application of dual internal standards for precise sizing of polymerase chain reaction products using capillary electrophresis. *Electrophoresis* **16**, 974–980.
34 Gelfi, C.; Leoncini, F.; Righetti, P.G.; Cremonesi, L.; di Blasio, A.M.; Carniti, C.; Vignali, M. (1995) Separation and quantitation of reverse transcriptase polymerase chain reaction fragments of basic fibroblast growth factor by capillary electrophoresis in polymer networks. *Electrophoresis* **16**, 780–783.
35 Lu, W.; Han, D.S.; Yuan, J.; Andrieu, J.M. (1994) Multi-target PCR analysis by capillary electrophoresis and laser-induced fluorescence. *Nature* **368**, 269–271.

36 Rossomando, E.F.; White, L.; Ulfelder, K.J. (1994) Capillary electrophoresis: separation and quantitation of reverse transcriptase polymerase chain reaction products from polio virus. *J. Chromatogr.* **656**, 159–168.
37 Stalbom, B.-M.; Torven, A.; Lundberg, L.G. (1994) Application of capillary electrophoresis to the post-polymerse chain reaction analysis of rat mRNA gastric H^+, K^+-ATPase. *Anal. Biochem.* **217**, 91–97.
38 Kuypers, A.W.H.M.; Meijerink, J.P.P.; Smetsers, T.F.C.M.; Linssen, P.C.M.; Mensink, E.J.B.M. (1994) Quantitative analysis of DNA aberrations amplified by competitive polymerase chain reaction using capillary electrophoresis. *J. Chromatogr.* **660**, 271–277.
39 Bates, S.R.E.; Knorr, D.A.; Weller, J.W.; Ziegle, J. (1996) Instrumentation for automated marker aquisition and data analysis. In: *The Impact of Plant Molecular Genetics*, Sobrall, B. (ed.), Birkenauer, Boston, pp. 239–255.
40 Huang, X.C.; Quesada, M.A.; Mathies, R.A. (1992) Capillary array electrophoresis using laser-excited confocal fluorescence detection. *Anal. Chem.* **64**, 967–972.
41 Ueno, K.; Yeung, E.W. (1994) Simultaneous monitoring of DNA fragments separated by electrophoresis in a multiplexed array of 100 capillaries. *Anal. Chem.* **66**, 1424–1431.
41a Woolley, A.T.; Mathies, R.A. (1995) Ultra-high-speed DNA sequencing using capillary electrophoresis chips. *Anal. Chem.* **67**, 3676–3680.
42 Kambara, H.; Takahashi, S. (1993) Multiple-sheathflow capillary array DNA analyser. *Nature* **361**, 565–566.
43 Bae, Y.C.; Soane, D. (1993) Polymeric separation media for electrophoresis: cross-linked systems or entangled solutions. *J. Chromatogr.* **652**, 17–22.
44 Fung, E.N.; Yeung, E.S. (1995) High-speed DNA sequencing by using mixed poly(ethylene oxide) solutions in uncoated capillary columns. *Anal. Chem.* **67**, 1913–1919.
45 Menchen, S.; Johnson, B.; Winnick, M.A.; Xu, B. (in press) Flowable networks as equilibrium DNA sequencing media in capillary columns. *Electrophoresis*.

31 Gene mutation and DNA sequencing
K. KITAGISHI

31.1 Introduction

Polymer sieving in capillary electrophoresis (CE) has been applied to DNA analysis because complementary double-stranded DNA or completely denatured single-stranded DNA molecules have a constant charge-to-mass ratio. These molecules migrate in sieving media, dependent only upon their molecular size, so that the migration velocity of the DNA of interest reflects its molecular size and secondary structure. Both gel-filled capillaries and replaceable linear polymer solutions can be used as sieving media. Polyacrylamide gel-filled capillaries can be prepared by procedures similar to conventional polyacrylamide slab gels, described in section 4.3. Gel-like molecular sieving electrophoresis is performed in replaceable linear polymer solutions such as non-crosslinked polyacrylamide, polysaccharides, poly(ethylene glycol), etc. Recently replaceable linear polymer solutions have been employed more frequently than gel-filled capillaries because of their easy operation and highly reproducible measurements [1–3].

In this chapter, gene mutation and DNA sequencing analysis in CE based on polymer sieving will be described. Although the concept of gene analysis probably includes gene damage under toxic conditions, the analysis of DNA adduct nucleotides damaged by exposure is not described here (see section 21.2). Recent publications on gene analysis are summarized in Table 31.1.

31.2 Polymerase chain reaction (PCR) product analysis

31.2.1 Size determination and quantitation of PCR-amplified DNA fragments

The molecular size of DNA fragments amplified by the polymerase chain reaction (PCR) in range of 20 to 1600 bp can be determined by sieving in CE with linear polymer solutions (LPS) in less than 20 min [4]. DNA separation with gel-filled capillaries generally takes longer than with LPS. A plot of migration time vs. DNA size is commonly employed for calibration. For an accurate determination of fragment size, the correct choice of DNA markers is required [2]. The molecular sizes of human mitochondrial

Table 31.1 Gene mutation and DNA sequencing

Methods	Compounds	Sample solution contents; sample origin	Refs
0.5% Hydroxyethyl cellulose, 100 mM Tris–borate pH 8.7, 0.1 mM EDTA, 1.27 µM ethidium bromide 37.8 µA, 25°C, 70 cm ×100 µm (i.d.) DB-17 capillary 260 nm	pBR322/Hae III; PCR products (D1S80, SE33)	Desalting solution by dialysis Commercial product by Sigma; human chromosome	1
Commercial polyacrylamide gel-filled capillary; two kinds of commercial sieving buffer and capillary, 10 µM ethidium bromide 210 V cm^{-1}, 30°C, 40 or 50 cm to detector × 100 µm (i.d.), 260 nm	ϕX174/Hinf I; ϕX174/Hae III; DNA marker XI; HUMTHO1 allelic ladder; PCR products (Williams and Jackson F2 soybean)	Desalting solution by dialysis Commercial products by Gibco BRL, Boeringer Mannheim and Promega; human mitochondria	2
0.8% Hydroxyethyl cellulose (M_n = 438k), 45 mM Tris, 45 mM borate, 1 mM EDTA, pH 8.3 80 V cm^{-1}, 22°C, 50 cm × 75 µm (i.d.) polyacrylamide-coated capillary, capillary array LIF, ex. 488 nm (Ar laser), em. bandpass filter of 520 nm and longpass filter of 590 nm	HUMTHO1 PCR products (primers are labeled with fluorescent dyes)	Desalting by dialysis Human chromosome	3
0.5% Hydroxyethyl cellulose, 10 mM Tris–borate buffer, 0.1 mM EDTA, pH 8.7, 25 mM NaCl, 1.27 µM ethidium bromide 15 kV, 35°C, 72 cm (50 cm to detector) × 50 µm (i.d.), 260 nm	PCR products of androgen receptor mRNA	10 mM Tris–HCl, 1 mM EDTA, pH 8.0 Human fibroblast	4
0.5% Hydroxypropyl methyl cellulose, 89 mM Tris, 89 mM borate, 2 mM EDTA, pH 8.5 175–350 V cm^{-1}, 25°C, 57 cm (50 cm to detector) × 100 µm (i.d.) dimethylpolysiloxane-coated capillary, 254 nm	PCR-amplified rDNA internal transcribed spacer and intergenic spacer	PCR amplified solution *Laccaria laccata* and *Laccaria bicolor*	5
1.0% Hydroxyethylcellulose, 100 mM Tris, 100 mM borate, 0.1 mM EDTA, pH 8.7, 1.27 µM ethidium bromide, 5 ng ml^{-1} YO-PRO-1 38 µA, 25°C, 60 cm to detector, DB-17 capillary LIF, ex. 488 nm (Ar laser)	HUMTHO1 alleles; vWA alleles; MBP alleles; pBR322/Hae III	Diluted with YO-PRO-1 solution Human chromosome; human myelin	6

Table 31.1 Continued

Methods	Compounds	Sample solution contents; sample origin	Refs
89 mM Tris, 89 mM borate, 2 mM EDTA, pH 8.5, 1/15–2/15 µM TOTO 0.5% Methylcellulose, 40–50 cm ×50 or 100 µm (i.d.) DB-17 capillary 8–15 kV, 25°C, 40 cm × 50 µm (i.d.) polyacrylamide gel-filled capillary (3% T, 3% C) LIF, ex. 488 nm (Ar laser), em. 530 nm bandpass filter	φX174/Hae III; PCR products (apolipoprotein B; VNTR locus D1S80; mitochondrial DNA)	Prestained with TOTO solution Commercial product by BRL laboratories; human hair roots	7
1% Hydroxyethyl cellulose, 89 mM Tris–89 mM borate buffer, 2 mM EDTA, 630 ng ml^{-1} YO-PRO-1 7 kV (260 V cm^{-1}), 25°C, 20 cm to detector × 50 µm (i.d.) DB-17 coated capillary LIF, ex. 488 nm (Ar laser), em. 510 nm	RT-PCR products of HCV RNA	10 mM Tris, 50 mM KCl, 1.5 mM MgCl$_2$, pH 8.3 Human serum of hepatitis C patient	8, 26
1.0% Hydroxyethyl cellulose, 100 mM Tris–borate buffer, 1 mM EDTA, pH 8.7 or 8.9, 1.27 µM ethidium bromide, 50 ng ml^{-1} YO-PRO-1 15 kV, 20°C or 25°C, 57 cm (50 cm to detector) × 50 or 100 µm (i.d.) DB-17 coated capillary LIF, ex. 488 nm (Ar laser), em. 520 nm	PCR products of human mitochondrial DNA: HV1A, HV1B and HV2	Desalting solution Human hair and blood	9
LIFluor dsDNA 1000 Kit (Beckman) 7.4 kV, 37 cm × 100 µm (i.d.) LIF, ex. 488 nm (Ar laser)	PCR and RT-PCR products of HIV DNA and RNA	Aqueous solution Human peripheral blood mononuclear cells	10
0.5% Hydroxypropyl methyl cellulose, 89 mM Tris, 89 mM borate, 2 mM EDTA, pH 8.5 175 V cm^{-1}, 25°C, 57 cm (50 cm to detector) × 100 µm (i.d.) DB-17 capillary, 260 nm	PCR amplified RFLP samples of ERBB2 oncogene; φX174/Hae III	Desalting solution of PCR-RFLP samples; diluted φX174/Hae III marker with water Oncogen locus on human chromosome; commercial product by New England Biolabs	11

Conditions	Sample	Others	Ref.
0.5% Hydroxypropyl methyl cellulose, 90 mM Tris, 90 mM borate, 2 mM EDTA, pH 8.5, 0–7.5% glycerol 228 V cm^{-1}, 22°C, 57 cm (50 cm to detector) × 100 μm (i.d.) DB-17 capillary, 260 nm	φX174/Hae III	Diluted with water	12
25 mM Tris–borate, 25 mM EDTA, pH 8.0 300 V cm^{-1}, 45 cm (25 cm to detector) × 75 μm (i.d.), polyacrylamide gel-filled capillary (9% T, 0% C) UV at 260 nm and LIF, ex. 488 nm (Ar laser), em. 520 nm	Fluorescence-tagged oligonucleotide	10 mM TBE Synthesized compound	13
Linear polyacrylamide (6% T, 0% C), 0.1 M Tris, 0.1 M borate, 2 mM EDTA, pH 8.3 140 or 100 V cm^{-1}, 37 cm (30 cm to detector) × 100 μm (i.d.) polyacrylamide-coated capillary, 254 nm	PCR products of Δ508 mutation in exon 10 in cystic fibrosis	Cystic fibrosis patients	15
Linear polyacrylamide (6% T, 0% C), 0.1 M Tris, 0.1 m borate, 2 mM EDTA, pH 8.3, 10 μM ethidium bromide 140 or 100 V cm^{-1}, 37 cm (30 cm to detector) × 100 μm (i.d.) polyacrylamide-coated capillary, 254 nm	PCR products of tetranucleotide repeats at the junction of intron IVS6a and exon 6b in cystic fibrosis	Cystic fibrosis patients	16
0.7% Methylcellulose, 50 mM Tris–borate, 2.5 mM EDTA, pH 8.3, 1 μg ml^{-1} ethidium bromide 200 V cm^{-1}, 30°C, 50 cm (30 cm to detector) × 100 μm (i.d.) DB-17 coated capillary, 260 nm	PCR products of VNTR locus in the human apolipoprotein B gene	Diluted to water Human leukocytes (heart disease)	17
100 mM Tris–borate, 2 mM EDTA, pH 8.3, 0.1 μg ml^{-1} thiazole orange 7.4 kV, 30°C, 37 cm (30 cm to detector) × 100 μm (i.d.), polyacrylamide gel-filled capillary LIF, ex. 488 nm (Ar laser), em. 530 nm bandpass filter	DNA fragments amplified with allele specific primers for medium-chain acyl-coenzyme A dehydrogenase deficiency	Dried blood spots of medium-chain acyl-coenzyme A dehydrogenase deficiency	18

Table 31.1 Continued

Methods	Compounds	Sample solution contents; sample origin	Refs
100mM Tris–borate, 2mM EDTA, pH 8.3, 0.1 μg ml^{-1} thiazole orange 10kV, 30°C, 37cm (30cm to detector) × 100μm (i.d.), polyacrylamide gel-filled capillary (3% T, 0.5% C) UV at 254nm and LIF, ex. 488nm (Ar laser), em. 530nm bandpass filter	PCR products containing Arg^{413} → Pro^{413} at the mutation site of exon 12 in a hepatic phenylalanine hydroxylase gene	Dried blood spots of phenylketonuria	19
0.5% Hydroxylpropyl methylcellulose, 89mM Tris–borate, 2mM EDTA, pH 8.5 160 V cm^{-1}, 50cm × 50μm (i.d.) preassembled coated capillary (Bio-Rad), 260nm	PCR products from DXS 164 locus of the dystrophin gene	Extracted with phenol–chloroform (3:1, v/v) and dissolved Human venous blood	20
100mM Tris–borate, 2mM EDTA, pH 8.5, in the absence or presence of 7M urea 400 V cm^{-1}, 25°C, 47cm (40cm to detector) × 100μm (i.d.), polyacrylamide gel-filled capillary, 254nm	Homo- and heterooligonucleotides in the 10–20-mer range	Diluted with water Commercial products by Pharmacia	21
Polyacrylamide gel-filled capillary (eCAP gel U100P by Beckman), 7M urea 11.1 kV, 30°C, 37cm × 100μm (i.d.), 254nm	Heterogeneous oligo-DNAs in 20–61-mer range	Gift from Takara; synthesized with Milligen DNA synthesizer	22
4% Linear polyacrylamide, 40mM Tris-acetate, 2mM EDTA, pH 8.3 250 V cm^{-1}, 25°C, 31 cm × 75μm (i.d.) polyacrylamide-coated capillary, 260nm	PCR products of p53 gene	Precipitated with ethanol and dissolved TBE buffer Normal white blood cells derived from normal persons and from bone marrow cells of multiple myeloma patients	24
7–10% Linear polyacrylamide, 89mM Tris–borate buffer containing EDTA, 5% glycerol 200 V cm^{-1}, 5–50°C, 20cm (14cm to detector) × 100μm (i.d.) polyacrylamide-coated capillary LIF, ex. 488nm (Ar laser), em. two bandpass filters of 520nm and 560nm	PCR products of insulin-like growth factor 1-binding protein 3 gene (primers are labeled with fluorescent dyes)	Diluted with water and formamide Human plasmid	25

Conditions	Analyte	Sample	Ref.
Polymer network (Dionex), 40 mM Tris–acetate, 2 mM EDTA, pH 8.3, 10 μg ml^{-1} ethidium bromide 150 V cm^{-1}, 18°C, 50 cm to detector × 50 μm (i.d.), 254 nm	Specific amplified DNA fragment of rpoB gene from *Mycobacterium tuberculosis*	Suspended in water and diluted in 10 mM Tris, pH 8.3, 50 mM KCl, 1.5 mM MgCl$_2$ Clinical isolate of *Mycobacterium tuberculosis*	26
89 mM Tris–borate, 2 mM EDTA, pH 8.3 150 V cm^{-1}, 50 cm to detector × 50 μm (i.d.), polyacrylamide gel-filled capillary (3% T, 3% C) LIF, ex. 488 nm (Ar laser), em. 520 nm	Specific amplified DNA fragment of rpoB gene from *Mycobacterium tuberculosis*, labeled with fluorescein at 5′ end	Precipitated with alcohol and suspended in deionized water Clinical isolate of *Mycobacterium tuberculosis*	26
0.5% Hydroxypropyl methylcellulose, 90 mM Tris–borate, 2 mM EDTA, pH 8.5, 3 μM ethidium bromide, 4.8% glycerol 228 V cm^{-1}, 23°C, 57 cm (50 cm to detector) × 100 μm (i.d.) DB-17 coated capillary, 260 nm	PCR products from the ITS2 region of *Heterobasidion annosum* rRNA gene	PCR-amplified solution European *Heterobasidion annosum*	27
6% Linear polyacrylamide, 89 mM Tris–borate, 1 mM EDTA, pH 8.3, 3.3 M urea, 20% formamide 250 V cm^{-1}, 36°C, 75 μm (i.d.), polyacrylamide-coated capillary LIF, ex. 488 nm (Ar laser), em 520 nm	PCR products of 206 bp fragments from bp 10011 of the human mitochondrial genome, labeled with fluorescein at 5′ end	Water or dilute buffer Human male TK-6 lymphoblasts	29
90 mM Tris–borate, 0.2 mM EDTA, pH 8.3 20 kV, 50 cm (25 cm to detector) × 50 μm (i.d.), polyacrylamide gel-filled capillary (4% T, 5% C) containing 8.3 M urea LIF, ex. 488 nm (Ar laser), em. 530 nm	M13mp18 single-stranded DNA labeled with fluorescein at 5′ end	Dissolved in formamide	31
0.1 M Tris–borate, 2.5 mM EDTA, pH 8.0, 7 M urea 310–350 V cm^{-1}, 65–92 cm (50–75 cm to detector) × 75 μm (i.d.), polyacrylamide gel-filled capillary (3% T, 5% C) containing 7 M urea LIF, ex. 488 nm (Ar laser), em. 520 nm	M13mp18 single-stranded DNA labeled with JOE at 5′ end	Precipitated with ethanol and resuspended in 80% formamide, 8 mM EDTA	32

Table 31.1 *Continued*

Methods	Compounds	Sample solution contents; sample origin	Refs
75 mM Tris–phosphate, 10 mM EDTA, pH 7.5 50–400 V cm^{-1}, 40 cm to detector × 50 μm (i.d.), polyacrylamide gel-filled capillary (4% T, 5% C) containing 8.3 M urea LIF, ex. 488 nm (Ar laser), em. 4 bandpass filters (540 nm, 560 nm, 580 nm, 610 nm)	M13mp18 single-stranded DNA labeled with four fluorescent dyes	5:1 formamide: 0.01 × running buffer (or 50 mM EDTA, pH 8.0)	33, 38
0.1 M Tris–borate, pH 8.6, 7 M urea 200 V cm^{-1}, 50 cm (36.3 cm to detector) × 100 μm (i.d.), polyacrylamide gel-filled capillary (5% T, 5% C) LIF, ex. 488 nm (Ar laser), em. 4 bandpass filters (540 nm, 560 nm, 580 nm, 610 nm)	M13mp18 single-stranded DNA labeled with four fluorescent dyes	Precipitated and resuspended in deionized formamide	34
1 × TBE 150 V cm^{-1}, 41 cm (27 cm to detector) × 50 μm (i.d.), polyacrylamide gel-filled capillary (6% T, 5% C) LIF, ex. 488 nm (Ar laser) and 543.5 nm (He–Ne laser), em. 4 bandpass filters (540 nm, 560 nm, 580 nm, 610 nm)	M13mp18 single-stranded DNA labeled with four fluorescent dyes	No description	35
1 × TBE 465 V cm^{-1}, 34 cm × 50 μm (i.d.), polyacrylamide gel-filled capillary (4% T, 5% C) LIF, ex. 488 nm (Ar laser), em. 515 nm and 535 nm	M13mp18 single-stranded DNA labeled with succinylfluorescein	Precipitated with ethanol and resuspended in 49:1 mixture of formamide:0.5 M EDTA at pH 8.0	35
1 × TBE 150–200 V cm^{-1}, 30–50 cm × 50 μm (i.d.), polyacrylamide gel-filled capillary (6% T, 5% C) containing 7–8 M urea and 20–30% formamide LIF, ex. 543.5 nm (He–Ne laser), em. 590 nm	M13mp18 single-stranded DNA fragments labeled with one dye (ROX, TAMRA or fluorescein)	Precipitated with ethanol and resuspended in formamide containing trace EDTA	35–37, 39, 40

Conditions	Sample	Sample preparation	Ref.
6% T linear polyacrylamide, 1 × TBE, 3.5 M urea, 30% formamide 250 V cm^{-1}, 32°C, 33 cm (18 cm to detector) × 75 μm (i.d.), polyacrylamide-coated capillary LIF, ex. 488 nm (Ar laser), em. 515 nm and 550 nm	M13mp18 single-stranded DNA fragments labeled with FAM and JOE	Precipitated and resuspended in 95% formamide:5% 50 mM EDTA	41
5%, 7%, 9% T linear polyacrylamide, 100 mM Tris-borate, 7 M urea, pH 8.3 82–200 V cm^{-1}, 20–100 cm to detector × 75 μm (i.d.), polyacrylamide-coated capillary LIF, ex. 488 nm (Ar laser), em. 520 nm	M13mp18 single-stranded DNA fragments labeled with fluorescein	Mixed with formamide (80% v/v)	42, 43
5% T linear polyacrylamide, 1 × TBE, 7 M urea 150 V cm^{-1}, 60°C, 39 cm × 50 μm (i.d.), polyacrylamide-coated capillary LIF, ex. 488 nm (Ar laser) and 543.5 nm (He-Ne laser), em. 4 bandpass filters (540 nm, 560 nm, 580 nm, 610 nm)	M13mp18 single-stranded DNA labeled with four fluorescent dyes	Precipitated with methanol and resuspended in formamide	44
Mixed solution of 1.5% PEO ($M_n = 8000k$) and 1.4% PEO ($M_n = 600k$), 89 mM Tris, 89 mM borate, 2 mM EDTA, 3.5 M urea 12 kV, 45 cm (35 cm to detector) × 75 μm (i.d.) polyacrylamide-coated capillary or bare capillary LIF, ex. 543.6 nm (He-Ne laser), em. two RG610 long-pass filters	DNA fragments from standard Sanger reactions, labeled with four fluorescent dyes	5:1 Formamide-50 mM aqueous EDTA solution	45
1 × TBE, 7 M urea 240 V cm^{-1}, 40 cm (25 cm to detector) × 100 μm (i.d.), polyacrylamide gel-filled capillary (9% T, 0% C), 4 to 24-capillary array LIF, ex. 488 nm (Ar laser), em. 530 nm	M13mp18 single-stranded DNA fragments labeled with fluorescein	Precipitated with ethanol and resuspended in 80% formamide	46
0.5% methyl cellulose, 100 mM Tris-borate, 2 mM EDTA, pH 8.2 50 V cm^{-1}, 50 cm (35 cm to detector) × 75 μm (i.d.) DB-1 coated capillary, 100-capillary array LIF, ex. 488 nm and 514 nm (Ar laser), em. 550 nm detected by CCD camera	φX174/Hae III mixed to TOTO-1 solution	2 μl Mixed with 8 μl of TOTO-1 (1×10^{-6} M) aqueous solution Commercial product by United States Biochemical	48

Table 31.1 *Continued*

Methods	Compounds	Sample solution contents; sample origin	Refs
89 mM Tris–borate, 2 mM EDTA, pH 8.3, 7 M urea ~230 V cm^{-1}, 30 cm to detector ×100 µm (i.d.), polyacrylamide gel-filled capillary (9% T, 0% C or 4% T, 5% C) containing 7 M urea, 20-capillary array LIF, ex. 488 nm (Ar laser) and 532 nm (YAG laser), em. 4 bandpass filters (540 nm, 560 nm, 580 nm, 610 nm), detected by CCD camera	M13mp18 single-stranded DNA fragments labeled with four fluorescent dyes	Precipitated with ethanol and resuspended in formamide	49
10% T Non-crosslinked polyacrylamide, 100 mM Tris–borate, 2 mM EDTA, pH 8.5, 7 M urea 2000–2300 V cm^{-1}, separation distance of 3.8 cm of micromachined capillary channel of 50 µm × 12 µm LIF, ex. 488 nm (Ar laser), em. 514 nm	Fluorescent oligonucleotides from 10 to 25 bases	Synthesized using Applied Biosystems 380B automated DNA synthesizer	50
0.75% Hydroxyethyl cellulose, 40 mM Tris, 40 mM acetate, 1 mM EDTA, pH 8.2, 1 µM TO or 0.1 µM TO6 180 V cm^{-1}, separation distance of 3.5 cm of micromachined capillary channel of 30, 50, 70 µm-wide coated with polyacrylamide LIF, ex. 488 nm (Ar laser), em. 530 nm	φX174/Hae III; PCR products of a hypervariable region in the second exon of the HLA-DQα locus	1 mM Tris, 0.1 mM EDTA, pH 8.2 Commercial product by New England Biolabs; provided by University of California, Berkeley	51
9% T, 0% C Polyacrylamide, 45 mM Tris, 45 mM borate, 1 mM EDTA, pH 8.3, 8.3 M urea 200 V cm^{-1}, separation distance of 3.5 cm of micromachined capillary channel of 50 µm × 8 µm LIF, ex. 488 nm (Ar laser), em. 510–540 nm, 545–570 nm, 570–590 nm, 590–660 nm	M13mp18 single-stranded DNA fragments labeled with four fluorescent dyes	Precipitated with ethanol and resuspended in 95% formamide/2.5 mM EDTA	52

TOTO: 1,1′-(4,4,7,7-tetramethyl-4,7-diazundecamethylene)-bis-4-[3-methyl-2,3-dihydro-(benzo-1,3-thiazole)-2-methylidene]-quinolinium tetraiodide, HCV: hepatitis C virus, TBE: Tris–borate–EDTA, JOE: 2′,7′-dimethyoxy-4′,5′-dichloro-6-carboxyfluorescein, FAM: 6-carboxyfluorescein, TAMRA: N,N,N′,N′-tetramethyl-6-carboxyrhodamine, ROX: 6-carboxy-X-rhodamine.

dinucleotide repeat, R-136, and human short tandem repeat (STR), HUMTHO1, were determined correctly. An example of genotyping was shown for a simple sequence repeat of the F2 generation of a Williams and Jackson soybean genotype at the locus [2].

The amplified internal transcribed spacer (ITS) and intergenic spacer (IGS) of RNA from *Laccaria laccata* and *Laccaria bicolor* could be sized. Multiple amplified IGS fragments of heterogenous size were detected in several strains [5].

Three loci containing a variable number tandem repeats (VNTR) with repeat units of 4 bp, being HUMTHO1, the Von Willenbrand Factor gene and a myelin basic protein gene of human origin, were able to be sized. This means that CE enables genetic typing of loci of interest [6]. An allelic mixture of VNTR with repeat units of 2 bp was completely separated by McCord *et al.* [1]. DNA typing was achieved for PCR-amplified DNA fragments from three different genetic loci, i.e. apolipoprotein B, VNTR locus D1S80 and mitochondrial DNA for heterozygous and homozygous individuals [7].

Molecular sizing by polymer sieving in CE is applied to a PCR-amplified DNA fragment derived from RNA, combining with reverse transcription (RT). Felmlee *et al.* [8] reported DNA sizing of HCV-specific DNA products in human serum by LIF–CE, compared with a conventional method, Southern blot. Analysis for 39 patients by the two methods showed a 100% correlation for detection of the 308 bp fragment. It suggests that CE is a potential technique for clinical assay for hepatitis C viral infection [8].

The possibility of quantitative analysis from peak areas/heights of PCR-amplified DNA fragments was investigated for androgen receptor mRNA [4], hypervariable areas in the control region of human mtDNA [9] and HIV DNA or RNA [10], following the CE separation during polymer sieving. Quantitation by CE shows advantages due to its ability to separate unincorporated primers and PCR byproducts from the targeted PCR product [9]. The plots of peak height vs. number of DNA templates or number of RNA virions of HIV-1 show good linearity [10] and a good peak area vs. concentration of PCR products relationship [4]. It admits the feasibility of diagnostic screening and therapy monitoring by this technique.

A high salt concentration in PCR-amplified solution reduces separation efficiency and reproducibility of DNA sizing in some cases, especially in electrokinetic injection. Desalting processes such as dialysis [1, 2, 3, 6] and DNA precipitation by organic solvents [5] were performed prior to sample injection. The processes enable DNA to be concentrated in sample solution and electropherograms to be simplified, ignoring the untargeted peaks of small molecules or proteins. On the other hand, part of the DNA of interest was lost during the desalting processes, causing a reduction in the accuracy of quantitation [9]. PCR products were injected without desalting in some studies [8, 10]. Even in electrokinetic injection of PCR products with high ionic strength, electrokinetic preinjection with water prior to sample injection avoids the sample diffusion effects due to high salt concentration in the sample [8].

Detection systems for DNA sizing are based on UV absorbance or on laser-induced fluorescence (LIF). The sensitivity of UV detection in the CE system is relatively low but sometimes enough for PCR products near 260 nm due to the characteristic absorption peak of nucleic acids [1, 2, 4, 5]. For more sensitive detection, LIF is applied by using intercalating dyes [6–10] or fluorescent labeling [3]. An intercalating dye, ethidium bromide, is often added to separation buffers in UV absorption detection, because its addition enhances the separation efficiency [1, 4, 11] as does glycerol [12].

Southern blotting was achieved using fluorescence-tagged oligonucleotides as probes. DNA molecules were hybridized with the probes prior to electrophoresis and the hybridized species could be detected by both UV absorbance and LIF [13].

Rapid genetic typing of STR alleles was developed using energy-transfer (ET) fluorescent labeled primers and capillary array electrophoresis. Unknown alleles of interest that were amplified by an ET primer with a fluorescein donor and a rhodamine acceptor were separated electrophoretically. The sizes of the alleles were identified by migration together with a standard ladder amplified by primers with only fluorescein. The detection system consists of a two-color confocal fluorescence scanner. Moreover, the design of the five-capillary array provides an increment in the throughput of STR typing [3].

31.2.2 DNA mutational analysis

Double-stranded DNA (dsDNA) can be separated by size difference in CE with a gel-filled capillary or a sieving polymer buffer. This technique can be applied to a DNA diagnosis like detection of DNA mutation, which is one of several new potential fields in molecular biology and biological medicine. The separation of dsDNA by difference in size has been achieved conventionally with slab-gel electrophoresis combined with PCR-amplification of the target gene. They include restriction fragment length polymorphism (RFLP), single-strand conformation polymorphism (SSCP), heteroduplex polymorphism assay (HPA), denaturing gradient gel electrophoresis (DGGE) and ligase chain reaction (LCR), in which conventional gel electrophoresis would possibly be substituted for CE. A recent review of DNA mutation detections in CE introduces some examples of various analytical techniques for DNA polymorphism [14].

Two different gene mutations in cystic fibrosis (CF) were detected by Righetti's group [15]. One is a 3 bp-deletion in exon 10, ΔF508 mutation, accounting for ~70% of the molecular defects in the CF patient. The molecular sizes of PCR products are 98 bp and 95 bp for normal and mutation, respectively [15]. Another is the variable number of tandem repeats of tetranucleotide (GATT) repeat polymorphism at the junction of intron IVS6a and exon 6b. Two main forms are six- and seven-repeats and the six-repeat allele is strongly linked to ΔF508 mutation. CE provides identification of individuals by DNA sizing of PCR products, with 111 bp for a six-repeat, 115 bp for a seven-repeat and their heteroduplex [16].

The number of the repeat for the VNTR locus in the human apolipoprotein B gene was determined by Baba et al. for genotyping of coronary heart disease [17].

Arakawa et al. [18] reported PCR amplification with two sets of allele specific oligonucleotide primers, followed by sizing with LIF–CE. The mutant and normal alleles produced 175 bp and 202 bp DNA fragments in medium-chain acyl-coenzyme A dehydrogenase (MCAD) deficiency [18].

Restriction fragment length polymorphism (RFLP) analyses for PCR-amplified DNA fragments have been achieved for ERBB2 oncogen [11], hepatic phenylalanine hydroxylase (PAH) gene with $Arg^{413} \rightarrow Pro^{413}$ at the mutation site [19] and DXS 164 locus of the dystrophin gene [20]. The analysis of ERBB2 oncogen showed homo- or heterozygosity based on the presence of 550 and/or 520 bp DNA fragments [11]. Single base substitutions could be detected both for 245 bp DNA fragments of PAH gene [19] and for 740 bp DNA fragments of dystrophin gene [20].

The electrophoretic migration properties of single-stranded oligonucleotides in CE were found to be highly dependent upon their base composition both in denaturing and non-denaturing gels [21]. The existence of a secondary structure in single-stranded oligo-DNA was strongly suggested even in denaturing gel, i.e. in the presence of 7 M urea, by analyzing the migration time of heterogeneous oligo-DNAs [22]. These studies revealed the feasibility of point-mutation detection in CE as single-strand conformation polymorphism following PCR amplification (PCR–SSCP) which was reported on a conventional slab gel [23]. SSCP in CE has advantages in non-radioactive detection, from the automated system, small amount of sample needed, short analysis time and accurate temperature control during electrophoresis, compared to PCR–SSCP using slab gel. SSCP analysis in CE was done for the tumor suppressor, p53 gene on chromosome 17. PCR-amplified 372 bp DNA samples were electrophoretically separated using linear polyacrylamide solution as a sieving medium in UV detection, following heating to 90°C to denature double-stranded DNA and snap-cooling in iced water to prevent reannealing [24]. LIF detection was applied to SSCP–CE with two-dye labeling for mutation detections of standard lac I gene of 255 bp and insulin-like growth factor 1-binding protein 3 gene of 276 bp [25]. Recently this technique was employed to detect single mutations within *Mycobacterium tuberculosis*-specific amplified DNA fragments by Felmlee *et al.* They described dideoxy fingerprinting (ddF), in which fragments can be analyzed based on the combination of chain terminating differences similar to DNA sequencing and secondary structural differences similar to SSCP [26].

Heteroduplex polymorphism analysis involves reannealing single-stranded DNA in a mixture of the wild type double-stranded DNA and a mutant one. HPA of 125 bp from the ITS 2 region of *Heterobasidion annosum* rRNA gene was reported. The reannealing was carried out by heating at 94°C and cooling slowly to room temperature. Heteroduplex with single-base deletion could be detected in CE by linear polymer sieving. Ethidium bromide and glycerol were added to the polymer solution in order to improve the peak resolution [27].

Denaturing gradient gel electrophoresis was introduced as a point-mutation detection technique using a slab gel containing a concentration gradient of one or more denaturants. A partially melted DNA molecule migrates much slower than an unmelted one [28]. In CE, DGGE is partially modified form of constant denaturant gel electrophoresis (CDGE). The heteroduplex of single-base substitution of 206 bp DNA from human TK-6 lymphoblasts was resolved by CDGE under constant denaturant conditions, combining a column temperature with a denaturant buffer [29].

Ligase chain reaction (LCR) can be applied to the routine analysis of a known point mutation. Two sets of adjacent oligonucleotides, complementary to each target strand, are used and thermostable DNA ligase combines the two oligonucleotides which are perfectly base-paired at the junction to the target. Repeated thermal cycling can amplify the complementary DNA fragment but no amplification is shown for single-base mismatch DNA [30]. An example of CE separation of LCR products has been reported [14].

31.3 DNA sequencing

31.3.1 DNA sequencing in CE with gel-filled capillaries

DNA sequencing in CE was developed initially with gel-filled capillaries, i.e. capillary gel electrophoresis (CGE), in LIF detection to analyze fluorescently labeled DNA fragments generated in the enzymatic sequencing reaction, the Sanger method [31]. One of the major advantages of CE for DNA sequencing is that much higher electric fields can be applied than in a conventional slab gel due to the high efficiency of heat dissipation. It results in a remarkable reduction in separation time. However, multilane electrophoresis on a slab gel brings not only high throughput but also parallel migration of DNA fragments terminated at different bases but labeled with the same compounds such as fluorescein. The sequencing products labeled with a single fluorophore were analyzed by CGE in four independent runs, one for each of the four base-specific reactions [32]. A strategy of multiple fluorophore was applied to DNA sequencing with a single capillary in a single run. This approach has been widely used for a DNA sequencer with a slab gel, available commercially from Perkin Elmer, Applied Biosystems Division. Four sequencing reactions are performed using different fluorophore-conjugated primers and the reaction products are loaded together onto a denaturing polyacrylamide gel. An optical system for multiple fluorophore detection in CE originally consisted of an argon ion laser as a light source and four photomultipliers as detectors. The fluorescence emitted is collected at right angles with a microscope objective, spatially filtered to eliminate the scattering light and split into four equal parts with beam splitters. The fluorescence in four different wavelength regions of interest can be observed with photomultipliers through four kinds of bandpass filter [33]. Tomisaki et al. developed an electrophoretic system based upon the same concept but simplified to use instrumentation not equipped with beam splitters [34].

Dovichi and co-workers demonstrated three different detection systems for DNA sequencing products. Their aim is the development of a CE system for accurate, low-cost DNA sequencing with high sensitivity. Their instrumentation which consists of a LIF detector for four-spectral-channel sequencing for multifluorophore DNA sequencing has two lasers, an argon ion laser and a He–Ne laser, for excitation, and a rotating filter wheel consisting of four sectors of four bandpass interference filters and one photomultiplier for emitted fluorescence. A detection limit of 200 zmol was obtained [35]. A postcolumn LIF detector based on the sheath flow cuvette shows a remarkable reduction in the background signal deriving from scattering light, and consequently the detection limit is lowered for DNA sequencing [36]. A two-spectral-channel detector for the DNA products prepared according to the Du Pont sequencing technique and a one-spectral-channel detector for DNA sequencing using the strategy of Richardson and Tabor, both of which contain a sheath flow cuvette, produced detection limits of 20 and 2 zmol, respectively [35]. Particularly, a low-cost, 0.75 mW He–Ne laser is used for one-spectral-channel detection in order to analyze tetramethylrhodamine isothiocyanate-labeled DNA fragments [37].

Although a high electric field applied during electrophoresis reduces separation time, it also leads to a loss of separation mainly due to thermal gradient peak broadening. The resulting trade-off between speed and resolution should be considered. The optimum electric field strength for the best fragment resolution depends upon the length of DNA fragment to be separated; a longer DNA fragment shows a lower optimum electric field at peak resolution [38].

31.3.2 DNA sequencing in CE with soluble linear polymers

In order to maintain high resolving power in automated reproducible DNA sequencing in CE, the stability and optimization of capillary gels were investigated [39, 40]. Replaceable linear polymers showed promise for use in DNA sequencing, since difficulties associated with gel stability could be solved in replaceable polymers. A linear polyacrylamide solution as a sieving medium has the advantage of easy operation in capillaries and good stability during repeated runs. DNA sequencing for 350 nucleotides in close to 30 min was demonstrated by Ruiz-Martinez et al. [41] Chen et al. reported that DNA fragments longer than 500 bases could be sequenced in linear polyacrylamide [42].

Optimization of the electrophoretic conditions was studied for the performance of DNA sequencing with linear polyacrylamide in LIF detection. Using 9% T linear polyacrylamide and an electric field of 100 V cm^{-1}, single-base resolution for M13mp10 DNA fragments labeled with fluorescein isothiocyanate (FITC) up to 520 nucleotides has been obtained [43].

At elevated temperature, the viscosity of linear polyacrylamide decreases, which leads to reduction in the separation time. Moreover, separation at high temperature improves the sequencing accuracy, because the deviation in migration time is lowered, as the secondary structure of single-stranded DNA diminishes as the temperature increases. Dovichi's group reported the separation of M13mp18 DNA fragments labeled with four-color dyes of up to 640 bases in less than 2 h using 5% T linear polyacrylamide in 7 M urea at 60°C and in an electric field of 150 V cm^{-1} [44].

Fung and Yeung [45] performed DNA sequencing using polymer solutions of linear polyacrylamide and poly(ethylene oxide), both of which are available commercially. A mixed poly(ethylene oxide) in two size ranges with a lower viscosity than linear polyacrylamide solution provides a comparable separation and increases the separation speed. Using bare fused-silica capillaries preconditioned with 0.1 N HCl, DNA fragments from the Sanger reaction up to 420 bases can be separated within 26 min with an average rate of 25 bases min^{-1} [45].

31.4 Enhancement of throughput in gene analysis in CE

31.4.1 Parallel runs in an array of capillaries

Sequencing of the human genome (Human Genome Initiative) requires an automatic DNA sequencer with a high throughput. DNA analysis in CE shows rapid separation even with gel-filled capillaries, compared to conventional gel electrophoresis or high performance liquid chromatography

(HPLC). In order to analyze the base composition of DNA at higher speed, molecular sieving with polymer solutions and separations at higher temperatures have been attempted as described in the previous section. However, total sample throughput is not enough high, because only one capillary can be analyzed at a time although DNA sequencing with conventional slab gel allows parallel multi-lane electrophoresis. For DNA sequencing or diagnosis in CE, an increase in sample throughput is an urgent subject to be solved. Parallel separation of an array of capillaries with simultaneous monitoring yields enhancement of the sample throughput in CE. The technical features of capillary array electrophoresis are high, relatively uniform detection sensitivity and high-speed data acquisition traceable to the separation efficiency covered over all capillaries.

Huang et al. [46] designed a laser-excited confocal fluorescence capillary array scanner. Laser light of 1 mW at 488 nm is focused within a capillary and the resulting fluorescence is collected, filtered and detected by a cooled photomultiplier. During electrophoresis, the computer-controlled stage translates the capillary array holder past the optical system continuously at $20\,\text{mm s}^{-1}$ with 1 s repeat cycles. Satisfactory sequencing can be read from the electropherogram image with a high signal-to-noise ratio out to greater than 500 bases for four capillaries in less than 2 h. By using an array of 24 capillaries, similar resolution and sensitivity can be obtained [46]. This instrumental design was applied to genetic typing of short tandem repeat (STR) polymorphism. The feasibility of high-speed, high-throughput STR typing of double-stranded DNA is proposed [3].

Taylor and Yeung [47] developed multiplexed detection with axial beam excitation and a charge-coupled device (CCD) camera as the method of detection for simultaneously monitoring ten capillaries. Laser power of 5 mW at 488 nm was divided among 10–12 optical fibers, each of which is inserted into a different capillary for excitation. The capillary array in the detection region was imaged onto a CCD camera. A frame of data was taken every 0.9 s [47]. Ueno and Yeung [48] evaluated the modified design with an array of 100 capillaries for DNA sequencing, where the excitation laser beam was directed through two mirrors and then focused into a line by a planoconvex cylindrical lens. An aluminum block with 100 triangle-shaped grooves machined onto its surface was used to mount the 100 capillaries. Fluorescence signals excited by a laser can be detected over all capillaries by a CCD camera at the rate of 0.6 frames s^{-1}. The crosstalk in the capillary array is small enough for DNA sequencing and a satisfactory level of reproducibility among all capillaries is obtained [48].

Takahashi et al. [49] developed capillary array electrophoresis of 20 capillaries with a multiple sheath-flow cell. The sheath-flow cell can solve the serious problem of high background light scattered at the capillary surface and in polyacrylamide gel in a gel-filled capillary array LIF–CE. The detection system consists of an image-splitting prism with four color filters and a cooled CCD camera coupled to a cooled image intensifier. Twenty gel-filled capillaries are aligned at a 0.35 mm pitch in an optical cell (26 mm × 26 mm × 4 mm). The end of each capillary faces an open capillary 1 mm away. Two lasers (Ar ion laser 488 nm and YAG laser 532 nm) used for excitation can irradiate all migration lanes simultaneously from one side of the capillary array like commercial DNA sequencers with a slab gel. Fluorescence labeled DNA with four different fluorophores could be detected at the minimum concentration of 2×10^{-12} M with a base reading speed of 200 bases/h/capillary [49].

31.4.2 High speed separations on a micromachined CE chip

The use of microfabrication to produce electrophoretic separation capillaries enables additional miniaturization and ultra-high speed analysis in CE. Effenhauser et al. [50] reported a successful application of polyacrylamide gel electrophoresis in micromachined planar glass structures for the size separation of fluorescent oligonucleotides ranging from 10 to 25 bases. A channel system was formed in the surface of polished glass plate by a standard photolithographical process, thermally bonded to a second glass slide. By application of an electric field up to 2300 V cm^{-1}, size separation of the oligonucleotides could be achieved in less than 45 s with a separation distance of 3.8 cm. Fast repetitive sample injection and analysis shows excellent reproducibility for migration time (<0.06%) and peak height (<1.7%) [50].

Woolley and Mathies [51] designed a capillary array fabricated on planar glass chips. This capillary array had 15 CE devices on each chip and was developed for optimization of sizes (widths) of separation and injection channels (ranging from 30 to 120 µm), but not for parallel migration, although the possibility was suggested. Molecular sizing of DNA fragments from 70 to 1000 bp is complete in only 120 s using hydroxyethyl cellulose as a sieving matrix and with a separation distance of 3.5 cm [51]. Woolley and Mathies [52] also performed DNA sequencing on microfabricated capillary electrophoresis chips by using a denaturing 9% T, 0% C polyacrylamide sieving medium. For one-color DNA sequencing, the peak corresponding to 433 bases is detected in 10 min with a separation channel 50 µm wide, 8 µm deep and 3.5 cm effective length. Four-color DNA sequencing shows 97% accuracy out to 147 bases in a separation time of 9 min with a separation channel of the same dimensions. A resolution of greater than 0.5 was obtained out to 200 bases for both the one- and four-color separations [52]. Integration of microfabricated capillary electrophoresis as a reliable system will lead to gene analysis with ultra-high speed and ultra-high throughput.

31.5 Conclusion

Gene analyses in CE have been achieved based on molecular sieving with a gel-filled capillary or a replaceable linear polymer solution. The latter has become more popular recently mainly due to its ease of operation. CE-based sizing of double-stranded DNA fragments amplified by PCR reaction enables genetic typing of STR alleles and VNTR alleles. Some electrophoretic modifications to sample pretreatments with commercial CE systems can produce gene mutational analysis, such as RFLP, SSCP, HPA, etc., which often resolves even single-base mutations. DNA sequencing in CE has the advantage of short separation time due to the high electric field applied. In order to increase the total throughput in gene analysis, novel instrumental designs are in progress, e.g. an array of multicapillaries in which parallel electrophoretic runs are possible, and a micromachined CE chip.

References

1. McCord, B.R.; Jung, J.M.; Holleran, E.A. (1993) High resolution capillary electrophoresis of forensic DNA using a non-gel sieving buffer. *J. Liq. Chromatogr.* **16**, 1963–1981.
2. Williams, P.E.; Marino, M.A.; Del Rio, S.A.; Turni, L.A.; Devaney, J.M. (1994) Analysis of DNA restriction fragments and polymerase chain reaction products by capillary electrophoresis. *J. Chromatogr.* **680**, 525–540.
3. Wang, Y.; Ju, J.; Carpenter, B.A.; Atherton, J.M.; Sensabaugh, G.F.; Mathies, R.A. (1995) Rapid sizing of short tandem repeat alleles using capillary array electrophoresis and energy-transfer fluorescent primers. *Anal. Chem.* **67**, 1197–1203.
4. Nathakarnkitkool, S.; Oefner, P.J.; Bartsch, G; Chin, M.A.; Bonn, G.K. (1992) High-resolution capillary electrophoretic analysis of DNA in free solution. *Electrophoresis* **13**, 18–31.
5. Martin, F.; Vairelles, D.; Henrion, B. (1993) Automated ribosomal DNA fingerprinting by capillary electrophoresis of PCR products. *Anal. Biochem.* **214**, 182–189.
6. McCord, B.R.; McClure, D.L.; Jung, J.M. (1993) Capillary electrophoresis of polymerase chain reaction-amplified DNA using fluorescence detection with an intercalating dye. *J. Chromatogr.* **652**, 75–82.
7. Srinivasan, K.; Girard, J.E.; Williams, P.; Roby, R.K.; Weedn, V.W.; Morris, S.C.; Kline, M.C.; Reeder, D.J. (1993) Electrophoretic separations of polymerase chain reaction-amplified DNA fragments in DNA typing using a capillary electrophoresis–laser induced fluorescence system. *J. Chromatogr.* **652**, 83–91.
8. Felmlee, T.A.; Mitchell, P.S.; Ulfelder, K.J.; Persing, D.H.; Landers, J.P. (1995) Capillary electrophoresis for the post-amplification detection of a hepatitis C virus-specific DNA product in human serum. *J. Capillary Electrophoresis* **2**, 125–130.
9. Butler, J.M.; McCord, B.R.; Jung, J.M.; Wilson, M.R.; Budowle, B.; Allen, R.O. (1994) Quantitation of polymerase chain reaction products by capillary electrophoresis using laser fluorescence. *J. Chromatogr.* **658**, 271–280.
10. Lu, W.; Han, D.-S.; Yuan, J.; Andrieu, J.-M. (1994) Multi-target PCR analysis by capillary electrophoresis and laser-induced fluorescence. *Nature* **368**, 269–271.
11. Ulfelder, K.J.; Schwartz, H.E.; Hall, J.M.; Sunzeri, F.J. (1992) Restriction fragment length polymorphism analysis of ERBB2 oncogene by capillary electrophoresis. *Anal. Biochem.* **200**, 260–267.
12. Cheng, J.; Mitchelson, K.R. (1994) Glycerol-enhanced separation of DNA fragments in entangled solution capillary electrophoresis. *Anal. Chem.* **66**, 4210–4214.

13. Chen, J.W.; Cohen, A.S.; Karger, B.L. (1991) Identification of DNA molecules by pre-column hybridization using capillary electrophoresis. *J. Chromatogr.* **559**, 295–305.
14. Mitchelson, K.R.; Cheng, J. (1995) Point mutation screening by high-performance capillary electrophoresis. *J. Capillary Electrophoresis* **2**, 137–143.
15. Gelfi, C.; Righetti, P.G.; Brancolini, V.; Cremonesi, L.; Ferrari, M. (1994) Capillary electrophoresis in polymer networks for analysis of PCR products: Detection of ΔF508 mutation in cystic fibrosis. *Clin. Chem.* **40**, 1603–1605.
16. Gelfi, C.; Orsi, A.; Righetti, P.G.; Brancolini, V.; Cremonesi, L.; Ferrari, M. (1995) Capillary zone electrophoresis of polymerase chain reaction-amplified DNA fragments in polymer networks: The case of GATT microsatellites in cystic fibrosis. *Electrophoresis* **15**, 640–643.
17. Baba, Y.; Tomisaki, R.; Sumita, C.; Morimoto, I.; Sugita, S.; Tsuhako, M.; Miki, T.; Ogihara, T. (1995) Rapid typing of variable number of tandem repeat locus in the human apolipoprotein B gene for DNA diagnosis of heart disease by polymerase chain reaction and capillary electrophoresis. *Electrophoresis* **16**, 1437–1440.
18. Arakawa, H.; Uetanaka, K.; Maeda, M.; Tsuji, A.; Matsubara, Y.; Narisawa, K. (1994) Analysis of polymerase chain reaction-product by capillary electrophoresis with laser-induced fluorescence detection and its application to the diagnosis of medium-chain acyl-coenzyme A dehydrogenase deficiency. *J. Chromatogr.* **680**, 517–523.
19. Arakawa, H.; Uetanaka, K.; Maeda, M.; Tsuji, A. (1994) Analysis of polymerase chain reaction product by capillary electrophoresis and its application to the detection of single base substitution in genes. *J. Chromatogr.* **664**, 89–98.
20. Del Principe, D.; Iampieri, M.P.; Germani, D.; Menichelli, A.; Novelli, G.; Dallapiccola, B. (1993) Detection by capillary electrophoresis of restriction fragment length polymorphism. Analysis of a polymerase chain reaction-amplified product of the DXS 164 locus in the dystrophin gene. *J. Chromatogr.* **638**, 277–281.
21. Guttman, A.; Nelson, R.J.; Cooke, N. (1992) Prediction of migration behavior of oligonucleotides in capillary gel electrophoresis. *J. Chromatogr.* **593**, 297–303.
22. Satow, T.; Akiyama, T.; Machida, A.; Utagawa, Y.; Kobayashi, H. (1993) Simultaneous determination of the migration coefficient of each base in heterogeneous oligo-DNA by gel filled capillary electrophoresis. *J. Chromatogr.* **652**, 23–30.
23. Orita, M.; Iwahana, H.; Kanazawa, H.; Hayashi, K.; Sekiya, T. (1989) Detection of polymorphisms of human DNA by gel electrophoresis as single-strand conformation polymorphisms. *Proc. Natl Acad. Sci. USA* **86**, 2766–2770.
24. Kuypers, A.W.H.M.; Willems, P.M.W.; van der Schans, M.J.; Linssen, P.C.M.; Wessels, H.M.C.; de Bruijn, C.H.M.M.; Everaerts, F.M.; Mensink, E.J.B.M. (1993) Detection of point mutations in DNA using capillary electrophoresis in a polymer network. *J. Chromatogr.* **621**, 149–156.
25. Hebenbrock, K.; Williams, P.M.; Karger, B.L. (1995) Single strand conformational polymorphism using capillary electrophoresis with two-dye laser-induced fluorescence detection. *Electrophoresis* **16**, 1429–1436.
26. Felmlee, T.A.; Oda, R.P.; Persing, D.A.; Landers, J.P. (1995) Capillary electrophoresis of DNA. Potential utility for clinical diagnoses. *J. Chromatogr.* **717**, 127–137.
27. Cheng, J.; Kasuga, T.; Mitchelson, K.R.; Lightly, E.R.T.; Watson, N.D.; Martin, W.J.; Atkinson, D. (1994) Polymerase chain reaction heteroduplex polymorphism analysis by entangled solution capillary electrophoresis. *J. Chromatogr.* **677**, 169–177.
28. Abrams, E.S.; Murdaugh, S.E.; Lerman, L.S. (1990) Comprehensive detection of single base changes in human genomic DNA using denaturing gradient gel electrophoresis and a GC clamp. *Genomics* **7**, 463–475.
29. Khrapko, K.; Hanekamp, J.S.; Thilly, W.G.; Belenkii, A.; Foret, F.; Karger, B.L. (1994) Constant denaturant capillary electrophoresis (CDCE): a high resolution approach to mutational analysis. *Nucleic Acids Res.* **22**, 364–369.
30. Barany, F. (1991) Genetic disease detection and DNA amplification using cloned thermostable ligase. *Proc. Natl Acad. Sci. USA* **88**, 189–193.
31. Drossman, H.; Luckey, J.A.; Kostichka, A.J.; D'Cunha, J.; Smith, L.M. (1990) High-speed separations of DNA sequencing reactions by capillary electrophoresis. *Anal. Chem.* **62**, 900–903.
32. Cohen, A.S.; Najarian, D.R.; Karger, B.L. (1990) Separation and analysis of DNA sequence reaction products by capillary gel electrophoresis. *J. Chromatogr.* **516**, 49–60.

33. Luckey, J.A.; Drossman, H.; Kostichka, A.J.; Mead, D.A.; D'Cunha, J.; Norris, T.B.; Smith, L.M. (1990) High speed DNA sequencing by capillary electrophoresis. *Nucleic Acids Res.* **18**, 4417–4421.
34. Tomisaki, R.; Baba, Y.; Tsuhako, M.; Takahashi, S.; Murakami, K.; Anazawa, T.; Kambara, H. (1994) High-speed DNA sequencer using capillary gel electrophoresis with a laser-induced four-color fluorescent DNA detector. *Anal. Sci.* **10**, 817–820.
35. Swerdlow, H.; Zhang, J.Z.; Chen, D.Y.; Harke, H.R.; Grey, R.; Wu, S.; Dovichi, N.J.; Fuller, C. (1991) Three DNA sequencing methods using capillary gel electrophoresis and laser-induced fluorescence. *Anal. Chem.* **63**, 2835–2841.
36. Swerdlow, H.; Wu, S.; Harke, H.R.; Dovichi, N.J. (1990) Capillary gel electrophoresis for DNA sequencing. Laser-induced fluorescence detection with the sheath flow cuvette. *J. Chromatogr.* **516**, 61–67.
37. Chen, D.Y.; Swerdlow, H.P.; Harke, H.R.; Zhang, J.Z.; Dovichi, N.J. (1991) Low-cost, high-sensitivity laser-induced fluorescence detection for DNA sequencing by capillary gel electrophoresis. *J. Chromatogr.* **559**, 237–246.
38. Luckey, J.A.; Smith, L.M. (1993) Optimization of electric field strength for DNA sequencing in capillary gel electrophoresis. *Anal Chem.* **65**, 2841–2850.
39. Swerdlow, H.; Dew-Jager, K.E.; Brady, K.; Grey, R.; Dovichi, N.J.; Gesteland, R. (1992) Stability of capillary gels for automated sequencing of DNA. *Electrophoresis* **13**, 475–483.
40. Rocheleau, M.J.; Grey, R.J.; Chen, D.Y.; Harke, H.R.; Dovichi, N.J. (1992) Formamide modified polyacrylamide gels for DNA sequencing by capillary gel electrophoresis. *Electrophoresis* **13**, 484–486.
41. Ruiz-Martinez, M.C.; Berka, J.; Belenkii, A.; Foret, F.; Miller, A.W.; Karger, B.L. (1993) DNA sequencing by capillary electrophoresis with replaceable linear polyacrylamide and laser-induced fluorescence detection. *Anal. Chem.* **65**, 2851–2858.
42. Chen, N.; Manabe, T.; Terabe, S.; Yohda, M.; Endo, I. (1994) High-resolution separation of oligonucleotides and DNA sequencing reaction products by capillary electrophoresis with linear polyacrylamide and laser-induced fluorescence detection. *J. Microcol. Sep.* **6**, 539–543.
43. Manabe, T.; Chen, N.; Terabe, S.; Yohda, M.; Endo, I. (1994) Effects of linear polyacrylamide concentrations and applied voltages on the separation of oligonucleotides and DNA sequencing fragments by capillary electrophoresis. *Anal. Chem.* **66**, 4243–4252.
44. Zhang, J.Z.; Fang, Y.; Hou, J.Y.; Ren, H.J.; Jiang, R.; Roos, P.; Dovichi, N.J. (1995) Use of non-cross-linked polyacrylamide for four-color DNA sequencing by capillary electrophoresis separation of fragments up to 640 bases in length in two hours. *Anal. Chem.* **67**, 4587–4593.
45. Fung, E.N.; Yeung, E.S. (1995) High-speed DNA sequencing by using mixed poly(ethylene oxide) solutions in uncoated capillary columns. *Anal. Chem.* **67**, 1913–1919.
46. Huang, X.C.; Quesada, M.A.; Mathies, R.A. (1992) Capillary array electrophoresis using laser-excited confocal fluorescence detection. *Anal. Chem.* **64**, 967–972.
47. Taylor, J.A.; Yeung, E.A. (1993) Multiplexed fluorescence detector for capillary electrophoresis using axial optical fiber illumination. *Anal. Chem.* **65**, 956–960.
48. Ueno, K.; Yeung, E.S. (1994) Simultaneous monitoring of DNA fragments separated by electrophoresis in a multiplexed array of 100 capillaries. *Anal. Chem.* **66**, 1424–1431.
49. Takahashi, S.; Murakami, K.; Anazawa, T.; Kambara, H. (1994) Multiple sheath-flow gel capillary-array electrophoresis for multicolor fluorescent DNA detection. *Anal. Chem.* **66**, 1021–1026.
50. Effenhauser, C.S.; Paulus, A.; Manz, A.; Widmer, H.M. (1994) High-speed separation of antisense oligonucleotides on a micromachined capillary electrophoresis device. *Anal. Chem.* **66**, 2949–2953.
51. Woolley, A.T.; Mathies, R.A. (1994) Ultra-high-speed DNA fragment separations using microfabricated capillary array electrophoresis chips. *Proc. Natl Acad. Sci. USA* **91**, 11348–11352.
52. Woolley, A.T.; Mathies, R.A. (1995) Ultra-high-speed DNA sequencing using capillary electrophoresis chips. *Anal. Chem.* **67**, 3676–3680.

32 Analysis of body fluids: urine, blood, saliva and tears

L.A. COLON

32.1 Introduction

The analysis of body fluids plays a critical role in the diagnosis and prognosis of a disease. A change in concentration or composition of a particular biochemical constituent in body fluids is used as an indicator of a physiological or pathological condition. A particular component in body fluids can thus be considered as a marker for the detection of a disease. Detection of such markers at an early stage can lead to a prompt diagnosis, which can result in an appropriate therapeutic treatment. The monitoring of a particular chemical species during the course of an illness, either an endogenous or exogenous compound, can also give an insight into the effectiveness of a particular therapy. Health evaluations and therapeutic screenings are also performed by analyzing body fluids. Many medical decisions rely on the results provided by the clinical laboratory on the analysis of body fluids. The clinical laboratory thus has the task of providing reliable information for the detection, diagnosis, prognosis, prevention and/or treatment of human diseases. It is extremely important then that the clinical laboratory is properly equipped with the latest and most appropriate analytical methodology to measure chemical species in biological fluids.

In this chapter, we take a look at the use of capillary electrophoresis (CE), a relatively new analytical technique for the analysis of body fluids; particular interest focuses on urine, blood (serum and plasma) and other less frequently analyzed fluids (e.g. tears and saliva). The high separation efficiencies, reduced analysis time and low cost of operation of CE place the technique as a viable alternative to HPLC. CE methods are not yet commonly used for clinical diagnosis; however, the great potential of CE for the determination of species of interest in a biological matrix and the profiling of metabolites of physiological interest has already been demonstrated [1–5]. The potential of CE for biomedical analysis and its prospects for the future have been pointed out very clearly [6]. Many compounds of biological importance can be determined in body fluids using several modes of CE, including CE in free solution, gel CE, capillary isotachophoresis (CITP), capillary isoelectric focusing (CIEF) and micellar electrokinetic capillary chromatography (MECC). MECC has gained popularity because it permits the separation and determination of analytes with discrimination based not only on electrophoretic mobilities but on the hydrophobic properties of the molecules. MECC methods have selectivities comparable to HPLC, requiring less time for analysis, less sample quantities and less costly procedures. The selectivity in MECC can be adjusted by controlling the nature of the micelles, the pH of the separation electrolyte and the use of other additives.

Here we recollect and illustrate a series of methods that employ CE for the analysis of body fluids, keeping in perspective their potential for clinical diagnosis. We emphasize methods that have been employed in the analysis of human fluids; therefore, little (or nothing) will be covered on fluids in research animals or on methods that have not yet been used on biological fluids. The determination of proteins in body fluids is used for clinical diagnosis; however, we will cover very little (if any) about protein determination by CE because the subject is covered in Chapter 14.

32.2 Capillary electrophoresis of urine samples

The analysis of urine specimens is performed on samples collected in a predetermined interval of time, which depends on the type of chemical determination to be conducted [7]. The time of collection is important because of the effect that dietary components can impart to the analysis. Sample conditions prior to analysis can also influence the determination of a specific analyte. Guzman et al., using a CE method, showed that the concentration of a particular species in urine can vary when a sample is frozen and then thawed for analysis [1]. It is, therefore, very important to handle the samples with care.

One of the most attractive features of CE methods is the possibility of analyzing biological fluids without any sample pretreatment. This approach simplifies the analytical method, avoids possible errors in sample manipulation and is economical. Samples of urine, as well as serum, from patients with medical drug intoxication have been analyzed by CE, CITP and MECC [8]. All methods were shown to be suitable for rapid screening and confirmation of drugs in urine and serum. The body fluids were injected directly or after dilution, providing results within 30 min. A comparison of HPLC and CE methods, using caffeine as a probe compound in the determination of N-acetylator phenotype, has also shown that CE is more advantageous because no sample extraction was necessary, resulting in short analysis time and removal of a potential source of error [9]. Nitrate and nitrite have also been determined in urine by CE without sample pretreatment (other than dilution) [10]. No interferences were observed in the separation performed in polyacrylamide-coated capillaries with a modified buffer at pH 3, using UV detection at 214 nm. Other CE methods have been developed to determine analytes in urine without sample pretreatment [8, 11–16]; these methods are also reproducible, linear and compare favorably with other procedures in the literature.

The screening of therapeutic drugs, as well as drugs of abuse and their metabolites is mostly performed by analyzing urine samples. The speed of analysis, resolution power and relatively low cost of operation of CE methods provide an alternative to the costly HPLC procedures. Several approaches have been developed to determine drugs and/or their metabolites in urine samples [15, 17–22]. Urine specimens have been analyzed by CE for the determination of diuretics, sport-banned drugs, narcotics–analgesics and β-blockers [17–19]. The CE procedures included the use of UV or laser-induced fluorescence (LIF) detection schemes. The separation power of CE also permits the determination of several species in the same analytical procedure within a relatively short period of time. For example, Chen and Evangelista reported a method for simultaneous quantification of multiple analytes in urine, based on immunochemical binding with CE separation [23]. They focused on competitive immunoassays with fluorescent-labeled antigen. The free and antibody-bound labeled antigen can be readily separated and detected by CE with LIF detection.

The analysis of urine samples also provides a useful means of monitoring the state of a patient with a known diagnosis. A particular case is the monitoring of pathological metabolites from phenylketonuric individuals. Phenylketonuria is a genetic disease resulting from deficiency in the enzyme phenylalanine hydroxylase; this deficiency alters the phenylalanine (Phe) metabolism. Phe accumulates in blood and the pathological metabolites are manifested in the urine fluid; the treatment is a low Phe diet beginning at an early age [24]. A simple, rapid and relatively inexpensive method to determine the pathological metabolites in urine can facilitate the monitoring of the state of a phenylketonuric patient. Dolnik showed that a CE method is convenient and appropriate for the determination of the Phe metabolites from the urine of phenylketonuric individuals [25].

Many other chemical species have been determined by CE in urine samples [26–33]. These include biochemical markers of malignancy (e.g. putrescine, cadaverine, spermidine and spermine) in monitoring transplantation [29], and sulfur-containing compounds (e.g. glutathion and cysteine) using electrochemical detection (ECD) [30–32], among others. ECD is applicable to the compounds with a relatively lower oxidation–reduction potential such as aromatic OH, aromatic SH, catecholamine etc. A selected list of applications of CE methods for the analysis of the urine fluid is presented in Table 32.1.

32.3 Capillary electrophoresis of blood samples

The determination of biochemical constituents in blood continues to be the most commonly used procedure for clinical diagnosis and for the monitor-

Table 32.1 Selected applications of capillary electrophoresis methods for the analysis of body fluids

Method Separation conditions Detection	Compounds	Matrix	Ref.
CE 50 mM borax, pH 10.0, 100 cm × 75 μm i.d., 20 kV UV 185 nm and indirect UV 254 nm	Oxalic, formic, malonic, fumaric, succinic, citric, acetic, pyruvic, lactic, isovaleric, hippuric acids	Urine	20
CE 100 mM imidazole, 0.1% HPMC, pH 7.4, 70 cm × 50 μm i.d., 15 kV; or 5 mM copper sulfate, 52 cm × 75 μm i.d., 30 kV UV 214 nm or indirect UV 214 nm	K^+, Na^+, Ca^{2+}, Mg^{2+}, Li^+	Serum	37, 38
MECC 100 mM borate, 100 mM NaH_2PO_4, pH 8.43, 70 mM SDS UV 280 nm	5-Acetylamino-6-formylamino-3-methyluracil, 1-methylxanthine	Urine	9
CE 10 mM MES, pH 5.5 or 10 mM TES pH 7.0, column i.d. = 25 μm, 30 kV Electrochemical (ECD)	Cysteine, glutathione, hydrazine, thioguanine, ascorbic acid, glucose	Urine	30
CE 60 mM CAPS, pH 10.6; or 70 mM acetate/500 mM betaine, pH 4.5, 67 cm × 50 μm i.d. UV 220 nm	Diuretics	Serum	19
CE 50 mM phosphate buffer, pH 6.8 column i.d. = 50 μm, 15 or 30 kV UV 280 nm	Cefixime and its metabolites	Urine	14
CZE, MECC 75 mM SDS, 16 mM borate and 10 mM phosphate, acetonitrile or propanol, pH 8.6 UV (multiwavelength, 195–320 nm)	Metabolites of dextromethorphan, mephenytoin and caffeine	Urine	13
CE 10 mM sodium phosphate, 10 mM HCl, pH 2.8 Electrochemical	Homocystine, cystine, GSSG	Urine	31
MECC 5 mM borate buffer, pH 9.0, 0.1% ethylenediamine, 2% SDS and 5% ethylene glycol, 50 cm × 50 μm i.d. fluorescence detection	Polyamines–putrescine, cadaverine, spermidine, spermine	Urine	29
CE MECC CE: 12 mM borate, 20 mM phosphate (pH 8.2–10.2); MECC: 75 mM SDS, 6 mM borate and 10 mM phosphate (pH ~9.2), 10% AcN UV multiwavelength (195–320 nm)	Metabolites of heroin and codeine	Urine	22

Table 32.1 Continued

Method / Separation conditions / Detection	Compounds	Matrix	Ref.
CE 0.01–0.75 M borate, 72 cm × 50 μm i.d., 17 kV UV 200 nm	Dextromethorphan, dextrorphan	Urine	13
CE (1) 50:50 methanol/10 mM borate–50 mM boric acid, pH 8.5, 30 kV; (2) 10 mM borate–50 mM boric acid, pH 8.5, 18 kV; (3) 4 mM copper(II) sulfate–4 mM formic acid (3:6), 20 kV (1) and (2) UV @ 200 nm; (3) indirect UV 215 nm	(1) Cocoamphocarboxyglycinate, (2) N-acetylcysteine, (3) inorganic ions	Tear	75
CE 50 mM phosphate, 1% acetonitrile, pH 9.5, 1 m × 75 μm, 10 kV UV 210 nm	Norephedrine, ephedrine	Urine	15
CE 1.5 mM phosphate–15.4 mM sodium 4-hydroxybenzoate, 1.3 mM cetrimide, 25 mM lithium hydroxide and 2.5% (v/v) methanol (pH 11.4), 15 kV UV 254 nm	Fosfomycin	Tear, serum	58
CE 20 mM citrate acid, pH 2.5, 20 kV, 122 cm × 50 μm i.d. UV 200 nm	Histidine-rich polypeptide	Saliva	66–69
CE 220 mM boric acid, pH 8.8 UV 254 nm	Iohexol	Serum	51
CE Phosphate, pH 7.4, with and without CD modification, 41 cm × 50 μm i.d., 10 kV UV 200 nm	Verapamil, norverapamil	Serum	45, 59
CE 25 mM sodium sulfate UV 210 nm	Nitrate	Plasma	39
CE 200 mM borate buffer, pH 10.2 LIF	Morphine and phencyclidine derivatives	Urine	23
CE 40 mM SDS–0.012 mM BSA, 18 kV LIF	Bilirubin	Serum	62
CE 0.1 M Tris–borate, pH 8.0, 16 kV UV 22 nm, LIF	Ibuprofen, warfarin, ketoprofen, verapamil, norverapamil, fluoxetine	Serum	41

Table 32.1 *Continued*

Method Separation conditions Detection	Compounds	Matrix	Ref.
CE 100 mM boric acid containing SDS, pH 8.4, 50 cm ×50 μm, 16.5 kV UV 214 nm	Felbamate	Serum	48
CE 30 mM Tris–citrate, pH 6.4, with 50% DMF and 0.1% HPMC LIF, He–Cd laser 325 nm	Methylmalonic acid	Serum	46
CE 100 mM phosphate, 2 mM to 20 mM [CD], 50 cm × 50 μm i.d. UV 310 nm	Warfarin enantiomers	Plasma	44
MECC 100 mM borate, pH 8.6, 25 mM SDS, 10% AcN, 75 μm i.d. UV 214 nm	Cicletanine	Plasma	49
MECC 10 mM sodium tetraborate, 50 mM boric acid, 50 mM SDS, pH 8.2, 20 kV UV 242 nm	Antipyrine	Serum, plasma	63, 57
CE 300 mM borate, pH 8.5, 13 kV UV 254 nm	Iohexol, theophylline	Serum	55
CZE 0.1 M aminocaproic acid, 0.1 M adipic acid, pH 4.3, 40 cm × 50 μm i.d., 11 kV UV 260 nm	Phenylalanine metabolites	Urine	25
CE 0.025 M borate, pH 10, 65 cm × 50 μm i.d., 20 kV UV 214 nm	Phenylalanine	Serum	60
CE 20 mM TAPS/AMPD, 30 kV, pH 8.8 UV 214 nm, LIF (ex. 488 nm/em. 520 nm)	Anti-cortisol antibody, cortisol	Serum	77

BSA, bovine serum albumin; DMF, dimethyl formamide; AMPD, 2-amino-2-methyl-1,3-propanediol; HPMC, hydroxypropyl methyl cellulose; TES (*N*-(tris(hydroxymetyl)methyl)-2-aminoethanesulfonic acid.

ing of patients undergoing clinical therapy. CE has been employed for the determination of a variety of compounds in blood derived samples (i.e. whole blood, serum and plasma). Of the several approaches developed, the analysis of serum samples by CE has received the most attention [1, 8, 21,

33–64]. Methodology developed to analyze urine samples can also be employed for the analysis of blood samples [25, 33–35]. The samples have been analyzed for the determination of endogenous and exogenous compounds (e.g. metal ions, antibiotics, anti-inflammatories and others) which can be used for diagnosis, monitoring the state of a disease and to determine the levels of a therapeutic drug (or its metabolites) in a patient. Most of the procedures involve sample pretreatment, although direct injection is possible [8, 42, 43].

Most of the CE methods developed to analyze blood samples have focused on the monitoring of therapeutic drugs or metabolites [8, 44–58], including enantiomeric separations [41, 44, 45, 56]. The samples are typically pretreated before analysis. Deproteinization with acetonitrile before analysis seems to be a very simple and effective pretreatment procedure [47, 48, 51, 53, 55]. The simplicity of CE methods and their speed of analysis facilitates the frequent monitoring of certain species of interest. For example, sumarin, an antiprostate-tumor drug, has a narrow therapeutic range and a long half life. Frequent monitoring of this drug during therapy is extremely important to avoid toxicity. Garcia and Shihabi [52] have developed a rapid CE method to determine sumarin in serum samples. The samples are deproteinized with acetonitrile and then analyzed by CE in less than 3 min, using 3-isobutyl-1-methylxanthine as internal standard and UV detection (254 nm).

Analysis of serum samples for the determination of Phe is a current procedure to confirm the diagnosis of the genetic disease phenylketonuria. Elevated concentrations are indicative of the disease. Follow-up Phe screening for patients undergoing therapeutic treatment of the disease can also be performed on serum samples. Tagliaro $et\ al.$ [60] developed a CE method for the rapid determination of Phe in serum which provides an alternative procedure to the routine laboratory diagnosis of phenylketonuria and can also be used to monitor patients undergoing therapy. Using an internal standard, their procedure was shown to be linear in the range of 5–175 $\mu g\ ml^{-1}$ with detection limits around 3 $\mu g\ ml^{-1}$ ($S/N = 3$). Intra- and inter-day reproducibilities were 4% and 7%, respectively. The method was used to determine Phe in sera from phenylketonuria patients.

The determination of bilirubin, a metabolic breakdown product of blood heme, is another example of the potential of CE in analyzing sera for diagnosis. Bilirubin plays an important role in the diagnosis and treatment of diseases associated with liver dysfunctions [61]. The four major bilirubin species in serum have been determined by CE using LIF as the detection scheme at a physiological pH [62]. The limit of detection of the method (30–150 nM) was shown to be two orders of magnitude lower than other procedures; this, in turn, allows the quantitation of bilirubin species in normal sera without sample pretreatment. Antipyrine, a marker of hepatic oxidative function and total body water, has also been determined in serum

[57, 63], as well as creatinine, an indicator of kidney disfunction [64]. Selected applications are presented in Table 32.1. They show the various possibilities of using CE methods to analyze blood-derived fluids.

32.4 Capillary electrophoresis of saliva and tear samples

The chemical/clinical analysis of saliva and tear samples represents a non-invasive approach for gathering information about the biochemical constituents present in the human body. This approach offers the possibility of monitoring drug therapies and pathological disorders while at the same time avoiding invasive sampling procedures, which can be painful or uncomfortable to the patient. The realization of non-invasive procedures for clinical analysis could encourage better adherence to therapies that require frequent monitoring during treatment (e.g. glucose monitoring in diabetics). The many advantages of CE combined with non-invasive sampling procedures can constitute a very powerful and indispensable methodology for the clinical laboratory.

The analysis of saliva can provide information revealing the manifestation of a disease. For example, the saliva can be analyzed to detect HIV antibodies [65], or to monitor the production of elevated salivary amylase quantities in patients with cystic fibrosis [65]. The CE methods developed to analyze salivary fluids have focused on the determination of proteins [66–70]. For example, Lal *et al.* developed CE methods to quantify the salivary cationic protein concentration in healthy adults and AIDS patients [66], and to examine histidine-rich polypeptides [67]. The potential for monitoring small molecules in saliva has been demonstrated using a capillary ultrafiltration probe to eliminate large biological molecules (e.g. proteins) to determine therapeutic drugs [71]. The collected sample can then be analyzed by a CE method. The concept was demonstrated in the analysis of theophylline, a common bronchial dilator; however, many other possibilities have been suggested [71]. The determination of caffeine and paraxanthine by MECC is another example [72]. Because of the low quantities required, CE may be useful in studying individual gland secretions to isolate the source of a particular secreted component.

Tear fluid also has the potential to be a non-invasive sampling medium for clinical diagnosis. The capability to analyze minute quantities of sample makes CE an excellent analytical tool to study tear fluid. Samples can be collected without induced tearing, preserving the most natural state of the fluid, which can then be analyzed directly by CE. Proteins as well as other components in tears have been determined by CE methods [6, 40, 73–75]. A CE method to analyze tear fluid before and after wounding of the cornea of the eye has shown differences before and after wounding [73]. A distinct pattern in peak intensities before and after wounding was observed, revealing a decrease or increase in certain tear components (i.e. proteins). Glucose in tear fluid can also be determined by CE [76]. The glucose in the tear sample (~3 µl) was derivatized with dansylhydrazine and then analyzed by CE using LIF as the detection system.

32.5 Concluding remarks

All the examples presented here demonstrate the great potential of CE for the analysis of body fluids. Although the use of CE for the analysis of biological fluids is indeed promising, more work still remains to be performed in order to use the technique for routine analysis in the clinical laboratory. Because of the small quantities examined, detectability can be an issue for improvement. More CE methods are being developed for routine analysis using LIF as the detection scheme (see applications in Table 32.1), which improves mass sensitivity. This progress is possible because of the availability of commercial instrumentation that can easily be implemented in the laboratory. However, there still is room for improvement; for example, the potential of ECD cannot be exploited until a well-designed commercial system becomes available. The combination of CE with immunoassays will indeed enhance selectivity and sensitivity. In principle, the separation power of CE can provide simultaneous determination of multiple analytes in a single capillary column. The principle has been demonstrated [23, 77, 78] and holds great promise. Another important aspect in the development of CE methods is the validation and optimization of the methodology. A method will gain acceptance and ultimately be implemented after rigorous validation with control samples, samples from patients and external quality samples; the results should also be comparable with well established procedures, when possible. Nevertheless, the future of CE as an analytical tool for the analysis of biological fluids is indeed encouraging; without any doubt, it will play a very important role in the clinical laboratory in future.

References

1. Guzman, N.A.; Gonzalez, C.L.; Hernandez, L.; Berck, C.M.; Trebilcock, M.A.; Advis, J.P. (1993) The use of CE in clinical diagnosis. In *Capillary Electrophoresis Technology*, Guzman, N.A. (ed.), Marcel Dekker, New York, pp. 643–672.
2. Thormann, W. (1993) The use of CE in clinical diagnosis. In *Capillary Electrophoresis Technology*, Guzman, N.A. (ed.), Marcel Dekker, New York, pp. 693–704.
3. Thorman, W.; Molleni, S.; Caslavska, J.; Schmutz, A. (1994) Clinical and forensic applications of capillary electrophoresis. *Electrophoresis* **15**, 3–12.
4. Shihabi, Z.K. (1992) Clinical applications of capillary electrophoresis. *Ann. Clin. Lab. Sci.* **22**, 398–405.
5. Xu, Y. (1995) Capillary electrophoresis. *Anal. Chem.* **67**, 463R–473R.

6. Deyl, Z.; Tagliaro, F.; Miksik, I. (1994) Biomedical applications of capillary electrophoresis. *J. Chromatogr.* **656**, 3–27.
7. Young, D.S.; Bermes, Jr, E.W. (1986) Specimen collection and processing: Sources of biological variation. In *Textbook of Clinical Chemistry*, Tietz, N.W. (ed.), Saunders Company, Philadelphia, pp. 478–518.
8. Caslavska, J.; Lienhard, S.; Thormann, W. (1993) Comparative use of three electrokinetic capillary methods for the determination of drugs in body fluids: prospects for rapid determination of intoxicants. *J. Chromatogr.* **638**, 335–342.
9. Lloyd, D.K.; Fried, K.; Wainer, I.W. (1992) Determination of N-acetylator phenotype using caffeine as a probe compound: a comparison of high-performance liquid chromatography and capillary electrophoresis methods. *J. Chromatogr.* **578**, 283–291.
10. Janini, G.M.; Chan, K.C.; Muschik, G.M.; Issaq, H.J. (1994) Analysis of nitrate and nitrite in water and urine by capillary zone electrophoresis. *J. Chromatogr.* **657**, 419–423.
11. Zhang, C.H.; Sun, Z.P.; Ling, D.K.; Zheng, J.; Guo, J.S.; Li, X.Y. (1993) Determination of 3-methylflavone-8-carboxylic acid, the main metabolite of flavoxate, in human urine by capillary electrophoresis with direct injection. *J. Chromatogr.* **612**, 287–294.
12. Li, S.; Fried, K.; Wainer, I.W.; Lloyd, D.K. (1993) Determination of dextromethorphan and dextrorphan in urine by capillary zone electrophoresis: application to the determination of debrisoquin-oxidation metabolic phenotype. *Chromatographia* **35**, 216–222.
13. Caslavska, J.; Hufschmid, E.; Theurillat, R.; Desiderio, C.; Wolfisberg, H.; Thormann, W. (1994) Screening for hydroxylation and acetylation polymorphisms in man via simultaneous analysis of urinary metabolites of mephenytoin, dextromethorphan and caffeine by capillary electrophoretic procedures. *J. Chromatogr.* **583**, 105–110.
14. Honda, S.; Taga, A.; Kakehi, K.; Koda, S.; Okamoto, Y. (1992) Determination of cefixime and its metabolites by high-performance capillary electrophoresis. *J. Chromatogr.* **590**, 364–368.
15. Chicharro, M.; Zapardiel, A.; Bermejo, E.; Perez, J.A.; Hernandez, L. (1993) Direct determination of ephedrine and norephedrine in human urine by capillary electrophoresis. *J. Chromatogr.* **622**, 103–108.
16. Jansen, E.H.J.M.; de Fluiter, P. (1994) Determination of n-methylnicotinamide in urine with capillary zone electrophoresis. *J. Liq. Chromatogr.* **17**, 1929–1939.
17. Gonzalez, E.; Montes, R.; Laserna, J.J. (1993) Pulsed-laser fluorescence detection in capillary zone electrophoresis of some banned substances in sport. *Anal. Chim. Acta* **282**, 687–693.
18. Gonzalez, E.; Laserna, J.J. (1994) Capillary zone electrophoresis for the rapid screening of banned drugs in sport. *Electrophoresis* **15**, 240–243.
19. Jumppanen, J.; SirÇn, H.; Riekkola, M.L. (1993) Screening for diuretics in urine and blood serum by capillary zone electrophoresis. *J. Chromatogr.* **652**, 441–450.
20. Shirao, M.; Furuta, R.; Sumiko, S.; Nakazawa, H.; Fujita, S.; Maruyama, T. (1994) Determination of organic acids in urine by capillary zone electrophoresis. *J. Chromatogr.* **680**, 247–251.
21. Perrett, D.; Ross, G. (1992) Capillary electrophoresis of drugs in biological fluids. *Methodol. Surv. Biochem. Anal.* **22**, 269–278.
22. Wernly, P.; Thormann, W.; Bourquin, D.; Brenneisen, R. (1993) Determination of morphine-3-glucuronide in human urine by capillary zone electrophoresis and micellar electrokinetic capillary chromatography. *J. Chromatogr.* **616**, 305–310.
23. Chen, A.; Evangelista, R.A. (1994) Feasibility studies for simultaneous immunochemical multianalyte drug assay by capillary electrophoresis with laser-induced fluorescence. *Clin. Chem.* **40**, 1819–1822.
24. Kelly, T.E. (1989) Diseases of genetic origin. In *Clinical Chemistry*, Kaplan, L.A.; Pesce, A.J. (eds), C.V. Mosby Company, St. Louis, 2nd edn, p. 698–699.
25. Dolnik, V. (1994) Capillary zone electrophoresis of pathological metabolites in phenylketonuria. *J. Microcol. Sep.* **6**, 63–67.
26. Aumatell, A.; Wells, R.J. (1993) Chiral differentiation of the optical isomers of racemethorphan and racemorphan in urine by capillary zone electrophoresis. *J. Chromatogr. Sci.* **31**, 502–508.
27. Schafroth, M.; Thormann, W.; Allemann, D. (1994) Micellar electrokinetic capillary chromatography of benzodiazepines in human urine. *Electrophoresis* **15**, 72–78.

28 Heuermann, M.; Blaschke, G. (1994) Simultaneous enantioselective determination and quantification of dimethindene and its metabolite N-dimethyl-dimethindene in human urine using cyclodextrins as chiral additives in capillary electrophoresis. *J. Pharm. Biomed. Anal.* **12**, 753–760.

29 Hori, A.; Matsumoto, T.; Nimura, Y.; Tsuda, T. (1992) Capillary electrophoresis on polyamine analysis. In *Progress in Clinical Biochemistry*, Elsevier Science B.V., Amsterdam, pp. 149–150.

30 O'Shea, T.J.; Lunte, S.M. (1994) Chemically modified microelectrodes for capillary electrophoresis/electrochemistry. *Anal. Chem.* **66**, 307–311.

31 Zhou, J.; O'Shea, T.J.; Lunte, S.M. (1994) Simultaneous detection of thiols and disulfides by capillary electrophoresis–electrochemical detection using a mixed-valence ruthenium cyanide-modified microelectrode. *J. Chromatogr.* **680**, 271–277.

32 Lin, B.L.; Colon, L.A.; Zare, R.N. (1994) Dual electrochemical detection of cysteine and cystine in capillary electrophoresis. *J. Chromatogr.* **680**, 263–270.

33 Meuleamans, A.; Delsenne, F. (1994) Measurement of nitrite and nitrate levels in biological samples by capillary electrophoresis. *J. Chromatogr.* **660**, 401–404.

34 Bogan, D.; Deasy, B.; O'Kennedy, R.; Smyth, M.; Fuhr, U. (1995) Determination of free and total 7-hydroxycoumarin in urine and serum by capillary electrophoresis. *J. Chromatogr.* **663**, 371–378.

35 Rhemrev-Boom, M.M. (1994) Determination of anions with capillary electrophoresis and indirect ultraviolet detection. *J. Chromatogr.* **680**, 675–680.

36 Nann, A.; Silvestri, I.; Simon, W. (1993) Quantitative analysis in capillary zone electrophoresis using ion-selective microelectrodes as on-column detectors. *Anal. Chem.* **65**, 1662–1667.

37 Zhang, R.; Shi, H.; Ma, Y. (1994) Quantitative determination of ionized calcium and total calcium in human serum by capillary zone electrophoresis with indirect photometric detection. *J. Microcol. Sep.* **6**, 217–221.

38 Buchberger, W.; Winna, K.; Turner, M. (1994) Applications of capillary zone electrophoresis in clinical chemistry: determination of low-molecular-mass ions in body fluids. *J. Chromatogr.* **671**, 375–382.

39 Leone, A.M.; Francis, P.L.; Rhodes, P.; Moncada, S. (1994) A rapid and simple method for the measurement of nitrate in plasma by high performance capillary electrophoresis. *Biochem. Biophys. Res. Commun.* **200**, 951–957.

40 Leveque, D.; Gallion, C.; Tarral, E.; Monteil, H.; Jehl, F. (1994) Determination of fosfomycin in biological fluids by capillary electrophoresis. *J. Chromatogr.* **655**, 320–324.

41 Soini, H.; Stefansson, M.; Riekkola, M.L.; Novotny, M.V. (1994) Maltooligosaccharides as chiral selectors for the separation of pharmaceuticals by capillary electrophoresis. *Anal. Chem.* **66**, 3477–3484.

42 Schmutz, A.; Thormann, W. (1994) Factors affecting the determination of drugs and endogenous low molecular mass compounds in human serum by micellar electrokinetic capillary chromatography with direct sample injection. *Electrophoresis* **15**, 51–61.

43 Morita, I.; Sawada, J. (1993) Capillary electrophoresis with online sample pretreatment for the analysis of biological samples with direct injection. *J. Chromatogr.* **641**, 375–381.

44 Gareil, P.; Gramond, J.P.; Guyon, F. (1993) Separation and determination of warfarin enantiomers in human plasma samples by capillary zone electrophoresis using a methylated-beta-cyclodextrin-containing electrolyte. *J. Chromatogr.* **615**, 317–325.

45 Dethy, J.M.; De Broux, S.; Lesne, M.; Longstreth, J.; Gilbert, P. (1994) Stereoselective determination of verapamil and norverapamil by capillary electrophoresis. *J. Chromatogr.* **654**, 121–127.

46 Schneede, J.; Ueland, P. (1995) Application of capillary electrophoresis with laser-induced fluorescence detection for routine determination of methylmalonic acid in human serum. *Anal. Chem.* **67**, 812–819.

47 Shihabi, Z.K. (1993) Serum pentobarbital assay by capillary electrophoresis. *J. Liq. Chromatogr.* **16**, 2059–2068.

48 Shihabi, Z.K.; Oles, K.S. (1994) Felbamate measured in serum by two methods: HPLC and capillary electrophoresis. *Clin. Chem.* **40**, 1904–1908.

49 Prunonosa, J.; Obach, R.; Diez-Cascon, A.; Gouesclou, L. (1992) Comparison of high-performance liquid chromatography and high-performance capillary electrophoresis for the determination of cicletanine in plasma. *J. Chromatogr.* **581**, 219–226.

50 Chen, F.A.; Pentoney, S.L. (1994) Characterization of digoxigenin-labeled B-phycoerythrin by capillary electrophoresis with laser-induced fluorescence: application to homogeneous digoxin immunoassay. *J. Chromatogr.* **680**, 425–430.
51 Shihabi, Z.K.; Constantinescu, M.S. (1992) Iohexol in serum determined by capillary electrophoresis. *Clin. Chem.* **38**, 2117–2120.
52 Garcia, L.L.; Shihabi, Z.K. (1993) Suramin determination by capillary electrophoresis. *J. Liq. Chromatogr.* **16**, 2049–2057.
53 Johansson, I.M.; Groen-Rydberg, M.B.; Schmekel, B. (1993) Determination of theophylline in plasma using different capillary electrophoretic systems. *J. Chromatogr.* **652**, 487–493.
54 Garcia, L.L.; Shihabi, Z.K. (1993) Sample matrix effects in capillary electrophoresis. I. Basic considerations. *J. Chromatogr.* **652**, 465–469.
55 Shihabi, Z.K. (1993) Sample matrix effects in capillary electrophoresis. II. Acetonitrile deproteinization. *J. Chromatogr.* **652**, 471–475.
56 Shibukawa, A.; Lloyd, D.K.; Wainer, I.W. (1993) Simultaneous chiral separation of leucovorin and its major metabolite 5-methyltetrahydrofolate by capillary electrophoresis using cyclodextrins as chiral selectors: estimation of the formation constant and mobility of the solute–CD complexes. *Chromatographia* **37**, 419–429.
57 Wolfisberg, H.; Schmutz, A.; Stotzer, R.; Thormann, W. (1993) Assessment of automated capillary electrophoresis for therapeutic and diagnostic drug monitoring: determination of bupivacaine in drain fluid and antipyrine in plasma. *J. Chromatogr.* **652**, 407–416.
58 Baillet, A.; Pianetti, G.A.; Taverna, M.; Mahuzier, G.; Baylocq-Ferrier, D. (1993) Fosfomycin determination in serum by capillary zone electrophoresis with indirect ultraviolet detection. *J. Chromatogr.* **616**, 311–316.
59 Shibukawa, A.; Yoshimoto, Y.; Ohara, T.; Nakagawa, T. (1994) High-performance capillary electrophoresis/frontal analysis for the study of protein binding of a basic drug. *J. Pharm. Sci.* **83**, 616–619.
60 Tagliaro, F.; Moretto, S.; Valentini, R.; Gambaro, G.; Antonioli, C.; Moffa, M.; Tato, L. (1994) Capillary zone electrophoresis determination of phenylalanine in serum: a rapid, inexpensive and simple method for the diagnosis of phenylketonuria. *Electrophoresis* **15**, 94–97.
61 Balistreri, W.F.; Shawn, L.M. (1986) Liver function. In *Textbook of Clinical Chemistry*, Tietz, N.W. (ed.), Saunders Company, Philadelphia, pp. 1373–1433.
62 Wu, N.; Sweedler, J.V.; Lin, M. (1994) Enhanced separation and detection of serum bilirubin species by capillary electrophoresis using an anionic surfactant–protein buffer system with laser-induced fluorescence detection. *J. Chromatogr.* **654**, 185–191.
63 Brunner, L.J.; Dipiro, J.T.; Feldman, S. (1993) Serum antipyrine concentrations determined by micellar electrokinetic capillary chromatography. *J. Chromatogr.* **622**, 98–102.
64 Lee, K.; Heo, G.S.; Doh, H.J. (1992) Determination of creatinine in serum by capillary electrophoresis. *Clin. Chem.* **38**, 2322–2323.
65 Beeley, J.A. (1991) Clinical applications of electrophoresis of human salivary proteins. *J. Chromatogr.* **569**, 261–280.
66 Lal, K.; Pollock, J.J.; Santarpia, P.R.; Heller, H.M.; Kaufman, W.H.; Fuhrer, J.; Steigbiegel, R.T. (1992) Pilot study comparing the salivary cationic protein concentrations in healthy adults and AIDS patients: correlation with antifungal activity *J. Acquired Immune Defic. Syndr.* **5**, 904.
67 Lal, K.; Santarpia, R.P.; Xu, L.; Manssuri, F.; Pollock, J.J. (1992) One-step purification of histidine-rich polypeptides from human parotid saliva and determination of anti-candidal activity. *Oral Microbiol. Immunol.* **7**, 44–50.
68 Xu, L.; Lal, K.; Pollock, J.J. (1992) Histatins 2 and 4 are autoproteolytic degradation products of human parotid saliva, *Oral Microbiol. Immunol.* **7**, 127–128.
69 Lal, K.; Xu, L.; Colburn, J.; Hong, Q.L.; Pollock, J.J. (1992) The use of capillary electrophoresis to identify cationic proteins in human parotid saliva. *Arch. Oral Biol.* **37**, 7–13.
70 Tie, J.; Tsukamoto, S.; Oshida, S. (1992) Analysis of human saliva protein by capillary electrophoresis. *Nihon Univ. J. Med.* **34**, 315–323.
71 Linhares, M.C.; Kissinger, D.T. (1992) Capillary ultrafiltration: in-vivo sampling probes for small molecules. *Anal. Chem.* **64**, 2831–2835.
72 Thormann, W.; Minger, A.; Molteni, S.; Caslavska, J.; Gebauer, P. (1992) Determination

of substituted purines in body fluids by micellar electrokinetic capillary chromatography with direct sample injection. *J. Chromatogr.* **593**, 275–288.
73 Varnell, R.J.; Maitchocek, D.; Beuerman, R.W. (1995) Capillary electrophoretic analysis of proteins in tear fluid before and after wounding of the cornea of the eye. *2nd Annual Symposium on Biomedical, Biopharmaceutical, and Clinical Applications of Capillary Electrophoresis*, Paper No. 14, East Rutherford, New Jersey, August.
74 Phillip, T.M.; Chmielinska, J.J. (1994) Immunoaffinity capillary electrophoresis analysis of cyclosporin in tear. *Biomed. Chromatogr.* **8**, 242–246.
75 Chadwick, R.R.; Hsieh, J.C.; Resham, K.S.; Nelson, R.B. (1994) Applications of capillary electrophoresis in the eye-care pharmaceutical industry. *J. Chromatogr.* **671**, 403–410.
76 Perez, S.A.; Colon, L.A. (1996) Determination of carbohydrates as their dansylhydrazine derivatives by capillary electrophoresis with laser-induced fluorescence detection. *Electrophoresis* **17**, 352–358.
77 Schmalzing, D.; Nashabeh, W.; Yao, X.W.; Mhatre, R.; Regnier, F.E.; Afeyan, N.B.; Fuchs, M. (1995) Capillary electrophoresis-based immunoassay for cortisol in serum. *Anal. Chem.* **67**, 606–612.
78 Chen, F.-T.A.; Sternberg, J.C. (1994) Characterization of proteins by capillary electrophoresis in fused silica columns: Review on serum protein analysis and application to immunoassays. *Electrophoresis* **15**, 13–21.

33 Toxins associated mainly with uremia and cancer

H. SHINTANI

33.1 Introduction

When someone gets a disease, newly occurring marker substances for that disease may be observed in the person's blood or urine, or a change is seen in the level of inherent components. The determination of such substances allows us to take preventive measures against the disease, to estimate at what stage the disease may be, and to monitor the patient's response to the treatment prescribed.

In renal disease, substances that would otherwise be excreted through the kidney accumulate in the blood. The patient will develop uremia and die unless the blood can be cleared of these substances by artificial dialysis or filtration. Although the principal compounds responsible for uremia, called uremic toxins, are still to be defined, compounds such as urea, uric acid, creatinine, guanidine compounds (methylguanidine, dimethylguanidine, guanidinosuccinic acid, guanidinoacetic acid, etc.), phenolic and hydroxyphenolic acids (hippurate, hydroxyhippurate, indoxysulfate, aromatic amines and indoles), polyamines (putrescine, spermidine, spermine and cadaverine), conjugated amino acids, peptides, aluminum, and inorganic and organic ions are suggested to be responsible for uremia. For example, differential analysis of selenium in blood is an important diagnosis tool for uremia [1, 2].

In patients with renal disorders, adequate control of the variation of electrolyte concentration in body fluids is indispensable. Unless properly managed, compromised renal function will lead to the accumulation of uremic toxins in blood leading to kidney failure. In modern medicine, individuals who suffer from kidney failure can live a moderately normal life by means of pleurtoneau dialysis or external dialysis with an artificial kidney. For these individuals, analytical techniques that can evaluate the extent of dialysis conveniently and precisely are invaluable. For diagnostic renal functions, biochemical measurements of inorganic cations, such as sodium, potassium and aluminum, are required. The simultaneous determination of these parameters is highly desirable, however simultaneous determination of these uremic toxins and electrolytes is difficult to attain by existing techniques, although it can be attained by micellar electrokinetic chromatography (MECC or MEKC), which is required for accurate diagnosis. Although many individual types of sensor for individual compound detection have been developed, simultaneous determination of these markers is still required. MECC will provide an effective technique for this purpose. There have been some interesting attempts to determine markers for various types of cancers and to monitor the course of certain diseases by using high performance capillary electrophoresis (HPCE).

For pretreatment of body fluid (e.g. blood or urine) prior to HPCE application, simple procedures such as dialysis or ultrafiltration are useful to remove protein.

Conventionally, the ion-selective electrode method currently used for anion analysis cannot be used for simultaneous multianion analysis and cannot always be used for differential analysis. Coexisting ions interfere with the ion-selective electrode method and low sensitivity is indicated.

Colorimetry, the ion-selective electrode method, atomic absorption spectrometry and flame spectrometry which are currently used for cation analysis cannot be used for differential analysis and simultaneous multi-ion analysis. An ammonium ion-selective electrode is used for urea analysis, but this method cannot differentiate urea and endogenous ammonium as both were determined as ammonium, so-called BUN (blood urea nitrogen), by immobilized urease attached to the electrode.

Comparison of HPCE with atomic absorption spectroscopy was reported [3]. Comparison of HPLC or ion chromatography (IC) with HPCE was also reported in several papers including that by Perett [4]. In Chapter 11, comparison of other analytical methods with HPCE is described. The significant differences between HPLC and HPCE are sample amount, theoretical plate numbers and analysis time. In the author's opinion, reproducible quantitative analysis is essential in analytical chemistry and with that in mind HPCE is still inferior to HPLC. IC is one type of HPLC and is mainly focused on ionic compound analysis, which normally involves difficulties in analyzing non-ionic compounds by IC. Inductively coupled plasma atomic emission spectrometry (ICP), which is capable of simultaneous multication analysis, cannot be used for differential analysis or anion analysis. Colorimetry has a poor reproducibility and a complicated handling procedure. Immunoassays (RIA (radio immunoassay), ELISA (enzyme-linked immunosorbent assay)) are expensive and multicompound analysis with a single immunoassay is impossible and has the potential for error due to crossreactivity because of the similar chemical structure of the compounds of interest.

By using MECC, these deficiencies are improved. Selective and simultaneous multi-ion analysis becomes possible. The anion gap is successfully estimated. Anion gap estimation can be attained for sodium, chloride and some bicarbonate, a clinically important factor in estimating ion balance in body fluids.

33.2 Analysis of toxic compounds from uremia or cancer patients

Since 1988, when CE became commercially available, the number of papers on CE has increased exponentially. Uremic toxin analysis using HPCE is reviewed here from 1992 to date (Table 33.1). Papers on uremic toxin analysis before 1992 are intentionally omitted because have been reviewed elsewhere [4–10]. Data obtained after 1988 can be evaluated interlaboratory using identical instruments, buffer conditions etc., thus requiring more reproducible data when publishing papers. Creatinine [11–13], uric acid [12, 14] and urea [12] were successfully separated using HPCE excepting urea and uric acid as the elution of urea and uric acid were overlapped with body fluid admixtures according to the author's experiment [in preparation for submission]. Uric acid in biological fluids was often overlapped with admixtures in MECC analysis depending on the mixing status of uric acid with added micelle [12]. Analysis of most of uremic toxin compounds has not yet been reported by HPCE, indicating HPCE is still a developing instrument and technology compared with HPLC.

In the current status, application of HPCE to clinical chemistry is inferior to HPLC mostly in terms of reproducibility, selectivity and separation efficiency [5]. Urea and uric acid were not successfully separated from blood admixtures [12]. In the author's experiment, uric acid and creatinine in blood were successfully separated by HPLC rather than HPCE without any pretreatment [15]. Urea analysis required column switching combined with immobilized urease column [15]. Using a pretreatment of solid phase extraction (SPE) or ultrafiltration for the analysis of body fluid compounds, this deficiency will be overcome in HPCE as it was in HPLC [16, 17]. This was confirmed in the analysis of other toxic compounds [18–20]. Among several toxic compounds, oxides and halogenated compounds were found to be quite hazardous. The author reported toxicity of oxides due to their prompt reactivity [19–21]. The analysis of polyhaloganated compounds and polycyclic aromatic hydrocarbons will be discussed in Part Seven of this book on Environmental Science. One example of HPCE separation of polycyclic aromatic amine is given by Perett [4].

As HPCE has several superior features, so that the technique has been improved and the deficiences overcome, HPCE will be superior to HPLC especially in terms of the small amount of sample needed, the fast analysis time and the capability of analyzing simultaneously charged and uncharged ions. Several significant papers on optimization of HPCE have been published [22, 23].

The detection of shellfish toxins by HPCE with MS was also reported [24–26], but only one paper on bacterial endotoxins, like aflatoxin, was reported using HPCE before 1992 [7], so in the current status HPLC application is still superior to HPCE.

Table 33.1 CE conditions for toxin analysis

Methods Separation conditions, voltage applied, current Capillary length with i.d. Detection	Compounds	Matrix	Reference
CZE 10 mmol l^{-1} Citric acid–sodium citrate, pH 5.0 with MeOH, CH$_3$CN or SDS modifier addition, 30 kV 100 cm (63.5 cm to detector) × 50 μm (i.d.) 215 nm	Creatinine	Serum	11
MECC 6 mM Na$_2$B$_4$O$_7$ and 10 mM Na$_2$HPO$_4$, pH 9.2 with 75 mM SDS, 10 mM hydrogenphosphate, 6 mM tetraborate, 25 kV, 70 μA 68 cm, 75 μm i.d. 215 nm	Creatinine, urea, uric acid	Human serum	12
CE 20 nM Borate buffer, pH 9.4, 37°C, 20 kV 44 cm, 75 μm i.d. 254 nm	Purine bases, uric acid, necleosides	Human cord plasma	13
MECC 6 mM Na$_2$B$_4$O$_7$ and 10 mM Na$_2$HPO$_4$, pH 9.2 with 75 mM SDS, 20 kV, 70 μA 70 cm, 75 μm i.d. 195–320 nm	Uric acid	Body fluids	14
CE Trisma buffer (pH 7.2), 24.4 kV 90 cm, 50 μm i.d. MS	Paralytic shellfish poisons		24
CE Buffer of 100 mM morpholine to pH 4.0 with formic acid, 30 kV 50 or 90 cm, 50 μm i.d. MS	Paralytic shellfish poisons		25
Beckman CE 2100 Buffer of Trisma^R (pH 7.2), 30 kV 90 cm, 50 μm i.d. MS	Paralytic shellfish poisons		26
CE 30 kV 60 cm × 25 μm i.d. 236 nm indirect	Putrescine, spermidine, spermine	Tumor cells	27

Table 33.1 *Continued*

Methods Separation conditions, voltage applied, current Capillary length with i.d. Detection	Compounds	Matrix	Reference
CE 30 kV 60 cm × 75 µm i.d. 236 nm indirect	Putrescine, spermidine, spermine	Biological samples	28
CE 150 mM Borate buffer, pH 9.0, 50 mbar pressure injection for 5 s, 22 kV, 32.5–34.5 µA, 25°C 64 cm × 50 µm i.d. 185–300 nm diode-array UV detection	Hypoxanthine, pseudouridine, hippuric acid	Serum	32

Polyamines such as putrescine, spermidine, spermine and cadaverine are involved in the function of DNA replication, gene expression, protein synthesis through m-RNA and cell surface receptor. These compounds are essential in a variety of cell functions, however overproduction of polyamines is toxic to cells and facilitates cell death by oxidative mechanisms. Tumor cells contain a much higher concentration of polyamines than normal cells and patients with cancer have an enhanced urinary excretion of polyamines, thus urinary polyamines are regarded as a marker compounds for tumor, so that an accurate and reproducible determination method for these compounds is required. HPCE is one analytical procedure that can be used and some papers have been published concerning the analysis of these compounds [27–31]. The recent paper dealing with clinical analysis indicated that some organic ions such as hypoxanthine, pseudouridine and hippuric acid tended to increase in uremic serum. These ions may or may not be a result of uremia, but it is certain that these compounds including uric acid are an indicator of uremia if they increase significantly compared with the normal range [32].

Major toxic compounds of phenol including phenol metabolite *in vitro* were analyzed by MECC [33]. Phenol is also discussed in Chapter 37, Organic Acids and Organic Ions.

33.3 Conclusion

As described, HPCE has several superior features, but this methodology is still a developing technique especially in terms of reproducibility which is essential in analytical chemistry. Additionally neutral compounds applied to MECC analysis are not detected by mass spectrometry (MS) under atmosphere pressure chemical ionization (APCI), so analysis of neutral compounds is done by CE–MS with MECC. Compared with HPLC, CE has several advantages including the small amount of sample needed, the fast analysis time and the capability of analyzing charged and uncharged ions simultaneously. When several problems including reproducibility are overcome, HPCE will become an excellent analytical method to deal with day-to-day innovation in science.

References

1. Vandael, P.; Deschuyt, A.; Robberec, H.; Vancaill, M.; Lamand, M.; Deelstra, H. (1994) Capillary whole-blood selenium determination in assessing selenium status of children. *J. Tr. Elem. E.* **8**, 225–228.
2. Bonomini, M.; Albertaz, A. (1995) Selenium in uremia. *Artif. Organ.* **19**, 443–448.
3. Pretswel, E.L.; Morrisson, A.R. (1995) The comparison of capillary zone electrophoresis and atomic spectroscopy for the determination of the cation content of a standard reference material IAEA-A-11 milk powder. *Talanta* **42**, 283–289.
4. Perett, D. (1993) Capillary electrophoresis in biomedical and pharmaceutical research. In *Capillary Electrophoresis Theory and Practice*, Camilleri, P. (ed.), CRC Press, Boca Raton, pp. 371–408.
5. Thormann, W.; Molteni, S.; Caslavska, J.; Schmutz, A. (1994) Clinical and forensic applications of capillary electrophoresis. *Electrophoresis* **15**, 3–12.
6. Deyl, Z.; Tagliaro, F.; Miksik, I. (1994) Biomedical applications of capillary electrophoresis. *J. Chromatogr.* **656**, 3–27.
7. Waetzig, H.; Dette, C. (1994) Capillary electrophoreis (CE)- a review strategies for method development and applications related to pharmaceutical and biological sciences. *Pharmazie* **49**, 83–96.
8. Brunner, L.J.; DiPiro, J.T.; Feldman, S. (1995) Science for clinicians, high-performance capillary electrophoresis in the pharmaceutical sciences. *Pharmacotherapy* **15**, 1–22.
9. Landers, J.P. (1995) Clinical capillary electrophoresis. *Clin. Chem.* **41**, 495–509.
10. Xu, Y. (1995) Application of CE in clinical diagnosis: capillary electrophoresis. *Anal. Chem.* **67**, 463R–473R.
11. Lee, K.J.; Heo, G.S.; Doh, H.J. (1992) Determination of creatinine in serum by capillary zone electrophoresis. *Clin. Chem.* **38**, 2322–2323.
12. Schmutz, A.; Thormann, W. (1994) Factors affecting the determination of drugs and endogenous low molecular mass compounds in human serum by micellar electrokinetic capillary chromatography with direct sample injection. *Electrophoresis* **15**, 51–61.
13. Grune, T.; Ross, G.A.; Schmidt, H.; Siems, W.; Perrett, D. (1993) Optimized separation of purine bases and nucleosides in human cord plasma by capillary zone electrophoresis. *J. Chromatogr.* **636**, 105–111.
14. Thormann, W.; Minger, A.; Molteni, S.; Caslavska, J.; Gebauer, P. (1992) Determination of substituted purines in body fluids by micellar electrokinetic capillary chromatography with direct sample injection. *J. Chromatogr.* **593**, 275–288.
15. Shintani, H.; Wojcik, A.B.; Tawa, R.; Uchiyama, S. (1994) Uremic toxin analysis with pre- and post-column immobilized enzyme reactors. In *Analytical Application of Immobilized Enzyme Reactors*, Lam, S.; Malikin, G. (eds), Blackie Academic & Professional, Glasgow, p. 131.
16. Shintani, H. (1996) Comparison of solid phase extraction and dialysis on pre-treatment efficiency of blood urea analysis. *J. Chromatogr. Sci.* **34**, 92–94.
17. Shintani, H. (1995) Solid phase extraction (SPE) of blood urea compared with liquid–liquid extraction regarding artifact formation. *J. Liq. Chromatogr.* **18**, 2167–2174.
18. Shintani, H. (1992) Solid-phase extraction (SPE) and HPLC analysis of toxic compounds and comparison of SPE and liquid–liquid extraction. *J. Liq. Chromatogr.* **15**, 1315–1336.

REFERENCES

19a Shintani, H. (1993) Solid-phase extraction and HPLC analysis of toxic compounds eluted from methyl methacrylate dental materials. *J. Anal. Toxicol.* **17**, 73–78.
19b Shintani, H. (1995) HPLC analysis of toxic additives and residual monomer from dental plate. *J. Liq. Chromatogr. Clin. Anal.* **18**, 613–626.
19c Shintani, H. (1994) Elution of toxic compounds from dental composite resin material. *Jpn. J. Med. Instr.* **64**, 345–349.
19d Shintani, H. (1992) Solid-phase extraction and high-performance liquid chromatographic analysis of a toxic compound from gamma-irradiated polyurethane. *J. Chromatogr.* **600**, 93–97.
20 Shintani, H. (1996) Analysis of newly found toxic compounds from dental material. *Jpn. J. Med. Instr.* **65**, 486–488.
21 Vanrolli, M.; Knapp, H.R. (1995) Identification of arachidonate epoxides diols by capillary chromatography mass-spectrometry. *J. Lipid Res.* **36**, 952–966.
22 Altria, K.D.; Fabre, H. (1995) Approaches to optimization of precision in capillary electrophoresis. *Chromatographia* **40**, 313–320.
23 Pyell, U.; Butehorn, U. (1995) Optimization strategies in micellar electrokinetic capillary chromatography- one-parameter optimizations of the concentration of sodium dodecylsulfate and the concentration of modifiers (urea and glucose). *Chromatographia* **40**, 175–184.
24 Pleasance, S.; Thibault, P.; Kelly, J. (1992) Comparison of liquid-junction and coaxial interfaces for capillary electrophoresis–mass spectrometry with application to compounds of concern to the aquaculture industry. *J. Chromatogr.* **591**, 325–339.
25 Buzy, A.; Thibault, P.; Laycock, M.V. (1994) Development of a capillary electrophoresis method for the characterization of enzymatic products arising from the carbamoylase digestion of paralytic shellfish poisoning toxins. *J. Chromatogr.* **688**, 301–316.
26 Pleasance, S.; Ayer, S.W.; Laycock, M.V.; Thibault, P. (1992) *Rapid Commun. Mass Spectrom.* **6**, 14.
27 Zhang, R.; Cooper, C.L.; Ma, Y. (1993) Determination of total polyamines in tumor cells by high-performance capillary zone electrophoresis with indirect photometric detection. *Anal. Chem.* **65**, 704–706.
28 Ma, Y.; Zhang, R.; Cooper, C.L. (1992) Indirect photometric detection of polyamines in biological samples separated by high-performance capillary electrophoresis. *J. Chromatogr.* **608**, 93–96.
29 Li, S.F.Y. (1993) *Capillary Electrophoresis Principles, Practice and Applications*, Journal of Chromatography Library, Volume 52, Elsevier, Amsterdam, pp. 453.
30 Rabel, S.R. (1993) Applications of CE in pharmaceutical analysis. *Pharm. Res.* **10**, 171–186.
31 Lunte, S.M.; O'Shea, T.J. (1994) Pharmaceutical and biomedical applications of capillary electrophoresis/electrochemistry. *Electrophoresis* **15**, 79–86.
32 Petucci, C.J.; Kantes, H.L.; Strein, T.G.; Veening, H. (1995) Capillary electrophoresis as a clinical tool determination of organic anions in normal and uremic serum using photodiode-array detection. *J. Chromatogr.* **668**, 241–251.
33 Davies, M.I.; Lunte, C.E.; Smyth, M.R. (1995) Use of micellar electrokinetic capillary chromatography in the study of *in-vitro* metabolism of phenol by human liver-microsomes. *J. Pharm. Biol.* **13**, 893–897.

Part Five

Ion Analysis Applications

34 Capillary electrophoresis of metal complexes
W. BUCHBERGER

34.1 General strategies for metal ion analysis after complexation

Currently, capillary zone electrophoresis (CZE) is attracting considerable interest as an alternative technique for the determination of metal ions. However, a range of metal ions such as transition and lanthanide metal ions exhibit nearly identical charge and size, so that a separation solely based on differences in the electrophoretic mobilities presents a serious problem. The most successful approach to optimization of separation selectivity for metal ions is the employment of complex-forming reactions that can take place in the capillary during the separation or in the sample before injection. The mobility of metal cations can be manipulated by addition of weak complexing agents such as organic acids to the carrier electrolyte. The higher the respective complex-formation constant, the lower will be the electrophoretic mobility due to the decreased effective charge (the effective charge is still positive, as the complexing properties of organic acids are not very strong). This strategy and various applications have been discussed in Chapter 36 on Inorganic Ions. Another possibility for improving the separation selectivity is the use of strong complexing reagents that allow the full conversion of metal ions into positively charged, negatively charged or neutral complexes. In this case, the complexation reaction can be carried out in the sample before injection (precapillary complexation), if the thermodynamic and kinetic stability of the resulting complexes is high enough. Precapillary complexation can improve the detectability of metal ions considerably when reagents are chosen that lead to UV-absorbing complexes. In some cases, the reagent can be added to the carrier electrolyte, so that the complex formation takes place during injection of the sample metal ions. The presence of the complexing reagent in the carrier electrolyte is also advisable to avoid dissociation of the complexes during the separation process if the stability is not sufficiently high. In the same way, the separation and detectability of free ligands can be improved after precapillary or on-capillary conversion into metal complexes. A detailed discussion of separation of metal ions after complexation is given by Timerbaev [1].

This chapter will focus on various reagents suitable for metal ion complexation and on appropriate separation conditions. In addition, some aspects of speciation analysis are included such as the analysis of metals occurring in the form of different complexes or in different oxidation states.

34.2 Precapillary and on-capillary complexation of metal ions

Among inorganic reagents, cyanide is a versatile ligand that forms complexes with a wide range of metal ions, including precious metals, which otherwise can be difficult to analyze by CZE [2]. Addition of cyanide to the carrier electrolyte is recommended to avoid the dissociation of less stable metallocyanide complexes (such as copper(I) cyanide). Chloride is another useful inorganic ligand for the analysis of platinum group metals, which can be separated as their chloro complexes in chloride-containing carrier electrolytes [3].

The majority of suitable reagents for metal complexation in capillary electrophoresis (CE) are organic ligands. Aminopolycarboxylic acids can be used in various applications to analyze a wide range of metal ions after derivatization into metal chelates. EDTA (ethylene diaminetetraacetic acid) has been employed for the determination of alkaline earth metals in the presence of alkali metals [4, 5]. Cyclohexane-1,2-diaminetetraacetic acid (CDTA) allows the separation of the lanthanides [6] and several transition metal ions [7]. N,N'-di(2-hydroxybenzyl)ethylenediamine-N,N'-diacetic acid (HBED) has been suggested instead of EDTA for improved detectability [8]. Derivatives of simple aminopolycarboxylic acids such as 2,6-bis[N,N-bis(carboxymethyl)aminomethyl]-4-benzoylphenol or 4-(phenylethynyl)-2,6-bis[N,N-bis(carboxymethyl)aminomethyl]pyridine can form complexes with europium(III) that are suited for luminescence detection [9]. One should be aware of the fact that some metal ions such as Fe(III) form sufficiently stable complexes with aminopolycarboxylic acids to allow a precapillary complexation, whereas others, depending on the pH, require the presence of the reagent in the carrier electrolyte. In addition, the kinetics of the complex formation reaction can be so slow that a sole on-capillary complexation would be unsuccessful. For these reasons, most metal ion separations with complexation by aminopolycarboxylic acids were based on both the precapillary derivatization reaction and on-capillary complex formation conditions.

Another organic ligand that has attracted wider interest is 4-(2-pyridylazo)resorcinol (PAR) [10–13]. The comments given above for precapillary and on-capillary complexation with aminopolycarboxylic acids are also valid for PAR.

Other complexing agents so far reported for metal ion analysis by CE include arsenazo III [12], 2, 2′-dihydroxyazobenzene-4,4′-disulfonate [14], 8-hydroxyquinoline-5-sulfonic acid (HQS) [15], 2-(5-bromo-2-pyridylazo)-5-(N-propyl-N-sulfopropylamino)phenol (5-Br-PAPS) [16],

2,6-diacetylpyridine-bis(N-methylenepyridiniohydrazone) [17], 4-chloro-3-(2,4-dihydroxyphenylazo)-2-hydroxybenzene-1-sulfonic acid [18] or N-methyl-N-2-sulfoethyldithiocarbamate [19].

The determination of free ligands after precapillary and on-capillary derivatization with metal ions has been described for various aminopolycarboxylic acids [20–22].

34.3 Modes for CE separation of metal complexes

In free solution CZE, the electrophoretic mobility of an ion generally depends on its charge and size. Difficulties with the separation of free metal ions of nearly identical charge and size are not necessarily avoided by converting them into more or less stable complexes. The ligand chosen for a certain application might form the same type of complex with a range of different metal ions so that once again the analytes exhibit the same charge and size (in this case, the only benefit of complexation would be the improved detectability). Fortunately, the careful selection of an appropriate reagent can overcome these problems in different ways. In the simplest case, a ligand is chosen that can form complexes containing different numbers of ligands for different metal ions. This case is represented by metal cyanides. Alternatively, a ligand is chosen that contains at least one acidic group that expresses the characteristics of the central metal ion by means of the magnitude of the acid dissociation constant (pK_a). Thereby, the effective charge can be manipulated as has been discussed for PAR complexes [10]. Working in a certain pH range can also favor the formation of mixed chelates containing OH⁻ as an additional ligand so that separations are made possible according to the different stabilities of different mixed metal chelates. Such effects have been observed for metal complexes with aminopolycarboxylic acids [4].

Another approach to the optimization of separation selectivity of metal complexes is the addition of ion-pairing reagents to the carrier electrolyte [14, 17]. The term 'ion-association capillary electrophoresis' has been used recently for this technique [14].

Finally, the employment of micelles in the carrier electrolyte (micellar electrokinetic chromatography, MEKC or MECC) can be a generic way of improving the separation of metal complexes that exhibit sufficient hydrophobicity [12].

Depending on the type of metal complex, coelectroosmotic or counterelectroosmotic migration modes are possible, whereby the detection side can be the anode or the cathode. Anionic complexes with bulky organic ligands are best separated in a conventional configuration with the detector at the cathodic side. Highly mobile anionic complexes such as several metallocyanide complexes might overcome EOF (electroosmotic

Table 34.1 Configurations for the separation of metal complexes

Metal complex	Detection side	Direction of electroosmotic and electrophoretic mobility
Anionic complexes	Cathode	Opposite direction
	Anode	Opposite direction (reduced EOF)
	Anode	Same direction (reversed EOF)
Anionic complexes in the presence of micelles	Cathode	Opposite direction
Cationic complexes	Cathode	Same direction
	Anode	Opposite direction (reversed EOF)
Cationic complexes in the presence of micelles	Anode	Opposite direction (reversed EOF)
Neutral complexes in the presence of micelles	Cathode	No electrophoretic mobility of the complex

Table 34.2 Summary of applications of CZE to separation of metal complexes

Carrier electrolyte composition Voltage Detection	Sample complexes	Reference
Alkaline earth metal complexes 20 mM Sodium borate, 2–3 mM EDTA 20 kV Direct UV 200 nm	Water and serum samples; EDTA complexes	4
10 mM Pyridine, 0.8 mM EDTA, pH 5 7.5 kV Indirect UV 254 nm	River water, urine, calcium carbonate; EDTA complexes	5
20 mM Sodium borate, 0.2 mM HBED 25 kV Direct UV 200–294 nm	Water and serum samples; HBED complexes	8
Transition metal complexes 20 mM Phosphate, 2 mM cyanide, pH 9.4 −4.5 to −7 kV Direct UV 214 nm	Aqueous standards; cyanide complexes	2
10 mM Carbonate buffer, pH 9.6 25 kV Direct UV 214 nm	Au and Ag in ores; cyanide complexes	24
30 mM Phosphate buffer, 15% acetonitrile, pH 7 −25 kV Direct UV 214 nm	Samples from electrolytic Cu refinery; cyanide complexes	25

50 mM HCl–KCl, 0.2 mM cetyl-
  trimethylammonium bromide,
  pH 3
−17 kV
Direct UV 214 nm
	Aqueous standards;
	chloro complexes
		3

Table 34.2 *Continued*

Carrier electrolyte composition Voltage Detection	Sample complexes	Reference
10 mM Borate buffer, 1 mM CDTA, pH 9 15 kV Direct UV 214 nm	Pharmaceuticals; CDTA complexes	7
10 mM TAPS, 0.1 mM PAR, pH 8.4 30 kV Direct vis 500 nm	Aqueous standards, vitamin tablets, water samples; PAR complexes	11
10 mM Ammonium phosphate, 75 mM SDS, 0.1 mM PAR, pH 8 15 kV Direct UV 254 nm	Aqueous standards; PAR complexes	12
10 mM Phosphate buffer, 0.1 mM arsenazo III, 0–50 mM SDS, pH 8 20–28 kV Direct UV 254 nm	Aqueous standards, zirconium ceramic; arsenazo III complexes	12
20 mM Ammonia/ammonium buffer, 0.3 mM PAR, pH 8.5 15 kV Direct UV 520 nm; (modified capillaries containing COOH functional groups)	Aqueous standards; PAR complexes	13
20 mM Phosphate buffer, 25 mM tetrabutylammonium bromide, pH 7 21.7 kV Direct vis 490 nm	Aqueous standards; 2,2′-dihydroxyazo- benzene-4,4′-disulfonate complexes	14
10 mM Borate buffer, 0.1 mM HQS, pH 9 –15 kV Direct UV 254 nm	Aqueous standards, tap water; HQS complexes	15
24 mM Acetate buffer, 0.12 mM 5-Br-PAPS, pH 4.9 30 kV Direct vis 550 nm	Aqueous standards, impurities in nickel salts; 5-Br-PAPS complexes	16
10 mM Borate buffer, 75–150 mM tetradecyltrimethylammonium bromide, 10 mM sodium octanesulfonate, pH 9 –15 kV Direct UV 254 nm	Aqueous standards; 2,6-diacetylpyridine-bis(N- methylenepyridiniohydrazone) complexes	17
5 mM EDTA, 0.01 mM fluorescein, pH 7.5 30 kV Fluorescence detection	Aqueous standards; EDTA complexes	26

Table 34.2 *Continued*

Carrier electrolyte composition Voltage Detection	Sample complexes	Reference
Rare earth metal complexes 20 mM Borate buffer, 1 mM CDTA, pH 11.1 15 kV Direct UV 214 nm	Aqueous standards; CDTA complexes	6
10 mM Borate buffer, 150 mM tetradecyltrimethylammonium bromide, 10 mM sodium octanesulfonate, pH 9 −15 kV Direct UV 254 nm	Aqueous standards; 2,6-diacetylpyridine-bis(*N*-methylenepyridiniohydrazone) complexes	17
Speciation analysis 10 mM Formate buffer, 1 mM CDTA, pH 3.8 −20 kV Direct UV 214 nm	Cr(III)/Cr(VI) in electroplating solutions; CDTA complex	7
10 mM Borate buffer, 0.2 mM tetradecyltrimethylammonium bromide, pH 9.3 −20 kV Direct UV 254 nm	Waste water; free DTPA and metal–DTPA complexes	21
5 mM Imidazole, pH 3.5 16 kV Indirect UV 214 nm	Aqueous standards; aluminum ions, fluoro- and oxalatoaluminum complexes	27
5 mM Phosphate/triethanolamine buffer, 0.8 mM hexamethonium bromide, pH 8.5 −20 kV Direct UV 214 nm	Aqueous standards; cyanide present in the form of metallocyanide complexes	28
Miscellaneous applications 100 mM Acetate buffer, pH 4 15 kV Fluorescence detection	Serum; Al after complexation with lumogallion	18

DTPA, diethylenetriaminepentaacetic acid; TAPS, *N*-tris(hydroxymethyl)methyl-3-aminopropanesulfonic acid; SDS, sodium dodecyl sulfate.

flow) and move away from the detector. For these species, one might simply reverse the polarity. Unfortunately, in this case samples containing both species with high and low mobility would cause problems as the latter would escape detection. Therefore, a much better approach is the reduction or reversal of EOF by adding cationic surfactants and a detection at the anodic side. Cationic metal complexes can be analyzed in a coelectroosmotic mode with the cathode being the detection side, although in this case the time window for the separation is limited. Sometimes, better separations can be obtained for cationic metal complexes after reversing EOF and reversing the polarity. Table 34.1 summarizes the configurations of CZE that have been used so far for separation of metal complexes.

Apart from CZE, ITP (isotachophoresis) has also been used for the separation of metal complexes [23], although its importance is limited when compared with CZE separations described in recent years.

34.4 Applications of metal ion analysis after complexation

Table 34.2 lists recent applications of CE for alkaline earth metal ions, transition metal ions and rare earth metal ions after complexation. In addition, examples of the separation of metal ions and metal complexes in speciation analysis are given (applications dealing with samples containing both free metal ions and organometallic compounds are excluded. For these compounds, refer to Chapter 35 on Metal Chelation).

References

1 Timerbaev, A.R. (1995) Separation of metal ions by capillary electrophoresis: An understanding of the basic principles. *J. Capillary Electrophoresis* **2**, 165–174.
2 Buchberger, W.; Semenova, O.P.; Timerbaev, A.R. (1993) Metal ion capillary zone electrophoresis with direct UV detection: Separation of metal cyanide complexes. *J. High Resolut. Chromatogr.* **16**, 153–156.
3 Zhang, H.; Jia, L.; Hu, Z. (1995) Determination of palladium(II) as a chloro complex by capillary zone electrophoresis. *J. Chromatogr.* **704**, 242–246.
4 Motomizu, S.; Oshima, M.; Matsuda, S.; Obata, Y.; Tanaka, H. (1992) Separation and determination of alkaline-earth metal ions as UV-absorbing chelates with EDTA by capillary electrophoresis. Determination of calcium and magnesium in water and serum samples. *Anal. Sci.* **8**, 619–625.
5 Wang, T.; Li, S.F.Y. (1995) Migration behaviour of alkali and alkaline-earth metal ion–EDTA complexes and quantitative analysis of magnesium in real samples by capillary electrophoresis with indirect ultraviolet detection. *J. Chromatogr.* **707**, 343–353.
6 Timerbaev, A.R.; Semenova, O.P.; Bonn, G.K. (1994) Capillary zone electrophoresis of lanthanoid elements after complexation with aminopolycarboxylic acids. *Analyst* **119**, 2795–2799.
7 Timerbaev, A.R.; Semenova, O.P.; Buchberger, W.; Bonn, G.K. (1996) Speciation studies by capillary electrophoresis – Simultaneous determination of chromium(III) and chromium (VI), *Fresenius J. Anal. Chem.* **354**, 414–419.
8 Motomizu, S.; Morimoto, K.; Kuwabara, M.; Obata, Y.; Izumi, K. (1993) Separation of metal ions and the determination of calcium and magnesium ions with N,N'-di(2-hydroxybenzyl)ethylenediamine-N,N'-diacetic acid by high-performance capillary electrophoresis. *Bunseki Kagaku* **42**, 873–880.
9 Latva, M.; Ala-Kleme, T.; Bjennes, H.; Kankare, J.; Haapakka, K. (1995) Time-resolved luminescence detection of europium(III) chelates in capillary electrophoresis. *Analyst* **120**, 367–372.
10 Iki, N.; Hoshino, H.; Yotsuyanagi, T. (1993) High-performance separation and determination of Co(III) and Ni(II) chelates of 4-(2-pyridylazo)-resorcinol at femtomole levels by capillary electrophoresis. *Chem. Lett.* 701–704.
11 Regan, F.B.; Meaney, M.P.; Lunte, S.M. (1994) Determination of metal ions by capillary electrophoresis using on-column complexation with 4-(2-pyridylazo)resorcinol following trace enrichment by peak stacking. *J. Chromatogr.* **657**, 409–417.

12. Timerbaev, A.R.; Semenova, O.P.; Jandik, P.; Bonn, G.K. (1994) Metal ion capillary electrophoresis with direct UV detection. Effect of a charged surfactant on the migration behaviour of metal chelates. *J. Chromatogr.* **671**, 419–427.
13. Chen, G.J.; Lee, N.M.; Hu, C.C.; Liu, C.Y. (1995) Chemical modification of capillary column for electrophoretic separations of transition metal ions. *J. Chromatogr.* **699**, 343–351.
14. Iki, N.; Hoshino, H.; Yotsuyanagi, T. (1993) Ion association capillary electrophoresis. New separation mode for equally and highly charged metal chelates. *J. Chromatogr.* **652**, 539–546.
15. Timerbaev, A.R.; Buchberger, W.; Semenova, O.P.; Bonn, G.K. (1993) Metal ion capillary zone electrophoresis with direct UV detection: determination of transition metals using an 8-hydroxyquinoline-5-sulphonic acid chelating system. *J. Chromatogr.* **630**, 379–389.
16. Motomizu, S.; Oshima, M.; Kuwabara, M.; Obata, Y. (1994) Separation and sensitive determination of metal ions by capillary zone electrophoresis with 2-(5-bromo-2-pyridylazo)-5-(*N*-propyl-*N*-sulfopropylamino)phenol. *Analyst* **119**, 1787–1792.
17. Timerbaev, A.R.; Semenova, O.P.; Bonn, G.K.; Fritz, J.S. (1994) Determination of metal ions complexed with 2,6-diacetylpyridine-bis (*N*-methylenepyridiniohydrazone) by capillary electrophoresis. *Anal. Chim. Acta* **296**, 119–128.
18. Takatsu, A.; Eyama, S.; Uchiumi, A. (1995) Determination of aluminium in serum by capillary zone electrophoresis with laser-induced fluorescence detection. *Chromatographia* **40**, 125–128.
19. Haiböck, U. (1994) *Diploma thesis*, Johannes-Kepler-University, Linz, Austria.
20. Wiley, J.P. (1995) Determination of polycarboxylic acids by capillary electrophoresis with copper complexation. *J. Chromatogr.* **692**, 267–274.
21. Buchberger, W.; Mülleder, S. (1995) Determination of chelating agents and metal chelates by capillary zone electrophoresis. *Mikrochim. Acta* **119**, 103–111.
22. O'Keeffe, M.; Dunemann, L.; Theobald, A.; Svehla, G. (1995) Capillary electrophoresis in speciation analysis. Investigations of metal–polyaminopolycarboxylate complexes and effects of metals on proteins. *Anal. Chim. Acta* **306**, 91–97.
23. Hirokawa, T.; Ohta, T.; Nakamura, K.; Nishimoto, K.; Nishiyama, F. (1995) Bidirectional isotachophoretic separation of metal cations using EDTA as a chelating agent. *J. Chromatogr.* **709**, 171–180.
24. Aguilar, M.; Farran, A.; Martinez, M. (1993) Determination of gold(I) and silver(I) cyanide in ores by capillary zone electrophoresis. *J. Chromatogr.* **635**, 127–131.
25. Lee, H.J.; Lee, S.H.; Chung, K.S. (1994) Determination of Pd(II) and Pt(II) metal cyano complexes using capillary electrophoresis. *Bull. Korean Chem. Soc.* **15**, 945–949.
26. Desbene, P.L.; Morin, C.J.; Desbene-Monvernay, A.M.; Groult, R.S. (1995) Utilization of fluorescein sodium salt in laser-induced indirect fluorimetric detection of ions separated by capillary zone electrophoresis. *J. Chromatogr.* **689**, 135–148.
27. Wu, N.; Horvath, W.J.; Sun, P.; Huie, C.W. (1993) Speciation of aluminium using capillary zone electrophoresis with indirect UV detection. *J. Chromatogr.* **635**, 307–312.
28. Buchberger, W.; Haddad, P.R. (1994) Separation of metallo-cyanide complexes by capillary zone electrophoresis. *J. Chromatogr.* **687**, 343–349.

35 Metal chelation
F.B. ERIM

35.1 Introduction

Complexing agents are used in capillary electrophoresis to separate analytes which show similar electrophoretic behavior. Various ligands serving this purpose have been reported and these are either strong chelating reagents or weak complexing agents. Generally, complex equilibrium occurs in one step if the complex forming ligand forms a strong chelate and in a step-wise fashion when a weak ligand is used.

In the case of partially complexing analytes, on-column complexation is preferred by adding excess amounts of complexing agent to the running buffer. Precolumn complexation is widely used when complexing occurs with a strong chelating agent. Even so, almost all of the literature on precolumn complexation recommends adding the complexing agent in low concentrations to the electrophoretic buffer solution. Otherwise dissociation of complexes and multiple peaks can occur.

Optimal separation conditions are usually determined in terms of buffer, pH, complexing agent concentration, type of counterion and, where applicable, type and concentration of ion pairing reagent and type and concentration of surfactant.

35.2 Metal chelation in capillary electrophoresis

Iki *et al.* [1] separated 4-(2-pyridylazo)resorcinol (PAR) and its complexes with Co(III), Fe(II), Ni(II) and V(V) in 6 min by capillary electrophoresis (CE). Metal–(PAR) and arsenazo(III) complexes were investigated by Timerbaev *et al.* [2] using micellar solutions of sodium dodecyl sulphate (SDS). They observed no interaction between doubly charged anionic metal–PAR complexes and SDS micelles in either borate or sodium phosphate buffer because of a strong repulsion between the anionic solutes and the negatively charged end groups of the micelles. However when these buffers were replaced with ammonium phosphate buffer, improved resolution of nine metal complexes was obtained. This was attributed to the ion pairing effect of ammonium ion which decreased the negative charge of the complexes through ion association and thereby overcame the electrostatic repulsion between anionic complexes and the negatively charged SDS micelles. The arsenazo(III) complexes did not show any observable interaction with SDS micelles because of their higher charge. Using this ligand non-micellar electrophoresis of seven metal ions was achieved. The applicability of this method to the zirconium ceramic samples containing high levels of acids was demonstrated.

Regan et al. [3] used PAR to separate Co(II), Fe(II), Cu(II) and Zn(II) ions. They used both precolumn complexation and an on-column complexation method combined with peak stacking. For both methods, N-tris(hydroxymethyl)methyl-3-aminopropanesulfonic acid (TAPS) containing a small amount of PAR was used as a running buffer. Their on-column peak stacking method involved three step sequences: in the first step, a high concentration of PAR plug was injected. A sample plug containing metal ion in water was then injected. The third step in the stacking program was a pause time, during which the run buffer was introduced to the capillary. After this step, the running voltage was applied. Since the positively charged metal ions moved more rapidly toward the cathode than did PAR, they mixed with the PAR plug, thus bringing about on-column complexation and stacking of the metal ions. The method resulted in a 100-fold reduction in detection limits for Co(II), Fe(II) and Zn(II) and 10-fold reduction for Cu(II), compared with precolumn complexation. The singularly charged Au(I) and Ag(I) cyanide complexes in carbonate buffer containing free cyanide were separated by Aguilar et al. [4]. The dicyanoargentate came off before the dicyonoaurate. Aguilar et al. attributed the mobility difference between these complexes, which have similar charge, to the difference in their anionic volume, as $Au(CN)_2$ is a much less hydrated anion than $Ag(CN)_2$. Because the other metal cyano complexes have between two and four negative charges and consequently large ionic mobilities, they did not reach the cathodic detection point under the conditions used in this method. Ag(I) and Au(I) complexes were separated without interacting with the ion matrix of these metal ions. It was suggested that this method was appropriate for the determination of Au and Ag in ore samples.

Separation of metal cyanides differing in charge and number of coordinated cyano ligands was however achieved by Buchberger et al. [5] using the cathodic injection method. The larger electrophoretic mobilities of the metal ions than of EOF in the opposite direction resulted in their detection at the anodic side.

Separation of negatively charged 8-hydroxyquinoline-5-sulfonic acid (HQS) complexes was performed by Timerbaev and co-workers [6, 7]. They investigated several possible migration modes and compared these modes with respect to analysis time, resolution efficiency and detectability. In their 'decelerated movement mode' method complexes were injected from the cathodic side to the capillary which was pretreated with a tetraalkylammonium salt. This resulted in such a decrease in electroosmotic flow (EOF) that the complexes moved in the opposite direction. However in their 'reversed movement mode' method, negatively charged complexes injected from the anodic side moved with EOF in the direction opposite to their electrophoretic movement. They found the second method to be superior achieving the separation of eight metal ions in less than half the time of the first method and with better resolution. The addition of a micelle to the electrophoretic buffer in the second migration mode was also investigated but no improvement was observed.

Separation of the metal chelates of Al(III), Co(III), Cr(III) and Fe(III) with 2,2'-dihydroxyazobenzene-5,5'-disulfonate (DHABS) was performed by Iki et al. [8]. They termed this 'ion-association CE'. While they were not able to resolve these metal ions by capillary zone electrophoresis (CZE), because of their identical and high negative charges (−5), adding tetrabutylammonium bromide (TBAB) led to a complete resolution of four metal chelates. Iki et al. deduced that the ion association reaction between the chelates and ammonium ions played the major role in the separation system. They also examined the effect of the sizes and concentrations of the ion association agents for five different tetraalkylammonium salts and a tetraphenylphosphonium salt concluding that hydrophobic interaction between solute anion and countercation contributed to the formation of ion associates as well as electrostatic attraction. Timerbaev et al. [9] used 2,6-diacetylpyridinebis(N-methylenepyridiniohydrozone) (H_2 dapmp) which formed positively charged metal complexes because of the quaternary ammonium groups in the molecule. Since this ligand formed very stable chelates, there was no need for capillary electrolyte containing a chelating agent. This separation process was performed using 75 mM tetradecyltrimethylammonium bromide (TDTAB) micellar media. The bulk solution of cationic micelle affected both the electrophoretic mobilities of analytes and caused the reversal of EOF. Resolution of 14 metal ions was achieved by adding a strong ion pairing anion, sodium n-octanesulfonate (SOS) to the micellar buffer. The formation of ion pairs between positively charged quaternary ammonium groups in the dapmp ligand and ion pairing anion enhanced the resolution of metal ion peaks.

The determination of Al complexed with desferrioxamine (DFO) in the presence of alkali and alkaline earth ions was performed by Baechmann et al. [10] using 4-methyaminophenol sulfate (metol) as a buffer for the indirect detection of alkali and alkaline earth metal ions. The Al(H-DFO)$^+$ complex was detected as a positive peak in simultaneous determination with other ions. The area of complex peak was examined for pH ranging from 2 to 10 and found to be independent of pH in this range. The buffer solution did not contain DFO because it showed absorption at the same wavelength as the metal ion and also interacted with the surface of the fused silica.

Motomizu et al. [11] used vis-absorbing chelating agent, 2-(5-bromo-2 pyridyazo)-5-(N-propyl-N-sulfopropylamino)phenol(5-Br-PAPS) for the

CE of 22 metal ions using the precolumn complexation method. The carrier solution also contained cheating agent. They applied their method to the determination of trace amounts of Co contained as impurities in Ni salts.

Motomizu et al. [12] also reported 2-(5-nitro-2 pyridyazo)-5-(N-propyl-N-sulfopropylamino) phenol as a chelating agent for the separation of ten metal ions. Ba(II), Sr(II), Ca(II) and Mg(II) were separated using ethylenediaminetetraacetic acid (EDTA) as a chelating agent by Motomizu et al. [13]. The method was applied to the determination of Ca(II) and Mg(II) in river, tap and underground water samples and serum sample. Several aminopolycarboxylic reagents, namely cyclohexanediaminetetraacetic acid (CDTA), EDTA, diethylenetriaminepentaacetic acid (DTPA) and triethylenetetraaminehexaacetic acid (TTHA) were used to separate and detect rare earth metal ions by Timerbaev et al. [14]. Of these, the most satisfactory chelating agent was found to be CDTA, which provides the best resolution and the shortest analysis time. The pH and borate buffer concentration were obtained for optimal CE conditions for this ligand. Using CDTA, lanthanoids(III) and also scandium(III) and yttrium(III) were separated in 12 min. The applicability of this method to the determination of lanthanoid elements in nuclear fuel waste was demonstrated.

The most common complexing agent used for on-column complexation is hydroxyisobutyric acid (HIBA). There are a number of studies in the literature on the use of HIBA to separate alkali, alkaline earth, transition and rare earth metal ions. Weston and co-workers [15, 16] separated a mixture of alkaline, alkaline earth and transition metal ions in 8 min, and 19 different cations including lanthanides and alkali and alkaline metal ions in 2 min using HIBA as a complexing agent. They showed how this technique could be applied to the analysis of ions in tap water, orange juice and in acid-etching baths.

Simultaneous quantitation of sodium, potassium and magnesium in parenteral solutions using HIBA as a weak complexing agent has been reported by Koberda et al. [17]. An Al assay method proposed by Barger et al. [18] involved separating Al ion from other cations in an electrolyte solution containing HIBA and ephedrin. They chose ephedrine as a UV absorbing agent as it has a mobility close to that of Al ion.

Lin et al. [19] reported separation of six alkaline and alkaline earth cations using ten different mono-, di- and triprotic carboxylic and hydroxycarboxylic acids as complexing agents. The agents were acetic, glycolic, lactic, hydroxyisobutyric, oxalic, malonic, malic, tartaric, succinic and citric acids. They used a buffer containing imidazole as a visualizing agent. An interesting point reported in this study was that the type of acid used to adjust the pH of solutions influenced the resolution of the metal ions. When H_2SO_4 was used instead of HCl to adjust pH, it competed with the complexing agent resulting in a change of resolution and order of migration of metal ions. They found the sulfate ion influenced the migration of divalent ions more than the weak acid anions.

A method of separating of alkaline and alkaline earth cations using an imidazole–H_2SO_4 background electrolyte system was proposed by Beck and Engelhardt [20]. Yang et al. [21] applied this method to the quantitative analysis of pharmaceutical electrolyte solutions and beverages. Buchberger et al. [22] recommended copper sulfate as a carrier electrolyte salt and demonstrated its applicability to cation determination in serum samples.

Lee and Lin [23] separated 17 different metal ions including alkali, alkaline earth and transitions metal ions and Group IB, IIB and IVA metal ions with pyridine or imidazole as a background electrolyte and glycolic acid as a metal complexing agent. Lanthanide metal ions were separated by Chen and Cassidy [24, 25] utilizing C1 and C18 saturated hydrocarbon bonded capillaries and using HIBA as complexing agent and benzylamine as a chromophore. They reported reduced surface interaction and improved resolution for the bonded phases. The same authors separated 26 different metal ions using HIBA and dimethylbenzylamine in an electrolyte solution.

Desiderio et al. [26] used Cu(II) complexes with L-amino acid or aspartame ligands for the separation of optical isomers of some α-hydroxy acids, 2-, 3-phenylacetic acid, mandalic, p-hydroxy-, m-hydroxy- and 3,4-dihydroxymandalic acid.

Cai and El Rassi [27] developed the capillaries with surface bound metal chelating functions, i.e. iminodiacetic acid functions for the selective on-line preconcentration of dilute protein samples prior to capillary zone electrophoresis (CZE). Zn(II) was immobilized on the capillary surface and metal chelate capillary tubes were connected in series to a separation capillary. The method permitted the detection of samples 25 times less concentrated than by CZE alone. Although many studies have reported the use of chelating agents for the separation of metal ions, only two studies have provided a mathematical method for variable optimization. Jimidar et al. [28] developed a model for optimal separation conditions for rare earth elements providing resolution, analysis time and sensitivity in terms of the concentration of the complexing agent, HIBA, and the solution pH. Quang et al. [29] developed a simulation model which was based however on a small number of experimental data. They again determined optimum operating conditions in terms of HIBA concentration and solution pH, predicting conditions which resulted in the separation of 14 metal cations including alkali, alkaline earth and transition metal (II) cations in less than 4 min. They demonstrated the applicability of this method to tap water analysis.

As is well known, the separation of metal ions using complexing agents is based on the differences between the stability constants of the complexes formed. The overall stability constant β_n of a complex is given by,

$$\beta_n = \frac{[ML_n]}{[M][L]^n} \tag{35.1}$$

where M is the metal and L is the ligand and the average number of ligands, $\bar{n}$ is given by,

$$\bar{n} = \frac{\sum_{n=1}^{n=N} n\beta_n [L]^n}{1 + \sum_{n=1}^{n=N} \beta_n [L]^n} \tag{35.2}$$

It can be seen from equation 35.2 that, $\bar{n}$ depends only on the free ligand concentration [L]. [L] varies with total ligand concentration C_L and buffer pH. Optimum metal ion separation, therefore, occurs when there are maximum differences in $\bar{n}$.

A mathematical evaluation of separation conditions in terms of complex equilibrium data has been performed by only three groups of researchers. Vogt and Conradi [30] used the complex formation constants of rare earth elements to precalculate optimum ligand concentration. They selected HIBA, lactic acid and acetic acid as complexing agents. As all of these ligands are weak complexing agents, only partial complexation occurred in solution. After investigating the influence of pH on the migration time of rare earth complexes, Vogt and Conradi selected a pH value of 5 to enable sufficient ligand deprotonation, but to prevent hydroxide precipitation and large EOF. For this pH, they used stability constants of the complexes formed to calculate [L] and $\bar{n}$ values for the complexing agent concentrations varying from 0 to 10mM. They obtained the optimum [L] value by plotting $\bar{n}$ vs. log[L] (known as a complex formation curve), for different metal ions, to obtain the point of maximum difference in $\bar{n}$ between the curves where the degrees of complexation of each of the metal ions with the same ligand would differ the most. The theoretical plots of $\bar{n}$ above were used to determine the optimum ligand concentration of HIBA and lactic acid, but not of acetate since it had no such differences in its complex formation curves for any metal ion at any free ligand concentration. Therefore, no satisfactory separation using acetic acid was achieved. However, better separations were reported with mixed ligand complexes, e.g. those formed by HIBA and acetate together.

Shi and Fritz [31] reported successful separations of metal ions using phthalate, tartrate, lactate and HIBA as the complexing reagents. They separated eight metal ions with phthalate, 12 alkali, alkaline earth and transition metal ions with tartrate, 27 metal ions, including rare earths, with lactate and 13 lanthanides with HIBA. For the last of these systems, the optimum concentration of HIBA and the optimum pH were obtained by calculating the $\bar{n}$ values from published formation constants of complexes for the 13 lanthanides. An interesting finding was that the average ligand number of a complex of one particular rare earth ion increased linearly with the atomic number of the element. A plot of $\bar{n}$ vs. atomic number had a correlation coefficient of 0.9958 and a plot of $\bar{n}$ vs. migration time had a correlation coefficient of 0.9978.

Timerbaev et al. [32] proposed a theoretical model for the migration behavior of lanthanide metal ions complexed with aminopolycarboxyclic reagents. The electrophoretic mobilities of the complexes were measured in buffer solutions of varying pH and borate concentration. They observed that the migration times of the complexes increased in the order CDTA < EDTA < DTPA < TTHA and concluded that the net charge on the complex, which depends strictly on the number of carboxyl groups of the ligand, was the main parameter determining the mobility of the complex. While the charge on the complex mainly determined this order, they also found ligand size to be an additional parameter and this depended on the functional groups on the ligand. However, regardless of the nature of the ligand, they found that migration times of lanthanides increased with atomic number and explained this in terms of the effective charge on the metal ion. That is, the lower the e^- acceptor ability of the metal ion, the larger the negative charge on the complex and the faster the mobility would be. Consequently, complexes of lanthanide ions with higher atomic numbers migrated more slowly towards the cathode. Timerbaev et al. reasoned that since e^- acceptor ability is closely related to the stability constant of a complex, these constants should be the overall parameter determining the migration behavior of lanthanides ions. Above pH 9.5, a significant increase in the electrophoretic mobilities and an improvement in resolution due to their appearance as hydroxo forms of lanthanide complexes were observed. Formation of mixed ligand complexes increased the negative charge on the complexes and consequently the electrophoretic mobilities. On the basis of their findings Timerbaev et al. constructed a six term equation expressing the mobility of metal complexes in borate buffer in terms of the stability constants of mono and mixed ligand complexes, pH and number of carboxyclic, methylene and tertiary amine functional groups of each ligand.

A different approach to metal complexation in CE was recently taken by Erim et al. [33] who investigated the step-wise complexation of Cu(II) with 1,10-phenanthroline and with 2,2'-bipyridyl and determined the stability constants of complexes by CZE. The method they used was based on the direct or indirect measurements of the free ligand concentration and calculation of $\bar{n}$, and extraction of β_n values from equation 35.2. The CZE method exploited the difference in electrophoretic mobility of the ligand and complex to determine the total free and/or bound ligand concentration. The investigation showed the applicability of CZE to the determination of stability constants of metal complexes.

Table 35.1 Summary of applications of chelating or complexing agents in capillary electrophoresis

Chelating/ complexing agents	Separation conditions	Compounds	Applications to real samples	Ref.
PAR precolumn	10 mM NaH$_2$PO$_4$, 10 mM Na$_2$B$_4$O$_7$, pH = 8.4, 21 kV, 70 cm (55 cm to detector) × 50 μm i.d., 500 nm	Co(II), V(V), Cu(II), Fe(II), Ni(II)	–	1
PAR precolumn	10 mM Ammonium phosphate, 75 mM SDS, 1.10^{-4} M PAR, pH = 8, 15 kV, 50 cm (42 cm to detector) × 75 μm i.d., 254 nm	Cr(III), Co(II), Cu(II), Pb(II), Ni(II), Fe(II), Zn(II), Fe(III), Cd(II)	–	2
PAR (a) precolumn (b) on-column + peak stacking	(a) 10 mM TAPS, 1 × 10^{-4} M PAR, pH = 8.4, 30 kV, 85 cm (75 cm to detector) × 50 μm i.d., 500 nm (b) Stacking voltage, 10 kV inj.times: [PAR]: 1 × 10^{-3} M, 10 s; [M]: 8.10^{-7}, 10 s; [PAR]: 1 × 10^{-4} M, 5 s	Co(II), Cu(II), Fe(II), Zn(II)	Vitamin tablets, pond water	3
Arsenazo(III) precolumn	10 mM Ammonium phosphate. 1.10^{-4} M arsenazo(III), pH = 8, 20 kV, 50 cm (42 cm to detector) × 75 μm i.d., 254 nm	Ce(III), La(III), U(VI), Cu(II), Pb(II), Co(II), Fe(III)	Zirconium ceramics	2
Cyanide precolumn	1.10^{-2} M Carbonate buffer, pH = 9.6, 25 kV, 64 cm × 50 μm i.d., 214 nm	Ag(I), Au(I)	Ore samples	4
Cyanide precolumn	20 mM Phosphate, 2 mM sodium cyanide, pH = 9.4, 4.5 kV(a), 7 kV(b), 35 cm (to detector) × 75 μm i.d., 214 nm	(a) Fe(III), Cu(I), Ni(II), Cr(III), Ag(I), Co(III) (b) Fe(III), Cu(I), Fe(II), Pb(II), Ag(I)	–	5
HQS precolumn	10 mM Borate buffer, 0.1 mM HQS (a,b); +50 mM SDS(c), pH = 9, –15 kV(a), +15 kV(b,c) 42 cm (35 cm to detector) × 75 μm i.d., (a) decelarated movement (b) reversed movement (c) MEKC migration mode	(a,b,c) Ni(II), Co(II), Zn(II), Cd(II), Fe(III), Cu(II) (c) Mn(II), Cu(II), Al(III), Cd(II), Fe(III), Zn(II), Co(II), Ni(II)	Tap water	6, 7
DHABS precolumn	20 mM NaH$_2$PO$_4$, 25 mM TBABr, pH = 7, 21.7 kV, 65 cm (50 cm to detector) × 50 μm i.d., 490 or 494 nm	Al(III), Co(II), Cr(III), Fe(III)	–	8
H$_2$ dapmp precolumn	10 mM Borate buffer, 75 mM TDTAB, 10 mM SDS, pH = 9, 15 kV, 50 cm × 75 μm i.d., 254 nm	Mo(VI), Sc(III), Fe(III), Y(III), Zn(II), Cd(II), Zr(IV), Co(II), U(VI), Cu(II), Sn(IV), Ta(V), Hg(II)	–	9
DFO	4 mM Metol, pH = 2–10, 60 cm × 75 μm i.d., 30 kV.	Cs(I), K(I), Na(I), Ca(II), Mg(II), Li(I), Al(III), Ce(III)	–	10
5-Br-PAPS precolumn	2.4 × 10^{-2} M Acetate buffer 1.2 × 10^{-4} M 5-Br-PAPS, pH = 4.9, 30 kV, 70 cm (50 cm to detector) × 50 μm i.d., 550 nm	Cd(II), Zn(II), Pb(II), V(IV), Hg(II), Cu(II), Co(II), Ni(II), Fe(II) (apparent electrophoretic mobilities of 22 metal ions were reported)	Trace amount of Co in nickel salts	11

Table 35.1 *Continued*

Chelating/ complexing agents	Separation conditions	Compounds	Applications to real samples	Ref.
5-Nitro-PAPS precolumn	2.4×10^{-2} M Acetate buffer, 1.2×10^{-4} M 5-nitro-PAPS, pH = 4.1, 30 kV, 72 cm (50 cm to detector) × 50 μm i.d., 560 nm	Cd(II), Zn(II), Pb(II), V(IV), Cu(II), Co(II), Ni(II), Fe(II)	Trace amount of Co in nickel salts	12
EDTA	2×10^{-3} M Borax, 2×10^{-3} EDTA, pH = 9.2, 30 kV, 75 cm (to detector) × 50 μm i.d., 200 nm	Ba(II), Sr(II), Ca(II), Mg(II)	River, tap, underground water, serum sample	13
CDTA precolumn	20 mM Borate, 1×10^{-3} M CDTA, pH = 11.1, 50 cm × 75 μm i.d., 214 nm	Ce(III), Pr(III), Y(III), Nd(III), Sm(III), Eu(III), Gd(III), Tb(III), Dy(III), Ho(III), Er(III), Tm(III), Yb(III), Lu(III), Sc(III)	Nuclear fuel waste	14
HIBA on-column	(a) 5 mM Waters UV Cat-1, 6.5 mM HIBA, pH = 4.4, 20 kV, 60 cm × 75 μm i.d., 214 nm (b) 10 mM waters UV Cat-1, 4 mM HIBA, pH = 4.4, 30 kV, 36.5 cm × 75 μm i.d., 214 nm	(a) K(I), Ba(II), Sr(II), Ca(II), Na(I), Mg(II), Mn(II), Cd(II), Fe(II), Co(II), Pb(II), Ni(II), Li(I), Zn(II) (b) Ru(I), K(I), Ca(II), Na(I) Mg(II), Li(I), lanthanides	Tap water, orange juice, acid etching baths	15, 16
HIBA on-column	5 mM Waters UV Cat-1, 6.5 mM HIBA, pH = 4.4, 20 kV, 60 cm × 75 μm i.d., 214 nm	K(I), Ca(II), Mg(II), Na(I)	–	17
HIBA on-column	5.2 mM Ephedrine, 4.7 mM HIBA, pH = 2.8, 15 kV, 50 cm × 50 μm i.d., 204 nm	Li(I), K(I), Ca(II), Cr(III), Zn(II), Al(III), Cu(II)	–	18
Acetic, glycolic, lactic, hydroxyisobutyric, oxalic, malonic, malic, tartaric, succinic, citric on-column	5 mM Imidazole, 0.1–6.4 mM complexing acids, pH = 3–6, 25 kV, 42 cm (35 cm to detector) × 50 μm i.d., 215 nm	K(I), Ba(II), Ca(II), Na(I), Mg(II), Li(I)	–	19
H_2SO_4–imidazole	5 mM Imidazole, pH = 4.5, 25 kV, 63 cm (55 cm to detector) × 75 μm i.d., 214 nm	K(I), Na(I), Ba(II), Ca(II), Mg(II), Li(I)	Apple and orange juice, pharmaceutical electrolyte solutions	20, 21
$CuSO_4$ on-column	5 mM $CuSO_4$, pH = 4.5, 30 kV, 52 cm (to detector) × 75 μm i.d., 214 nm	K(I), Na(I), Ca(II), Mg(II), Li(I) in serum sample	Serum	22
Glycolic acid on-column	13 mM Glycolic acid, 10 mM imidazole or 12 mM glycolic acid, 10 mM pyridine, pH = 4, 25 kV, 90 cm (80 cm to detector) × 75 μm i.d., 210 nm (for imidazole), 254 nm (for pyridine)	Cs(I), K(I), Ba(II), Sr(II), Na(I), Ca(II), Mg(II), Mn(II), Cr(III), Fe(II), Cd(II), Li(I), Co(II), Ni(II), Pb(II), Zn(II), Cu(II)	–	23
HIBA, acetic acid on-column	(a) 4 mM HIBA, 9 mM benzylamine, 20 mM acetic acid, C1 and C18 saturated. H.K. bonded capillaries 70 cm (60 cm to detector) × 75 μm i.d., 214 nm	(a) Lanthanides (b) Lanthanides	–	24, 25

Table 35.1 *Continued*

Chelating/complexing agents	Separation conditions	Compounds	Applications to real samples	Ref.
	(b) 4.2 mM HIBA, 6 mM N,N-dimethyl-benzylamine (DBA), pH = 5 (adj: acetic acid), 0.2 mM Triton × 100, 30 kV, unbonded capillary 60 cm × 75 μm i.d., 214 nm			
Copper(II)-L-aminoacid or aspartame	(a) 0.02 M Sodium dihydrogen phosphate, Cu(II)-L-proline (8 mM/16 mM) or (6 mM/12 mM), pH = 4.4, 8 kV, 20 cm × 25 μm i.d. (b) 0.02 M acetate Cu(II)-aspartame(8 mM/16 mM), 12 kV, 30 cm × 50 μm i.d.	(a) Racemic 3-phenyllactic acid (b) Racemic α-hydroxyacids	–	26
Zn–iminodiacetic acid(IDA) metal–chelate capillaries	Preconcentration capillary: Zn(II)-IDA, etched at 250°C 20 cm × 75 μm i.d., separation capillary I-200, 60 cm (30 cm to detector) × 50 μm i.d. Binding electrolyte: 10 mM sodium phosphate, pH = 6 Debinding electrolyte: 10 mM sodium phosphate 30 mM EDTA, pH = 3.8, 20 kV, 210 nm	Preconcentration and detection of dilute carbonic anhydrase	–	27
HIBA on-column mathematical method for variable optimization based on experimental data	10 mM Creatinine, acetate buffer, 2 mM HIBA, pH = 4.1, 30 kV, 52 cm (to detector) × 75 μm i.d., 214 nm	La(III), Ce(III), Gd(III), Tb(III)	–	28
HIBA on-column mathematical method for variable optimization based on experimental data	12 mM HIBA, 6 mM imidazole, pH = 3.95, 25 kV, 62 cm (50 cm to detector) × 50 μm i.d., 214 nm	K(I), Ba(II), Sr(II), Ca(II), Na(I), Mg(II), Mn(II), Cd(II), Fe(II), Co(II), Li(I) Ni(II), Zn(II), Cu(II)	Tap water	29
HIBA, lactic acid, acetic acid (optimization with equilibrium data)	(a) 20 mM Lactic acid, 30 mM creatinine (b) 7 mM HIBA, 30 mM creatinine, pH = 4.8, 6 kV, 40 cm = 100 μm i.d., 214 nm	K(I), Na(I), Mg(II), lanthanides	–	30
Phthalate, tartrate, lactate, HIBA (optimization with equilibrium data)	(a) 2.5 mM Tartaric acid, 6 mM p-toluidine, 20% methanol, pH = 4.8, 30 kV (b) 2 mM Phthalic acid, UV-Cat-1, 20% methanol, pH = 3.3, 15 kV (c) 15 mM Lactic acid, 8 mM 4-methylbenzylamine, 5% methanol, pH = 4.25, 30 kV (d) 4 mM HIBA, 5 mM UV Cat 1, pH = 4.3, 30 kV, 60 cm × 75 μm i.d., 214 nm	(a) K(I), Na(I), Li(I), Mg(II), Ba(II), Sr(II), Mn(II), Ca(II), Cd(II), Co(II), Ni(II), Zn(II) (b) K(I), Na(I), Pb(II), Mn(II), Co(II), Ni(II), Zn(II), Cd(II) (c) K(I), Ba(II), Sr(II), Na(I), Ca(II), Mg(II), Mn(II), Li(I), Co(II), Pb(II), Ni(II), Zn(II), lanthanides (d) Lanthanides	–	31

Table 35.1 *Continued*

Chelating/complexing agents	Separation conditions	Compounds	Applications to real samples	Ref.
CDTA theoretical estimation	–	–	–	32, 14
1,10-Phenanthroline, 2,2′-bipyridyl determination of stability constants	Phenanthroline, chloroacetic acid buffer, pH = 1.96, bipyridyl, chloroacetic acid, pH = 2.85 (a) Frontal analysis method: 20 kV, inj: 60 s, 10 kV (b) Hummel Dreyer method: 12 kV, inj: 2.4 s, 10 kV (c) Vacany peak method 12 kV, inj: 2.4 s, 10 kV 75 cm × 50 µm i.d., 273 or 302 nm	Stability constants of Cu(II) – 1,10-phenanthroline, Cu(II) – 2,2′-bipyridyl complexes	–	33
18-Crown-6 ether on-column	4 mM Formic acid, 4 mM cupric sulfate, 4 mM 18-crown-6, pH = 3, 20 kV, 50 cm × 50 µm i.d., 215 nm	$NH_4(I)$, K(I), Na(I), Ca(II), Mg(II), Sr(II), Li(I), Ba(II)	Drinking water	34
18-Crown-6 ether on-column	11 mM Lactic acid, 2.6 mM 18-crown-6, 7.5 mM 4-methylbenzylamine, 8% methanol, pH = 4.3, 30 kV, 60 cm (52.5 cm to detector) × 75 µm i.d., 214 or 254 nm	$NH_4(I)$, K(I), Na(I), Ca(II), Sr(II), Mg(II), Mn(II), Ba(II), Cd(II), Fe(II), Li(I), Co(II), Ni(II), Zn(II), Pb(II), Cu(II)	–	35
18-Crown-6 ether on-column	5 mM Collidine, tartrate, 0.1% HEC (w/v), 40 mM 18-crown-6, pH = 5.2, 25 cm (to detector) × 300 µm i.d., 254 nm	K(I), Ba(II), Sr(II), Li(I), Ca(II), Mg(II), Na(I), Rb(I), $NH_4(I)$, Cs(I)	Rain, tap, mineral water	36
18-Crown-6 ether on-column	500 µM Cerium(III) sulfate, 2.5 mM 18-crown-6, 30 kV, 55 cm × 75 µm i.d., 251 nm	Cs(I), $NH_4(I)$, K(I), Ca(II), Na(I), Mg(II), Sr(II), Ba(II), Li(I)	Rain sample, cola beverage	37
Borate on-column	100 mM Borate buffer, PH = 8.4, 25 kV, 57 × 50 µm i.d., 200 nm	Biologically active molecules differing only by a single hydroxyl group	–	
Borate on-column	175 mM Borate buffer, pH = 10.5, 25 kV, 72 cm (50 cm to detector) × 50 µm i.d., 305 nm	Mono and oligo saccharides	–	
Borate on-column	0.2 M Boric acid, pH = 10.5, 20 kV, 70 cm × 50 µm i.d., 270 nm	Flavonoid-3*O*-glycosides	–	
Borate on-column	0.5 mM Borate, 12.5 mM β-CD, 2% tetrahydrofuran, pH = 8.2, 25 kV, 100 cm (85 cm to detector) × 50 µm i.d., 375 nm	Monosaccharide enantiomers	–	

Because of its ability to form complexes with alkaline and alkaline earth metal ions, 18-crown-6 ether has been used as a complexing agent for the separation of these metal ions by several researchers. Riviello and Harrold [34] used 18-crown-6 in a Cu-based electrolyte system together with formic acid and found that besides causing the selectivity of the metal ions to change it also had an obvious effect on the resolution of NH_4^+ and K^+ peaks. Shi and Fritz [35] added 18-crown-6 to an electrolyte buffer containing lactic acid to resolve NH_4^+ and K^+ peaks from one another in the presence of 16 metal ions. Simunicova et al. [36] used 18-crown-6 with tartrate to separate alkali, alkaline earth metal and ammonium cations in one run. Baechmann et al. [37] reported the separation of ammonium and alkaline earth metal ions in a Ce(III) sulfate electrolyte system. They added 18-crown-6 to the electrolyte solution to achieve a resolution of NH_4^+, K^+, Sr and Ca peaks.

Borate complexation with neutral compounds containing diol groups results in the formation of negatively charged complexes and enables their separation by CZE on the basis of their charge-to-mass ratio. Landers et al. [38] used borate complexation to separate several biologically active molecules differing only by a single hydroxyl group. Vorndran et al. [39] separated mono and oligosaccharide borate complexes after a precolumn derivatization.

Morin et al. [40] separated a mixture of flavonoid 3-O-glycosides which differed in their sugar moiety. The separation was based on the structural preference for the formation of the borate complex, i.e. the disposition of hydroxyl groups in the molecule. Stefansson et al. [41] reported analytical resolution of fluorescently labeled monosaccharide enantiomers through complexation with borate and dextrins.

The applications of chelating or complexing agents in CE are summarized in Table 35.1.

35.3 Conclusion

Complexing agents have been used for the separation and detection of metal ions, carbohydrates and chiral compounds in CE. In addition, metal chelate capillaries have been used for the separation of proteins and CE has been applied to the investigation of metal complex equilibria. Taking into consideration the huge number of complexing agents used in chemistry until now, there are numerous possibilities for the application of metal chelation in CE.

References

1. Iki, N.; Hoshino, H.; Yotsuyanagi, T. (1993) High-performance separation and determination of Co(III) and Ni(II) chelates of 4-(2-pyridylazo)resorcinol at femtomole levels by capillary electrophoresis. *Chem. Lett.* **4**, 701–704.
2. Timerbaev, A.R.; Semenova, O.P.; Jandik, P.; Bonn, G.K. (1994) Metal ion capillary electrophoresis with direct UV detection. Effect of a charged surfactant on the migration behaviour of metal chelates. *J. Chromatogr.* **671**, 419–427.

3   Regan, F.B.; Meaney, M.P.; Lunte, S.M. (1994) Determination of metal ions by capillary electrophoresis using on-column complexation with 4-(2-pyridylazo)resorcinol following trace enrichment by peak stacking. *J. Chromatogr.* **657**, 409–417.
4   Aguilar, M.; Farran, A.; Martinez, M. (1993) Determination of gold(I) and silver(I) cyanide in ores by capillary zone electrophoresis. *J. Chromatogr.* **635**, 127–131.
5   Buchberger, W.; Semenova, O.P.; Timerbaev, A.R. (1993) Metal ion capillary zone electrophoresis with direct UV detection. Separation of metal cyanide complexes. *J. High Resolut. Chromatogr.* **16**, 153–156.
6   Timerbaev, A.R.; Buchberger, W.; Semenova, O.P.; Bonn, G.K. (1993) Metal ion capillary zone electrophoresis with direct UV detection. Determination of transition metals using an 8-hydroxyquinoline-5-sulphonic acid chelating system. *J. Chromatogr.* **630**, 379–389.
7   Timerbaev, A.; Semenova, O.; Bonn, G. (1993) Metal ion capillary zone electrophoresis with direct UV detection: Comparison of different migration modes for negatively charged chelates. *Chromatographia* **37**, 497–500.
8   Iki, N.; Hoshino, H.; Yotsuyanagi, T. (1993) Ion-association capillary electrophoresis. New separation mode for equally and highly charged metal chelates. *J. Chromatogr.* **652**, 539–546.
9   Timerbaev, A.R.; Semenova, O.P.; Bonn, G.K.; Fritz, J.S. (1994) Determination of metal ions complexed with 2,6-diacetylpyridine bis(*N*-methylenepyridiniohydrazone) by capillary electrophoresis. *Anal. Chim. Acta* **296**, 119–128.
10 Baechmann, K.; Ehmann, Th.; Haumann, I. (1994) pH-Independent determination of aluminum as a cationic complex using capillary electrophoresis. *J. Chromatogr.* **662**, 434–436.
11 Motomizu, S.; Oshima, M.; Kuwabara, M. (1994) Separation and sensitive determination of metal ions by capillary zone electrophoresis with 2-(5-bromo-2-pyridylazo)-5-(*N*-propyl-*N*-sulfopropylamino)phenol. *Analyst* **119**, 1787–1792.
12 Motomizu, S.; Mori, N.; Kuwabara, M.; Oshima, M. (1994) Separation and sensitive determination of metal ions by capillary zone electrophoresis with 2-(5-nitro-2 pyridylazo)-5-(*N*-propyl-*N*-sulfopropylamino)phenol. *Anal. Sci.* **10**, 101–103.
13 Motomizu, S.; Oshima, M.; Matsuda, S.; Obata, Y. (1992) Separation and determination of alkaline-earth metal ions as UV-absorbing chelates with EDTA by capillary electrophoresis. Determination of calcium and magnesium in water and serum samples. *Anal. Sci.* **8**, 619–625.
14 Timerbaev, A.R.; Semenova, O.P.; Bonn, G.K. (1994) Capillary zone electrophoresis of lanthanoid elements after complexation with aminopolycarboxylic acids. *Analyst* **119**, 2795–2799.
15 Weston, A.; Brown, P.R.; Jandik, P.; Jones, W.R.; Heckenberg, A.L. (1992) Factors affecting the separation inorganic metal cations by capillary electrophoresis. *J. Chromatogr.* **593**, 289–295.
16 Weston, A.; Brown, P.R; Heckenberg, A.L.; Jandik, P.; Williams, R.J. (1992) Effect of electrolyte composition on the separation of inorganic metal cations by capillary ion electrophoresis. *J. Chromatogr.* **602**, 249–256.
17 Koberda, M.; Konkowski, M.; Youngberg, P. (1992) Capillary electrophoretic determination of alkali and alkaline-earth cations in various multiple electrolyte solutions for parenteral use. *J. Chromatogr.* **602**, 235–240.
18 Barger, W.R.; Mowery, R.L.; Wyatt, J.R. (1994) Separation and indirect detection by capillary zone electrophoresis of ppb (w/w) levels of aluminum ions in solutions of multiple cations. *J. Chromatogr.* **680**, 659–665.
19 Lin, T.-I.; Lee, Y.-H.; Chen, Y.-C. (1993) Capillary electrophoretic analysis of inorganic cations. Role of complexing agent and buffer pH. *J. Chromatogr.* **654**, 167–176.
20 Beck, W.; Engelhardt, H. (1992) Capillary electrophoresis of organic and inorganic cations with indirect UV detection. *Chromatographia* **33**, 313–316.
21 Yang, Q.; Jimidar, M.; Hamoir, T.P.; Smeyers-Verbeke, J.; Massart, D.L. (1994) Determination of alkali and alkaline earth metals in real samples by capillary ion analysis. *J. Chromatogr.* **673**, 275–285.
22 Buchberger, W.; Winna, K.; Turner, M. (1994) Applications of capillary zone electrophoresis in clinical chemistry. Determination of low-molecular-mass ions in body fluids. *J. Chromatogr.* **671**, 375–382.

23 Lee, Y.-H.; Lin, T.-I. (1994) Determination of metal cations by capillary electrophoresis. Effect of background carrier and complexing agents. *J. Chromatogr.* **675**, 227–236.
24 Chen, M.; Cassidy, R.M. (1992) Bonded-phase capillaries and the separation of inorganic ions by high-voltage capillary electrophoresis. *J. Chromatogr.* **602**, 227–234.
25 Chen, M.; Cassidy, R.M. (1993) Separation of metal ions by capillary electrophoresis. *J. Chromatogr.* **640**, 425–431.
26 Desiderio, C.; Aturki, Z.; Fanali, S. (1994) Separation of some α-hydroxy acid enantiomers by high performance capillary electrophoresis using copper(II)–L-amino acid and copper(II)–aspartame complexes as chiral selectors in the background electrolyte. *Electrophoresis* **15**, 864–869.
27 Cai, J.; El Rassi, Z. (1993) Selective on-line preconcentration of proteins by tandem metal chelate capillaries–capillary zone electrophoresis. *J. Liq. Chromatogr.* **16**, 2007–2024.
28 Jimidar, M.; Hamoir, T.; Degezelle, W.; Massart, D.L.; Soykenç, S.; Van de Winkel, P. (1993) Method development and optimization for the determination of rare earth metal ions by capillary zone electrophoresis. *Anal. Chim. Acta.* **284**, 217–225.
29 Quang, C.; Khaledi, M.G. (1994) Prediction and optimization of the separation of metal cations by capillary electrophoresis with indirect UV detection. *J. Chromatogr.* **659**, 459–466.
30 Vogt, C.; Conradi, S. (1994) Complex equilibria in capillary zone electrophoresis and their use for the separation of rare earth metal ions. *Anal. Chim. Acta.* **294**, 145–153.
31 Shi, Y.; Fritz, J.S. (1993) Separation of metal ions by capillary electrophoresis with a complexing electrolyte. *J. Chromatogr.* **640**, 473–479.
32 Timerbaev, A.R.; Semenova, O.P. (1995) Theoretical estimation of capillary zone electrophoresis behaviour of metal complexes using multivariate regression analysis. *J. Chromatogr.* **690**, 141–148.
33 Erim, F.B.; Boelens, H.F.M.; Kraak, J.C. (1994) Applicability of capillary zone electrophoresis to study metal complexation in solution. *Anal. Chim. Acta.* **294**, 155–163.
34 Riviello, J.M.; Harrold, M.P. (1993) Capillary electrophoresis of inorganic cations and low-molecular-mass amines using a copper-based electrolyte with indirect UV detection. *J. Chromatogr.* **652**, 385–392.
35 Shi, Y.; Fritz, J.S. (1994) New electrolyte systems for the determination of metal cations by capillary zone electrophoresis. *J. Chromatogr.* **671**, 429–435.
36 Simunicová, E.; Kaniansky, D.; Loksiková, K. (1994) Separation of alkali and alkaline earth metal and ammonium cations by capillary zone electrophoresis with indirect UV absorbance detection. *J. Chromatogr.* **665**, 203–209.
37 Bachmann, K.; Boden, J.; Haumann, I. (1992) Indirect fluorimetric detection of alkali and alkaline earth metal ions in capillary zone electrophoresis with cerium(III) as carrier electrolyte. *J. Chromatogr.* **626**, 259–265.
38 Landers, J.P.; Oda, R.P.; Schuchard, M.D. (1992) Separation of boron complexed diol compounds using high-performance capillary electrophoresis. *Anal. Chem.* **64**, 2846–2851.
39 Vorndran, A.E.; Grill, E.; Huber, C.; Oefner, P.J.; Bonn, G.K. (1992) Capillary zone electrophoresis of aldoses, ketoses and uronic acids derivatized with ethyl *p*-aminobenzoate. *Chromatographia* **34**, 109–114.
40 Morin, Ph.; Villard, F.; Dreux, M. (1993) Borate complexation of flavonoid-*O*-glycosides in capillary electrophoresis. II. Separation of flavonoid-3-*O*-glycosides differing in their sugar moiety. *J. Chromatogr.* **628**, 161–169.
41 Stefansson, M.; Novotny, M. (1993) Electrophoretic resolution of monosaccharide enantiomers in borate–oligosaccharide complexation media. *J. Am. Chem. Soc.* **115**, 11573–11580.

36 Inorganic ions
W. BUCHBERGER

36.1 Introduction

During the last few years, capillary zone electrophoresis (CZE) has become a rapidly developing analytical technique for the separation of inorganic ions. Thereby, it has attracted considerable attention as an alternative to the well-established technique of ion chromatography (IC). The growing importance of CZE in the field of inorganic ion analysis is emphasized by the appearance of a textbook [1] as well as review articles [2, 3] focusing on both inorganic anions and cations. To utilize the potential power of CZE for inorganic ion analysis, several variables of this technique need careful consideration and optimization that might be less important in other application areas of CZE. Among other requirements, a way to control the direction and rate of the electroosmotic flow (EOF) must be found, manipulation of the separation selectivity of CZE for a range of ions that might have similar mobilities is necessary and adequate detection techniques for ions that do not absorb in the UV–vis range must be available.

This chapter will focus on various possibilities for meeting the requirements of CZE for inorganic ions, will discuss suitable carrier electrolytes for inorganic anion and cation separations and will list applications so far reported for real samples in the fields of water analysis, environmental analysis, food analysis, industrial analysis, pharmaceutical analysis and clinical analysis.

Other modes of capillary electrophoresis (CE) such as isotachophoresis (ITP) are not discussed in this review on inorganic ions. ITP investigations have been published for inorganic ions but are currently only of minor importance when compared with CZE. Nevertheless, some aspects of ITP for preconcentration in combination with CZE techniques are included.

36.2 Requirements for inorganic ion analysis by CZE

When CZE separations are carried out in untreated fused-silica capillaries with the cathode at the detection side, EOF will speed up the migration of cations to the detector and will also carry neutral species to the detector. Anions will reach the detection side if their electrophoretic mobility (directed away from the detector) is smaller than the electroosmotic mobility.

In this case, the observed mobility of an anion (that is the vector sum of electrophoretic and electroosmotic mobility) is still directed to the detector. Unfortunately, many inorganic anions have relatively high electrophoretic mobilities so that the observed mobility is small or even directed to the injection side. Therefore, excessively long migration times would hamper the analysis, provided that the analyte anions reach the detector at all. This problem can be circumvented by changing the polarity (anode at the detection side) and at the same time reversing the direction of EOF or at least reducing its magnitude.

Various additives (often called EOF modifiers) to the carrier electrolyte have been suggested in order to reverse the EOF. Many of these EOF modifiers belong to the class of cationic surfactants as will be outlined in detail later in this review. Chemical modification of the fused-silica surface is another way to decrease EOF for inorganic anion analysis [4, 5].

The manipulation of EOF is of minor importance for separation of inorganic cations, as both electroosmotic and electrophoretic mobility point in the same direction. A considerably more serious problem is the fact that many inorganic cations exhibit very similar electrophoretic mobilities due to nearly identical charge and size so that the separation performance of CZE might not be satisfactory. In this case, separation selectivity can be enhanced by incorporation of a weak complexing agent into the carrier electrolyte. Obviously, the higher the respective complex formation constant, the lower will be the electrophoretic mobility due to the decreased positive charge. Other additives in the carrier electrolyte that convert inorganic cations into negatively charged anionic species are not included in this chapter but are discussed in Chapter 34. The same is true for precapillary derivatization procedures of metal ions with ligands that form thermodynamically and kinetically stable complexes in the sample.

Adequate detection is a particularly difficult problem in CZE of inorganic cations and anions. Many inorganic ions (except for a few anions like nitrite, nitrate, iodide and some others) have negligible absorbance at useful wavelengths and this precludes the use of direct absorption detection. A common alternative to detect UV transparent inorganic ions is indirect UV absorption detection. In this form of detection, a UV absorbing species having the same charge as the sample ion is used as carrier electrolyte. Displacement of these species by the migrating sample creates a region of decreased concentration of UV absorbing ions, so that the sample ions are monitored as a decrease in the background absorbance. Sensitivity in indirect UV detection is governed by the molar absorptivity of the carrier electrolyte and its charge. One might expect that the displacement of the probe ion by the migrating sample ion would occur on an equivalent-per-equivalent basis. This is true only if the sample ion has the same electrophoretic mobility as the background ion. Otherwise, the sensitivity may be decreased or increased, depending whether the mobility of the background ion is lower or higher than the sample ion. On the other hand, the mobility of the sample ion must be closely matched by that of the carrier electrolyte ion to avoid deformed peak shapes and decreased resolution. These requirements must be considered when choosing an appropriate carrier electrolyte for separation of inorganic ions. Details of indirect UV detection in inorganic anion analysis are given by Buchberger et al. [6].

Considerable efforts have been made to develop other detection techniques in addition to the commonly used indirect UV detection for UV transparent inorganic ions. Indirect photometric detection in the visible range has been reported [7] as well as indirect fluorescence detection [8–12]. Amperometric detection [13–15] and potentiometric detection with ion-selective electrodes [16–19] shows some promising features, although their robustness might need some further improvement for routine applications. Conductometric detection is commercially available [20]. Suppressed conductivity detection (CD) is normally utilized for the detection in IC, but application to CE is a breakthrough in thinking since tiny current must be detected in a large current [21–23]. Suppressed CD is becoming commercially available from Dionex Co. Ltd. A refractive index (RI) detector has been described for separations of metal ions [24]. Combination of CE with mass spectrometry (MS) [25], inductively coupled plasma optical emission spectrometry [26] or inductively coupled plasma mass spectrometry [26, 27] holds great potential for the trace analysis of metal ions in complex samples, speciation studies and other applications, although the expense of the instrumentation still limits its usefulness for routine analytical work.

36.3 Carrier electrolytes for separation of inorganic anions

The selection of the carrier electrolyte strongly depends on the detection technique chosen. For direct UV detection, phosphate buffers or simple electrolytes like sodium sulfate are appropriate. Borate buffers are compatible with suppressed conductivity detection [21].

Commonly used carrier electrolytes for indirect UV detection of inorganic anions are chromate [28] and pyromellitic acid [29]; alternatively, benzoic acid, 2-sulfobenzoic acid or o-benzylbenzoic acid [30], phthalic acid [30, 15], naphthalenesulfonates [31], molybdate [32], p-aminobenzoate [33], ribonucleotides [34], vanadate or iodide[5] can be used. Dihydroxybenzoic acid has been reported for indirect fluorescence detection [8].

As mentioned above, the addition of EOF modifiers to the carrier electrolyte is of considerable importance in order to obtain a fast separation of inorganic anions. Besides commercially available proprietary chemicals [28], C12 to C18 alkyltrimethylammonium salts have been widely used [35–37] as well as C6 alkyl diquaternary ammonium salts such as hexamethonium hydroxide [29]. EOF can also be reversed by cationic polymers such as hexadimethrine bromide [37], poly(1,1-dimethyl-3,5-dimethylenepiperidinium) or poly(1,1-dimethyl-3,5-dimethylenepyrrolidinium salts [38] and Praestol 185K [36]. Often it is sufficient to purge the capillary with a solution of these polymers before the analytical run. A summary of other additives that have been applied for suppression of EOF such as diethylenetriamine or hydroxyethylcellulose is given by Schomburg et al. [36].

It should be noted that the addition of EOF modifiers to the carrier electrolyte not only affects the EOF but can also be used to manipulate the separation selectivity for anions due to ion-pairing or ion-exchange interactions of the analytes with EOF modifiers [35, 38, 39] or due to partitioning into micelles if EOF modifiers are used above the critical micelle concentration [40]. These effects could be used in empirical and theoretical models in order to predict and optimize anion separations in carrier electrolytes containing chromate and cetyltrimethylammonium bromide [41, 42]. Separation selectivity for inorganic anions can also be changed by using organic solvents in the carrier electrolyte [35, 15] or by adding metal ions (e.g. lead ions [43]) or cyclodextrins (CDs) [44] to the carrier electrolyte, thereby affecting the size of the charge on the analyte anions.

36.4 Carrier electrolytes for separation of inorganic cations

Carrier electrolytes suitable for indirect UV detection of inorganic cations include imidazole [45], aromatic amines such as benzylamine [46], 4-methyl-2-benzylamine [47] or N,N-dimethylbenzylamine [48], pyridine [49] and its derivatives such as p-aminopyridine [50], 1,1'-di-n-heptyl-4,4'-bipyridine [51] or nicotinamide [52], copper sulfate [53], ephedrine [49], creatinine [54, 55] and some commercially available proprietary chemicals [56]. Some of these carrier electrolytes are also compatible with amperometric detection [13, 14]. Indirect fluorescence detection can be accomplished in carrier electrolytes like cerium(III) sulfate [8, 9], 6-aminoquinoline [10] or 2-aminopyridine [12]. A magnesium acetate buffer has been described for detection with ion selective electrodes [17].

As mentioned above, the success in CE of inorganic cations strongly depends on the addition of suitable weak complexing agents to the carrier electrolyte. These include organic acids such as citric acid [57], α-hydroxyisobutyric acid [57, 58], phthalic acid, tartaric acid and lactic acid [59], glycolic acid [49] or glycine [51] as well as tropolone [56]. Addition of crown ethers [44, 60] or organic solvents [52, 61] to the carrier electrolyte yields a further enhancement of separation selectivity. Modeling and computer-assisted prediction of electrophoretic mobilites of inorganic cations in carrier electrolytes containing complexing agents can be a powerful approach to optimization of separation selectivity [54, 55, 61–63].

In a few cases, the use of coated capillaries has been suggested to reduce interactions between the silica surface and the positively charged ions and to optimize EOF [46, 64].

36.5 Applications of capillary zone electrophoresis in inorganic ion analysis

Applications of CZE to the analysis of inorganic anions and cations in various fields are summarized in Table 36.1. Inorganic ion analysis by CZE is still a developing technique, which explains why a considerable percentage of papers published in this field deal with fundamental studies carried out with standard mixtures of ions. The predominant goal of Table 36.1 is a comprehensive presentation of recent applications to real samples, although several separations of standard mixtures are included (applications that require the use of not yet commonly available instrumentation are generally not included in Table 36.1). Many of the applications given in Table 36.1 deal with mixtures of inorganic ions and other low molecular mass organic ions. These organic ions are not listed in the table, but are described in Chapter 37 on Organic Acids and Organic Ions.

The analysis of inorganic ions in real samples by CZE requires a careful optimization of sensitivity. Apart from the selection of an appropriate detection technique, the use of on-capillary preconcentration procedures is highly recommended. The injection of samples into a matrix of considerably lower ionic strength than the carrier electrolytes leads to a preconcentration by sample stacking (electrostacking) due to the increased electric field across the sample segment. Preconcentration is also possible by on-column transient ITP if a matrix ion in the sample acts as a leading or terminating electrolyte (a typical example would be a body fluid with large amounts of sodium and chloride ions). In principle, both techniques can be used with hydrostatic as well as electromigration injection. Detection limits in the low ppb range can be achieved in combination with direct or indirect UV detection. Practical aspects of on-capillary preconcentration for inorganic ions have been demonstrated by Boden *et al.* [32], Jackson and Haddad [65], Janini *et al.* [66] and Wojtusik and Harrold [67]; a theoretical description of transient on-column ITP focusing is given by Gebauer *et al.* [68]. Another interesting approach to sensitivity optimization is the coupling of a short capillary wherein an ITP preseparation and

Table 36.1 Summary of applications of capillary zone electrophoresis to inorganic ion analysis

Carrier electrolyte composition Voltage Detection	Sample analytes	Reference
Water analysis and environmental analysis		
4 mM Chromate, 0.3 mM OFM Anion-BT, pH 8.1 −15 kV Indirect UV 254 nm	Tap water, ground water, waste water; chloride, sulfate, nitrate, fluoride, phosphate	70
25 mM Phosphate buffer, pH 3, 0.5% 3-(*N*,*N*-dimethylmyristylammonio) propanesulfonate, 1% Brij 35 −15 kV Direct UV 214 nm	Well water; nitrite, nitrate	4
3 mM *p*-Aminobenzoic acid, 4.5 mM sodium *p*-aminobenzoate, 0.76 mM barium hydroxide, 0.055 mM tetradecyltrimethylammonium hydroxide −30 kV Indirect UV 264 nm	Rain water; inorganic anions	33
10 mM Tris, acetic acid, pH 6.6, 0.5 mM chlorphenol red −10 kV Indirect vis 578 nm	Drinking water; chloride, nitrate, sulfate	7
Phosphate/borate buffers 20 kV Direct UV 195 nm	Tap water, spring water; selenium and arsenic species	71
5 mM Chromate, 0.2 mM tetradecyltrimethylammonium bromide, pH 8.2 −30 kV Indirect UV 275 nm	Soil extracts; chloride, sulfate, nitrate, fluoride, carbonate	72
2.25 mM Pyromellitic acid, 0.75 mM hexamethonium hydroxide, 6.5 mM NaOH, 1.6 mM triethanolamine, pH 7.7 −30 kV Indirect UV 254 nm	Atmospheric aerosols; chloride, sulfate, nitrate	73, 74
6 mM Chromate, pH 10, OFM Anion-BT −28 kV Indirect UV 274 nm	Coal fly ash; chloride, arsenate vanadate	75
5 mM UVCat-1, 6.5 mM hydroxyisobutyric acid, pH 4.4 20 kV Indirect UV 214 nm	Tap water; potassium, calcium, sodium, magnesium, copper	58

Table 36.1 *Continued*

Carrier electrolyte composition Voltage Detection	Sample analytes	Reference
0.5 mM Cerium(III) sulfate, 2.5 mM 18-crown-6 30 kV Indirect fluorescence	Rain water; ammonium, potassium, calcium, sodium, magnesium	9
4 mM copper sulfate, 4 mM formic acid, 4 mM 18-crown-6, pH 3 20 kV Indirect UV 215 nm	Drinking water; sodium, calcium, magnesium	53
1 mM Tris–2 mM acetate buffer, pH 4.2, 0.12 mM methyl green 10 kV Indirect vis 635 nm	Drinking water; potassium, calcium, magnesium, sodium	7
5 mM 1,1'-di-*n*-heptyl-4-4'- bipyridinium hydroxide, 6 mM glycine, 2 mM 18-crown-6- ether, 2% methanol, pH 6.5 25 kV Indirect UV 280 nm	Aerosol extracts; inorganic cations	51
Food analysis Chromate, OFM Anion-BT −20 kV Indirect UV 254 nm	Ultrafiltrate of milk; chloride, sulfate, phosphate, carbonate	76
10 mM Chromate, pH 11.5, 2.3 mM cetyltrimethyl- ammonium bromide −20 kV Indirect UV 254 nm	Vegetables; chloride, bromide, sulfate, nitrate, nitrite phosphate, carbonate	77
0.5 mM Cerium(III) sulfate, 2.5 mM 18-crown-6 30 kV Indirect fluorescence	Beverages; ammonium, potassium, calcium, sodium	9
5 mM UVCat-1, 8 mM hydroxyisobutyric acid, pH 4.4 20 kV Indirect UV 185 nm	Orange juice; potassium, calcium, sodium, magnesium	58
5 mM Imidazole, pH 4.5 20 kV Indirect UV 214 nm	Fruit juices; potassium, sodium, calcium, magnesium	78
5 mM Imidazole, pH 4.2 20 kV Indirect UV 214 nm	Apple vinegar, mineral waters; potassium, sodium, calcium, magnesium	45
5 mM UVCat-1, 6.5 mM hydroxyisobutyric acid, pH 4.4 20 kV Indirect UV 214 nm	Ultrafiltrate of milk; potassium, sodium, calcium, magnesium	76

Table 36.1 *Continued*

Carrier electrolyte composition Voltage Detection	Sample analytes	Reference

5 mM Imidazole, 6.5 mM
  hydroxyisobutyric acid,
  20% methanol, 0.53 mM
  18-crown-6, pH 4.5
20 kV
Indirect UV 214 nm
	Chinese green tea
	  infusion;
	  sodium, calcium,
	  magnesium, manganese
		61

5 mM Imidazole, 6.5 mM
  hydroxyisobutyric acid, 0.55 mM
  18-crown-6-ether, 20% methanol,
  pH 4.5
20 kV
Indirect UV 214 nm
	Tea;
	  potassium, calcium,
	  magnesium, manganese,
	  sodium
		79

Industrial analysis
10 mM Sodium chromate,
  0.5 mM OFM Anion-BT, pH 8
−15 kV
Indirect UV 254 nm
	Primary water and
	  secondary water from
	  nuclear pressurized
	  water reactors;
	  inorganic anions
		80

5 mM Sodium chromate,
  0.5 mM OFM Anion-BT
−20 kV
Indirect UV 254 nm
	Bayer liquor;
	  chloride, sulfate
		81

5 mM Chromate, pH 9.1,
  dodecyl- / tetradecyl-
  trimethylammonium bromide
−20 kV
Indirect UV 254 nm
	Bayer liquor;
	  chloride, sulfate,
	  fluoride, phosphate,
	  carbonate
		82

5 mM Sodium chromate,
  0.5 mM OFM Anion-BT
−20 kV
Indirect UV 254, 214 nm
	Kraft pulping process
	  liquors;
	  chloride, sulfate, sulfite,
	  thiosulfate, hydrosulfide,
	  carbonate
		83

10 mM Sodium chromate,
  0.05 mM OFM Anion-BT, pH 7.9
−20 kV
Indirect UV 254 nm
	Thin films on
	  silicon substrates;
	  boron, phosphorus
		84

6 mM Chromate,
  0.07 mM tetradecyltrimethyl-
  ammonium hydroxide, pH 8.1
−30 kV
Indirect UV 254 nm
	Silicon wafers;
	  bromide, chloride,
	  sulfate, nitrite, nitrate
		32

2.25 mM Pyromellitic acid, 1.6 mM
  triethanolamine, 6.5 mM NaOH,
  0.75 hexamethonium hydroxide
−25 kV
Indirect UV 254 nm
	Detergents;
	  sulfate
		85

Table 36.1 *Continued*

Carrier electrolyte composition Voltage Detection	Sample analytes	Reference

5 mM Chromate,
  0.5 mM OFM Anion-BT
  −20 kV
  Indirect UV 254 nm
	Detergent products;
	sulfate
		86

5 mM Chromate,
  0.5 mM OFM Anion-BT
  −25 kV
  Indirect UV 254 nm
	Toothpaste;
	fluoride, monofluoro-
	phosphate, phosphate
		87

5 mM Adenosine monophosphate,
  2 mM diethylenetriamine,
  100 mM boric acid, pH 6.8
  −30 kV
  Indirect UV 259 nm
	Toothpaste;
	sulfate, chloride,
	fluoride, phosphate,
	diphosphate, fluorophosphate,
	carbonate
		34

5 mM UVCat-1, 8 mM
  hydroxyisobutyric acid, pH 4.4
  20 kV
  Indirect UV 185 nm
	Acid etching bath;
	sodium, nickel, zinc
		58

1.2 mM UVCat-2, 3 mM
  tropolone
  20 kV
  Indirect UV 185 nm
	Semiconductor
	grade hydrogen
	peroxide;
	inorganic cations
		47

30 mM Creatinine, 8 mM
  hydroxyisobutyric acid, pH 4.8
  27 kV
  Atmospheric pressure,
  ionization mass spectrometry
	Aircraft engine oil;
	nickel, chromium,
	lead
		25

Pharmaceutical and clinical analysis
5 mM Chromate,
  0.5 mM OFM Anion-BT
  −20 kV
  Indirect UV 254 nm
	Drugs;
	bromide, fluoride,
	sulfate, chloride
		88

5 mM Chromate,
  0.4 mM OFM Anion-BT, pH 8
  −20 kV
  Indirect UV 254 nm
	Prenatal vitamin
	formulation;
	chloride, sulfate,
	nitrate, phosphate,
	carbonate
		89

25 mM Phosphate buffer, pH 3,
  0.5% 3-(*N*,*N*-dimethylmyristyl-
  ammonio) propanesulfonate,
  1% Brij 35
  −15 kV
  Direct UV 200, 214 nm
	Urine, serum;
	chloride, nitrate
	nitrite
		66, 4

10 mM Sodium sulfate,
  OFM Anion-BT
  −15 kV
  Direct UV 214 nm
	Urine, brain ultrafiltrate;
	nitrite, nitrate
		90

Table 36.1 *Continued*

Carrier electrolyte composition Voltage Detection	Sample analytes	Reference
2.25 mM Pyromellitic acid, 6.5 mM NaOH, 1.6 mM triethanolamine, 0.75 mM hexamethonium bromide, pH 7.7 −20 kV Indirect UV, 250 nm	Urine, serum; chloride, sulfate, nitrite, nitrate	91
10 mM Chromate, 1 mM 5,5-diethylbarbiturate, pH 8 −30 kV Indirect UV 254 nm (capillary coated with 0.0003% hexadimethrine bromide)	Human seminal plasma; chloride, phosphate, carbonate	37
750 mM Sodium chloride, 5% OFM Anion-BT −20 kV Direct UV 214 nm	Blood plasma; nitrite, nitrate	92
5 mM Chromate, 0.5% (hydroxypropyl)-methyl-cellulose, pH 7 −20 kV Indirect UV 273 nm	Airway surface fluid, chloride	93
5 mM UVCat-1, 6.5 mM hydroxyisobutyric acid, pH 4.2 20 kV Indirect UV 214 nm	Cough syrup; potassium, calcium, sodium, magnesium	57
5 mM UVCat-1, 6.5 mM hydroxyisobutyric acid, pH 4.4 20 kV Indirect UV 185 nm	Prenatal vitamin formulation; calcium, sodium, iron, zinc	89
6 mM Imidazole, 4 mM formic acid 5–12 kV Indirect UV 214 nm	Drugs; potassium, sodium	94, 95
5 mM Imidazole, pH 4.5 20 kV Indirect UV 214 nm	Electrolyte solutions for parenteral use; potassium, sodium, calcium, magnesium	78
0.5 mM 6-Aminoquinoline, 0.2 mM cetyltrimethyl-ammonium bromide, pH 3.8 20 kV Indirect fluorescence	Single erythrocyte; sodium, potassium	10
0.25 mM 2-Aminopyridine, 1.5% glucose, pH 4.8 20 kV Indirect fluorescence	Single erythrocyte; potassium, sodium	12

Table 36.1 *Continued*

Carrier electrolyte composition Voltage Detection	Sample analytes	Reference
5 mM Copper sulfate 30 kV Indirect UV 214 nm	Serum; potassium, sodium, calcium, magnesium	96
4 mM Copper sulfate–4 mM formic acid (3:6) 20 kV Indirect UV 215 nm	Tear fluid; potassium, sodium, calcium, magnesium	97

20 mM Imidazole, 0.1%
  hydroxypropyl methyl cellulose,
  pH 6
15 kV
Indirect UV 214 nm
	Ocular lenses;
	potassium, sodium,
	calcium, magnesium
		98

5 mM Pyridine, 3.6 mM tartaric acid, 2 mM 18-crown-6, pH 4.05 25 kV Indirect UV 255 nm	Urine; ammonium, sodium, potassium, magnesium, calcium	99
10 mM Imidazole, 8% 2-propanol, pH 3.5 20 kV Indirect UV 214 nm	Airway surface fluid; sodium, potassium, calcium, magnesium	93

Miscellaneous applications
5 mM Chromate,
  quaternary ammonium salt,
  1–8% butanol
−30 kV
Indirect UV 254 nm
	Aqueous standards;
	inorganic anions
		100

25 mM Phosphate buffer pH 6.8 25 kV Direct UV 190 nm	Aqueous standards; arsenic species	101

5 mM Chromate, pH 10,
  0.25 mM hexadecytrimethyl-
  ammonium bromide
−15 kV
Indirect UV 254 nm
	Aqueous standards;
	arsenic and selenium
	species
		102

25 mM Phosphate buffer, pH 5.6–8 25 kV Direct UV detection 190 nm	Aqueous standards; arsenic species	103, 104

80 mM Phosphate buffer, pH 5–8.5,
  2 mM tetradecyltrimethyl-
  ammonium bromide
−12 kV
Direct UV 200 nm
	Aqueous standards;
	selenium species
		104

2.3 mM Pyromellitic acid,
  0.75 mM hexamethonium
  hydroxide, 6.5 mM NaOH,
  1.6 mM triethanolamine, pH 7.7
−30 kV
ICP–MS
	Aqueous standards;
	arsenic and selenium
	species
		27

Table 36.1 *Continued*

Carrier electrolyte composition Voltage Detection	Sample analytes	Reference
0.1 mM Fluorescein, pH 10 25 kV Indirect fluorescence	Aqueous standards; cyanide, cyanate, thiocyanate, nitrate, chloride, sulfate	105
2 mM Borate, 40 mM boric acid, 1.8 mM dichromate, 1 mM diethylenetriamine −20 kV Indirect UV detection 280, 265, 205 nm	Explosive residues; inorganic anions	106
5 mM Chromate, 3 mM borate, pH 9.2 20 kV Indirect UV 254 nm	Aqueous standards; chloride isotopes	107
5 mM Chromate, 0.5 mM cetyltrimethylammonium bromide, pH 8 −15 kV Indirect UV 254 nm	Caustic solution, technical grade sodium carbonate, HCl digest of paper coating; inorganic anions; (sample clean-up by membrane-based solid-phase extraction)	108
6 mM N,N-Dimethylbenzylamine, 4.2 mM hydroxyisobutyric acid, pH 5.2, 0.2 mM Triton X-100 30 kV Indirect UV 214 nm	Aqueous standards; alkali, alkaline earth, transition metal ions, lanthanides	48
8 mM 4-Methylbenzylamine, 15 mM lactic acid, 5% methanol, pH 4.25 30 kV Indirect UV 214 nm	Aqueous standards; alkali, alkaline earth, transition metal ions, lanthanides	59
8 mM Nicotinamide, 0.6 mM 18-crown-6, pH 3.2 25 kV Indirect UV	Aqueous standards; alkali, alkaline earth ions, Cu, Al, V	52
5.2 mM Ephedrine, 4.7 mM hydroxyisobutyric acid, pH 2.8 15 kV Indirect UV 204 nm	Aqueous standards; aluminum ions in the presence of other metal ions	109
10 mM Ammonium nitrate, 5 mM phenanthroline, 50 µg l^{-1} cesium 25 kV ICP–MS	Aqueous standards; inorganic cations	27
5 mM UVCat-1, 6.5 mM hydroxyisobutyric acid, pH 4.2 20 kV Indirect UV 214 nm	Fermentation broth; potassium, sodium, magnesium, manganese, zinc	57

Table 36.1 *Continued*

Carrier electrolyte composition Voltage Detection	Sample analytes	Reference
10 mM 4-Aminopyridine, acetic acid, pH 4.5 22 kV Indirect UV 254 nm	Cell suspension; ammonium, potassium, sodium	50
5 mM Imidazole, 6.5 mM hydroxyisobutyric acid, 0.55 mM 18-crown-6-ether, 20% methanol, pH 4.5 20 kV Indirect UV 214 nm	Standard reference materials (bovine liver, total diet, oyster tissue, fish tissue, pine needles, citrus leaves); inorganic cations	79

ICP–MS, inductively coupled plasma mass spectrometry; OFM Anion-BT, trademark for a proprietary EOF modifier (Waters); UVCat-1, UVCat-2, trademarks for proprietary carrier electrolytes (Waters).

preconcentration is carried out with a second capillary for the analytical CZE separation [69].

36.6 Conclusions

CZE for the analysis of inorganic ions is rapidly evolving from a technique used by a few specialists into a tool for routine applications and validated analytical methods. Comparison of CZE with IC, photometric methods, flame atomic spectroscopy or wet chemical methods generally shows excellent agreement [51, 74, 78, 85, 86]. CZE analysis of inorganic ions in reference materials and comparison of the results with certified values also proves the satisfactory accuracy of this analytical technique [79].

It seems fair to mention that many of the applications described so far for inorganic ions do not yet take advantage of all benefits of CZE. Although CZE is a microtechnique requiring just a few nanoliters of sample, most sample preparation procedures are still done on the macroscale. It is obvious that there is a mismatch between the sampling technique and the analytical technique. Nevertheless, there are some new approaches that do take advantage of CZE as a microtechnique. Gases that form inorganic anions in solutions can be sampled in a liquid droplet or liquid film at the end of a CE capillary; after collection, the absorbed species can be subjected to electrophoretic separation and detection [110, 111]. Only a small area of possible applications of CZE in inorganic ion analysis has been exploited so far and interest in this technique is likely to increase over the next few years.

References

1. Jandik, P.; Bonn, G. (1993) *Capillary Electrophoresis of Small Molecules and Ions.* VCH, New York.
2. Jackson, P.E.; Haddad, P.R. (1993) Capillary electrophoresis of inorganic ions and low-molecular-mass ionic solutes. *Trends Anal. Chem.* **12**, 231–238.
3. Timerbaev, A.R. (1995) Metal ion analysis by capillary electrophoresis: New possibilities in separation and detection. *J. Capillary Electrophoresis* **2**, 14–23.
4. Janini, G.M.; Chan, K.C.; Muschik, G.M.; Issaq, H.J. (1994) Analysis of nitrate and nitrite in water and urine by capillary zone electrophoresis. *J. Chromatogr.* **657**, 419–423.
5. Tindall, G.W.; Perry, R.L. (1995) Separation of fast anions by capillary electrophoresis without flow reversal. *J. Chromatogr.* **696**, 349–352.
6. Buchberger, W.; Cousins, S.M.; Haddad, P.R. (1994) Optimisation of indirect UV detection in capillary zone electrophoresis of low-molecular-mass anions. *Trends Anal. Chem.* **13**, 313–319.
7. Mala, Z.; Vespalec, R.; Bocek, P. (1994) Capillary zone electrophoresis with indirect photometric detection in the visible range. *Electrophoresis* **15**, 1526–1530.
8. Bächmann, K.; Haumann, I.; Groh, T. (1992) Simultaneous determination of inorganic cations and anions in capillary zone electrophoresis (CZE) with indirect fluorescence detection. *Fresenius J. Anal. Chem.* **343**, 901–902.
9. Bächmann, K.; Boden, J.; Haumann, I. (1992) Indirect fluorimetric detection of alkali and alkaline earth metal ions in capillary zone electrophoresis with cerium (III) as carrier electrolyte. *J. Chromatogr.* **626**, 259–265.
10. Hogan, B.L.; Yeung, E.S. (1992) Determination of intracellular species at the level of a single erythrocyte via capillary electrophoresis with direct and indirect fluorescence detection. *Anal. Chem.* **64**, 2841–2845.
11. Hogan, B.L.; Yeung, E.S. (1993) Single-cell analysis at the level of a single human erythrocyte. *Trends Anal. Chem.* **12**, 4–9.
12. Li, Q.; Yeung, E.S. (1994) Contamination control in capillary electrophoresis and quantitative measurement of potassium and sodium in single human erythrocytes. *J. Capillary Electrophoresis* **1**, 55–61.
13. Lu, W.; Cassidy, R.M. (1993) Evaluation of ultramicroelectrodes for the detection of metal ions separated by capillary electrophoresis. *Anal. Chem.* **65**, 1649–1653.
14. Lu, W.; Cassidy, R.M.; Baranski, A.S. (1993) End-column electrochemical detection for inorganic and organic species in high-voltage capillary electrophoresis. *J. Chromatogr.* **640**, 433–440.
15. Salimi-Moosavi, H.; Cassidy, R.M. (1995) Capillary electrophoresis of inorganic anions in nonaqueous media with electrochemical and indirect UV detection. *Anal. Chem.* **67**, 1067–1073.
16. Nann, A.; Simon, W. (1993) On-column detection in capillary zone electrophoresis with ion-selective microelectrodes in conical capillary apertures. *J. Chromatogr.* **633**, 207–211.
17. Nann, A.; Silvestri, I.; Simon, W. (1993) Quantitative analysis in capillary zone electrophoresis using ion-selective microelectrodes as on-column detectors. *Anal. Chem.* **65**, 1662–1667.
18. Hauser, P.C.; Renner, N.D.; Hong, A.P.C. (1994) Anion detection in capillary electrophoresis with ion-selective microelectrodes. *Anal. Chim. Acta* **295**, 181–186.
19. Nann, A.; Pretsch, E. (1994) Potentiometric detection of anions separated by capillary electrophoresis using an ion-selective microelectrode. *J. Chromatogr.* **676**, 437–442.
20. Jones, W.R.; Haber, C.; Reineck, J. (1994) Small molecular weight ion analysis using capillary electrophoresis and an open architecture conductivity detector. Paper No. 56 presented at the *International Ion Chromatography Symposium*, Turin. Century International, Medfield, MA.
21. Dasgupta, P.K.; Bao, L. (1993) Suppressed conductometric capillary electrophoresis separation systems. *Anal. Chem.* **65**, 1003–1011.

22 Avdalovic, N.; Pohl, C.A.; Rocklin, R.D.; Stillian, J.R. (1993) Determination of cations and anions by capillary electrophoresis combined with suppressed conductivity detection. *Anal. Chem.* **65**, 1470–1475.
23 Kar, S.; Dasgupta, P.K.; Liu, H.; Hwang, H. (1994) Computer-interfaced bipolar pulse conductivity detector for capillary systems. *Anal. Chem.* **66**, 2537–2543.
24 Krattiger, B.; Bruin, G.J.M.; Bruno, A.E. (1994) Hologram-based refractive index detector for capillary electrophoresis: Separation of metal ions. *Anal. Chem.* **66**, 1–8.
25 Huggins, T.G.; Henion, J.D. (1993) Capillary electrophoresis/mass spectrometry determination of inorganic ions using an ion spray–sheath flow interface. *Electrophoresis* **14**, 531–539.
26 Olesik, J.W.; Kinzer, J.A.; Olesik, S.V. (1995) Capillary electrophoresis inductively coupled plasma spectrometry for rapid elemental speciation. *Anal. Chem.* **67**, 1–12.
27 Liu, Y.; Lopez-Avila, V.; Zhu, J.J.; Wiederin, D.R.; Beckert, W.F. (1995) Capillary electrophoresis coupled on-line with inductively coupled plasma mass spectrometry for elemental speciation. *Anal. Chem.* **67**, 2020–2025.
28 Jones, W.R. (1993) Method development approaches for capillary ion electrophoresis. *J. Chromatogr.* **640**, 387–395.
29 Harrold, M.P.; Wojtusik, M.J.; Riviello, J.; Henson, P. (1993) Parameters influencing separation and detection of anions by capillary electrophoresis. *J. Chromatogr.* **640**, 463–471.
30 Ma, Y.; Zhang, R. (1992) Optimization of indirect photometric detection of anions in high-performance capillary electrophoresis. *J. Chromatogr.* **625**, 341–348.
31 Shamsi, S.A.; Danielson, N.D. (1994) Naphthalenesulfonates as electrolytes for capillary electrophoresis of inorganic anions, organic acids, and surfactants with indirect photometric detection. *Anal. Chem.* **66**, 3757–3764.
32 Boden, J.; Bächmann, K.; Kotz, L.; Fabry, L.; Pahlke, S. (1995) Application of capillary zone electrophoresis with an isotachophoretic initial state to determine anionic impurities on as-polished silicon wafer surfaces. *J. Chromatogr.* **696**, 321–332.
33 Röder, A.; Bächmann, K. (1995) Simultaneous determination of organic and inorganic anions in the sub-µmol/l range in rain water by capillary zone electrophoresis. *J. Chromatogr.* **689**, 305–311.
34 Shamsi, S.A.; Danielson, N.D. (1995) Ribonucleotide electrolytes for capillary electrophoresis of polyphosphates and polyphosphonates with indirect photometric detection. *Anal. Chem.* **67**, 1845–1852.
35 Buchberger, W.; Haddad, P.R. (1992) Effects of carrier electrolyte composition on separation selectivity in capillary zone electrophoresis of low-molecular-mass anions. *J. Chromatogr.* **608**, 59–64.
36 Schomburg, G.; Belder, D.; Gilges, M.; Motsch, S. (1994) Ionic and nonionic polymers as wall modifiers in capillary electrophoresis. *J. Capillary Electrophoresis* **1**, 219–230.
37 Oefner, P.J. (1995) Surface-charge reversed capillary zone electrophoresis of inorganic and organic anions. *Electrophoresis* **16**, 46–56.
38 Stathakis, C.; Cassidy, R.M. (1994) Cationic polymers for selectivity control in the capillary electrophoretic separation of inorganic anions. *Anal. Chem.* **66**, 2110–2115.
39 Stathakis, C.; Cassidy, R.M. (1995) Effect of electrolyte composition in the capillary electrophoretic separation of inorganic/organic anions in the presence of cationic polymers. *J. Chromatogr.* **699**, 353–361.
40 Kaneta, T.; Tanaka, S.; Taga, M.; Yoshida, H. (1992) Migration behavior of inorganic anions in micellar electrokinetic capillary chromatography using a cationic surfactant. *Anal. Chem.* **64**, 798–801.
41 Jimidar, M.; Massart, D.L. (1994) Prediction of the migration behaviour of anions in capillary ion analysis. *Anal. Chim. Acta* **294**, 165–176.
42 Jimidar, M.; Bourguignon, B.; Massart, D.L. (1995) Selectivity optimization after prediction of the migration behaviour of anions in capillary ion analysis in the presence of micelles. *Anal. Chim. Acta* **310**, 27–42.
43 Groh, T.; Bächmann, K. (1992) Selective influencing of the migration time of SO_4^{2-} by forming neutral species with Pb^{2+} in capillary zone electrophoresis. *Electrophoresis* **13**, 458–461.

44 Boden, J.; Ehmann, T.; Groh, T.; Haumann, I.; Bächmann, K. (1994) Influencing of the migration time in CZE. *Fresenius J. Anal. Chem.* **348**, 572–575.
45 Beck, W.; Engelhardt, H. (1992) Capillary electrophoresis of organic and inorganic cations with indirect UV detection. *Chromatographia* **33**, 313–316.
46 Chen, M.; Cassidy, R.M. (1992) Bonded-phase capillaries and the separation of inorganic ions by high-voltage capillary electrophoresis. *J. Chromatogr.* **602**, 227–234.
47 Carpio, R.A.; Jandik, P.; Fallon, E. (1993) Capillary electrophoresis methods for the trace cation analysis of semiconductor grades of hydrogen peroxide. *J. Chromatogr.* **657**, 185–191.
48 Chen, M.; Cassidy, R.M. (1993) Separation of metal ions by capillary electrophoresis. *J. Chromatogr.* **640**, 425–431.
49 Lee, Y.H.; Lin, T.I. (1994) Determination of metal cations by capillary electrophoresis. Effect of background carrier and complexing agents. *J. Chromatogr.* **675**, 227–236.
50 Beck, W.; Engelhardt, H. (1993) Separation of non UV-absorbing cations by capillary electrophoresis. *Fresenius J. Anal. Chem.* **346**, 618–621.
51 Dabek-Zlotorzynska, E.; Dlouhy, J.F. (1995) Application of capillary electrophoresis in atmospheric aerosol analysis: determination of cations. *J. Chromatogr.* **706**, 527–534.
52 Shi, Y.; Fritz, J.S. (1994) New electrolyte systems for the determination of metal cations by capillary zone electrophoresis. *J. Chromatogr.* **671**, 429–435.
53 Riviello, J.M.; Harrold, M.P. (1993) Capillary electrophoresis of inorganic cations and low-molecular-mass amines using a copper-based electrolyte with indirect UV detection. *J. Chromatogr.* **652**, 385–392.
54 Jimidar, M.; Hamoir, T.; Degezelle, W.; Massart, D.L.; Soykenc, S.; Van de Winkel, P. (1993) Method development and optimization for the determination of rare earth metal ions by capillary electrophoresis. *Anal. Chim. Acta* **284**, 217–225.
55 Vogt, C.; Conradi, S. (1994) Complex equilibria in capillary zone electrophoresis and their use for the separation of rare earth metal ions. *Anal. Chim. Acta* **294**, 145–153.
56 Weston, A.; Brown, P.R.; Jandik, P.; Heckenberg, A.L.; Jones, W.R. (1992) Optimization of detection sensitivity in the analysis of inorganic cations by capillary ion electrophoresis using indirect photometric detection. *J. Chromatogr.* **608**, 395–402.
57 Weston, A.; Brown, P.R.; Jandik, P.; Jones, W.R.; Heckenberg, A.L. (1992) Factors affecting the separation of inorganic metal cations by capillary electrophoresis. *J. Chromatogr.* **593**, 289–295.
58 Weston, A.; Brown, P.R.; Heckenberg, A.L.; Jandik, P.; Jones, W.R. (1992) Effect of electrolyte composition on the separation of inorganic metal cations by capillary ion electrophoresis. *J. Chromatogr.* **602**, 249–256.
59 Shi, Y.; Fritz, J.S. (1993) Separation of metal ions by capillary electrophoresis with a complexing electrolyte. *J. Chromatogr.* **640**, 473–479.
60 Francois, C.; Morin, P.; Dreux, M. (1995) Effect of the concentration of 18-crown-6 added to the electrolyte upon the separation of ammonium, alkali and alkaline-earth cations by capillary electrophoresis. *J. Chromatogr.* **706**, 535–553.
61 Yang, Q.; Smeyers-Verbeke, J.; Wu, W.; Khots, M.S.; Massart, D.L. (1994) Simultaneous separation of ammonium and alkali, alkaline earth and transition metal ions in aqueous-organic media by capillary ion analysis. *J. Chromatogr.* **688**, 339–349.
62 Quang, C.; Khaledi, M.G. (1994) Prediction and optimization of the separation of metal cations by capillary electrophoresis with indirect UV detection. *J. Chromatogr.* **659**, 459–466.
63 Yang, Q.; Zhuang, Y.; Smeyers-Verbeke, J.; Massart, D.L. (1995) Interpretation of migration behaviour of inorganic cations in capillary ion electrophoresis based on an equilibrium model. *J. Chromatogr.* **706**, 503–515.
64 Cheng, K.; Zhao, Z.; Garrick, R.; Nordmeyer, F.R.; Lee, M.L.; Lamb, J.D. (1995) Separation of metal cations by electrophoresis in a positively charged coated capillary. *J. Chromatogr.* **706**, 517–526.
65 Jackson, P.E.; Haddad, P.R. (1993) Optimization of injection technique in capillary ion electrophoresis for the determination of trace level anions in environmental samples. *J. Chromatogr.* **640**, 481–487.

66. Janini, G.M.; Muschik, G.M.; Issaq, H.J. (1994) Sample matrix effects in capillary zone electrophoresis. Effect of chloride ion on nitrate and nitrite. *J. Capillary Electrophoresis* **1**, 116–120.
67. Wojtusik, M.J.; Harrold, M.P. (1994) Factors influencing trace ion analysis with preconcentration by electrostacking. *J. Chromatogr.* **671**, 411–417.
68. Gebauer, P.; Thormann, W.; Bocek, P. (1992) Sample self-stacking in zone electrophoresis. Theoretical description of the zone electrophoretic separation of minor compounds in the presence of bulk amounts of a sample component with high mobility and like charge. *J. Chromatogr.* **608**, 47–57.
69. Kaniansky, D.; Zelensky, I.; Hybenova, A.; Onuska, F.I. (1994) Determination of chloride, nitrate, sulfate, nitrite, fluoride, and phosphate by on-line coupled capillary isotachophoresis–capillary zone electrophoresis with conductivity detection. *Anal. Chem.* **66**, 4258–4264.
70. Romano, J.P.; Krol, J. (1993) Capillary ion electrophoresis, an environmental method for the determination of anions in water. *J. Chromatogr.* **640**, 403–412.
71. Li, K.; Li, S.F.Y. (1995) Speciation of selenium and arsenic compounds in natural waters by capillary zone electrophoresis after on-column preconcentration with field-amplified injection. *Analyst* **120**, 361–366.
72. Stahl, R. (1994) Determination of anions by capillary electrophoresis and ion chromatography. *J. Chromatogr.* **686**, 143–148.
73. Dabek-Zlotorzynska, E.; Dlouhy, J.F. (1994) Application of capillary electrophoresis in atmospheric aerosols analysis: determination of inorganic and organic anions. *J. Chromatogr.* **671**, 389–395.
74. Dabek-Zlotorzynska, E.; Dlouhy, J.F.; Houle, N.; Piechowski, M.; Ritchie, S. (1995) Comparison of capillary zone electrophoresis with ion chromatography and standard photometric methods for the determination of inorganic anions in atmospheric aerosols. *J. Chromatogr.* **706**, 469–478.
75. Lin, L.; Wang, J.; Caruso, J. (1995) Arsenic speciation using capillary zone electrophoresis with indirect ultraviolet detection. *J. Chromatogr. Sci.* **33**, 177–180.
76. Schmitt, M.; Saulnier, F.; Malhautier, L.; Linden, G. (1993) Effect of temperature on the salt balance of milk studied by capillary ion electrophoresis. *J. Chromatogr.* **640**, 419–424.
77. Jimidar, M.; Hartmann, C.; Cousement, N.; Massart, D.L. (1995) Determination of nitrate and nitrite in vegetables by capillary electrophoresis with indirect detection. *J. Chromatogr.* **706**, 479–492.
78. Yang, Q.; Jimidar, M.; Hamoir, T.P.; Smeyers-Verbeke, J.; Massart, D.L. (1994) Determination of alkali and alkaline earth metals in real samples by capillary ion analysis. *J. Chromatogr.* **673**, 275–285.
79. Yang, Q.; Hartmann, C.; Smeyers-Verbeke, J.; Massart, D.L. (1995) Methods development and validation for the determination of mineral elements in food and botanical materials by capillary electrophoresis. *J. Chromtogr. A* **717**, 415–425.
80. Bondoux, G.; Jandik, P.; Jones, W.R. (1992) New approach to the analysis of low levels of anions in water. *J. Chromatogr.* **602**, 79–88.
81. Grocott, S.C.; Jefferies, L.P.; Bowser, T.; Carnevale, J.; Jackson, P.E. (1992) Applications of ion chromatography and capillary ion electrophoresis in the alumina and aluminium industry. *J. Chromatogr.* **602**, 257–264.
82. Haddad, P.R.; Harakuwe, A.H.; Buchberger, W. (1995) Separation of inorganic and organic anionic components of Bayer liquor by capillary zone electrophoresis: I. Optimisation of resolution with electrolyte-containing surfactant mixtures. *J. Chromatogr.* **706**, 571–578.
83. Salomon, D.R.; Romano, J. (1992) Applications of capillary ion electrophoresis in the pulp and paper industry. *J. Chromatogr.* **602**, 219–225.
84. Carpio, R.A.; Mariscal, R.; Welch, J. (1992) Determination of boron and phosphorus in borophosphosilicate thin films on silicon substrates by capillary electrophoresis. *Anal. Chem.* **64**, 2123–2129.
85. Pretswell, E.L.; Morrisson, A.R.; Park, J.S. (1993) Comparison of capillary zone electrophoresis with standard gravimetric analysis and ion chromatography for the determination of inorganic anions in detergent matrices. *Analyst* **118**, 1265–1267.

86. Jordan, J.M.; Moese, R.L.; Johnson-Watts, R.; Burton, D.E. (1994) Determination of inorganic sulfate in detergent products by capillary electrophoresis. *J. Chromatogr.* **671**, 445–451.
87. Skocir, E.; Pecavar, A.; Krasnja, A.; Prosek, M. (1993) Quantitative determination of fluorine in toothpastes. *J. High Resolut. Chromatogr.* **16**, 243–246.
88. Nair, J.B.; Izzo, C.G. (1993) Anion screening for drugs and intermediates by capillary ion electrophoresis. *J. Chromatogr.* **640**, 445–461.
89. Swartz, M.E. (1993) Capillary electrophoretic determination of inorganic ions in a prenatal vitamin formulation. *J. Chromatogr.* **640**, 441–444.
90. Meulemans, A.; Delsenne, F. (1994) Measurement of nitrite and nitrate levels in biological samples by capillary electrophoresis. *J. Chromatogr.* **660**, 401–404.
91. Rhemrev-Boom, M.M. (1994) Determination of anions with capillary electrophoresis and indirect ultraviolet detection. *J. Chromatogr.* **680**, 675–684.
92. Ueda, T.; Maekawa, T.; Sadamitsu, D.; Oshita, S.; Ogino, K.; Nakamura, K. (1995) The determination of nitrite and nitrate in human blood plasma by capillary zone electrophoresis. *Electrophoresis* **16**, 1002–1004.
93. Transfiguracion, J.C.; Dolman, C.; Eidelman, D.H.; Lloyd, D.K. (1995) Determination of the inorganic ion composition of rat airway surface fluid by capillary electrophoresis: Direct sample injection to allow multiple analyses from nanoliter volumes. *Anal. Chem.* **67**, 2937–2942.
94. Filbey, S.D.; Altria, K.D. (1994) Robustness testing of a capillary electrophoresis method for the determination of potassium content in the potassium salt of an acidic drug. *J. Capillary Electrophoresis* **1**, 190–195.
95. Altria, K.D.; Clayton, N.G.; Harden, R.C.; Makwana, J.V.; Portsmouth, M.J. (1995) Inter-company cross validation exercise on capillary electrophoresis. Quantitative determination of drug counter-ion level. *Chromatographia* **40**, 47–50.
96. Buchberger, W.; Winna, K.; Turner, M. (1994) Applications of capillary zone electrophoresis in clinical chemistry: Determination of low-molecular-mass ions in body fluids. *J. Chromatogr.* **671**, 375–382.
97. Chadwick, R.R.; Hsieh, J.C.; Resham, K.S.; Nelson, R.B. (1994) Applications of capillary electrophoresis in the eye-care pharmaceutical industry. *J. Chromatogr.* **671**, 403–410.
98. Shi, H.; Zhang, R.; Chandrasekher, G.; Ma, Y. (1994) Simultaneous detection and quantitation of sodium, potassium, calcium and magnesium in ocular lenses by high-performance capillary electrophoresis with indirect photometric detection. *J. Chromatogr.* **680**, 653–658.
99. Xu, X.; Kok, W.T.; Kraak, J.C.; Poppe, H. (1994) Simultaneous determination of urinary creatinine, calcium and other inorganic cations by capillary zone electrophoresis with indirect UV detection. *J. Chromatogr.* **661**, 35–45.
100. Benz, N.J.; Fritz, J.S. (1994) Studies on the determination of inorganic anions by capillary electrophoresis. *J. Chromatogr.* **671**, 437–443.
101. Lopez-Sanchez, J.F.; Amram, M.B.; Lakkis, M.D.; Lagarde, F.; Rauret, G.; Leroy, M.J.F. (1994) Quantitative aspects of the separation of arsenical species by free solution capillary electrophoresis. *Fresenius J. Anal. Chem.* **348**, 810–814.
102. Vogt, C.; Werner, G. (1994) Speciation of heavy metals by capillary electrophoresis. *J. Chromatogr.* **686**, 325–332.
103. Morin, P.; Amran, M.B.; Favier, S.; Heimburger, R.; Leroy, M. (1992) Separation of arsenic anions by capillary zone electrophoresis with UV detection. *Fresenius J. Anal. Chem.* **342**, 357–362.
104. Albert, M.; Demesmay, C.; Rocca, J.L. (1995) Analysis of organic and non-organic arsenious or selenious compounds by capillary electrophoresis. *Fresenius J. Anal. Chem.* **351**, 426–432.
105. Marti, V.; Aguilar, M.; Yeung, E.S. (1995) Indirect fluorescence detection of free cyanide and related compounds by capillary electrophoresis. *J. Chromatogr.* **709**, 367–374.
106. Hargadon, K.A.; McCord, B.R. (1992) Explosive residue analysis by capillary electrophoresis and ion chromatography. *J. Chromatogr.* **602**, 241–247.
107. Lucy, C.A.; McDonald, T.L. (1995) Separation of chloride isotopes by capillary electrophoresis based on the isotope effect on ion mobility. *Anal. Chem.* **67**, 1074–1078.

108 Saari-Nordhaus, R.; Anderson, J.M. (1995) Membrane-based solid-phase extraction as a sample clean-up technique for anion analysis by capillary electrophoresis. *J. Chromatogr.* **706**, 563–569.
109 Barger, W.R.; Mowery, R.L.; Wyatt, J.R. (1994) Separation and indirect detection by capillary zone electrophoresis of ppb (w/w) levels of aluminium ions in solutions of multiple cations. *J. Chromatogr.* **680**, 659–665.
110 Liu, S.; Dasgupta, P.K. (1995) Liquid droplet. A renewable gas sampling interface. *Anal. Chem.* **67**, 2042–2049.
111 Dasgupta, P.K. (1995) Analysis of the atmospheric environment: the next challenge for ion analysis techniques. Paper No. 2 presented at the *International Ion Chromatography Symposium, Dallas*. Century International, Medfield, MA.

37 Organic acids and organic ions
F.S. STOVER

37.1 Introduction

The separation and quantitation of organic ions and organic acids is an important application area of capillary electrophoresis (CE). The earliest commercial form of CE, capillary isotachophoresis (ITP), has been used for determining numerous organic ions [1]. Conventional ITP, performed with discontinuous buffers in 0.2–0.5 mm i.d. PTFE tubes using conductivity detection, has found applications in analyses for organic ions in food, environmental, biological and industrial chemical samples [2, 3]. Extensive applications of ITP were found from its development in the mid-1970s through the 1980s.

With the development of modern capillary zone electrophoresis (CZE) in the early 1980s [4], a new high-efficiency approach to separations of small molecules became available. Modern CE, performed in continuous buffer systems and sub-100 µm i.d. fused-silica capillaries, allows electrophoretic separations analogous to elution chromatography to be performed based on differences in effective mobilities. With theoretical plate numbers in the 10^5–10^6 range, CE offers separation efficiencies approaching capillary gas chromatography (GC) for non-volatile solutes in (mostly) aqueous solution. The introduction of micelle-containing buffer systems (micellar electrokinetic chromatography, MEKC or MECC) added a new dimension to selectivity in CE, namely hydrophobic partitioning into micelles [5]. Initial CZE and MEKC separations dealt with fluorescent or UV absorbing ions.

The growth in CE for small ion analysis in the 1990s is due in part to a breakthrough in the use of indirect UV detection [6]. This approach of using the displacement of a UV absorbing co-ion for detection, combined with reversal or reduction of electroosmotic flow using buffer additives, resulted in generally useful methods to analyze small, non-absorbing organic acids. In addition, the availability and promotion of instruments from numerous commercial manufacturers has spurred the growth of CE. Kits with buffer systems 'optimized' for a variety of small ion analyses currently are available from CE instrument manufacturers.

The results of a literature search on 'capillary electrophoresis of carboxylates' illustrate the rapid increase in CE applications. In the period 1990–1992, the number of papers on this topic averaged about eight per year, while for the period 1993–1995, the number increased to about 30 per year. This trend is expected to continue due to further application-oriented research in an expanded user base, and due to continuing instrumental developments (e.g. conductivity and mass spectrometry detection).

The growth of CE for organic acid/ion applications is occurring despite the existence of well-developed, competing techniques such as liquid (HPLC), ion (IC) and gas chromatography. A major advantage of CE with respect to other techniques is the speed of both the separations themselves and of CE method developments. Many analyses can be completed in 4–10 min, and column re-equilibrations, common in gradient elution HPLC, can largely be avoided. CE also offers a unique selectivity mechanism, which can be tuned through simple buffer modifications. Rapid, semi-automated buffer changes, along with a good understanding of the physicochemical basis of the separations, can result in fast method developments. Other advantages of CE include small injection volumes and low solvent/buffer consumption. Finally, the general chromatographic problem of late-eluting strongly retained components is avoided since the separation compartment is prepared for the next injection by flushing with fresh electrolyte/buffer.

CE is not without its disadvantages. CE methods often exhibit a poorer concentration-based sensitivity, more limited dynamic range (while maintaining resolution), more limited load capacity for highly ionic matrices and poorer precision and reproducibility compared to HPLC, IC and GC. Nevertheless, the quantitative capabilities of CE are often sufficient for solving important analytical problems, and work continues to improve its performance [7–10]. At present, CE is viewed as a technique complementary to other chromatographic methods, and is often the method of choice for analysis of organic acids and organic ions. Applications are summarized in Table 37.1.

37.2 Aspects of CE separations of organic ions and organic acids

In its simplest form, capillary zone electrophoresis separates ions based on differences in electrophoretic mobilities, which are proportional to effective charge and an inverse function of hydrodynamic radius or ionic size. While the hydrodynamic radius is fairly constant (reflected in the fully ionized or absolute mobility), many organic ions are weak acids or bases, with pK_a values of 2–11, in the pH range accessible to CE buffer systems. Thus, the degree of ionization (and hence the effective charge) of analytes can be modified by simple pH changes. Of great importance is the fact that these two fundamental parameters of CE separations, pK_a and absolute mobility, are independently measurable using potentiometric titration and electrolytic conductance measurements. Extensive listings of pK_a values and mobilities of common organic ions are available [11–14], allowing simple calculation of effective mobility vs. pH to facilitate buffer selection. While optimizing pH can be the main concern in developing a CE method for organic ions, numerous other factors must be taken into account. Non-diffusional peak broadening can occur due to electric field inhomogeneity when analyte concentrations approach those of the buffer [15]. This is particularly problematic for indirect detection systems, where low concentration buffers are used to maximize sensitivity. Thus, a second important parameter in selecting buffers for ion separations is the mobility of the buffer co-ion [16, 17]. Choosing a buffer with an effective mobility close to those of the analyte ions will minimize electrodispersive peak broadening.

A third factor to consider is the need to control the electroosmotic flow (EOF) present in bare fused-silica capillaries during electrophoresis. Controlling EOF is important from a reproducibility standpoint, as well as for minimizing analysis times. For small organic anions, electrophoretic velocities often are of the same magnitude, but of opposite sign, as the EOF. This can lead to excessive time prior to analyte appearance at the detector. Addition of polyvalent and/or hydrophobic cations to the buffer can reduce or reverse the zeta potential of the silica surface, resulting in reduced or reversed EOF. Addition of EOF modifiers has an additional value in suppressing wall interactions, often a problem for cationic separations. Use of coated capillaries is also common for controlling EOF and wall interaction. Finally, tuning of the selectivity of the separation can involve the use of a variety of other buffer additives. Addition of organic solvents to buffer systems, often used to suppress EOF or improve solubility, can have a beneficial effect on separation selectivity through shifts in protolytic equilibria or solvation. Simple changes in buffer concentration can affect mobilities differentially through ionic strength effects. The EOF modifiers noted above can change the selectivity through ion pairing interactions with anionic analytes. For acidic compounds, metal additives can improve separations through metal–ligand interactions. Complexation by compounds such as cyclodextrins offers a way of enhancing molecular shape and/or chiral selectivity. Addition of entangled polymer solutions to buffers provides a sieving medium to give size- or molecular weight-based separations. In the special case of MEKC, inclusion of micelles in the buffer system can switch the selectivity to one predominated by analyte hydrophobicity. Partitioning of analyte molecules into charged micelles allows separation of non-ionic compounds, as well as altered selectivity toward ionic solutes.

As electrolyte/buffer systems become more complex, optimization requires a more empirical approach. Nevertheless, numerous papers show that modeling CE systems for organic ion separations can be successful. Models have been developed which include peak broadening [18, 19], surfactant type [20], solvent gradients [21] and ionic strength [22]. Kenndler

[23] used cluster analysis for the classification of buffers and the optimization of CE separations based on the mobilities of 55 anions. Prediction of optimized separations has been shown for benzoates [18, 24–26], phenols [27–29] and chloroanilines [30] using CZE, and for alkylphenols [31] and aromatic amines [32, 33] using MEKC.

37.3 Applications of CE to organic acids and organic ions

Many applications of capillary electrophoresis to organic acids and ions can be found elsewhere in this book covering specific application areas (e.g. food, environmental, pharmaceutical, clinical). The applications presented in this chapter are not intended to be comprehensive for all organic ions studied by CE. Rather, the focus is on separation/detection systems that are widely applicable to classes of organic ions, and on applications which fall outside the scope of other chapters. Also, applications presented here focus on recent developments in CE, since several texts [34–40] and review articles [41, 42] cover previous applications to organic ion analysis.

37.3.1 Organic acids and anions

37.3.1.1 Carboxylic acids. Small carboxylic acids without strong UV-absorbing chromophores usually are determined using indirect UV detection and EOF modifiers to ensure rapid migration toward the detector. The widely used system for inorganic anions, chromate with the proprietary Anion-BT osmotic flow modifier [17], has been applied to carboxylate separations. Determinations with this electrolyte include oxalate in Bayer liquors [43], and the results have been compared to HPLC [44]. Other applications of this buffer include determining anions in pulp and paper processing [45], food [46], plasma [47] and determining carboxylates in morpholine with ITP stacking to enhance sensitivity [48]. Chromate electrolytes have been used with other EOF modifiers. Buchberger showed carboxylate separations with tetradecyltrimethylammonium bromide (TTAB) and ethylene glycol [49]. Chromate with mixed dodecyltrimethylammonium bromide (DTAB)/TTAB modifiers was proposed for improved Bayer liquor analysis [50, 51].

In addition to chromate, a wide range of other UV-absorbing buffer co-ions have been used for indirect detection of carboxylates. Phthalate has been used in a number of applications, with TTAB for organic acids in coffee and wine [52] and with Anion-BT for acids in cigarette smoke [53], wine [54] and sugar juices [55]. The latter application used the addition of calcium to improve formate/tartrate resolution.

Benzoic acid and its derivatives are also popular background electrolytes. Benzoate was used for carboxylate separations with Triton added [16], and for propionate determination in bread [56]. Devevre et al. [57] used p-hydroxybenzoate with calcium to obtain high resolution carboxylate separation. Acetate and methanesulfonate were separated using dihydroxybenzoate with lead addition in an amine-coated capillary [58]. Aminobenzoate buffers modified with tetradecyltrimethylammonium hydroxide (TTAH) and barium were used to determine organic acids in rain water [59]. Tindall and Perry [60] employed nitrophthalic acid with TTAB to determine aliphatic acids in cellulose esters, and Heiger and Weinberger [61] analyzed caboxylates in beer and dairy products using dinitrobenzoates with cetyltrimethylammonium bromide (CTAB). Gutnikov et al. [62] used dinitrobenzoate with acetone/CTAB to determine C2–C14 fatty acids in fat hydrolysates. The same paper showed the separation of C3–C18 fatty acids using trinitrobenzene sulfonate with acetonitrile. Roldan-Assad and Gareil [63] developed separations of C2–C18 fatty acids using Tris/p-anisate buffer with cyclodextrin and methanol.

Naphthalene carboxylates and sulfonates are useful as electrolytes for indirect detection. Dabek-Zlotorzynska and Dlouhy [64, 65] used naphthalene dicarboxylate (NDC) to quantitate carboxylates in air. Tindall et al. demonstrated that by including β-cyclodextrin in buffers, the mobility of NDC electrolytes could be optimized to match carboxylate mobility [66]. Both naphthalene mono- and disulfonates have utility as co-ions for carboxylate and sulfonate analyses [67] with diethylenetriamine (DETA) added to suppress EOF. A buffer mixture of 1-naphthylacetate and benzenetricarboxylate was used as a mixed background electrolyte for hexanoate, malonate and tartrate separations [68].

While CZE is the major mode used for carboxylate separations, several reports of using MEKC with indirect UV detection have appeared. Sodium dodecylbenzenesulfonate (SDBS) served as both the micellar component and visualizing agent for the separation of C4–C7 carboxylates [69]. A similar buffer system [70] was employed for determining C8–C20 fatty acids in butter. In addition to indirect UV, several other detection schemes have been used for the CE of carboxylates. Mala et al. [71] used a dye-containing buffer with a coated capillary for indirect visible detection of carboxylates. Direct detection of aminopolycarboxylates including nitrilotriacetic acid (NTA), ethylenediamine tetraacetic acid (EDTA) and diethylenetriamine pentaacetic acid (DTPA) as copper complexes was demonstrated by Wiley [72] using copper sulfate electrolyte containing TTAB. Aminopolycarboxylates and their copper and nickel complexes were measured with direct UV detection at 200 nm. Bullock used CZE with 200 nm UV detection to separate free DTPA from its mono- and diamides as zinc complexes [73]. Direct UV detection of carboxylates in urine was accomplished at 185 nm [74]. Several UV-absorbing carboxylates were analyzed by direct detection, including maleate using a phosphate buffer containing CTAB [75] and ascorbate in fruit juices [76] and other beverages [77].

Unique buffer additives have been employed to enhance carboxylate separations. Stathakis and Cassidy [78] explored the use of polymeric cations as pseudo ion-exchangers, and Kaniansky et al. [79] employed polyethylene glycol for the analysis of sorbate in foods. Derivatization of carboxylates with aminonaphthalene disulfonic acid was used for sugar acids in a borate buffer [80], and derivatives of dicarboxylates with pyrenyldiazomethane were determined in serum [81] using an acetate/phosphate buffer with organic modifiers.

Polymeric carboxylates have been analysed by CE. Dolnik and Novotny [82] used a gel filled column to separate oligomers of poly-L-aspartic acid. CZE was used by Rigol et al. [83] to separate humic acids into two fractions, and by Garrison et al. [84] to characterize both humic and fulvic acids from different sources. Ryder [85] separated molecular weight 900–14000 acrylic acid copolymers using a borate buffer.

Conductivity detection shows great promise for improved analysis of carboxylates. Huang et al. separated C1–C7 carboxylates in morpholinoethanesulfonic acid (MES)/histidine buffers with TTAB and hydroxyethylcellulose (HEC) [86], and determined carboxylates in wine using a similar buffer in 30% methanol [87]. Borate buffers have been used in suppressed conductivity detection systems for separating malonate from inorganic ions [88] and for determining carboxylates in grape juice. ITP recently was applied to the determination of carboxylates in fermentation media and broths [89, 90].

37.3.1.2 Benzoates and aromatic acids. The UV-absorbing properties of benzoates and other aromatic acids simplify buffer selection for this class of analytes. The main concern is to use the proper pH and appropriate buffer additives to obtain analyte separation. Simple buffers, or those containing EOF modifiers, organic solvents, micelles or chiral selectors, have been used for separations of aromatic acids. Also, this class of analytes is often used as test solutes for other detection schemes.

Phosphate buffers have been used to analyse soft drinks for benzoate [91] and for separation of *meta-* and *para-*aminobenzoate [92]. Other simple buffers used include borate for determining acids in citrus pulp wash [93], acetate for aromatic acids in urine [94] and for phenoxyacid herbicides [95, 96], hexanesulfonate for determining impurities in terephthalic acid [97] and aminomethylpropanediol (AMPD)–bicine for hippurate and orotate in whey [98].

Additives to minimize EOF commonly are used in aromatic acid separations. Ng [99] employed phosphate-borate buffer with tetrabutylammonium (TBA) bisulfate to determine benzoate in oyster sauce. Kaneta et al. [100] studied the effect of added cetyltrimethylammonium chloride (CTAC) on the separation of benzoates. Birrell et al. [101] used phytate as a low conductivity EOF modifier in borate buffer, and Esaka et al. [102] used polyethylene glycol in phosphate buffer, both for separations of substituted benzoates. Gassner et al. [103] used a coated capillary with sodium acetate buffer to explore the effect of adsorbed Fe and calcium on benzoate separations. Yeo and co-workers [104, 105] used cyclodextrins to optimize separations of phenoxyacetate plant growth regulators. Garrison et al. [96] showed that cyclodextrins allow chiral separations of phenoxyacid herbicides.

Organic solvents with or without EOF modifiers have long been used for improving CE separation of aromatics. Fujiwara and Honda [106] used phosphate/acetonitrile to study the effect of organic solvent on positional isomer separations. Benz and Fritz [107] used butanol in borate buffer with Anion-BT to separate benzoates. In an elegant separation, polyethylene glycols with phthalic acid end groups were analyzed using borate buffer with acetonitrile and diaminopropane added [108]. In an extreme case of using organic solvents, Jansson and Roeraade [109] showed that unbuffered, 100% N-methylformamide could be used as a CE solvent for the separation of pyridine carboxylates and salicylate.

MEKC also plays a prominent role in separation schemes for aromatic acids. Wu et al. [110] used (1) phosphate–borate buffers with sodium dodecyl sulfate (SDS) and bovine serum albumin or cholate to separate porphyrin carboxylates, and (2) phosphate buffer with SDS, methanol and the detergent Brij to separate chlorophenoxy acid herbicides. Chiari et al. [111] showed that Tris acetate buffer containing SDS could be used to separate perdeuterated and normal benzoic acids. Iso-alpha acids in beer have been determined using phosphate/SDS buffer [112], and phosphate–borate buffers with SDS were used to determine aromatic carboxylates in plant extracts [113]. Brumley and Brownrigg [114] used borate with sodium cholate to analyze for aromatic carboxylates in soil and water. Bile salts have also been used as micelles in phosphate–borate buffer to determine acetylsalicylate and salicylate in analgesic tablets [115]. *In situ* charged micelles formed by alkylglucosides in butylboronate buffers were used by Smith and El Rassi [116] to separate phenoxyacid herbicides. Cationic micelles also have utility for separating aromatic acids, as shown by Janini et al. [117] for pyridine carboxylates, by Bjergegaard et al. [118] for benzoate in dietary fiber and by Lee et al. [119] for methylhippurates in urine. Chiral separations are a major application of CE for aromatic acid analysis. Cyclodextrins have been predominantly used for chiral analyses, being applied to such compounds as phenylacetates [120, 121], phenoxyacid herbicides [122], profens [123, 124] and other aromatic acids [125, 126]. Both coated and uncoated capillaries have been used. Other chiral selectors for aromatic acids include albumin [127] and antibiotics [128, 129].

While UV detection is by far the most popular method for aromatic acids, several other systems have been described. Conductivity detection was applied to anions including benzoate [130]; electrochemical detection was used for anthraquinone carboxylates [131]; impurities in the dye Eosin Y were detected using mass spectrometry [132]; laser-induced fluorescence was employed for separations of porphyrin carboxylates [133] and for phenoxyacid herbicides derivatized with a fluorescein agent [134].

37.3.1.3 Phenols. Like benzoates and aromatic acids, phenols are easily detected using direct UV at 200–260 nm. Due to solubility and selectivity considerations, buffer systems often contain organic solvents and/or use pH > 7. MEKC has also been extensively employed for optimization of phenolic separations.

In the CZE mode, phosphate buffer was used to separate chlorophenols [135]. The optimum pH for separation of dichlorophenols was found to be 7.72 [136]. Phosphate–borate buffers were used to analyze chloro- and nitrophenols [137], phenolic lignin degradation products [138] and to separate a variety of phenolic compounds in biomass degradation products [139]. Phenolic acids in wines were determined using borate [140] or bicarbonate [141] buffers. Dissolved kraft lignins were characterized using a glycine buffer at pH 10 [142]. Step changes in voltage [143] or buffer ion [144] have been used to improve the separation of substituted phenols. Addition of methanol [145, 146] or acetonitrile [147] has been used to separate chloro-/nitrophenols and alkylphenols, respectively. Harms and Schwedt [148] used a coated capillary with cyclohexanediamine tetraacetic acid (CDTA) added to analyze polyphenols in tea. Masselter and coworkers [149–151] used 40% isopropanol and hexadimethrine bromide (HDB) as EOF modifiers to optimize the separation of positional isomers of substituted phenols. Shohat and Grushka [152] added a calixarene sulfonate to phosphate buffer to improve the separation of chlorophenols. Bao and Dasgupta [153] used a silicone rubber membrane to preconcentrate phenols prior to separation in a borate buffer.

SDS micelles in the MEKC mode have been used widely for the separation of a variety of phenolics [154–156]. Use of a capillary with an anionic coating improved the pH range for phenol analysis by MEKC using SDS [157]. Brumley and Jones [158] used sodium cholate with 10% acetone for the determination of phenols in soils and water. The addition of longer chain alcohols to borate/SDS buffers gave improved separations of phenols by MEKC [159, 160]. A unique additive for CE, dendrimers with carboxylate end groups, offers an alternative selectivity vs. SDS for phenolic separations [161]. Micellar emulsion electrokinetic chromatography (MEEKC) with heptane/SDS mobile phase was developed by Terabe *et al.* [162] for its unique selectivity for various compounds, including phenols.

Non-UV detection schemes for phenols include indirect fluorescence for priority phenols [163], and electrochemical detection for chlorophenols in wastewater [164] and phenolic acids in apple juice [165]. Praus and Dombek [166–169] have shown the utility of ITP for determining chlorophenols in wastewater.

37.3.1.4 Organic sulfates and sulfonates. Many of the electrolyte/EOF modifier systems used for indirect detection of carboxylates can be used for separations of alkylsulfates and sulfonates. These include naphthalene dicarboxylate [65], naphthalene disulfonate [67], hydroxybenzoate [170], chromate with borate as a buffer [171] and sorbate [172]. Harrold *et al.* [173] used pyromellitic acid with hexamethonium bromide to separate alkylsulfonates and haloacetates. Alkylsulfates and sulfonates were separated using a salicylate/phosphate buffer with magnesium [174]. Other systems for alkylsulfonates include a veronal buffer [175] and a naphthalene monosulfonate buffer with acetonitrile [176]. Benzoate/borate/DETA electrolytes were used for separating alkyl ether sulfates [177]. Thompson *et al.* [178] employed a benzoate buffer to determine cyclamate in beverages, while Gibbons and Hoke [179] used dihydroxybenzoate in 5% methanol to determine SDS in simulated waters. Aromatic sulfonate separations commonly use direct UV detection in typical CZE buffers [180], with a variety of additives. Phosphate and/or borate buffers have been used for aromatic sulfonates in soil leachates [181], heparin disaccharides [182], anthraquinone sulfonates [183], sulfonated reactive dyes [184] and vinyl sulfonate monomer in polymers [185]. Citrate buffer in 10% acetonitrile was used to separate reactive textile dyes [186]. Altria *et al.* [187] fully validated a CZE method for determining trace residues of SDBS. Borate buffer with acetone and SDS added gave impressive separations of linear alkylbenzene sulfonates [188, 189] in commercial mixtures. Other applications include Tris/borate with added polyethylene glycol for analysis of the polysulfonate, quinobene [190], borate plus magnesium for separating sulfonated calixarenes [191], phosphate with triethylamine for separating aminonaphthalene sulfonate derivatives of saccharides [192, 193] and phosphate with a polymeric cation for the separation of naphthalene sulfonates [194]. Addition of bisulfite to a phosphate buffer allowed *in situ* formation and separation of the bisulfite addition product of benzaldehyde [195]. Polystyrene sulfonate polymers have been separated in phosphate buffers containing hydroxyethylcellulose as a soluble sieving medium using both constant [196] and pulsed [197] voltage. These latter two applications were performed in coated capillaries.

MEKC has been employed for aromatic sulfate/sulfonate analyses. Borate/SDS buffers were used for dye intermediates [198], sulfated disaccharides [199], reactive dyes [170] and for toluenesulfonic acid in a drug intermediate [200]. Borate/cholate with acetone was employed for determining dyes in soils [201], and borate/SDS with 25% methanol was used to separate UV-absorbing, anionic surfactants [177]. Sun et al. [202] showed that a capillary with a tunable coating for EOF can be employed to separate naphthalene- and benzenesulfonates at low pH.

A variety of non-UV detectors have been used for sulfate and sulfonate determinations. CZE–MS using ammonium acetate buffers in 30–60% acetonitrile has been used for naphthalene and benzene sulfonates [203] and for sulfonated azo dyes [204]. Aminonaphthalene sulfonates were separated in coated capillaries using fluorescence detection [205, 206]. Stefansson and Novotny [207] used CZE–fluorescence to separate heparin as a fluorescent derivative in an ethylenediamine/MES buffer. Suppressed conductivity detection was demonstrated for alkylsulfonates [208]. In a unique application, intact SDS micelles could be detected using indirect fluorescence when a fluorophore was added to the buffer [209]. Tribet et al. [210] used ITP with conductivity detection to separate C8–C20 alkyl sulfonates.

37.3.1.5 Sulfonamides, sulfonyl ureas and thiols. Miscellaneous organics capable of forming anions in solution include these important classes of antibiotics and herbicides. Sulfonamides are covered more completely in the chapter on pharmaceutical applications; a brief overview is presented here.

The initial sulfonamide separations described by Wainwright [211], and the separation of 18 sulfonamides by Ricci and Cross [212], used neutral phosphate buffers. Ng and co-workers [213–215] showed that sulfonamide separations could be optimized using CZE with cyclodextrins. Seven sulfonamides were separated by MEKC using SDS with TBA bromide added [216]. Sulfonamides have also been separated and detected by mass spectrometry [217, 218]. Ackermans et al. [219] described the determination of 12 sulfonamides in pork using CZE in phosphate–borate buffer. Benzothiazole sulfonamides, accelerators used in rubber manufacture, were separated using MEKC with urea added to the buffer [220].

Sulfonyl urea herbicides have been analyzed in soils using MEKC with mixed organic solvents [221, 222], in grains using MEKC with SDS/phosphate buffer [223] and using CZE–UV/MS in ammonium acetate/acetonitrile buffer [224]. Thiols generally are not analyzed as ions, due to their high pK_a, and often are detected electrochemically [225, 226]. Applications have dealt mainly with biologically important thiols in blood [227, 228].

37.3.1.6 Organic phosphates and phosphonates. Organic phosphates and phosphonates are as equally amenable to CE analysis as organosulfates and sulfonates, although applications in the literature are more limited. Shamsi and Danielson [229] have shown that buffers containing adenosine-5-monophosphate (AMP) with added Mg and/or diethylenetriamine are excellent electrolytes for indirect UV detection of aminopolyphosphonate compounds. This electrolyte system was used to determine the amount of N-phosphonomethyl glycine (NPMG) and aminomethyl phosphonic acid (AMPA) in a commercial herbicide formulation. Pianetti et al. [230] used phenylphosphonate as a buffer co-ion for separations of alkylphosphonates. Inositol phosphates and phytate have been analyzed using phthalate buffer with TTAB [231], in soybeans using benzoate buffer [232] and in fermentation broths using naphthalene disulfonate with added hydroxypropylmethylcellulose (HPMC) [233].

Direct UV detection has been used for phospholipids in MEKC using borate/cholate/isopropanol buffers [234]. Robins and Wright [235] showed that borate buffers could form UV-absorbing complexes for the separation of ethyl- and methylphosphonate. NPMG and AMPA were determined in serum after derivatization with p-toluenesulfonyl chloride using borate buffers with 10% methanol [236]. Enantiomeric binaphthyl phosphates were separated by CE using anionic cyclodextrins [237]. Indirect fluorescence detection was employed in the separation of isoprenyl pyrophosphate using salicylate buffer [238]. Other unique detection schemes for organophosphorus compounds include mass spectrometry [239] and phosphorus-specific flame photometric detection [240]. Velasquez [241] used ITP to separate mono- and diethylhexylphosphate.

37.3.2 Organic bases and cations

37.3.2.1 Simple amines. In the original description of CE [4], Jorgenson and Lukacs showed the separation of fluorescent derivatives of alkylamines. Analysis of amines remains an important application area of CE. Separation of derivatized amines can be accomplished by MEKC with SDS micelles [242], with isopropanol step gradients [243] and with cholate micelles [24]. Mattusch et al. [245] used a fluorescein isothiocyanate derivatization method for the fluorescent detection of diamines in pine needles. Since many amine derivatization methods are directed toward amino acid applications, the reader is referred to the amino acid chapter for more complete details. CZE with direct UV detection has employed buffers with UV-absorbing co-ions such as imidazole [246], copper [247] or quinine [248]. Quang and Khaledi [249] separated small amines and metals using imidazole with hydroxyisobutyric acid. Hydroxylamine and di- and triethanolamine were determined using a commercial cation buffer [250]. Direct UV detection at 200 nm was employed in the determination of alanylglutamine in a personal wash formulation by MEKC [251]. Substituted amines have also been determined by indirect fluorescence using a quinine sulfate buffer [252] and by electrogenerated chemiluminescence [253].

37.3.2.2 Aromatic amines. Since many aromatic amines find important uses in pharmaceutical applications, consultation of Part Two of this book on Biochemistry and Part Three on Pharmaceutical Science is suggested for additional references. Early CZE work by Tsuda *et al.* [180] showed the separation of pyridines in phosphate buffer. Lithium phosphate buffer at pH 2.5 has been used for the separation of alkylpyridines [254] and isomers of pentenyl pyridine [255]. Improved phosphate buffers have been obtained by adding cyclodextrins for separating xylidines [256] and chiral imidazoles [257], and by adding a crown ether for aminotetralin separations [258]. Wren [259] showed that CZE with TTAB to suppress EOF could be used to separate methylpyridines. The tendency of aromatic amines to adsorb to fused-silica capillaries has led to the extensive use of coated capillaries and/or MEKC to separate these analytes. Borate buffers with SDS were used to determine diphenylamines in gunshot residue [260] and to analyze aromatic quaternary amines used as dye intermediates [198]. SDS and amphoteric surfactants were used as mixed micelles for the MEKC separation of anilines in explosive residues [261]. Addition of TBA salts improved the separation of imidazoles by MEKC [262]. MEKC with γ-cyclodextrin was employed to separate aniline derivatives [263]. Another optimized MEKC separation of aniline derivatives used urea as an organic modifier [264]. Brumley and Brownrigg [265] determined benzidines in soil using borate buffer with sodium cholate micelles in 5% acetone.

Silica capillaries coated with cationic polyethyleneimine [87] and polypropylene capillaries coated with polyacrylamide [266] have been used to separate substituted pyridines. Nielen [267] determined phenylenediamines in saline solutions using CZE in a commercially coated capillary. By adding a non-ionic detergent to Tris–acetate buffer, Chiari *et al.* [111] separated perdeuterated pyridine and aniline from the hydrogen forms.

37.3.2.3 Cationic surfactants, herbicides, phosphonium compounds and cationic polymers. Several quaternary amine surfactants have bene analyzed using capillary electrophoresis. Weiss *et al.* [268] used indirect UV detection with a buffer containing SDS, tetrahydrofuran (THF) and an aromatic surfactant as a co-ion for quaternary alkylamines. Borate buffer with CTAC added [117] and a TBA phosphate electrolyte in a polypropylene capillary with added cationic surfactant [269] have been used to separate aromatic quaternary amine surfactants with direct UV detection. Mass spectrometry has also been used for detection of similar compounds [203]. ITP was used by Tribet *et al.* [210] to separate C8–C20

Table 37.1 Selected separation schemes for CE of organic acids and organic ions*

Mode	Buffer (pH)†	Analytes	Matrix	Ref.
Carboxylates				
CZE IUV 254	5 mM Chromate + H_2SO_4 (8), 0.5 mM Anion-BT–OFM	Carboxylates, alkyl sulfonates	Standards	17
CZE IUV 214	5 mM Chromate + H_2SO_4 (10.6), 0.5 mM Anion-BT–OFM	Oxalate, formate	Pulp/paper	45
CZE IUV 254	5 mM Chromate + H_2SO_4 (8), 2.5 mM Anion-BT–2.5 mM OFM, 1 M Z1-Methyl, 5% MeOH	Oxalate	Bayer liqors	43
CZE IUV 254	10 mM Chromate, 1 mM veronal + H_2SO_4 (8), HDB pre-coat	Carboxylates	Seminal plasma	47
CZE IUV 254	5 mM Chromate, 0.5 mM TTAB, 5% ethylene glycol, 0.78 mM Ca	Carboxylates	Standards	49
CZE IUV 254	6.5 mM Chromate + KOH (8.8), 5 mM TTAB, 1 mM DTAB	Carboxylates	Bayer liqors	50
CZE IUV 205	5 mM KHP, 50 mM MES (5.2), 0.5 mM TTAB	Carboxylates	Coffee, wine	52
CZE IUV 254	5 mM Phthalate (5.6), 1 mM Anion-BT	Anions	Cigarette smoke	53
CZE IUV 214	5 mM Tris + benzoic acid (4.6)	Propionate	Bread	56
CZE IUV 254	5 mM *p*-Hydroxybenzoate (4.5), 0.4 mM Ca, 2.5% Anion-BT	Organic acids	Standards	57
CZE IUV 264	7.5 mM *p*-Aminobenzoate + NH_4OH (9.4), 0.055 mM TTAH, 0.76 mM BaOH	Carboxylates	Rain	59
CZE IUV 254	6 mM Sorbate (5)	Sugar acids	Standards	286
CZE IUV 220	7 mM Nitrophthalate (7), 0.5 mM TTAB	C1–C5 carboxylates	Cellulose esters	60
CZE IUV 214	5 mM 3,5-Dinitro-benzoic acid (9), 0.5 mM CTAB, 40% acetone	C2–C14 fatty acids	Fat hydrolysates	62
CZE IUV 254	10 mM Trinitro-benzenesulfonate (9), 60% acetonitrile	C3–C18 fatty acids	Standards	62
CZE IUV 270	20 mM Tris + 10 mM *p*-anisic acid, 1 mM trimethyl-β-cyclodextrin, 50% methanol (8)	C2–C18 fatty acids	Cocoa oil	63
CZE IUV 283	1 mM Naphthalene dicarboxylate (9), 2 mM β-cyclodextrin	C2–C4 aliphatic acids	Standards	66
CZE IUV 280	2 mM Naphthalene dicarboxylate (11), 0.5 mM TTAB	Carboxylates	Air	65
CZE IUV 288	4 mM Naphthalene disulfonate 5 mM borate, 100 mM boric acid (8), 2 mM DETA	Carboxylates	Standards	67
MEKC IUV 214	4 mM SDBS, 50 mM SDS, 12.5 mM borax	C4–C7 carboxylates	Standards	69
MEKC IUV 198	5 mM Tris/HCl (8.3), 50% AcCN, 10 mM SDBS, 30 mM Brij	C8–C20 fatty acids	Butter	70
CZE UV 254	100 mM MES, 10 mM bisTris (5.2), 0.2% PEG	Sorbate	Foods	79

Table 37.1 *Continued*

Mode	Buffer (pH)†	Analytes	Matrix	Ref.
CZE IUV 254	5 mM Chromate + H_2SO_4 (8), 0.15% PDDPiCl	Carboxylates	Wine	78
CZE UV 185	50 mM Borate (10), 0.5 mM Anion-BT	Organic acids	Urine	74
CZE UV 254	10 mM Tricine (8.8)	Ascorbate	Fruits, urine	76
CZE UV 247	50 mM Phosphate (10)	ANDSA derivs. of sugar acids	Standards	80
CZE UV 254	5 mM $CuSO_4$, 1 mM TTAB	ODS, NTA, EDTA DTPA, EHDP	Liquid hand soap	72
Benzoates and aromatic acids				
CZE UV 214	25 mM Phosphate (11)	Benzoate	Soft drinks	91
CZE UV 185	25 mM Hexanesulfonate + LiOH (10), 0.5 mM Anion-BT	Benzoate, toluate	Terephthalic acid	97
CZE UV 210	8 mM Borate, 2 mM phosphate (9.6), 10 mM DTAC, 30% AcCN	8 Aromatic acids	Standards	287
CZE UV 214	300 mM Borate (9.2), 30 mM phytic acid	Substituted benzoates	Standards	101
CZE UV 210	10 mM Phosphate (7.8), 5% polyethylene glycol-400	Substituted benzoates	Standards	102
CZE UV 254	20 mM Na_2HPO_4/20 mM KH_2PO_4 (7), 50% AcCN	Substituted benzoates	Standards	106
CZE UV 254	1 mM Borate (8.5) 30 µM Anion-BT, 4% *n*-butanol	Substituted benzoates	Standards	107
CZE UV 215	5 mM Phosphate (6.8) 0.2% PDDPiCl	Substituted benzoates	Standards	78
CZE UV 205	57 mM Borate (9.7), 35 mM diaminopropane, AcCN	Phthalic anhydride derivatives	$n = 30–70$ polyethylene glycol	108
MEKC UV 254	30 mM Phosphate (7.6), 40 mM SDS	iso-α Acids	Beer	112
MEKC UV 214	50 mM Borate, 100 mM cholate (8.3)	Aromatic carboxylates	Water, soil	114
MEKC UV 214	20 mM Borate (7), 25 mM Na cholate, 50 mM Na deoxycholate	Salicylate acetylsalicylate	Tablets	115
MEKC UV 254	10 mM Phosphate (2.7/7.5), 30 mM CTAB	Pyridine carboxylates	Standards	117
MEKC UV 224	20 mM Phosphate (8), 100 mM DTAB, 4 M urea	Methyl-hippurates	Urine	119
MEKC UV 231	200 mM Butyl-boronate (11), 28 mM alkylglycoside	Phenoxy acid herbicides	Standards	116
CZE UV 206	50 mM Phosphate (7), 40 mM γ-cyclodextrin	Mandelic acid isomers	Standards {chiral}	120
CZE UV 200	50 mM Li Acetate (4.8), 10 mM α-cyclodextrin	Phenoxy acid herbicides	Production samples	122
Phenols				
CZE UV 206	20 mM Borate, 10 mM phosphate, 15% methanol (10)	Chloro- and nitrophenols	Standards	145
CZE UV 254	1 mM Borate, 40% AcCN (11.2)	Alkylphenols	Standards	147
CZE UV 215	50 mM Phosphate, 100 mM borate (7.72)	6 Dichloro-phenols	Standards	136

Table 37.1 *Continued*

Mode	Buffer (pH)†	Analytes	Matrix	Ref.
CZE UV 254	15 mM Phosphate, 1.25 mM borate (11), 0.001% HDB, 40% isopropanol	Substituted phenols	Standards	149
MEKC UV 214	50 mM Borate (8.35), 100 mM Na cholate, 10% acetone	Phenols	Soil, water	158
MEKC UV 214	50 mM Phosphate, 25 mM borate (9), 50 mM SDS	22 Phenolics	Standards	155
MEEKC UV 214	100 mM Borate, 50 mM phosphate (7), 60 mM SDS, 79 mM heptane, 874 mM butanol	Cresols, xylenols, alkylphenols	Standards	162
Organic sulfates and sulfonates				
CZE IUV 254	0.5 mM Anion BT-OFM 10 mM Trinitrobenzene sulfonate (8), 60% acetonitrile	Alkyl sulfonates	Air	65
CZE IUV 254	5 mM *p*-Hydroxybenzoate (6) Anion-BT	Alkyl sulfonates	Brine	170
CZE IUV 265	1 mM K Dichromate, 1 mM borate + boric acid (8), 30% AcCN	Ethoxylated alcohol and alkyl sulfates	Commercial products	171
CZE IUV 254	5 mM K sorbate + H_2SO_4 (7.6), 0.5 mM Anion-BT	Alkyl sufonates	Standards	172
CZE IUV 254	12 mM Borate, 10 mM benzoate (6.9), 4 mM diethylenetriamine	Alkyl ether sulfates	Standards	177
CZE IUV 288	5 mM NMS, 100 mM borate, 5 mM boric acid (8)	Alkyl sulfonates alkyl sulfates	Standards	67
CZE IUV 230	5 mM Phosphate, 5 mM salicylate, 1 mM Mg	Alkyl sulfates and sulfonates	Standards	174
CZE UV 220	10 mM Acetate (9) 40% acetonitrile, 3 mM Mg	LAS	Commercial mixtures	174
CZE IUV 254	10 mM Benzoate (6.6) 1 mM CTAH	Cyclamate	Beverages	178
CZE UV 214	50 mM Borate (8.3)	Aromatic sulfonates	Leachate	181
CZE UV 235	25 mM Phosphate, 1% HEC (5) coated capillary	Polystyrene sulfonate	Standards	196
CZE UV 214	6.25 mM Borate (9) 50 mM SDS, 30% AcCN	LAS	Commercial mixtures	189
MECK UV 254	10 mM Borate, 40 mM SDS	Dye intermediates	Products	198
MEKC	50 mM Borate (8.35) 100 mM Na cholate 5% acetone	Dyes	Soils	201
Sulfonamides, sulfonyl ureas and thiols				
CZE UV 254	20 mM Phosphate, 20 mM borate (7)	12 Sulfonamides	Pork	219
CZE UV 210	50 mM Phosphate, 50 mM borate (7), 10 mM β-cyclodextrin	7 Sulfonamides	Standards	214

Table 37.1 *Continued*

Mode	Buffer (pH)†	Analytes	Matrix	Ref.
MEKC UV 254	10 mM Phosphate–borate (8.7), 50 mM SDS, 6 M urea	Benzothiazole, sulfonamides	Standards	220
MEKC UV 214	30 mM Borate, 80 mM SDS (7), 14% methanol, 20% isopropanol	Sulfonyl ureas	Soil	222
CZE UV 220 or fluorescence, ex. = 380, em. = 510	100 mM Phosphate (2.5), coated capillary	Thiol derivatives	Blood	227
Organic phosphates and phosphonates				
CZE IUV 259	5 mM AMP, 100 mM borate (7.8), 2 mM diethylenetriamine	Aminoalkyl phosphonates	Commercial products	229
CZE IUV 214	0.5 mM Naphthol-disulfonic acid, 30 mM HOAc, 0.01% HPMC	Inositol phosphates	Fermentation broths	233
MEKC UV 200	10 mM Phosphate, 6 mM borate (8.5), 75 mM cholate, 30% 1-propanol	Phospholipids	Lecithins	234
ITP CD	L = 10 mM HCl + β-alanine (3.3), 0.25% HPMC, T = 10 mM hexanoic acid	Mono-, di-alkyl phosphates	Standards	241
Simple amines				
CZE IUV 214	6 mM Imidazole, 12 mM HIBA (3.95)	Small amines	Standards	249
CZE IUV 215	4 mM CuSO$_4$, 4 mM formic acid (3), 4 mM 18-crown-6	Small amines	Standards	247
CZE IUV 236	5 mM Quinine sulfate + HCl (3), 19% ethanol	Diamines	Tumor cells	288
Aromatic amines				
CZE UV 254	40 mM Lithium phosphate + H$_3$PO$_4$ (2.5)	Alkyl pyridines	Standards	254
CZE UV 210	50 mM KH$_2$PO$_4$/NaH$_2$PO$_4$ + H$_3$PO$_4$ (3), 100 mM SBE-β-cyclodextrin, 10% methanol	Imidazole derivatives	Standards (chiral)	257
CZE UV 225	600 mM Serine acetate (4.5) coated capillary	Phenylene diamines	Saline	267
MEKC UV 214	25 mM Phosphate–borate (8.7), 100 mM SDS, 10 mM TBA bisulfate	Imidazole derivatives	Commercial products	262
MECK UV 254	10 mM Borate, 10 mM Na$_2$HPO$_4$, 100 mM SDS, 2 M urea	Aniline derivatives	Standards	264
Cationic surfactants, herbicides and polymers				
CZE IUV 210	8 mM NaH$_2$PO$_4$, 3 mM benzyldimethylammonium bromide, 3 mM SDS, 57.5% THF	Quaternary ammonium surfactants	Disinfectant	268
MEKC UV 254	60 mM Borate (8.5), 0.25% CTAC	Aromatic quaternary amines	Standards	177
ITP CD	L = 10 mM KOAc, 10 mM HOAc (5.64), tetraethylene glycol, 50% MeOh, T = 10 mM creatinine, 50% MeOH	C8–C20 alkyl trimethylammonium surfactants	Commercial products	210
CZE UV310	9.8 mM Tris–citrate (6.1), 0.15% polyethylene glycol	Diquat, paraquat	Lake water	271

Table 37.1 *Continued*

Mode	Buffer (pH)†	Analytes	Matrix	Ref.
CZE UV 254	20 mM Tris + TCA (3), 30% ethanol	Triazines	Standards	276
MEKC UV 200–300	5 mM Borate, 30 mM SDS (8)	Triazines	Standards	278
CZE IUV 220	30 mM Creatinine acetate (4.8), 0.1% polyethylene oxide	Polyethylene oxide diamines	Polymer samples	108

*Buffer anions are as sodium salts, and/or pH adjusted with NaOH, unless otherwise specified.
†Numbers in parentheses are pH values.
Anion-BT is a proprietary osmotic flow modifier from Waters Corp. AcCN = acetonitrile, AMP = adenosine-5-monophosphate, ANDSA = 7-aminonaphthalene-1,3-disulfonic acid, bicine = N,N-bis(2-hydroxyethyl)glycine, bistris = 2,2-bis(hydroxymethyl)-2,2′,2″-nitrilotriethanol, CD = conductivity detection, CTAH = cetyltrimethylammonium hydroxide, DTAB = dodecyltrimethylammonium bromide, DTAC = dodecyltrimethylammonium chloride, EHDP = ethylhydroxydiphosphonate, HDB = hexadimethrine bromide (1,5-dimethyl-1,5-diazaundecamethylene polymethobromide), HIBA = hydroxyisobutyric acid, HOAc = acetic acid, IUV = indirect UV detection, KHP = potassium hydrogen phthalate, KOAc = potassium acetate, LAS = linear alkylbenzene sulfonates, NMS = naphthalenemonosulfonate, ODS = oxydisuccinic acid, OFM = osmotic flow modifier, PDDPiCl = poly(1,1-dimethyl-3,5-dimethylenepiperidinium chloride), PEG = polyethylene glycol, PTFE = poly(tetrafluoroethylene), SBE = sulfobutyl ether, THF = tetrahydrofuran, tricine = N-tris(hydroxymethyl)methylglycine, Tris = tris(hydroxymethyl)aminomethane, Z1-Methyl is a proprietary CE additive from Waters Corp.

alkyltrimethylammonium cations, and by Kolecek *et al.* [270] to determine 2-chloroethyltrimethylammonium cations.

Cationic herbicides have also been analyzed by CE. Paraquat and diquat were determined in lake water in Tris–citrate buffers with added polyethylene glycol and diethylenetriamine to lower adsorption [271]. The same herbicides were determined in potatoes using a coated capillary [272]. Paraquat, diquat and difenzoquat were separated using acetate buffer with salt [273], and the method was compared with HPLC analysis [274]. Cationic triazine herbicides have been separated in simple acetate buffers [275] and in those containing ethanol [276] or acetonitrile [277]. MEKC with SDS [278], or alkylglucoside–boronate micelles [279], has also been used for triazine determinations. Imidazolinone herbicides were separated using MEKC with DTAB addition [280].

Phosphonium cations rarely have been studied by CE. Smith and coworkers [281, 282] used mass spectrometric detection for the CZE and ITP separation of quaternary phosphonium compounds. Tetraphenylphosphonium ions were determined using ITP with conductivity detection [283]. Diphenyldimethylphosphonium has been used as a visualizing co-ion in buffers for cation separations [284].

In a paper on the CE of polymers, Bullock [108] used creatinine buffers containing polyethylene oxide for the separation and indirect UV detection of oligomers of polyoxyethylene/propylene diamines and polyethylene oxide diamines. Amankwa *et al.* [285] also separated the former polymer types after derivatization with naphthalenedialdehyde using fluorescence detection.

References

1. Everaerts, F.M.; Beckers, J.L.; Verheggen, Th.P.E.M. (1976) *Isotachophoresis – Theory, Instrumentation and Applications*, Elsevier, Amsterdam.
2. Baldesten, A. (1981) Capillary isotachophoresis. *Crit. Rev. Anal. Chem.* **11**, 261–352.
3. Stover, F.S. (1990) Applications of capillary electrophoresis for industrial analysis. *Electrophoresis* **11**, 750–756.
4. Jorgenson, J.W.; Lukacs, K.D. (1981) Zone electrophoresis in open-tubular glass capillaries. *Anal. Chem.* **53**, 1295–1302.
5. Terabe, S.; Otsuka, K.; Ando, T. (1985) Electrokinetic chromatography with micellar solution and open-tubular capillary. *Anal. Chem.* **57**, 834–841.
6. Jandik, P.; Jones, W.R. (1991) Optimization of detection sensitivity in the capillary electrophoreis of inorganic anions. *J. Chromatogr.* **546**, 431–443.
7. Altria, K.D.; Fabre, H. (1991) Approaches to optimisation of precision in capillary electrophoresis. *Chromatographia.* **40**, 313–320.
8. Leube, J.; Roeckel, O. (1994) Quantification in capillary zone electrophoresis for samples differing in composition from the electrophoresis buffer. *Anal. Chem.* **66**, 1090–1096.
9. Smith, S.C.; Strasters, J.K.; Khaledi, M.G. (1991) Influence of operating parameters on reproducibility in capillary electrophoresis. *J. Chromatogr.* **559**, 57–68.
10. Jumppanen, J.H.; Riekkola, M. (1995) Marker techniques for high-accuracy identification in CZE. *Anal. Chem.* **67**, 1060–1066.
11. Hirokawa, T.; Nishino, M.; Aoki, N.; Kiso, Y.; Sawamoto, Y.; Yagi, T.; Akiyama, J. (1983) Table of isotachophoretic indices. I. Simulated qualitative and quantitative indices of 287 anionic substances in the range pH 3–10. *J. Chromatogr.* **271**, D1–D106.
12. Pospichal, J.; Gebauer, P.; Bocek, P. (1989) Measurement of mobilities and dissociation constants by capillary isotachophoresis. *Chem. Rev.* **89**, 419–430.
13. Wronski, M. (1993) Concept of effective mass and hidden mass for calculation of mobility of organic anions and peptides. *J. Chromatogr.* **657**, 165–173.
14. Cross, R.F.; Ricci, M.C. (1995) pH and pK_a limitations in the CZE analysis of sulfonamides. *LC–GC* **13**, 132–142.
15. Sustacek, V.; Foret, F.; Bocek, P. (1991) Selection of the background electrolyte composition with respect of electromigration dispersion and detection of weakly absorbing substances in capillary zone electrophoresis. *J. Chromatogr.* **545**, 239–248.
16. Foret, F.; Fanali, S.; Ossicini, L.; Bocek, P. (1989) Indirect photometric detection in capillary zone electrophoresis. *J. Chromatogr.* **470**, 299–308.
17. Jones, W.R.; Jandik, P. (1991) Controlled changes of selectivity in the separation of ions by capillary electrophoresis. *J. Chromatogr.* **546**, 445–458.
18. Reijenga, J.C.; Kenndler, E. (1994) Computational simulation of migration and dispersion in free capillary zone electrophoresis. II. Results of simulation and comparison with measurements. *J. Chromatogr.* **659**, 417–426.
19. Gas, B.; Stedry, M.; Rizzi, A.; Kenndler, E. (1995) Dynamics of peak dispersion in capillary zone electrophoresis including wall adsorption. I. Theoretical model and results of simulation. *Electrophoresis* **16**, 958–967.
20. Yang, S.; Khaledi, M.G. (1995) Chemical selectivity in micellar electrokinetic chromatography: characterization of solute–micelle interactions for classification of surfactants. *Anal. Chem.* **67**, 499–510.
21. Powell, A.C.; Sepaniak, M.J. (1990) Development of a model for predicting retention times in solvent-gradient micellar electrokinetic capillary chromatography. *J. Microcol. Sep.* **2**, 278–284.
22. Friedl, W.; Reijenga, J.C.; Kenndler, E. (1995) Ionic strength and charge number correction for mobilities of multivalent organic anions in capillary electrophoresis. *J. Chromatogr.* **709**, 163–170.
23. Kenndler, E.; Gassner, B. (1990) Cluster analysis applied to the selection and combination of buffering electrolyte systems used for capillary electrophoresis of anions with water or methanol as solvents. *Anal. Chem.* **62**, 431–436.
24. Jumppanen, J.H.; Siren, H.; Riekkola, M.; Soderman, O. (1993) Correlation of resolution with frictional coefficients and pK_a values in capillary electrophoresis of four diuretics:

Determination of electric field strength and electroosmotic velocity. *J. Microcol. Sep.* **5**, 451–457.
25. Friedl, W.; Kenndler, E. (1993) Resolution as a function of the pH of the buffer based on the analyte charge number for multivalent ions in capillary zone electrophoresis without electroosmotic flow: Theoretical prediction and experimental evaluation. *Anal. Chem.* **65**, 2003–2009.
26. Friedl, W.; Kenndler, E. (1994) Limitations of the optimization of the resolution by the buffer pH in capillary zone electrophoresis. *Fresenius J. Anal. Chem.* **348**, 576–582.
27. Sahota, R.S.; Khaledi, M.G. (1994) Target factor modeling of migration behavior in capillary electrophoresis. *Anal. Chem.* **66**, 2374–2381.
28. Smith, S.C.; Khaledi, M.G. (1993) Optimization of pH for the separation of organic acids in capillary zone electrophoresis. *Anal. Chem.* **65**, 193–198.
29. Reijenga, J.C.; Hutta, M. (1995) MECCSIM, training software for micellar electrokinetic capillary chromatography. *J. Chromatogr.* **709**, 21–29.
30. Jacquier, J.C.; Rony, C.; Desbene, P.L. (1993) Computer-assisted pH optimization for the separation of geometric isomers in capillary zone electrophoresis. *J. Chromatogr.* **652**, 337–345.
31. Smith, S.C.; Khaledi, M.G. (1993) Prediction of the migration behavior of organic acids in micellar electrokinetic chromatography. *J. Chromatogr.* **632**, 177–184.
32. Quang, C.; Strasters, J.K.; Khaledi, M.G. (1994) Computer-assisted modeling, prediction, and multifactor optimization in micellar electrokinetic chromatography of ionizable compounds. *Anal. Chem.* **66**, 1646–1653.
33. Strasters, J.K.; Khaledi, M.G. (1991) Migration behavior of cationic solutes in micellar electrokinetic capillary chromatography. *Anal. Chem.* **63**, 2503–2508.
34. Li, S.F.Y. (1992) *Capillary Electrophoresis: Principles, Practice and Applications*, Elsevier, Amsterdam.
35. Demarest, C.W.; Monnot-Chase, E.A.; Jiu, J.; Weinberger, R. (1992) Separation of small molecules by high-performance capillary electrophoresis. In *Capillary Electrophoresis Theory & Practice*, Grossman, P.D.; Colburn, J.C. (eds), Academic Press, San Diego, pp. 301–341.
36. Jones, W.R. (1993) Electrophoretic capillary ion analysis. In *Handbook of Capillary Electrophoresis*, Landers, J.P. (ed.), CRC Press, Boca Raton, pp. 209–232.
37. Kuhr, W.G. (1993) Separation of small organic molecules by capillary electrophoresis. In *Capillary Electrophoresis Theory and Practice*, Camilleri, P. (ed.), CRC Press, Boca Raton, pp. 65–115.
38. Guzman, N.A. (1993) *Capillary Electrophoresis Technology*, Marcel Dekker, New York.
39. Kuhn, R.; Hofstetter-Kuhn, S. (1993) *Capillary Electrophoresis: Principles and Practice*, Springer-Verlag, Berlin.
40. Jandik, P.; Bonn, G. (1993) *Capillary Electrophoresis of Small Molecules and Ions*, VCH, New York.
41. Kuhr, W.G.; Monning, C.A. (1992) Capillary electrophoresis. *Anal. Chem.* **64**, 389R–407R.
42. Monning, C.A.; Kennedy, R.T. (1994) Capillary electrophoresis. *Anal. Chem.* **66**, 280R–314R.
43. Gorcott, S.C.; Jefferies, L.P.; Bowser, T.; Carnevale, J.; Jackson, P.E. (1992) Applications of ion chromatography and capillary ion electrophoresis in the alumina and aluminium industry. *J. Chromatogr.* **602**, 257–264.
44. Jackson, P.E. (1995) Analysis of oxalate in Bayer liquors: A comparison of ion chromatography and capillary electrophoresis. *J. Chromatogr.* **693**, 155–161.
45. Salomon, D.R.; Romano, J. (1992) Applications of capillary ion electrophoresis in the pulp and paper industry. *J. Chromatogr.* **602**, 219–225.
46. Kenney, B.F. (1991) Determination of organic acids in food samples by capillary electrophoresis. *J. Chromatogr.* **546**, 423–430.
47. Oefner, P.J. (1995) Surface-charge reversed capillary zone electrophoresis of inorganic and organic anions. *Electrophoresis* **16**, 46–56.
48. Bondoux, G.; Jandik, P.; Jones, W.R. (1992) New approach to the analysis of low levels of anions in water. *J. Chromatogr.* **602**, 79–88.

REFERENCES

49 Buchberger, W.; Cousins, S.M.; Haddad, P.R. (1994) Optimisation of indirect UV detection in capillary zone electrophoresis of low-molecular-mass anions. *Trends Anal. Chem.* **13**, 313–319.

50 Harakuwe, A.H.; Haddad, P.R; Buchberger, W. (1994) Optimisation of separation selectivity in capillary zone electrophoresis of inorganic anions using binary cationic surfactant mixtures. *J. Chromatogr.* **685**, 161–165.

51 Haddad, P.R.; Harakuwe, A.H.; Buchberger, W. (1995) Separation of inorganic and organic anionic components of Bayer liquor by capillary zone electrophoresis. I. Optimisation of resolution with electrolyte-containing surfactant mixtures. *J. Chromatogr.* **706**, 571–578.

52 Kelly, L.; Nelson, R.J. (1993) Capillary zone electrophoresis of organic acids and anions. *J. Liq. Chromatogr.* **16**, 2103–2112.

53 Lagoutte, D.; Lombard, G.; Nisseron, S.; Papet, M.P.; Saint-Jalm, Y. (1994) Determination of organic acids in cigarette smoke by high-performance liquid chromatography and capillary electrophoresis. *J. Chromatogr.* **684**, 251–257.

54 Levi, V.; Wehr, T.; Talmadge, K.; Zhu, M. (1993) Analysis of organic acids in wines by capillary electrophoresis and HPLC. *Am. Lab.* **25**, 29–32.

55 Lalljie, S.P.D.; Vindevogel, J.; Sandra, P. (1993) Quantitation of organic acids in sugar refinery juices with capillary zone electrophoresis and indirect UV detection. *J. Chromatogr.* **652**, 563–569.

56 Ackermans, M.T.; Ackermans-Loonen, J.C.; Beckers, J.L. (1992) Determination of propionate in bread using capillary zone electrophoresis. *J. Chromatogr.* **627**, 273–279.

57 Devevre, O.; Putra, D.P.; Botton, B.; Garbaye, J. (1994) Sensitive and selective method for the separation of organic acids by capillary zone electrophoresis. *J. Chromatogr.* **679**, 349–357.

58 Bachmann, K.; Steeg, K.; Groh, T.; Haumann, I.; Boden, J.; Holtheus, H. (1992) Analysis of cations and anions in small volumes using micro-LC and CE. *J. Microcol. Sep.* **4**, 431–438.

59 Roder, A.; Bachmann, K. (1995) Simultaneous determination of organic and inorganic anions in the sub-micromol/l range in rain water by capillary zone electrophoresis. *J. Chromatogr.* **689**, 305–311.

60 Tindall, G.W.; Perry, R.L. (1993) Determination of ester subtituents in cellulose esters. *J. Chromatogr.* **633**, 227–233.

61 Heiger, D.; Weinberger, R. (1994) Determination of small ions by capillary zone electrophoresis with indirect photometric detection. *Hewlett Packard Application Note*, Publication No. 12-5963-1138E.

62 Gutnikov, G.; Beck, W.; Engelhardt, H. (1994) Separation of homologous fatty acids by capillary electrophoresis. *J. Microcol. Sep.* **6**, 565–570.

63 Roldan-Assad, R.; Gareil, P. (1995) Capillary zone electrophoretic determination of C2–C18 linear saturated free fatty acids with indirect absorbance detection. *J. Chromatogr.* **708**, 339–350.

64 Dabek-Zlotorszynska, E.; Dlouhy, J.F. (1994) Application of capillary electrophoresis in atmospheric aerosols analysis: Determination of inorganic and organic anions. *J. Chromatogr.* **671**, 389–395.

65 Debek-Zlotorzynska, E.; Dlouhy, J.F. (1994) Capillary zone electrophoresis with indirect UV detection of organic anions using 2,6-naphthalenedicarboxylic acid. *J. Chromatogr.* **685**, 145–153.

66 Tindall, G.W.; Wilder, D.R.; Perry, R.L. (1993) Optimizing dynamic range for the analysis of small ions by capillary zone electrophoresis. *J. Chromatogr.* **641**, 163–167.

67 Shamsi, S.A.; Danielson, N.D. (1994) Naphthalenesulfonates as electrolytes for capillary electrophoresis of inorganic anions, organic acids, and surfactants with indirect photometric detection. *Anal. Chem.* **66**, 3757–3764.

68 Wang, T.; Hartwick, R.A. (1992) Binary buffers for indirect absorption detection in capillary zone electrophoresis. *J. Chromatogr.* **589**, 307–313.

69 Szucs, R.; Vindevogel, J.; Sandra, P. (1991) Micellar electrokinetic chromatography of aliphatic compounds with indirect UV detection. *J. High Resolut. Chromatogr.* **14**, 692–693.

70 Erim, F.B.; Xu, X.; Kraak, J.C. (1995) Application of micellar electrokinetic chromatography and indirect UV detection for the analysis of fatty acids. *J. Chromatogr.* **694**, 471–479.

71 Mala, Z.; Vespalec, R.; Bocek, P. (1994) Capillary zone electrophoresis with indirect photometric detection in the visible range. *Electrophoresis* **15**, 1526–1530.

72 Wiley, J.P. (1995) Determination of polycarboxylic acids by capillary electrophoresis with copper complexation. *J. Chromatogr.* 692, 267–274.

73 Bullock, J. (1995) Assay for the dianhydride of diethylenetriamine pentaacetic acid and its major degradation products by capillary electrophoresis. *J. Chromatogr.* **669**, 149–155.

74 Shirao, M.; Furuta, R.; Suzuki, S.; Nakazawa, H.; Fujita, S.; Maruyama, T. (1994) Determination of organic acids in urine by capillary zone electrophoresis. *J. Chromatogr.* **680**, 247–251.

75 Chang, H.; Yeung, E.S. (1992) Optimization of selectivity in capillary zone electrophoresis via dynamic pH gradient and dynamic flow gradient. *J. Chromatogr.* **608**, 65–67.

76 Koh, E.V.; Bissell, M.G.; Ito, R.K. (1993) Measurement of vitamin C by capillary electrophoresis in biological fluids and fruit beverages using a stereoisomer as an internal standard. *J. Chromatogr.* **633**, 245–250.

77 Marshall, P.A.; Trenerry, C.; Thompson, C.O. (1995) The determination of total ascorbic acid in beers, wines, and fruit drinks by micellar electrokinetic capillary chromatography. *J. Chromatogr. Sci.* **33**, 426–432.

78 Stathakis, C.; Cassidy, R.M. (1995) Effect of electrolyte composition in the capillary electrophoretic separation of inorganic/organic anions in the presence of cationic polymers. *J. Chromatogr.* **699**, 353–361.

79 Kaniansky, D.; Masar, M.; Madajova, V.; Marak, J. (1994) Determination of sorbic acid in food products by capillary zone electrophoresis in a hydrodynamically closed separation compartment. *J. Chromatogr.* **677**, 179–185.

80 Mechref, Y.; El Rassi, Z. (1994) Capillary zone electrophoresis of derivatized acidic monosaccharides. *Electrophoresis* **15**, 627–634.

81 Schneede, J.; Mortensen, J.H.; Kvalheim, G.; Ueland, P.M. (1994) Capillary zone electrophoresis with laser-induced fluoresence detection for analysis of methylmalonic acid and other short-chain dicarboxylic acids derivatized with 1-pyrenyldiazomethane. *J. Chromatogr.* **669**, 185–193.

82 Dolnik, V.; Novotny, M.V. (1993) Separation of amino acid homopolymers by capillary gel electrophoresis. *Anal. Chem.* **65**, 563–567.

83 Rigol, A.; Lopez-Sanchez, J.F.; Rauret, G. (1994) Capillary zone electrophoresis of humic acids. *J. Chromatogr.* **664**, 301–305.

84 Garrison, A.W.; Schmitt, P.; Kettrup, A. (1995) Capillary electrophoresis for the characterization of humic substances. *Water Res.* **29**, 2149–2159.

85 Ryder, D.S. (1993) The use of capillary zone electrophoresis for the analysis of polymeric water treatment additives and comparison with liquid chromatography techniques. In *Ion Exchange Processes: Advances and Applications*, Dyer, A.; Hudson, M.J.; Williams, P.A. (eds), Royal Society of Chemistry, London, pp. 101–110.

86 Huang, X.; Gordon, M.J.; Zare, R.N. (1989) Effect of electrolyte and sample concentration on the relationship between sensitivity and resolution in capillary zone electrophoresis using conductivity detection. *J. Chromatogr.* **480**, 285–288.

87 Huang, M.; Vorkink, W.P.; Lee, M.L. (1992) Evaluation of surface-bonded polyethylene glycol and polyethylene imine in capillary electreophoresis. *J. Microcol. Sep.* **4**, 135–143.

88 Kar, S.; Dasgupta, P.K.; Liu, H.; Hwang, H. (1994) Computer-interfaced bipolar pulse conductivity detector for capillary systems. *Anal. Chem.* **66**, 2537–2543.

89 Futschik, K.; Ammann, M.; Bachmayer, S.; Kenndler, E. (1993) Determination of ionic species formed during growth of *Escherichia coli* by capillary isotachophoresis. *J. Chromatogr.* **644**, 389–395.

90 Karovicova, J.; Polonsky, J.; Drdak, M.; Simko, P.; Vollek, V. (1993) Capillary isotachophoresis of organic acids produced by selected microorganisms during lactic acid fermentation. *J. Chromatogr.* **638**, 241–246.

91 Jimidar, M.; Hamoir, T.P.; Foriers, A.; Massart, D.L. (1993) Comparison of capillary zone electrophoresis with high-performance liquid chromatography for the determination of additives in foodstuffs. *J. Chromatogr.* **636**, 179–186.
92 Nielen, M.W.F. (1991) Impact of experimental parameters on the resolution of positional isomers of aminobenzoic acid in capillary zone electrophoresis. *J. Chromatogr.* **542**, 173–183.
93 Cancalon, P.F. (1993) Rapid monitoring of fruit juice adulteration by capillary electrophoresis. *LC–GC* **11**, 748–751.
94 Isaaq, H.J.; Delviks, K.; Janini, G.M.; Muschik, G.M. (1992) Capillary zone electrophoretic separation of homovanillic and vanillylmandelic acids. *J. Liq. Chromatogr.* **15**, 3193–3201.
95 Nielen, M.W.F. (1993) Trace enrichment of environmental samples in capillary zone electrophoresis. *Trends Anal. Chem.* **12**, 345–356.
96 Garrison, A.W.; Schmitt, P.; Kettrup, A. (1994) Separation of phenoxy acid herbicides and their enantiomers by high-performance capillary electrophoresis. *J. Chromatogr.* **688**, 317–327.
97 Jones, W.R.; Jandik, P. (1992) Various approaches to analysis of difficult sample matrices of anions using capillary ion electrophoresis. *J. Chromatogr.* **608**, 385–393.
98 Tienstra, P.A.; van Riel, J.A.M.; Mingorance, M.D.; Olieman, C. (1992) Assessment of the capabilities of capillary zone electrophoresis for the determination of hippuric and orotic acid in whey. *J. Chromatogr.* **608**, 357–361.
99 Ng, C.L.; Lee, H.K.; Li, S.F.Y. (1992) Analysis of food additives by ion-pairing electrokinetic chromatography. *J. Chromatogr.* **30**, 167–170.
100 Kaneta, T.; Tanaka, S.; Taga, M. (1993) Effect of cetyltrimethylammonium chloride on electroosmotic and electrophoretic mobilities in capillary zone electrophoresis. *J. Chromatogr.* **653**, 313–319.
101 Birrell, H.C.; Camilleri, P.; Okafo, G.N. (1994) Phytic acid can greatly enhance resolution in capillary electrophoresis. *J. Chem. Soc., Chem. Commun.* 43–44.
102 Esaka, Y.; Yamaguchi, Y.; Kano, K.; Goto, M.; Haraguchi, H.; Takahashi, J. (1994) Separation of hydrogen-bonding donors in capillary electrophoresis using polyethers as matrix. *Anal. Chem.* **66**, 2441–2445.
103 Gassner, B.; Friedl, W.; Kenndler, E. (1994) Wall adsorption of small anions in capillary zone electrophoresis induced by cationic trace constituents of the buffer. *J. Chromatogr.* **680**, 25–31.
104 Yeo, S.K.; Lee, H.K.; Li, S.F.Y. (1992) Separation of plant growth regulators by capillary electrophoresis. *J. Chromatogr.* **594**, 335–340.
105 Yeo, S.K.; Ong, C.P.; Li, S.F.Y. (1991) Optimization of high-performance capillary electrophoresis of plant growth regulators using the overlapping resolution mapping scheme. *Anal. Chem.* **63**, 2222–2225.
106 Fujiwara, S.; Honda, S. (1987) Effect of addition of organic solvent on the separation of positional isomers in high-voltage capillary zone electrophoresis. *Anal. Chem.* **59**, 487–490.
107 Benz, N.J.; Fritz, J.S. (1994) Studies on the determination of inorganic anions by capillary electrophoresis. *J. Chromatogr.* **671**, 437–443.
108 Bullock, J. (1993) Application of capillary electrophoresis to the analysis of the oligomeric distribution of polydisperse polymers. *J. Chromatogr.* **645**, 169–177.
109 Jansson, M.; Roeraade, J. (1995) *N*-methylformamide as a separation medium in capillary electrophoresis. *Chromatographia* **40**, 163–169.
110 Wu, N.; Barker, G.E.; Huie, C.W. (1994) Separation of porphyrins and prophyrin isomers in capillary electrophoresis using mixed ionic surfactant–bovine serum albumin buffer systems. *J. Chromatogr.* **659**, 435–442.
111 Chiari, M; Nesi, M.; Ottolina, G.; Righetti, P.G. (1994) Separation of charged and neutral isotopic molecules by micellar electrokinetic chromatography in coated capillaries. *J. Chromatogr.* **680**, 571–577.
112 Szucs, R.; Vindevogel, J.; Sandra, P.; Verhagen, L.C. (1993) Sample stacking effects and large injection volumes in micellar electrokinetic chromatography of ionic compounds: direct determination of iso-alpha-acids in beer. *Chromatographia* **36**, 323–329.
113 Morin, Ph.; Dreux, M. (1993) Factors influencing the separation of ionic and non-ionic

chemical natural compounds in plant extracts by capillary electrophoresis. *J. Liq. Chromatogr.* **16**, 3735–3755.
114 Brumley, W.C.; Brownrigg, C.M. (1993) Electrophoretic behavior of aromatic-containing organic acids and the determination of selected compounds in water and soil by capillary electrophoresis. *J. Chromatogr.* **646**, 377–389.
115 Boonkerd, S.; Lauwers, M.; Detaevernier, M.R.; Michotte, Y. (1995) Separation and simultaneous determination of the components in an analgesic tablet formulation by micellar electrokinetic chromatography. *J. Chromatogr.* **695**, 97–102.
116 Smith, J.T.; El Rassi, Z. (1994) Micellar electrokinetic capillary chromatography with *in situ* charged micelles: 3. Evaluation of alkylglucoside surfactants as anionic butylboronate complexes. *Electrophoresis* **15**, 1248–1259.
117 Janini, G.M.; Chan, K.C.; Barnes, J.A.; Muschik, G.M.; Issaq, H.J. (1993) Separation of pyridinecarboxylic acid isomers and related compounds by capillary zone electrophoresis. Effect of cetyltrimethylammonium bromide on electroosmotic flow and resolution. *J. Chromatogr.* **653**, 321–327.
118 Bjergegaard, C.; Michaelsen, S.; Sorensen, H. (1992) Determination of phenolic carboxylic acids by micellar electrokinetic capillary chromatography and evaluation of factors affecting the method. *J. Chromatogr.* **608**, 403–411.
119 Lee, K.; Lee, J.J.; Moon, D.C. (1994) Application of micellar electrokinetic capillary chromatography for monitoring of hippuric and methylhippuric acid in human urine. *Electrophoresis* **15**, 98–102.
120 Valko, I.E.; Billiet, H.A.H.; Frank, J.; Luyben, K.Ch.A.M. (1994) Factors affecting the separation of mandelic acid enantiomers by capillary electrophoresis. *Chromatographia* **38**, 730–736.
121 Valko, I.E.; Billiet, H.A.H.; Frank, J.; Luyben, K.Ch.A.M. (1994) Effect of the degree of substitution of (2-hydroxy)propyl-β-cyclodextrin on the enantioseparation of organic acids by capillary electrophoresis. *J. Chromatogr.* **678**, 139–144.
122 Nielen, M.W.F. (1993) (Enantio-) separation of phenoxy acid herbicides using capillary zone electrophoresis. *J. Chromatogr.* **637**, 81–90.
123 Rawjee, Y.Y.; Staerk, D.U.; Vigh, G. (1993) Capillary electrophoretic chiral separations with cyclodextrin additives. I. Acids: chiral selectivity as a function of pH and the concentration of beta-cyclodextrin for fenoprofen and ibuprofen. *J. Chromatogr.* **635**, 291–306.
124 Fanali, S.; Aturki, Z. (1995) Use of cyclodextrins in capillary electrophoresis for the chiral resolution of some 2-arylpropionic acid non-steroidal anti-inflammatory drugs. *J. Chromatogr.* **694**, 297–305.
125 Belder, D.; Gerhard, S. (1994) Chiral separations of basic and acidic compounds in modified capillaries using cyclodextrin-modified capillary zone electrophoresis. *J. Chromatogr.* **666**, 351–365.
126 Nardi, A.; Eliseev, A.; Bocek, P.; Fanali, S. (1993) Use of charged and neutral cyclodextrins in capillary zone electrophoresis: enantiomeric resdolution of some 2-hydroxy acids. *J. Chromatogr.* **638**, 247–253.
127 Vespalec, R.; Sustacek, V.; Bocek, P. (1993) Prospects of dissolved albumin as a chiral selector in capillary zone electrophoresis. *J. Chromatogr.* **638**, 255–261.
128 Armstrong, D.W.; Gasper, M.P.; Rundlett, K.L. (1995) Highly enantioselective capillary electrophoretic separations with dilute solutions of the macrocyclic antibiotic ristocetin A. *J. Chromatogr.* **689**, 285–304.
129 Rundlett, K.L.; Armstrong, D.W. (1995) Effect of micelles and mixed micelles on efficiency and selectivity of antibiotic-based capillary electrophoretic enantioseparations. *Anal. Chem.* **67**, 2088–2095.
130 Dasgupta, P.K.; Bao, L. (1993) Suppressed conductometric capillary electrophoresis separation systems. *Anal. Chem.* **65**, 1003–1011.
131 Malone, M.Z.; Weber, P.L.; Smyth, M.R.; Lunte, S.M. (1994) Reductive electrochemical detection for capillary electrophoresis. *Anal. Chem.* **66**, 3782–3787.
132 Varghese, J.; Cole, R.B. (1993) Optimization of capillary zone electrophoresis-electrospray mass spectrometry for cationic and anionic laser dye analysis employing opposite polarities at injector and interface. *J. Chromatogr.* **639**, 301–316.
133 Wu, Q.; Claessens, H.A.; Cramers, C.A. (1992) The separation of herbicides by micellar electrokinetic capillary chromatography. *Chromatographia* **34**, 25–30.

134 Brumley, W.C. (1995) Environmental applications of capillary electrophoresis for organic pollutant determination. *LC–GC* **13**, 556–568.
135 Gonnord, M.F.; Collet, J. (1993) Optimized capillary zone electrophoretic separation of chlorinated phenols. *J. Chromatagr.* **645**, 327–336.
136 Lin, C.; Lin, W.; Chiou, W. (1995) Migration behaviour and optimization of selectivity of dichlorophenols in capillary zone electrophoresis. *J. Chromatogr.* **705**, 325–333.
137 Li, G.; Locke, D.C. (1995) Separation of the eleven priority pollutant phenols by capillary zone electrophoresis. *J. Chromatogr.* **669**, 93–102.
138 Masselter, S.; Zemann, A.; Bobleter, O. (1995) Analysis of lignin degradation products by capillary electrophoresis. *Chromatographia* **40**, 51–57.
139 Zemann, A.J.; Bobleter, O. (1993) Separation of biomass degradation products by capillary electrophoresis. In *Advances in Thermochemical Biomass Conversion*, Bridgwater, A.V. (ed.), Blackie Academic & Professional, Glasgow, pp. 953–965.
140 Gil, M.I.; Garcia-Viguera, C.; Bridle, P.; Tomas-Barberan, F.A. (1995) Analysis of phenolic compounds in Spanish red wines by capillary zone electrophoresis. *Z. Lebensm-Unters. Forsch.* **200**, 278–281.
141 Cartoni, G.; Coccioli, F.; Jasionowska, R. (1995) Capillary electrophoresis separation of phenolic acids. *J. Chromatogr.* **709**, 209–214.
142 Sjoholm, E.; Nilvebrant, N. (1993) Characterization of dissolved kraft lignin by capillary zone electrophoresis. *J. Wood Chem. Tech.* **13**, 529–544.
143 Chang, H.; Yeung, E.S. (1993) Voltage programming in capillary zone electrophoresis. *J. Chromatogr.* **632**, 149–155.
144 Sudor, J.; Pospichal, J.; Deml, M.; Bocek, P. (1991) Step change of co-ion, a new option in capillary zone electrophoresis. *J. Chromatogr.* **545**, 331–336.
145 Bachmann, K.; Gottlicher, B.; Haag, I.; Hannina, M.; Hensel, W. (1994) Sample stacking for charged phenol derivatives in capillary zone electrophoresis. *Fresenius J. Anal. Chem.* **350**, 368–371.
146 Bachmann, K.; Gottlicher, B.; Haag, I.; Hensel, W. (1994) Improvement of reproducibility in capillary zone electrophoresis by sequential application of high voltage and pressure. *Fresenius J. Anal. Chem.* **350**, 716–717.
147 Benz, N.J.; Fritz, J.S. (1995) Optimization of separations of alkyl-substituted phenolate anions by capillary zone electrophoresis. *J. High Resolut. Chromatogr.* **18**, 175–178.
148 Harms, J.; Schwedt, G. (1994) Applications of capillary electrophoresis in element speciation. Analysis of plant and food extracts. *Fresenius J. Anal. Chem.* **350**, 93–100.
149 Masselter, S.; Zemann, A.J. (1995) Influence of organic solvents in coelectroosmotic capillary electrophoresis of phenols. *Anal. Chem.* **67**, 1047–1053.
150 Masselter, S.; Zemann, A.J. (1995) Influence of buffer electrolyte pH on the migration behavior of phenolic compounds in co-electroosmotic capillary electrophoresis. *J. Chromatogr.* **693**, 359–365.
151 Masselter, S.M.; Zemann, A.J.; Bobleter, O. (1993) Separation of cresols using coelectroosmotic capillary electrophoresis. *Electrophoresis* **14**, 36–39.
152 Shohat, D.; Grushka, E. (1994) Use of calixarenes to modify selectivities in capillary electrophoresis. *Anal. Chem.* **66**, 747–750.
153 Bao, L.; Dasgupta, P.K. (1992) Membrane interfaces for sample introduction in capillary zone electrophoresis. *Anal. Chem.* **64**, 991–996.
154 Otsuka, K.; Terabe, S.; Ando, T. (1985) Electrokinetic chromatography with micellar solutions. Retention behavior and separation of chlorinated phenols. *J. Chromatogr.* **348**, 39–47.
155 Rony, C.; Jacquier, J.C.; Desbene, P.L. (1994) Analytical study of biomass pyrolysis oils. II. Optimization of analytical conditions for the phenolic fraction using micellar electrokinetic chromatography. *J. Chromatogr.* **669**, 195–204.
156 Khaledi, M.G.; Smith, S.C.; Strasters, J.K. (1991) Micellar electrokinetic capillary chromatography of acidic solutes: Migration behavior and optimization strategies. *Anal. Chem.* **63**, 1820–1830.
157 Landman, A.; Sun, P.; Hartwick, R.A. (1994) Enhanced micellar electrokinetic capillary chromatography separations on anionic polymer-coated capillary with pH-independent electroosmotic flow. *J. Chromatogr.* **669**, 259–262.
158 Brumley, W.C.; Jones, W.J. (1994) Comparison of micellar electrokinetic chromatography (MEKC) with capillary gas chromatography in the separation of phenols, anilines

and polynuclear aromatics. Potential field-screening applications of MEKC. *J. Chromatogr.* **680**, 163–173.
159 Katsuta, S.; Tsumura, T.; Saitoh, K.; Teramae, N. (1995) Control of selectivity in micellar electrokinetic chromatography by modification of sodium dodecyl sulfate micelles with organic hydroxy compounds. *J. Chromatogr.* **705**, 319–324.
160 Aiken, J.H.; Huie, C.W. (1993) Effects of 1-alkanols on separation performance in micellar electrokinetic capillary chromatography. *J. Microcol. Sep.* **5**, 95–99.
161 Muijselaar, P.G.H.M.; Claessens, H.A.; Cramers, C.A.; Jansen, J.F.G.A.; Meijer, E.W.; de Brabander-van den Berg; E.M.M.; van der Wal, S. (1995) Dendrimers as pseudo-stationary phases in electrokinetic chromatography. *J. High Resolut. Chromatogr.* **18**, 121–123.
162 Terabe, S.; Matsubara, N.; Ishihama, Y.; Okada, Y. (1992) Microemulsion electrokinetic chromatography: comparison with micellar electrokinetic chromatography. *J. Chromatogr.* **608**, 23–29.
163 Chao, Y.; Whang, C. (1994) Capillary zone electrophoresis of eleven priority phenols with indirect fluorescence detection. *J. Chromatogr.* **663**, 229–237.
164 Gaitonde, C.; Pathak, P.V. (1990) Capillary zone electrophoretic separation of chlorophenols in industrial waste water with on-column electrochemical detection. *J. Chromatogr.* **514**, 389–393.
165 O'Shea, T.J.; Greenhagen, R.D.; Lunte, S.M.; Lunte, C.E.; Smyth, M.R.; Radzik, D.M.; Watanabe, N. (1992) Capillary electrophoresis with electrochemical detection employing an on-column nafion joint. *J. Chromatogr.* **593**, 305–312.
166 Praus, P.; Dombek, V. (1993) Separation of chlorophenols by capillary isotachophoresis. *Anal. Chim. Acta* **277**, 97–101.
167 Praus, P.; Dombek, V. (1993) β-cyclodextrin in the capillary isotachophoretic separation of chlorophenols. *Anal. Chim. Acta* **281**, 397–400.
168 Praus, P.; Dombek, V. (1993) Utilization of polyethylene glycol for the separation of chlorophenols by capillary isotachophoresis. *Anal. Chim. Acta* **283**, 917–921.
169 Praus, P. (1995) Possibilities of simple steam distillation in combination with capillary isotachophoresis for the determination of chlorophenols in river and industrial waste waters. *Anal. Chim. Acta* **302**, 39–44.
170 Evans, K.P.; Beaumont, G.L. (1993) Role of capillary electrophoresis in specialty chemical research. *J. Chromatogr.* **636**, 153–169.
171 Goebel, L.K.; McNair, H.M. (1993) Separation of ethoxylated alcohol sulfates by capillary electrophoresis using indirect UV detection. *J. Microcol. Sep.* **5**, 47–50.
172 Jones, W.R. (1993) Method development approaches for capillary ion electrophoresis. *J. Chromatogr.* **640**, 387–395.
173 Harrold, M.P.; Wojtusik, M.J.; Riviello, J.; Henson, P. (1993) Parameters influencing separation and detection of anions by capillary electrophoresis. *J. Chromatogr.* **640**, 463–471.
174 Chen, S.; Pietrzyk, D.J. (1993) Separation of sulfonate and sulfate surfactants by capillary electrophoresis: Effect of buffer cation. *Anal. Chem.* **65**, 2770–2775.
175 Nielen, M.W.F. (1991) Quantitative aspects of indirect UV detection in capillary zone electrophoresis. *J. Chromatogr.* **588**, 321–326.
176 Romano, J.; Jandik, P.; Jones, W.R.; Jackson, P.E. (1991) Optimization of inorganic capillary electrophoresis for the analysis of anionic solutes in real samples. *J. Chromatogr.* **546**, 411–421.
177 Dionex Corporation (1991) *Surfactant Analysis by Capillary Electrophoresis*. Applications Note 74.
178 Thompson, C.O.; Trenerry, V.C.; Kemmery, B. (1995) Determination of cyclamate in low joule foods by capillary zone electrophoresis with indirect ultraviolet detection. *J. Chromatogr.* **704**, 203–210.
179 Gibbons, J.M.; Hoke, S.H. (1994) Capillary zone electrophoresis with indirect UV detection: Determination of sodium dodecyl sulfate in simulated stream water. *J. High Resolut. Chromatogr.* **17**, 665–667.
180 Tsuda, T.; Nomura, K.; Nakagawa, G. (1983) Separation of organic and metal ions by high-voltage capillary electrophoresis. *J. Chromatogr.* **264**, 385–392.
181 Brumley, W.C. (1992) Qualitative analysis of environmental samples for aromatic

sulfonic acids by high-performance capillary electrophoresis. *J. Chromatogr.* **603**, 267–272.
182 Damm, J.B.L.; Overklift, G.T.; Vermeulen, B.W.M. (1992) Separation of natural and synthetic heparin fragments by high-performance capillary electrophoresis. *J. Chromatogr.* **608**, 297–309.
183 Williams, S.J.; Goodall, D.M.; Evans, K.P. (1993) Analysis of anthraquinone sulphonates. Comparison of capillary electrophoresis with high-performance liquid chromatography. *J. Chromatogr.* **629**, 379–384.
184 Croft, S.N.; Lewis, D.M. (1992) Analysis of reactive dyes and related derivatives using high-performance capillary electrophoresis. *Dyes Pigm.* **18**, 309–317.
185 Ryder, D.S. (1992) Determination of sodium vinyl sulphonate in watersoluble polymers using capillary zone electrophoresis. *J. Chromatogr.* **605**, 143–147.
186 Oxspring, D.A.; Smyth, W.F.; Marchant, R. (1995) Comparison of reversed-polarity capillary electrophoresis and adsorptive stripping voltammetry for detection and determination of reactive textile dyes. *Analyst* **120**, 1995–2000.
187 Altria, K.D.; Gill, I.; Howells, J.S.; Luscombe, C.N.; Williams, R.Z. (1995) Trace analysis of detergent residues by capillary electrophoresis. *Chromatographia.* **40**, 527–531.
188 Desbene, P.L.; Rony, C.; Desmazieres, B.; Jacquier, J.C. (1992) Analysis of alkylaromatic sulphonates by high-performance capillary electrophoresis. *J. Chromatogr.* **608**, 375–383.
189 Desbene, P.L.; Rony, C.M. (1995) Determination by high-performance capillary electrophoresis of alkylaromatics used as bases of sulfonation in the preparation of industrial surfactants. *J. Chromatogr.* **689**, 107–121.
190 Cheung, A.P.; Nguyenle, T.; Hettiarachchi, K. (1993) Liquid chromatography and capillary electrophoresis analysis of polyanionic quinobene. *J. Pharm. Biomed. Anal.* **11**, 1261–1267.
191 Zhang, Y.; Warner, I.M. (1994) Separation of water soluble *p*-sulfonated calixarenes 4, 6, and 8 and 4-hydroxybenzene sulfonate by use of capillary zone electrophoresis. *J. Chromatogr.* **688**, 293–300.
192 Chiesa, C.; Horvath, C. (1993) Capillary zone electrophoresis of malto-oligosaccharides derivatized with 8-aminonaphthalene-1,3,6-trisulfonic acid. *J. Chromatogr.* **645**, 337–352.
193 Chiesa, C.; O'Neill, R.A. (1994) Capillary zone electrophoresis of oligosaccharides derivatized with various aminonaphthalene sulfonic acids. *Electrophoresis* **15**, 1132–1140.
194 Terabe, S.; Isemura, T. (1990) Ion-exchange electrokinetic chromatography with polymer ions for the separation of isomeric ions having identical electrophoretic mobilities. *Anal. Chem.* **62**, 650–652.
195 Lores Aguin, M.; Vindevogel, J.; Sandra, P. (1993) Utilisation of the bisulfite addition reaction for the separation of neutral aldehydes by capillary electrophoresis. *Chromatographia* **37**, 451–454.
196 Poli, J.B.; Schure, M.R. (1992) Separation of poly (styrenesulfonates) by capillary electrophoresis with polymeric additives. *Anal. Chem.* **64**, 896–904.
197 Sudor, J.; Novotny, M. (1994) Pulsed-field capillary electrophoresis: optimizing separation parameters with model mixtures of sulfonated polystyrenes. *Anal. Chem.* **66**, 2139–2147.
198 Burkinshaw, S.M.; Hinks, D.; Lewis, D.M. (1993) Capillary zone electrophoresis in the analysis of dyes and other compounds employed in the dye-manufacturing and dye-using industries. *J. Chromatogr.* **640**, 413–417.
199 Kerns, R.J.; Vlahov, I.R.; Linhardt, R.J. (1995) Capillary electrophoresis for monitoring chemical reactions: sulfation and synthetic manipulation of sulfated carbohydrates. *Carbohydr. Res.* **267**, 143–152.
200 Shah, P.A.; Quinones, L. (1995) Validation of a micellar electrokinetic capillary chromatography (MECC) method for the determination of *p*-toluenesulfonic acid impurity in a pharmaceutical intermediate. *J. Liq. Chromatogr.* **18**, 1349–1362.
201 Brumley, W.C.; Brownrigg, C.M.; Grange, A.H. (1994) Capillary liquid chromatography-mass spectrometry and micellar electrokinetic chromatography as complementary techniques in environmental analysis. *J. Chromatogr.* **680**, 635–644.

202 Sun, P.; Landman, A.; Barker, G.E.; Hartwick, R.A. (1994) Synthesis and evaluation of anionic polymer-coated capillaries with pH-independent electroosmotic flows for capillary electrophoresis. *J. Chromatogr.* **685**, 303–312.
203 Nichols, W.; Zweigenbaum, J.; Garcia, M.J.; Henion, J. (1992) CE–MS for industrial applications using a liquid juntion with ion-spray and CF–FAB mass spectrometry. *LC–GC* **10**, 676–686.
204 Lee, E.D.; Muck, W.; Henion, J.D. (1989) Capillary zone electrophoresis/tandem mass spectrometry for the determination of sulfonated azo dyes. *Biomed. Environ. Mass Spectrom.* **18**, 253–257.
205 Pfeffer, W.D.; Yeung, E.S. (1991) Electroosmotically driven electrochromatography of anions having similar electrophoretic mobilities by ion pairing. *J. Chromatogr.* **577**, 125–136.
206 Garner, T.W.; Yeung, E.S. (1993) Increased selectivity for electrochromatography by dynamic ion exchange. *J. Chromatogr.* **640**, 397–402.
207 Stefansson, M.; Novotny, M. (1994) Modification of the electrophoretic mobility of neutral and charged polysaccharides. *Anal. Chem.* **66**, 3466–3471.
208 Avdalovic, N.; Pohl, C.A.; Rocklin, R.D.; Stillian, J.R. (1993) Determination of cations and anions by capillary electrophoresis combined with suppressed conductivity detection. *Anal. Chem.* **65**, 1470–1475.
209 Swaile, D.F.; Copper, C.L.; Sepaniak, M.J.; Burton, D.E.; Powell, L.L. (1994) The use of fluorescent probe molecules for detection in capillary electrokinetic separations. *Talanta* **41**, 1499–1505.
210 Tribet, C.; Gaboriaud, R.; Gareil, P. (1992) Determination of C8–C20 saturated anionic and cationic surfactant mixtures by capillary isotachophoresis with conductivity detection. *J. Chromatogr.* **609**, 381–390.
211 Wainwright, A. (1990) Capillary electrophoresis applied to the analysis of pharmaceutical compounds. *J. Microcol. Sep.* **2**, 166–175.
212 Ricci, M.C.; Cross, R.F. (1993) Capillary electrophoresis separation of sulphonamides and dihydrofolate reductase inhibitors. *J. Microcol. Sep.* **5**, 207–215.
213 Ng, C.L.; Lee, H.K.; Li, S.F.Y. (1992) Systematic optimization of capillary electrophoretic separation of sulphonamides. *J. Chromatogr.* **598**, 133–138.
214 Ng, C.L.; Lee, H.K.; Li, S.F.Y. (1993) Determination of sulphonamides in pharmaceuticals by capillary electrophoresis. *J. Chromatogr.* **632**, 165–170.
215 Ng, C.L.; Ong, C.P.; Lee, H.K.; Li, S.F.Y. (1993) Systematic optimization of capillary electrophoretic separations using the overlapping resolution mapping scheme. *J. Microcol. Sep.* **5**, 191–197.
216 Dang, Q.; Sun, Z.; Ling, D. (1992) Separation of sulphonamides and determination of the active ingredients in tablets by micellar electrokinetic capillary chromatography. *J. Chromatogr.* **603**, 259–266.
217 Pleasance, S.; Thibault, P.; Kelly, J. (1992) Comparison of liquid-junction and coaxial interfaces for capillary electrophoresis–mass spectrometry with application to compounds of concern to the aquaculture industry. *J. Chromatogr.* **591**, 325–339.
218 Perkins, J.R.; Parker, C.E.; Tomer, K.B. (1992) Nanoscale separations combined with electrospray ionization mass spectrometry: sulfonamide determination. *J. Am. Soc. Mass Spectrom.* **3**, 139–149.
219 Ackermans, M.T.; Beckers, J.L.; Everaerts, F.M.; Hoogland, H.; Tomassen, M.J. (1992) Determination of sulphonamides in pork meat extracts by capillary zone electrophoresis. *J. Chromatogr.* **596**, 101–109.
220 Nielen, M.W.F.; Mensink, M.J.A. (1991) Separation of benzothiazole sulfenamides using micellar electrokinetic capillary chromatography. *J. High Resolut. Chromatogr.* **14**, 417–419.
221 Dinelli, G.; Vicari, A.; Bonetti, A. (1995) Separation of sulfonylurea metabolites in water by capillary electrophoresis. *J. Chromatogr.* **700**, 195–200.
222 Dinelli, G.; Vicari, A.; Brandolini, V. (1995) Detection and quantitation of sulfonylurea herbicides in soil at the ppb level by capillary electrophoresis. *J. Chromatogr.* **700**, 201–207.
223 Krynitsky, A.J.; Swineford, D.M. (1995) Determination of sulfonylurea herbicides in grains by capillary electrophoresis. *J. Assoc. Off. Anal. Chem. Int.* **78**, 1091–1096.

224 Garcia, F.; Henion, J. (1992) Fast capillary electrophoresis–ion spray mass spectrometric determination of sulfonylureas. *J. Chromatogr.* **606**, 237–247.
225 O'Shea, T.J.; Lunte, S.M. (1993) Selective detection of free thiols by capillary electrophoresis–electrochemistry using a gold/mercury amalgam microelectrode. *Anal. Chem.* **65**, 247–250.
226 Zhou, J.; O'Shea, T.J.; Lunte, S.L. (1994) Simultaneous detection of thiols and disulfides by capillary electrophoresis–electrochemical detection using a mixed valence ruthenium cyanide-modified microelectrode. *J. Chromatogr.* **680**, 271–277.
227 Ling, B.L.; Baeyens, W.R.G.; Dewaele, C. (1991) Capillary zone electrophoresis with ultraviolet and fluorescence detection for the analysis of thiols. Application to mixtures and blood. *Anal. Chim. Acta* **255**, 283–288.
228 Ling, B.L.; Baeyens, W.R.G.; Dewaele, C. (1991) Comparison of micro-LC and capillary zone electrophoresis for the analysis of thiols. *J. High Resolut. Chromatogr.* **14**, 169–173.
229 Shamsi, S.A.; Danielson, N.D. (1995) Ribonucleotide electrolytes for capillary electrophoresis of polyphosphates and polyphosphonates with indirect photometric detection. *Anal. Chem.* **67**, 1845–1852.
230 Pianetti, G.A.; Taverna, M.; Mahuzier, G.; Baylocq-Ferrier, D. (1993) Determination of alkylphosphonic acids by capillary zone electrophoresis using indirect UV detection. *J. Chromatogr.* **630**, 371–377.
231 Henshall, A.; Harrold, M.P.; Tso, J.M.Y. (1992) Separation of inositol phosphates by capillary electrophoresis. *J. Chromatogr.* **608**, 413–419.
232 Nardi, A.; Cristalli, M.; Desiderio, C.; Ossicini, L.; Shukla, S.K.; Fanali, S. (1992) Indirect UV photometric detection in capillary zone electrophoresis for the determination of phytate in soybeans. *J. Microcol. Sep.* **4**, 9–11.
233 Buscher, B.A.P.; Irth, H.; Andersson, E.; Tjaden, U.R.; van der Greef, J. (1994) Determination of inositol phosphates in fermentation broth using capillary zone electrophoresis with indirect UV detection. *J. Chromatogr.* **678**, 145–150.
234 Ingvardsen, L.; Michaelsen, S.; Sorensen, H. (1994) Analysis of individual phospholipids by high-performance capillary electrophoresis. *J. Am. Oil Chem. Soc.* **71**, 183–188.
235 Robins, W.H.; Wright, B.W. (1994) Capillary electrophoretic separation of organophosphonic acids using borate esterification and direct UV detection. *J. Chromatogr.* **680**, 667–673.
236 Tomita, M.; Okuyama, T.; Nigo, Y.; Uno, B.; Kawai, S. (1991) Determination of glyphosate and its metabolite, (aminomethyl) phosphonic acid, in serum using capillary electrophoresis. *J. Chromatogr.* **571**, 324–330.
237 Chankvetadze, B.; Endresz, G.; Blaschke, G. (1995) Enantiomeric resolution of anionic R/S-1,1'-binaphthyl-2,2'-diyl hydrogen phosphate by capillary electrophoresis using anionic cyclodextrin derivatives as chiral selectors. *J. Chromatogr.* **704**, 234–237.
238 Andersson, P.E.; Pfeffer, W.D.; Blomberg, L.G. (1995) Indirect detection in capillary electrophoresis. Comparison between indirect UV and indirect laser-induced fluorescence detection for the determination of isoprenyl pyrophosphates. *J. Chromatogr.* **699**, 323–330.
239 Kostiainen, R.; Bruins, A.P.; Hakkinen, V.M.A. (1993) Identification of degradation products of some chemical warfare agents by capillary electrophoresis–ion spray mass spectrometry. *J. Chromatogr.* **634**, 113–118.
240 Sanger-van de Griend, C.E.; Kientz, Ch.E.; Brinkman, U.A.Th. (1994) Capillary electrophoresis coupled on-line with flame photometric detection. *J. Chromatogr.* **673**, 299–302.
241 Velasquez, R.; Tsika, D.; Brunner, G. (1993) Analytical capillary isotachophoresis of bis(2-ethylhexyl)hydrogenphosphate and 2-ethylhexyl dihydrogenphosphate. *J. Chromatogr.* **639**, 317–321.
242 Sepaniak, M.J.; Burton, D.E.; Maskarinec, M.P. (1987) Micellar electrokinetic capillary chromatography. In *Ordered Media in Chemical Separations*, Hinze, W.L.; Armstrong, D.W. (eds), American Chemical Society, Washington.
243 Balchunas, A.T.; Sepaniak, M.J. (1988) Gradient elution for micellar electrokinetic capillary chromatography. *Anal. Chem.* **60**, 617–621.
244 Cole, R.; Sepaniak, M. (1992) The use of bile salt surfactants in micellar electrokinetic capillary chromatography. *LC–GC* **10**, 380–385.

245 Mattusch, J.; Huhn, G.; Wennrich, R. (1995) Sensitive laser induced fluorescence detection of polyamine-fluoresceinisothiocyanate-derivatives after capillary zone electrophoretic separation. *Fresenius J. Anal. Chem.* **351**, 732—738.
246 Beck, W.; Engelhardt, H. (1992) Capillary electrophoresis of organic and inorganic cations with indirect UV detection. *Chromatographia* **33**, 313–316.
247 Riviello, J.M.; Harrold, M.P. (1993) Capillary electrophoresis of inorganic cations and low-molecular-mass amines using a copper-based electrolyte with indirect UV detection. *J. Chromatogr.* **652**, 385–392.
248 Ma, Y.; Zhang, R.; Cooper, C.L. (1992) Indirect photometric detection of polyamines in biological samples separated by high-performance capillary electrophoresis. *J. Chromatogr.* **608**, 93–96.
249 Quang, Q.; Khaledi, M.G. (1994) Prediction and optimization of metal cations by capillary electrophoresis with indirect UV detection. *J. Chromatogr.* **659**, 459–466.
250 Oehrle, S.A. (1994) Analysis of cationic ingredients and degradation products in liquid propellants by capillary ion electrophoresis (CIE). *J. Energ. Mater.* **12**, 197–209.
251 Jones, D.; Scarborough, A.; Tier, C.M. (1994) Quantitative determination of a dipeptide in personal wash liquid by capillary electrophoresis. *J. Chromatogr.* **661**, 1–6.
252 Gross, L.; Yeung, E.S. (1990) Indirect fluorometric detection of cations in capillary zone electrophoresis. *Anal. Chem.* **62**, 427–431.
253 Gilman, S.D.; Silverman, C.E.; Ewing, A.G. (1994) Electrogenerated chemiluminescence detection for capillary electrophoresis. *J. Microcol. Sep.* **6**, 97–106.
254 Rowe, R.C.; Wren, S.A.; McKillop, A.G. (1994) Molecular size/shape effects in the separation of the monosubstituted alkyl pyridines using capillary electrophoresis. *Electrophoresis* **15**, 635–639.
255 McKillop, A.G.; Smith, R.M.; Rowe, R.C.; Wren, S.A.C. (1995) Separation and identification of the Z and E isomers of 2-(3-pentenyl) pyridine by capillary electrophoresis and nuclear magnetic resonance spectroscopy. *J. Chromatogr.* **700**, 69–72.
256 Terabe, S.; Ozaki, H.; Otsuka, K.; Ando, T. (1985) Electrokinetic chromatography with 2-*O*-carboxymethyl-beta-cyclodextrin as a moving 'stationary' phase. *J. Chromatogr.* **332**, 211–217.
257 Chankvetadze, B.; Endresz, G.; Blaschke, G. (1995) Enantiomeric resolution of chiral imidazole derivatives using capillary electrophoresis with cyclodextrin-type buffer modifiers. *J. Chromatogr.* **700**, 43–49.
258 Walbroehl, Y.; Wagner, J. (1994) Enantiomeric resolution of primary amines by capillary electrophoresis and high-performance liquid chromatography using chiral crown ethers. *J. Chromatogr.* **680**, 253–261.
259 Wren, S. (1991) Optimization of pH in the electrophoretic separation of 2-, 3-, and 4-methyl pyridines. *J. Microcol. Sep.* **3**, 147–154.
260 Northrop, D.M.; Martire, D.E.; MacCrehan, W.A. (1991) Separation and identification of organic gunshot and explosive constituents by micellar electrokinetic capillary electrophoresis. *Anal. Chem.* **63**, 1038–1042.
261 Mussenbrock, E.; Kleibohmer, W. (1995) Separation strategies for the determination of residues of explosives in soils using micellar electrokinetic capillary chromatography. *J. Microcol. Sep.* **7**, 107–116.
262 Ong, C.P.; Ng, C.L.; Lee, H.K.; Li, S.F.Y. (1994) Separation of imidazole and its derivatives by capillary electrophoresis. *J. Chromatogr.* **686**, 319–324.
263 Takeda, S.; Wakida, S.; Yamane, M.; Kawahara, A.; Higashi, K. (1993) Separation of aniline derivatives by micellar electrokinetic chromatography. *J. Chromatogr.* **653**, 109–114.
264 Pyell, U.; Butehorn, U. (1995) Optimization strategies in micellar electrokinetic chromatography. One-parameter optimizations of the concentration of sodium dodecyl sulfate and the concentration of modifiers (urea and glucose). Chromatographia **40**, 175–184.
265 Brumley, W.C.; Brownrigg, C.M. (1994) Applications of MEKC in the determination of benzidines following extraction from water, soil, sediment, and chromatographic adsorbents. *J. Chromatogr. Sci.* **32**, 69–75.
266 Liu, P.Z.; Malik, A.; Kuchar, M.C.J.; Lee, M.L. Polyacrylamide-modified polypropylene hollow fibers for capillary electrophoresis. *J. Microcol. Sep.* **6**, 581–589.
267 Nielen, M.W.F. (1992) Separation of phenylenediamine isomers by capillary zone electrophoresis. *J. Chromatogr.* **625**, 387–391.

268 Weiss, C.S.; Hazlett, J.S.; Datta, M.H.; Danzer, M.H. (1992) Determination of quaternary ammonium compounds by capillary electrophoresis using direct and indirect UV detection. *J. Chromatogr.* **608**, 325–332.
269 Nielen, M.W.F. (1993) Capillary zone electrophoresis using a hollow polypropylene fiber. *J. High Resolut. Chromatogr.* **16**, 62–64.
270 Kolecek, J.; Riha, V.; Vytras, K. (1993) Determination of 2-chloroethyltrimethylammonium chloride in Retacel by ion-selective electrode potentiometry and capillary isotachophoresis. *Anal. Chim. Acta* **273**, 431–433.
271 Kaniansky, D.; Ivanyl, F.; Onuska, F.I. (1994) On-line isotachophoretic sample pretreatment in ultratrace determination of paraquat and diquat in water by capillary zone electrophoresis. *Anal. Chem.* **66**, 1817–1824.
272 Wigfield, Y.Y.; McCormack, K.A.; Grant, R. (1993) Simultaneous determination of residues of paraquat and diquat in potatoes using high-performance capillary electrophoresis with ultraviolet detection. *J. Agric. Food Chem.* **41**, 2315–2318.
273 Galceran, M.T.; Carneiro, M.C.; Puignou, L. (1994) Capillary electrophoresis of quaternary ammonium ion herbicides: paraquat, diquat and difenzoquat. *Chromatographia* **39**, 581–586.
274 Carniero, M.C.; Puignou, L.; Galceran, M.T. (1994) Comparison of capillary electrophoresis and reversed-phase ion-pair high-performance liquid chromatography for the determination of paraquat, diquat and difenzoquat. *J. Chromatogr.* **669**, 217–224.
275 Schmitt, Ph.; Freitag, D.; Sanlaville, Y.; Lintelmann, J.; Kettrup, A. (1995) Capillary electrophoretic study of atrazine photolysis. *J. Chromatogr.* **709**, 215–225.
276 Foret, F.; Sustacek, V.; Bocek, P. (1990) Separation of some triazine herbicides and their solvolytic products by capillary zone electrophoresis. *Electrophoresis* **11**, 95–97.
277 Cai, J.; El Rassi, Z. (1992) On-line preconcentration of triazine herbicides with tandem octadecyl capillaries–capillary zone electrophoresis. *J. Liq. Chromatogr.* **15**, 1179–1192.
278 Susse, H.; Muller, H. (1995) Application of micellar electrokinetic capillary chromatography to the analysis of pesticides. *Fresenius J. Anal. Chem.* **352**, 470–473.
279 Smith, J.T.; El Rassi, Z. (1994) Micellar electrokinetic capillary chromatography with *in situ* charged micelles. II. Evaluation and comparison of octylmaltoside and octanoylsucrose surfactants as anionic borate complexes in the separation of herbicides. *J. Microcol. Sep.* **6**, 127–138.
280 Perkin Elmer Corporation (1994) Separation of imidazolinone herbicides by capillary electrophoresis. *Perkin Elmer Application Note* CEENV-3.
281 Smith, R.D.; Barinaga, C.J.; Udseth, H.R. (1988) Improved electrospray ionization interface for capillary zone electrophoresis–mass spectrometry. *Anal. Chem.* **60**, 1948–1952.
282 Udseth, H.R.; Lee, J.A.; Smith, R.D. (1988) Capillary isotachophoresis/mass spectrometry. *Anal. Chem.* **61**, 228–232.
283 Burgstaller, W.; Schinner, F. (1992) Isotachophoretic analysis of the lipophilic cation tetraphenylphosphoniumbromide. *Fresenius J. Anal. Chem.* **343**, 893–895.
284 Dionex Corporation (1995) *Ionphore electrolyte buffers for CE*. Product brochure No. LPN-034881-01.
285 Amankwa, L.N.; Scholl, J.; Kuhr, W.E. (1990) Characterization of the oligomeric dispersion of poly(oxyalkylene)diamine polymers by precolumn derivatization and capillary zone electrophoresis with fluorescence detection. *Anal. Chem.* **62**, 2189–2193.
286 Bergholdt, A.; Overgaard, J.; Colding, A. (1993) Separation of D-galactonic and D-gluconic acids by capillary zone electrophoresis. *J. Chromatogr.* **644**, 412–415.
287 Liu, Y.-M.; Sheu, S.-J. (1994) Separation of aromatic acids by reversed electroosmotic flow capillary electrophoresis. *J. Chromatogr.* **663**, 239–243.
288 Zhang, R.; Cooper, C.L.; Ma, Y. (1993) Determination of total polyamines in tumor cells by high-performance capillary electrophoresis with indirect photometric detection. *Anal. Chem.* **65**, 704–706.

Part Six

Food Analysis Applications

38 Food analysis by capillary electrophoresis
P.F. CANCALON

38.1 Introduction

Capillary electrophoresis (CE) is a recent technique that has become widely available only in the last few years. During most of the 1980s, CE was used mainly as a research instrument. During this period, different types of CE analysis were developed and analytical techniques were perfected (for a review of earlier work see Ewing *et al.* [1]). Many commercial instruments have become available; a description of the main features of most CE has been published by Oda *et al.* [2]. In the last few years, CE has been applied to more practical tasks, particularly in the areas of pharmaceutical and food analysis (summarized in Table 38.1). This has been reflected in the publication of validation reports for various CE procedures. Altria and co-workers [3, 4] reported the validation of a CE method to determine the level of potassium in acidic drugs and Taylor and Reid [5] for the analysis of antimalarial drugs. Such rapid development is due to several factors, but especially to speed and versatility. These are important in the monitoring of products where the repeated analysis of compounds having different chemical properties is required (for review of earlier work in the area of food analysis see Zeece [6]).

A rapid summary of the major CE techniques should help to understand food CE applications. Free solution capillary electrophoresis (FSCE) or capillary zone electrophoresis (CZE) is based on the early work by Tiselius [7]. CZE consists of a microbore capillary usually made of fused silica. The tube is filled with a buffer and subjected to an electric field. Under these conditions the positive charged species migrate toward the cathode at a rate determined by their mass-to-charge ratios. However, an electroosmotic flow (EOF) or net movement of the buffer toward the cathode is generated within the tube by the ionized capillary wall silanol groups. As a result, the neutral molecules migrate as a single group transported by the EOF, and each negative species moves toward the cathode at a rate determined by the difference between the EOF and its anodic attraction. Since any charged species can move in an electric field, CE has been used to separate widely different compounds ranging from lithium [8] to polymers with a molecular weight of 2 000 000 [9] and even viable bacteria [10].

The free solution electrophoresis described above is the most widely used but other procedures have been developed by modifying some factors

Table 38.1 Analytical conditions of food components

Methods Separation conditions, voltage applied, current, capillary length with i.d., detection	Compounds	Matrix	Reference
CE 8 mM Benzyltrimethylammonium chloride and 4 mM α-hydroxyisobutyric acid at pH 4.6 with NaOH, 30 kV, 60 cm to × 75 μm i.d., 214 nm	Metal ions	Standard	8
CE 6 mM N,N-Dimethylbenzylamine and 4 mM α-hydroxyisobutyric acid at pH 4.6 with NaOH, 30 kV, 60 cm × 75 μm i.d., 214 nm	Metal ions	Standard	8
CZE 5 mM Sodium chromate electrolyte containing OFM-BT (0.4 mM), 20 kV, 60 cm × 75 μm i.d., 214 nm (direct)	Anions	Orange juice, orange pulpwash	47
CZE 5 mM Sodium chromate electrolyte containing OFM-BT (0.4 mM) containing $Ca(OH)_2$ (50 μl of 10% solution), 20 kV, 60 cm × 75 μm i.d., 254 nm (indirect)	Anions	Orange juice, orange pulpwash	47
ITP Leading electrolyte: 10 mM HCl, counter ion, ε-aminocaproic acid, pH 4.5; additive, methylhydroxyethylcellulose (0.1%); terminating electrolyte, 5 mM caproic acid–5 mM histidine (pH 4–5), 50 μA	Organic acids	Green pea, carrot, celery, paprika	49
CZE 5 mM Phthalate at pH 5.6 with 0.2–0.6 mM Ca^{2+}, −20 kV, 53 cm × 75 μm i.d., 254 nm	Organic acids	Chicory root (Raftisweet)	54
CE 100 mM $Na_2B_4O_7$, pH 9.9, 10 kV, 50°C, 50 cm × 50 μm i.d., 195 nm	Carbohydrates	Mature seeds of pea lines	65
CE 20 mM Citrate–phosphate, pH 6–7, 20 kV, 25°C, 28 cm × 50 μm i.d., 560 nm	Potato starch, amaizo V, corn starch, maltodextrin, amylose, amylopectin	Saccharides	66
CZE 6 mM Sorbate, pH 12.2 amaizo V, 10 kV, 15°C, 83 cm × 50 μm i.d., 256 nm	Monosaccharide, N-acetyl-monosaccharide, uronic acid	Fruit juices	67
CZE 6 mM Sorbate at pH 5.0, 20 kV (1.6 μA), 92.4 cm × 50 μm i.d., 254 nm	D-galactonic acid, D-gluconic acids	Standards	68

INTRODUCTION

Table 38.1 *Continued*

Methods Separation conditions, voltage applied, current, capillary length with i.d., detection	Compounds	Matrix	Reference
CE 50 mM NaOH, 30 kV, 80 cm × 25 µm i.d., electrochemical detection (Cu disk + 0.6 V vs. Ag/AgCl)	Sugars, alditols Sugar acids	Standards	78
CE 35 mM Solution of borax pH 9.3, 100 cm × 75 µm i.d. 30°C, 21 kV for initial 12 min and 25 kV for the remainder of analysis, 200–360 nm (UV), 370–500 nm (vis)	Flavonoids, amino acids, ascorbic acid, feruloyl and sinapyl glucose	Juice samples	81
CE 23 cm × 23 µm i.d., 10 kV (52 µA), 0.5 M sodium phosphate, 4.0 M urea pH 7.0, 23°C, 200 nm	Casein, lactalbumin, lactoglobulin	Milk proteins	89
CZE 10 mM Sodium phosphate containing 6 M urea and 0.05% methylhydroxyethylcellulose 30 000 (mol. wt), 20 kV, 57 cm × 50 µm, 45°C, 214 nm	Casein, lactalbumin, lactoglobulin	Milk proteins	90
CE 40 mM Tris, 40 mM boric acid, 0.1% (w/v) SDS and 10% (w/v) PEG 8000 (mol. wt), pH 8.6, 16 kV (15 µA), 20 cm × 25 µm, 21°C, 214 nm	BSA, lactalbumin, lactoglobulin	Bovine whey, proteins	91
CZE 50 mM Sodium borate, pH 9.0, containing 0.1% ethanolamine and 0.1% Tween-20, 20 kV, 50 cm × 50 µm, 215 nm	BSA, lactalbumin, lactoglobulin	Bovine whey, proteins	96
CZE 50 mM 2-(N-Morpholino)ethanesulfonic acid pH 8.0 with 0.1% Tween-20, 0.1% ethanolamine, 20 kV, 50 cm × 50 µm, 215 nm	BSA, lactalbumin, lactoglobulin	Bovine whey, proteins	96
CZE 150 mM Sodium borate, pH 8.5 plus 0.05% Tween-20, 16 kV (15 µA), 50 cm × 50 µm, conductivity	BSA. lactalbumin, lactoglobulin, IgG	Bovine whey proteins	97
CE 100 mM Sodium borate, pH 8.0 plus 60 mM sodium borate, 10 kV, 30°C, 50 cm × 50 µm, 200 nm	Lysozyme, conalbumin, ovalbumin	Egg white proteins	98
CE 300 mM Borate buffer, pH 10.0, 12 kV (26 µA), 23°C, 25 cm × 20 µm, 200 nm	Lysozyme, conalbumin, ovalbumin, globulin, ovomucoid	Egg white proteins	99

Table 38.1 *Continued*

Methods Separation conditions, voltage applied, current, capillary length with i.d., detection	Compounds	Matrix	Reference
CZE 500 mM Phosphate buffer, pH 7.4, 10 kV, 20°C, 20 cm × 20 μm, 200 nm	Sarcoplasmic protein	Fish muscle proteins	108
CZE 100 mM Phosphate buffer, pH 2.44, 25°C, 214 nm, 57 cm × 75 μm, 0–10 kV for 0.17 min, 10–17 kV for 22 min, and 15–25 kV for 1 min of three- step gradient voltage	Lysozyme, protein	Fish proteins	109
MECC 5 mM Phosphate buffer, pH 7.0, 25°C, 214 nm, 57 cm × 75 μm, 10 kV	5-Hydroxy- methyl-furfural, glycine	Millard reaction mixtures	109
MECC 50 mM Phosphate buffer, pH 7.5, 25°C, 280 nm, 37 cm × 75 μm, 10 kV	5-Hydroxy- methyl-2-furaldehyde, and 2-furaldehyde and 2-furyl methyl ketone	Grapefruit juice	112
CZE 20 mM Phosphate buffer, pH 7.0, room temperature, 254 nm, 40 cm × 100 μm, 6 kV (60 μA)	Degradation products of L-ascorbic acid	Orange juice	117
FSCE or CE 200 H_3BO_3, pH 10.5, 40°C, 270 nm, 65 cm × 50 μm, 24.4 kV (80 μA)	Flavonoid-*o*- glycosides	Querceti-3-*o*- glycosides	131
CE 500 mM Acetate, pH 4.6: acetonitrile (1:1), 25°C, 280 nm, 50 cm × 50 μm, 15 kV	Alkaloids	*Phellodendri* *Cortex*	140
CE 18 mM SDS, 2 mM sodium cholate, 12.5 mM $Na_2B_4O_7$ and 10 mM NaH_2PO_4, 30°C, 254 nm, 70 cm × 75 μm, 20 kV	Anthraquinoids, flavones, saponin, carboxylic acids	Chinese herbal extracts	141
CZE 20 mM Imidazole–acetate at pH 7, 25°C, 254 nm, 50 cm × 75 μm, 10 kV	Sulfonamides	Standards	146
CE 100 mM Phosphate buffer pH 2.5, 20°C, 200 nm, 60 cm × 75 μm, 20 kV (84 μA)	Paraquat, diquat	Potatoes	153

Table 38.1 *Continued*

Methods Separation conditions, voltage applied, current, capillary length with i.d., detection	Compounds	Matrix	Reference
CZE 0.5 mM 1-Naphthol-3,6- disulfonic acid, 30 mM acetic acid, 0.01% hydroxypropyl methylcellulose, 24°C, 214 nm, 50 cm × 75 µm, −30 kV (10 µA)	Inositol, phosphates	Fermentation broth	155
ITP Leading electrolyte: 10 mM HCl + bis-tris-propane pH 6.1, terminating electrolyte, 5 mM MES, 90 × 0.3 mm, 30 µA, conductivity detection	Phytic acid, inositol pentaphosphate	Grains	156
CZE 65 mM Sodium borate, pH 9.5, 35°C, 50 cm × 75 µm, 250 µA, induced fluorescence detection	Fumonisins B1, fumonisins B2, hydrolyzed fumonisin B1	Horse serum	157
CE 100 mM 3-(Cyclohexylamino)- 1-propanesulfonic acid buffer, pH 11, 30°C, 22 cm × 50 µm, 9.2 kV (50 µA), 250 nm	Hypoxanthine	Fish extract	159
CZE 20 mM Citrate, pH 2.5, 35°C, 72 cm × 50 µm, 27 kV, 210 nm	Histamine	Tuna extract, mahi-mahi extract	160
CZE 40 mM AMPD–BICINE buffer, pH 8.8, 25°C, 50 cm × 75 µm, 25 kV, 254 nm	Hippuric acid, sorbic acid, orotic acid	Rennet whey	161
MEKC 50 mM Sodium deoxycholate, 10 mM potassium dihydrogen ortho- phosphate, 10 mM sodium borate, pH 8.6, 27°C, 50 cm × 75 µm, 20 kV, 220 nm	Caffeine dulcin, alitamine aspartame, dehydroacetic acid, sorbic acid, benzoic acid, saccharin acesulfamine-K	Low-joule soft drinks, tomato sauce, cordials, marmalade jam, sweeteners	165
CZE 1 mM Hexadecyltrimethylammonium, hydroxide and 10 mM sodium benzoate, pH 6.6, 28°C, 50 cm × 75 µm, −20 kV, 254 nm (indirect)	Cyclamate, sorbate, aspartame, alitame, saccharin, acesulphame-K	Soft drinks, cordials	166
MEKC 50 mM Sodium borate buffer, pH 9.5 with 50 mM SDS, 50 cm × 50 µm, 24 kV, 214 nm	PG, TBHQ, BHA, BHT	Food antioxidant	169

Table 38.1 *Continued*

Methods Separation conditions, voltage applied, current, capillary length with i.d., detection	Compounds	Matrix	Reference
MEKC 15% CH_3CN and 85% 50 mM SDS/5 mM potassium dihydrogen-orthophosphate/5 mM sodium borate buffer, 40 cm × 50 μm, 30 kV, 214 nm, 25°C	Green S, brilliant blue, erythrosine B, allura red, indigo carmine, sunset yellow, azorubine, amaranth, ponceau, tartrazine	Synthetic colors	173
MEKC 15% CH_3CN and 85% 50 mM sodium deoxycholate/5 mM potassium dihydrogen ortho-phosphate/5 mM sodium borate buffer, 40 cm × 50 μm, 30 kV, 214 nm, 25°C	Green S, brilliant blue, erythrosine B, allura red, indigo carmine, sunset yellow, azorubine, amaranth, ponceau, tartrazine	Synthetic colors	173
CE 50 mM NaOH, 80 cm × 25 μm, 30 kV, electrochemical detection (Cu disk +0.6 V vs. Ag/AgCl)	Glucitol, glucose, gluconic acid, glucuronic acid, glucaric acid, acesulphame-K	Standards	174

AMPD = 2-amino-2-methyl-1,3-propane diol.

contributing to the movement of the molecules. These modifications have given rise to different CE techniques that have been described in detail in several reviews [11–15] as well as several books [16–18]. In isotachophoresis (ITP), the sample is sandwiched between electrolytes of high and low mobility and the analytes migrate in order of decreasing mobility until a steady state is reached. Neutral molecules are not separated by conventional CE; however, in micellar electrokinetic chromatography (MEKC), separation can be achieved by adding an anionic surfactant that interacts with neutral species to form negatively charged micelles. Isoelectric focusing (IEF) has been developed from protein gel electrophoresis methods: charged species migrate in a pH gradient, stop when they reach their isoelectric point (pI) and accumulated as a sharp band that can be subsequently eluted.

Detection has been the major problem in CE (see Albin and co-workers [19, 20]). Since the tube itself is used as the detection cell, the lightpath is very small, and in early CE systems the lack of sensitivity was a major problem. Detection procedures have been largely based on conventional UV absorbance methods as well as visible absorbance and fluorescence. However, many types of molecules do not absorb under these conditions.

To remedy this problem, compounds such as ions, organic acids and sugars have been examined by indirect UV detection as follows: UV absorbing species are added to the electrolyte, during separation displacement of this probe by the analyte creates a zone of lower UV absorbance. As a result a negative peak is created that can be recorded and quantified.

Recent improvements in CE detection can be divided into two main categories, optical and chemical. Most of the optical modifications have been aimed at increasing the path length using bent capillaries (Z cells), mirrors or bubbles. Alternatively, the use of high intensity UV lasers and laser-induced fluorescence can produce a major increase in sensitivity. Conductivity and electrochemical detection have now been adapted to CE and should give a 2–3 fold increase in sensitivity compared with indirect UV detection. Mass spectrometry is also available with CE instruments. Chemical modifications, such as isotachophoresis and stacking, involve the concentration of the sample in narrow bands before electrophoresis *per se*. Tagging of the analytes with various types of chromophore has also been widely used (see Krull and Deyl [21]).

This chapter is dedicated to the use of CE in food analysis. However applications in other areas such as pharmaceutical and biomedical are also relevant and may provide useful information for the examination of food components. Many of these studies are examined in detail elsewhere in this book. Landers [17], Xu [22, 23] and Rabel and Stobaugh [12] reviewed early pharmaceutical applications of CE. Lunte and O'Shea [24] reviewed the biomedical applications of CE in the area of peptides and sugars. The use of CE in biomedical analysis was followed by Smith and Evans [25] and among the molecules reviewed were proteins, nucleic acids and flavonoids that have applications in food analysis; Sadecka *et al*. [26] evaluated CE analysis of pharmaceutical compounds. Watzig and Dette [14] examined method development and applications related to biological sciences, and the wide range of products which followed such as alkaloids, antibiotics, flavonoids, peptides, nucleotides, DNA or vitamins are also of interest in food analysis.

38.2 Analysis of food components

Until recently HPLC (high performance liquid chromatography) has been one of the main tools in food analysis [27, 28]. CE has proved to be effective in analyzing most of the molecules found in food [6, 29]. The main exception has been lipids, which because of their hydrophobicity and lack of charge, have not been extensively investigated. However recent progress in the use of non-aqueous solvents may lead to major developments in this area [30, 31]. Not all studies reviewed in this chapter deal directly with food components but they describe the analysis of molecules found in foods and therefore could be used in food analysis.

38.2.1 Small ions and organic acid analysis

Since its introduction, CE has been used to quantify small ions, but they do not absorb in the UV and an indirect detection method had to be used. Shi and Fritz [32, 33] described the separation of up to 27 metal ions using weak complexing agents and compared the effect of various conditions; lactic acid and crown ether provided the best results. The authors also explored in detail complexing agents that allow the separation of metal cations with similar mobility. Of the chemicals tried, Shi and Fritz were most satisfied with phenylethylamine, benzylamine, *p*-toluidine and 4-methylbenzylamine. Iki *et al.* [34] separated metal ions as chelates. By combining CE with suppressed conductivity detection, Avdalovic *et al.* [35] separated many inorganic anions at the ppm level. Indirect photometric detection with an imidazole–sulfuric acid buffer was also used by Yang *et al.* [36] to measure K, Na, Ca and Mg in apple and orange juice. Šimuničova *et al.* [37] examined alkali and alkaline earth metals in mineral water at a 0.1 mM level using tartrate and 18-crown-6 and benzimidazole as a visualization co-ion. Optimal conditions for the indirect UV analysis of metal ions have been examined by Chen and Cassidy [8]. They examined the separation of 26 metal ions under various parameters such as the UV adsorbing probe, the pH, the concentration of the complexing agents and the nature of the capillary column. Cationic polymers, as electrolyte additives, were examined by Stathakis and Cassidy [38] to improve the separation of inorganic anions. Cousins *et al.* [39] compared the effectiveness of several carrier electrolytes ranging from chromate to benzoate for the separation of small anions. A similar separation of metal cations was obtained by Morin *et al.* [40] using imidazole and benzylamine buffers. Inorganic anions were also separated in non-aqueous media containing methanol and dimethylformamide [41]. For small ions very significant concentration of the analyte (several orders of magnitude) can be obtained by stacking. In this method the sample in a low ionic strength buffer is injected using electromigration, where under the influence of the electric field, the ions move rapidly to the interface between the two buffers and stack at the boundary. Using this technique, Burgi [42] was able to examine chloride and sulfate at the 100 ppb level. Wojtusik and Harrold [43], using electrostacking, detected metal ions at the microgram level.

The use of CE for the analysis of small ions in food products has been reported by several groups. Baechmann *et al.* [44] used indirect fluorometric detection to examine alkaline earth ions in beverages, such as cola. Dedieu *et al.* [45] used CE to analyze low molecular weight ions in musts and wines. Nitrates were quantified by ITP in milk [46]. Swallow and Low [47] analyzed chloride, sulfate and nitrates by indirect UV detection at a level as low as 0.2 ppm. However, the authors were unable to use this technique to detect the addition of pulpwash (an orange juice byproduct) to orange juice.

Jimidar *et al.* [48] determined nitrate and nitrite in vegetables by CE with indirect UV detection. Calcium and magnesium in wheat flour were examined using CZE [170].

In CE, organic acids behave very much like inorganic ions. Organic acids are also molecules that do not absorb in the UV and detection had to be done indirectly with UV absorbent molecules. The organic acids (lactic, acetic, phosphoric, citric, propionic, butyric), generated during lactic fermentation of various vegetables, were determined by ITP and detected by a conductivity detector. This CE method compared favorably with other chromatographic procedures [49]. Isocitric, citric, malic and tartric acids were separated in 14 min from orange juice with inverse polarity and a buffer containing phthalic acid, cetyltrimethylammonium bromide (CTAB) and 20% methanol at pH 7. Hops bitter acids have been analyzed by various groups. Szucs *et al.* [50] were able to analyze directly iso-alpha acids in beer by sample stacking of large injected volumes. They obtained similar separation by microemulsion electrokinetic chromatography [51]. Organic acids in beer were also examined by CZE [52]. CE (phthalate buffer pH 5.6) and HPLC were compared by Levi *et al.* [53] in the study of organic acids in wine and similar results were found. Lalljie *et al.* [54] analyzed organic acids in juice produced during sugar refining and in chicory root extract. Polycarboxylic acids complexed with copper were analyzed by Wiley [55] with myristyltrimethylammonium bromide as an electroosmotic modifier; the author concluded that the CE method was as sensitive as HPLC. The advantage of CE can be seen in recent experiments where the analyses of organic acids and small ions were combined. ITP has been used by Kvasnicka *et al.* [56] to monitor metals, organic, inorganic acids and volatile fatty acids during sugar production. Oehrle [57, 58] examined organic acids and anions in less than 6 min at the mid ppb level. In wine, molecules ranging from chloride to lactic acid were separated in 10 min [59]. The development of CE conductivity detection allowed Jones *et al.* [60] to examine 37 anions at the low ppb level. Finally using simultaneous conductivity and UV detection, the same authors [61] quantified in a single analysis, molecules as diverse as bromide, sulfate, acetate or benzoate. We have been analyzing citrus organic acids, malic, citric and isocitric using 5 mM phthalate pH 3.5 and 30 kV. Under these conditions standards of these three acids are well separated. However, in orange juice isocitric acid is present in amounts about 125 times less than citric acid and the small isocitric peak can be covered by the other peaks. Separation of isocitric acid from the other acids was achieved under different conditions: 5 mM phthalate pH 3.6 with 3 mM $CaCl_2$ and 30 kV. The calcium chloride improves the separation between isocitric and malic acids and allows distinct peaks to be seen even with very different amounts of material. However, because of the differences in peak size two analyses had to be performed, one for malic and citric acids and the other for isocitric acid. More recently, the three organic acids, including isocitric acid, were quantified in a single analysis using conductivity detection with a buffer containing 130 mM histidine/MES, 0.7 mM TTAB, 0.03% Triton × 100, 100 ppm EDTA and 5% acetonitrile [61a]. Metals such as Na and Ca can be seen but with phthalate as a chromophore, the peaks are too small to allow quantification [62].

38.2.2 Saccharides

As reported by Oefner and Chiesa [63] in a comprehensive review, saccharides have been extensively analyzed by CE. Detailed protocols for the CE analysis of oligosaccharides have been reported by Linhardt [64]. Saccharides do not absorb readily in the UV and methods had to take into account peak visualization. The separation of borate complex was one of the first types of CE analysis developed for saccharides. Arentoft et al. [65] used this method to examine pea oligosaccharides. After extensive extraction of the oligosaccharides from the seeds, the polyol–borate complexes were separated and detected at 195 nm. Recently, a colored reaction generated by the formation of complexes between starch and iodine was used by Brewster et al. [66] to separate and detect in the visible. Indirect UV detection has also been used extensively. Klockow et al. [67] compared CZE with indirect UV detection and HPLC with pulsed amperometric detection (PAD) to the routine determination of sugars in fruit juices (orange, apple and grape). The authors concluded that there were no major differences between the two methods, CE being simple, inexpensive and rapid. Bergholdt et al. [68] used sorbic acid as an electrolyte anion and a chromophore to measure galacturonic and gluconic acids by CZE. We have developed a similar method to examine sugars in citrus products using sorbate at pH 11.5, sugars can be analyzed in 12 min and in 20 min if galacturonic acid is also monitored [62]. The method does not require any preparation besides a four-fold dilution of the juice and the results are similar to those obtained by HPLC [69]. The analysis of derivatized sugars has been an important development; derivatization improves the sensitivity, but increases the preparation time, since an extra step has to be introduced. The added tag can provide a strong charge and/or an easily detectable moiety. Stefansson and Novotny [70, 71] analyzed neutral and charged polysaccharides including modified celluloses on a capillary column coated with polyacrylamide after formation of fluorescent derivatives. Schwaiger et al. [72] separated 4-aminobenzonitrile carbohydrate derivatives by CZE as borate complexes or by MEKC, with this last method a concentration sensitivity of 0.3 µM was reached for glucose. Acidic monosaccharides were separated as sulfanilic acid or 7-aminonaphthalene-1,3-disulfonic acid [73]. This method was applied by the same group to the analysis of sialogangliosides [74]. *p*-Aminobenzoic acid derivatives of mostly aldoses, ketoses and uronic acids were separated by Grill et al. [75].

More recently electrochemical detection (ECD) has started to be used in CE, the equipment was usually developed in individual laboratories. With the commercial availability of such systems, a large increase in this type of analysis is likely to occur. Colon et al. [76] used a copper microelectrode to examine glucose and fructose in carbonated soft drinks. A similar method was used by Lu and Cassidy [77] to detect saccharides and polyols and by Ye and Baldwin [174] who determined the saccharide composition of apple juice with a detection limit at or below the femtomole level. The analysis of oligosaccharides was performed by Guttman and Starr [79] who compared CE with polyacryamide gel electrophoresis.

38.2.3 Amino acids, peptides and proteins

Proteinaceous compounds represent a main component of food products and they have been extensively analyzed by CE (for a review see Schoneich et al. [171]). A detailed review of CE analysis of amino acids has recently been published [80]. Only the three aromatic amino acids, namely phenylalanine (Phe), tyrosine (Tyr) and tryptophan (Trp) can be absorbed significantly in the UV. The authors were able to examine Phe, Trp and Tyr in citrus juice with a method developed for the general analysis of citrus juice by UV detection [81]. Cu chelates of amines, amino acids and peptides were analysed by ITP with detection at 580 nm [82]. Derivatization represents an effective method for amino acid detection. Using fluorescence detection, Higashijima et al. [83] detected amino acids at the 10 pmol level. A very sensitive analysis of fluorescamine labeled amino acids and peptides has been described in single nerve cells by Shippy et al. [84]. ECD should improve CE amino acid analysis. Ye and Baldwin [78] used CE with electrooxidation at a Cu electrode to examine amino acid mixtures without any derivatization procedure. The method was applied to the detection of aspartame in soft drinks. Amperometric detection was used by Guo et al. [85] to examine amino acids without derivatization with a strong alkaline buffer (pH 12). The authors used the method to analyze cytochrome c hydrolysates. CZE with a citrate buffer pH 2.5 was applied to the essential amino acid composition of dry cured ham [100].

Protein detection can usually be done in the UV since most proteins contain the three UV absorbing amino acids. A review on the use of CE for the analysis of proteins and nucleic acids has been published by Karger et al. [86] and the principles of peptide analysis by CZE were reviewed by Wheat [172]. Chen and Sternberg [87] separated proteins, with pI ranging from 4 to 11 mol. wt from 5000 to 77 000 in less than 10 min. Extensive studies of milk proteins by CE have been reported in recent years and were reviewed by Olieman [88]. Chen and Zang [89] used CE, with UV detection at 200 nm and a phosphate buffer pH 6, to differentiate proteins from fresh and reconstituted milk allowing adulterated milk to be detected. Milk proteins were studies by De Jong et al. [90] using a coated capillary and a buffer containing 6 M urea and 0.05% methylhydroxyethylcellulose. The authors found that the results compared with those obtained by reversed-phase HPLC and the method was used to compare milk from cow, goat and sheep and heat-damaged milks. Cifuentes et al. [91] used CE to analyze whey proteins with a Tris–boric acid, SDS (sodium dodecyl sulfate) and PEG (polyethylene glycol). The caseinomacropeptides from acid casein and rennet whey were analyzed at pH 2.5 in 11 min, by a method that could provide a tool in monitoring the kinetics of rennet coagulation [92]. Rennet whey solids in milk and buttermilk powder were estimated by CZE using 6 M urea with a 0.15 mM citrate buffer [93] and casein phosphopeptides were examined by Adamson and Reynolds [94]. Whey proteins were also analyzed by Recio et al. [95] using an uncoated capillary with a 100 mM borate buffer, pH 8.2 and 30 mM sodium sulfate. Paterson et al. [96] separated lactoglobulin by CZE and the same group [97] compared CZE with other methods of analyzing milk proteins. Egg white proteins were analyzed by McCulloch [98] using untreated and bonded columns. A more general procedure for the analysis of food proteins (egg white, egg yolk and milk proteins) has been developed by Chen and Tusak [99] using untreated silica with a borate buffer at pH 10. The method was applied to egg white and yolk as well as milk caseins and whey proteins. Separation of fermentation broth was achieved by Strege and Lagu [101] by MEKC with a C18-derivatized capillary (%RSD (relative standard deviation) = 0.2) as compared with 2.2 %RSD for bare silica columns. The authors [102] performed a similar study on fermentation broth using a polyacrylamide-coated capillary. Werner and co-workers [103–105] used CE to analyze gliadins and glutenin in wheat and determine varietal origins. The authors used a coated column with an aluminum buffer at pH 2.3 or a capillary filled with a polymer and a buffer containing glycerol and methanol. Wheat gliadins were analyzed by Bietz [106] with a phosphate buffer at pH 2.5 containing a linear hydrophilic polymer. Lookhart and Bean [107] analyzed oat and rice proteins by CZE with phosphate buffer at pH 2.5 containing 0.05% hydroxypropylmethylcellulose and concluded that CE performed as well or better than polyacrylamide electrophoresis or HPLC. CZE at pH 7.4 was used by Leblanc et al. [108] to examine fish sarcoplasmic proteins. Several fish could be differentiated in less than 10 min and the authors concluded that this CE method was superior to conventional electrophoretic and HPLC procedures for the screening and monitoring of fish products. A similar study was performed by Gallardo et al. [109] using CZE with a 30 mM phosphate buffer pH 2.4. Tomlison et al. [110] examined the reaction products of 5-hydroxymethyl furfural and glycine as a model of the Maillard reaction. They [111] examined the colored products from the Maillard reaction with a phosphate buffer pH 6.5 and showed a better separation of the products than with a HPLC method. 5-Hydroxymethyl-2-

furaldehyde and 2-furaldehyde were estimated in fruit juices by MEKC with direct sample injection [112].

38.2.4 Lipids and hormones

As a group, lipids are often neutral and hydrophobic and are not well suited to separation by CE. To overcome the charge and solubility problems two approaches have been pursued: MEKC and CE in non-aqueous solvent. A recent review [113] examined the use of CE in lipid analysis and described the separation of a mixture of isoprenylpyrophosphate with indirect laser-induced fluorescence detection. Potter et al. [114] separated estrogens with aqueous–methanolic buffers. Non-ionic aromatic compounds were examined by Shi and Fritz [30] with a borate buffer containing 50 mM tetraheptylammonium bromide and 42% acetonitrile. MEKC was used by Ingvardsen et al. [115] to separate phospholipids with a buffer containing cholate and propanol. Yoo et al. [116] used a phosphate–borate buffer pH 9.3 with cyclodextrin to examine a mixture of gangliosides. However, the use of CE in food lipid analysis has remained very limited. The development of non-aqueous buffers should improve the use of CE for hydrophobic compounds.

38.2.5 Vitamins

The quantitation of vitamins is also an important part of food analysis. In the USA, the introduction of the Nutrition Labeling and Education Act has made the determination of vitamins in food products a major requirement. Water-soluble UV absorbing vitamins can easily be monitored by CE. Chiari et al. [117] measured vitamin C in fruit by CZE, this method was sensitive from 1.6 to 480 mg l^{-1}. Ascorbic and isoascorbic acids were analyzed by CZE in 5 min with a phosphate buffer pH 5 in fruit juices and pharmaceutical products [118]. Total L-ascorbic acid was assessed by Thompson and Trenerry [119] in fruits and vegetables after extraction with metaphosphoric acid and stabilization with dithiothreitol by MEKC with a sodium deoxycholate buffer. The dithiothreitol reduces the dehydroascorbic acid to ascorbic acid and allows the estimation of the total amount of vitamin C. The authors used a similar method to measure ascorbic acid in beers, wines and fruit beverages [120]. Cahill and Wightman [121] also measured ascorbate and catecholamine by CE. We have measured vitamin C by CZE as part of a general analysis of citrus juices [81]. The value of CE was shown in recent studies. Huapalahti and Sunell [122] monitored simultaneously vitamins B1, B2, and B6 by CZE. Jegle [123] developed a CZE method with a phosphate buffer pH 7 for the analysis of up to nine water-soluble vitamins (vitamin C, vitamins B1, B3, B6, B12, folinate, orotic acid, pantothenate and nicotinamide) in about 15 min with an excellent repeatability. L- and D-folinic acid were separated by Cellia et al. [124] using cyclodextrin as a chiral active component. During the CE analysis of citrus juice, the free form of tetrahydrofolic acid represents a minor peak; however following enzymatic treatment to cleave the polyglutamate associated to the vitamin, a significant peak was found (Cancalon, unpublished observation).

38.2.6 Natural compounds

Natural compounds that absorb well in the UV such as flavonoids and polyphenols have been extensively examined by CE. A specific study of the flavonoids in two sugar cane cultivars was performed by CZE with a borate buffer, pH 9.5 and 20% methanol [125]. Pietta et al. reported several studies on the analysis of flavonoids. In 1992 they used MEKC [126] and UV photodiode array detection to separate flavonoids and the method was optimized in 1994 [127]. Pietta et al. [128] used MEKC with a 25 mM buffer pH 8.25 with 50 mM SDS to analyze components from the root of *Eleutherococcus senticosus* and to detect adulteration in *Arnica montana* and *A. chamissonis* [129]. Bjergegaard et al. [130] developed a similar method to study kaempferol and quercitin glycosides in cruciferous plants. Morin et al. [131, 132] separated flavonoid-7-o-glycosides differing in their flavonoid aglycone or flavonoid-3-o-glycosides differing in their sugar moiety after complexation with borate. A comparison of HPLC and CE separation of isoflavones was performed by Shihabi et al. [133]; the HPLC method required a complex isocratic elution while the CE separation was done on a simple capillary with a borate buffer at pH 8.6 and was complete in less than 8 min. A comparison between CE and HPLC for the analysis of honey flavonoids has been reported by Delgado et al. [134], and by Ferreres et al. [135] who used MEKC with a borate buffer pH 8 containing 50 mM sodium dodecyl sulfate (SDS) and 10% methanol to differentiate various types of honey by their flavonoid content. Shepherd and McGhie [136] measured chlorogenic acid, apigenin, isoorientin and quercitin by CE with a tetraborate buffer. The method was used to monitor the activity of polyphenol oxidase. The phenolic compounds in red wine were analyzed by Gil et al. [137] with a borate buffer at pH 9.5 in a fused-silica column. The electrophoretic profiles were used to differentiate wines from different areas. We routinely examine citrus juices for hesperidin, neohesperidin, narirutin and naringin to assess the purity of the products [81].

Many other natural compounds have also been examined by CE. Ferulic and coumaric acids and esters as well as feruloyl glucosides from wheat bran were analyzed by Donaghy and McKay [138] using CZE. Feruloyl and synapyl glucosides are routinely separated in our analysis of orange juice components [81]. Alkaloids can be responsible for various food poisoning and have been followed by CE. Ergovaline in seeds was measured by My et al. [139] with a phosphate buffer pH 3.5 containing 50% methanol. Liu and Sheu [140] used CE to compare alkaloid composition in samples of *Ephedrae Herba* and *Phellodendri Cortex*. The composition of Chinese herbal tea including the flavonoid constituents has been examined by Sheu and Chen [141] by HPLC and CE. Rhubarb ale-emodin, emodin and rhein were measured by MEKC with a buffer containing CAPS (3-cyclohexylamino-1-proparesulfonic acid), SDS and acetonitrile at pH 10.96 [142].

38.2.7 Drugs, additives, toxins and degradation products

The monitoring of food products requires constant examination of many drugs, food additives and degradation products which indicate that the product is fit for consumption. A review of the use of CE for the analysis of organic pollutants has been published by Brumley [143] and for antibiotics and related drugs by Gilpin and La Pachla [144]. Separation of nitrosamines was achieved with a cyclodextrin-modified MEKC and a combined packed and open-tubular column [145]. Ackermans et al. [146] quantified several sulfonamides in pork meat by CZE with a detection limit of 2–9 ppm. A CE immunofluorescence method was used by Blais et al. [147] to determine chloramphenicol in milk. Bound and unbound forms of the chloramphenicol conjugate were titrated with laser-induced fluorescence detection. Sorbic acid and benzoic acid in foods and beverages were separated by Pant and Trenerry [148] with an SDS buffer pH 9.2 (MEKC). We separate and monitor routinely sorbate and benzoate in citrus juice with a direct CZE analysis [81]. Food grade antioxidants were examined by Hall et al. [169]. Recently, Dinelli et al. [149] compared CE, HPLC and immunoassay for the detection of the herbicide terbutylazine in tap water. CE gave the best sensitivity and the shortest retention time; however CE and HPLC require prior extraction of the sample. Dinelli and co-workers [150, 151] also quantitated sulfonylurea herbicides at the ppb level. Nine herbicides were separated in 16 min by Farran and Ruiz [152] with MEKC. CZE on a silanized column was also used to analyze paraquat and diquat in potatoes, the method was shown to be superior to an HPLC method with 20 times more theoretical plates [153]. Polyphosphates and polyphosphonates were analyzed by Shamsi and Danielson [154] in toothpaste and herbicides with ribonucleotide monophosphates as electrolytes using indirect photometric detection. Using 1-naphthol-3,6-disulfonic acid as a chromophore, Buscher et al. [155] monitored the enzymatic hydrolysis of inositol phosphates (such as phytic acid from grain) during fermentation. Phytic acid was analyzed in plant material with a fluorinated ethylene–propylene copolymer capillary associated with two conductivity cells [156]. Maragos [157] examined the mycotoxins fumonisins by CZE after derivatization with fluorescein isocyanate. The two isomers of the neurotoxin 2-(N-oxalyl)-L-2,3-diaminopropionic acid (ODAP) were measured in grass peas with a phosphate buffer pH 7.8 with direct UV detection at 195 nm [158]. Degradation products have been followed to assess the degree of freshness of proteinaceous compounds. Fish freshness was estimated by Luong et al. [159] using CE and immobilized enzymes to follow nucleotide degradation products such as hypoxanthine. Histamine in fish was measured by Mopper and Sciacchitano [160] by CZE with a citrate buffer pH 2.5, who showed that the method was superior to the official AOAC fluorometric method, particularly at low concentration. The authors also reported distinct electropherograms for various fish species. Hippuric and orotic acids were quantified in whey by Tiensdra et al. [161] with an uncoated column, a N,N-bis(2-hydroxyethyl) glycine (BICINE) buffer pH 8.8 and UV detection. The quality of fish products has been monitored by Pleasance et al. [162] by CE–MS of various compounds such as toxins and antibiotics. This technique of detection of shellfish toxins was improved by two orders of magnitude with an ITP preconcentration step [163]. A similar method using 100 mM morpholine at pH 4 as a buffer was used by Buzy et al. [164] to estimate shellfish poisoning toxins. Artificial sweeteners can easily be detected in the UV and several analyses have been published. Recently, MEKC with a sodium deoxycholate buffer was used by Thompson et al. [165, 166] to measure many artificial sweeteners in soft drinks. Using the same technique, Thompson and Trenerry [173] also examined synthetic color in confectionery and cordials. Liu and Li [167] compared CE and HPLC for the analysis of Stevia sweeteners.

38.3 Multiple analyses

As shown in this review, a trend is appearing in the use of CE to monitor the production and/or quality of food products. Methods are being developed that cover the analysis of many compounds in a single procedure. A description of the use of ITP for the monitoring of sugar production has been published by Kvasnicka et al. [56]. Metals, organic acids, inorganic acids and volatile fatty acids were routinely analyzed by this method. An automated analysis of food proteins has been developed by Chen and Tusak [99]. Compounds such as egg white and yolk as well as milk caseins and whey proteins were examined in less than 10 min. Brumley [143] used CE to examine a wide range of organic pollutants.

We have tried to take advantage of these qualities of CE to develop a monitoring technique for citrus juices. With UV detection a procedure has been developed that provides information on the flavonoid content of the juice, the concentration of vitamin C and the presence of the preservatives sorbate and benzoate. Other compounds such as amines, amino acids and polyphenols are also seen, but are not monitored. The presence of pulpwash (PW) (an orange citrus juice byproduct whose addition to some types of juices is controlled) is determined by inputting some of the CE results into an artificial neural network [168]. The same method can be used to determine the dilution of a juice. Using indirect UV detection, organic acids and sugars could also be analyzed. The total analytical time has been cut in half when compared to the previous analytical process involving PW determination and HPLC measurement of flavonoids and preservatives. CE offers the possibility of examining simultaneously many chemicals in a single analytical procedure. This technique is therefore well suited for the quality monitoring of food products.

References

1. Ewing, A.; Wallingford, R.; Olefirowicz, T. (1989) Capillary electrophoresis. *Anal. Chem.* **61**, 292A–303A.
2. Oda, P.; Spelsberg, T.; Landers, J. (1994) Commercial capillary electrophoresis instrumentation. *LC–GC* **12**, 50–51.
3. Altria, K.; Wood, T.; Kitscha, R.; McIntosh, A. (1995) Validation of a capillary electrophoresis method for the determination of potassium counter-ion levels in an acidic drug salt. *J. Pharm. Biomed. Anal.* **13**, 33–38.
4. Altria, K.; Clayton, N.; Harden, R.; Makwana, J.; Portsmouth, M. (1995) Inter-company cross validation exercise on capillary electrophoresis. Quantitative determination of drug counter-ion level. *Chromatogr.* **40**, 47–50.
5. Taylor, R.; Reid, R. (1995) Analysis of basic antimalarial drugs by CZE: Part 2. Validation and application to bioanalysis. *J. Pharm. Biomed. Anal.* **13**, 21–26.
6. Zeece, M. (1992) Capillary electrophoresis a new analytical tool for food science. *Trends Food Sci. Technol.* **3**, 6–10.
7. Tiselius, A. (1937) A new apparatus for electrophoretic analysis of colloidal mixtures. *Trans. Faraday Soc.* **33**, 524–531.
8. Chen, M.; Cassidy, R. (1993) Separation of metal ions by capillary electrophoresis. *J. Chromatogr.* **640**, 425–431.
9. Richmond, M.; Yeung, E. (1993) Development of laser-excited indirect fluorescence detection for high-molecular-weight polysaccharide in capillary electrophoresis. *Anal. Biochem.* **210**, 245–248.
10. Ebersole, R.; McCormick, R. (1993) Separation and isolation of viable bacteria by capillary zone electrophoresis. *Biotechnology* **11**, 1278–1282.
11. Landers, J.; Oda, R.; Spelsberg, T.; Noman, J.; Ulfelder, K. (1993) Capillary electrophoresis – a powerful microanalytical technique for biologically active molecules. *Biotechniques* **14**, 98–111.
12. Rabel, S.; Stobaugh, J. (1993) Applications of capillary electrophoresis in pharmaceutical analysis. *Pharm. Res.* **10**, 171–186.
13. Weinberger, R. (1993) Separation science for the nineties and beyond: High-performance capillary electrophoresis. *Today's Chemist* **2**, 30–34.
14. Watzig, H.; Dette, C. (1994) Capillary electrophoresis (CE) – a review – Strategies for method development and applications related to pharmaceutical and biological sciences. *Pharmazie* **49**, 83–96.
15. Monnig, C.; Kennedy, R. (1994) Capillary electrophoresis. *Anal. Chem.* **66**, R280–R314.
16. Li, S. (1992) *Capillary Electrophoresis: Principles, Practice, Applications*, Elsevier, Amsterdam.
17. Landers, J. (1993) *Handbook of Capillary Electrophoresis*, CRC Press, Boca Raton.

18. Camilleri, P. (1993) *Capillary Electrophoresis. Theory and Practice*, CRC Press, Boca Raton.
19. Albin, M.; Grossman, P.; Moring, S. (1993) Sensitivity enhancement for capillary electrophoresis. *Anal. Chem.* **65**, 489A–497A.
20. Moring, S.; Pairaud, C.; Albin, M.; Locke, S.; Thibault, P.; Tindall, G. (1993) Enhancement of UV detection sensitivity for capillary electrophoresis, with a high-sensitivity optical cell. *Am. Lab.*, July, 32–39.
21. Krull, I.; Deyl, L. (1994) General strategies and selection of derivatization reactions for liquid chromatography and capillary electrophoresis. *J. Chromatogr.* **659**, 1–17.
22. Xu, Y. (1993) Capillary electrophoresis. *Anal. Chem.* **65**, R425–R433.
23. Xu, Y. (1995) Capillary electrophoresis. *Anal. Chem.* **67**, R463–R473.
24. Lunte, S.; O'Shea, T. (1994) Pharmaceutical and biomedical applications of capillary electrophoresis/electrochemistry. *Electrophoresis* **15**, 79–86.
25. Smith, N.; Evans, M. (1994) Capillary zone electrophoresis. *J. Pharm. Biomed. Anal.* **12**, 579–611.
26. Sadecka, J.; Polonsky, J.; Shintani, H. (1994) Current advancement of pharmaceutical analysis by capillary zone electrophoresis, micellar electrokinetic chromatography and iso-tachophoresis. *Pharmazie* **49**, 631–641.
27. Widmer, W.; Cancalon, P.; Nagy, S. (1992) Recent advances in the detection of adulteration in citrus juices. *Trends Food Sci. Technol.* **3**, 278–286.
28. Chang, S.; Holm, E.; Schwarz, J.; Rayasdurate, P. (1995) Food. *Anal. Chem.* **67**, R127–R153.
29. Cancalon, P. (1995) A new tool in food analysis: Capillary electrophoresis. *J. Assoc. Off. Anal. Chem. Int.* **78**, 12–15.
30. Shi, Y.; Fritz, J. (1994) Capillary zone electrophoresis of neutral organic molecules in organic–aqueous solution HRC. *J. High Resolut. Chromatogr.* **17**, 713–718.
31. Whitaker, K.; Yin, X.; Sepaniak, M. (1995) Separation of polycyclic aromatic hydrocarbons and fullerenes with non-aqueous capillary electrokinetic chromatography. *Pittsburgh Conference Abstracts*, 315P, New Orleans.
32. Shi, Y.; Fritz, J. (1993) Separation of metal ions by capillary electrophoresis with a complexing electrolyte. *J. Chromatogr.* **640**, 473–479.
33. Shi, Y.; Fritz, J. (1994) New electrolyte systems for the determination of metal cations by capillary zone electrophoresis. *J. Chromatogr.* **671**, 429–435.
34. Iki, N.; Hoshino, H.; Yotsuyanagi, T. (1993) Ion-association capillary electrophoresis. New separation method for equally and highly charged metal chelates. *J. Chromatogr.* **652**, 539–546.
35. Avdalovic, N.; Pohl, C.; Stillian, J. (1993) Determination of cations and anions by capillary electrophoresis combined with suppressed conductivity detection. *Anal. Chem.* **65**, 1470–1475.
36. Yang, Q.; Jimidar, M.; Hamoir, T.; Smeyers-Verbeke, J.; Massart, D. (1994) Determination of alkali and alkaline earth metals in real samples by capillary ion analysis. *J. Chromatogr.* **673**, 275–285.
37. Simunicova, E.; Kaniansky, D.; Loksikova, K. (1994) Separation of alkali and alkaline earth metal and ammonium cations by capillary zone electrophoresis with indirect UV absorbance detection. *J. Chromatogr.* **665**, 203–209.
38. Stathakis, C.; Cassidy, R. (1994) Cationic polymers for selectivity control in the capillary electrophoretic separation of inorganic anions. *Anal. Chem.* **66**, 2110–2115.
39. Cousins, S.; Haddad, P.; Buchberger, W. (1994) Evaluation of carrier electrolytes for capillary zone electrophoresis of low-molecular-mass anions with indirect UV detection. *J. Chromatogr.* **671**, 39–402.
40. Morin, P.; Francois, C.; Dreux, M. (1994) Capillary electrophoresis of alkali and alkaline earth cations with imidazole or benzylamine buffers. *J. Liq. Chromatogr.* **17**, 3869–3888.
41. Salimi-Moosavi, H.; Cassidy, R. (1995) Capillary electrophoresis of inorganic anions in nonaqueous media with electrochemical and indirect UV detection. *Anal. Chem.* **67**, 1067–1073.
42. Burgi, D. (1993) Large volume stacking of anions in capillary electrophoresis using an electroosmotic flow modifier as a pump. *Anal. Chem.* **65**, 3726–3729.

43 Wojtusik, M.; Harrold, M. (1994) Factors influencing trace ion analysis with pre-concentration by electrostacking. *J. Chromatogr.* **671**, 411–417.
44 Baechmann, K.; Boden, J.; Haumann, I. (1992) Indirect fluorimetric detection of alkali and alkaline earth metal ions in capillary zone electrophoresis with cerium (III) as carrier electrolyte. *J. Chromatogr.* **626**, 259–265.
45 Dedieu, F.; Nouadje, G.; Puig, P. (1994) Contribution of capillary electrophoresis to the analysis of low molecular weight ions in musts and wines. *Revue Oenol. Tech. Vitivinicoles Oenolog.* **72**, 7–10.
46 Stransky, Z.; Dostal, V.; Cvak, Z.; Dulova, V. (1991) Isotachophoretic determination of nitrates in milk. *Acta Univ. Palacki. Olomuc. Fac. Rerum. Nat.* **102**, 121–128.
47 Swallow, K.; Low, N. (1994) Capillary electrophoretic analysis of minor anions present in orange juice and orange pulpwash. *J. Agric. Food Chem.* **42**, 2808–2811.
48 Jimidar, M.; Hartmann, C.; Cousement, N.; Massart, D. (1995) Determination of nitrate and nitrite in vegetables by capillary electrophoresis with indirect detection. *J. Chromatogr.* **706**, 479–492.
49 Karovicova, J.; Polonsky, J.; Drdak, M.; Simko, P.; Vollek, V. (1993) Capillary isotachophoresis of organic acids produced by selected microorganisms during lactic acid fermentation. *J. Chromatogr.* **638**, 241–246.
50 Szucs, R.; Vindevogel, J.; Sandra, P.; Verhagen, L. (1993) Sample stacking effects and large injection volumes in micellar electrokinetic chromatography of ionic compounds: direct determination of iso-acids in beer. *Chromatographia* **36**, 323–329.
51 Szucs, R.; Vindevogel, J.; Everaert, E.; Decooman, L.; Sandra, P.; Dekeukeleire, D. (1994) Separation and quantification of all main hop acids in different hop cultivars by microemulsion. *Electrokin. Chromatogr. J. Int. Brewing* **100**, 293–296.
52 Vries, K. De (1993) Determination of organic acids in beer by capillary electrophoresis. *J. Am. Soc. Brewing Chem.* **51**, 155–157.
53 Levi, V.; Wehr, T.; Talmadge, K.; Zhu, M. (1993) Analysis of organic acids in wines by capillary electrophoresis and HPLC. *Am. Lab.* 29–31.
54 Lalljie, S.; Vindevogel, J.; Sandra, P. (1993) Quantitation of organic acids in sugar refinery juices with capillary zone electrophoresis and indirect UV detection. *J. Chromatogr.* **652**, 563–569.
55 Wiley, J. (1995) Determination of polycarboxylic acids by capillary electrophoresis with copper complexation. *J. Chromatogr.* **692**, 267–274.
56 Kvasnicka, F.; Parking, G.; Harvey, C. (1993) Capillary isotachophoresis as a new tool in sugar factory analysis. *Int. Sugar J.* **95**, 451–458.
57 Oehrle, S. (1994) Versatility of capillary electrophoresis of anions with a high-mobility chromate electrolyte. *J. Chromatogr.* **671**, 383–387.
58 Oehrle, S. (1995) Capillary ion analysis of ion and organic acids in circuit board extracts. *Pittsburgh Conference Abst.* 1368 New Orleans.
59 Cotter, R.; Benvenuti, M.; Krol, J. (1995) The use of capillary ion analysis in the analysis of organic acids inpotable liquids. *Pittsburgh Conference Abst.* 1365, New Orleans.
60 Jones, W.; Soglia, J.; McGlynn, M. (1995) Trace ion analysis of high purity and concentrated samples matrices using capillary electrophoresis and conductivity detection. *Pittsburgh Conference Abst.* 1366, New Orleans.
61 Jones, W.; Soglia, J.; McGlynn, M. (1995) Method development approaches for capillary ion electrophoresis using conductivity and UV detection. *Pittsburgh Conference Abst.* 706, New Orleans.
61a Cancalon, P. (1997) Analysis of organic acids in citrus juices by capillary electrophoresis. Pittsburgh Conference Abstract, in press.
62 Cancalon, P. (1994) Analysis of organic acids by capillary electrophoresis. *45th Ann. Citrus Proc. Meeting.*
63 Oefner, P.; Chiesa, C. (1994) Capillary electrophoresis of carbohydrates. *Glycobiology* **4**, 397–412.
64 Linhardt, R. (1994) Capillary electrophoresis of oligosaccharides. In *Guide to Techniques in Glycobiology*, Lennarz, W.J.; Hart, G.W. (eds), Academic Press Inc, San Diego, pp. 265–280.
65 Arentoft, A.; Michaelsen, S.; Sorensen, H. (1993) Determination of oligosaccharides by capillary zone electrophoresis. *J. Chromatogr.* **652**, 517–524.

66   Brewster, J.; Fishman, M. (1995) Capillary electrophoresis of plant starches as the iodine complexes. *J. Chromatogr.* **693**, 382–387.
67   Klockow, A.; Paulus, A.; Figueiredo, V.; Amado, R.; Widmer, H. (1994) Determination of carbohydrates in fruit juices by capillary electrophoresis and high performance liquid chromatography. *J. Chromatogr.* **680**, 187–200.
68   Bergholdt, A.; Overgaard, J.; Colding, A.; Frederiksen, R. (1993) Separation of D-galactonic and D-gluconic acids by capillary zone electrophoresis. *J. Chromatogr.* **644**, 412–415.
69   Cancalon, P. (1993) Rapid monitoring of fuit juice adulteration by capillary electrophoresis. *LC–GC* **11**, 748–751.
70   Stefansson, M.; Novotny, M. (1994) Separation of complex oligosaccharide mixtures by capillary electrophoresis in the open-tubular format. *Anal. Chem.* **66**, 1134–1140.
71   Stefansson, M.; Novotny, M. (1994) Modification of the electrophoretic mobility of neutral and charged polysaccharides. *Anal. Chem.* **66**, 3466–3471.
72   Schwaiger, H.; Oefner, P.; Huber, C.; Grill, E.; Bonn, G. (1994) Capillary zone electrophoresis and micellar electrokinetic chromatography of 4-aminobenzonitrile carbohydrate derivatives. *Electrophoresis* **15**, 941–952.
73   Mechref, Y.; El Rassi, Z. (1994) Capillary zone electrophoresis of derivatized acidic monosaccharide. *Electrophoresis* **15**, 627–634.
74   Mechref, Y.; Ostrander, G.; El Rassi, Z. (1995) Capillary electrophoresis of carboxylated carbohydrates. 1. Selective precolumn derivatization of gangliosides with UV absorbing and fluorescent tags. *J. Chromatogr.* **695**, 83–95.
75   Grill, E.; Huber, C.; Oefner, P.; Vorndran, A.; Bonn, G. (1993) Capillary zone electrophoresis of *p*-aminobenzoic acid derivatives of aldoses, ketoses, and uronic acid. *Electrophoresis* **14**, 1004–1010.
76   Colon, L.; Dadoo, R.; Zare, R. (1993) Determination of carbohydrates by capillary zone electrophoresis with amperometric detection at a copper microelectrode. *Anal. Chem.* **65**, 476–481.
77   Lu, W.; Cassidy, R. (1993) Pulsed amperometric detection of carbohydrates separated by capillary electrophoresis. *Anal. Chem.* **65**, 2878–2881.
78   Ye, J.; Baldwin, R. (1994) Determination of carbohydrates, sugar acids and alcohols by capillary electrophoresis and electrochemical detection at the copper electrode. *J. Chromatogr.* **687**, 141–148.
79   Guttman, A.; Starr, C. (1995) Capillary and slab gel electrophoresis profiling of oligosaccharides. *Electrophoresis* **16**, 993–997.
80   Issaq, H.; Chan, K. (1994) Separation and detection of amino acids and their enantiomers by capillary electrophoresis: A review. *Electrophoresis* **16**, 467–480.
81   Cancalon, P.; Bryan, C. (1993) Use of capillary electrophoresis for monitoring citrus juice composition. *J. Chromatogr.* **652**, 555–562.
82   Kaniansky, D.; Zelensky, I. (1993) Photometric detection of amino-containing compounds in capillary isotachophoresis based on reaction with copper(II) ions. *J. Chromatogr.* **638**, 225–229.
83   Higashijima, T.; Fuchiggami, T.; Imasaka, T.; Ishibashi, N. (1992) Determination of amino acids by capillary zone electrophoresis based on semiconductor laser fluorescence detection. *Anal. Chem.* **64**, 711–714.
84   Shippy, S.; Jankowsky, J.; Timperman, A.; Sweedler, J. (1995) Neuropeptide determination utilizing a novel laser induced fluorescence scheme for CE. *Pittsburgh Conference Abst.* 1060, New Orleans.
85   Guo, Y.; Colon, L.; Dadoo, R.; Zare, R. (1995) Analysis of underivatized amino acids by capillary electrophoresis using constant potential amperometric detection. *Electrophoresis* **16**, 493–497.
86   Karger, B.; Chu, Y.; Foret, F. (1995) Capillary electrophoresis of proteins and nucleic acids. *Ann. Rev. Biophysics* and *Biomolecular Structure* **24**, 579–610.
87   Chen, F.-T.; Sternberg, J. (1994) Characterization of proteins by capillary electrophoresis in fused-silica columns: Review on serum protein analysis and application to immunoassay. *Electrophoresis* **15**, 13–21.
88   Olieman, C. (1993) Capillary electrophoresis in dairy research. *Voedings Middelen Technol.* **26**, 10–11.

89 Chen, F.-T.; Zang, J. (1992) Determination of milk proteins by capillary electrophoresis. *J. Assoc. Off. Anal. Chem. Int.* **75**, 905–909.
90 De Jong, N.; Visser, S.; Olieman, C. (1993) Determination of milk proteins by capillary electrophoresis. *J. Chromatogr.* **652**, 207–213.
91 Cifuentes, A.; de Frutos, M.; Diez-Masa, J. (1993) Analysis of whey proteins by capillary electrophoresis using buffer-containing polymeric additives. *J. Dairy Sci.* **76**, 1870–1875.
92 Otte, J.; Midtgaard, L.; Qvist, K. (1995) Analysis of caseinomacropeptide(s) by free solution capillary electrophoresis. *Milchwissenschaft – Milk Sci. Internat.* **50**, 75–79.
93 Van Riel, J.; Olieman, C. (1995) Determination of caseinomacropeptide with capillary zone electrophoresis and its application to the detection and estimation of rennet whey solids in milk and buttermilk powder. *Electrophoresis* **16**, 529–533.
94 Adammson, N.; Reynolds, E. (1995) High performance capillary electrophoresis of casein phosphopeptides containing 2–5 phosphoseryl residues: Relationship between absolute electrophoretic mobility and peptide charge and size. *Electrophoresis* **16**, 525–528.
95 Recio, I.; Molina, E.; Ramos, M.; Defrutos, M. (1995) Quantitative analysis of major whey proteins by capillary electrophoresis using uncoated capillaries. *Electrophoresis* **16**, 654–658.
96 Paterson, G.; Hill, J.; Otter, D. (1995) Separation of beta-lactoglobulin A, B and C variants of bovine whey using capillary zone electrophoresis. *J. Chromatogr.* **700**, 105–110.
97 Kinghorn, N.; Norris, C.; Paterson, G.; Otter, D. (1995) Comparison of capillary electrophoresis with traditional methods to analyse bovine whey proteins. *J. Chromatogr.* **700**, 111–123.
98 McCulloch, J. (1993) Egg white protein separation development – Effect of buffers and untreated fused silica vs. neutral covalently bonded columns. *J. Liq. Chromatogr.* **16**, 2025–2038.
99 Chen, F.-T.; Tusak, A. (1994) Characterization of food proteins by capillary electrophoresis. *J. Chromatogr.* **685**, 331–337.
100 Toldra, F.; Aristoy, M. (1993) Availability of essential amino acids in dry-cured ham. *Int. J. Food Sci. Nutrition* **44**, 215–219.
101 Strege, M.; Lagu, A. (1993) Micellar electrokinetic capillary chromatography of proteins. *Anal. Biochem.* **210**, 402–410.
102 Strege, M.; Lagu, A. (1995) Capillary electrophoretic separations of biotechnology-derived proteins in *E-coli* fermentation broth. *Electrophoresis* **16**, 642–646.
103 Werner, W.; Wiktorowicz, J.; Kasarda, D. (1994) Wheat varietal identification by capillary electrophoresis of gliadins and high molecular weight glutenin subunits. *Cereal Chem.* **71**, 397–402.
104 Werner, W.; Wiktorowicz, J.; Kasarda, D. (1994) Wheat varietal identification by capillary electrophoresis. *Am. Lab.* **26**, 32NN–32OO.
105 Werner, W. (1995) Ferguson plot analysis of high molecular weight glutenin subunits by capillary electrophoresis. *Cereal Chem.* **72**, 248–251.
106 Bietz, J. (1994) Fractionation of wheat gluten proteins by capillary electrophoresis in gluten proteins. *Proc. 5th International Workshop. Assoc. Cereal Research*, Detmold, Germany pp. 404–413.
107 Lookhart, G.; Bean, S. (1995) Rapid differentiation of oat cultivars and of rice cultivars by capillary zone electrophoresis. *Cereal Chem.* **72**, 312–316.
108 Leblanc, E.; Singh, S.; Leblanc, R. (1994) Capillary zone electrophoresis of fish muscle sarcoplasmic proteins. *J. Food Sci.* **59**, 1267–1270.
109 Gallardo, J.; Sotelo, C.; Perez-Martin, P. (1995) Use of capillary zone electrophoresis for fish species identification. Differentiation of flatfish species. *J. Agric. Food Chem.* **43**, 1238–1244.
110 Tomlinson, A.; Landers, J.; Lewis, I.; Naylor, S. (1993) Buffer conditions affecting the separation of Maillard reaction products by capillary electrophoresis. *J. Chromatogr.* **652**, 171–177.
111 Tomlinson, A.; Mlotkiewicz, J.; Lewis, I. (1994) Application of capillary electrophoresis to the separation of coloured products of Maillard reactions. *Food Chem.* **49**, 219–223.

112 Corradini, D.; Corradini, C. (1992) Separation and determination of 5-hydroxymethyl-2-furaldehyde in fruit juice by micellar electrokinetic capillary chromatography with direct sample injection. *J. Chromatogr.* **624**, 503–509.
113 Blomberg, L.; Andersson, P. (1994) Capillary electrophoresis for lipid analysis. *Information* **5**, 1030–1037.
114 Potter, K.; Allington, R.; Algaier, J. (1993) Separation of estrogens and rodenticides using capillary electrophoresis with aqueous–methanolic buffers. *J. Chromatrogr.* **652**, 427–430.
115 Ingvardsen, L.; Michaelsen, S.; Sorensen, H. (1994) Analysis of individual phospholipids by high-performance capillary electrophoresis. *J. Am. Oil Chem. Soc.* **71**, 183.
116 Yoo, Y.; Kim, Y.; Jhon, G.; Park, J. (1993) Separation of gangliosides using cyclodextrin in capillary zone electrophoresis. *J. Chromatogr.* **652**, 431–439.
117 Chiari, M.; Nesi, M.; Carrea, G.; Righetti, P. (1993) Determination of total vitamin-C in fruits by capillary zone electrophoresis. *J. Chromatogr.* **645**, 197–200.
118 Ling, B.; Baeyens, W.; Vanacker, P.; Dewaele, C. (1992) Determination of ascorbic acid and isoascorbic acid by capillary zone electrophoresis – Application to fruit juices and to a pharmaceutical formulation. *J. Pharm. Biomed. Anal.* **10**, 717–721.
119 Thompson, C.; Trenerry, V. (1995) A rapid method for the determination of total L-ascorbic acid in fruits and vegetables by micellar electrokinetic capillary chromatography. *Food Chem.* **53**, 43–50.
120 Marshall, P.; Trenerry, V.; Thompson, C. (1995) The determination of total ascorbic acid in beers, wines, and fruit drinks by micellar electrokinetic capillary chromatography. *J. Chromatogr. Sci.* **33**, 426–432.
121 Cahill, P.; Wightman, R. (1995) Simultaneous amperometric measurement of ascorbate and catecholamine secretion from individual bovine adrenal medullary cells. *Anal. Chem.* **67**, 2599–2605.
122 Huopalahti, R.; Sunell, J. (1993) Use of capillary electrophoresis in the determination of B vitamins in pharmaceutical products. *J Chromatogr.* **636**, 133–135.
123 Jegle, U. (1993) Separation of water-soluble vitamins via high-performance capillary electrophoresis. *J. Chromatogr.* **652**, 495–501.
124 Cellai, L.; Desiderio, C.; Filippetti, R.; Fanali, S. (1993) Capillary electrophoresis quantitation of *l*-L-folinic acid in the presence of its inactive *d*-L-form. *Electrophoresis* **14**, 823–825.
125 McGhie, T. (1993) Analysis of sugar cane flavonoids by capillary zone electrophoresis. *J. Chromatogr.* **634**, 107–112.
126 Pietta, P.; Mauri, P.; Facino, R.; Carini, M. (1992) Analysis of flavonoids by MECC with ultraviolet diode array detection. *J. Pharm. Biomed. Anal.* **10**, 1041–1045.
127 Pietta, P.; Mauri, P.; Zini, L.; Gardana, C. (1994) Optimization of separation selectivity in capillary electrophoresis of flavonoids. *J. Chromatogr.* **680**, 175–179.
128 Pietta, P.; Mauri, P.; Gardana, C.; Zini, L. (1994) Micellar electrokinetic chromatographic/ultraviolet diode array analysis of *Eleutherococcus senticosus*. *Phytochem. Anal.* **5**, 305–308.
129 Pietta, P.; Mauri, P.; Bruno, A.; Merfort, I. (1994) MEKC as an improved method to detect falsifications in the flowers of *Arnica montana* and *A.chamissonis*. *Planta Medica* **60**, 369–372.
130 Bjergegaard, C.; Michaelsen, S.; Mortensen, K.; Sorensen, H. (1993) Determination of flavonoids by micellar electrokinetic capillary electrophoresis. *J. Chromatogr.* **652**, 477–485.
131 Morin, P.; Villard, F.; Dreux, M. (1993) Borate complexation of flavonoid-*o*-glycosides in capillary electrophoresis II. Separation of flavonoid-3-*o*-glycosides differing in their sugar moiety. *J. Chromatogr.* **628**, 161–169.
132 Morin, P.; Villard, F.; Druex, M. (1993) Borate complexation of flavonoid-*o*-glycosides in capillary electrophoresis I. Separation of flavonoid-7-*o*-glycosides differing in their flavonoid aglycone. *J. Chromatogr.* **628**, 153–160.
133 Shihabi, Z.; Kute, T.; Garcia, L.; Hinsdale, M. (1994) Analysis of isoflavones by capillary electrophoresis. *J. Chromatogr.* **680**, 181–185.
134 Delgado, C.; Tomasbarderan, F.; Talou, T.; Gaset, A. (1994) Capillary electrophoresis as an alternative to HPLC for determination of honey flavonoids. *Chromatographia* **38**, 71–78.

135 Ferreres, F.; Amparo-Blanquez, M.; Gil, M.; Tomas-Barberan, F. (1994) Separation of honey flavonoids by micellar electrokinetic capillary chromatography. *J. Chromatogr.* **669**, 264–274.
136 Shepherd, K., McGhie, T. (1995) Novel capillary electrophoresis technique for the study of plant phenolic enzymic oxidation mechanisms. *J. Agric. Food Chem.* **43**, 657–661.
137 Gil, M.; Garciaviguera, C.; Bridle, P.; Tomasbarberan, F. (1995) Analysis of phenolic compounds in Spanish red wines by capillary zone electrophoresis. *Z. Lebensm. Unters. Forsch.* **200**, 278–281.
138 Donaghy, J.; McKay, A. (1995) Measurement of feruloyl/*p*-coumaroyl esterase by capillary zone electrophoresis. *World J. Microbiol. Biotechnol.* **11**, 160–162.
139 Ma, Y.; Meyer, K.; Afzal, D. (1993) Isolation and quantification of ergovaline from *Festuca arundinacea* (tall fescue) infected with the fungus *Acremonium coenophilum* by high performance capillary electrophoresis. *J. Chromatrogr.* **652**, 535–538.
140 Liu, Y.; Sheu, S. (1993) Determination of quaternary alkaloids from *Phellodendri cortex* by capillary electrophoresis. *J. Chromatogr.* **634**, 329–333.
141 Sheu, S.; Chen, H. (1995) Simultaneous determination of twelve constituents of *l*-tzu-tang, a Chinese herbal preparation, by high-performance liquid chromatography and capillary electrophoresis. *J. Chromatogr.* **704**, 141–148.
142 Zong, Y.; Che, C. (1995) Analysis of rhubarb by micellar electrokinetic capillary chromatography. *J. Natural Products – Lloydia* **58**, 577–582.
143 Brumley, W. (1995) Enviromental applications of capillary electrophoresis for pollutant determination. *LC–GC* **13**, 556–566.
144 Gilpin, R.; Pachla, L. (1995) Pharmaceuticals and related drugs. *Anal. Chem.* **67**, R295–R313.
145 Ng, C.; Ong, C.; Lee, H.; Li, S. (1994) Capillary electrophoretic separation of nitrosamines using combined open-tubular and packed capillary columns. *J. Chromatogr. Sci.* **32**, 121–125.
146 Ackermans, M.; Beckers, J.; Everaerts, F.; Hoogland, H.; Tomassen, M. (1992) Determination of sulfamides in pork meat extracts by capillary zone electrophoresis. *J. Chromatogr.* **596**, 101–109.
147 Blais, B.; Cunningham, A.; Yamazaki, H. (1994) A novel immunofluorescence capillary electrophoresis assay system for the determination of chloramphenicol in milk. *Food Agric. Immunol.* **6**, 409–417.
148 Pant, I.; Trenerry, V. (1995) The determination of ascorbic acid and benzoic acid in foods and beverages by micellar electrokinetic capillary chromatography. *Food Chem.* **53**, 219–226.
149 Dinelli, G.; Vicari, A.; Bonetti, A.; Catizone, P. (1995) Comparison of capillary electrophoresis, HPLC and enzyme immunoassay for terbuthylazine detection in water. *J. Agric. Food Chem.* **43**, 951–955.
150 Dinelli, G.; Vicari, A.; Bonetti, A. (1995) Separation of sulfonylurea metabolites in water by capillary electrophoresis. *J. Chromatogr.* **700**, 195–200.
151 Dinelli, G.; Vicari, A.; Brandolini, V. (1995) Detection and quantitation of sulfonylurea herbicides in soil at the ppb level by capillary electrophoresis. *J. Chromatogr.* **700**, 201–207.
152 Farran, A.; Ruiz, S. (1995) Simultaneous separation of neutral and acidic herbicides by micellar electrokinetic capillary chromatography. *Int. J. Environ. Anal. Chem.* **58**, 121–129.
153 Wigfield, Y.; McCormack, K.; Grant, R. (1993) Simultaneous determination of residues of paraquat and diquat in potatoes using high-performance capillary electrophoresis with ultaviolet detection. *J. Agric. Food Chem.* **41**, 2315–2318.
154 Shamsi, S.; Danielson, N. (1995) Ribonucleotide electrolytes for capillary electrophoresis of polyphosphates and polyphosphonates with indirect photometric detection. *Anal. Chem.* **67**, 1845–1852.
155 Buscher, B.; Irth, H.; Andersson, E.; Tjaden, U.; van der Greef, J. (1994) Determination of inositol phosphates in fermentation broth using capillary electrophoresis with indirect UV detection. *J. Chromatogr.* **678**, 145–150.
156 Blatny, P.; Kvasnicka, F.; Kenndler, E. (1995) Determination of phytic acid in cereal

grains, legumes, and feeds by capillary isotachophoresis. *J. Agric. Food Chem.* **43**, 129–133.
157 Maragos, C. (1995) Capillary zone electrophoresis and HPLC for the analysis of fluorescein isothiocyanate-labeled fumonisin B1. *J. Agric. Food Chem.* **43**, 390–394.
158 Arentoft, A.; Greirson, B. (1995) Analysis of 3-(*N*-oxalyl)-L-2,3-diaminopropanoic acid and its alpha-isomer in grass pea (*Lathyrus sativus*) by capillary zone electrophoresis. *J. Agric. Food Chem.* **43**, 942–945.
159 Luong, J.; Male, K.; Masson, C.; Nguyen, A. (1992) Hypoxanthine ratio determination in fish extract using capillary electrophoresis. *J. Food Sci.* **57**, 77–81.
160 Mopper, B.; Sciacchitano, C. (1994) Capillary zone electrophoretic determination of histamine in fish. *J. Assoc. Off. Anal. Chem. Int.* **77**, 881–884.
161 Tienstra, P.; van Riel, J.; Mingorance, M.; Olieman, C. (1992) Assessment of the capabilities of capillary zone electrophoresis for the determination of hippuric and orotic acid in whey. *J. Chromatogr.* **608**, 357–361.
162 Pleasance, S.; Thibault, P.; Kelly, J. (1992) Comparison of liquid-junction and coaxial interfaces for capillary electrophoresis–mass spectrometry with application to compounds of concern to the aquaculture industry. *J. Chromatogr.* **591**, 325–339.
163 Locke, S.; Thibault, P. (1992) Improvement in detection limits for the determination of paralytic shellfish toxins in shellfish tissues using capillary electrophoresis/electrospray mass spectrometry and dicontinuous buffer sytems. *Anal. Chem.* **66**, 3436–3446.
164 Buzy, A.; Thibault, P.; Laycock, M. (1994) Development of a capillary electrophoresis method for the characterization of enzymatic products arising from the carbamoylase digestion of paralytic shellfish poisoning toxins. *J. Chromatogr.* **688**, 301–316.
165 Thompson, C.; Trenerry, V.; Kemmery, B. (1995) Micellar electrokinetic capillary chromatographic determination of artificial sweeteners in low-joule soft drinks and other foods. *J. Chromatogr.* **694**, 507–514.
166 Thompson, C.; Trenerry, V.; Kemmery, B. (1995) Determination of cyclamate in low joule foods by capillary zone electrophoresis with indirect ultraviolet detection. *J. Chromatogr.* **704**, 203–210.
167 Liu, J.; Li, S. (1995) Separation and determination of stevia sweeteners by capillary electrophoresis and high performance liquid chromatography. *J. Liq. Chromatogr.* **18**, 1703–1719.
168 Cancalon, P. (1995) Capillary electrophoresis and fruit juice monitoring. In *Methods to Detect Adulteration of Fruit Juice Beverages*, Nagy, S.; Wade, R.L. (eds), AgScience, Auburndale, FL, pp. 181–197.
169 Hall, C.; An, Zhu; Zeece, M. (1994) Comparison between capillary electrophoresis and high-performance chromatography separation of food grade antioxidants. *J. Agric. Food Chem.* **42**, 919–921.
170 Kajiwara, H.; Sato, A.; Kaneko, S. (1993) Analysis of calcium and magnesium ions in wheat flour by capillary zone electrophoresis. *Biosci. Biotech. Biochem.* **57**, 1010–1011.
171 Schoneich, C.; Huhmer, A.; Rabel, S.; Stobaugh, J.; Jois, S.; Larive, C.; Siahaan, T.; Squier, T.; Bigelow, D.; Williams, T. (1995) Separation and analysis of peptides and proteins. *Anal. Chem.* **67**, R155–R181.
172 Wheat, T. (1995) In *Principles and Practice of Peptide Analysis with Capillary Zone Electrophoresis. Peptide Analysis Protocols*, Dunn, B.M.; Pennington, M.W. (eds), Humana Press Inc, Totowa, NJ.
173 Thompson, C.; Trenerry, V. (1995) Determination of synthetic colours in confectionery and cordials by micellar electrokinetic capillary chromatography. *J. Chromatogr.* **704**, 195–201.
174 Ye, J.; Baldwin, R.P. (1994) Determination of carbohydrates, sugar acids and aldihols by capillary electrophoresis and electrochemical detection at the copper electrode. *J. Chromatogr.* **687**, 141–148.
175 Lopez Martinez, N.; Rodriguez Roldan, A. (1993) Application de la electrophoresis capilar micelar al analisis cuantitativo de acidos carboxilicos. *Alimentaria* **243**, 71–74.

39 Capillary zone electrophoresis analysis of additives in food samples
R. SCHUSTER

39.1 Introduction

Which analytical technique is used very often is determined by the physical properties of the compounds to be analyzed. Highly volatiles and slightly polar compounds can be examined by gas chromatography (GC) and non-volatile and polar compounds by high performance liquid chromatography (HPLC) or thin layer chromatography (TLC). The highly polar or low volatility compounds quite often require special conditions in HPLC, mobile phases (e.g. ion pairing) or stationary phases (ion exchange) which are troublesome to handle and result in poor reliability, efficiency and separation capability. Figure 39.1 shows compounds arranged according to polarity and volatility. The domain of GC is located on the left hand side, HPLC ideally fits into the middle and partially onto the right hand side depending on analytical modes or stationary phases, while capillary zone electrophoresis (CZE) is best suited to analysis of the non-volatile, hydrophilic analytes. From these facts the analysis of artificial sweeteners and synthetic dyes has been selected to demonstrate the capabilities of CZE.

39.2 Artificial sweeteners

Sweetening agents can be classified as belonging to one of two main groups: caloric or nutritive, and non-caloric or non-nutritive compounds. Nutritive sweeteners are carbohydrates (or their derivatives such as glucose, fructose and maltose) or products hydrolyzed from starch. Non-nutritive sweeteners do not belong to any particular chemical group. Synthetic sweeteners are steadily increasing in importance with increased public awareness of diabetes and its special dietary requirements, and more consumers becoming concerned about obesity and dental caries. The most frequently used synthetic sweeteners are saccharin, cyclamate, aspartame and acesulfame.

To date, artificial sweeteners have been determined by HPLC with reversed phase chromatography using different buffer systems, ion pairing

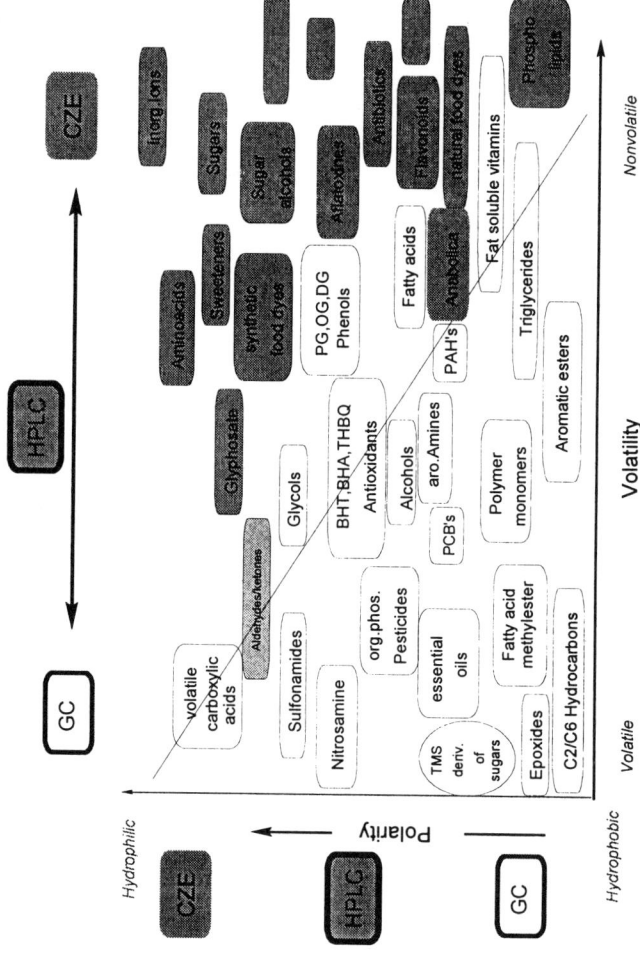

Figure 39.1 Diagram of different analytes arranged according polarity and volatility.

reagents and specific derivatization procedures. Derivatization improves detection limitations for these compounds, indicating absorbance only at lower UV wavelengths such as for acesulfame and cyclamate.

Toxicological data has led to the use of some artificial sweeteners being controlled, for example cyclamate is banned in the United States, the United Kingdom and Japan. Reliability means obtaining analytical data for original as well as degraded compounds. Such varied methods with their differing derivatization protocols make the analysis of artificial sweeteners time consuming and labor intensive.

Aspartame is metabolized to aspartic acid, methanol and phenylalanine. Simultaneous analysis of these compounds is thought to be difficult due to their different polarities and the lack of UV absorbance for aspartic acid and methanol, so that derivatization is required individually. Thus, CZE is an alternative method to HPLC. All compounds can be separated simultaneously.

39.3 Natural and synthetic dyes

Color is a vital constituent of foods and probably the first characteristic perceived by the human senses. Almost all foods, from raw agricultural commodities to finished products, have an associated color. Furthermore, many tests have shown that color can dominate the flavor of a food organoleptically.

Dyes have been added to foods and include chlorophylls, carotinoids, flavonoids and anthocyans extracted from different plants. Nowadays synthetic dyes have widely replaced natural dyes. These dyes are used to supplement and enhance the natural colors destroyed during processing or storage, and substantially increase the appeal and acceptability of foodstuffs where no natural colors exist, for example, soft drinks or ice cream. But synthetic dyes are also used to mask decay, to redye food, to mask aging effects or to disguise poor products.

Colors permitted for food use can be divided in three categories: synthetic dyes – described in this study, natural colors (for example, caramel or beetroot) and naturally identical colors (canthaxanthine), and inorganic pigments.

Synthetic colors in food products are predominantly azo and triarylmethane dyes. These are mostly acidic or anionic dyes containing carboxylic acid, sulfonic acid or hydroxy groups, which form negatively charged colored ions at basic pH ranges. CZE is an ideal tool because it can separate all the different functional groups in a single run. CZE separates compounds based on charge and size.

The range of color shades covered by the azo group includes red, orange, yellow, brown and blue-black. These dyes are prepared by coupling diazotized sulfanilic acid to a phenolic sulfonic acid moiety that often contains unwanted byproducts from corresponding impurities. Triarylmethane colors are distinguished by their brilliance of color and high tinctorial strength, but have poor light stability. The United States (US) and the European Community (EC) are the two major geographical areas where color regulations are enforced. The lists of permitted colors are updated continuously because of suspected carcinogenicity. For example:

- amaranth (E123, FD&C Red No. 2) – one of the most widely-used red colors – was delisted in the US in 1970;
- indigo carmine (E132, FD&C Blue No. 2) was delisted in 1980;
- tartrazine (E102, FD&C Yellow No. 5) was subjected to rigorous tests; and
- in 1990 the use of erythrosine (E127, FD&C Red No. 3) was discontinued.

39.4 Concrete CZE analytical procedure

CZE separations were performed combined with photodiode array detector and ChemStation software.

39.4.1 Sweeteners

Separations were carried out on fused-silica 50 µm i.d. capillaries with 64.5 cm total length and 56 cm effective length with an extended path length (bubble cell). All separations were performed at 25°C using a 20 mM sodium tetraborate buffer at pH 9.4. Capillaries were preconditioned by flushing first with 1 M sodium hydroxide for 3 min followed by a flush with running buffer for 10 min. Samples were introduced hydrodynamically at 50 mbar and an injection time of 2 s. They were analyzed with an applied voltage of 30 kV and detected at 192 nm/2 nm bandwidth. After each run the column was rinsed with the separation buffer for 2 min.

The separation of common artificial sweeteners can be attained by this method including phenylalanine, aspartame, p-hydroxybenzoic acid (PHB) propyl ester, PHB ethyl ester, PHB methyl ester, dehydroacetic acid, cyclamide, sorbic acid, benzoic acid, aspartic acid, saccharin and acesulfame in this order of elution in an electropherogram within 9 min from the injection. Aspartame contains the decomposition products of phenylalanine, aspartic acid and methanol.

Apart from several methods used for analyzing sweeteners, the lack of chromophores (cyclamate, acesulfame) means that wavelengths in the low UV range <200 nm are needed but which cannot be detected without using ion pairing reagents. The use of a photodiode array detector with the CE system as a monitoring unit allows detection of compounds at 192 nm and simultaneously obtains spectral information of the compound of interest. Using an accumulated spectra library, compounds can be analyzed not only simply by migration time, but also by spectral comparison of peak purity, which can be performed by overlaying spectra of the peak with standard spectra in the library. With the generation of new software, just one step enabled combination of all these capabilities, migration time, spectral identification and peak purity, to produce quantitative reports based on three-dimensional data.

39.4.2 Synthetic dyes

Separation of the dyes was performed at 30°C using 10 mM sodium phosphate with 5 mM sodium hydrogencarbonate buffer at pH 10.5. Capillaries were preconditioned by flushing first with 1 M sodium hydroxide for 3 min followed by a flush with running buffer for 10 min. A fused-silica capillary of 50 μm i.d. with 64.5 cm total length and 56 cm effective length was used. Samples were introduced hydrodynamically at 100 and 200 mbar s^{-1} followed by a 200 mbar s^{-1} buffer plug in each case. They were analyzed with an applied voltage of 30 kV and detected at 215/50 nm, 520/60 nm red, 410/60 nm yellow and 598/4 nm blue. After each run inlet and outlet, the vials were replenished and the column was rinsed with the separation buffer for 1 min. A voltage ramp from 0 to 30 kV over 0.5 min was performed to avoid possible thermal expansion and loss of sample.

Following this procedure, the separation of the common synthetic dyes patent blue E131, acid brilliant green E142, erythrosin E127, indigocarmine E132, E105, carminic acid, chryosin E103, sunset yellow E110, sunset yellow E111, scarlet red E125, quinoline yellow E104, azorubine E122, ponceau 4R E124, naphthol yellow, amaranth E123, brilliant black E151, tartrazine E102 and ponceau E126 was attained in this elution order in the electrogram within 12 min of injection.

Separation of colors as anions by CZE was chosen for method development according to the functional groups, mostly sulfonic acid groups, in the pH range of 8–11. Different buffers like borate, CAPS (3-cyclohexylamino-1-propane sulfonic acid) and phosphate were tested at different pH values and concentrations. The optimal buffer temporally used was a 10 mM phosphate at pH 10.5. Reliable data showed a clear trend to shorter migration times according to pH changes in the buffer even when replenishment of the two vials was used mainly due to the non-stable pH of this buffer system. At this pH, sodium hydrogen carbonate was found to be an alternative buffer system. Finally, a mix of 10 mM phosphate and 5 mM sodium hydrogencarbonate adjusted to pH 10.5 with sodium hydroxide gave optimal conditions for separation of all colors. Employing different pH values (e.g. 10.5–10.2) the migration order of two colors, azorubine E122 and quinoline yellow E104, was reversed while the migration times of other compounds were more or less stable.

39.5 Application

39.5.1 Sweeteners

The method has been applied to different matrices: beverages (Coke, Sprite, coffee) and tablets (Figure 39.2). All peaks were identified. In Figure 39.3 the electropherogram of a sweetener tablet overlaid with the electropherogram of directly injected coffee sweetened with tablets was shown.

39.5.2 Dyes

In Figure 39.4, an electropherogram of a carbonated drink containing colors and artificial sweeteners is presented. The method for colors has been applied to different matrices: beverages (carbonated drinks), 'spaetzle' (German noodles), candy cherries (sugar-coated cherries), 'taramasalata' extract and capsules. It has also been tested on the quality control of colors themselves, e.g. quinoline yellow. A carbonated drink, a 'woodruff-ade' (lemonade from plant extract), was injected directly and peaks were identified (Figure 39.4). The 'mint' impression (green color) here was given by a mix of quinoline yellow E104 monitored at 410 nm with patent blue E131

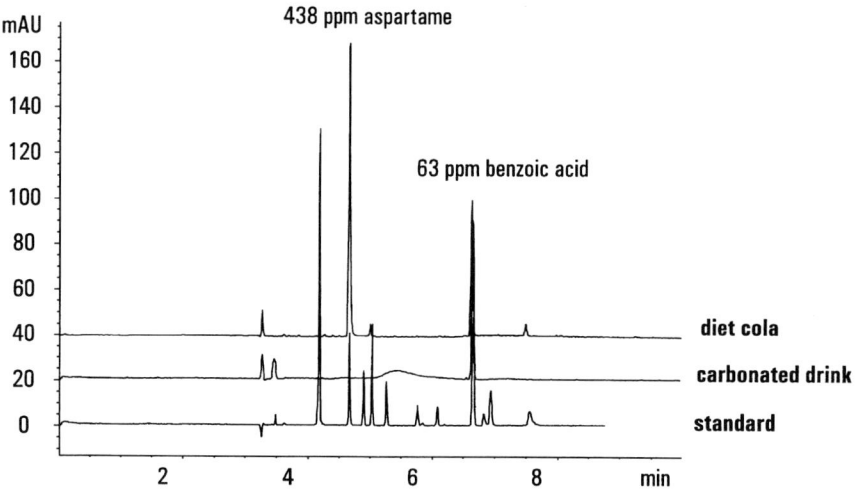

Figure 39.2 Electrogram of a diet cola containing aspartame, benzoic acid and a carbonated drink containing benzoic acid.

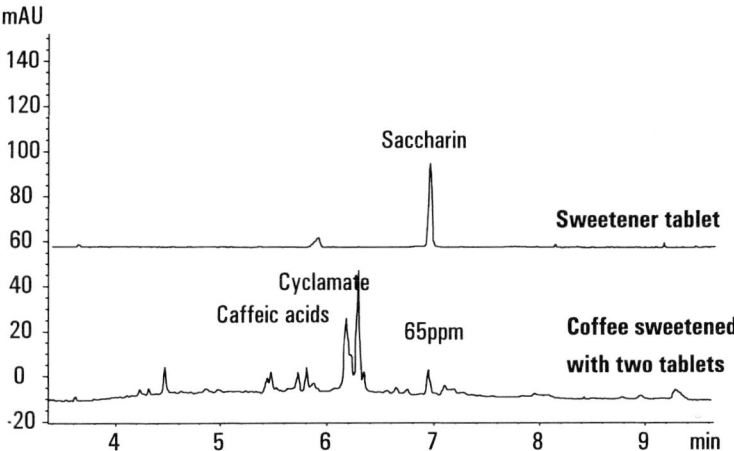

Figure 39.3 Electrogram of analysis of a sweetener tablet overlaid with the electropherogram of directly injected coffee sweetened with two such tablets.

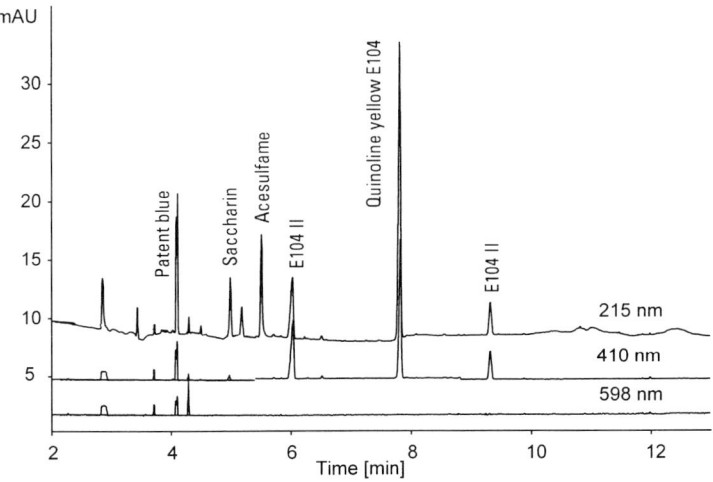

Figure 39.4 Electropherogram of a carbonated drink containing colors and artificial sweeteners.

at 598 nm. Other compounds labeled on a bottle only as 'sweeteners' could also be quantified as acesulfam and saccharin. Figure 39.5 shows an electropherogram of a pasta extract, with a sample prepared by the 'woolfiber' method. There the food sample 'spaetzle' containing quinoline yellow E104 with impurities and sunset yellow E110 to simulate the use of eggs is indicated. Both 'yellow' compounds were selectively detected at 410 nm.

Figure 39.6 shows the electropherogram of a capsule. This indicates the analysis of capsules, dissolved in water and injected directly. A red 'capsule' used is composed of two red colors, erythrosine E127 and amaranth E123 monitored at 520 nm and a yellow one, tartrazine E102 at 410 nm. The preservatives PHB propyl, PHB ethyl and PHB methyl esters, at 215 nm, could be identified. Under these conditions some of the sweeteners and preservatives could be determined by the method used for sweeteners.

Apart from classical food applications the quality control of individual colors plays an important role. The use of intermediates containing impurities and coupling reactions results in the formation of unwanted products. An example was the analysis of quinoline yellow E104 containing seven impurities. After spectral analysis two different 'yellow' types were identified, type I with a maximum at 422 nm and type II at 387 nm. All the other compounds could be associated with either of the two spectral types, a fact which can be seen in all samples containing quinoline yellow E104 (Figures 39.4 and 39.5). Similar problems arise for sunset yellow E110–

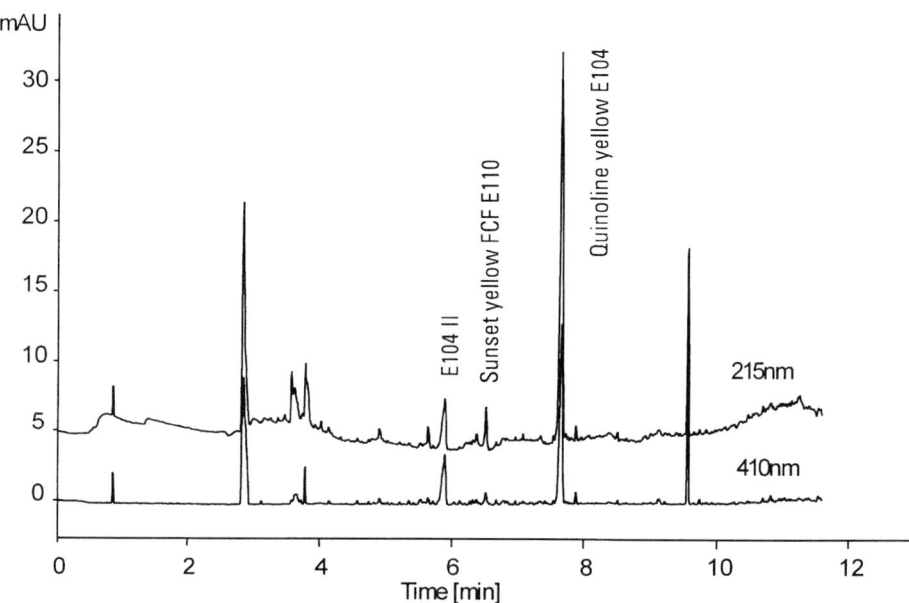

Figure 39.5 Electropherogram of pasta extract using a wool-fiber sample preparation method.

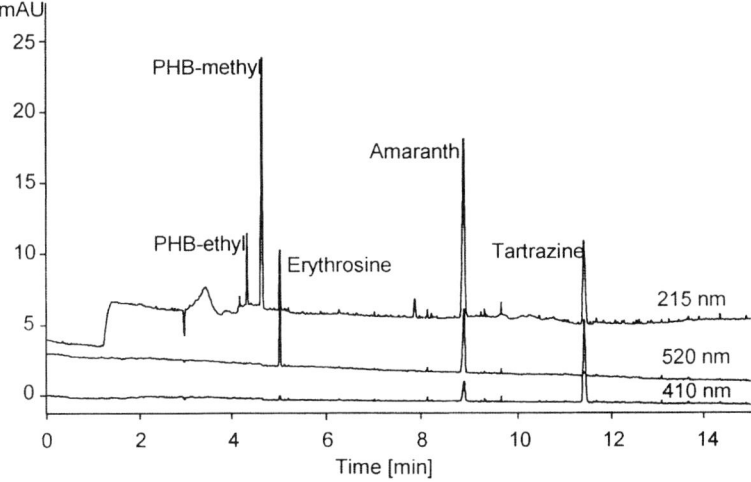

Figure 39.6 Electropherogram of a tablet capsule.

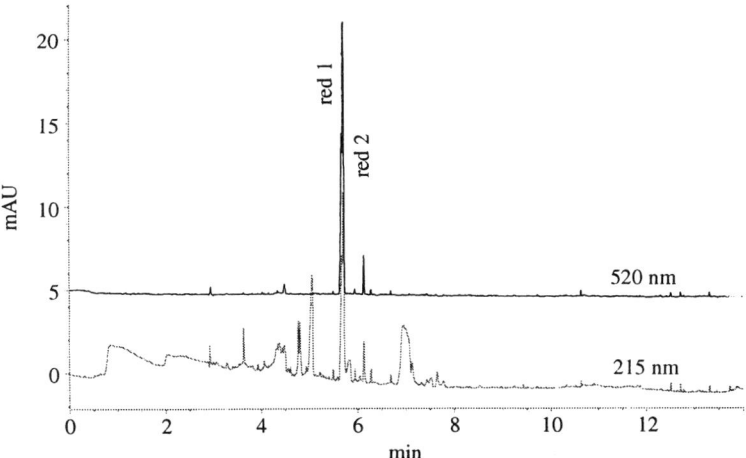

Figure 39.7 Electropherogram of an aqueous 'beetroot' extract.

FD&C Yellow No. 6, tartrazine E102–FD&C Yellow No. 5 and FD&C Red No. 40.

In recent years concern has been expressed about the safety of certain synthetic dyes and this has prompted an increased consideration of the use of natural colorants for food samples. Natural and nature-identical colors are usually mixtures of several colored as well as non-colored compounds. Figure 39.7 shows the analysis of an aqueous beetroot extract with two red compounds (520 nm).

39.6 Reproducibility and linearity

Repeatability for all compounds was better than 0.5% for retention time and between 2 and 4% for peak area. The calculation was based on ten runs with injected amounts of 60 to 260 ng with 100 mbar s^{-1} and 6–26 ng with 200 mbar s^{-1}, respectively. After each run, replenishment is necessary to attain satisfactory reproducibility. The correlation of linearity is 0.9992 and regression line is $Y = 6.1747\mathrm{e}^{-1} X + 1.0606$. X and Y indicate the amount and area, respectively.

39.7 Conclusion

CZE is found to be well suited to controlling synthetic food dyes in food samples, peak purity in sweeteners and preservatives and for peak identification and confirmation of compounds. Use of diode array detection allowed individual color groups to be monitored selectively at appropriate wavelengths. The detection limit for most compounds is in the low nanogram range.

40 Analysis of underivatized carbohydrates by capillary electrophoresis and electrochemical detection
R.P. BALDWIN

40.1 Introduction

The determination of carbohydrate compounds represents one of the most challenging analytical problems in biotechnology today. Compared with peptides and proteins, for example, carbohydrates present far greater difficulties for separation because of the large number of different monosaccharides that occur naturally, the large number of ways in which they can be linked together to form oligo- and polysaccharides, the very similar chemical reactivities that different carbohydrates exhibit and the variety of non-carbohydrate substituents with which they can be modified. Nevertheless, considerable progress has been made in carbohydrate analysis and a range of instrumental techniques are available for this purpose.

Until quite recently, methods utilizing high performance liquid chromatography (HPLC) have shown the greatest promise for carbohydrate analysis; and most of the different HPLC approaches, e.g. reverse-phase, ion-exchange, size-exclusion and affinity chromatography, have been exploited with some success to resolve relatively complex carbohydrate mixtures. However, over the past few years, capillary electrophoresis (CE) has begun to be used as an alternative to HPLC in these applications. The principal reason for this switch lies in the higher separation efficiencies, typically 100 000 plates m^{-1} or greater, generally available via CE. In addition, CE also provides separations that are relatively rapid and require only minute sample volumes. Two excellent reviews covering the broad topic of CE of carbohydrates have recently appeared [1, 2].

Unfortunately, the application of CE techniques to carbohydrate analysis is not completely straightforward. Rather, before CE can be rationally applied to carbohydrate analysis, two specific problems must be addressed. First, most carbohydrates do not possess readily ionizable functionalities and hence are uncharged under most solution conditions. Therefore, it is necessary to find a way to impart some fixed charge to any carbohydrate analyte before it can undergo the ionic migration fundamental to the CE separation. This can be accomplished through a variety of special procedures such as formation of a charged carbohydrate complex by adding borate to the electrophoresis buffer or formation of a charged carbohydrate derivative by covalent functionalization. Second, carbohydrates generally do not possess native chromophoric groups that absorb strongly at readily accessible UV–visible wavelengths. Therefore, detection by the usual spectrophotometric methods, involving either absorption or fluorescence, is attractive only if appropriate sample derivatization steps have first been carried out on the carbohydrate species. Several such procedures have been employed for CE detection, for example the formation of pyridylamino [3, 4], quinolinecarboxaldehyde [5, 6] and aminonaphthalene [7, 8] derivatives. However, these procedures all add time and complexity to the analysis and also compromise the ability to work with small sample volumes, which is one of the most useful characteristics of CE.

For these reasons, the development of direct CE separation and detection schemes, offering good analytical performance for carbohydrates but avoiding the need for sample derivatization, are of considerable practical interest. One such direct CE analysis approach, which has begun to be investigated in several research laboratories, is provided by electrochemical (EC) detection using metal electrodes. The basis for this 'CE–EC' approach is found in analogous HPLC work from the past 10–15 years which showed that underivatized carbohydrates and related compounds can be oxidized and detected directly at metal electrodes such as Pt, Au and Cu [9, 10]. One important restriction for the use of all of these metal electrodes for carbohydrate analysis is that the detection is carried out in highly alkaline media, typically containing OH⁻ concentrations of 0.10 M or higher. This requirement has been somewhat limiting for earlier HPLC applications because only specialized column packing materials which are stable at high pH are appropriate for use. However, for CE, this high pH requirement can actually be quite useful because most simple carbohydrates have pK_a values in the 12–13 range and thus are deprotonated anions and perfectly suited for electrophoresis under these conditions. In fact, for many carbohydrate compounds, the pH of the electrophoresis medium provides an important new parameter for CE that can be used to control the degree of ionization of high pK_a functionalities present and therefore the migration times and separation selectivities obtained [11].

Thus, the use of metal electrodes in combination with strongly alkaline solution conditions represents an attractive new CE approach for the separation and the detection of carbohydrates directly without the need for prior derivatization. In this report, we discuss the basis of this approach. In particular, we will consider the nature of the processes involved in the oxidation of carbohydrates at metal electrodes, the practical factors to be considered in adapting EC methods to CE systems, and most important, the nature and variety of carbohydrate separations that have so far been obtained by CE in high pH media.

40.2 Electrochemical detection of carbohydrates

The metals that have been employed for carbohydrate oxidation and detection in HPLC have included Pt, Au, Ni, Cu and Ru. Of these, only Au and Cu have so far been used in CE. The principal difference between these two electrode materials is that, with Au, the applied potential must be continuously pulsed (hence, the term 'pulsed amperometric detection' or PAD) for stable long-term response to be maintained while constant potential operation is feasible with Cu. In PAD, triply pulsed waveforms are usually employed where the detection step, an oxidation carried out at a low positive potential, is followed by application of a more positive potential to form an oxide layer on the Pt or Au that allows products of the carbohydrate oxidation to desorb and then by switching to a negative, reducing potential to remove the passivating oxide film [9]. With Cu, carbohydrate detection is also accomplished by oxidation at a modest positive potential where background currents are low and highly sensitive detection is therefore possible. However, unlike Au, the Cu electrode is not subject to fouling problems and even without pulsing the applied potential, nearly the same signal levels persist at Cu for periods of continuous use lasting at least 2 weeks [12]. For both electrodes, highly sensitive detection of carbohydrate species is possible in CE–EC, typically down to the femtomole level. However, the constant potential capability of the Cu electrode permits the use of somewhat less expensive and less complicated instrumentation. Consequently, this electrode material has been the preferred one in our laboratory.

In general, metal electrodes can provide detection for a wide variety of carbohydrate compounds in addition to simple mono- and disaccharides. With Cu electrodes, for example, the key structural feature required for detection is the presence of multiple hydroxyl groups in adjacent or, at least, nearby positions [13]. Thus, not only simple sugars but also many related compounds, including alditols, aminosugars and aldonic, aldaric and uronic acids, can be detected equally well at Cu in CE–EC. In addition, detection is not restricted to small molecular systems. Rather, relatively large polysaccharides such as 18 000 molecular weight dextrans also yield very good response [14]. Typical analytical performance obtained for CE–EC at Cu electrodes for a range of carbohydrate compounds is shown in Table 40.1.

40.3 CE separations at high pH

Nearly all CE work to date involving high pH separation of carbohydrate compounds has employed the technique of capillary zone electrophoresis

Table 40.1 Analytical performance of Cu electrode for CE–EC of carbohydrate compounds[a]

Carbohydrate compound	Detection limit (fmol)	Linear range
Glucose	1	1 μM–1 mM
Galactose	1	1 μM–1 mM
Ribose	1	1 μM–1 mM
Maltose	0.7	2 μM–5 mM
Glucitol	0.5	0.5 μM–1 mM
Inositol	0.4	0.4 μM–0.6 mM
Gluconic acid	1.5	1.5 μM–1 mM
Galactonic acid	2	2 μM–1 mM
Glucuronic acid	4	4 μM–2 mM
Galacturonic acid	3	3 μM–2 mM
Glucaric acid	6	6 μM–4 mM
Galactaric acid	8	8 μM–4 mM
Maltoheptaose	0.05	0.2 μM–1 mM
Maltopentaose	0.3	0.8 μM–1 mM

[a] From references 11, 14 and 20.

with the electrophoresis media consisting of simple aqueous NaOH solutions of 50 mM concentration or higher. These OH⁻ concentrations, which are required by Au and Cu electrodes for effective carbohydrate detection, are also sufficient to begin deprotonation and ionization of most simple sugars. (The pK_a values for glucose and galactose = 12.35, pK_a for ribose = 12.21, pK_a for maltose = 11.94 [15]). Under these conditions, most small carbohydrate compounds can be readily separated from one another and variation of the OH⁻ concentration can be used effectively to control migration times and resolution. For example, different members of the same carbohydrate family (e.g. glucitol, glucose, gluconic acid, glucuronic acid and glucaric acid) can be easily resolved from one another in 50 mM NaOH [11]. However, different alditols (e.g. glucitol, galactitol, mannitol and inositol), which have pK_a values in the 13–14 range [15] and are virtually uncharged under these conditions, all exhibit very short migration times and have completely overlapping peaks. Thus, alditols are not able to be resolved from one another unless OH⁻ levels of at least 0.25 M are employed [11].

Unfortunately, the simple OH⁻ variations useful for achieving separations with mono- and disaccharides have not been effective for oligo- and polysaccharides. Rather, for CE in aqueous OH⁻ media, the migration pattern of an oligosaccharide mixture is such that the larger components emerge first, with the smallest appearing a relatively short time later and all intermediate-sized species crowded in between [14]. The reason for this can be described as follows. In all CE using an open capillary format, a sample component's residence time in the capillary depends on the mutual interplay of electromigration (EM) and electroosmotic flow (EOF). At high pH, nearly all carbohydrate compounds are negatively charged and tend to migrate toward the electrophoresis anode. At the same time, EOF is generally very fast and in the direction of the cathode under these conditions. As a result, the first peak detected in a CE run at high pH generally corresponds to the most slowly migrating carbohydrate, whose electromigration is least able to offset the more rapid electroosmosis. Similarly, the last peaks represent the more mobile or highly charged species, whose EM rates are higher.

The net result is that the decreases in EM rates that occur as the length of the carbohydrate chain is increased are not large enough to allow a good CE separation to be achieved. Thus, although good separations of small carbohydrates can usually be obtained under the simple high pH conditions required for EC detection, an upper limit in size is rapidly reached above which a useful separation cannot be obtained because of crowding in the early part of the electropherogram. In our work, this limit is usually reached for carbohydrates containing only 2–3 sugar units.

A simple way around this difficulty is just to reverse the direction of EOF by addition of a cationic surfactant to the alkaline electrophoresis buffer [14]. The idea is that, by interaction with the negative siloxy functionalities in the inner capillary wall, the positive surfactant adsorbs and alters the charge on the inside of the capillary from its usual negative value to an overall positive state. If this is accomplished and the polarity of the electrophoresis voltage is reversed (so that detection is performed at the anode end of the capillary), then the oligosaccharide elution order can be reversed as well. In this situation, the smallest carbohydrate systems exit the capillary first and no *a priori* upper limit on oligomer size applies. Table 40.2 compares the CE separations obtained for a mixture of linear maltoses from glucose to maltoheptaose both with no surfactant present and with the surfactant cetyltrimethylammoniumbromide (CTAB) added. The potential

Table 40.2 CE separations for linear maltoses[a]

Compound	Migration time (min)	
	0.10 M NaOH[b]	0.10 M NaOH/10 mM CTAB[c]
Glucose	24.0	18.5
Maltose	23.0	19.0
Maltotriose	22.8	19.5
Maltotetraose	22.8	20.0
Maltopentaose	22.8	21.0
Maltohexaose	22.8	22.6
Maltoheptaose	22.8	24.0

[a] From reference 14.
[b] Separation voltage, −10 kV.
[c] Separation voltage, +20 kV.

of the latter approach for application to samples containing larger oligo- and polysaccharides is clear. In fact, extension of the approach to the separation of much larger carbohydrates, such as 18 000 molecular weight dextrans, has been reported [14].

40.4 Applications

The application of CE–EC techniques for the analysis of carbohydrates represents a development which was first reported only in 1993. Consequently, the number of such applications reported to date is not large. Those that have appeared since then are summarized in Table 40.3. Despite the brevity of this listing, it is clear that CE–EC using metal electrodes and high-pH electrophoresis conditions possesses great potential for a wide range of carbohydrate samples.

The different cell configurations used in CE–EC have been described in detail in an informative review by Ewing *et al.* [16]. All of these have been employed in the carbohydrate analyses reported thus far and it is appropriate to provide a few comments on CE–EC cell design here. In all cases, the cell configuration employed must address two specific detection difficulties common to all CE–EC applications, whether the system is intended for carbohydrates or some other analyte. First, there is the need to adapt the size and placement of the detector electrode to the micron dimensions of the capillary. Because of this, microelectrodes consisting of wires 5–50 µm in diameter and 100–500 µm in length have usually been employed. These are either inserted a short distance into the capillary itself or are left in solution but are carefully aligned with the exit of the capillary tube by means of a micropositioning device. Second, there is the need to isolate the high voltages (typically 10–30 kV) used for the CE separation from the much smaller EC potentials (typically 0–1 V) used to control analyte oxidation and reduction. Such electrical 'decoupling' can be accomplished by placing a small fracture, usually coated with Nafion, in the capillary a short distance before the exit. When this is done correctly, ion flow and electrical conductivity, but not bulk solution flow, are possible through the fracture. As a result, the CE voltage is dropped across the capillary only up to the point of the fracture but not beyond and an electrode placed at the capillary exit is effectively electrically isolated. This arrangement is usually termed 'off-column' detection.

Alternatively, when the capillaries in use possess sufficiently small internal diameters, the electric field due to the CE voltage decays rapidly enough in the electrolyte solution outside the capillary that special decoupling procedures are not required as long as the detection electrode is kept outside the capillary channel. This configuration, in which the electrode is carefully placed at the end of the capillary without the use of decoupling, is

Table 40.3 Summary of applications employing CE–EC for carbohydrate analysis

Analyte	Electrode	EC Mode	CE Conditions	Ref.
(1) 15 Mono-, di- and trisaccharides in soft drinks	25 μm Cu wire	End column, +0.6 V vs. Ag/AgCl	100 mM NaOH	17
(2) Glucose and glucosamines in human blood	50 μm Au wire	Off column, PAD	10 mM NaOH, 8 mM Na_2CO_3	18
(3) 8 Monosaccharides and alditols	10 μm Au disk	End column, PAD	100 mM NaOH	19
(4) 5 Mono- and disaccharides	127 and 320 μm Cu disks	Wall-jet, +0.6 V vs. Ag/AgCl	100 mM NaOH	20
(5) 7 Alditols and mono- and disaccharides; glucose in human blood	25 μm Au wire	Optimized end column, PAD	80 mM NaOH	21
(6) Alditols, monosaccharides and sugar acids of glucose; mixtures of 8 alditols	100 μm Cu disk	Wall-jet, +0.6 V vs. Ag/AgCl	25–200 mM NaOH	11
(7) 13 Alditols, mono- and disaccharides, and amino acids in human urine	127 μm Cu disk	Wall-jet, +0.6 V vs. Ag/AgCl	100 mM NaOH	12
(8) 7 Malto-oligo-saccharides; starch hydrolysates; 1500–18 300 Da dextrans	127 μm Cu disk	Wall-jet, +0.6 V vs. Ag/AgCl	100–200 mM NaOH, 10 mM CTAB	14

termed 'end-column' detection. The obvious disadvantage of this simpler approach is the considerable difficulty involved in placing the microelectrode at the optimum position and then maintaining it in this position stably and reproducibly. Two approaches are available to address this problem. In the first, a short distance up the exit end of the capillary is enlarged by etching with HF in order to provide a relatively large cavity into which the electrode can be inserted and thereby fixed in position; this is termed 'optimized end column' detection. In the second approach, a disk-shaped electrode possessing a diameter much larger than the capillary opening can be placed at the end of the capillary in a 'wall-jet' configuration. Because of its relatively large size, such an electrode is easier to work with and initial electrode placement with respect to the capillary is far less critical and far easier to reproduce over the course of an extended series of experiments.

Most of the CE–EC applications with carbohydrates reported to date have been concerned with the separation and determination of simple sugars. The first such application was reported by Colon et al. [17] who used a 25 µm diameter Cu wire to detect 15 mono-, di- and trisaccharides following separation by CE in 0.10 M NaOH. A constant potential of +0.60 V vs. Ag/AgCl was applied to a Cu electrode set up in an end-column mode. Detection limits of approximately 50 fmol, separation efficiencies of up to 200 000 and stability over several hundred runs were obtained. O'Shea et al. [18] were the first to employ PAD in CE–EC. In this work, a 50 µm diameter, 350 µm long Au wire electrode in an off-column configuration was able to achieve slightly improved detection limits (22.5 fmol for glucose). A triply pulsed potential waveform similar to that used in HPLC–PAD was employed, and the suitability of the approach for the determination of glucose in human blood was demonstrated. Subsequently, Lu and Cassidy [19] also reported the use of PAD at an Au electrode, this time a 10 µm diameter disk operated in an end-column configuration, and were able to obtain detection limits in the low femtomole range for eight simple sugars and alditols. Ye and Baldwin [20] demonstrated that highly sensitive detection of these compounds could also be carried out at a large electrode operated in a wall-jet arrangement. In this work, Cu disks, 127 µm and 320 µm in diameter, gave detection limits of 1–2 fmol but with much simpler and more reproducible electrode placement. With experience, such electrodes could be positioned acceptably with respect to the capillary exit in a matter of minutes with a variation of only 10–15% from one alignment to the next. Very recently, Roberts and Johnson [21] completed a systematic evaluation of the important waveform parameters for PAD at an Au electrode in the detection of several alditols and simple carbohydrates.

In addition to the above reports, a few other studies have focused on extending CE–EC applications to a wider range of carbohydrate compounds than just the common mono- and disaccharides. For example, Ye and Baldwin [11] showed that the Cu electrode can give just as effective detection for alditols and sugar acids (i.e. aldonic, uronic and aldaric acids) as it does for simple sugars, with detection limits at the low femtomole level and linear ranges extending from micromole to millimole concentrations. Furthermore, CE in highly basic media was particularly well suited to the separation of such species because of their different charges at pH 12 and above. In order to demonstrate this, it was shown that all of the members of the glucose and galactose families could be easily resolved from one another in 50 mM NaOH. However, a mixture of eight alditols, which were virtually inseparable at this [OH$^-$], were able to be resolved completely when 250 mM OH$^-$ was employed as the electrophoresis buffer. Thus, it was possible to use pH as a means of optimizing the CE separation to best suit the kind of carbohydrate compounds present in a given sample. Subsequently, Voegel and Baldwin [12] used this approach to separate and detect 13 different alditols, mono- and disaccharides and amino acids in human urine in a single CE run. Most recently, Zhou and Baldwin [14] described how CE–EC with Cu electrodes can be applied to mixtures of complex polysaccharides. In this work, the positively charged surfactant CTAB was added to the strongly alkaline electrophoresis medium in order to reverse the EOF direction and permit elution to be in order of increasing molecular size. Carbohydrate systems analyzed by this approach included linear maltoses, enzymatically hydrolyzed starches and commercial dextran samples with nominal molecular weight from 1500 up to 18 300. The electropherograms obtained for these samples were virtually identical to those that had been reported previously when the carbohydrates were first derivatized with an aminonaphthalene trisulfonic acid reagent and the CE was carried out with fluorescence detection [7, 8]. A significant advantage of the CE–EC approach was that it required no derivatization and therefore was simpler and much faster.

40.5 Conclusions

The use of high-pH CE–EC methods for carbohydrate analysis represents an extremely recent development for which relatively few applications have been reported at this point. However, enough work with this approach has been performed to demonstrate its comparative sensitivity (at the femtomole level) and simplicity (no precolumn derivatization required). While it is not likely that the CE–EC methods discussed will soon displace the more longstanding approaches that rely on the formation of highly charged and easily detected carbohydrate derivatives before CE, we believe that, for many applications, CE–EC can certainly be a valuable addition to these techniques.

References

1. Oefner, P.; Chiesa, C.; Bonn, G.; Horvath, C. (1994) Developments in capillary electrophoresis of carbohydrates. *J. Capillary Electrophoresis.* **1**, 5–26.
2. El Rassi, Z.; Nashabeh, W. (1995) High performance capillary electrophoresis of carbohydrates and glycoconjugates. In *Carbohydrate Analysis*, El Rassi, Z. (ed.), Elsevier, Amsterdam, pp. 267–360.
3. Honda, S.; Iwase, S.; Makino, A.; Fujiwara, S. (1989) Simultaneous determination of reducing monosaccharides by capillary zone electrophoresis as the borate complexes of N-2-pyridylglycamines. *Anal. Biochem.* **176**, 72–77.
4. Nashabeh, W.; El Rassi, Z. (1990) Capillary zone electrophoresis of pyridylamino derivatives of maltooligosaccharides. *J. Chromatogr.* **514**, 57–64.
5. Liu, J.; Shirota, O.; Wiesler, D.; Novotny, M. (1991) Ultrasensitive fluorometric detection of carbohydrates as derivatives in mixtures separated by capillary electrophoresis. *Proc. Natl Acad. Sci. USA* **88**, 2302–2306.
6. Liu, J.; Dolnik, V.; Hsieh, Y.-Z.; Novotny, M. (1992) Experimental evaluation of the separation efficiency in capillary electrophoresis using open tubular and gel-filled columns. *Anal. Chem.* **64**, 1328–1336.
7. Chiesa, C.; Horvath, C. (1993) Capillary zone electrophoresis of malto-oligosaccharides derivatized with 8-aminonaphthalene-1,3,6-trisulfonic acid. *J. Chromatogr.* **645**, 337–352.
8. Stefansson, M.; Novotny, M. (1994) Separation of complex oligosaccharide mixtures by capillary electrophoresis in the open-tubular format. *Anal. Chem.* **66**, 1134–1140.
9. Johnson, D.C.; LaCourse, W.R. (1990) Liquid chromatography with pulsed electrochemical detection at gold and platinum electrodes. *Anal. Chem.* **62**, 589A–597A.
10. Luo, P.; Prabhu, S.V.; Baldwin, R.P. (1990) Constant potential amperometric detection at a copper-based electrode: electrode formation and operation. *Anal. Chem.* **62**, 752–755.
11. Ye, J.; Baldwin, R.P. (1994) Determination of carbohydrates, sugar acids, and alditols by capillary electrophoresis and electrochemical detection at a copper electrode. *J. Chromatogr. A.* **687**, 141–148.
12. Voegel, P.D.; Baldwin, R.P. (1996) Electrochemical detection with copper electrodes in liquid chromatography and capillary electrophoresis. *Am. Lab.*, January 39–45.
13. Luo, M.Z.; Baldwin, R.P. (1995) Characterization of carbohydrate oxidation at copper electrodes. *J. Electroanal. Chem.* **387**, 87–94.
14. Zhou, W.; Baldwin, R.P. (1996) Capillary electrophoresis and electrochemical detection of underivatized oligo- and polysaccharides with surfactant-controlled electroosmotic flow. *Electrophoresis*, **17**, 319–324.
15. Andrews, A.T. (1981) *Electrophoresis: Theory, Techniques, and Biomedical and Clinical Applications*, Oxford University Press, New York, p. 7.
16. Ewing, A.G.; Mesaros, J.M.; Gavin, P.F. (1994) Electrochemical detection in microcolumn separations. *Anal. Chem.* **66**, 527A–536A.
17. Colon, L.A.; Dadoo, R.; Zare, R.N. (1993) Determination of carbohydrates by capillary zone electrophoresis with amperometric detection at a copper microelectrode. *Anal. Chem.* **65**, 476–481.
18. O'Shea, T.J.; Lunte, S.M.; LaCourse, W.R. (1993) Detection of carbohydrates by capillary electrophoresis with pulsed amperometric detection. *Anal. Chem.* **65**, 948–951.
19. Lu, W.; Cassidy, R.M. (1993) Pulsed amperometric detection of carbohydrates separated by capillary electrophoresis. *Anal. Chem.* **65**, 2878–2881.
20. Ye, J.; Baldwin, R.P. (1993) Amperometric detection in capillary electrophoresis with normal size electrodes. *Anal. Chem.* **65**, 3525–3527.
21. Roberts, R.E.; Johnson, D.C. (1995) Variation in PED response at a gold microelectrode as a function of waveform parameters when applied to alditols and carbohdrates separated by capillary electrophoresis. *Electroanalysis* **7**, 1015–1019.

Part Seven

Environmental Science Applications

41 Hydrocarbons
E. DABEK-ZLOTORZYNSKA

41.1 Introduction

Capillary electrophoresis (CE) with its high resolving power has attracted great interest for the analysis of different classes of compounds including neutral hydrocarbons. In principle, neutral substances like aliphatic and aromatic hydrocarbons cannot be separated electrophoretically unless they are either given a charge by complexation or solvophobic association, or micellar electrokinetic capillary chromatography (MEKC) is used for the separation of the uncharged molecules.

41.2 Application of micellar electrokinetic capillary chromatography to analysis of hydrocarbons

When employing the MEKC technique, it is generally necessary to manipulate migration range and capacity factor, k', values in order to resolve structurally similar and/or highly hydrophobic hydrocarbons. Several studies have been focused on the separation of this class of compounds by MEKC using various surfactants and organic modifiers. Yang and Khaledi [1, 2] studied the influence of surfactant type on migration behavior and chemical selectivity in MEKC through a linear solvation energy relationship. High correlation between the logarithm of solutes' capacity factors and their solvatochromic parameters was observed for a group of 26 uncharged substituted aromatic compounds and polynuclear aromatic hydrocarbons (PAHs) with sodium dodecyl sulfate (SDS) and sodium cholate micelles. It was found that solutes' size and basicity are the two dominant factors that influence the migration behavior in MEKC. Yik et al. [3] studied three types of cyclodextrins (CDs), α-, β- and γ-, as modifiers for the separation of PAHs. Yik reported that the migration behavior of PAHs is related to their size and the cavity sizes of the CDs. The addition of γ-CD to the migration buffer was found to improve selectivity for seven PAHs tested. Jinno and Sawada [4] observed a correlation between the logarithmic capacity factor and the hydrophobicity of 16 PAHs and several nonplanar hydrophobic aromatic compounds. Very good linearity was found particularly when γ-CD was used. Jinno reported an improvement in the reproducibility of separation of the selected analytes after addition of dimethyl sulfoxide (DMSO) to the buffer, due to better solubility of hydrophobic compounds in the buffer solution. Terabe et al. [5] reported the separation of nine isomeric dimethylnaphthalenes, and 16 PAHs on the U.S. EPA priority pollutant list using γ-CD-modified MEKC. Two pairs of PAHs, fluoranthene and benz[a]anthracene, and benzo[k]fluoaranthene and dibenz[a,h]anthracene were not resolved. Imasaka et al. [6] determined six anthracene derivatives with CD–MEKC (cyclodextrin-modified micellar electrokinetic capillary chromatography). The detection limit was 7×10^{-9} M for 9,10-dimethylanthracene, when He–Cd laser fluorimetry (325 nm) was used. The method was applied to heavy oil and solvent-refined coal. Cooper et al. [7] utilized CD–MEKC for analyses of mixture of PAHs, including those isolated from complex samples such as shale oil. Efficient separations of alkyl-substituted benzopyrene (BaP) isomers were reported [8]. A comparison of SDS and sodium cholate surfactant systems indicated that CD–SDS mobile phases were more favorable for separation of BaP isomers. Otsuka et al. [9] investigated the applicability of MEKC using SDS surfactant with organic modifiers for the separation of PAHs. By using a SDS solution containing DMSO or acetone, eight PAHs and 13 PAHs were separated, respectively. Use of bile salts as the micellar phases for the separation of hydrophobic compounds such as PAHs proved successful. Bile salts are relatively polar and consequently provide a reduction in capacity factor values for most non-ionic analytes due to weaker solvation capability compared to SDS. Kaneta et al. [10] reported the separation of six PAHs with relatively large molecular weights by using a bile salt, sodium deoxycholate, as a surfactant and DMF (N,N-dimethylformamide) as an organic modifier. The detection limit for perylene was 1.4×10^{-7} M under He–Cd laser excitation or 3.4×10^{-7} under semiconductor laser excitation. Dabek-Zlotorzynska and Lai [11] investigated the usefulness of sodium taurodeoxycholate as a surfactant and four organic solvents to separate of 16 PAHs on the U.S. EPA pollutant list. Excellent resolution was obtained with 30% acetone. One pair, indeno[1,2,3-cd]pyrene and benzo[g,h,i]perylene, could not be separated. The capability of the developed separation system was demonstrated in the analysis of ambient air extracts. Brumley and Jones [12] used the CD–MEKC with sodium cholate surfactant for the separation of PAHs and other semivolatile analytes of interest to environmental analysis. The MEKC method was compared to GC.

Several groups investigated the possibility of replacing the micelle with alternative phases. Palmer et al. [13] used an oligomer of sodium 10-undecylenate as a pseudostationary phase in MEKC for separations of neutral organic compounds in 20–40% methanol. Excellent separations of PAHs were reported but some difficulties were encounted in reproducing migration times as a result of occasional precipitation of the oligomer. Tanaka and co-workers [14, 15] reported SBDs (starburst dendrimers polyamidoamines) as pseudostationary phases in electrokinetic capillary chromatography (EKC) of hydrophobic compounds. The SBDs provided similar selectivity as polymer gel packing materials in reversed-phase liquid chromatography, showing little selectivity for alkyl groups and clear preference for aromatic compounds, especially for rigid, planar PAHs. The separation of neutral aromatic compounds was found to be influenced by the size and charge state of the SBDs and the organic solvent content of the buffer. Bachmann et al. [16] reported successful separations of 12 PAHs with resorcarenes in EKC. This novel pseudostationary phase showed several advantages over traditionally used surfactants in MEKC such as high stability in the presence of higher content of organic modifiers. Yang et al. [17] reported the use of an ionic polymer, poly(methyl methacrylate/ethyl acrylate/methacrylic acid), and high concentrations of organic modifiers in MEKC for the separation of a wide range of hydrophobic compounds from straight n-alkylphenone homologs to a complex mixture of PAHs, as well as a mixture of fullerenes.

Sepaniak et al. [18] described a dual-CD phase variant EKC for separations of non-ionizable solutes. Separations were based on a differential distribution between charged CM-β-CD (carboxymethyl-β-cyclodextrin) and neutral CD. Efficiency, selectivity and system retention were evaluated on the basis of separations of PAHs. Szolar et al. [19] illustrated a similar approach for separation of some PAHs using CE with a buffer containing a mixture of a neutral and anionic β-CD derivatives. Methods were developed for separation of a mixture of PAHs, containing phenanthrene, anthracene, pyrene, chrysene, Ba[a]P and Ba[e]P, using various concentration ratios of the neutral and charged CDs.

Nie et al. [20] employed CE with UV-laser-excited native fluorescence for ultrasensitive determination of some PAHs. The separation was based on solvophobic association of the analytes with tetraalkylammonium (TAA$^+$) ions in mixed acetonitrile/water solvent. The detection limits were in the range of $(6-30) \times 10^{-11}$ M. Shi and Fritz [21] reported very good separation of some PAHs and other non-ionic organic compounds by CE using a THA (tetraheptylammonium) salt as an additive in aqueous acetonitrile as solvent. A systematic study was undertaken to determine the effect of experimental parameters on electroosmotic mobility and electrophoretic mobility. It was found that pH, concentration of acetonitrile and type and concentration of quaternary ammonium salts were important variables. Excellent separation of a broad range of non-ionic aromatic compounds, including PAHs and fairly hydrophilic compounds was obtained by Shi and Fritz [22] with DOSS (sodium dioctyl sulfosuccinate) as a solvophobic additive. Separation was based on differences in the strength of analyte–DOSS association 'complexes' in solution, which resulted in differences in effective electrophoretic mobility.

Yan et al. [23] reported a successful separation of 16 PAHs using capillary electrochromatography (CEC) with a fused-silica capillary packed with

3 μm octadecylsilica particles. The detection limits for individual PAHs using laser-induced fluorescence (LIF) ranged between 10^{-9} and 10^{-11} M. Guo and Colon [24] described the preparation of a silica glass coating material, based on the sol–gel method for chromatographic applications. PAHs were used as a probe mixture to test the sol–gel coated capillary columns under liquid chromatography (LC) and CEC conditions.

Kaneta et al. [25] demonstrated separation of several hydrophilic aromatic compounds with indirect semiconductor laser fluometry. Tetradecyltrimethylammonium chloride (TTAC) was used as a surfactant and oxazine 750 as a visualizing agent with indirect fluorescence detection. Tanaka et al. [26] characterized a new surfactant with two ionic groups and two lipophilic chains in MEKC. Compared with widely used SDS, this new surfactant exhibited different selectivity for several substituted naphthalene and benzene derivatives and gave a wider migration time window. Use of butyl acrylate–butyl acid copolymers sodium salts (BBMA) for naphthalene derivatives was reported by Ozaki et al. [27]. BBMA showed significantly different selectivity for the tested non-ionic solutes in comparison with SDS. Smith and co-workers [28, 29] evaluated a series of N-D-gluco-N-methylalkanamide (MEGA) surfactants in MEKC of neutral and charged species. The MEGA–borate micellar phases were useful in the separation of a number of polyaromatics, herbicides, barbiturates and dansyl amino acids. Katsuta et al. [30] studied the effects of organic hydroxy modifiers (1-hexanol, cyclohexanol and phenol) on the MEKC separation of various neutral compounds. The effects of these modifiers were mainly explained in terms of the saturation of the micellar palisade layer with the modifiers and the hydrogen-bond interaction between the modifier and analyte molecules in the micellar phase. Terabe et al. [31] studied the fundamental characteristics of microemulsion electrokinetic capillary chromatography (MEEKC) in comparison with MEKC. The microemulsion showed a stronger affinity for non-polar compounds than the SDS micelle. MEEKC was used by Ishihama et al. [32] to evaluate the hydrophobicity of various neutral aromatic compounds. Ishihama observed that the logarithm of the capacity factors, which is proportional to the free energy of transfer into the microemulsion, was highly correlated with the logarithm of the octanol–water partition coefficients rather than with that of the SDS micelle/water partition coefficients measured by MEKC.

Nielsen and Foley [33] and Ahuja and Foley [34] examined the effects of using different counterions, various TAA^+, Li^+ and K^+ ions, with the SDS micelles in order to ascertain the influence of the counterion on efficiency, selectivity, migration, elution range and resolution of very hydrophobic compounds in MEKC. It was reported that the nature of the counterion can significantly influence separation by MEKC. The buffer systems that used symmetrical TAA^+ ions showed reductions in retention specific for the carbonyl functional groups. The potassium dodecyl sulfate (KDS) micelles provided the largest elution range followed by SDS and then lithium dodecyl sulfate (LiDS). Ahuja and Foley [35] introduced hydrophobic interaction electrokinetic capillary chromatography (HI–EKC) for the separation of some very hydrophobic alkyl aryl ketone homologs. Through hydrophobic interactions between the free cetyltrimethylammonium bromide (CTAB) or SDS monomers and the analytes, separation of C20, C22, C24 alkyl aryl ketones was obtained. Three homologous series; alkyl aryl ketones, 1-nitroalkalenes and alkylbenzenes, were studied for use as retention index standards [36]. The authors recommended the use of 1-nitroalkalenes as retention index standards, especially with aqueous to 10% organic modified micellar systems. Alkyl aryl ketones were found to have an effective retention index for more hydrophobic analytes than acetophenone. Separation of five n-alkylphenone homologs using a nonionic/anionic mixed micellar system was reported by Ahuja et al. [37]. Retention characteristics suggested a nearly infinite elution range. Nielsen and Foley [38] utilized zone sharpening for a homologous series of alkylphenones using electrokinetic injection in MEKC. As a result of the increased amount of analyte loaded into the capillary without significant loss in efficiency, the detection limits were reduced in some cases by a factor of ten. Fujimoto [39] utilized the polyacrylamide gels in CEC to separate ketone homologs. Some selected applications are summarized in Table 41.1.

Table 41.1 Summary of applications of CE for the analysis of hydrocarbons

Methods Separation conditions, voltage applied, current, capillary length with i.d., temperature, detection	Compounds	Reference
PAHs		1–24
CD–MEKC 100 mM Borate buffer, pH = 9.0, 100 mM SDS, 5 M urea, 20 mM γ-CD, 25 kV, 80 cm (60 cm to detector), 50 μm i.d., 25°C, UV 254 nm	16 U.S. EPA priority PAHs	5
MEKC Borate–phosphate buffer, pH = 7.0, 25 mM SDS, 30% acetone, 20 kV, 30 cm to detector, 52 μm i.d., 30°C, pressure injection 1 s, UV 200 nm	p-Quinone, quinoline, benzene, benzoin, naphthalene, benzanthrone, phenanthrene, anthracene, pyrene, 1,2-benzanthraquinone, 2,3-benzofluorene, benz[a]anthracene	9
MEKC 10 mM Phosphate–20 mM Tris buffer, 70 mM sodium deoxycholate, 20% DMF, 20 kV, 60 cm (50 cm to detector), 50 μm i.d., hydrostatic injection 4 cm/10 s, He–Cd laser fluorimetry 325 nm	Fluoranthene, pyrene, perylene, Ba[a]P, 2,3-benztriphenylene, dibenz[a,h]anthracene	10

Table 41.1 *Continued*

Methods Separation conditions, voltage applied, current, capillary length with i.d., temperature, detection	Compounds	Reference
MEKC 10 mM Phosphate–6 mM borate buffer, pH = 9.0, 50 mM sodium taurodeoxycholate, 30% acetone, 25 kV, 57 cm (50 cm to detector), 50 μm i.d., 25°C, pressure injection 1 s, UV 214 nm	16 U.S. EPA priority PAHs	11
MEKC 10 mM Phosphate–12 mM borate buffer, pH = 8.2, 5 mM sodium undecylenate, 30% acetonitrile, 10 kV, 56 cm to detector, 100 μm i.d., vacuum injection 7 kPa s, UV 275 nm	Naphthalene, acenaphthene, fluorene, phenanthrene, pyrene, chrysene, benzo[b]fluoranthene, Ba[a]P, benzo[ghi]perylene	13
EKC 12 mM Borate buffer, pH = 10.6, 5 mM SBD, 40% methanol, 48 cm (33 cm to detector), 50 μm i.d., UV 210 nm	Diphenylmethane, o-terphenyl, naphthalene, anthracene, pyrene, triphenylene	15
EKC 5.8 mM Resorcarene, pH = 13.25, 6 M urea, 50% acetonitrile, 20 kV, 6 μA, 76 cm (62 cm to detector), 50 μm i.d., hydrostatic injection 10 cm/30 s, UV 260 nm	Naphthalene, phenanthrene, anthracene, fluoranthene, pyrene, triphenylene, chrysene, benzo[k]fluoranthene, Ba[a]P, indeno[1,2,3-cd]pyrene, anthanthrene, benzo[ghi]perylene	16
CD–EKC 10 mM Phosphate–6 mM borate buffer, pH = 6.0, 8 mM β-CD, 1 mM γ-CD: 10 mM CM-β-CD with 30% methanol, 20 kV, 50 cm to detector, 50 μm i.d., hydrostatic injection 10 cm/10 s, He–Cd laser fluorimetry 325 nm	Anthracene, pyrene, chrysene, Ba[a]P	18
CE 25 mM THA perchlorate in 3:1 (v/v) mixed acetonitrile/water, 20 kV, 100 cm, (80 cm to detector), 25 μm and 150 μm i.d., hydrostatic injection 15 cm/20 s, UV–LIF 325 nm	Perylene, benz[a]anthracene, pyrene, anthracene	20

Table 41.1 *Continued*

Methods Separation conditions, voltage applied, current, capillary length with i.d., temperature, detection	Compounds	Reference
Substituted hydrocarbons		1, 2, 4, 15, 21, 22, 25–32, 34
MEKC 1 mM Phosphate–0.5 mM Tris buffer, pH = 6.0, 10 µM oxazine 750, 50 mM TTAC, 4% methanol, 20 kV, 12 µA, 50 cm capillary (40 cm to detector), 50 µm i.d., hydrostatic injection 4 cm/5–10 s, indirect semiconductor laser fluorometry 670 nm	Aniline, *p*-nitrobenzaldehyde, nitrobenzene, *m*-dinitrobenzene, *o*-dinitroaniline, 2,4-dinitrotoluene	25
MEKC 100 mM Borate–50 mM phosphate buffer, pH = 8.0, 2% BBMA, 20 kV, 36.5 cm (32 cm to detector), 50 µm i.d., 30°C, pressure injection, UV 210 nm	1-naphthalenemethanol, 1,6-dihydroxynaphthalene, 1-naphthylamine, 1-naphthaleneethanol, 2-naphthol, 1-naphthol	27
MEEKC Borate–phosphate buffer, pH = 7.0, 79 mM heptane–60 mM SDS–874 mM butanol, 13 kV, 33 µA, 48.2 cm, (28.2 cm to detector), 52 µm i.d., 25°C, pressure injection, UV 214 nm	*o*-cresol, *p*-cresol, 2,6- 2,3-, 3,4-, 2,4-xylenol, *p*-propylphenol, *p*-butylphenol, *p*-amylphenol	31
Ketone homolog		17, 29, 33–39
MEKC 200 mM Borate buffer, pH = 10.0, 100 mM MEGA, 15 kV, 64 cm (56 cm to detector), 50 µm i.d., 30°C, pressure injection, UV 254 nm	Acetophenone, propiophenone, butyrophenone, valerophenone, hexanophenone, heptanophenone	29
HI–EKC 5 mM Phosphate buffer, pH = 7.0 or pH = 2.8, 50 mM SDS or 20 mM CTAB, 50% acetonitrile, 30 kV, <40 µA, 37.5 cm to 42.5 cm (30 cm or 35 cm to detector), 50 µm i.d., 25°C, hydrostatic injection 1 s, UV 254 nm or 214 nm	Dodecanophenone, tetradecanophenone, hexadecanophenone, octadecanophenone	35

Table 41.1 *Continued*

Methods Separation conditions, voltage applied, current, capillary length with i.d., temperature, detection	Compounds	Reference
MEKC 10 mM Phosphate buffer, pH = 6.2, 20 mM SDS, 14 mM Brij 35, 25 kV, 35 µA, 37.5 cm (30 cm to detector), 50 µm i.d., hydrostatic injection 1 s, UV 254 nm	Acetophenone, propiophenone, butyrophenone, valerophenone, hexanophenone, heptanophenone, octanophenone, decanophenone, dodecanophenone	37
CEC 100 mM Tris–150 mM boric acid, pH = 8.1, UV 254 nm	Dimethyl-, methyl ethyl-, methyl n-propyl-, methyl n-butyl-, methyl n-pentyl-, methyl phenyl-, ethyl phenyl-, n-propyl phenyl-, n-butyl phenyl ketone	39

Brij 35 = polyoxyethylene lauryl ether, DMF = dimethyl formamide, Tris = tris(hydroxymethyl) amino methane, U.S. EPA = United States Environmental Protection Agency.

References

1. Yang, S.; Khaledi, M.G. (1995) Chemical selectivity in micellar electrokinetic chromatography: characterization of solute–micelle interactions for classification of surfactants. *Anal. Chem.* **67**, 499–510.
2. Yang, S.; Khaledi, M.G. (1995) Linear solvation energy relationships in micellar liquid chromatography and micellar electrokinetic capillary chromatography. *J. Chromatogr.* **692**, 301–310.
3. Yik, Y.F.; Ong, C.P.; Khoo, S.B.; Lee, H.K.; Li, S.F.Y. (1992) Separation of polycyclic aromatic hydrocarbons by micellar electrokinetic chromatography with cyclodextrins as modifiers. *J. Chromatogr.* **589**, 333–338.
4. Jinno, K.; Sawada, Y. (1994) Relationship between capacity factors and hydrophobicity of polycyclic aromatic hydrocarbons in cyclodextrin-modified micellar electrokinetic capillary chromatography. *J. Capillary Electrophoresis* **1**, 106–111.
5. Terabe, S.; Miyashita, Y.; Ishihama, Y.; Shibata, O. (1993) Cyclodextrin-modified micellar electrokinetic chromatography: separation of hydrophobic and enantiomeric compounds. *J. Chromatogr.* **636**, 47–55.
6. Imasaka, T.; Nishitani, K.; Ishibashi, N. (1992) Determination of anthracene derivatives by cyclodextrin-modified micellar electrokinetic chromatography combined with helium–cadmium laser fluorimetry. *Anal. Chim. Acta* **256**, 3–7.
7. Cooper, C.L.; Staller, T.D.; Sepaniak, M.J. (1993) Characterization of polyaromatic hydrocarbons mixtures by micellar electrokinetic capillary chromatography. *Polycycl. Arom. Comp.* **3**, 121–135.
8. Cooper, C.L.; Sepaniak, M.J. (1994) Cyclodextrin-modified micellar electrokinetic capillary chromatography separations of benzopyrene isomers: correlation with computationally derived host–guest energies. *Anal. Chem.* **66**, 147–154.
9. Otsuka, K.; Higashimori, M.; Koike, R.; Karuhaka, K.; Okada, Y.; Terabe, S. (1994) Separation of lipophilic compounds by micellar electrokinetic chromatography with organic modifiers. *Electrophoresis* **15**, 1280–1283.
10. Kaneta, T.; Yamashita, T.; Imasaka, T. (1995) Separation of polyclic aromatic hydrocar-

bons by micellar electrokinetic chromatography with laser fluorescence detection. *Anal. Chim. Acta* **299**, 371–375.
11. Dabek-Zlotorzynska, E.; Lai, E.P.C. (1996) Separation of polynuclear aromatic hydrocarbons by micellar electrokinetic capillary chromatography using sodium taurodeoxycholate modified with organic solvents. *J. Capillary Electrophoresis*, **3**, 31–35.
12. Brumley, W.C.; Jones, W.J. (1994) Comparison of micellar electrokinetic chromatography (MEKC) with capillary gas chromatography in the separation of phenols, anilines and polynuclear aromatics. Potential field-screening applications of MEKC. *J. Chromatogr.* **680**, 163–173.
13. Palmer, C.P.; Khaledi, M.Y.; McNair, H.M. (1992) A monomolecular pseudostationary phase for micellar electrokinetic capillary chromatography. *J. High Resolut. Chromatogr.* **15**, 756–762.
14. Tanaka, N.; Tanigawa, T.; Hosoya, K.; Kimata, K.; Araki, T.; Terabe, S. (1992) Starburst dendrimers as carriers in electrokinetic chromatography. *Chem. Lett.* **6**, 959–962.
15. Tanaka, N.; Fukutome, T.; Tanigawa, T.; Hosoya, K.; Kimata, K.; Araki, T.; Unger, K.K. (1995) Structural selectivity provided by starburst dendrimers as pseudostationary phase in electrokinetic chromatography. *J. Chromatogr.* **699**, 331–341.
16. Bachmann, K.; Bazzanella, A.; Haag, I.; Han, K.-Y.; Arnecke, R.; Bohmer, V.; Vogt, W. (1995) Resorcarenes as pseudostationary phases with selectivity for electrokinetic chromatography. *Anal. Chem.* **67**, 1722–1726.
17. Yang, S.; Bumgarner, J.G.; Khaledi, M.G. (1995) Separation of highly hydrophobic compounds with an ionic polymer. *J. High Resolut. Chromatogr.* **18**, 443–445.
18. Sepaniak, M.J.; Copper, C.L.; Whitaker, K.W.; Anigbogu, V.C. (1995) Evaluation of a dual-cyclodextrin phase variant of capillary electrokinetic chromatography for separations of nonionizable solutes. *Anal. Chem.* **67**, 2037–2041.
19. Szolar, O.H.J.; Brown, S.R.; Luong, J.H. (1995) Separation of PAHs by capillary electrophoresis with laser-induced fluorescence detection using mixtures of neutral and anionic β-cyclodextrins. *Anal. Chem.* **67**, 3004–3010.
20. Nie, S.; Dadoo, R.; Zare, R.N. (1993) Ultrasensitive fluorescence detection of polycyclic aromatic hydrocarbons in capillary electrophoresis. *Anal. Chem.* **65**, 3571–3575.
21. Shi, Y.; Fritz, J.S. (1994) Capillary zone electrophoresis of neutral organic molecules in organic–aqueous solution. *J. High Resolut. Chromatogr.* **17**, 713–718.
22. Shi, Y.; Fritz, J.S. (1995) HPCZE of nonionic compounds using a novel anionic surfactant additive. *Anal. Chem.* **67**, 3023–3027.
23. Yan, C.; Dadoo, R.; Zhao, H.; Zare, R.N.; Rakestraw, D.J. (1995) Capillary electrochromatography: analysis of polycyclic aromatic hydrocarbons. *Anal. Chem.* **67**, 2026–2029.
24. Guo, Y.; Colon, L.A. (1995) A stationary phase for open tubular liquid chromatography and electrochromatography using sol-gel technology. *Anal. Chem.* **67**, 2511–2516.
25. Kaneta, T.; Yamashita, T.; Imasaka, T. (1995) Separation of polycyclic aromatic hydrocarbons by micellar electrokinetic chromatography with laser fluorescence detection. *Anal. Chim. Acta* **299**, 371–375.
26. Tanaka, M.; Ishida, T.; Araki, T.; Masuyama, A.; Nakatsuji, Y.; Okahara, M.; Terabe, S. (1993) Double-chain surfactant as a new and useful micelle-forming reagent for micellar electrokinetic chromatography. *J. Chromatogr.* **648**, 469–473.
27. Ozaki, H.; Terabe, S.; Ichihara, A. (1994) Micellar electrokinetic chromatography using high-molecular surfactants. Use of butyl acrylate–butyl methacrylate–methacrylic acid copolymers sodium salts as pseudo-stationary phases. *J. Chromatogr.* **680**, 117–123.
28. Smith, J.T.; Nashabeh, W.; Rassi, Z. (1994) Micellar electrokinetic capillary chromatography with *in situ* charged micelles. I. Evaluation of N-D-gluco-N-methylalkanamide surfactants as anionic borate complexes. *Anal. Chem.* **66**, 1119–1133.
29. Smith, J.T.; Rassi, Z. (1994) Micellar electrokinetic capillary chromatography with *in situ* charged micelles. IV. Influence of the nature of the alkylglycoside surfactant. *J. Chromatogr.* **685**, 131–143.
30. Katsuta, S.; Tsumura, T.; Saitoh, K.; Terame, N. (1995) Control of selectivity in micellar electrokinetic chromatography by modification of sodium dodecyl sulfate micelles with organic hydroxy compounds. *J. Chromatogr.* **705**, 319–324.

31. Terabe, S.; Matsubara, N.; Ishihama, Y.; Okada, Y. (1992) Microemulsion electrokinetic chromatography: comparison with micellar electrokinetic chromatography. *J. Chromatogr.* **608**, 23–29.
32. Ishihama, Y.; Oda, Y.; Uchikawa, K.; Asakawa, N. (1995) Evaluation of solute hydrophobicity by microemulsion electrokinetic chromatography. *Anal. Chem.* **67**, 1588–1595.
33. Nielsen, K.R.; Foley, J.P. (1994) Effect of the dodecyl sulfate counterion on selectivity and resolution in micellar electokinetic chromatography: II. Organic counterions. *J. Microcol. Sep.* **6**, 139–149.
34. Ahuja, E.S.; Foley, J.P. (1995) Influence of dodecyl sulfate counterion on efficiency, selectivity, retention, elution range, and resolution in micellar electrokinetic chromatography. *Anal. Chem.* **67**, 2315–2324.
35. Ahuja, E.S.; Foley, J.P. (1994) Separation of very hydrophobic compounds by hydrophobic interaction electrokinetic chromatography. *J. Chromatogr.* **680**, 73–83.
36. Ahuja, E.S.; Foley, J.P. (1994) A retention index for micellar electrokinetic chromatography. *Analyst* **119**, 353–360.
37. Ahuja, E.S.; Little, E.L.; Nielsen, K.R.; Foley, J.P. (1995) Infinite elution range in micellar electrokinetic capillary chromatography using a nonionic/anionic mixed micellar system. *Anal. Chem.* **67**, 26–33.
38. Nielsen, K.R.; Foley, J.P. (1994) Zone sharpening of neutral solutes in micellar electrokinetic chromatography with electrokinetic injection. *J. Chromatogr.* **686**, 283–291.
39. Fujimoto, C. (1995) Charged polyacrylamide gels for capillary electrochromatographic separations of uncharged, low molecular weight compounds. *Anal. Chem.* **67**, 2050–2053.

42 Environmental pollutants
E. DABEK-ZLOTORZYNSKA

42.1 Introduction

There has been in the recent years an increasing number of application of capillary electrophoresis (CE) to the determination of organic pollutants in environmental samples [1]. Groups of compounds which have received much attention include polynuclear aromatic hydrocarbons (PAHs), phenols, phthalate esters, amines, alkyl and aryl sulfonates, dyes, organochlorine and other pesticides, and organometallics. The need to monitor these pollutants in air, soil and sludge and water is increasingly being recognized, and novel procedures for their determination are being published continuously. However, several application areas are still relatively undeveloped for CE. Some selected applications are summarized in Table 42.1.

42.2 Phenol and its derivatives

Some of the earliest environmental applications of CE include phenols which are of great environmental concern due to their common use and toxicity. Gonnord *et al.* [2] demonstrated the suitability of capillary zone electrophoresis (CZE) for the separation of 13 chlorinated phenols within 16 min. The migration order was a function of both charge and size of the analytes. Bachmann *et al.* [3] reported improvement in the resolution and selectivity of ten selected phenol derivatives by adding methanol as an organic modifier to the CZE buffer. The detection limits were found to be in the low microgram per liter range using a specially optimized electrostacking procedure. Aguilar *et al.* [4] utilized CZE system for the determination of two chlorophenols in conjunction with two phenoxyalkyl acid herbicides in water. The detection limit was lower than 6 pg. Chao and Whang [5] and Chen and Wheng [6] descirbed a scheme for the separation and detection of 11 United States Environmental Protection Agency (U.S. EPA) priority phenols using CZE coupled with indirect fluorescence detection and amperometric detection, respectively. Linearity over two orders of magnitude of concentration was generally obtained and limits of detection for the priority phenols were in the microgram per liter range. The methods were successfully applied to the determination of priority phenols in industrial waste waters. Li and Locke [7] reported the CZE

Table 42.1 Summary of applications of CE to the determination of environmental pollutants

Methods Separation conditions, voltage applied, current, capillary length with i.d., temperature, detection	Compounds	Reference
Phenols		1–22
CZE 10 mM Phosphate buffer, pH = 6.9, 30 kV, 65 µA, 57 cm, (50 cm to detector), 75 µm i.d., pressure injection 5 s, 22°C, UV 214 nm	13 Chlorinated phenols	2
CZE 20 mM Borate–10 mM phosphate buffer, pH = 10.0, 15% methanol, 30 kV, 68 cm (51 cm to detector), 50 µm i.d., hydrodynamic injection, 50 mbar/12 s, UV 206 nm	10 Phenol derivatives	3
CZE 15 mM Borate buffer, pH = 9.9, 1 mM fluorescein, 9 kV, 2.8 µA, 50 cm (45 cm to detector), 20 µm i.d., electrokinetic injection 2 s, indirect fluorimetry 520 nm	11 U.S. EPA priority phenols	5
CZE 20 mM CHES, pH = 10.1, 9 kV, 62.5 cm, 50 µm i.d., amperometric +1.10 V (vs. Ag/AgCl reference)	11 U.S. EPA priority phenols	6
CZE 1 mM Borate buffer, pH = 11.2, 30% acetonitrile, 30 kV, 30 µA, 70 cm (62.5 cm to detector), 75 µm i.d., hydrostatic injection 10 cm/20 s, UV 254 nm	15 Chloro-, nitro- and alkyl-substituted phenols	9
CE 50 mM Phosphate buffer, pH = 7.0, 4 mM *p*-sulfonic calix, 80 cm (40 cm to detector), 50 µm i.d., UV 220 nm	*o*-, *p*-, *m*-Chlorophenols, *p*-, *m*-, *o*-benzenediols	10
ITP LE: 1 mM HCl, Tris, 0.1% HEC, pH = 6.3, TE: 5 mM asparagine, Ba(OH)$_2$, pH 10.5–11.0, conductometric detection	Pentachlorophenol, 2,4,6-trichlorophenol, 2,4-dichlorophenol, 2-chlorophenol	14
MEKC 10 mM Phosphate buffer, pH = 9.18, 50 mM SDS, 18 kV, anionic coated capillary, 75 µm i.d., hydrodynamic injection 10 cm/5 s, UV 210 nm	5 Substituted phenols	18
CE 15 mM Phosphate–1.25 mM borate buffer, pH 11, 0.001% (w/v) HDB, 40% (v/v) organic solvent, 30 kv, 32 cm, (24.5 cm to detector), 50 µm i.d., hydrostatic injection 10 cm/5 s, UV 254 nm	12 Substituted phenols	19

Table 42.1 *Continued*

Methods Separation conditions, voltage applied, current, capillary length with i.d., temperature, detection	Compounds	Reference
Anilines and benzidines		23–26
CD–MEKC 20 mM Borate–phosphate buffer, pH = 6.3, 50 mM SDS, 8 mM γ-CD, 20 kV, 20 μA, 72 cm, (50 cm to detector), 50 μm i.d., 30°C, UV 210 nm	Aniline, *m*-anisidine, *o*-chloroaniline, *o*-anisidine, *m*-chloroaniline, *p*-chloroaniline, *N*-methylaniline, *p*-anisidine, diphenylamine *N*-nitrosodiphenylamine	23
CE or MEKC 20 mM Phosphate buffer or 20 mM phosphate buffer with 20 mM SDS, pH = 6.6, 10 kV, 50 cm, 50 μm i.d., hydrodynamic injection 10 cm/10 s, MS	Aniline, *p*-nitroaniline, 1-naphthylamine	24
MEKC 50 mM Boric acid–sodium borate buffer, pH = 8.3, 100 mM sodium cholate, 5% acetone, 25 kV, 57 cm (50 cm to detector), 50 μm i.d., pressure injection 1–2.5 s, UV 214 nm	4,4′-Diaminobiphenyl, 3,3′-dimethoxybenzidine, 3,3′-dimethylbenzidine, 3,3′-dichlorobenzidine	25
Aliphatic amines		27–30
CE 5 mM Imidazole, pH = 4.5, 25 kV, 9 μA, 63 cm (55 cm to detector), 75 μm i.d., hydrostatic injection 10 cm/30 s, indirect UV 214 nm	Dimethylamine, trimethylamine, diethylamine, triethylamine, diethanolamine, triethanoamine	27
CE 5 mM 1,1′-di-*n*-heptyl-4,4′-bipyridinium hydroxide, pH = 6.5, 6 mM glycine, 2 mM 18-crown-6 ether, 2% methanol, 25 kV, 12.5 μA, 57 cm (50 cm to detector), 75 μm i.d., indirect UV 280 cm	Methylamine, dimethylamine, trimethylamine, ethylamine, triethylamine	29
CE 5 mM Phosphoric acid, 6 mM β-alanine, 25 kV, 50 cm, 75 μm i.d., suppressed conductivity	Tetrabutylammonium, tributylamine, tripropylamine, dibutylamine	30
Nitrosoamines		31, 32
MEKC 50 mM Phosphate–borate buffer, pH 6.6, 40 mM SDS, 15 kV, 50 cm, 75 μm i.d., UV 254 nm	*N*-nitrosodiethanolamine, *N*-nitrosodimethylamine, *N*-nitrosomorpholine	31
Carbonyls (aldehydes and ketones)		33, 34
CE 5 mM Phosphate–10 mM borate buffer, pH = 7.1, 350 V cm^{-1}, 85 cm (65 cm to detector), 50 μm i.d., hydrostatic injection 10 cm/30 s, UV 218 nm	Propionaldehyde, acetaldehyde, *p*-toluylaldehyde, benzaldehyde, acrolein, formaldehyde, methylglyoxal	33

Table 42.1 *Continued*

Methods Separation conditions, voltage applied, current, capillary length with i.d., temperature, detection	Compounds	Reference
CE 12.5 mM Borate buffer, pH = 7.3, 30 mM sodium bisulfite, 20 kV, 60 cm (53 cm to detector), 75 μm i.d., hydrodynamic injection 10 cm/5 s, UV 280 nm	13 Substituted benzaldehydes	34
Phthalate esters MEKC 20 mM Borate–phosphate buffer, pH = 9.0, 50 mM SDS–methanol (80:20, v/v) 20 kV, 72 cm, 50 μm i.d., 38.9°C, UV 210 nm	10 Phthalate esters	35, 36 35
MEKC 10 mM Phosphate – 12 mM borate buffer, pH = 8.2, 50%(v/v) methanol, 5 mM sodium 10-undecylenate, 240 V cm^{-1}, 38 cm to detector, 100 μm i.d., hydrodynamic injection 3.8 cm/10 s, UV 250 nm	Alkyl phthalates	36
Explosives MEKC 12.5 mM Boric acid–2.5 mM sodium borate, pH = 8.5, 25 mM SDS, 20 kV, 57 cm (51 cm to detector), 75 μm i.d., gravity injection 3 cm/5 s, UV 230 nm	1,3,5-Trinitro-1,3,5-triazocyclohexane, 2,4,6-trinitrotoluene-4-amino, 2,4,6-dinitrotoluene, 2,4,6-trinitrobenzene, 2,6-dinitrotoluene, 2-nitrotoluene, 4-nitrotoluene, 2,4,6,*N*-tetranitro-*N*-methylaniline, 1,3,5,7-tetranitro-1,3,5,7-tetraazocyclooctane, nitrobenzene	37 37
Surfactants CE 8 mM Phosphate buffer, 3 mM C$_{12}$ benzyl, 3 mM SDS–THF (57.5:42.5 v/v), 18 kV, 24 cm, 50 μm i.d., gravity injection 10 cm, indirect UV 210 nm	C12, C14, C16, C18 alkylbenzyldimethylammonium quaternary compounds	38–44 38
CE 10 mM Acetate buffer, pH = 6.0, 3 mM Mg^{2+}, 2:3 acetonitrile–water, 30 kV, 60 cm, 50 μm i.d., UV 230 nm	C10–C14 LAS homologs	39
CE 12.5 mM Borate–boric acid buffer, pH = 9.0, 30% acetonitrile, 30 kV, 57 cm, 50 μm i.d., hydrodynamic injection, 2 s, UV 214 nm	C2–C12 LAS homologs	40
Dyes MEKC 50 mM Boric acid–sodium borate buffer, pH = 8.35, 100 mM sodium cholate, 10% acetone, 25 kV, 57 cm (50 cm to detector), 75 μm i.d., UV 214 nm	Cresol red, orange II, acid blue 40, acid orange 8, nuclear fast red, acid red 151, tropaeolin	45–52 49

Table 42.1 *Continued*

Methods Separation conditions, voltage applied, current, capillary length with i.d., temperature, detection	Compounds	Reference
CE 50 mM Citric acid, pH = 3.25, 10% acetonitrile, 25 kV, 75 cm (68 cm to detector), 75 μm i.d., hydrodynamic injection, 1 s, DAD 400–610 nm	8 Hydrolyzed dyes	52
Pesticides CE 10.5 mM Borate buffer, pH = 7.2, 30 kV, 67 cm (57 cm to detector), 75 μm i.d., 50°C, UV 254 nm	Alkylphosphonic acids	53–79 55
MEKC 5 mM Borate buffer, pH = 8.0, 30 mM SDS, 20 kV, 70 cm (63 cm to detector), 75 μm i.d., vacuum injection 1–2 s, UV scanning 200–300 nm	Propoxur, carbofuran, simazine, prophame, despetryn, atrazine, parathion methyl, parathion ethyl, chlorfenvinyphos	57
MEKC 200 mM Borate buffer, pH = 10.0, 50 mM octyl-β-D-maltopyranoside, 15 kV, 80 cm, (50 cm to detector), 50 μm i.d., UV 210 nm	Monuron, fluometuron, metobromuron, siduron, diuron, linuron, neburon, chloroxuron	61
CD–CZE 30 ml Lithium acetate buffer, pH = 4.80, 20 g l^{-1} heptakis (2,6-di-*O*-methyl)-β-CD, 30 kV, 79.5 cm (63.1 cm to detector), 50 μm i.d., 30°C, UV 200 nm	7 Phenoxy acid herbicides	63
CZE 50 mM Acetate buffer, pH = 4.5, 20 kV, 50 cm to detector, 75 μm i.d., 30°C, UV 230 nm	Fenoprop, mecoprop, dichloroprop, 2,4-dichlorophenoxyacetic acid	65
MEKC 30 mM Sodium borate buffer, pH = 7.0, 80 mM SDS, 14% methanol, 20% isopropanol, 25 kV, 50 cm, 75 μm i.d., 35°C, UV 214 nm	Metsulforon, chlorimuron, chlorsulforon	71
MEKC 20 mM Phosphate–borate buffer, pH = 7.0, 70 mM SDS, 15 kV, 55 cm (40 cm to detector), 50 μm i.d., manual injection (1.5 nl), UV 205 nm	9 Herbicides	73
CE Acetate buffer, pH = 4.0, 100 mM NaCl, 15 kV, 72 cm (50 cm to detector), 50 μm i.d., 30°C, UV 210 nm	Paraquat, diquat, difenzoquat	77

Table 42.1 *Continued*

Methods Separation conditions, voltage applied, current, capillary length with i.d., temperature, detection	Compounds	Reference
ITP–CZE LE: 10 mM Potassium citrate, pH = 6.08, 0.2% MHEC TE: 5 mM Tris–citric, pH = 7.80 LE: 9.8 mM Tris–citric, pH = 6.08, 0.15% PEG, 200 µA, 35 cm (25 cm to detector) UV 310 nm	Paraquat, diquat	79
Organometallics CE 100 mM Borate buffer, pH = 8.35, 10% methanol, 15 kV, 50 cm (45 cm to detector), 75 µm i.d., pressure injection 8.6 s, 30°C, UV 200 nm	Methyl-, ethyl-, phenyl-, inorganic- mercury	80–89 80
MEKC 25 mM Phosphate–borate buffer, pH = 6.0, 50 mM SDS, 5 mM β-CD, 15 kV, 44 cm 50 µm i.d., UV 210 nm	Trimethyllead, triethyllead, diphenyl selenide, phenylselenyl compounds	82
CZE Phosphate–borate buffer, pH = 8.65, 20 kV, 50 cm (33 cm to detector), 50 µm i.d., UV 195 nm	Selenium and arsenic compounds	86
CE 5 mM Pyridine, pH = 2.65, 20 kV, 60 cm (53 cm to detector), 75 µm i.d., hydrostatic injection 10 cm/20 s, UV 254 nm	Organotin compounds	87

*C12 benzyl = benzyldimethylstearylammonium chloride, CD-MEKC = cyclodextrin-modified micellar electrokinetic capillary chromatography, CHES = 2-(cyclohexylamine)ethane-2-sulfonic acid, DAD = diode array detection, HDB = hexadimethrine bromide, HEC = hydroxyethyl cellulose, HI-EKC = hydrophobic interaction EKC, LE = leading electrolyte, MHEC = methylhydroxyethylcellulose, TE = terminating electrolyte, THF = tetrahydrofuran, Tris = tris(hydroxymethyl)aminomethane.

separation of 11 U.S. EPA priority pollutant phenols in less than 15 min using a pH 9.8 phosphate buffer. Good calibration data were obtained for phenol concentrations up to 50 mg l^{-1}. Detection limits for all phenols were less than 1 mg l^{-1}. Lin *et al.* [8] investigated the migration behavior and selectivity of six isomeric dichlorophenols using CZE. Baseline separation of these dichlorophenols was achieved at 10 kV with a phosphate–borate buffer with pH in the range of 6.7–9.6. Benz and Fritz [9] reported the separation of a mixture of chloro-, nitro- and alkyl-substituted phenols by CZE. By including an aqueous acetonitrile electrolyte solution and optimizing conditions such as the apparent pH, the separation of phenols with very similar pK_a values and chemical structures was possible. Shohat and Grushka [10] investigated migration behavior of chlorinated phenols, benzenediols and toluidines in the presence of *p*-sulfonate calixarene [6] as a selectivity modifier in CE. Praus and Dombek [11–13] studied conditions for the isotachophoresis (ITP) separation and determination of chlorophenols. The separation was improved by addition of β-CD (cyclodextrin) or poly(ethylene glycol) (PEG) into the leading electrolytes. Praus [14] reported the promising conjunction of ITP and simple steam distillation, as a preconcentration step for the determination of chlorophenols in river and industrial wastewaters. The method presented permitted the detection of 10–15 μg l^{-1} chlorophenols in water. The application of ITP and micellar electrokinetic capillary chromatography (MEKC) for the analysis of chlorophenols and other organic analytes in water was reported by Praus and Dombek [15] and Takeda et al. [16], respectively. Brumley and Jones [17] compared the MEKC separation of phenols and other organic pollutants with those obtained by capillary gas chromatography (GC). Landman et al. [18] used the anionic polymer-coated capillary with pH-independent flow to separate phenols in MEKC. Masselter and Zemann [19, 20] investigated the separation behavior of phenolic compounds with various modifiers with coelectroosmotic CE (migration of the analytes in the same direction as the electroosmotic flow, EOF). The influence of various organic modifiers and buffer electrolyte pH on EOF, electrophoretic mobility of the selected phenols and theoretical plate numbers in CE was reported. It was shown that a separation of the chosen phenols was possible under coelectroosmotic conditions by buffer pH optimization. The addition of organic solvents significantly improved peak shape of analytes and enabled the separation of positional isomers of phenols with both short times and high separation efficiencies. Aiken and Huie [21] examined the effect of alcohols on the MEKC separation of a variety of substituted phenols. The enhancement in resolution was found to be a function of the carbon chain length of alcohols as well the ratio of the 1-alkanols. Cresols were studied by Masselter et al. [22]. The separations were carried out at high pH values (pH 12) to obtain complete dissociation of the cresols and short time of analysis (less than 5 min) through the fast migration of the analytes towards the anode. Reversion of the EOF was carried out by a polycationic ammonium compound.

Another group of compounds which received much attention are PAHs. Several significant studies have focused on optimizing CE for PAHs. The application of CE for the separation of these pollutants is discussed in Chapter 41 on Hydrocarbons.

42.3 Aromatic and aliphatic amines

Aromatic amines including benzidines have been the subject of several studies. Takeda et al. [23] studied the migration behavior of the aniline derivatives under various analytical MEKC conditions. The complete separation of ten aniline derivatives, including these aniline derivatives, which have been found in environmental water was achieved. Analysis of aromatic amines by MEKC–MS (mass spectrometry) using an electrospray–chemical ionization (ECI) source was demonstrated by Takeda et al. [24]. Brumley and Jones [17] examined the separation of anilines by MEKC using a cholic acid as the micellar agent and the prospects of short capillaries for screening analyses. Comparison of MEKC with column GC separation indicated relatively high efficiency and selectivity for both techniques. Brumley and Brownrigg [25] utilized MEKC with UV detection to determine benzidines in water, soil and sediment. The use of short capillaries made possible the determination of benzidines in soil extracts in about 2 min after extraction and clean-up. Nielen et al. [26] studied CZE for the separation of phenylenediamine isomers.

Several groups investigated the possibility of using the CE with indirect UV detection for the separation of aliphatic amines. Beck and Engelhardt [27], Riviello and Harrold [28] and Dabek-Zlotorzynska and Dlouhy [29] utilized different background carrier electrolytes for the separation of low molecular mass amines using indirect UV. The amines were well separated from the inorganic cations. Avdalovic et al. [30] reported separation of amine cations by CE with the suppressed conductivity detection.

42.4 Nitrosoamines

Ng et al. [31] reported the separation of three nitrosoamines by CE using a combined packed and open-tubular capillary column. Improvement in selectivity in the separation of nitrosoamines was observed by using different composition and types of packing materials and by changing the concentration of the electrophoretic buffer. Janini et al. [32] investigated the separation of a selected group of naturally occurring, heterocyclic nitrosoamino compounds by CE. Subambient column temperatures improved the resolution of these compounds and resulted in the separation of their *syn* and *anti* conformers.

42.5 Carbonyl compounds

Carbonyls which are compounds abundant in the environment with concentrations depending on anthropogeneous pollution were analyzed by Bachmann et al. [33]. The carbonyls were derivatized with 5-(dimethylamino)naphthalene-1-sulfonehydrazide (DNSH, dansylhydrazide) to form negative detectable (absorbance or fluorescence) compounds. Eight carbonyls were separated in less than 8 min. Aguin et al. [34]

utilized in-column reaction of neutral aldehydes with the bisulfite in their separation under electrophoretic conditions.

42.6 Phthalate esters

The phthalate esters, commonly used as plasticizers, are ubiquitous in the environment. Although their biological effects are not certain (most probably mutagen and/or carcinogen according to the published papers, reviews and books), due to their widespread use and potential harmful effects, some of them are included in the U.S. EPA priority pollutant list. Takeda et al. [35] studied the MEKC separation and migration behavior of selected phthalate esters including priority pollutants. Migration times were found to correlate well with octanol–water partition coefficients. The enthalpy and entropy changes associated with partitioning into the micelle were also examined. Palmer et al. [36] utilized the oligomerized sodium 10-undecylenate to form a stable, monomolecular pseudostationary phase in MEKC. The highly stable structure allowed high concentrations of organic modifier to be used in the migration buffer, which in turn facilitated the separation of highly hydrophobic substances such as alkyl phthalates.

42.7 Explosive compounds

Kleibohmer et al. [37] separated 26 constituents of gunshot and explosives using MEKC. The highly toxic nature of many of these substances, coupled with their persistence in the environment, required thorough characterization of contaminated areas. The MEKC method was compared with an established high performance liguid chromatography (HPLC) method in terms of separation efficiency, sensitivity, specificity, analysis time and calibration linearity.

42.8 Surfactants

Several studies have focused on applying CE to various surfactants. Surfactants have a toxic impact on aquatic organisms and therefore the determination of them in environmental samples is of great importance. Weiss et al. [38] used CZE with indirect UV absorbance detection to analyze alkyl and alkylbenzylquaternary ammonium compounds. It was necessary to modify the migration buffer with organic solvents to prevent formation of micelles by the longer chain surfactants. Chen and Pietrzyk [39] reported that the separation of alkanesulfonate (RSO_3^-), alkyl sulphate

(ROSO$_3^-$) and linear alkylbenzene sulfonates (LAS) surfactants by CE was dependent on the buffer cation and its concentration. Divalent cations increased migration time and resolution to a greater extent than did monovalent cations. Resolution was markedly improved by using Mg^{2+} as the buffer additive. Desbene et al. [40] compared CZE and MEKC separations of LAS with chain length C2–C12. CZE and MEKC appeared to be complementary techniques. CZE allowed not only an efficient sorting according to the number of the sulfonic groups, but also allowed better resolution of the structural homologs of the alkylbenzenesulfonates than did MEKC. However, MEKC allowed the separation of alkybenzenesulfonate isomers. Shamshi and Danielson [41] used naphthalene mono, di- and trisulfonates (NMS, NDS, NTS) as electrolytes for CE with indirect photometric detection for a wide range of inorganic anions, organic acids and aliphatic anionic surfactants. The rapid separation in 6 min of C4–C14 SO$_4^-$ or SO$_3^-$-type surfactants was obtained with NMS-based electrolyte. Altria et al. [42] described the application of CE to monitor levels of dodecylbenzenesulfonate quantitatively. The method employed borate–phosphate buffer at pH = 9.5 with detection at 200 nm. Vogt et al. [43] reported the use of HPLC and CZE with UV detection for the separation of homologs and structural isomers of LAS. The optimized CZE technique was applied to the analysis of various samples, including household detergents as well as river waters. Goebel et al. [44] analyzed ethoxylated alcohol sufates without derivatization by CE using a borate buffer with 1 mM potassium dichromate and 30% (v/v) acetonitrile.

42.9 Dyes and their hydrolyzed compounds

Several studies have focused on applying CE to dyes and their hydrolyzed products. Hydrolyzed dye is found in dye effluent and, hence, this fact is important when carrying out environmental analysis. Croft and co-workers [45, 46] utilized CE for the analysis of disperse and reactive dyes. Croft reported that the CZE technique has proved extremely useful in the analysis of water-soluble reactive and related non-reactive triazine dyes. Lord et al. [47] used CEC–ES–MS (capillary electrochromatography–electrospray mass spectrometry) for the analysis of non-ionic disperse textile dyes. Tapley [48] studied the hydrolysis reactions of a selection of model dyes under various pH and temperature conditions with the aid of CE analysis. Brumley et al. [49] used MEKC for the analysis of environmental matrices for synthetic dyes and compared MEKC with capillary LC (liquid chromatography). Brumley reported that MEKC provided a powerful screening and determinative technique, while LC–MS provides a confirmatory tool. Burkinshaw et al. [50] studied several dyes and related intermediates by MEKC. On-line CZE–ES–MS was used to investigate the purity of laser dyes by Varghese and Cole [51]. Oxspring *et al.* [52] compared CE and adsorptive stripping voltammetry (AdSV) using a hanging mercury drop electrode for the separation and determination of six reactive textile dyes. Optimum CE separation of the hydrolyzed dyes was achieved using 50 mM citic acid (pH 3.25) with 10% acetonitrile. AdSV was shown to be a more sensitive technique for monitoring individual dyes but lacked the selectivity to deal with a complex sample.

42.10 Pesticides and herbicides

Interest in CE in pesticides, herbicides and other aromatic containing organic acids analysis has increased dramatically since 1990. Brumley [53] investigated CE as a qualitative tool in the analysis of environmental samples for aromatic sulfonic acids and related compounds. Brumley and Brownrigg [54] presented the behavior of 56 aromatic containing organic acids and herbicides under CZE and under MEKC using cholic acid as the micellar agent. Seven compounds were studied in detail with respect to extraction and clean-up from spiked water and soil. Robins and Wright [55] used sodium borate as both a buffer and derivatization agent for the separation and direct detection of organophosphonic acids by CE. Shamsi and Danielson [56] investigated the use of ribonulcleotide electrolytes CE with indirect photometric detection of polyphosphates and polyphosphonates. The quantitation of glyphosphate and aminomethylphosphonic acid in a commercial herbicide was performed with an average relative error of 1.6%.

Susse and Miller [57] developed a method for the simultaneous separation and determination of selected triazines, carbamates and phosphorganic compounds by MEKC. A separation of nine pesticides has been successfully achieved within 13 min. The detection limits were determined between 8 and 13 µg l^{-1} for the investigated pesticides. Crosby and El Rassi [58] evaluated a series of alkyltrimethylammonium chloride and bromide surfactants in MEKC of urea herbicides, alkylbenzenes and phenylalkyl alcohols. An increase in the migration window was observed as the size of the alkyl tail of the surfactant decreased. The separation of a mixture of six urea herbicides was best achieved when a MEKC system of low hydrophobic phase ratio and wide migration window, such as dodecyl- or decyltrimethylammonium chloride, was used as the micellar phase. Cai and El Rassi [59] and Smith and co-workers [60–62] introduced a novel micelle with adjustable surface charge density for MEKC of neutral and charged herbicides. High separation efficiencies were obtained over a wide range of migration conditions, and consequently the detection limit for the herbicides was in the range of 18–52 fmol using UV detection. Aguilar [4] used CZE for the determination of two phenoxyalkyl acid herbicides.

Nielen [63] described the CZE separation of several phenoxy acid herbicides and related impurities originating from the production process, as well as chiral separation of some phenoxypropionic acid herbicides using CD chiral selectors. Farran and Hernandez [64] utilized a CZE method for the analysis of phenylurea and phenoxyalkyl acid herbicides. Good results were obtained when the method was applied to spiked river water samples. Garrison et al. [65] established CZE conditions for the separation and detection of 2,4-dichlorophenoxyacetic acid and three optically active phenoxy acid herbicides. Baseline separation of the two enantiomers of each of the three active herbicides was accomplished by the addition of tri-O-methyl-β-CD. Wu et al. [66] developed a MEKC method for the separation of a number of herbicides consisting of chlorophenoxy acids. Sodium dodecyl sulfate (SDS) combined with Brij 35 (polyoxyethylene lauryl ether) as the micellar agent was found to provide the best overall separation of these compounds. Nielen [67] descirbed various preconcentration methods for environmental trace analysis by CE. Phenoxy acid herbicides were used as model compounds in drinking and river water at the sub-μg l^{-1} level. Dinelli and co-workers [68, 69] described determination of herbicides in tap water by CE and MEKC. The methods were suitable for mono- and multiresidue analysis of herbicides in tap water at the μg l^{-1} level, using a 1000-fold condensation step by solid phase extraction (SPE). Dinelli et al. [70] evaluated the potential of CE for the separation and detection of the metabolites of nine sulfonylurea herbicides in water. CE was confirmed to be a very efficient separation technique, suitable for the determination of sulfonylurea herbicides and their metabolites formed during hydrolysis. Dinelli et al. [71] developed a multiresidue analytical method based on SPE enrichment with MEKC to quantitate three sulfonylurea herbicides in soil samples. The recovery of each herbicide was >80% and the detection limit was 10μg l^{-1}. Kolecek et al. [72] determined 2-chloroethyltrimethylammonium chloride (chlormequat) by capillary isotachophoresis using sodium acetate as the leading electrolyte and 6-aminocaproic acid as the terminator in a medium buffered to pH 4.5 with citric acid. Farran and Ruitz [73] utilized MEKC for the simultaneous determination of mixtures of phenylurea and phenoxyalkyl acid herbicides. Total separation of a mixture of nine different herbicides in 16 min has been achieved. The usefulness of electrokinetic capillary chromatography (EKC) for the determination of phenylureas and chlorosulfuron herbicides was investigated by Song et al. [74]. Micellar, mixed micellar and microemulsion EKC were examined for this purpose and compared systematically. The separation efficiencies in mixed micellar and microemulsion micellar electrokinetic capillary chromatography (MEEKC) were higher than that in MEKC. Dinelli et al. [75] compared CE, HPLC and enzyme immunoassay for terbutylazine determination in tap and ground waters at the microgram per liter level. Over the range of concentration tested, the results obtained by the different methods were highly correlated. Compared to enzyme immunoassay, CE and HPLC needed sample extraction and concentration before analysis but showed higher accuracy and lower variation.

Paraquat and diquat have been the subject of several studies. Cai and Rassi [76] investigated the potential of CZE for the determination of these compounds. The limits of detection for paraquat and diquat were 15.4 and 16.8 fmol, respectively, with a UV detector. Galceran et al. [77] established the optimum CE conditions for the simultaneous separation of the quaternary ammonium ion herbicides. Resolution of paraquat and diquat was critical and required well established conditions. Carneiro et al. [78] compared CE and HPLC for determination of cationic pesticides. Detection limits ranged from 2.9 to 5.5 µg l^{-1} and were similar for both techniques. On-line isotachophoretic sample pretreatment in the CZE determination of paraquat and diquat in water was described by Kaniansky et al. [79]. Using this technique, the herbicides can be analyzed with a detection limit of 10^{-9} M.

42.11 Organic heavy metal compounds

Several studies have focused on applying CE to organic forms of tin, lead, mercury, selenium and arsenic, among other heavy metal ions. Medina et al. [80] reported a rapid method for speciation and determination of organomercury compounds in a biological sample of marine origin using CE. Results were compared to that obtained by GC. Lai and Dabek-Zlotorzynska [81] developed a method for the speciation of mercury based on CE with amperometric detection. Electrophoretic separation of four inorganic and organomercury species was achieved in less than 10 min. Ng et al. [82] developed a MEKC method for the determination of organolead and organoselenium in environmental samples. High separation efficiencies and detection limits comparable to those obtained by GC were achieved. Lopez-Sanchez et al. [83] and Amran et al. [84] applied CZE to separate arsenical species. Albert et al. [85] and Li and Li [86] reported CE as an efficient tool for the separation of organic and non-organic arsenious or selenious compounds. On-column enrichment of these compounds was performed by stacking large volumes of samples injected onto the column with the field-amplified injection technique [86]. Liu et al. [87] reported the use of CE with an inductively coupled plasma mass spectrometric detector to determine various inorganic and organic arsenic species and other heavy metal ions at nanogram per liter to microgram per liter levels. The separation of organotin species using CE with indirect photometric detection was investigated by Han et al. [88]. Pyridine was used as the UV absorption additive at 254 nm and separation was achieved in 6 min. Li and Li [89] demonstrated the combined subcritical fluid extraction–MEKC method as a promising technique for the extraction, separation and determination of alkyllead and alkyltin compounds in solid samples.

References

1. Brumley, W.C. (1995) Environmental applications of capillary electrophoresis for organic pollutant determination. *LC–GC* **13**, 556–568.
2. Gonnord, M.F.; Collet, J. (1993) Optimized capillary zone electrophoretic separation of chlorinated phenols. *J. Chromatogr.* **645**, 327–336.
3. Bachmann, K.; Gottlicher, B.; Haag, L.; Hannina, M.; Hensel, W. (1994) Sample stacking for charged phenol derivatives in capillary zone electrophoresis. *Fresenius J. Anal. Chem.* **350**, 368–371.
4. Aguilar, M.; Farran, A.; Marti, V. (1993) Analysis of phenoxyalkyl acid herbicides and chlorophenols by capillary zone electrophoresis. *Sci. Total Environ.* **132**, 133–140.
5. Chao, Y.-C.; Whang, C.-W. (1994) Capillary zone electrophoresis of eleven priority phenols with indirect fluorescence detection. *J. Chromatogr.* **663**, 229–237.
6. Chen, I.C.; Whang, C.W. (1994) Capillary zone electrophoresis of 11 priority phenols with amperometric detection. *J. Chin. Chem. Soc.* **41**, 419–424.
7. Li, G.; Locke, D. (1995) Separation of the eleven priority phenols by capillary zone electrophoresis. *J. Chromatogr.* **669**, 93–102.
8. Lin, C.; Lin, W.; Chiou, W. (1995) Migration behavior and optimization of selectivity of dichlorophenols in capillary zone electrophoresis. *J. Chromatogr.* **705**, 325–333.
9. Benz, N.; Fritz, J.S. (1995) Optimization of separation of alkyl-substituted phenolate anions by capillary zone electrophoresis. *J. High Resolut. Chromatogr.* **18**, 175–178.
10. Shohat, D.; Grushka, E. (1994) Use of calixarenes to modify selectivities in capillary electrophoresis. *Anal. Chem.* **66**, 747–750.
11. Praus, P.; Dombek, V. (1993) Separation of chlorophenols by capillary isotachophoresis. *Anal. Chim. Acta* **277**, 97–101.
12. Praus, P.; Dombek, V. (1993) Utlilization of polyethylene glycol for the separation of chlorophenols by capillary isotachophoresis. *Anal. Chim. Acta* **283**, 917–921.
13. Praus, P.; Dombek, V. (1993) β-Cyclodextrin in the capillary isotaphoretic separation of chlorophenols. *Anal. Chim. Acta* **281**, 397–400.
14. Praus, P. (1995) Possibilities of simple steam distillation in combination with capillary isotachophoresis for the determination of chlorophenols in river and industrial waste waters. *Anal. Chim. Acta* **302**, 39–44.
15. Praus, P.; Dombek, V. (1993) The utilization of capillary isotachophoresis in the analysis of water. *Chem. Listy* **87**, 101–109.
16. Takeda, S.; Wakida, S.; Kunishige, H. (1992) Water analysis by electrokinetic chromatography. *Kogyo Yosui* **410**, 74–78.
17. Brumley, W.C.; Jones, W.J. (1994) Comparison of micellar electrokinetic chromatography (MEKC) with capillary gas chromatography in the separation of phenols, anilines and polynuclear aromatics. Potential field-screening applications of MEKC. *J. Chromatogr.* **680**, 163–173.
18. Landman, A.; Sun, P.; Hartwick, R.A. (1994) Enhanced micellar electrokinetic capillary chromatography separations on anionic polymer-coated capillary with pH-independent electroosmotic flow. *J. Chromatogr.* **669**, 259–262.
19. Masselter, S.M.; Zemann, A.J. (1995) Influence of organic solvents in coelectroosmotic capillary electrophoresis of phenols. *Anal. Chem.* **67**, 1047–1053.
20. Masselter, S.M.; Zemann, A.J. (1995) Influence of buffer electrolyte pH on the migration behavior of phenolic compounds in co-electroosmotic capillary electrophoresis. *J. Chromatogr.* **693**, 359–365.
21. Aiken, J.H.; Huie, C.W. (1993) Effects of 1-alkanols on separation performance in micellar electrokinetic capillary chromatography. *J. Microcol. Sep.* **5**, 95–99.
22. Masselter, S.M.; Zemann, A.J.; Bobleter, O. (1993) Separation of cresols using coelectroosmotic capillary electrophoresis. *Electrophoresis* **14**, 36–39.

23. Takeda, S.; Wakida, S.; Yamane, M.; Kawahara, A.; Higashi, K. (1993) Separation of aniline derivatives by micellar electrokinetic chromatography. *J. Chromatogr.* **653**, 109–114.
24. Takeda, Y.; Sakairi, M.; Koizumi, H. (1995) On-line combination of micellar electrokinetic chromatography and mass spectrometry using an electrospray–chemical ionization interface. *Rapid Commun. Mass Spectrom.* 488–490.
25. Brumley, W.C.; Brownrigg, C.M. (1994) Application of MEKC in the determination of benzidines following extraction from water, soil, sediment, and chromatographic adsorbents. *J. Chromatogr. Sci.* **32**, 69–75.
26. Nielen, M.W.F. (1992) Separation of phenylenediamine isomers by capillary zone electrophoresis. *J. Chromatogr.* **625**, 387–391.
27. Beck, W.; Engelhardt, H. (1992) Capillary electrophoresis of organic and inorganic cations with indirect UV detection. *Chromatographia.* **33**, 313–316.
28. Riviello, J.M.; Harrold, M.P. (1993) Capillary electrophoresis of inorganic cations and low-molecular-mass amines using a copper-based electrolyte with indirect UV detection. *J. Chromatogr.* **652**, 385–392.
29. Dabek-Zlotorzynska, E.; Dlouhy, J.F. (1995) Application of capillary electrophoresis in atmospheric aerosol analysis: determination of cations. *J. Chromatogr.* **706**, 527–534.
30. Avdalovic, N.; Pohl, C.A.; Rocklin, R.D.; Stillian, J.R. (1993) Determination of cations and anions by capillary electrophoresis combined with suppressed conductivity detection. *Anal. Chem.* **65**, 1470–1475.
31. Ng, C.L.; Ong, C.P.; Lee, H.K.; Li, S.F.Y. (1994) Capillary electrophoretic separation of nitrosoamines using combined open-tubular and packed capillary columns. *J. Chromatogr. Sci.* **32**, 121–125.
32. Janini, G.M.; Muschik, G.M.; Issaq, H.J. (1994) Capillary electrophoresis separation of heterocyclic nitrosoamino acid-conformers at sub-ambient temperatures. *J. High Resolut. Chromatogr.* **17**, 753–755.
33. Bachmann, K.; Haag, I.; Schmitzer, R.Q. (1993) Determination of carbonyl compounds in the low ppb-range by capillary electrophoresis. *Fresenius J. Anal. Chem.* **346**, 786–788.
34. Aguin, M.L.; Vindevogel, J.; Sandra, P. (1993) Utilisation of the bisulfite addition reaction for the separation of neutral aldehydes by capillary electrophoresis. *Chromatographia* **37**, 451–454.
35. Takeda, S.; Wakida, S.; Yamane, M.; Kawahara, A.; Higashi, K. (1993) Migration behavior of phthalate esters in micellar electrokinetic chromatography with or without added methanol. *Anal. Chem.* **65**, 2489–2492.
36. Palmer, C.P.; Khaledi, M.Y.; McNair, H.M. (1992) A monomolecular pseudostationary phase for micellar electrokinetic capillary chromatography. *J. High Resolut. Chromatogr.* **15**, 756–762.
37. Kleibohmer, W.; Cammann, K.; Robert, J.; Mussenbrock, E. (1993) Determination of explosives residues in soils by micellar electrokinetic capillary chromatography and high-performance liquid chromatography. A comparative study. *J. Chromatogr.* **638**, 349–356.
38. Weiss, C.S.; Hazlett, J.S.; Datta, M.H.; Danzer, M.H. (1992) Determination of quaternary ammonium compounds by capillary electrophoresis using direct and indirect UV detection. *J. Chromatogr.* **608**, 325–332.
39. Chen, S.; Pietrzyk, D.J. (1993) Separation of sulfonate and sulfate surfactants by capillary electrophoresis: effect of buffer cation. *Anal. Chem.* **65**, 2770–2775.
40. Desbene, P.L.; Rony, C.; Desmazieres, B.; Jacquier, J.C. (1992) Analysis of alkylaromatic sulphonates by high-performance capillary electrophoresis. *J. Chromatogr.* **608**, 375–383.
41. Shamsi, S.A.; Danielson, N.D. (1994) Naphthalenesulfonates as electrolytes for capillary electrophoresis of inorganic anions, organic acids, and surfactants with indirect photometric detection. *Anal. Chem.* **66**, 3757–3764.
42. Altria, K.D.; Gill, I.; Howells, J.S.; Luscombe, C.N.; Williams, R.Z. (1995) Trace analysis of detergent residues by capillary electrophoresis. *Chromatographia* **40**, 527–531.
43. Vogt, C.; Heinig, K.; Longer, B.; Mattusch, J.; Werner, G. (1995) Determination of linear alkylbenzenesulfonates by high-performance liquid chromatography and capillary zone electrophoresis. *Fresenius J. Anal. Chem.* **352**, 508–514.

44. Goebel, L.K.; McNair, H.M.; Rasmussen, H.K.; McPherson, B.P. (1993) Separation of ethoxylated alcohol sulfates by capillary electrophoresis using indirect UV detection. *J. Microcol. Sep.* **5**, 47–50.
45. Croft, S.N.; Lewis, D.M. (1992) Analysis of reactive dyes and related derivatives using high-performance capillary electrophoresis. *Dyes Pigm.* **18**, 309–317.
46. Croft, S.N.; Hinks, D. (1993) Analysis of dyes by capillary electrophoresis. *Textile Chem. Colours* **25**, 47–51.
47. Lord, G.A.; Gordon, D.B.; Tetler, L.W.; Carr, C.M. (1995) Electrochromatography electrospray mass spectrometry of textile dyes. *J. Chromatogr.* **700**, 27–33.
48. Tapley, K.N. (1995) Capillary electrophoretic analysis of the reactions of bifunctional reactive dyes under various conditions including of the analysis of the traditionally difficult to analyse phthalocyanine dyes. *J. Chromatogr.* **706**, 555–562.
49. Brumley, W.C.; Brownrigg, C.M.; Grange, A.H. (1994) Capillary liquid chromatography–mass spectrometry and micellar electrokinetic chromatography as complementary techniques in environmental analysis. *J. Chromatogr.* **680**, 635–644.
50. Burkinshaw, S.M.; Hinks, D.; Lewis, D.M. (1993) Capillary zone electrophoresis in the analysis of dyes and other compounds in the dye-manufacturing and dye-using industries. *J. Chromatogr.* **640**, 413–417.
51. Varghese, J.; Cole, R.B. (1993) Optimization of capillary zone electrophoresis–electrospray mass spectrometry for cationic and anionic laser dye analysis employing opposite polarities at the injector and interface. *J. Chromatogr.* **639**, 303–316.
52. Oxspring, D.; Smyth, W.F.; Marchant, R. (1995) Comparison of reversed-polarity capillary electrophoresis and adsorptive stripping voltammetry for the detection and determination of reactive textile dyes. *Analyst* **120**, 1995–2000.
53. Brumley, W.C. (1992) Qualitative analysis of environmental samples for aromatic sulfonic acids by high-performance capillary electrophoresis. *J. Chromatogr.* **603**, 267–272.
54. Brumley, W.C.; Brownrigg, C.M. (1993) Electrophoretic behavior of aromatic-containing organic acids and the determination of selected compounds in water and soil by capillary electrophoresis. *J. Chromatogr.* **646**, 377–389.
55. Robins, W.H.; Wright, B.W. (1994) Capillary electrophoretic separation of organophosphonic acids using borate estrification and direct UV detection. *J. Chromatogr.* **680**, 667–673.
56. Shamsi, S.A.; Danielson, N.D. (1995) Ribonucleotide electrolytes for capillary electrophoresis of polyphosphates and polyphosphonates with indirect detection. *Anal. Chem.* **67**, 1845–1852.
57. Susse, H.; Miller, H. (1995) Application of micellar electrokinetic capillary chromatography to the analysis of pesticides. *Fresenius J. Anal. Chem.* **352**, 470–473.
58. Crosby, D.; Rassi, Z. (1993) Micellar electrokinetic capillary chromatography with cationic surfactants. *J. Liq. Chromatogr.* **16**, 2161–2187.
59. Cai, J.; Rassi, Z. (1992) Micellar electrokinetic capillary chromatography of neutral solutes with micelles of adjustable surface charge density. *J. Chromatogr.* **608**, 31–45.
60. Smith, J.T.; Nashabeh, W.; Rassi, Z. (1994) Micellar electrokinetic capillary chromatography with *in situ* charged micelles. I. Evaluation of N-D-gluco-N-methylalkanamide surfactants as anionic borate complexes. *Anal. Chem.* **66**, 1119–1133.
61. Smith, J.T.; Rassi, Z. (1994) Micellar electrokinetic capillary chromatography with *in situ* charged micelles: II. Evaluation and comparison of octylmaltoside and octylsucrose surfactants as anionic borate complexes in the separation of herbicides. *J. Microcol. Sep.* **6**, 127–138.
62. Smith, J.T.; Rassi, Z. (1994) Micellar electrokinetic capillary chromatography with *in situ* charged micelles. IV. Influence of the nature of the alkylglycoside surfactant. *J. Chromatogr.* **685**, 131–143.
63. Nielen, M.W.F. (1993) (Enantio-) separation of phenoxy acid herbicides using capillary zone electrophoresis. *J. Chromatogr.* **637**, 81–90.
64. Farran, A.; Hernandez, O. (1993) Separation of mixtures of phenylurea and phenoxyalkyl acid herbicides by capillary zone electrophoresis. *Quim. Anal.* **12**, 205–208.
65. Garrison, A.W.; Schmott, P.; Kettrup, A. (1994) Separation of phenoxy acid herbicides and their enantiomers by high-performance capillary electrophoresis. *J. Chromatogr.* **688**, 317–327.

66. Wu, Q.; Claessens, H.A.; Cramers, C.A. (1992) The separation of herbicides by micellar electrokinetic capillary chromatography. *Chromatographia* **34**, 25–30.
67. Nielen, M.W.F. (1993) Trace enrichment of environmental samples in capillary zone electrophoresis. *Trends Anal. Chem.* **12**, 345–356.
68. Dinelli, G.; Vicari, A.; Catizone, P. (1993) Use of capillary electrophoresis for detection of metsulfuron and chlorosulfuron in tap water. *J. Agric. Food Chem.* **41**, 742–746.
69. Dinelli, G.; Bonetti, A.; Catizone, P.; Galletti, G.C. (1994) Separation and detection of herbicides in water by micellar electrokinetic chromatography. *J. Chromatogr.* **656**, 275–280.
70. Dinelli, G.; Vicari, A.; Bonetti, A. (1995) Separation of sulfonylurea metabolites in water by capillary electrophoresis. *J. Chromatogr.* **700**, 195–200.
71. Dinelli, G.; Vicari, A.; Brandolini, V. (1995) Detection and quantitation of sulfonylurea herbicides in soil at the ppb level by capillary electrophoresis. *J. Chromatogr.* **700**, 201–207.
72. Kolecek, J.; Riha, V.; Vytras, K. (1993) Determination of 2-chloroethyltrimethyl-ammonium chloride in Retacel by ion-selective electrode potentiometry and capillary isotachophoresis. *Anal. Chim. Acta.* **273**, 431–433.
73. Farran, A.; Ruitz, S. (1995) Simultaneous separation of neutral and acidic herbicides by micellar electrokinetic capillary chromatography. *Int. J. Environ. Anal. Chem.* **58**, 121–129.
74. Song, L.; Ou, Q.; Yu, W.; Li, G. (1995) Separation of six phenylureas and chlorosulfuron standards by micellar, mixed micellar and microemulsion electrokinetic chromatography. *J. Chromatogr.* **699**, 371–382.
75. Dinelli, G.; Vicari, A.; Bonetti, A.; Catizone, P. (1995) Comparison of capillary electrophoresis, HPLC, and enzyme immunoassay for terbutylazine detection in water. *J. Agric. Food Chem.* **43**, 951–955.
76. Cai, J.; Rassi, Z. (1992) Capillary zone electrophoresis of two cationic herbicides, paraquat and diquat. *J. Liq. Chromatogr.* **15**, 1193–1200.
77. Galceran, M.T.; Carneiro, M.C.; Puignou, L. (1994) Capillary electrophoresis of quaternary ammonium ion herbicides: paraquat, diquat and difenzoquat. *Chromatographia.* **39**, 581–586.
78. Carneiro, M.C.; Puignou, L.; Galceran, M.T. (1994) Comparison of capillary electrophoresis and reversed-phase ion-pair high performance liquid chromatography for the determination of paraquat, diquat and difenzoquat. *J. Chromatogr.* **669**, 217–224.
79. Kaniansky, D.; Ivanyi, F.; Onuska, F.I. (1994) On-line isotachophoresis sample pretreatment in ultratrace determination of paraquat and diquat in water by capillary electrophoresis. *Anal. Chem.* **66**, 1817–1824.
80. Medina, I.; Rubi, E.; Mejuto, M.C.; Cela, R. (1993) Speciation of organomercurials in marine samples using capillary electrophoresis. *Talanta* **40**, 1631–1636.
81. Lai, E.P.C.; Dabek-Zlotorzynska, E. (1996) Capillary electrophoresis with amperometric detection for mercury speciation. *Am. Env. Lab.* **6**, 6–7.
82. Ng, C.L.; Lee, H.K.; Li, S.F.Y. (1993) Determination of organolead and organoselenium compounds by micellar electrokinetic chromatography. *J. Chromatogr.* **652**, 547–553.
83. Lopez-Sanchez, J.F.; Amram, M.B.; Lakkis, M.D.; Lagarde, F.; Rauret, G.; Leroy, M.J.F. (1994) Quantitative aspects of the separation of arsenical species by free solution capillary electrophoresis. *Fresenius J. Anal. Chem.* **348**, 810–814.
84. Amran, M.B.; Hagege, A.; Legarde, F.; Leroy, M. (1995) Improvement of detection sensitivity of arsenic species using capillary zone electrophoresis. *Chemia Analityczna* **40**, 309–318.
85. Albert, M.M.; Demesmay, C.; Rocca, J.L. (1995) Analysis of organic and non-organic arsenious or selenium compounds by capillary electrophoresis. *Fresenius J. Anal. Chem.* **351**, 426–432.
86. Li, K.; Li, S.F.Y. (1995) Speciation of selenium and arsenic compounds in natural waters by capillary zone electrophoresis after on-column preconcentration with field-amplified injection. *Analyst* **120**, 361–366.
87. Liu, Y.; Lopez-Avila, V.; Zhu, J.J.; Wiederin, D.R.; Beckert, W.F. (1995) Capillary electrophoresis coupled on-line with inductively coupled plasma mass spectrometry for elemental speciation. *Anal. Chem.* **67**, 2020–2025.

88. Han, F.; Fashing, J.L.; Brown, P.R. (1995) Speciation of organotin compounds by capillary electrophoresis using indirect ultraviolet absorbance detection. *J. Chromatogr.* **669**, 103–112.
89. Li, K.; Li, S.F.Y. (1995) Determination of alkyllead and alkyltin compounds in solid samples by supercritical and subcritical fluid extraction and MEKC. *J. Chromatogr. Sci.* **33**, 309–314.

43 Analysis of synthetic polymers by capillary electrophoretic techniques

J.A. BULLOCK

43.1 Introduction

As a class of compounds, synthetic polymers present some of the more difficult challenges for the separation chemist. In the context of this chapter, synthetic polymers are to be differentiated from biopolymers. The former are synthetic materials produced from one or more monomeric starting materials for which there is no intended biological activity. This is in contrast to biopolymers, which include peptides, proteins and oligonucleotides, either naturally occurring or synthesized, for which there is an intended biological activity. By nature, synthetic polymers are heterogeneous materials. This heterogeneity derives not only from the molecular weight distribution, but can also result from the chemical composition (copolymers), structural sequence and structural motif (branching, grafted, etc.) as well. The techniques applied to polymer separations span a broad range of chromatographic modalities but are typically based on size or adsorption (normal phase, reversed phase, ion exchange or GPC). Classical size exclusion or gel permeation chromatographic techniques (GPC) are firmly established for evaluating molecular weight information of polymers while adsorption techniques provide the ability to probe chemical composition and, in some cases, molecular weight information as well.

Previously a refractive index (RI) detector was used to determine the molecular weight (mol. wt) of polymers, but the mol. wt obtained is relative. In order to obtain an exact mol. wt of the polymer, a light scattering detector is required. Compared with relative and exact mol. wt, the former was often greater (editor's data).

Each of these techniques has its own advantages and disadvantages. In general, the large size of polymers places limitations on the efficiency of chromatographic separations. The discriminating power of size exclusion chromatography in terms of separation of individual oligomers drops off quickly with increasing molecular weight. The slow mass transport phenomena characteristic of adsorption chromatography of polymers imposes restrictions on the efficiency and recovery achievable with this class of separations.

Electrophoretic separations composing capillary zone electrophoresis (CZE), micellar electrokinetic chromatography (MEKC or MECC) and capillary isotachophoresis (CITP) present some interesting alternative separation mechanisms for synthetic polymers. In addition to providing a different mode of separation, each of these techniques can achieve higher efficiencies than traditional column chromatographic methods. With CZE, in fact, separation efficiency is predicted to increase with increasing polymer molecular weight. For these reasons, among others, these CE techniques are an attractive alternative or compliment to column chromatographic methods.

In spite of the potential advantages of these CE techniques, the development of methods for synthetic polymers has lagged behind developments for biopolymers. There are perhaps several reasons for this difference in the rate of development of this technology for these two different classes of polymers. First of all, electrophoretic techniques in general have traditionally been the domain of biopolymer analyses. The characteristic charge properties and high water solubility of biopolymers makes them ideal candidates for electrophoretic separation. By contrast, many synthetic polymers carry no charge and are insoluble in aqueous systems which are obvious obstacles to electrophoretic analysis. In general, the physical properties of synthetic polymers as a whole are more diverse in terms of their charge and solubilities. In addition, the size heterogeneity of synthetic polymers is typically much greater than that of biopolymers. Because of the greater variety of physical and chemical properties compared to biopolymers, the development of 'standard' separation protocols for synthetic polymers are, perhaps, not possible. This is in contrast to biopolymers for which standard protocols are well established, facilitating the development and acceptance of CE techniques for this class of compounds.

While the development of standard CE methods for synthetic polymers may not be possible, these techniques are, nonetheless, gaining importance for polymer characterization. Among the different types of applications that have been investigated for polymers are purity analysis, oligomer separation, size analysis, kinetic analysis and composition analysis. Separations of charged as well as uncharged polymers, water soluble as well as insoluble polymers and low as well as high molecular weight polymers have been demonstrated. It is clear that for most polymer separations, the electrophoretic conditions will need to be uniquely tailored to that particular polymer. In this paper, a summary of some of the different types of materials analyzed and variety of different applications developed for synthetic polymers will be presented. This covers a broad range of polymer sizes, chemistries and applications which should provide the reader interested in polymer separations with a starting point to develop methods suitable for other types of synthetic polymers. Table 43.1 categorizes the different applications by type of material analyzed and provides some details on the experimental conditions.

INTRODUCTION

Table 43.1 Summary of applications of CE to analysis of synthetic polymers

Methods Separation conditions Voltage, current Capillary length, i.d. Detection	Compounds	Ref.
Impurities and monomers CZE 20 mM Borate buffer, pH 8.0 with H_3PO_4 20 kV, 50 μA 27 cm, 50 μm UV at 109 nm	Vinyl sulfonate	1

CZE
pH 4 Acetate or pH 7 phosphate
  buffer with or without 2 mM
    CTAB
20 kV, 23 μA
50 cm, 50 μm
	Acrylic acid	2

pH 2.5 Phosphate buffer with 1% Ficoll-400 25 kV, 74 μA 50 cm, 50 μm UV at 214 and 254 nm	Polymer impurities	2
MECC 200 mM Tris–borate, pH 9.0, 50 mM SDS 12 kV, 60 μA 57 cm, 75 μm UV at 254 nm	Acrylamide monomer	3
CZE 0.1 M pH 9 Borate buffer 15 kV, 50 μA 50 cm, 75 μm UV at 254 nm	Acrylamido monomers	4
CITP LE: 10 mM K^+, 13.2 mM acetate pH 5.5, 2% TEG additive TE: 10 mM β-alanine Electrical resistance	Cationic polymers	5
CZE 100 mM Phosphate, pH 2.0 30 kV, 132 μA 72 cm, 50 μm UV 200 nm	Synthetic polylysine	6
CZE 20 mM Tetraborate, pH 9.0 20 kV 64.5 cm, 75 μm UV 200 nm	Polyacrylate polymer	7
CZE 50 mM Borate, pH 9.0 25 kV, 22 μA 57 cm, 50 μm UV at 200 nm	Residual DTPA	8

Table 43.1 *Continued*

Methods Separation conditions Voltage, current Capillary length, i.d. Detection	Compounds	Ref.
CZE 20 mM Borate, pH 8.0, 20% methanol 30 kV 62 cm, 50 μm UV at 190 nm	Residual monomers	9
Insoluble polymers and particles CZE 1 mM ACES, pH 5.8 or 6.46, CTAB treated capillary 30 kV 47.6 cm, 50 μm UV at 254 nm	Polystyrene nanoparticles	10, 11
CZE 10 mM Phosphate, pH 7.4 30 kV 27 cm, 50 μm UV at 225 nm	Polystyrene nanoparticles	12
CZE Phosphate buffers, pH 6.6–10.7 30 kV 55 cm, 75 μm UV at 254 nm	Latex particles	13
CZE 5 mM Phosphate, pH 10.7 25 kV 46 cm, 75 μm UV at 254 nm	Polystyrene latex particles	14
CGE Tris–borate–EDTA, pH 8.3 with polyacrylamide of mol. wt 5–6×10^6 10 kV, 60 μA 37 cm, 150 μm, polyacrylamide coated capillary Fluorescence detection	Polystyrene carboxylate particles	15
CZE 2.5 mM NH_4OH, 4.65 mM NH_4Cl, pH 9.0 buffer 30 kV 101 cm, 50 μm Detection by light scattering at 190 nm	Silica sols	16
Oligomer separations CZE 77 mM Boric acid, 19.4 mM citric acid, 11.3 mM Na_3PO_4, pH 4.2/20% methanol 30 kV, 3.1 μA 75 cm, 25 μm LIF detection	NDA poly(oxyalkylene)diamines	17

Table 43.1 *Continued*

Methods Separation conditions Voltage, current Capillary length, i.d. Detection	Compounds	Ref.
CZE 0.1 M Sodium borate, pH 9.2 50 cm, 50 μm UV at 200 nm	Poly(β-malic acid) oligomers	18
CZE 20 mM Borate buffer, pH 8.0 30 kV 120 cm, 50 μm UV at 190 nm	Acrylic acid copolymers	9
CZE 30 mM Creatinine/acetate, pH 4.8, 1 mg ml^{-1} PEO – 86 kDa 25 kV, 13 μA 44 cm, 50 μm Indirect UV at 220 nm	Poly(alkyl oxide) diamines	19
CZE 25 mM Boric acid, 50 mM SDS, pH 8.6, 35% acetonitrile 25 kV, 27 μA 67 cm, 50 μm UV at 200 nm	Triton-X oligomers	19
CZE 57 mM Boric acid, 35 mM 1,3-diaminopropane, pH 9.7 – 30% acetonitrile 25 kV, 14 μA 44 cm, 50 μm UV at 205 nm	Derivatized poly(ethylene glycol)	19
CGE Micro-Gel 100 capillary, 100 mM Tris–borate/10% methanol −30 kV 50 cm, 75 μm Mass spectrometric detection	Poly(acrylic acid) oligomers	20
CGE Hydrolink gel-filled capillary, 0.1–0.2 M Tris–tricine 9.2 kV 49 cm, 75 μm UV at 220 nm or LIF	Homo poly(amino acid) oligomers	21
CGE Hydrolink gel-filled capillary, 0.1 M Tris–tricine buffer 6 kV 30 cm, 75 μm UV at 220 nm	Homo poly(amino acid) oligomers	22

Table 43.1 *Continued*

Methods Separation conditions Voltage, current Capillary length, i.d. Detection	Compounds	Ref.
CZE 12.5 mM Borate, pH 9 – 30% acetonitrile 30 kV 57 cm, 50 μm UV at 214 nm	Alkylbenzene sulfonate oligomers	23
MECC 6.25 mM Borate, pH 9, 50 mM SDS 25 kV 57 cm, 50 μm UV at 214 nm	Alkylbenzene sulfonate oligomers	23

CZE
1 mM Potassium dichromate,
  1 mM sodium tetraborate,
  30 mM boric acid
10 kV
100 cm, 100 μm
Indirect UV at 265 nm
	Ethoxylated alcohol sulfate oligomers	24

CZE 50 mM NH$_4$Cl, 50 mM Et$_3$N in methanol 20 kV, 35 μA 44.6 cm, 50 μm UV at 250 nm	Polyether oligomers	25

High molecular weight polymers
CZE
Coated capillary,
  25 mM sodium phosphate, pH 5,
  5–10 mg ml^{-1} HEC
20 kV
50 cm, 50 μm
UV at 235 nm
	Poly(styrenesulfonates)	26

CZE
Coated capillary,
  25 mM sodium phosphate, pH 3,
  0.6% HEC (mol. wt = 100 000) or
  polyacrylamide (mol. wt = 500 000)
75–300 V cm^{-1}
40 cm, 75–100 μm
UV at 227 nm
	Poly(styenesulfonates)	27

Inorganic polymers CZE 5 mM AMP, 100 mM boric acid, 2 mM EDTA, pH 6.8 −30 kV 75 cm, 50 μm Indirect UV at 259 nm	Polyphosphates and polyphosphonates	28

Table 43.1 *Continued*

Methods Separation conditions Voltage, current Capillary length, i.d. Detection	Compounds	Ref.
CZE Coated capillary, 0.1 M sodium phosphate, pH 7, 30% acetonitrile 3 kV 24 cm, 25 µm UV at 210 nm	Polyoxometalates	29
Polymer composition analysis CZE 75 mM Sodium borate pH 9.2 or 25 mM sodium phosphate pH 8.2 10 kV, 40 µA 61 cm, 75 µm UV at 254 nm	Polyimide monomers	30
CZE 7 mM 3-Nitrophthalic acid, 0.5 mM myristytrimethylammonium bromide, pH 7.0 −20 kV 60 cm, 75 µm Indirect UV at 254 nm	Cellulose ester substituents	31
Kinetics of polymerization MECC 100 mM Tris–borate, 50 mM SDS, pH 9.0 4 kV, 4 µA 47 cm, 100 µm UV at 214 nm	Polyacrylamide gels	32
MECC 50 mM Sodium phosphate, 50 mM SDS, pH 7, 10% methanol 10 kV, 66 µA 50 cm, 75 µm UV at 214 nm	Acrylamide weak acids and bases	33
MECC 100 mM Borate, 100 mM SDS, pH 9.0 4 kV, 31 µA 50 cm, 100 µm UV at 214 nm	Polyacrylamide gels	34
MECC 100 mM Tris–borate, 50 mM SDS, pH 9.0 5 kV, 35 µA 50 cm, 100 µm UV at 214 nm	Substituted polyacrylamides	35

Table 43.1 *Continued*

Methods Separation conditions Voltage, current Capillary length, i.d. Detection	Compounds	Ref.
MECC 100 mM Borate, 100 mM SDS, pH 9 4 kV, 31 µA 50 cm, 75 µm UV at 214 nm	Polyacrylamide gels	36

ACES = (2-[2-amino-2-oxoethyl)amino]ethanesulfonic acid, AMP = adenosine monophosphate, CTAB = cetyltrimethylammonium bromide, DTPA = diethylenetriaminetetraacetic acid, HEC = hydroxyethyl cellulose, HPCE = high performance capillary electrophoresis, LE = leading electrolyte, LIF = laser-induced fluorescence, NDA = 2,3-naphthalenedialdehyde, PEO = polyethylene oxide, TE = trailing electrolyte, TEG = tetraethylene glycol, Tris = tris(hydroxymethyl)aminomethane.

43.2 Analysis of impurities and residual monomers in polymers

Ryder [1] used CZE to quantitate residual vinyl sulfonate monomer in water-soluble polymer samples. A borate–phosphate buffer at pH 8.0 was used combined with UV detection at 190 nm. The method was highly precise and accurate. Comparisons were drawn with existing high performance liquid chromatography (HPLC) methods. Advantages of the CZE method over HPLC include accuracy and cost of analysis. The properties of CZE are well suited to analyzing hydrophilic charged monomers and the technique should find use for other types of monomer/polymer systems. Righetti *et al.* [2] utilized CZE to detect acrylic acid and polymeric impurities in immobiline compounds used for isoelectric focusing. A soluble polymeric buffer additive was used to achieve separation of the polymeric impurities. Without the polymeric additive Ficoll-400, separation of the polymers from the monomer was impossible. Chiari *et al.* [3] used CZE to monitor the scavenging of unreacted acrylamide monomer in polyacrylamide matrices used for capillary gel sieving electrophoresis of oligonucleotides and DNA. The unreacted acrylamide monomer is reacted with cysteine and the resulting adduct monitored by CE in a Tris–borate buffer (Tris = tris(hydroxymethyl)amino methane) containing 50 mM sodium dodecyl sulfate (SDS) to determine the percentage incorporation. In a 10% T matrix up to 20% unreacted monomer was found by CE. Gelfi *et al.* [4] used CZE to measure the percentage incorporation of a series of novel acrylamido monomers used to prepare polymeric matrices for gel electrophoresis. Unreacted monomers were first extracted from the polymers before being analyzed by CZE using a pH 9.0 borate buffer.

Tribet et al. [5] used capillary isotachophoresis to monitor the synthesis of cationic polymers formed by γ-irradiation of various dodecyl and hexadecylallydimethylammonium and dodecylvinylimidazolium salts. Separations of polymer from unreacted monomer and polymer oxidation products were achieved allowing investigation of polymer kinetics and optimization of polymerization conditions. Grimm [6] compared HPLC with CZE for determination of the purity of a synthetic polylysine preparation. Whereas reversed phase HPLC gave a single peak for the sample, CZE in a 100 mM phosphate buffer at pH 2.0 revealed at least ten minor impurity peaks which were presumably low molecular weight oligomers. Ross [7] demonstrated the potential for CZE to monitor the purity of a high molecular weight polyacrylate polymer. The 3 000 000 Da polymer was eluted in a borate buffer at pH 9.0 and detected at 200 nm. CZE was used to quantitate low levels of residual monomers in copolymeric magnetic resonance imaging contrast agents [8]. A series of different metal chelator monomers were conveniently separated from the polymer in borate buffer at pH 9.0. Ryder [9] reported on several applications involving the CZE analysis of polymeric water treatment additives. In most instances, advantages were noted over corresponding HPLC methods. CZE was shown to be well suited for the analysis of monomers from a copolymer in which the monomers differed significantly in their polarities. In another case [1], addition of methanol to the standard borate buffer was shown to improve the separation of vinyl sulfonate from a particular copolymer. CZE using indirect UV detection was used to detect trace levels of sulfate in a water-soluble polymer.

43.3 Analysis of polymeric particulates

CZE analysis of particulates involves an assessment of polymer particle size. Thus, analyses are typically conducted on water-insoluble polymers. VanOrman and McIntire [10, 11] reported on the capillary electrophoretic analysis of submicron polystyrene latex particles. Particles in the size range of 39 nm to 683 nm were separated in a 50 µm i.d. capillary and detected at 254 nm. Pretreatment of the capillary with a cationic surfactant was used to effect separation of the different sized particles. A type of particle–capillary wall interaction was proposed to account for the resulting separations. Although the exact mechanism of interaction was not well known, the separation was thought to be based on particle size. In a more recent report [12], the separation of these polystyrene nanoparticles in an untreated silica capillary was explained by the particle electrophoretic mobility and, thus, is related to the zeta potential. Satisfactory correlation was obtained between electrophoretic mobilities determined by CE and those determined by laser Doppler electrophoresis for negatively charged particles. For positively charged particles, however, the correlation was not good due to apparent particle–capillary wall interactions. Jones and Ballou [13] optimized CE separations of mixtures of chemically different latex particles with different numbers of attached carboxylate or sulfate groups. Separations were optimized with respect to capillary inside diameter and length, buffer concentration, pH and applied voltage. Particles in the size range from 30 nm to 1.16 µm could be separated. Size range distribution must be considered as one of factors affecting separation. Petersen and Ballou [14] investigated the effect of capillary temperature control and electrophoretic heterogeneity on the separation parameters for the CZE analysis of latex particles. Temperature is also considered in GPC because temperature affects the viscosity of the polymers as well as that of the eluent (editor's data). Electrophoretic heterogeneity in the particle samples was determined to be the major source of zone broadening when the system is adequately thermostated. Under these conditions, efficiency and resolution are largely independent of applied voltage.

Radko *et al.* [15] investigated using CE in un-crosslinked polyacrylamide solutions to separate polystyrene carboxylate particles up to 10 µm in diameter. A Tris–borate–EDTA (ethylene diaminetetraacetic acid) buffer was used with polyacrylamide of molecular weight of 500 000 Da. Detection was by fluorescence. McCormick [16] used free solution CE to separate colloidal silica particles with diameters in the range of 5 to 500 nm. Particles were detected by turbidity measurements at 190 nm. Resolution of sub-100 nm silica sols differing by a factor of two could be accomplished in 20 min. The ionic strength of the separation buffer was found to have a significant effect on the resolution and electrophoretic mobilities of the silica sols. With appropriate software and calibration standards, precise particle size distribution measurements should be possible.

43.4 Separation of oligomers of homopolymers and copolymers

The separation of oligomers by CE presents an opportunity to take advantage of the high efficiency and resolving power of the technique to achieve separations not attainable or difficult by column chromatography. Amankwa *et al.* [17] reported oligomer separation by CZE. The oligomer distribution of several poly(oxyalkylene) diamine polymers, ranging in molecular weight from 600 to 2000 Da, was characterized by precolumn derivatization and free zone CE with UV and fluorescence detection. Optimal separations were obtained with a pH 4.2 buffer containing 20% methanol using a 25 µm i.d. capillary and fluorescence detection. The method was capable of separating 30 oligomers with an average molecular weight of 600 which differed in size by a methylene unit. Braud and Vert [18] utilized CZE to aid in monitoring the degradation of poly(β-malic acid). A borate buffer at pH 9.2 was used to separate the first 15 oligomers in a degraded sample of poly(β-malic acid). A plot of logarithm of molecular weight of the oligomers versus migration time was linear demonstrating the potential of CZE for molecular weight analysis. This is consistent with GPC in which there is a linear relationship between the logarithm of molecular weight and elution time, although often a higher order function better describes this relationship. Compared with size exclusion chromatography, CZE was faster and provided superior resolution of oligomers.

Ryder [9] separated oligomers of acrylic copolymers in a borate buffer at pH 8.0 using a relatively long capillary of 120 cm. Although the apparent mass-to-charge ratio is constant throughout the copolymer, the stearic configuration change with molecular weight changes the ionization of the carboxylic acid groups, thus increasing the effective mass-to-charge ratio. Copolymers in the molecular weight range 900 to 14000 Da were separated although oligomers were separated only in the molecular weight range of 900 Da. Bullock [19] investigated CE to characterize the oligomeric distribution of some model ionic and non-ionic water soluble polymers. All separations were conducted in the open tube format without the aid of a sieving medium. Ionic poly(alkyl oxide) oligomers in the molecular weight range from a few hundred to over 4000 Da were separated into as many as 60 individual oligomers with an indirect UV detection. Neutral Triton X series oligomers from $n = 1$ to 46 were separated using a SDS matrix with high levels of acetonitrile. A solvophobic association separation mechanism was proposed whereby the hydrophobic portions of the Tritons associate with the hydrophobic tail of SDS in a non-micellar medium. Neutral PEG (polyethylene glycol) oligomers could be separated and detected by UV after derivatization with phthalic anhydride. The migration time/molecular weight data could be fitted to an appropriate function demonstrating the potential to determine the molecular weight properties of unknowns. In order to obtain the exact molecular weight, a light scattering detector can be recommended, otherwise only relative molecular weight data will be obtained.

Separation of oligomeric species using gel filled CE has also been reported. Garcia and Henion [20] described the separation of acrylic acid oligomers using a commercially available gel filled capillary. On-line MS was accomplished using a liquid junction–ion spray interface. One of the advantages noted for the gel filled capillary was that it effectively decouples the CE separation from the mass spectrometer allowing relatively high concentrations of buffers. Dolnik and Novotny [21] utilized highly concentrated and moderately crosslinked acrylamide-type gels to separate oligomers of some homopoly(amino acid) polymers. Separation of these model compounds serves to demonstrate the capabilities of gel filled CE for separating other types of ionic polymers (both positively and negatively charged). High concentrations (15–20%) of the various gel matrices were required to effect resolution of individual oligomers. For synthetic homopolymers in the mass range of 6000 Da, over 50 oligomers could be separated. Although impressive separations were achieved, it was noted that there was a need for additional, less UV-absorbing gel matrices and gels with less absorptive characteristics toward cationic polymers. Dolnik *et al.* [22] examined gel filled CE to study conformational properties of some model homopoly(amino acid) polymers. This work demonstrated the ability of CE to characterize the molecular weight dependency of polymer conformation applicable to other types of synthetic polymers.

CE has also been shown to be applicable to the separation of oligomers in various types of polymeric surfactants. As an example, Desbène *et al.* [23] used both CZE and MEKC to separate the oligomers of linear alkylbenzene sulfonates. Using CZE, baseline separation of oligomers with alkyl chain lengths between 2 and 12 could be achieved in 4 min. Although the corresponding MEKC separations were less efficient than CZE, MEKC had the advantage over CZE of being able to separate alkylbenzene sulfonate isomers. Goebel *et al.* [24] applied CZE to the analysis of ethoxylated alcohol sulfate surfactants. Oligomers differing in alkyl chain and degree of ethoxylation could be separated in a borate buffer using indirect UV detection based on dichromate as the UV absorbing buffer component.

All of the separations reviewed so far have involved aqueous or aqueous/organic buffer matrices. Okada [25] reported on the use of totally non-aqueous media for the separation of non-ionic polyethers. Methanol was used as the separation medium with various alkylamine salts added as the electrolyte. This work was used to determine the complexation constants of various polyethers with cations. This work demonstrates that it is possible to use totally non-aqueous media in the CE analysis of polymers. This will be important in extending CE technology to the analysis of polymers that are insoluble in aqueous media, since synthetic polymers are mostly hydrophobic. Additional work will be needed to investigate the use of other solvents because even in the use of 100% methanol, for example, polyurethane is still insoluble.

43.5 Capillary electrophoresis of high molecular weight polymers

For high molecular weight, soluble polymers where the degree of polymerization is high (>100) resolution of individual oligomers is perhaps not practical. However, separation of high molecular weight polymers with different average molecular weights are possible by CE analogous to size-exclusion chromatography. Poli and Schure [26] used CE with a polymeric sieving medium to separate polystyrene sulfonates encompassing a wide range of average molecular weights. A wall-coated capillary was used to reduce electroosmotic flow (EOF) in conjunction with a buffer to which is added hydroxyethyl cellulose to separate polystyrene sulfonates in the range from approximately 2000 to over 1 000 000 Da. The CE method was compared to size-exclusion chromatography and found to be superior in terms of analysis time, applicable molecular weight range and fractionating power. The work of Poli and Schure demonstrated that it was possible to perform molecular weight determinations analogous to size-exclusion chromatography with the advantages enumerated above. Sudor and Novotny [27] studied the electrophoretic transport of sulfonated polystyrenes under constant-field and pulsed-field conditions. A sodium phosphate buffer was used with hydroxyethyl cellulose added as a sieving medium. Under pulsed-field conditions, the influence of different pulse shapes, frequencies and amplitudes was studied. The results of these studies contributed to an understanding of the electrophoretic transport mechanism of polymers in a gel sieving matrix under constant and pulsed-field conditions.

43.6 Capillary electrophoresis of inorganic polymers

Shamsi and Danielson [28] investigated CE with indirect UV detection for separating various polyphosphates and polyphosphonates, which are relatively low molecular weight polymers of phosphoric acid and phosphorous acid, respectively. Various ribonucleotide monophosphates were used as the UV absorbing buffer species. Separations were optimized by adding different amine additives and Mg^{2+}. The methods were applied to the analysis of soaps, toothpaste and herbicides. Hettiarachchi *et al.* [29] applied CZE to the analysis of polyoxometalates. Polyoxometalates are polymers of oxides of early transition metal ions (i.e. tungsten and molybdenum). A pH 7 phosphate buffer was used with 30% acetonitrile added to stabilize and solubilize these inorganic polymers. The relative merits of the CZE technique were compared to those of the HPLC method.

43.7 Polymer composition analysis by capillary electrophoresis

In polymer composition analysis, the polymer is broken up into its constituent monomers or degraded to suitable derivatives which can then be analyzed revealing information about the chemical composition of the parent polymer. McNair and Sun [30] employed CZE to characterize the chemical composition of polyimide. The polyimide was decomposed to its corresponding aromatic diamine and aromatic acid monomers by an alkali fusion reaction. The low molecular weight reaction products were then separated in a pH 9.2 borate buffer. An advantage of the CZE method over existing gas chromatography (GC) and HPLC methods are a much simpler sample preparation. Tindall and Perry [31] developed a CZE method to determine most ester substituents in cellulose esters. The esters are hydrolyzed in a methanolic sodium hydroxide solution after dissolving in dimethyl sulfoxide (DMSO). The aliphatic acids released are separated and detected by an indirect UV.

43.8 Investigation of polymerization kinetics by capillary electrophoresis

Another application of CZE to polymer analysis is the investigation of polymerization reaction kinetics. Usually, this entails monitoring the disappearance of monomer over time. Therefore, the separation schemes used here share similarities with those used for purity analysis for residual monomer. CE seems to be well suited for this type of analysis due to the speed of analysis, attractive separation mechanisms, simple sample preparations and ability to elute polymer quickly or wash it out after each separation without contaminating the column. Several papers have been published on the use of CE to examine the kinetics of polyacrylamide polymerization under different conditions. Cagilo and Righetti [32] used CE to investigate the conversion efficiency of five different catalyst systems for polyacrylamide gel polymerization as a function of pH. Gel samples were extracted with methanol prior to MEKC, which uses a 100mM borate buffer with 50mM SDS for unreacted acrylamide and bisacrylamide. Chiari *et al.* [33] used MEKC to monitor the reaction kinetics of polyacrylamide and the formation of oxidation products of amino buffers in the mixture. Righetti and Caglio [34] used MEKC to study the kinetics of monomer incorporation into a polyacrylamide gel using a photopolymerization system. Righetti *et al.* [35] investigated the polymerization kinetics of polyacrylamide gels using a series of mono- and di-substituted acrylamide monomers. Caglio *et al.* [36] assessed the conversion efficiency of monomers into a polyacrylamide gel matrix in the presence of different types of detergents commonly used in biochemical analysis. In each of the above citations the authors used a MEKC procedure in place of CZE to separate and quantitate unreacted monomer.

43.9 Conclusions

The applications developed to date for polymers, while somewhat limited in number, have demonstrated the broad range of capabilities of CE and related techniques in polymer analysis. In order to expand upon this capability, advances in several areas will be needed. In terms of column technology, new gel matrices with the desired properties of chemical inertness, stability in different solvents and compatibility with different detection schemes are necessary. The use of electrokinetically driven chromatography with traditional chromatographic supports used in polymer analysis should also be explored.

More work in the area of non-aqueous CE will be needed to expand the applicability of the technique to hydrophobic polymers. Different detection schemes, especially refractive index, would make the technique more amenable to the analysis of polymers with poor or no UV chromophore even though sensitivity of RI is low. It is anticipated that as advances are made in these areas and as the application base expands, CE techniques will assume a major role in the analysis and characterization of synthetic polymers.

References

1. Ryder, D.S. (1992) Determination of sodium vinyl sulphonate in water soluble polymers using capillary electrophoresis. *J. Chromatogr.* **605**, 143–147.
2. Righetti, P.G.; Ettori, C.; Chiari, M. (1991) Analysis of acrylamide-buffers for isoelectric focusing by capillary zone electrophoresis. *Electrophoresis* **12**, 55–58.
3. Chiari, M.; Nesi, M.; Fazio, M.; Righetti, P.G. (1992) Capillary electrophoresis of macromolecules in 'syrupy' solutions: Facts and misfacts. *Electrophoresis* **13**, 690–697.
4. Gelfi, C.; de Besi, P.; Alloni, A.; Righetti, P.G. (1992) Investigation of the properties of novel acrylamido monomers by capillary electrophoresis. *J. Chromatogr.* **608**, 333–341.
5. Tribet, C.; Gaboriaud, R.; Gareil, P. (1992) Analogy between micelles and polymers of ionic surfactants: A capillary isotachophoretic study of small ionic aggregates in water–organic solutions. *J. Chromatogr.* **608**, 131–141.
6. Grimm, R. (1994) Comparison of capillary electrophoresis and HPLC for checking the purity of a synthetic polylysine preparation. *Hewlett-Packard Application Publication* Number 12-5962-7230E.
7. Ross, G. (1995) Applications of the HP 3D capillary electrophoresis system. *Hewlett-Packard Publication* number 12-5963-7140E.
8. Bullock, J. unpublished results.
9. Ryder, D.S. (1993) The use of capillary zone electrophoresis for the analysis of polymeric water treatment additives and comparison with liquid chromatography techniques. *Spec. Publ. Roy. Soc. Chem.* **122**, 101–110.
10. VanOrman, B.; McIntire, G.L. (1989) Analytical separations of polystyrene nanospheres by capillary electrophoresis. *J. Microcol. Sep.* **1**, 289–293.
11. VanOrman, B.B.; McIntire, G.L. (1990) Size-based separation of polystyrene nanoparticles by capillary electrophoresis. *Am. Lab.* **22**, 66–67.
12. Huff, B.V.; McIntire, G.L. (1994) Determination of the electrophoretic mobility of polystyrene particles by capillary electrophoresis. *J. Microcol. Sep.* **6**, 591–594.
13. Jones, H.K.; Ballou, N. (1990) Separation of chemically different particles by capillary electrophoresis. *Anal. Chem.* **62**, 2484–2490.
14. Petersen, S.L.; Ballou, N.E. (1992) Effects of capillary temperature control and electrophoretic heterogeneity on parameters characterizing separations of particles by capillary zone electrophoresis. *Anal. Chem.* **64**, 1676–1681.
15. Radko, S.P.; Garner, M.M.; Caiafi, G.; Chrambach, A. (1994) Molecular sieving of polystyrene carboxylate of a diameter up to 10µm in solutions of uncrosslinked polyacrylamide of Mw 5×10^6 using capillary zone electrophoresis. *Anal. Biochem.* **223**, 82–87.
16. McCormick, R.M. (1991) Characterization of silica sols using capillary zone electrophoresis. *J. Liq. Chromatogr.* **14**, 939–952.
17. Amankwa, L.N.; Scholl, J.; Kuhr, W.G. (1990) Characterization of the oligomeric dispersion of poly(oxyalkylene) diamine polymers by precolumn derivatization and capillary zone electrophoresis with fluorescence detection. *Anal. Chem.* **62**, 2189–2193.

18 Braud, C.; Vert, M. (1992) Degradation of poly(β-malic acid)-monitoring of oligomers formation by aqueous SEC and HPCE. *Polymer Bull.* **29**, 177–183.
19 Bullock, J. (1993) Application of capillary electrophoresis to the analysis of the oligomeric distribution of polydisperse polymers. *J. Chromatogr.* **645**, 169–177.
20 Garcia, F.; Henion, J.D. (1992) Gel-filled capillary electrophoresis/mass spectrometry using a liquid junction–ion spray interface. *Anal. Chem.* **64**, 985–990.
21 Dolnik, V.; Novotny, M. (1993) Separation of amino acid homopolymers by capillary gel electrophoresis. *Anal. Chem.* **65**, 563–567.
22 Dolnik, V.; Novotny, M.; Chmelik, J. (1993) Electromigration behavior of poly-(L-glutamate) conformers in concentrated polyacrylamide gels. *Biopolymers* **33**, 1299–1306.
23 Desbène, P.L.; Rony, C.; Desmazieres, B.; Jacquier, J.C. (1992) Analysis of alkylaromatic sulphonates by high performance capillary electrophoresis. *J. Chromatogr.* **608**, 375–383.
24 Goebel, L.K.; McNair, H.M.; Rasmussen, H.T.; McPherson, B.P. (1993) Separation of ethoxylated alcohol sulfates by capillary electrophoresis using indirect UV detection. *J. Microcol. Sep.* **5**, 47–50.
25 Okada, T. (1995) Non-aqueous capillary electrophoretic separation of polyethers and evaluation of weak complex formation. *J. Chromatogr.* **695**, 309–317.
26 Poli, J.B.; Schure, M.R. (1992) Separation of poly(styrenesulfonates) by capillary electrophoresis with polymeric additives. *Anal. Chem.* **64**, 896–904.
27 Sudor, J.; Novotny, M.V. (1994) Pulsed-field capillary electrophoresis: Optimizing separation parameters with model mixtures of sulfonated polystyrenes. *Anal. Chem.* **66**, 2139–2147.
28 Shamsi, S.A.; Danielson, N.D. (1995) Ribonucleotide electrolytes for capillary eletrophoresis of polyphosphates and polyphosphonates with indirect photometric detection. *Anal. Chem.* **67**, 1845–1952.
29 Hettiarachchi, K.; Ha, Y.; Tran, T.; Cheung, A.P. (1995) Application of HPLC and CZE to the analysis of polyoxometalates. *J. Pharm. Biomed. Anal.* **13**, 515–523.
30 McNair, H.M.; Sun, X. (1995) Capillary zone electrophoresis for polyimide composition analysis. *J. High Resolut. Chromatogr.* **18**, 115–116.
31 Tindall, G.W.; Perry, R.L. (1993) Determination of ester substituents in cellulose esters. *J. Chromatogr.* **633**, 227–233.
32 Caglio, S.; Righetti, P.G. (1993) On the pH dependence of polymerization efficiency as investigated by capillary zone electrophoresis. *Electrophoresis* **14**, 554–558.
33 Chiari, M.; Micheletti, C.; Righetti, P.G.; Poli, G. (1992) Polyacrylamide gel polymerization under non-oxidizing conditions, as monitored by capillary zone electrophoresis. *J. Chromatogr.* **598**, 287–297.
34 Righetti, P.G.; Caglio, S. (1993) On the kinetics of monomer incorporation into polyacrylamide gels, as investigated by capillary zone electrophoresis. *Electrophoresis* **14**, 573–582.
35 Righetti, P.G.; Chiari, M.; Nesi, M.; Caglio, S. (1993) Towards new formulations for polyacrylamide matrices, as investigated by capillary zone elctrophoresis. *J. Chromatogr.* **638**, 165–178.
36 Caglio, S.; Chiari, M.; Righetti, P.G. (1994) Gel polymerization in detergents: Conversion efficiency of methylene blue vs. persulfate catalysis, as investigated by capillary zone electrophoresis. *Electrophoresis* **15**, 209–214.

Appendix A
Commercially available instrumentation for capillary electrophoresis

Appendix A1 Overview of commercial instruments
T. WEHR AND M. ZHU

A1.1 Introduction

Commercial capillary electrophoresis (CE) instruments were first introduced in 1988 and at the time of writing there are at least 14 vendors offering CE systems or components. During the preceding decade many researchers had contructed experimental manual systems consisting of a power supply, capillary, detector and injection device. In contrast to these simple homemade systems, many of the first commercial units were automated instruments with integrated detectors, autosamplers with multiple injection modes and provision for temperature control. In the intervening period, additional systems have been introduced onto the market and earlier systems have evolved with the addition of new or more sophisticated detectors, more advance sample injection and liquid handling features and improved control and data processing software. This chapter summarizes the features of currently available CE instrumentation with a focus on automated systems (Table A1.1).

A1.2 Power supply

Power supplies capable of delivering constant voltage at high precision up to 30 kV are standard throughout the industry. Most systems offer, in addition, constant current operation at up to 300 µA; constant current operation may be desirable in systems without adequate temperature control. Some instruments are also capable of operation in constant power mode up to 6 W; use of constant power operation permits separation time to be minimized without excess heat generation. Early in the evolution of CE it was considered that efficiency in a CE separation was primarily diffusion-limited and that operation at very high voltages (up to 60 kV) could yield higher theoretical plate numbers and improved resolution. In practice, thermal effects at high voltage compromise efficiency gains, and today satisfactory separations are achieved below 30 kV (typically 10–25 kV).

Table A1.1 Capillary electrophoresis commercial instruments

Manufacturer	Advanced Molecular Systems	Applied BioSystems–Perkin Elmer	ATI/Unicam
Product	Model 2000	270-HT/HS	Crystal CE 300
High voltage supply			
Operation models	Constant voltage or current	Constant voltage	Constant voltage or current
Voltage range	0–30 kV	0–30 kV	0–30 kV
Current range	0–330 μA	0–330 μA	0–200 μA
Power range	Up to 9 W	–	–
Polarity reversal	Manual	Through software	Through software
Injection modes	Electrokinetic, vacuum	Electrokinetic, vacuum	Electrokinetic, pressure
Capillary temperature control			
Mode	Liquid	Forced air	Forced air
Range	2–75°C	Ambient–60°C	10°C below ambient–60°C
Liquid handling			
Capillary purge	Up to 800 mbar in 20 mbar increments	Vacuum	Vacuum
Buffer replenishment	Yes, inlet and outlet automatically	Yes	Yes
Autosampler			
Inlet positions	4, 8, 96-well tray	Fifty 0.5 ml positions and eight 4.0 ml positions	4 (Model CE 300), 30 (Models CE 310 and 310C)
Outlet positions	One	One 15 ml, replenishable	Up to 8
Temperature control	Liquid	Optional external bath	Optional external bath
Range	2–75°C	5–60°C	4–40°C
Detectors			
Absorbance	Photodiode array, 190–288 nm	Variable λ, 190–700 nm	Selectable λ (filter) or diode array
Fluorescence	190–288 nm excitation, 450–628 nm emission	–	Dual monochromator, UV–vis excitation and emission
Mass spectrometer coupling	–	–	Available for ESI
Other	–	–	Conductivity
Fraction collection	–	Automatic	Manual

*48 positions total, each position can be stationed at capillary inlet or outlet.

Table A1.1 *Continued*

Beckman Instruments	Bio-Rad Laboratories	Dionex	GTI/Spectro-Vision
P/ACE 5000 Series	BioFocus Series	CES 1	Modular
Constant or programmable voltage, constant current or power	Constant or programmable voltage, constant current	Constant or programmable voltage, current or power	Constant voltage or current
1–30 kV 1–250 µA 0–6 W Manual	0–30 kV 1–300 µA – Through software	0–30 kV 0–400 µA 0–6 W Manual	0–30 kV 0–250 µA – Manual
Electrokinetic, pressure, or both	Electrokinetic, pressure	Electrokinetic, pressure, gravity	Electrokinetic
Liquid 5°C below ambient to 50°C	Liquid 15–65°C, programmable	Forced air – –	– –
Pressure, 20 psi	Pressure, 5 or 100 psi	Pressure, 20 psi	Manual
No	No	Yes	–
24	32	40	–
10	32	1 with automatic purging	–
Optional external bath –	Optional integrated peltier 4–40°C	– –	– –
Selectable λ (filter) or diode array (190–600 nm)	Variable λ (190–800 nm) or fast-scanning (190–800 nm)	Variable λ (190–800 nm)	–
Argon ion laser source	–	Deuterium and tungsten sources	–
Available for ESI –	Available for ESI –	– Enhanced fluorescence kit	– –
Automatic	Automatic	–	–

Table A1.1 *Continued*

Manufacturer	Hewlett Packard	Kontron	Otsuka
Product	HP3D	Eureka 2100	CAP I-3100
High voltage supply Operation models	Constant or programmable voltage, current or power; CEC	Constant or programmable voltage or current	Constant or programmable voltage or current
Voltage range	0–30 kV	0–30 kV	0–30 kV
Current range	0–300 μA	0–300 μA	1–300 μA
Power range	0–6 W	–	–
Polarity reversal	Through software	Yes	Automatic
Injection modes	Electrokinetic, pressure, vacuum	Electrokinetic, pressure, gravity	Electrokinetic, gravity
Capillary temperature control Mode Range	Forced air 10°C below ambient–60°C	Forced air 5–50°C	Forced air 5–45°C
Liquid handling Capillary purge	Pressure, up 12 bar	20 psi	20 psi
Buffer replenishment	Yes	No	No
Autosampler Inlet positions	48*	20 6 buffer (4 ml)	45 sample (0.5 ml)
Outlet positions	*	10 fraction	7 buffer (15 ml) 6
Temperature control Range	Optional external bath 10–40°C	Optional external bath 4–45°C	Optional external bath 4–45°C
Detectors Absorbance	Diode array (190–600 nm)	Diode array (190–800 nm)	Diode array (190–600 nm)
Fluorescence	–	–	–
Mass spectrometer coupling	Adapter kit for all MS platforms	–	–
Other	–	–	–
Fraction collection	Automatic	No	Yes

Table A1.1 *Continued*

Thermo Separation Products	Waters Associates	Zeta Technology
SpectraPhoresis Series	Quanta 4000E	modular
Constant or programmable (optional) voltage, constant current	Constant voltage, constant current, or both ('programmable isomigration')	Constant voltage or current
0–30 kV	0–30 kV	0–30 kV
0–300 μA	0–250 μA	0–300 μA
–	–	–
Through software	Manual	Manual
Electrokinetic, vacuum	Electrokinetic, hydrostatic	Electrokinetic, vacuum
Forced air	Peltier	–
15–60°C, programmable with optional software	10–45°C	–
Vacuum	Vacuum	Yes
Yes	–	–
12 (Model 100), 82 (Models 500, 1000, 2000)	20	–
1	5	–
–	Integrated peltier	–
–	10–45°C	–
Selectable λ (Model 100), variable λ, (Model 500), fast scanning 190–800 nm (Models 1000, 2000)	Selectable λ (lamp, filter)	Argon ion laser source
–	–	–
–	–	–
–	–	–
		–
–	Automatic	–

High voltage is applied at the capillary inlet, with the outlet (detector) at ground potential. In many applications, capillary zone electrophoresis (CZE) is performed in uncoated fused-silica capillaries and electroosmotic flow (EOF) carries all analytes, regardless of charge, towards the cathode. In these cases, the inlet (high voltage) electrode is the anode ('normal' polarity). In an increasing variety of applications, however, 'reversed' polarity is used with the inlet electrode as cathode. These cases include analysis of anionic analytes in the absence of EOF (e.g. capillary gel electrophoresis and CZE using neutral coated capillaries) and analysis with reversed EOF (e.g. CZE using capillaries with positively charged coatings or with osmotic flow modifying additives in the background electrolyte). All commerical instruments have reversible polarity, although for frequent switching among different applications, polarity reversal through software is more convenient than manual reversal.

Six commercial instruments (Beckman, Bio-Rad Dionex, Hewlett-Packard, Kontron and TSP) offer time-programmable high voltage; this feature may be useful in preventing artifacts during application of high voltage, in reducing analysis time for samples containing low mobility components and in modulating migration rates in special applications (e.g. fraction collection and resolution of temperature-sensitive conformers of biopolymers).

A1.3 Injection

Sample injection in CE requires the introduction of very small amounts of analyte at the capillary inlet with high precision. All commercial instruments offer electromigration injection and at least one type of displacement injection.

Electromigration is the simplest injection method in CE; the capillary inlet is immersed in the sample solution and high voltage is applied for a brief period (typically a few seconds). If no EOF is present, sample ions enter the capillary by electophoretic mobility alone. If EOF is present, sample ions will be introduced by a combination of electrophoretic mobility and EOF; this mode is generally termed electrokinetic injection.

Electrophoretic injection offers two advantages. First, since only species of like charge will enter the capillary, it enables discrimination against compounds of opposite charge, simplifying the separation problem. Second, by varying the ionic strength of the sample relative to the background electrolyte, a focusing or stacking effect is created which allows the formation of very sharp sample zones with high analyte concentration. Unfortunately, these advantages are countered by two major limitations. Since sample ions enter the capillary based on mobility, low-mobility ions will be present in lower concentrations, decreasing detector response. More importantly, presence of non-analyte ions in the sample will reduce injection efficiency, so electrophoretic injection is very sensitive to the presence of salts or buffers in the sample matrix. The disadvantages of electrophoretic injection argue against its use in routine analysis except in cases where displacement injection is not possible, e.g. in capillary gel electrophoresis (CGE). Electrokinetic injection suffers from the further disadvantage that many sample matrices contain components such as proteins which adsorb to the capillary wall and change the magnitude of the EOF.

Displacement injection is usually the preferred method since analyte ions are present in the sample zone in proportion to their concentration in the bulk sample, and injection efficiency is less sensitive to variations in sample ionic strength. However, it should be noted that the presence of high salt can affect detector response with displacement injection, and variations in the sample viscosity (due to temperature variations or the presence of viscosity-modifying components) can affect displacement injection efficiency.

Two modes of displacement injection have been employed in commercial CE instruments: application of positive pressure at the capillary inlet and application of vacuum at the capillary outlet. The former method can employ pressurization of the sample headspace by gas (pressure injection) or application of hydrostatic pressure by elevating the capillary inlet relative to the capillary outlet (gravity injection). Gravity injection has been reported to provide reproducible injection of very small sample zones. On the other hand, pressure injection can allow the flexibility of introducing larger sample zones; this can be an advantage when using injection pneumatics to introduce a chiral selector or to perform on-column concentration. Moderate injection pressures can be an advantage when injecting samples into capillaries containing analysis buffers with viscous agents such as sieving polymers. At least two commerical instruments integrate the pressure signal during the injection process and inject to a constant time–pressure product; this compensates for variations in seal compliance and sample-to-sample headspace.

A1.4 Capillary temperature control

Electrophoretic mobility is strongly dependent on temperature, with migration rate varying by 2.5% for each degree of temperature rise. Fused-silica capillaries are very efficient in dissipation of Joule heat generated by the electrophoretic process, allowing very high resolution to be achieved. However, temperature control of the capillary environment is essential to attain satisfactory reproducibility. Inadequate temperature control results in variable migration times and in variable run-to-run peak areas since peak area is migration-rate dependent. Therefore, it is not surprising that all commercial automated CE systems include some form of temperature control. Operation at elevated temperatures may be useful in nucleic acid separations to help maintain species in single-stranded states, while subambient operation may be useful for achieving separations in some CE modes (e.g. MEKC) or for performing kinetic studies.

Two manufacturers, Beckman Instruments and Bio-Rad laboratories, consider circulating-liquid temperature control to be the most effective. This requires enclosure of the capillary in a sealed cartridge to maintain coolant flow; both manufacturers supply cartridges which can be user-assembled and accept empty or gel-filled capillaries from all commercial suppliers. The Beckman P/ACE systems employ Fluorinert® (Dow Chemical Co.), a non-conductive fluorocarbon, as coolant while the Bio-Rad BioFocus systems use deionized water as coolant for low voltage operation and Fluorinert at high voltages (>20 kV). The latter also employs a vacuum pump to draw coolant through the system, eliminating the risk of coolant leakage into the instrument. Both systems operate at subambient as well as at elevated temperatures. The cartridge format provides for automatic alignment of the capillary in the detector light path and reduces time required for capillary replacement when changing methods.

All other automated CE instruments employ forced air–nitrogen convection for capillary temperature control. Two systems (Applied Biosystems–Perkin Elmer and Dionex) use only forced-air heating, while the remaining systems can cool to subambient temperatures as well as operate at elevated temperatures. All of these systems use free-hanging capillaries except the TSP Spectraphoresis system in which the capillary is mounted in an open cartridge. The latter manufacturer offers temperature gradient programming as a software option.

The effectiveness of capillary thermostatting can be determined by variation in current as a function of voltage. According to Ohm's law, this should be a linear relationship, and deviation from linearity in an Ohm's law plot is indicative of poor efficiency in heat dissipation by the capillary temperature control system.

A1.5 Liquid handling

Replenishment of the electrolyte in the capillary after each analysis is done in all modes of CE except for CGE. Prior to replenishing the electrolyte, it is often necessary to purge the capillary with one or a series of solvents to remove adsorbed sample components, recondition the fused-silica surface or to replace a dynamic capillary coating. The advantage of an automated CE instrument is that complicated capillary wash and conditioning protocols can be programmed as part of the method and automatically executed between each analysis. Six commercial systems (Advanced Molecular Devices, Applied Biosystems–Perkin Elmer, ATI, Dionex, Hewlett Packard, TSP) offer automatic buffer replenishment systems which empty and refill the outlet electrode vessel. This feature can be advantageous when using electrolytes with low buffering capacity in which electrode reactions can cause pH drift after a few analyses. With the increased popularity of polymer sieving systems for separations of nucleic acids and SDS–protein complexes, the ability to handle viscous electrolytes is advantagous, and some commercial systems (e.g. the Bio-Rad BioFocus and the Hewlett Packard HP3D) provide high pressure purging capabilities for such applications. Simultaneous application of pressure and high voltage is desirable for techniques such as electrochromatography and zone mobilization in capillary isoelectric focusing (CIEF); several commercial systems offer this capability (e.g. Beckman, Bio-Rad, Hewlett Packard).

A1.6 Autosamplers

Rapid analysis and high sample throughput are two key advantages offered by CE compared to alternative separation techniques such as gel electrophoresis and high performance liquid chromatography (HPLC). Therefore automatic sampling devices with multisample capacity are integrated into almost all commercial CE systems. Sample capacities range from 32 to 80 (the AMD instrument uses a 96-position microtiter plate format). Four systems (ABI–Perkin Elmer, Beckman, Bio-Rad and Hewlett-Packard) allow for multiple outlet buffer positions, which enable multiple separation chemistries to be scheduled in automated runs.

Autosampler temperature control may be important in some cases, e.g. when analyzing heat-labile biological samples, when collecting fractions or when performing kinetic studies at elevated temperatures. All but two commercial CE systems offer optional temperature control of samples and buffers using circulating coolant from an external bath; one system (Bio-Rad) contains an integrated autosampler temperature control. The ABI–Perkin Elmer and Beckman systems permit selected cooling of sample vials without the need to cool buffer vials as well.

A1.7 Detectors

A1.7.1 UV–vis absorbance

As in HPLC, absorbance detection is used in the vast majority of CE applications and all commercial CE systems employ UV or UV–vis absorbance as the primary mode of detection. The simplest approach is the use of line-source lamps or continuum-source lamps with wavelength selection using filters. The Beckman P/ACE system employs a deuterium source with wavelength selection by a four-position filter wheel. The Waters Quanta system uses line-source lamps with filters. Most other systems use a continuum (e.g. deuterium lamp) source with wavelength selection using a grating monochromater. Since the low output of a deuterium lamp above 360 nm limits sensitivity in the visible range, several manufacturers (Bio-Rad, Dionex, TSP, Waters) provide additional tungsten sources for visible wavelength detection.

Several manufacturers offer UV or UV–vis scanning detection. A scanning diode-array detector is standard on the Hewlett-Packard HP3D system, while ATI, Beckman, Otsuka and Waters offer optional diode-array detectors. Bio-Rad and TSP offer optional fast-scanning detectors based on forward-optics systems.

Scanning detection enables on-the-fly acquisition of spectra as analytes migrate through the detection point; this information can assist in the identification of peaks based on spectral patterns, in detection of peak impurities by variation in spectral profiles across a peak, or in determination of the absorbance maximum of an unknown compound. Scanning detectors are usually equipped with specialized software packages for storage, retrieval and comparison of spectra and mathematical analysis of spectral data for determination of peak purity and identity. Of course, as in HPLC, spectra can be solvent dependent, requiring that analyte spectra and the compared library spectra should be acquired under the same CE analysis conditions.

All commercial CE absorbance detectors employ on-tube detection: a section of the capillary itself is used as the detection cell. This permits detection of separated zones with no loss in resolution. Most capillaries used for CE are coated with a polymer (usually polyimide) which protects the fused-silica capillary and provides it with mechanical stability. Since the polymer is not optically transparent, it must be removed from the detection segment to form a 'window' and this window must be accurately positioned in the optical path to achieve good sensitivity. This segment of bare capillary is very fragile and is subject to breakage during manipulation and installation of the capillary. Capillaries with a UV-transparent coating are available from Polymicro Technologies which eliminates this problem; however, the coating is not resistant to some coolants (e.g. Fluorinert) used in a liquid-cooled CE system.

In on-tube detection, the internal diameter of the capillary forms the detection light path. In accordance with Beer's law, the sensitivity of a concentration-sensitive detector is a direct function of the length of the light path. Therefore, in comparison to an HPLC detector with a 1 cm path length, the detector signal strength should be reduced 200-fold in a CE system equipped with a 50 μm i.d. capillary. Concentration sensitivity can be improved by employing focusing lenses to collect light at the capillary lumen, by detecting at low wavelengths (where most analytes have greater absorbance) and by using sample-focusing techniques during the injection process. However, even under ideal conditions the concentration limit of detection (CLOD) is about 10^{-7} M.

Two manufacturers supply enhanced-pathlength devices to increase absorbance detection sensitivity. ABI–Perkin Elmer offers an optional Z-cell for the Model 270HT CE system. As the name implies, the flowcell in this device contains two bends such that light is directed down the axis of the capillary, creating a 3 mm light path. The manufacturer has demonstrated a 30-fold increase in sensitivity using the Z-cell. Hewlett-Packard supplies a capillary with conventional geometry except for an enhanced inner diameter (typically by 3×) at the point of detection. Through a combination of increased light path and reduced field strength at the detection point, this 'bubble cell' has been shown to increase sensitivity three-fold. The drawbacks of these sensitivity-enhancing devices include high cost and reduced resolution.

Absorbance detection is not limited only to compounds with UV absorbance. In the technique of indirect absorbance detection, a UV-absorbing ion with the same charge as the analyte ions is incorporated into the background electrolyte. Displacement of the background chromophore ion in the sample zone forms a negative peak which can be inverted electronically (indirect absorbance detection method) and quantitated as a conventional positive peak. This technique has been widely employed in CE analysis of inorganic anions and cations, organic acids and bases, and carbohydrates. However, CLOD values are usually in the 10^{-4}–10^{-5} M range.

A1.7.2 Fluorescence

Additional information concerning superiority and inferiority of fluorescence detection can be referred to Chapter 6 on Detection.

Fluorescence detection offers the possibility of high sensitivity, and in the case of complex samples, improved selectivity. However, this mode of detection requires the analyte to exhibit native fluorescence or to contain a group to which a fluorophore can be attached by chemical derivatization. The number of compounds that fall into the former category are small and while many analytes contain derivatizable groups (e.g. amino, carboxyl, hydroxyl functional groups), most derivatization chemistries are limited by one or more disadvantages (slow reaction kinetics, complicated reaction or clean-up conditions, poor yields, interference by matrix components, derivative instability, interference by reaction side products or unreacted derivatizing agent). Because of these limitations, fluorescence detection, as in HPLC, has found few applications in CE. However, in the case of nucleic acids and carbohydrates, suitable derivatization procedures are available and fluorescence detection is widely used for these applications. Both covalent and intercalating fluorescent dyes are used for DNA separations, and there are at least two fluorescent tags to attach to amino-containing sugars.

When compared to fluorescence detectors for HPLC, the design of a fluorescence detector for CE presents some technical problems. In order to achieve acceptable sensitivity, it is necessary to focus sufficient excitation light on the capillary lumen, which is difficult to do with a conventional light source. One answer to this problem is the use of a laser as the excitation source. Two manufacturers (Beckman and Zeta Technology) currently offer laser-induced fluorescence (LIF) detectors. These detectors employ an argon ion laser providing excitation at 488 nm, which is close to the desired excitation wavelength for several DNA and carbohydrate dyes. The CLOD for a laser-based fluorescence detector can be as low as 10^{-12} M.

Three manufacturers offer fluorescence detectors with conventional excitation sources. The detectors manufactured by ATI and SpectroVision use xenon sources, while the Dionex system uses a tungsten source. A conventional-source fluorescence detector should be able to achieve CLODs of 10^{-9} M.

A1.7.3 Other detectors

ATI offers a conductivity monitor for CE. This detector is useful for analysis of inorganic ions and provides an order of magnitude improvement in sensitivity over indirect absorbance detection.

A1.7.4 On-line coupling with mass spectrometry (MS)

With the increasing need to obtain absolute identification of separated components and the gradual price reduction of mass spectormeters, there is a growing demand for direct coupling of CE with MS instruments. The most frequent configuration is introduction of the capillary outlet into an electrospray interface (ESI) coupled to a single or double quadrupole instrument. In this configuration, the outlet electrode of the CE is eliminated and the MS becomes the ground. Since the volumetric flow out of the capillary is neglible or nil, separated components are transported from the capillary to the electrospray using a liquid sheath flow. The major limitation in CE–ESI/MS is the requirement for volatile buffers. Ionic buffers are not appropriate. This narrows the choice of CE separation modes and resolving power.

CE couplings for ESI/MS are available from Beckman, Bio-Rad, and ATI. The Beckman system can be interfaced to the Finnegan MAT quadrupole instruments, while the Bio-Rad can be interfaced to the Finnegan systems and the VG Platform 2 system. The latter CE interface allows the capillary to be liquid cooled from the CE directly to the ESI.

A1.8 Fraction collection

Because of its high resolving power CE is often considered for micropreparative isolation of compounds. Seven commercial instruments offer fraction collection capability. The Waters and Isco systems provide for manual fraction collection; the Waters Quanta system permits replacement of the outlet vial holder with a circular polyvinylidene difluoride (PVDF) disk onto which separated components can be recovered for subsequent operations such as protein sequencing or Western blotting. For sample recovery using the Perkin Elmer–ABI system, sample is injected from the outlet position and fractions are collected into inlet carousel positions. The Beckman, Bio-Rad and Hewlett-Packard instruments have automatic fraction collection software permitting recovery of separated components in multiple outlet vials.

The desirability of using CE as a micropreparative tool has to be carefully weighed against the problems encountered in fraction collection. When using narrow-bore (e.g. 50 μm i.d.) capillaries, the volume injected into the capillary is quite small (typically a few nanoliters). Unless the analyte is in very high concentration, recovery of sufficient material will require repetitive injections of the same sample. In this case, the run-to-run migration times must be highly reproducible to ensure accurate collection of the analyte peak. Also, the recovered analyte must be stable under the collection conditions for the time required to collect the desired amount of material (often several hours). An alternative strategy is the use of larger diameter capillaries (≥ 75 μm). However, thermal effects may compromise resolution and low voltages or low conductivity buffers may be necessary to prevent excessive heating.

Appendix A2   The CAPI series of instruments
K. KITAGISHI

Otsuka Electronics Co. Ltd has two kinds of capillary electrophoresis which are system available commercially and are called the CAPI series. One is the CAPI-3100, a high-end of the market version with 3D-UV detection, controlled by a personal computer and the other is the CAPI-1000, a computer-free system at lower cost.

A2.1   CAPI-3100

A2.1.1   *Specifications*

Autosampler

- Left carousel (inlet end): 45 vials (500 µl) for samples and 6 vials (16 ml) for buffers. For sample vials, commercially available tubes made from polypropylene or polyethylene of 500 µl are acceptable.
- Right carousel (outlet end): 7 vials (16 ml) for buffers and 6 vials for fraction collection.
- Sample temperature control: sample temperature can be maintained at a constant level in the range 4°C to 45°C with a sample temperature controller and a circulating bath optionally installed to the system.

Electrophoresis

- Power supply: ±30 kV, 300 µA. Cathode/anode convertible by computer control. Constant voltage/constant current or gradient voltage/gradient current (9 steps each) system is available.
- Capillary: the standard has a 75 µm i.d. and is made of synthetic silica. The synthetic silica capillary has the advantage over fused silica of transparency, which causes a low background. Capillaries with i.d. of 20, 50 and 100 µm can be adapted. Minimal total length is 40 cm (27.5 cm minimal effective length from inlet end to detection window). A capillary replacement is easy with a one-touch cassette system.
- Capillary temperature control: Peltier cooling with a range of 5°C to 45°C. The high accuracy temperature controlling function (±0.2°C) inhibits temperature elevation in the capillary, thus ensuring high reproducibility.

- Injection: gravity and electrokinetic methods are available. Pressure injection is available optionally by regulating gas pressure.

Optics

- Detection mode: UV–vis absorbance.
- Detection element: photodiode array 512ch.
- Measurable wavelength range: 190 to 600 nm.
- Wavelength accuracy: ±1.0 nm.
- Wavelength resolution: 1.3 to 20 nm.
- Noise level: 5×10^{-5} AU (absorbance units) or less (at 254 nm).
- Time drift: 1×10^{-3} AU h^{-1} or less (at 254 nm).
- Minimum scale: 0.0001 AU.

Software

- Measurement: random access is possible for both sample and buffer vials on the carousels. Automatic measurements with a multimethod table consisting of up to 50 lines under different analytical conditions. An analytical condition defined in each line can be applied sequentially to 45 different sample vials, at maximum, of autosampler. A multitask function allows automatic data analysis while under run ('Run and Report').
- Analysis: both autoanalysis succeeding measurement and post-analysis under voluntary analytical conditions are available.
- 2-D analysis: calculations on pherograms such as addition, subtraction, division and differential can be achieved. Migration time, peak height, peak area and peak area % are calculated quantitatively from pherograms.
- 3-D analysis: three-dimensional data are displayed as a 3-D plot, a contour plot and a spectral index plot. Multiwavelength pherograms and 3-D cross-sectional view can be drawn. For the chromatographic analysis of 3-D data, peak purity analysis, index analysis, calculations on spectra (sum, difference, ratio and differential) and library search are available.

Application kits. Application kits as α-, β- and γ-series are available. The α-series are mainly ready-to-use electrolytes for common applications. They include linear polymer solutions for DNA and sodium dodecyl sulfate (SDS)–proteins, cationic surfactant or zwitterions in order to suppress protein adsorption onto the capillary wall and suitable buffers containing additives for cation or anion analysis. The β-series contains modified capillaries: neutral hydrophilically coated capillaries, gel-filled capillaries, highly sensitive cell capillaries with an extended light path in a bubble shape. For the γ-series, several kits are available: a kit for molecular weight analysis of SDS–proteins consisting of a coated capillary, a standard protein mixture and a linear polymer solution, a kit for isoelectric focusing consisting of a coated capillary, a standard protein mixture and ampholyte and a kit for chiral separation.

A2.1.2 Features

Versatile functionality and high quality operation ability are features of the CAPI-3100. The autosampler affords automatic measurement of up to 45 samples and six buffers. Each line of a multimethod table represents measurement conditions like rinse of capillary, filling of electrolyte, sample injection, electrophoresis, capillary temperature, detection, the number of repeated runs, threshold for safety operation and data processing conditions. Voluntary values can be input for these conditions. Measurements with a multimethod table randomly accessible to 45 samples and six buffers are capable of routine analysis for a batch of samples or investigation to optimize analytical conditions with little labor. Automatic analysis can be performed during or after a run in the multimethod table.

CAPI-3100 has a detection system consisting of a photodiode array as the detecting element with a remarkably low noise level. Spectral information obtained with this detection system enables purity analysis and library search to be undertaken. The maximum wavelength as well as the molar absorption coefficient at that wavelength are obtained from a spectrum. At this wavelength quantitative analysis can be made with higher sensitivity. Simultaneous quantitative of more than two components in a sample solution can be carried out by multiple peak quantitative analysis. Subtraction or division between pherograms at two different wavelengths in the same run improves detection sensitivity.

CAPI-3100 is equipped with all the devices necessary to obtain accuracy and reproducibility, such as a capillary temperature controller, various injection methods, autosampler, high-quality detecting element and fully automated operating system. CAPI-3100 is considered to be suitable both for method development and for routine analysis with high reproducibility.

A2.2 CAPI-1000

A2.2.1 Specifications

Sampler

- Sampler: six sample vials (250 µl made of polypropylene) manually replaceable.

Electrophoresis

- Power supply: +25 kV, 200 µA as standard; –25 kV, 200 µA as optional. Constant voltage (1 kV step) or constant current (1.0 µA step) is applied. Voltage at constant current and current at constant voltage are displayed.
- Capillary: the standard has a 75 µm i.d. capillaries with an i.d. of 20, 50 and 100 µm can be adapted. The minimal total length is 60 cm (40 cm minimal effective length from inlet end to detection window). The capillary is easily replaceable in the front of the instrument.
- Capillary temperature control: forced air circulation system. Control range is from room temperature plus 3 to 49°C by digital steps with an accuracy of ±0.5°C.
- Injection: vacuuming by electromotive pump. Injection time by digital set-up. An electrokinetic method is available optionally.

Optics

- Detection mode: double beam UV absorbance.
- Detection element: twin-type silicon photodiode.
- Measurable wavelength range: 195 to 380 nm (380 to 700 nm, if a filter is switched).
- Wavelength accuracy: ±1 nm.
- Wavelength resolution: 6 nm.
- Noise level: 5×10^{-5} AU or less (at 254 nm).
- Time drift: 1×10^{-3} AU h^{-1} or less (at 254 nm).
- Minimum scale: 0.001 AU.
- Output: 0.5, 1.0 V/AU 10 mV full scale.

Data output

1. CAPI-1000 connected with Chromatocorder 21 by System Instruments affords data output immediately after analysis. The memory has a capacity of up to 80 chromatograms each of about 10 min.
2. CAPI-1000 connected with Integrator I or Integrator II as Mac-Integrator provides storage, analysis and reporting data functions. This set-up makes it possible to display voltage and current, to store, reanalyze and print out data. The functions of addition and subtraction of chromatograms, ratio chromatogram, baseline correction and overlay of up to 10 chromatograms are optionally available. Integrator II allows data storage in the PICT file and transfer into drawing or word processor software.

A2.2.2 Features

CAPI-1000 features a simple design for ease of use and is available equipped with a function to ensure data accuracy. The system configuration is as simple as the principle of the block diagram and is easy to use with a one-touch system on the panel switches. Its performance-centered design provides reliable data, i.e. a well-controlled vacuum injection system, use of a high-performance double beam UV detector for low drift and low noise and use of a large buffer vial for long lasting stable measurements.

Safety is regarded as the most important for the configuration. The high voltage power supply is automatically turned off when the door is opened, and the ground remains fixed even if it is switched to the negative high voltage power supply optionally available.

A wide variety of application kits in CAPI-3100 can be commonly used in CAPI-1000, which is suitable both for beginners and experts of CE to develop new analytical methods.

Appendix B Troubleshooting
K. KITAGISHI

Table B.1 Possible causes of and solutions to problems in capillary electrophoresis instrumentation

Problems	Possible causes	Solutions
No current	A. Broken capillary	Replace capillary
	B. No capillary	Set capillary
	C. No buffer in capillary	Fill buffer in capillary
	D. Low conductivity buffer in capillary	Current monitoring system cannot detect very low (but not zero) current. Accordingly measurement can continue as normal.
	E. No potential field applied	1. Check high voltage
		2. Confirm voltage setting
	F. Electrodes or capillary not immersed in electrolyte	1. Confirm electrolyte volume in buffer reservoirs
		2. Confirm electrodes or capillary immersed in electrolyte
	G. Plug in capillary	
	G1: Dry electrolyte at edge of capillary	1. Trim capillary end
		2. Minimize the period when capillary ends are not immersed in solution.
	G2: Precipitation at isoelectric pH	Dilute sample
	G3: Insoluble product with mixing of sample and electrolyte	Change electrolyte components
	G4: Obstructive particles in sample or buffer	Filtrate sample or buffer
		(When plug in capillary cannot be removed by flushing the liquid, replace capillary)
	H. Bubble in capillary	
	H1: High conductivity of sample	Dilute sample with a low conductivity buffer
	H2: High conductivity of electrolyte	1. Dilute electrolyte
		2. Use narrower-bore capillary
		3. Apply lower electric potential
	H3: Organic solvent in sample	Add aqueous solution to sample
	H4: Organic solvent in electrolyte	Degass electrolyte
Current fluctuation	A. Cracked capillary	Replace capillary
	B. Contamination on capillary surface; inhomogeneity of capillary wall	Wash capillary thoroughly

Table B.1 *Continued*

Problems		Possible causes	Solutions
	C.	High conductivity of sample	Measurement can be continued. However, the current should be stabilized: 1. Reverse EOF with additive 2. Dilute sample
	D.	Inhomogeneity of medium in capillary (in gel or viscous polymer)	Precondition separation medium by applying potential field (there are some cases where the problem cannot be solved)
	E.	Different components between medium in capillary and electrolyte in buffer reservoir	Use the same components
	F.	Temperature fluctuation	Thermostat capillary
Abnormal current elevation	A.	High ionic strength of electrolyte	1. Dilute electrolyte 2. Use narrower-bore capillary 3. Lower applied voltage 4. Operate with constant current mode instead of constant potential mode
	B.	Different components between filling medium in capillary and electrolyte in buffer container	Use the same components
No peaks	A.	Insufficient sensitivity for detection	1. Raise sample concentration 2. Increase injection volume of sample solution 3. Increase optical path (with broader-bore or extended optical path capillary)
	B.	No injection of sample	Confirm injection method
	C.	No sample vial at injection position	Set sample vial correctly
	D.	No migration of sample	Reverse applied electric field
	E.	Sample adsorption on to capillary wall	Confirm separation method
	F.	Inappropriate wavelength for detection	Change detection wavelength
	G.	Insufficient conditioning of capillary inner wall	Wash capillary sufficiently
	H.	Detection lamp not turned on	Switch on lamp
	I.	Dead detection lamp	Replace lamp
Noisy baseline	A.	Smudge on slit	Clean slit
	B.	Detection window of capillary not clear	Clean detection window
	C.	Aging lamp	Replace lamp
	D.	Light source not turned on	Switch on lamp
	E.	Dead detection lamp	Replace lamp
	F.	High background (note especially in indirect detection)	1. Change electrolyte components 2. Change detection wavelength 3. Subtract or divide between absorbances at two wavelengths

Table B.1 *Continued*

Problems	Possible causes	Solutions
	G. Incorrect alignment of optics	Check optics and align correctly
	H. Inappropriate detection factor	
	H1. Inappropriate wavelength for detection	Change detection wavelength
	H2. Short data interval	Input longer interval
	H3. Small detection volume	1. Increase detection volume (extend detection window, use broader-bore capillary)
		2. Subtract or divide between absorbances at two wavelengths
Drift of baseline	Insufficient preconditioning of lamp	Precondition lamp sufficiently
Stepwise baseline	Current fluctuations	See 'current fluctuation'
Sharp spikes on baseline	A. Obstructive particles in sample or buffer	Filter sample or buffer
	B. Bubble in capillary	1. Degass electrolyte
		2. Lower applied electric potential
		3. Trim capillary end
Insufficient separation		
a. Tailing peaks	Adsorption of analyte on capillary inner surface	1. Use coating of capillary inner surface
		2. Confirm separation system
b. Overlapping peaks	Inappropriate separation system	Confirm separation system
c. Separation with poor efficiency	Separation system not sufficiently established	Improve separation conditions (detection interval, capillary length, detection volume, etc.)
Poor reproducibility of migration time	A. Fluctuation of electroosmotic flow	1. Preconditioning of capillary
		2. Use of internal standard
		3. Check capillary temperature during migration
	B. pH drift of electrolyte	1. Check capillary temperature during migration
		2. Minimize the number of assays for each vial
		3. Increase electrolyte volume in buffer reservoirs
		4. Minimize the period when capillary ends are not immersed in solution
	C. Undesirable gravity flow	Maintain a constant fluid level for vials
Poor reproducibility of peak area/height	A. Poor reproducibility of injection	1. Correct area/height with migration time
		2. Check injection method
		3. Use internal standard

Table B.1 *Continued*

Problems	Possible causes	Solutions
	B. Adsorption of analyte on capillary inner surface	1. Use coating of capillary inner surface
		2. Confirm separation system.
	C. Undesirable gravity flow during sample injection	Maintain a constant fluid level for vials
Impossible to flush buffer	A. Insufficient power of pump for flushing (especially with viscous polymer solution)	1. Increase injection pressure
		2. Prolong injection period
	B. Plug in capillary	Replace capillary
Unexpected pherogram with normal current value	Stale electrolyte	Freshly prepare electrolyte

Appendix C
Commercially available buffer reagents

H. SHINTANI

In the following, typical buffer systems normally used in high performance capillary electrophoresis (HPCE) analysis and available on the market from Fluka reagent company are listed.

The rapidly growing capabilities of HPCE have created a great demand for reagents of an appropriate quality. Ready-to-use buffers covering the pH range from 2.5 to 11 for HPCE are as follows:

- Buffer solution pH 2.5
  20 mM sodium citrate
- Buffer solution pH 3.0
  20 mM sodium citrate
- Buffer solution pH 3.5
  20 mM sodium citrate
- Buffer solution pH 4.0
  20 mM sodium citrate
- Buffer solution pH 4.5
  20 mM sodium citrate
- Buffer solution pH 5.0
  20 mM sodium citrate
- Buffer solution pH 5.5
  20 mM sodium citrate
- Buffer solution pH 6.0
  20 mM sodium citrate
- Buffer solution pH 6.5
  20 mM sodium phosphate
- Buffer solution pH 7.0
  20 mM sodium phosphate
- Buffer solution pH 7.5
  20 mM sodium phosphate
- Buffer solution pH 8.0
  20 mM sodium phosphate
- Buffer solution pH 8.0
  20 mM sodium tetraborate
- Buffer solution pH 8.5
  20 mM sodium phosphate

- Buffer solution pH 8.5
  20 mM sodium tetraborate
- Buffer solution pH 9.0
  20 mM sodium phosphate
- Buffer solution pH 9.0
  20 mM sodium tetraborate
- Buffer solution pH 9.5
  20 mM sodium phosphate
- Buffer solution pH 10.0
  20 mM CAPS (3-cyclohexylamino-1-propane sulfonic acid)
- Buffer solution pH 10.5
  20 mM CAPS
- Buffer solution pH 11.0
  20 mM CAPS

Capillary wash solution:

- Sodium hydroxide solution for HPCE
  0.1 M NaOH
- Hydrochloric acid solution for HPCE
  0.1 M HCl

Index

A_1c 440
Ab–Ag binding 220
Absolute
 mobility 551
 quantification 271
 recovery 54
Absorbance 86, 88, 681
 detection 681, 683
Absorption 6, 618
 detection 86
 imaging 123
Abuse 315
Accelerator 559
Acceptability 609
Acceptor, rhodamine 293
Accuracy 26, 34, 53, 314, 334
Accurate temperature control 477
Acesulfame 607, 609, 610, 614
Acetaminophen 392, 397, 404
Acetate 554, 555, 591
Acetate/phosphate buffer 555
Acetate-based buffer 199
Acetic acid 520, 522, 591
Acetone 180
Acetone/CTAB 554
Acetonitrile 386, 554, 556
 deproteinization 391
Acetophenone 633
Acetyl chloramphenicol 437
Acetylation 347
Acetylcystein 347, 368
Acetylsalicylate 556
Acetylsalicylic acid 313
Acetylthebaol 315
Achiral 372
 surfactant 367
Acid
 casein 594
 dissociation constant 511
Acid–base equilibrium constant 311

Acid-etching bath 520
Acidic
 amino acid enantiomer 378
 buffer 262–3
 compound 552
 drug 583
 monosaccharide 592
Acridinium
 chemiluminescence detection 89
Acrylamide 61–2, 449, 670
 bis-acrylamide mixture 181
 coating 61
 monomer 61–2, 76, 141, 664, 670
 polymerization 62
Acrylamide monomer 664
Acrylic
 acid 664, 667
 copolymer 667
ACTH 255, 263
Activation energy 207
Additive 6, 71, 74, 174, 379, 487, 532, 583, 597, 607, 669, 678
 polymer 72
Adenine 89
Adenosine 34
 deaminase 63
 diphosphate moiety 208
 diphosphate-glucose 208
 diphosphate-glucose pyrophosphorylase 208
 triphosphate 208
Adenosine-5-
 monophosphate 560
Adhesion 223
Adrenal
 cortex 255
 gland 418
 medullary cell 420
Adrenocorticotropin 255
Adsorption 57, 139, 199
 chromatography 657

effect 78
Adulterated milk 594
Adulteration 180, 596
Adult 265
Affinity 219, 224, 234, 291, 349, 632
 analysis 270, 275
 assay 250
 CE 25, 31
 chromatography 617
 complex 270
 ligand 250, 291
 parameter 273
 probe 250
 probe capillary electrophoresis 222
 system 278
Affinity-based interaction 271
Aflatoxin 501
African sleeping sickness 399
Ag 427, 442
 $(CN)_2$ 518
 (I) 518
 (I) cyanide complex 518
 staining 142
Agarose 61, 141, 174, 450
 polyacrylamide gel 173
 electrophoresis 425
 gel 141
 gel electrophoresis 177, 179, 427, 442, 445
 polyacrylamide-gel 8
Agarose electrophoresis 388
Age 150
Agenergic agent 345, 347, 352
Agglutinated particle 419
Aggregate 317
 phase 458
Aggregation 178, 199
 number 373
Agricultural
 application 219
Agriculture 456
Agrochemical 334

AIDS patient 493
Air cooling 34
Al 499
    (III) 519
    assay method 520
    ion 73, 520
Alanine 149–50, 368, 416, 420
Alanylglutamine 560
Albumin 425, 428, 431, 556, 557
Alcohol 224, 356, 645
    dehydrogenase 22
    precipitation 434
Aldaric 619
    acid 625
Aldehyde 647
Alditol 619–20, 624
Aldonic 619, 625
Aldose 592
Alendronate 318
Alendronate–Cu^{2+}
    chromophore 318
Aliphatic
    acid 72, 554
    amine 639, 646
    aromatic hydrocarbon 629
    anionic surfactant 648
Alkali 520
    buffer 263–4, 621
    earth ion 73, 510, 515, 520, 528, 590
    fusion reaction 670
    media 618
    metal ion 510, 520, 590
    metal salt 175
    phosphatase 214, 216, 437, 440
    sulfite treatment 149
    transition metal ion 521
Alkaloid 315, 589, 596
Alkanesulphonate 647
Alkybenzenesulphonate 648
Alkyl
    aryl ketone 633
    aryl sulphonates 639
    benzylquaternary ammonium 647
    chain length 668
    diquaternary ammonium 534
    ether sulfate 558
    phthalate 647
    sulfonate 6
    sulphate 647
Alkylamine 560, 668

Alkylaryl ketone 31
Alkylbenzene 31, 75, 633, 649
    sulphonate isomer 668
Alkylbenzenesulphonate 648
Alkyldisaccharaide 75
Alkylglucoside 556
Alkylglucoside-boronate
    micelle 566
Alkyllead 652
Alkylphenol 553, 557
Alkylphenone 631, 633
Alkylphosphonate 560
Alkylpyridine 561
Alkylsilane-derivatized
    amino acid 112
Alkylsulfate 558–9
Alkyltin 652
Alkyltrimethylammonium 534, 649
    cation 566
Allele 455, 477
    specific oligonucleotide
        primer 477
Allelic ladder 455
Allyl dextran 179
Alpha1-antitrypsin 430
Alpha1-globulin 428
Alprenolol 351
Aluminium, see Al
Alzheimerí dimentia 150
Amaranth 610–11, 614
Ambient temperature-tolerant SSCP
    analysis 451
Ambucetamide 360
Ameliorate 79
Amide bond 221
Amine 79, 416, 550, 554, 560, 593, 599, 639
    additive 43
    analogue 233
    cation 646
    nitrogen 224
Amine-containing
    compound 224
Amine-reactive
    group 224
    labelling reagent 233
    reagent 233
Amino 232, 345
    acid 87, 112, 119, 124, 149, 198, 264, 339, 345, 368, 416, 420, 445, 521, 583, 593, 599, 625
    analysis 369
    content 240

derivative 88
enantiomer 369
naphthylamide 124
neurotransmitter 156
buffer 670
group 221
nitrogen 227
polycarboxylic
    reagent 520
pyridine 534
sugar 112
Aminobenzoate 533, 555, 592
buffer 554
Aminocaproic acid 399, 428
Aminocyan 63
Aminoglutethimide 372
Aminoglycoside
    antibiotic 311
    monitoring 387
Aminomethyl phosphonic
    acid 560, 649
Aminomethyl
    propanediol 555
Aminonaphthalene 618
    disulfonic acid 555
    sulfonate 558, 559
    trisulfonic acid 625
Aminopeptidase 124
Aminopolycarboxyclic
    reagent 523
Aminopolycarboxylate 554
Aminopolycarboxylic
    acid 510–11
Aminopolyphosphonate 560
Aminoquinoline 534
Aminosalicyclic acid 74
Aminosugar 619
Aminotetralin 561
Amiodarone 397
Amitriptyline 393
Amlodipine 314
Ammonium 73, 528
    acetate buffer 394
    acetate/acetonitrile
        buffer 559
    ion-selective electrode 500
    peroxydisulfate 61, 141
Amobarbital 394
Amoxicilla 309
Amoxicillin 394
Amperometric detection 92, 95, 533, 534, 593, 639
Amperometry 2, 86, 91, 92

Amphetamine 315, 352–3, 368, 403–4, 420
Amphipathic polymer 57, 317
Ampholyte 12, 141–2, 183–5
Amphoteric surfactant 561
Ampicillin 309, 393
Amplification refractory mutation system 456
Amylose 78, 355
Anaesthetic 345, 353
Analgesics 301, 313, 386, 397, 556
Analogue 255
Analysis time 308, 647, 669
Analyte 12, 108, 265, 347
  diffusion 123
  ion 138
  molecule 457
  zone 78, 86, 118
Analyte–wall interaction 33
Analytical
  condition 26
  methodology 308
Androgen receptor mRNA 475
Anesthetic drug 402
Angiotensin 418, 438
  converting enzyme 314
Alpha-NH$_2$ terminal 174
Aniline 561, 639, 645
  derivative 646
Animal husbandry 456
  anion gap 500
Anion-BT 309, 553, 556
Anion-BT osmotic flow modifier 553
Anion 309
Anionic 559
  borate 75
  complex 517
  detergent 181
  dye 609
  solute 345, 360
  surfactant 179, 310, 584
Anode 179, 678
Anodic side 514
Ansamycin 78
Anterior
  pituitary gland 263
  pituitary lobe 255
Anthocyan 609
Anthracene 630, 631
  derivative 630

Anthraquinone
  carboxylate 557
  sulfonate 558
Anthropogeneous
  pollution 646
Anticonformer 646
Antiarrhythmic 386
  drug 397
Antiasthmatic 301, 309, 393
Antibacterial agent 399
Antibiotic 301, 309, 311, 347, 392–3, 492, 557, 559, 589, 597–8
Antibody 177, 220, 224, 240, 247, 270, 392–3, 401, 417, 419, 431
  binding 227, 233
  characterization 243
  complex 224, 229
  fab fragment 243
  fragment 223, 250
  heavy and light chains 247
  migration 223
  production 249
  quantitation 249
Antibody–antigen 219
  binding 219
  complex 221–2, 227, 229, 236
  reaction 219, 223
Antibody-bound
  antigen 219
  antigen binding 220
  labeled antigen 488
Antibody-digoxin-Cy5
  complex 229
Antibody-labeled antigen complex 221
Anticancer drug 110, 301, 311
Anti-CEA monoclonal antibody 245, 252
Anticoagulant 345, 355
Anticonvective media 271
Anticonvulsant drug 371
Anti-cyclosporin A monoclonal antibody 220
Antidepressant 301, 312, 345, 355, 393
Anti-digoxigenin Fab 223
Antiepileptic 386, 394
  drug 387, 390, 394
Antifungal agent 360
Antigen 224, 270
  binding 243, 451
  peak 220

Antihistamine 78
Antihistaminic 358, 360
Anti-human GH IgG 263
Antihypertensive 386, 397
  agent 313, 397
  drug 401
  furopyridine drug 402
Antiinflammatory drug 318, 401, 492
Anti-insulin antibody 221
Antileukemic activity 311
Antimalarial 360
  agent 360
  drug 78, 598
Anti-migraine 301, 312
  polyclonal antibody 245
Antineoplastic 421
Antioxidant 597
Anti-phosphotyrosine monoclonal antibody 245
Anti-prostate-tumor drug 492
Antipyrine 401, 492
Antispasmodic 357, 360, 399
Anti-t$_4$ Mab 227
Anti-theophylline Mab 229
Antitumor 311
  drug 399
  necrosis factor antibody 252
Antitussive 402
Anti-ulcer drug 301, 312, 399
Antiviral
  agent 212
  compound 318, 399
AOAC 598
Apex 45
Apigenin 596
Aplatic anemai patient 198
Aplysia californica 420
Apnea 393
Apolipoprotein 436, 475
Apparent mobility 42
Apple 587
  orange juice 590
  juice 558, 593
Applied
  potential 619
  voltage 3, 72, 610–11
Apricot seed 416
APTS-oligosaccharide 204
Aqueous buffer 271
Arg[413]–Pro[413] 477

INDEX

Arginine 150, 174, 345
Argon ion
   laser 86–9, 107, 110, 112, 123, 125, 222–3, 293, 479, 482, 684
   lazer beam 22
Arnica montana 596
   A. chamissonis 596
Aromatic 555–6
   acid 550, 555–6, 670
   amine 499, 534, 550, 553, 561, 645–6
   amino acid 221, 593
   carboxylate 556
   diamine 670
   moiety 347
   quaternary amine surfactant 561
   sulphonic acid 558, 649
Aromatic-containing organic acid 649
Arsenazo 510, 517
Arsenic species 651
Arsenious 651
Artifact 667
Artificial
   cerebrospinal fluid 155
   dialysis 499
   sweetener 607, 610, 612
Arylsulfonamide *d*-camphor-10-sulphonate 370
Ascite 177
   fluid 175, 177, 247
Ascorbate 555, 595
Ascorbic
   acid 420, 595
   isoascorbic acid 595
Asparagine 200, 345
Aspartame 361, 593, 607, 610
   ligand 521
Aspartate 149, 416, 420
Aspartic acid 149, 345, 609–10
Assay 310, 390
   consideration 233
   time 140
Association 75
   constant 280
Atenolol 78, 314, 347, 371
Atherosclerosis 436
Atomic
   absorption spectrometry 500
   absorption spectroscopy 500
   number 523

ATP 208
Au 2, 518
Auromatic micropositioning 19
Autoimmune disease 219
Automated
   analysis 121, 123
   CE instrument 680
   reproducible DNA sequencing 480
   sequencing of DNA 61
   system 477
   velocity programming 89
Automatic
   buffer replenishment 680
   DNA sequencer 480
   sampling device 681
Automation 271
Autoradiograph 455
Autosampler 2, 19, 124, 673, 681, 686
Autosampling 140
Average molecular weight 668–9
Avidin 213, 355–6, 360
Avidity 232
Axial
   beam 87
   beam excitation 456, 481
Axial-beam lazer 110
Azo 609
Azorubin 611

β-Agonist 347
β-Antagonist 347, 351–2
   adrenergic agent 347
β-Cyclodextran 156, 345, 347, 361, 631, 645
   cavity 345
β-Galactosidase 212–13, 278
β-Phycoerythrin 223
Baby hamster kidney 199
Baccatia 311
Background 74
   absorbent 90
   electrolyte 72, 74, 678, 683
   interference 96
   noise 105
   signal 104
Bacterial endotoxin 501
Bag cell 420
Baicalein 314
Ball lens 66

Band broadening 22, 275, 437
   bandpass filter 89, 114, 479
Barbital 440
   buffer 177
Barbiturate 370–1, 386, 390, 394, 404, 632
Bare
   capillary 682
   fused-silica capillary 480
Barium 554
Bark 311
Base 283, 294, 550, 560
   composition 292, 481
   pair mismatch 452
Baseline
   resolution 311, 376
   separation 264, 644
   shift 49
Base-specific detection 278
Basic
   compound 78
   drug 387
   fibroblast growth factor 455
   pH range 609
   protein 78, 79, 178
Basicity 231
Bayer liquor 553
Beam splitter 107, 479
Beer 554, 591, 595
Beer's law 682
Beetroot 609
β-Endorphin 255
Benz [*a*] anthracene 630
Benzaldehyde 558
Benzendiol 644
Benzene
   derivative 119, 632
   sulfonate 559
Benzenesulfonamide 32
Benzenetricarboxylate 554
Benzidine 561, 639, 645–6
Benzo [*a*] pyrene 111
   guanosinmonophosphate 290
Benzoate 6, 72–3, 550, 553, 555–7, 590, 591, 597, 598
Benzo [*g,h,i*] perylene 630
Benzo [*k*] fluoaranthene 630
Benzopyrene isomer 630
   buffer 560
Benzoate/borate electrolyte 558

INDEX

Benzodiazepine 393, 404
Benzoic
  acid 71, 533, 553, 610
  acid derivative 77
Benzoin 78, 356, 378
  enantiomer 379
Benzothiazole sulfonamide 559
Benzoylecgonine 230–1
Benzoylecgonine-Cy5.5 231
Benzylamine 6, 73, 521, 534
  buffer 590
Benzylbenzoic acid 533
Benzylpenicillin 394
Beverage 521, 555, 558, 590, 596, 612
Biantennary
  oligosaccharide
    structure 198
  sugar chain 205
Bi- tri-tetra-antennary 202
Biased reptation regime 292
Bicarbonate 500
  buffer 557
BICINE 555
  buffer 598
Bifunctional reagent 61
Big chap 402
Bile
  salt 74, 313, 334, 372–3, 376, 378, 556, 630
  salt monomer 373
  salt surfactant 372–3
Bilirubin 492
Binaphthyl compound 373
Binary mixture 457
Binding 270, 387
  constant 31, 32, 270, 273, 275
  parameter 270
  site 233
  system 273
Bioactive component 314
Bioassay application 342
Biochemical 61
  application 219
  constituent 486
  reaction 207
Biochemistry 140, 207
Biodegradation kinetic
  study 318
Biofluid 52, 53, 342
Biological 550
  activity 198, 255

effect 401
fluid 224, 399, 411, 442, 486
matrix 486
medicine 476
molecule 219
process 451
sample 220, 651
science 585
Biomass degradation
  product 557
Biomedical 589
  analysis 589
  application 589
  field 61
  science 124
Biomedicine 140
Biomolecule 89, 199, 277, 449
Biomolecule analysis 243
Biopolymer 9, 119, 140
Biopsy sample 418
Biosample 339, 342
Biosynthesis 198, 265
Biosynthetic process 204
Biot number 453
Biotechnology 617
Biotin 213
Bipyridine 534
Bisacrylamide 670
Bisulfate 556
Bisulfite 558, 647
Blocking agent 185
Blood 486, 499
  admixture 501
  pressure 262
  urea nitrogen 500
Blue
  diode laser 112
  laser diode 88
  semiconductor laser 88
Blue-black 609
Body fluid 114, 283, 486, 535
Bombardment 96
Bombesin 264
Bone 255
  disease 318
Borate 75, 428, 558, 618
  buffer 149, 177–8, 227, 264, 315, 371, 401, 438, 442, 533, 555, 561, 664
  complex 528
Borate-carbonate buffer 442
Borate: glycerol: cellulose
  derivative 293

Borate/phosphate buffer 648, 664
Borate/SDS buffer 150, 558
Borax buffer 436
Boric acid 443
Bound (complexed) ligand 270
Boundary 106
  material 109
Bovine 264
  hemoglobin 438
  serum albumin 78, 233, 556
  serum albumin-warfarin
    affinity system 273
Brain 263, 411
  dialysate 155
  tissue 411, 416
Branching 657
Bread 554
Breakdown product 492
Breast carcinoma 438
Bridging group 278
Brij 35, 314, 650
Brilliance 610
Broad gamma-globulin 177
Broad-band collisional
  Activation 96
Broadening 437
Broadness 229
  peak 233
Bromide 6, 92, 309, 591
Bromobimane 442
Bromocresol green 90
Bronchial dilator 493
Broth 555
Brown 609
BSA-warfarin 219
Bubble cell 683
Bubble-shaped cell 66
Buffer 33, 139, 247, 427, 390, 552
  additive 44, 45, 139
  aluminum 594
  ammonium acetate 559
  benzoate 558
  borate 555, 557–8
  borate/cholate/
    isopropanol 560
  borate/SDS 557
  butylboronate 556
  capacity 36
  CAPSO 399
  cation 560
  chiral additive 347
  composition 45, 317

creatinine 566
electrolyte 352
ethylenediamine/MES 559
glycine 557
modification 551
modifier 310
naphthalene sulfonate 558
pH 313, 522
phosphate 556, 558, 594
phosphate/borate 556
phosphate/SDS 556
phthalate 560
plug 271
quinine sulfate 561
replenishment 314
reservoir 118
salicylate 560
salicylate/phosphate 558
SDS/phosphate 559
sodium acetate 556
solution 29, 57
specific conductivity 453
tris/acetate 556, 561
tris/borate 558
tris/citrate 566
tris/p-anisate 554
veronal 558
Buffering 394
  capacity 45
Bulk
  phase 263, 265
  sample 679
BUN 500
Bupivacaine 402
Burn 434
Butalbital 394
Butanol 556
Byproduct 313, 610

C-6 Alkyl diquaternary ammonium 534
C-12 to C-18 Alkyltrimethylammonium 534
C1-C7 Carboxylate 555
C4-C7 Carboxylate 554
C18-Derivatized capillary 594
C2-C14 Fatty acid 554
C2-C18 Fatty acid 554
C8-C20
  alkyl sulfonate 559
  fatty acid 554
Caboxylate 554
Cadaverine 488, 499, 503
Caffeine 283, 313, 390, 393

Calcium 92, 528, 553, 554, 556, 590, 591, 592
  chloride 592
Calcium-free perfusate 417
Calibration 53
  linearity 647
  plot 293, 372
Calibrator 227
Calixarene 558
  sulfonate 557
Cancer 499, 503
Candy cherry 612
Cannabinoid 404
Canthaxanthine 609
Capacity factor 29, 402
Capillary 104, 427, 550, 673
  affinity gel electrophoresis 62, 291
  array 456, 476, 481
  array approach 456
  array electrophoresis 476
  axis 86
  bore 92
  cell 17
  channel 622
  coating 178, 244
  column 590
  diameter 33, 72, 84, 453
  displacement 86
  electrochromatography 12
  electrophoresis 1, 104, 112, 207, 219, 386, 411, 425, 486, 509, 517, 531, 550, 583, 617, 673
  electrophoretic immunosubtraction 221
  gel electrophoresis 3, 8, 18, 20, 61, 76, 174, 181, 214, 240, 291, 478, 678, 679
  gel sieving electrophoresis 664
  inlet 667
  ion electrophoresis 309
  isoelectric focusing 3, 12, 87, 174, 183, 222, 240, 425, 430, 486, 681
  isotachophoresis 12, 240, 404, 437, 486, 550, 553, 665
  length 3, 62
  lumen 682, 684
  scanning device 87
  separation tube 104
  surface 78, 444
  temperature control 679
  thermostatting 46, 680
  tube 109, 622
  wall 45, 57, 87, 107, 109, 199, 221, 265, 449
    charge 43
    effect 402
    zone 583
    electrophoresis 3, 4, 32, 63, 174–5, 240, 391, 411, 509, 521, 531, 550–1, 620, 673
Capillary-end geometry 50
Capillary-to-capillary
  correction 20
  difference 19
CAPSO 611
  buffer 393
Captopril 314
Caramel 609
Carbamate 649
Carbamazepine 387, 394
Carbohydrate 95, 198, 199, 202, 607, 617, 624, 683
  chain 621
  detection 620
  group 244
  moiety 212
  oxidation 619
  portion 184
  structure 198
Carbon fiber 92
Carbonate buffer 518
Carbonated
  drink 612
  soft drink 588
Carbonic
  anhydrase 32, 265, 275, 419, 437
Carbonyl 639, 646
  functional group 632
  group 232
Carboxylate 550, 553–5, 666
Carboxylic
  acid 550, 553, 609, 667
Carboxypeptidase 208–11, 438
Carcinogenic stain 450
Cardiotonic drug digoxin 223
Cardiovascular
  disease 313
  drug 301, 313–14
Carminic acid 611
Carotinoid 609

INDEX 703

Carousel 686
Carrier
　detection 453
　electrolyte 72, 509, 590
　protein 233
Carry-over 50
Carvedilol 355
Casein 213, 594
　phosphopeptide 594
Caseinomacropeptide 594
Catalyst 61, 141, 670
　solution 89
Catalytic
　activity 207, 436, 437
　conversion 214
Catechol 95
　compound 92
Catecholamine 95, 442, 445, 595
Cathepsin
　buffer 444
　D 425, 438, 443
Cathode 678
Cathodic side 180
Catholyte 185
Cation 6, 74, 143, 531, 550, 560
　analysis 500
　exchange
　　chromatography 241, 245
Cationic
　anionic zwitterionic neutral micelle 263
　drug 312, 345, 357
　fluorosurfactant 79
　herbicide 566
　metal complex 514
　micelle 556
　mixed micelle 21
　polyethyleneimine 561
　polymer 78, 550, 561, 590, 668
　surfactant 6, 57, 74, 78, 178, 309, 311, 550, 561, 621, 665
　triazine herbicide 566
Caudate nucleus 421
Cavity 345, 402, 629
CBI-derivatized amino acid 371
CBQCA 89, 112, 150
CCD 2, 87, 123, 430, 481
　camera 481
CD 6, 71, 72, 345, 369, 402, 534
　concentration 72
　derivative 78

CD/electrokinetic
　chromatography 6,8
CD/MEKC 6, 8
CD-CE Method 347
CD-CZE Separation 347
CD-modified MEKC 369, 371
CDGE 478
CDNA 198
CDTA 509–10, 557
CE 1, 4, 17, 87, 104, 207, 211–12, 219, 291, 308, 345, 366, 386, 411, 425, 486, 509, 531, 550, 583, 597, 617, 673
　assay 314
　column 1
　lipid analysis 595
　non-aqueous solvent 595
　method 345
　separation 8, 673
Cerium (III)
　sulphate 528, 534
CE/LIF 223, 227
　competitive
　　immunoassay 223, 230
CE/LIF-based
　immunoassay 224, 232, 237
　non-competitive
　　immunoassay 222
CE-buffer 347
CEC 12
CE-EC 618
Cefixime 394
CE-FL 107
CE-ion spray MS 36
Celebrated therapeutic
　drug 198
　cell 124, 264, 411
　culture medium 150, 420
　function 503
　injection 124
　surface receptor 503
Cell-line B8-300 204
Cellobiohydrolase 351
Cellulose 174, 588
　acetate 180
　acetate film 92
　derivative 9, 76, 77, 141
　ester 554, 670
CE-MS 21, 36, 598
　interface 95
CE–MS–MS 96
Central nervous system
　stimulant 352

Cephalomannine 311
Cephalosporin 310
　antibiotic 355, 394
Cephradine 310
Cerebellum 417
Cerebrospinal
　fluid 150, 177, 220, 411, 425, 434
　fluid (CSF) protein 425, 434
Cetrimide micelle 312
Cetyltrimethylammonium 534
　bromide 75, 78, 534, 554, 591, 621
　chloride 96, 556
CF 476
CGE 3, 8, 18, 20, 61, 76, 174, 181, 291, 478, 679
CGE/SDS 181
Change mobility 280
Channel 125
CHAPS 263, 402
Characterization 118
Charge 609
　density 31
　mass 438
Charge-coupled
　device 2, 87, 123, 430, 481
Charged 658
　benzenesulfonamide 275
　carbohydrate
　　complex 617
　carbohydrate derivative 618
　CD additive 360
　charged-β-CD
　　sulphobutyl ether IV 315
　ligand 31
　species 632
Charge-to-mass ratio 200, 207, 449, 528
Chelate 517, 519, 590
Chelating
　agent 6, 73–4, 90, 517, 519
　reagent 517
CHEM-ELISA 418
Chemical
　affinity 270
　composition 657, 669
　derivatization 419
　inertness 671
　modification 589
　reaction 22, 207
　synthesis 233

Chemically
    immobilized
        ribonuclease T$_1$ 213
    modified fused-silica
        capillary 179
Chemiluminescence 84, 89
    detection 293
Chemiluminescence-
    enhanced enzyme-
    linked immunoassay
    418
Chemotherapeutic agent
    309
CHES buffer 175
Chicory root extract 591
Childhood neurological
    disorder 150
Chimeric (human/mouse)
    monoclonal
    antibody 245
    monoclonal antibody-
    cytoxin conjugate
    246
Chinese
    herbal 352
    herbal drug 301, 314
    herbal preparation 315
    herbal tea 597
    medicinal formulation
        314
    tea 74
Chinolingel b 611–12, 614
Chiral 372, 388
    achiral component 367
    active component 596
    additive 345, 347, 352–3,
        355
    amino acid 368
    analysis 51, 301
    buffer additive 345
    CE 32
    CE analyses 51
    CE model 355
    compound 8
    electrolyte additive 212
    electrophoresis 270
    immobilised A$_1$-acid
        glycoprotein 351
    mercaptan derivative
        368
    method 345
    modifier 355
    pharmaceutical 334
    phase 366
    purity 72
    recognition 8, 352
    resolution 352, 355
    selectant 366, 367, 382

selectivity 72
selector 8, 71, 72, 78,
    345, 352, 355–7, 555,
    679
separation 7, 72, 78, 109,
    137, 339, 341, 349,
    351–2, 355, 386, 401,
    556
separation efficiency 32
stationary phase 366
surfactant 78, 367
Chirality 376
Chirasil-B-Dex 361
Chitosan 313
Chloramphenicol 597
    acetyltransferase 208,
    437
Chloride 309, 500, 510,
    535, 590, 591
Chlorinated phenol 639
Chlormequat 650
Chloro nitrophenol 557
Chloro-, nitro-, alkyl-
    substituted phenol
    644
Chloroacetaldehyde 89
Chloroaniline 553
Chlorodisopyramide 397
Chloroethyltrimethyl
    ammonium 566
    chloride 650
Chloroform-isopropanol
    403
Chlorogenic acid 596
Chloropheniramine 313,
    371
Chlorophenol 557–8, 639,
    645
Chlorophenoxy 556
    acid 650
Chlorophyl 112, 609
Chlorosulfuron 650
    herbicide 650
Chlorothiazide 314
Chlorowarfarin 355
CHO 198
Cholate 247, 556, 560, 595
Cholecystokinin 264
Cholesterol 371
    lowering agent 339
Cholic acid 646, 649
Choriogonadotropin 231
Chromate 6, 72, 73, 74,
    533–4, 553, 558, 590
Chromatic beam splitter
    107
Chromatographic
    analysis 25

mode 6
preconcentration 21
Chromatographic
    electrophoretic
    method 219
Chromogenic 394
Chromophore 26, 74, 86,
    111–12, 405, 521,
    589, 592, 597, 610,
    683
Chromophoric group 618
Chromosome 17, 455, 477
Chronic
    liver disorder 431
    obstructive pulmonary
        disease. 393
    use 404
Chryosin 611
Chrysene 631
Chylomicron VLDL 436
Cicletanine 372, 401, 402
CIEF 3, 12, 25, 31, 97, 123,
    174, 222, 240, 425,
    430, 437, 486, 681
Cigarette smoke 553
Cimetidine 312
Ciprofloxaci 311
Cis-trans-dothiepin 355
Citic acid 649
CITP 12, 97, 240, 486, 665
Citrate 442
    buffer 180, 438, 593
Citric 591, 591
    acid 534, 591
Citrus 591
    juice 593, 595, 596, 597,
        598
    pulp wash 555
Cleanup 649
    step 432
Clenbuterol 338
    enantiomer 340
Clinical 219
    analysis 207, 411, 425,
        531
    chemistry 387, 501
    diagnosis 150, 219, 445,
        486, 487
    examination 124
    forensic analysis 334,
        342
    improvement 387
    interest 438
    laboratory 486, 494
    sample 334
    significance 427
    test 425, 445
    therapy 492

ClO$_4^-$ 92
Cluster 553
  analysis 553
Carboxymethyl-β-CD 631
CMC 75, 367, 379
CN 416
Co 517–9
Coated capillary 139, 179
Coating
  procedure 179
  technique 57
Coaxial
  flow chamber 96
  sheath flow 95
Cocaine 315–16, 404
  metabolite 230
Cochenillerot 611
Codeine 313, 403
Coefficient
  variation 452
Coelectroosmotic 511, 514, 645
  mode 514
Coenzyme 437
  NAD 214
Coffee 553, 612
Coherent
  laser beam 110
  light 97
  UV 86
Coiling 62
Coke 612
Cola 590
Collection
  capillary 118
  vial 118
Collagen 443
  synthesis 443
Collinear 87
  geometry 87
  confocal optical configuration 107
Colloidal silica particle 666
Color 609, 612
Colored reaction 592
Colorimetric assay 438
Colorimetry 500
Column 33, 56, 596
  connection 63
  efficiency 62
  lifetime 181
  packing 387
  re-equilibration 551
  technology 16, 56, 671
Commercial
  dextran sample 625

DNA sequencer 482
  instrument 583
Common variant 440
Compartmental
  release 420
Compatibility 671
Competing species 223, 231, 233
Competition 227
  reaction 401
Competitive
  binding 224
  immunoassay 220, 222–4, 227, 231–2, 234, 236
  Rt-PCR assay 455
  PCR product 142
Complementary
  DNA fragment 478
  strand 452
Complex 233, 270, 273, 316, 523
  biological fluid 224
  biological mixture 208
  entangled polymer system 271
  equilibrium 517, 522
  formation 509, 522
    curve 522
  forming ligand 517
Complexation 31, 75, 273, 509–10, 515, 552, 629
  borate 596
  constant 668
Complexed ligand 275
Complex-formation constant 509
Complexing 517
  agent 72–3, 312, 517, 520, 590
  reagent 509
Complimentary tool 317
Composition 311, 669
  analysis 658
Compton derived semi-empirical model 240
Computer
  simulation 6, 119
Computer-controlled analysis 140
Conalbumin-arylaminona-phthalene sulfonate 219
Concentration 95, 340, 432, 434, 651
  gradient 478
  limit detection 683

sensitivity 84
Concious rat 150
Condensed zone 12
Conductive medium 57
Conductivity 2, 4, 12, 90–1, 179, 453, 551
  cell 56, 91
  detection 91, 550, 555, 557, 566
  detector 91, 591
  monitor 684
  suppression 533
Conductometric
  detection 533
  detector 533
Cone-shape 8
Confectionery 598
Configuration 106, 108, 271, 684
Confocal
  optical configuration 110
  sequencing detection 456
Conformation 77, 255, 355
Conformer 667
Conical aperture 92
Conjugate 233, 243, 283
Conjugated
  amino acid 499
  antibody 247–8
Conjugation 227
Constant
  denaturant gel electrophoresis 478
  velocity 25
  voltage operation 73
Constant time-pressure product 679
Constant-field 669
Constituent monomer 669
Constitute 308
Contaminant 219
Continuous
  pressure control 19
  variation 114
  zone electrophoretic separation 125
Continuous-flow fast atom bombardment 96
Contraction 262, 264
Contrast agent 386, 397
Conventional
  gel electrophoresis 61
  laser 112
  OPA post-column derivatization detection 112

slab gel 76, 477
Cooled
  CCD camera 125
  image intensifier 125
Cooling
  efficiency 33–4
  system 34
Coomassie
  blue 173
  brilliant blue 142
Coordinate bond 4
Copolymeric
  magnetic resonance imaging contrast agent 665
Copolymer 657, 665–6
Copolymerization 179
Copper
  (II)-S-amino acid complexe 360, 554
  microelectrode 416, 588
  sulfate 318, 521, 534, 554
  see also Cu
Copper (II) aspartame complex 361
Copper (I) cyanide 510
Coptidis rhizoma 315
CORD blood sample 440
Cordial 598
Cornea 493
Coronary heart disease 436, 477
Corpus striatum 150, 417
Correlation 255, 616
  coefficient 393, 523
Cortex 417
Corticotropin releasing hormone 272
Cortisol 223
Cough syrup 342
Coumarin 399
Counter
  cation 519
  flow 21
  ion 178, 517
Counterelectroosmotic 511
  migration mode 511
Counterion 2, 632
Coupling reaction 614
Covalent
  attachment 227
  bond 57
  capillary coating 449
  functionalization 618
Covalent-bond
  coating procedure 180
Covalently
  bonded 57

bonded modification 3
Cow 594
Cr (III) 519
Creatinine 6, 73
  kinase 215, 440
Creatinine 397, 493, 499, 501, 534
  clearance 397
Cresol 645
CRH 263
Criterion 308
Critical
  angle 105
  factor 26
  micelle concentration 6, 75, 183, 347, 367, 390, 534
  parameter 340
Cromakalin 379
Cross
  correlation 309, 312
  section 4
Cross-linked
  gel 181
  polyacrylamide 457
  polyacrylamide-filled capillary 449
Cross-linker 61, 62
Cross-linking 179
  reagent 76
Cross-reactivity 224, 231, 500
Cross-talk 481
Cross-validation 334, 339
Crown
  ether 6, 334, 345, 347, 357, 360, 534, 561, 590
Cryoglobulin 425, 431
  analysis 431
Cryopreciptate 431
Cryptand 74
CSF 434, 443
  protein electrophoresis 434
CTAB 78, 264, 311, 554–5, 591, 621
  micelle 315
CTAC 556
Cu 560, 591
  (II) 370, 518
  chelate 593
  electrode 593
  see also Copper
Cultivation media 251
Culture broth 317
Current flow 449
Curved DNA 451

CV 452, 455
Cy3-labeled
  angiotensin 211
Cy5 227, 229–31
  cyanine dye 223
  diacid internal standard 227
  label 224
  N-hydroxy-succinimide ester 224
  reactive dye 224, 227
Cy5-labeled
  antigen 227
  competing species 224
  drug 229
  primer 212
Cyanide 416, 510
Cyanine
  (Cy3) labeled angiotensin 211
  dye 230
  moiety 224
Cyanine-labeled
  antigen 236
Cyano ligand 518
Cyclamate 558, 607, 609–10
Cyclamide 610
Cyclic
  aromatic hydrocarbon 76
  oligomer 369
  peptide 220
  structure 78
Cyclization
  product 314
Cyclodextrin 6, 334, 402, 416, 534, 552, 554, 556, 595, 596
  (CD) modified MEKC 369
  modified MEKC 369
β-Cyclodextrin 554
Cyclodextrin-modified MEKC 597
Cycloethylene diaminetetraacetic acid 6
Cyclohexane-1,2-diaminetetraacetic acid 509
Cyclohexanediamine tetraacetic acid 520, 557
Cyclohexanol 632
Cyclosporine 220, 221, 386–7, 392
Cylindrical
  cell 106

cross-section 106
electrode 90
flowcell 106
hollow tube 86
hollow tubing 122
Cysteine 345, 488, 664
Cysteine-based
  anionic chiral surfactant 379
Cystic fibrosis 397, 476, 493
Cytidine
  deaminase 212
  deamination-catalyzed reaction 212
Cytochrome 175
  C 175
  C hydrolyzate 593
Cytokin 317
Cytoplasm 418, 420
Cytosine-β-arabinoside 399
CZE 3, 4, 21, 32, 63, 97, 222, 240, 347, 393, 404, 411, 428, 442, 509, 531, 550, 583, 591, 607, 665, 673
CZE-electrospray MS spectrometry 243, 559
CZE-fluorescence 559
CZE-MS spectrometry 243, 559

DAC-β-CD 352
  cavity 345
  chiral selector 345
Dairy product 554
Dansyl 376
  amino acid 75, 632
Dansylated
  amino acid 89, 109, 149, 370, 376
  amino acid enantiomer 371–3, 379
Dansylhydrazide 646
Dansylhydrazine 494
DAPMP ligand 519
DAT amino acid derivative 347
Data
  processing 26
  processing software 673
  processing time 456
  smoothing 26
Deamidation product 317
Decelerated movement mode method complex 518

Decomposition
  product 610
Decoupling procedure 622
Deep-red
  region 88, 112
  semiconductor laser 88
Deer antler 416
Deglucuronidated
  urine 371
Deglycosylation 198
Degradate 301, 308, 314
Degradation 246, 443, 667
  product 311, 597
Degree
  complexation 522
  ionization 551
Dehydroacetic acid 610
Dehydrogenase
  isozyme 419
Denaturant 453, 478
  buffer 478
  urea 62
Denaturation 452
Denatured strand 451
Denaturing 291, 477
  agent 452
  condition 294, 452
  gel 477
  gel-filled capillary 294
  gradient gel electrophoresis 476
  milieu 452
Denaturing gradient gel electrophoresis 478
Dendrimer 557
Densitometer 173
Densitometry 142
Density 17
Dental carry 607
Deoxycholate 630
Deoxynucleotide 283
Deoxynucleotidyl transferase 293
Deoxyribose 294
Deproteinization 386, 391, 394, 442, 492
  reagent 394
Deproteinizing reagent 444
Deprotonation 620
Derivatization 84, 87, 89, 112, 149–50, 204, 340, 345, 367, 369, 387, 394, 438, 442, 510, 555, 609, 667, 683
  agent 649
  carboxylate 555

microvial 124
procedure 609
Derivatized
  amino acid 112, 345
  aromatic amino acid 347
Derivatizing
  agent 683
  reagent 112, 369
Desalting 21
Desferrioxamine 519
Desialyated 204
Desialylation 198
Design 63
Designed 3-(4-caboxybenzoyl)-2-quinolinecarbox-aldehyde 112
Desipramine 393, 416
Detectability 369, 510
Detection 16, 84, 96, 111, 137, 311
  carbohydrate 618
  cell 588
  electrode 622
  limit 84, 112, 229, 308, 334, 371–2, 388, 535, 593, 624, 639
  point 110, 682
  segment 682
  sensitivity 84, 86, 88–9, 107, 138, 204
  system 84
  volume 32, 294
  window 25–6, 28, 50, 123, 125, 180, 338
  zone 25, 89
Detector 552, 673, 681
  electrode 622
  high sensitivity 84
  response 53
  window 29
Detergent 314, 648, 670
  Brij 556
Detoxicant 314
Detrimental effect 45
Deuterium
  lamp 86, 682
  source 681
  tungsten lamp 114
Deviation 19, 20, 26, 33, 480
Dextran 9, 77, 141, 182, 399, 619
  buffer 345
Dextran-based coating 174
Dextromethorphan 402
Dextropropoxyphene 313
Dextrose 355

Diabetes 493, 607
Diagnosis 124, 427, 434, 442, 481, 486, 488
Diagnostic significance 219
Dialysate 156
Dialysis 340, 418, 432, 434, 475
 tubing 21
Diamine 560
Diaminoalkane 179
 derivative 178
Diaminopropane 265, 556
Diastereomer 291, 367–9, 379
Diazepam 403, 404
Diazotized sulfanilic acid 610
Dibenz[a,h] anthracene 630
Dicarbocyanine succinimidyl ester 112
Dicarboxylate 555
Dichlorophenol 557, 644
Dichromate 668
Diclofenac 318
Dicyanoargentate 518
Dicyonoaurate 518
Didecyl-L-alanine 379
Dideoxy fingerprinting 478
Dideoxynucleoside 283
Dielectric constant 34, 73
Dietary
 component 487
 fiber 556
 requirement 607
Diethyl ether 401
Diethylenetriamine 534, 554, 560, 566
 pentaacetic acid 509, 520, 554
Difenzoquat 566
Differential analysis 500
Diffuse layer 2
Diffusion 18, 22, 32, 61, 123, 215, 294
 coefficient 32–3
 effect 476
Digestion 264
Digestive enzyme 264
Digital signal 26
Digitonin-sodium taurodeoxycholate 378
Digoxin 223, 227, 401
Dihydrodiazepam 355
Dihydrofolate reductase inhibitor 310

Dihydroxybenzoate 554, 558
Dihydroxybenzoic acid 533
Diketopiperazine 314
Diltiazem 314, 373
Diluted serum 223
Dilution 215, 487
Dimer 265
Dimeric 1:2 borate:di-diol complex 293
Dimethindene 358, 360
Dimethyl-β-cyclodextrin 212, 338
Dimethylbenzylamine 521, 534
Dimethylformamide 178, 590
Dimethylguanidine 499
Dimethylnaphthalenes 630
Dimethylsulfoxide 315
Diniconazole 371
Dinitrobenzoate 554
Diode
 array detection 616
 array detector 682
 laser 86, 110–12, 123, 224
Diol 63
Diphenylamine 561
Diphenyldimethyl-phosphonium 566
Diprophylline 442
Diquat 566, 597, 651
Direct
 chiral resolution 352
 detection 554, 649
 determination 51
 indirect UV detection 535
 injection 283
 method 367, 369
 separation 369, 373
 serum injection 390, 397
 UV detection 111, 533, 554, 558, 560
Discontinuous buffer system 21
Discrete zone 214, 215
Disease 150, 486, 492
Diseased state 451
Disk-shaped electrode 624
Disodium 5,12-bis (Dodecyloxymethyl)-4,7,10,13-tetraoxa-1,16-Hexadecanedisulfonate 75
Disopyramide 397

Disperse 648
 textile dye 648
Dispersion factor 32
Displacement 683
 electrophoresis 21
 injection 667, 679
Dissipation 679
 heat 1
Dissociation 227
 complex 234
 constant 234
Dissociation/association constant 25
Distribution
 carbohydrate 205
Disulfide
 bond 222, 247
 linkage 181
Disulfide-related instability 247
Diterpene amide 311
Dithiothreitol 180, 595
Diuretic 313, 314, 401, 488
 drug substance 314
Divalent 143
 cation 648
 cationic species 179
Diversity manipulation 56
DM-β-CD 347, 352–3, 361
DMF 178
DMSO 150, 670
DNA 199, 283, 294, 450, 455, 585, 664
 analysis 291, 449–50
 calibration marker 293
 denaturation 62
 diagnosis 476
 fragment 29, 76–7, 110, 119, 291, 293, 449–52, 466, 477, 479–80
 analysis 291
 Fx-174/Hae III 118
 hybridization 219
 marker 293, 466
 molecule 77, 476
 mutation 452, 476
 polymorphism 453, 476
 precipitation 475
 probe assay 207
 replication 503
 restriction fragment 291–2
 separation 684
 sequencer 479
 sequencing 62, 88, 114, 124–5, 143, 291, 293, 449, 451, 456–8, 478–82

INDEX 709

sizing 475, 477
technology 456
DNA-intercalator binding 219
DNB-amino acid 345
DNB-derivitization 156
DNP 345
DNS-amino acid 345
DNSH 646
DNS-Phe 347
Dodecyl 665
  decyltrimethylammonium chloride 649
Dodecyl-β-D-glucopyranoside 378
Dodecylbenzenesulphonate 648
Dodecyltrimethylammonium 57
  bromide 291, 404, 553
Dodecylvinylimidazolium 665
Dominating factor 74
Domipramine 312
Donor fluorescein 293
Dopamine 92, 124, 416, 420
  release 420
Dosage 301
Dothiepin 355
Double
  electrical layer 56
  layer 2
  stranded DNA 31, 212, 451, 476
Doubled diode laser source 111
Double-stranded 294
  DNA 76, 142, 212, 292–3, 449, 476–8, 481
Doxorubicin 110
Doxylamine 342
Driving force 2
Drug 114, 278, 334, 411, 583, 597
  abuse 230, 301, 315, 386–7
  attachment 110
  impurity 308
  intermediate 559
  metabolism 405
  receptor complex 270
Drug-fluorescent conjugate 401
Drum 118
Dry cured ham 593
Drying 76
Du Pont sequencing technique 479

Dual immunoassay 231
Duchenne muscular dystrophy 453
Duodenum 418
Durability 56
DXS 164 locus 477
Dye 558, 609, 612, 639, 648
  intermediate 558, 561
Dye-containing buffer 554
Dye-to-protein binding 173
Dynamic
  capillary coating 680
  coating 57
  elution 89
  equilibrium 379
Dyphylline 393
Dysfunction 492
Dystrophin gene 477

e⁻ acceptor ability 523
Eardrop 311
Easy-operation 88
ECD 84, 90, 95, 149, 314, 420, 488, 494, 593
Econazole 360
Edge shape 19
Edman sequence analysis 202
EDTA 6, 311, 509–10, 520, 554
Effective
  cell volume 109
  charge 551
  length 611
  mobility 550, 552
Efficiency 78, 315, 673
Egg
  white 183, 352, 355, 594, 598
  yolk 594
Egg-laying hormone 420
EI 95
EI-MS 318
Einstein-Nernst equation 32
Electophoretic mobility 678
Electric
  analog signal 26
  charge 4, 74, 137
  conductivity 33, 71, 75
  current 33
  density 292
  field 2, 4, 12, 22, 26, 33, 57, 62, 183, 361, 478, 480, 482, 545, 590

  force 3
  mobility 4, 292
Electrical
  decoupling 622
  double layer 34
Electrochemical 311
  cell 90
  detection 84, 90, 95, 149, 214, 310, 314, 411, 416, 420, 445, 488, 494, 557–8, 589, 593, 617–19
  property 416
Electrochromatography 681
Electrode 310, 678
Electrodispersive peak broadening 552
Electroelution-separation 220
Electroendoosmotic flow 199, 236
Electrogenerated chemiluminescence 561
Electrokinetic 312
  chromatography 76, 671
  flow 2
  injection 16–8, 20, 22, 36, 46, 53, 139, 450, 475, 476, 633, 678–9
  mode 6
  process 120
Electrolyte 6, 71, 74, 77, 145, 310, 340–1, 487, 588, 668, 680
  additive 74, 78, 590
  composition 96
  concentration 353
  ionic strength 72
  solution 402
  system 16
Electrolytic
  conductance 551
Electromigration 18, 119, 535, 590, 621, 667
  injection 667
  velocity 120
Electron spray
  ionization 262
Electroosmosis 119, 314, 621
Electroosmotic
  control 54, 56
  flow 1, 21, 42, 177, 245, 252, 275, 292, 376, 531, 550, 552, 583, 621, 645, 678
  mobility 34, 531–2, 631

modifier 591
property 56
velocity 33, 56
Electrooxidation 593
Electropherogram 25–6, 28, 34, 52, 119, 125, 173, 202, 221–2, 227, 232, 290, 397, 420, 428, 431, 443, 452, 598, 612, 614, 621, 625
Electrophoresis 111, 140–1, 220, 263, 418, 427, 552, 583
buffer 214, 394, 399, 625
medium 312, 620
Electrophoretic 6, 532
behavior 255
buffer 517
environment 271
EOF 3
heterogeneity 666
injection 678
migration 2, 233, 270, 291, 477
mobility 2, 3, 29, 31, 33–4, 42, 178, 180, 183, 200, 212, 251, 264, 273, 292, 310–11, 418, 450, 487, 509, 511, 523, 532, 551, 631, 665, 678–9
mode 4
process 679
run 229
separation 125, 181, 315
separation capillary 22, 482
transport 669
velocity 552
Electrophoretically 437
mediated
microanalysis 22
mediated microassay 214
Electrospray 96
interface 684
ionization 95
Electrospray-chemical ionization 646
Electrostacked band 63
Electrostacking 545
procedure 639
Electrostatic
force 4
interaction 57, 79
repulsion 78, 517

Eleutherococcus senticosus 596
ELISA 249, 316, 455, 500
Ellipsoidal mirror 87
Elution
order 314
range 283, 632
Emetine 313
Emission
band 229
light 107, 114
spectrometry 500
Emitted fluorescence 479
Emodin 597
Enalaprilat 314
Enantiomer 72, 75, 212, 334, 347, 351–2, 355–6, 367–9, 372–3, 376, 379, 402–3, 650
bamethan 355
chlorpheniramine 352
metoprolol 351
mixture 78
verapamil 355
warfarin 355
Enantiomeric 273
binaphthyl phosphate 560
compound 366–7
impurity 301, 338
mixture 366, 367
purity 334, 341
recognition 78, 367
separation 78, 334, 367, 492
Enantioselective 342, 372
MEKC technique 378
separation 367
Enantioselectivity 366, 376, 378, 382
Enalapril 314
Enatiomeric purity testing 51
End-column
amperometric detection 92
chemiluminescence detection 90
detection 91–2, 624
Endo
exopeptidase 211
Endo-β-N-acetyl-glucosaminidase F 198
Endocrine
gland 255
Endocrinology 219
Endogenous 486, 492

ammonium 500
compound 114, 391, 411
fluorescent compound 224
neurotoxin 242
opiate 262
peak 339
substance 386
Endonuclease digest 293
Endopeptidase 211
Endorphin 255, 262–3
Energy-transfer 476
Enilconazole 360
Enkephalin 255
Entangled
polymer 457
polymer solution 552
Entanglement 457
Enthalpy 647
Entropy 647
Environmental 219, 550
analysis 531
carcinogen 219
sample 639
water 646
Enzymatic
activity 417
analysis 207
desialylation 184
digest 317
engineering 124
immunoassay method 399
oxidation 22
sequencing reaction 478
transformation 212, 341
Enzymatically hydrolyzed starch 625
Enzyme 207, 264, 270, 425, 436–7, 488
activity 207–8, 219
assay 207, 419
catalysis 208
immunoassay 650
labeled alkaline phosphatase 242
profiling 124
reactor 63
substrate 88
Enzyme–antibody complex 419
Enzyme-catalyzed multiple phosphorylation site 208
Enzyme-labeled monoclonal antibody 242

Enzyme-linked
  immunoassay 207
  immunosorbent
    assay 316, 455
Enzyme-modified
  capillary reactor 22
EOF 2–3, 6–7, 22, 34, 42,
    56, 177, 179, 181,
    184, 199, 275, 347,
    376, 430, 440, 444,
    514, 531, 535, 552,
    583, 645, 669, 678
  mobility 63
  modifier 531
  reversal 347
  stabilization 42
  suppressor 185
  velocity 56
Eosin
  bis (2,4,6-
    trichlorophenyl)
    oxalate-H_2O_2
    system 89
  bovine serum albumin
    complex 89
  Y 557
Eperisone 352
Ephedra herba 315, 597
Ephedrine 73, 315, 341–2,
    347, 352, 372, 379,
    520, 534
  alkaloid 315
  compound 315
  enantiomer 352, 379
Epillumination
  confocal optics 110
  FL microscope 107, 110
Epinephrine 316, 341, 420
Epitope 231, 233
Equal concentration 271
Equilibrium 273, 275
  concentration 273
  condition 271
  constant 349
  dialysis 270
  reaction 270
  species 271
  stipulation 271
Equine insulin 264
Equipment system 15
$ERBB_2$ oncogen 477
Ergovaline 597
Erythrocyte 74, 418–19,
    422
Erythroid progenitor
  cell 198, 265
Erythropoietin 198, 265
Erythrosine 610–11, 614

Essential amino acid
  composition 593
Estrogen 595
Etched tip 420
Etching 92
Ethanolamine 315, 560
Ethidium
  bromide 76, 292–3, 476,
    478
Ethosuximide 394
Ethoxylated alcohol sulfate
  surfactant 668
Ethoxylation 668
Ethyl acetate 283
Ethylene
  glycol 553
Ethylenediamine 34, 79
  tetraacetic acid 6, 509,
    520, 554
Europium 510
  (III) 510
Evaporation 46, 391
Excimer 97
  laser 97
Excitation 125, 222, 293,
    482
  beam 106
  emission
    monochromator 109
  intensity 89
  light 106, 107, 114, 684
  source 108, 109, 223,
    224, 684
  wavelength 87, 88, 89,
    114
Exogenous compound 486,
    492
Expectorant 314
Experimental
  design 340
  optimization 86
Explosive 639, 647
External electric field 57
Extinction coefficient 221
Extracellular
  fluid 411
  level 149
Extraction 283, 387, 391,
    487, 649
  process 54
Extraneous injection 19

F_2 Generation 475
Fab 221
  antibody 222
  arm 247
  fragment 220, 222, 223,
    263

Factor VIII 204
Famotidine 399
Fast off rate 273
Fast atom bombardment
  interface 96
Fast-scanning
  detector 682
  multiwavelength UV
    detector 114
Fat 264
  hydrolysate 554
Fatty acid 554, 591, 598
FD&C
  red 615
  yellow 615
FDA 394
Fe 184, 373, 510, 517–19,
    556
Felbamate 386, 390, 394
Fenoldopam 372, 379
Fenoprofen 318, 356
Ferguson
  analysis 247
  plot 181
Fermentation 590, 597
  broth 74, 560, 594
  media 555
Fertility panel 231
Ferulic coumaric acid 596
Feruloyl
  synapyl glucoside 596
  glucoside 596
Fetus 198, 265
Fiber optic array 125
Fiber–optic-based UV
    detection cell 118
Field
  decoupler 92, 95, 118
  modulation 119
  programming CE 26
  strength 361, 683
Field-amplified injection
    138, 651
Filter 87, 107
Filtration 138, 499
Fish sarcoplasmic protein
    594
FITC 22, 88–9, 112, 417
  derivatized 150
FITC-labeled
  arginine 112
  insulin 419
Fk506 387
FL 110
  derivatization 112
  detection 104, 108, 111–
    12, 114
  emission 106–10

labeled amino acid 112
labeling reagent 112
microscopy 107
Flame
  atomic spectroscopy 543
  detector 387
  spectrometry 500
Flavonoid 314–15, 589, 596, 598, 609
  aglycone 596
  3-O-glycoside 528, 596
  7-O-glycoside 596
Flavor 609
Flavoxate 399
FLEC 368–9
Flexibility 271
Flow
  atom bombardment ionization 96
  counterbalanced CE 125
  direction 106
  profile 119
  rate 96
Flowcell 104, 106
  arrangement 107
Fluctuation 33
Fluor 227
  label 223, 232
Fluoranthene 630
Fluorescamine 87, 318, 394, 416, 420, 588
  labeled amino acid 593
Fluorescamine-derivatized sample 87
Fluorescamine-labeled peptide 207
Fluorescein 87, 114, 178, 293, 419
  agent 557
  donor 476
  isocyanate 598
  isothiocyanate 22, 88–9, 417
Fluorescein-di-β-galactopyranoside 212
Fluorescein-ethylenediamine 283
Fluorescein-ethylenediamine-deoxyadenosine-5′-monophosphate 283
Fluoresceinisothiocyanate 112
Fluorescein-labeled anti-IgA F(ab*)2 fragment 222

cortisol oxime derivative 223
insulin 221
myoglobin 178
Fluorescence 2, 74, 84, 87–8, 442, 479, 589, 618, 631, 683
  capillary array scanner 125
  detection 86–8, 104, 123, 207, 293, 369, 416, 419, 442, 450, 559, 566, 625, 666, 683
  detector 87, 419, 684
  labeled DNA 125, 482
  polarization 219
  quenching 270
  scanner 88
  signal 481
  spectrophotometry 293
  tagged oligonucleotide 476
Fluorescent 110, 416, 550
  analyte 221, 452
  antibody 418
  derivative 112, 264, 437
  derivatization 107
  dye 125, 237, 250, 293, 452, 684
  label 221, 224
  labeled amino acid 122
  labeled primer 476
  labeling 111–12, 476
  marker 125
  oligonucleotide 482
  protein 224
Fluorescent-labeled amino acid 22
  antigen 488
Fluoride 309
Fluorinated ethylene–propylene copolymer 597
Fluor-labeled antigen 221
  antibody 232
Fluorocarbon 680
Fluorogenic reagent 112
Fluorohydrocarbon resin 1
Fluorophore 125, 173, 283, 452, 454, 456, 482, 559, 683
Fluorophore-conjugated primer 479
Fluorospectrometer 87
Fluor-to-protein ratio 233
Fluoxetine 355

Fluparoxon 308
Flurazepam 393
Flurorescently-labeled sialyllactose 211
Flush 610–11
Flushing 36, 611
Flusting 610
FMOC 345
FMOC-derivatized amino acid 150
FMOC-peptide 275
Focused zone 185
Focusing 110
  lens 682
Folinate 595–6
Folinic acid 595–6
Follicle stimulating hormone 255, 263
Follitropin 231
Food 219, 550, 553, 583, 607
  additive 597
  analysis 531
  component 589
  dye 616
  lipid 595
  poisoning 597
  protein 593–4
  toxin 219
Foodstuff 609
Forced air 34
Forced-cooling function 34
Forensic
  analysis 386, 403
  application 403–4
  drug analysis 404
  drug testing 219
  identification 453–4
  sample 368
  science 315, 403
  significance 352
  testing 231
Formamide 62, 73
Formate 553
  buffer 444
Formate/tartarate 553
Formic acid 443, 528
Formulation 301
  stability testing 334, 341
Fosfomycin 399
Four sequencing reaction 479
Fourier transform ion cyclotron resonance MS 96
Fraction 681
  collection 63, 118, 418, 667, 685–6

INDEX 713

collector 118
dilution 118
Fractional
   factorial design 308
Fracture 22
Fragment
   analysis 449–50
   length 294
   mobility 452
   size 466
Frame photometric
   detection 97
Free 488
   antigen 219–20, 229
   cyanide 111
   energy 632
   galacturonic acid 92
   labeled antigen 224
   ligand 270, 273, 509
   unlabeled antigen 221
   zone capillary
      electrophoresis 4,
      283, 666
Free solution capillary
      electrophoresis 583
Freedom
   interference 334, 339
Free-zone 212
   capillary
      electrophoresis 1,
      199
Frequency
   doubling 86, 111
   tripled 88
Fresh
   reconstituted milk 593–4
Fresnel's law 105
Frictional
   coefficient 314, 449
   force 3
Frontal
   analysis 273
Fronting 275
Frontoparietal cortex 149
Fructose 593, 607
Fructuation 20
Fruit 595
   beverage 595
   juice 555, 591, 595
FSH 255, 263
Fucosyltransferase 211
Fullerenes 631
Full-scan collision 96
Fulvic acid 555
Functional
   group 523, 609, 611,
      632

Fused
   silica 1, 56, 104, 531,
      550, 583, 596, 611
   silica capillary 1, 2, 56,
      84, 179, 552, 561,
      682
   silica optical fiber 114
FZE 4, 29, 33, 283, 292

G Philadelphia 440
G-6P 215
G-6-PDH 213–15, 419
GABA 394, 416
Gabapentin 394
Galactitol 620
Galactose 620
Galactosyl residue 199
Galacturonic acid 592
Gal-β (1-3) Gal NAc 202
Gallbladder 264
Gamma
   CD 345, 361, 561, 629
   globulin 193, 425, 427–8
   GTP 214
   region 430, 434
Ganglioside 416, 590
Gas chromatography 95,
      386, 607
Gas-phase ion 96
Gastric acid 264
   inhibitory polypeptide
      264
   secretion 264
Gastrin 264
Gastrointestinal tract 255,
      264
GATT 477
Gaussian peak 26
GC 95, 386, 388, 551, 607
GC/MS 443
Ge 140
Gel
   electrophoresis 122, 140,
      291, 664, 681
   filled capillary 437
   filled CE 667
   filtration 232
   instability 61
   isoelectric focusing 430,
      440
   matrix 77, 668
   permeation 657
Gelborange 611, 614
Gel-filled
   capillary 54, 61–2, 181,
      291–2, 294, 476, 483
   column 56
Gene 449, 451, 477

analysis 451, 458, 480,
   483
expression 503
mutation 476
sequence 451
Genetic
   disease 453, 488, 492
   loci 475
   screening 452
   typing 475, 481
Genotyping 455, 475, 477
Geometry change 66
Gland 255, 419–20
   secretion 493
Glass 1
   chip 122
   coupler 92
Glaucoma 316
Gliadin 594
Globin 440
   chain 440
Glomerular filtration 397
Glucagon 264
Glucaric acid 620
Glucitol 620
Glucomannan 77
Gluconate 442
Gluconeogenolysis 264
Gluconic acid 592, 620
Glucopyranose 8
Glucose 264, 283, 369,
   493–4, 592, 607,
   620–1
   oligomer 355
Glucose-1-phosphate 208
Glucose-6-phosphate 215,
   437
   dehydrogenase 124,
      213–14, 419, 437
Glucuronic acid 620
Glutamate 114, 149, 416,
   420
Glutamic acid 149–50, 345
Glutamine 345
Glutamyltranspeptidase 214
Glutaraldehyde 63
Glutathione 124, 208, 419,
   437, 488
   peroxidase 208, 437
Glutenin 589
Glutethimide enantiomer
   378
Glycerol 292, 476, 478, 590
Glycine 112, 149, 345, 420,
   428, 534, 594
Glycoform 212, 265
   heterogeneity 198
Glycogen 264

Glycogenolysis 264
Glycol modified fused-
      silica capillary 179
Glycolic acid 73, 521, 534
Glycopeptide 200, 202, 205
Glycoprotein 178, 184,
      212, 247, 351, 353,
      356–7
   alpha-1-acid 353, 356
   hormone 198, 265
Glycosidic enzyme 212
Glycosylated 240, 244
   peptide 200
   rEPO 198
Glycyrrhizae radix 314
Glycyrrhizic acid 376
Gly-Gly-Phe 438
Glyphosphate 649
Goat 594
Gonadorelin 311
Good sensitivity 682
Gr50360a enantiomer 355
Gr57732a 358
GRA 376
Gradient
   conductivity 21
   elution 551
   LC assay 312
Grain 559, 597
Grape 592
   juice 555
Grass peas 598
Gravity 430
   flow 17
   injection 17, 679
Green
   color 612
   He-Ne laser 223
Ground state 207
Group
   Ib 521
Growth 198
Guanidine compound 499
Guanidinium 174
Guanidinoacetic acid 499
Guanidinosuccinic acid
      499
Guest-host complex 345,
      349, 356
Gunshot 561

H_2 dapmp 519
Hair analysis 386, 404
Half life 397, 492
Haloacetate 558
Halogen 86
   lamp 123
Haloperidol 399, 416

Harmful effect 647
Hb 440, 442, 443
Hb_A 440
Hb_{A1C} 440
Hb_{A2} 442
HBED 510
HbF 440
hCG 265
HCV-specific DNA
      product 475
HDL 436
Headspace 679
Heat 452
   dissipation 63, 125, 173,
      366
   generation 71, 119, 673
Heat-damaged milk 594
Heavy
   metal 651
   oil 630
He-Cd
   laser 87–9, 109–10, 112,
      114, 222
   laser detector 353
   laser excitation 630
   laser fluorimetry 630
HeLa cell 199
Helium gas 180
Helium-Neon 88
   green laser 110
   laser 87, 88, 97, 123,
      224, 479
Heme 492
Hemoglobin 184, 193, 419,
      425, 431, 434, 438,
      440, 445
   C trait patient 177
   variant 193, 440, 445
Hemolysis 434
Hene 111
Heparin 78, 402
   disaccharide 558
Hepatic
   oxidative function 492
   phenylalanine
      Hydroxylase 477
Hepatitis C 431, 475
Heptakis-β-CD 316
Heptane-SDS-butanol
      buffer 76
Herbal drug 314
Herbicide 550, 556, 559–
      61, 597, 639, 649,
      669
   terbutylazine 597
Heritable disease 453
Heroin 315, 316
Hesperidin 596

Heterobasidion annosum
      452
   rRNA gene 478
Heterocyclic
   nitrosoamino 646
Heteroduplex 452, 477–8
   analysis 452
   mobility assay 452
   molecule 452
   polymorphism
      analysis 478
   polymorphism assay 476
   single-base substitution
      478
Heterogeneity 223–4, 243,
      657, 658
Heterogeneous 202
   behavior 275
   immunoassay 219
   material 657
   oligo DNA 291, 477
   size 475
Hexadecylally-dimethylam-
      monium 665
Hexadimethrine
   bromide 534, 557
Hexamethonium
   bromide 265, 558
   hydroxide 534
Hexanesulfonate 555
Hexanoate 554
Hexanol 632
Hexobarbital 353
Hexokinase 63
HIBA 520
High
   background 87
   performance capillary
      electrophoresis 500
High-impedance circuit 95
High-molecular-weight
   polysaccharide 111
High-performance
   liquid chromatography
      137, 366, 417, 607,
      617
Hinge region 222
Hippocampus 416
Hippurate 499, 555
Hippuric
   acid 404, 442, 503
   and orotic acid 598
Histamine 92, 598
Histidine 112, 150, 174
   buffer 555
Histidine-rich polypeptides
      493
Histone 178

HIV
  antibody 493
  DNA 475
  subtype 452
HIV-1 455
HIV-positive patient 418
Homo
  heteroduplex 452–3
  heterooligonucleotide 291
  heterozygosity 477
Homocysteine 442
Homocystinuria 442
Homoduplex 452
Homogenate 417
Homogeneity 227
Homogeneous
  immunoassay 219
Homologous series 31
Homopolymer 666, 668
Honey 596
  flavonoid 596
Hops bitter acid 591
Hormone 223, 255, 583, 595
Host-guest
  complex 8, 349
  complex stability 8
  inclusion-complex
    stability 8
HP-β-CD 345, 347, 352–3, 361
HPCE 500
HPLC 25, 112, 137, 184, 232, 308, 366, 387–8, 399, 417, 425, 427, 437, 486, 500, 551, 589, 591, 597, 598, 607, 617, 681
  FL detector 109
  FL flowcell 109
HPMC 150, 357, 560
HQS 510, 518
$HSeO_4^-$ 96
Human
  apolipoprotein β
    gene 477
  blood 624
  cerebrospinal fluid 150
  chorionic gonadotropin 265
  cord plasma 442
  disease 270
  erythrocyte 411, 418
  genome initiative 480
  genome project 451
  growth hormone 220, 222, 317
  hemoglobin 438

IgG heavy and light
  chains 247
IgG-protein G 219
lymphoma cell 290
mitochondrial
  dinucleotide
    repeat 475
mtDNA 475
plasma 355
recombinant
  erythropoietin 178
red cell glucose
  transporter 193
serum 177
serum albumin 150
serum transferrin 193
short tandem repeat 475
tear fluid 221
tissue plamin activator 204
TK-6 lymphoblasts 478
urine 198, 360, 625
Humanized
  anti-Tac monoclonal
    antibody 245
  monoclonal antibody
    anti-Tac 317
HUMF13a 454
HUMFES 454
Humic acid 555
Hummel-dreyer method 271
HUMTHO1 454, 475
HUMvWA 454
HVA 442
Hybridization 293, 452
Hybridized species 476
Hybridoma medium 247
Hydrocarbon 629
Hydrochlorothiazide 314
Hydrocortison 311
Hydrodynamic 139
  flow 2, 18, 123, 141, 314
  injection 16, 17, 20, 36, 46, 53, 313, 318
  mode 34
  pump 9
  radius 551
Hydrogen
  bond 71, 77
  bonding 347, 352
  peroxide 90, 283
Hydrogen-bond
  interaction 632
Hydrolysis 309
  product 314
Hydrolyzed dye 648
Hydromorphone 224

Hydrophilic
  aromatic compound 632
  coated capillary 61
  compound 76
  head 373
  hydrophobic face 373
Hydrophilicity/lipophilicity 74
Hydrophobic
  bond 4
  compound 390, 595
  hydrocarbon 629
  interaction 367, 428, 519
  interior cavity 8
  olefinic residue 179
  polymer 671
  property 487
  region 181
  tail 373
Hydrophobicity 7, 56, 425
Hydrostatic 535
  pressure 679
Hydroxo form
  lanthanide complex 523
Hydroxy acid 361
  group 609
  hippurate 499
Hydroxybenzoate 554, 558
Hydroxyethyl
  methacrylate 179
Hydroxyethylcellulose 77, 122, 182, 482, 534, 555, 558, 669
  cluster 77
  molecule 77
  solution 77
Hydroxyindole
  acetic acid 442
Hydroxyisobutyric acid 520, 534, 560
Hydroxylamine 560
Hydroxylpropyl 353
  methylcellulose 77, 150, 178, 182–3, 357, 452, 560, 594
5-hydroxymethyl-2-
  furaldehyde
Hydroxyphenolic acid 499
Hydroxyphenylamine 347, 349, 351
  enantiomer 349
Hydroxyproline 443
Hydroxypropyl-β-CD 72, 315, 340, 371, 372
Hydroxypropylcellulose 77, 179

Hydroxy-substituted
 pyridinecarboxylic
  isomer 79
 steroid 373
Hyperkinetic behaviour
 disorder 352
Hyperlipidemia 401
Hypertension 313, 314
Hyphenated technique 243
Hypothalamus 255, 263
Hypoxanthine 425, 442,
 503, 593

I⁻ 92
Ibuprofen 356, 397, 405
IC 143, 500, 551
Ice cream 609
ICP 500
ICP-MS 531
Idazoxan 417
Identification 118, 220,
 310, 315, 352, 372,
 402, 418, 682
IDL 436
IEF 174, 178, 199, 245, 588
IgA 221, 222, 231, 430
 specific F(Ab)2 250
IgD 430
IgE 430
IgG 221, 230–1, 243, 247,
 430
 heterogeneity analysis
  240
IgM 231, 430
IgS 475
 fragment 475
ILe 345
Illicit
 drug 315, 352, 404
 heroin 315
Illness 486
Imafen 355
Image-splitting prism 125
Imaging
 detector 121, 123
Imidazole 6, 73, 174, 520–
 1, 534, 560–1, 590
Imidazole-H_2SO_4
 background
  electrolyte
  system 521
Imidazole–sulfuric acid
 buffer 590
Imidazolinone 566
 herbicide 566
Iminodiacetic
 acid 521
Imipramine 312, 393

Immersed flowcell 107–8
Immersing liquid 109
Immersion capillary 97
Immobiline compound 664
Immobilization 57, 278
Immobilized
 affinity analysis 278
 capillary enzyme reactor
  63
 carboxypeptidase Y 213
 enzyme 213, 598
 protease 213
 urease 500
 yeast cell 124
Immuno affinity
 capture 221
 CE 220–1, 240
Immunoassay 219–20, 251,
 263, 387–8, 393, 401,
 404, 436, 438, 488,
 494, 500, 597
 detection 419
 method 393
 procedure 419
 technology 237
Immunochemical 315, 488
 binding 231, 233
 CE analysis 222
Immunocomplex 417
Immunoconcentration 221
Immunodiffusion 270
Immuno-fixation 425,
 430–1
Immunofluorescence 597
Immunoglobulin 425, 428,
 430–1
Immunologically bound
 antigen 219
Immunometric
 direct light absorbancy
  method 436
Immunoprotein panel 231
Immunoreaction 234
Immunosubtraction 221
Immunosuppressant 392
Implement 92
Imprecision 402
Impurity 301, 610, 614,
 664
 analysis 301, 308
Incident
 angle 106
 light 87, 106
Incubation 437
Indeno[1,2,3-cd] pyrene
 630
Indigo carmine 610
Indigotin 611

Indirect 111
 absorbance detection
  683, 684
 detection 45, 53, 90,
  111–12, 345, 351,
  367, 369, 399, 405,
  419, 442, 519, 552,
  553
 fluorescence 558, 559
 fluorescence detection
  74, 111–12, 419,
  533–4, 560, 632, 639
 fluorometric detection
  590
 laser-induced
  fluorescence
  detection 111, 595
 method 367, 369
 photometric detection
  150, 590, 597, 648–9,
  651
 semiconductor laser
  fluometry 632
 separation 368
 spectroscopy 78
 UV 6, 71, 72, 111, 309,
  311, 531–3, 550, 554,
  560, 566, 589, 590,
  591, 646, 647, 665,
  667–8, 670
 visible detection 554
Indole 499
Indoxy sulfate 499
Inductively
 coupled plasma atomic
  emission 97, 500
 coupled plasma mass
  spectrometry 531,
  533, 651
 coupled plasma optical
  emission
  spectrometry 533
Industrial
 analysis 531
 chemical sample 550
 waste water 645
Inert
 gas 184
 gas pressure 185
Infection 309, 427, 434
Infectious disease 219
Influx 416
Ingredient 309–10, 313,
 436
Inherited metabolic
 disorder 443
Inhibitor 270, 314, 437
Inhibitor–antibody 248

INDEX

Inhibitor–protease
   association
   complex 248
Initiator 61
Injection 278, 294, 611, 667
   amount 138
   artifact 46
   channel 482
   device 673
   efficiency 678–9
   error 278
   length 32
   precision 36, 338
   pressure 679
   process 19, 679
   reproducibility 18, 20, 34, 36
   time 19, 72
   volume 17, 19–20, 402, 551
Injector linearity 51
Inlet 678
   electrode 678
   end 2, 19
Inner
   diameter 1, 92
   radius 17
   surface 56, 61, 119
   wall 2, 56
Inorganic 499, 590
   anion 72, 531–2, 535, 590, 648
   cation 6, 96, 531–2, 535, 646
   ligand 510
   pigment 609
Inositol 620
   hexaphosphate 178–9
   phosphate 74, 560, 593
Insolubility 137
Insoluble
   polymer 658
Installation 682
Instrumental
   technique 617
   variable 119
Instrumentation 456
Insulin 264, 317, 419
   secretion 419
Insulin-like growth factor 1-binding protein 477
Integral 278
Integrated detector 673
Integration 26
Intensity 227
Interaction 75, 174, 270, 275, 347, 445

Intercalating 684
   agent 76
   dye 142, 292–3, 476
Intercalator dye 212
Intercellular variation 419
Intercept 309
Inter-company
   cross validation 37, 308, 338
Interface 22, 95
   condition 96
Interfacial pressure
   difference 19–20
Interference 339, 683
Interfering substance 138
Interferon 317
Interferon-related 2–5$_A$ system 437
Intergenic spacer 475
Interleukin-1 317
Interleukin-4 317
Internal
   diameter 104, 682
   radius 18
   reflection 105
   size standard 452
   standard 20, 36, 50, 51, 139, 278, 293, 309, 338, 341, 392, 394, 402, 416, 430, 442, 450, 492
   transcribed spacer 475
Intersection point 123
Intoxication 487
Intra-assay reproducibility 229
Intra-strand conformation 451
Invertebrate neuron 420
Iodide 532, 533
Iodinated compound 399
Iodine 227, 587
Iodoacetamide 222
Iohexol 399, 405
Ion 583, 590
   association agent 519
   charge 3
   chromatography 143, 500, 531, 551
   exchange 607
   pairing 607
   paring effect 517
   pairing interaction 552
   pairing reagent 517, 609
   spray-sheath flow interface 96

Ion-association capillary
   electrophoresis 511, 519
Ion-exchange 232, 617
   interaction 534
   membrane 91
Ion-exchanger 555
Ionic
   micelle 7
   migration 617
   polymer 631, 668
   size 551
   strength 4, 33, 73, 119, 243, 262, 313, 318, 391, 394, 427, 444, 535, 552, 666, 678, 679
   strength buffer 139, 175, 590
   substance 4
   surfactant 179
Ionizable functionality 617
Ionization 227, 438, 551, 620, 667
   constant 29
   equilibrium 29
   MS 310
Ionized air 57
Ion-pair HPLC 140
Ion-pairing 534
   reagent 310, 511, 610
Ion-selective
   electrode 92, 500, 533
Ionspray MS detection 318
Ion-spray
   atmospheric pressure 310
Iopiperidol 309
Iothalamic acid 399, 444
Iron, see Fe
Irreversible adsorption 45, 139
Ischemic heart disease 313
Isis 2922 291
Islets of Langerhans 222
Isocitric acid 591
Isoelectric
   focusing 140–2, 174, 178, 183, 199, 263, 427, 588, 664
   mode 12
   point 12, 25, 31, 123, 141, 198, 427, 588
Isoenzyme 438
Isoflavone 596
Isofocusing 430
Isoform 175, 183, 200, 212, 243, 265, 445

Isolectric focusing 430
Isomer 349, 352, 357, 369, 401, 402, 403, 598
Isomeric dichlorophenol 644
Isomerization 301
Isoorientin 596
Isoprenyl pyrophosphate 111, 560, 595
Isopropanol 557, 560
Isoproterenol 352
Isotachophoresis 21, 404, 430, 436, 531, 588
Isotachophoretic
  sample concentration 16
  zone sharpening 144
Isothiocyanate
  derivatization method 560
  group 114
Isovue 399, 405
Isozyme 142, 215
Ithium 583
ITP 21, 309, 515, 531, 535, 550, 588, 591
ITS 475
IVa metal ion 521

JA 125
Joint 124
Joule (heat) 45, 73, 75, 97, 149, 173, 181, 450, 679
Jugular vein 421
Juice 591

K 431, 528, 590
Kaempferol 596
K_b 31
Ketamine 353
Ketoconazole 360
Ketone 629
Ketose 592
Keyhole limpet
  hemocyanin 233
Kidney 198, 255, 265, 318, 397, 431
  disfunction 493
Kinetic 208
  analysis 658
  constant 273
  parameter 310
  rate 270
  stability 509
  study 680, 681
Kinetic 208
Km 213

Known point-mutation 478
Kraft lignin 557
Krypton fluoride laser 97
Kynurenic acid 416
Kynurenine 411

L isomer 402
L-Menthoxyacetic acid 370
L/D-Marfey reagent 368
LA2 114
Labeled
  antigen 223–4, 227
  competing species 233
  FITC 480
  T_4 227
  theophylline-antibody complex 229
Labeling reagent 232
Laccaria
  bicolor 475
  laccata 475
Lactate 6, 74, 419, 522
  dehydrogenase 216
Lactic
  acid 522, 528, 534, 590, 591
  fermentation 590
Lactococcus lactis 438
Lactoglobulin 175, 594
Lambda
  chain 231
  light chain monoclonal gammopathy 221
Laminar 2
  flow 84, 119
Lamotrigine 394
Langerhans 420
Lanthanide 510, 520, 522
  metal ion 509, 521
Lanthanoid (III) 520
LAS 648
  surfactant 648
Laser 87, 684
  beam 86, 109
  diode 88, 90, 97
  doppler
    electrophoresis 665
  dye 649
  excitation 110, 630
  fluorimetry 630
  induced fluorescence 263, 311, 401, 434, 557
Laser-based
  detection 452

fluorescence detector 684
Laser-excited confocal
  fluorescence
    capillary array
      scanner 481
Laser-induced
  detection 371
  fluorescence 87, 112, 118, 123, 137–8, 202, 214, 216, 221, 263, 283, 399, 405, 443, 454–6, 476, 488, 589
  fluorescence (LIF)
    detection 109, 183, 211, 251, 684
  fluorescence detection 109, 369, 420, 597
  vibration 97
Latex particle 219, 666
Laudanosine 379
Lauric acid 428
LDH 438
LDL 436
Lead 651
  ion 534
Leading 12, 404, 535
  electrolyte 645, 650
    terminating buffer 21, 436
    terminating and running electrolyte 21
Least square regression 53
Lectin-sugar 219
Length 4, 621
  variation 453
Lens effect 86
Lentivirus 212
Leucine 345
  aminopeptidase 214, 437
Leucinostatin 317
Leukemia 399
Leukocyte-A 317
LHRH 263
LIF 87, 118, 202, 214, 216, 221, 263, 283, 455, 476, 488
  detection 456
LIFD 109, 112
Ligand 32, 270, 273, 275, 511, 520
  complex 273
Ligase chain reaction 476, 478
Light
  absorbency 436

INDEX 719

beam 110
chain 430–1
emmission 150
path 17, 581, 589, 682
scatter 88
scattering 77, 97, 104, 219, 419
source 86–8, 123
Light-control feedback 86
Lignin 557
Lilly drug 339
Limit
  detection 111, 112, 312, 372, 379
  individual 631
Limitation 678
Linear
  acid 361
  alkylbenzene sulfonate 558, 668
  amine 3
  CCD 123
  chain 450
  glycosaminoglycan 78
  hydrophilic polymer 594
  limit 234
  oligosaccharide 355
  polyacrylamide 76, 141, 174, 477, 480
  polymer 452
  polymer sieving 3, 9, 29, 122, 466
  polymer solution 466
  range 51
  regression analysis 53
  relationship 31, 53
Linearity 53, 250, 334, 338, 372, 403, 444, 616
  detector response 334
Lineweaver-Burk analysis 278
Lipid 583, 589, 595
Lipophilic
  chain 75
  dye 436
Lipoprotein 425, 436, 445
Liquid
  chromatography 95, 111
  cooling 34
  dosage formulation 313
  handling 680
  ion-exchange resin 114
  junction-ion spray interface 667
  junction interface 95
  sheath flow 684

thermostated capillary 34
Liquid-junction coupling 96
Lissamine 20, 434
Lithium 583
  chloride 317
  phosphate buffer 561
Liver 198, 265, 430
  oxidative function 401
Living cell 207
LOD 111–12, 372, 379
Longer
  chain surfactant 647
  DNA fragment 479
Longer-chain
  alcohol 557
Longitudinal
  coordinate 33
  diffusion 32
Lorazepam 404
Lot variation 373
Low
  concentration 237
  detection limit 20
  molecular weight ion 591
  pH buffer 347
  solvent/buffer consumption 551
  viscosity 18, 139
Lower
  directivity 88
  optimum electric field 479
Low-melting agarose 77
Low-mobility component 667
Low-molecular-mass organic ion 535
L-Phenylalanine 124
L-Pseudoepinephrine 341
Lubricating layer 458
Luciferase
  bioluminescence 90
Luminescence 90
  detection 510
Luminol 90
  chemiluminescence 90
Lumino-labeled compound 89
Lung 418
Luteinizing
  hormone 255, 263
  hormone releasing hormone 263, 272, 417
Lutropin 231

Lymphoma cell 455
Lyophilized human hemoglobin 443
Lysate 418
Lysine 150, 174, 345
Lysosomal enzyme 438
Lysozyme 175, 318

M13 sequencing primer 114
M13mp18 DNA fragment 480
Macrocyclic
  antibiotic 78, 349
Magnesium 520, 558, 560, 590, 591
  acetate buffer 534
  chloride 183
Magnetizable bead 219
Maillard reaction 594
Major hemoglobin variant 440
Malachite green 90
Malate 554
Malic acid 591
Malignancy 488
Malonate 554, 555
Maltodextrin 78, 355, 356
Maltoheptaose 204, 621
Maltose 607, 620, 621
Mammalian cell 124, 149
Mammary gland 255
Mandalic acid 521
Manipulation 291, 421, 682
Mannitol 620
Manual reversal 678
Mapping 22
Maprotiline 312
Marijuana metabolite 230
Marine origin 651
Mask 180
Mass
  charge ratio 317, 583
  sensitivity 450
  spectrometry 202, 310, 387, 533, 551, 557, 559–61, 589, 684
  spectroscopy 2, 95
  transfer rate 379
  transport 657
  transport equation 120
Matching
  fluid 97
  glue 97
Mathematical
  analysis 682
  model 56

Matrix-assisted laser
    adsorption
    ionization 96
  assisted laser desorption
    MS 247
Matrix 62
  assisted laser desorption
    ionization 202
  component 683
  effect 390
Maturation 198
Measurable difference 270
Meat extract 310
Me-β-CD 345, 355
MECC 174, 183, 240, 247,
    390, 393, 404, 417,
    434, 486, 550
Mechanical stability 682
Mechanism hydrolysis 310
Medicarpin 376
Medium 3
Medium-chain acyl-
    coenzyme A
    dehydrogenase 477
MEEKC 557, 632
Mefloquine 360
MEGA 373
MEGA-borate micellar
    phase 632
MEKC 3, 6, 111, 263–4,
    367–9, 373, 376, 550,
    588, 632
Membrane 118
  collection 214
  interface 22
Mephenytoin 371, 372
Mephobarbitol 357
Mercaptoethanol 180
Mercury 651
  lamp 89
Mesh
  formation 77
  size 77
Metabolism 342, 371
Metabolite 264, 360, 421,
    445, 488, 492
Metacerebral,
    B1 420
Metal 517, 590–1
  additive 552
  cation 72
  chelate 510, 519, 521
  chelate capillary
    tube 521
  chelation 517
  chelator monomer 665
  complex 509, 515, 521
  cyanide 511, 518

cyano complex 518
  electrode 618
  ion 74, 492, 509, 515,
    531, 534, 590
  mask 122
Metal–(PAR) 517
Metal–ligand
  exchange
    chromatography 379
  interaction 552
Metallo-cyanide
    complex 510, 511
Metallothionein 183
  isoform 418
Metaphosphoric acid 591
Metaproternol 352
Metastasizing 438
Methacryl residue 179
Methamphetamine 368
Methanesulfonate 554
Methanol 183, 554, 590,
    595, 609, 610
Methanolic sodium
    hydroxide solution
    670
Methaqualone 403, 404
Methionyl recombinant
    human growth
    hormone 222
Method 393
  development 32, 250,
    316, 382
  development kit 250
  development time 250
  robustness 334
  transfer 334, 340
  validation 301, 334
Methotrexate 399
Methoxylated agarose 77,
    182
Methydimethoxy
    amphetamine 352
Methyl
  β-CD 72
  amphetamine 352
  cellulose 243
Methyl-2-benzylamine 534
Methylcellulose 77, 179,
    184, 185
Methyldimethoxyethyl-
    amphetamine 352–
    3
Methylene 62
  bisacrylamide 61, 141,
    179
  blue 112
  group 31
  unit 667

Methylflavone carboxylic
    acid 399
Methylglucamine 428
Methylguanidine 499
Methylhippuric acid 404
Methylhydroxyethyl-
    cellulose 180, 594
Methylmalonic acid 425,
    443
Methylpyridine 561
Methymalonic aciduria 443
Metol 519
Metomidate 353
Met-recombinant human
    growth hormone
    antigen 234
MFD11 455
m-hydroxy 3,4-
    dihydroxymandalic
    acid 521
Mianserin 355
Micellar
  agent 646, 649, 650
  concentration 72
  electrokinetic capillary
    chromatography 3,
    6, 111, 174, 240, 367,
    390, 417, 486, 511,
    550, 588
  emulsion electrokinetic
    chromatography 557,
    632
  phase 310, 632
  pseudo-phase 317
Micelle 7, 71, 74, 373, 487,
    555
Micelle-forming
    fluorocarbon tail
    457
Micelle-like
    pseudostationary phase
    76
Micelle-mediated capillary
    electrokinetic
    chromatography
    366
MEKC 367
Micellular 263
Michaelis–Menten
    constant 213
Michaelis–Menten kinetic
    278
Miconazole 360
Micro
  liquid chromatography
    95
Micro-bore
  capillary 583

INDEX 721

Microchip 122–3
  CE 123
  electrophoresis 122
Microcolumn
  separation 22
Microconcentric
  column 66
Microdialysate 411, 416
Microdialysis 22, 150, 411
  probe 149
Microdrop injector 19
Microelectrode 92, 624
Microemulsion 76, 379,
  632
  electrokinetic
    chromatography 591
Microfabricated
  capillary electrophoresis
    chip 482
  channel 216
  system 252
Microfabrication 20, 482
Microfluidic circuit 252
Micro-handling 22
Microheterogeneity 198,
  200, 202, 212
Microinjection 22
Microinjector 419
Micromachined
  CE chip 482–3
  glass chip 122
  planar glass structure
    482
Micromachining 121, 123
Micropreparative 685
  CE 66
  tool 685
Microreactor 213, 437
Microsatellite 454
Microscope 107
  charge-coupled device
    66
  objective 87
  optic 2
Microstructure 122
Microtechnique 543
Migrated zone 125
Migration 89
  behavior 62, 71, 118,
    291, 523
  buffer 317, 629
  data 34
  dispersion 119
  index 371
  order 179, 357, 369, 611,
    639
  order reversal 373, 376
  peak 20

property 291
rate 679
reproducibility 34
time 20, 25, 29, 31, 33,
  41, 45, 51, 54, 119,
  181, 275, 291, 308,
  338, 402, 427, 445,
  452, 480, 611, 620,
  647, 667
time reproducibility 338
time window 75
velocity 3, 7, 20, 25, 29,
  33, 63, 137, 138, 142
zone 89
Migration-rate 679
Milk 183, 255, 591
  casein 593–4, 598
  protein 180, 183, 593–4
  secretion 262
Mimic biologic function
  436
Mineral water 590
Miniaturization 90, 122,
  390, 482
Minute sample volume
  271
Miotic agent 301, 316
Mirror 86, 87
Mitochondrial DNA 475
Mixed buffer 403
Mobile
  cation 2
  phase heptane/SDS 557
Mobility 18, 73, 273, 310,
  449, 523, 533, 678
  difference 275, 451
Mobilization 123, 185
  step 193
Mode 95
Modeling 72
  migration 531
Modifier 57, 532
Modulation depth 119
Moiety 347
Molar
  absorption coefficient
    688
  absorptivity 86, 532
Molecular
  absorptivity 86
  biology 207, 476
  mass 198, 247
  medicine 270
  recognition 277
  sieving 291, 292, 449,
    466, 481
  sieving gel 76
  sieving media 141

sieving mode 25, 71
size 8, 181, 449, 625
sizing 475, 482
  DNA fragment 482
weight 25, 31, 95, 180,
  181, 240, 270, 657
Molecule 224, 227
Molybdate 533
Molybdenum 669
Mono
  diamide 554
  di-ethylhexylphosphate
    560
  di, tri-protic carboxylic
    hydroxy carboxykic
    acid 520
  disaccharide 620
  disialyl residue 202
  oligo saccharide borate
    complexe 528
Monochromater 109, 682
Monoclonal 431
  antibody 221–3, 229,
    243, 248, 392
  cryoglobulin 431
  gammapathy 427
  immunoassay 393
  nature 177
Monokis-6-O-
  octamethylene-
  permethyl-β-CD
  356
Monomer 61, 664
Monosaccharide 592, 617
  enantiomer 528
Monovalent 143, 275
  cation 648
  ion 71
Morbidity 387
Morphine 223–4, 229, 231,
  315, 316, 404
  antibody 224
Morphine-Cy5 229, 231
Morpholine 3, 553, 598
Morpholinoethanesulfonic
  acid 555
Motilin 264
m-RNA 455, 503
MS 84, 95, 202, 310, 318,
  405, 501, 684
  CE 84
  detection sensitivity 137
Multianalyte immunoassay
  229, 231
Multi-anion analysis 500
Multi-capillary 88, 143,
  216, 445, 483
Multi-ion analysis 500

Multi-lane
    electrophoresis 478
Multimatrix
    adaptability 250
Multiple
    capillary 140
    capillary detection 110
    capillary system 463
    cation 73
    component 184
    fluorophore 294, 479
    injection 53, 673
    mutation 452
    reflection 107
    sclerosis 434
Multiplexed detection 481
Multiplexing 455
Multiplication 455
Multi-point detection 97
Multi-sample capacity 681
Multiwavelength 316
Murine monoclonal IgG 245
Muscle 255
Musts 590
Mutant 178, 478
Mutated fragment 452
Mutation 451–2
    detection 451, 453, 456, 477
    site 477
Mutation-specific
    primer 456
Mycobacterium
    tuberculosis-specific
    amplified DNA
    fragment 478
Mycotoxin fumonisin 597
Myelin basic protein gene 475
Myoglobin 175, 178, 431, 434
Myristyltrimethyl-
    ammonium
    bromide 591

N-*o*-linked sugar 199
N,N,N′,N′-
    tetramethylethylene-
    diamine 61–2, 141, 185
N,N′-di (2-hydroxybenzyl)
    ethylenediamine-N,
    N′-diacetic acid
    509–10
$Na_2B_4O_7$ 394
N-acetylator 487
N-acetylprocainamide 397

NAD 215, 419
NADH 214–15, 419
Nadolol 397
NADPH 216
Nafion tubing 92
NaH_2PO_4 394
Naphthalene 559, 632
    carboxylate 554
    dicarboxylate 554, 558
    disulfonate 558, 560
    sulfonate 554, 558
Naphthalene-2,3-
    dicarboxaldehyde 416
    derivatized amino acid
    112
Naphthalenedialdehyde 566
Naphthalene-mono, di, tri
    sulphonate 648
Naphthalenesulfonate 533
Naphthol yellow 611
Naphthylamide 345
Naproxen 318, 356, 401
Narcotic
    analgesic 402, 488
Naringin 596
Narirutin 596
Narrow inner diameter
    92
Narrow bore hollow
    tube 1
Native 452
    amino acid 111
    double-stranded 291
    FL detection 110–11
    peptide 88
    protein 88, 199
Natural
    color 609
    colorant 615
    compound 583, 596
    dye 609
    epo 198
    product additive 376
    product chiral micelle
    373, 376
Nd:YLF laser 88
NDA 112, 416
    derivative several amine
    369
    derivatization 114
NDA-derivatized 149
    glutamate 416, 417
    rat brain
    homogenate 416
NDC 554
    electrolyte 554
N-dodecoxy-
    carbonylproline 378

N-dodecoxy-
    carbonylvaline
    378–9
NDS 648
Near
    infrared indirect
    absorbance 90
Near-infrared 88
    diode Laser 111
Near-IR
    dye 114
    fluorescent dye 114
Needle 311
Nefion ionomer 57
Nefopam 355
Negative charge 2
    peak 271, 589, 683
Negatively
    charged 509
    charged colored ion 609
    charged complex 519
    charged micelle 588
    charged particle 665
    charged SDS
    micelle 517
Neohesperidin 596
Neomycin 311
Nephelometry 249
Net
    charge 78
    charge density 4
Neuraminidase 200, 204, 212
Neurochemical 416
Neuroleptic drug 399
Neuron 418
Neuropeptide Y 417
    β-endorphine 417
Neuroscience community 422
Neurotensin 264, 418
Neurotin 394
Neurotoxin 593
Neurotransmitter 92, 155, 156, 394, 416
Neutral 390, 632
    amino acid derivative
    enantiomer 378
    analyte 21
    complex 509
    compound 632
    hydrocarbon 629
    hydrophilic surface 179
    marker 56
    molecule 588
PAC 356
PEG oligomer 667
pH 263

INDEX 723

phosphate buffer 559
species 588
surfactant 75
triton X 667
New
born 393, 443
drug application 308
N-gluco-N-
methylalkanamides 75
N-glycanase 198, 202
N-hydroxy-succinimide
ester 221
Nicardipien 314
Nickel 517
complex 551
pinhole 86
Nicotinamide 534, 595
Nicotinic acid 386, 401
Nifedipin 314
Nitrate 6, 92, 417, 487, 532, 590–1
Nitrilotriacetic acid 554
Nitrite 6, 90, 92, 417, 487, 532, 591
Nitrophenol 268, 557
β-galactoside 213
Nitrophthalic acid 554
Nitrosoamine 63, 597, 639, 646
N-linked
glycopeptide 202
oligosaccharide 199, 202, 204
sugar chain 198
N-methylformamide 556
N-methyl-N-2-sulfoethyl
dithiocarbamate 511
N-methylpseudoephedrine
enantiomer 379
NMR, 178, 345, 352, 357, 371
Noble metal electrode 95
Noise 104
coefficient 90
Nomifensine 355
Nonaqueous
buffer 595
CE 671
media 73
Nonaromatic amino acid 345, 347
Noncaloric 607
Non-carbohydrate
substituent 617
Non-competitive
method 222

Non-covalent biospecific
interaction 219
Non-covalently bonded 57
Non-crosslinked
polyacrylamide 179
Non-crosslinked
hydrophilic polymer 181
Non-crosslinked linear
polymer network 449
Non-denaturing
condition 451–2
gel 477
polyacrylamide capillary
gel 291
Non-fluorescent amide
derivative 437
Non-gel polymer solution 292
Nonglycopeptide 200
Non-immunoreactive
UV-absorbing
species 220
Non-ionic
detergent 561
polyether 668
surfactant 75, 149, 178
Non-nutritive 607
sweetener 607
Nonpolar compound 137, 632
Non-radioactive detection 477
Non-reducing end 199, 212
Nonstable pH 611
Nonsteroidal
antiinflammatory 345, 355–6
aromatase inhibitor
enantiomer 371
Nonstoichiometric binding 455
Noradrenaline 416
Nordoxepin 312
Norephedrine 78, 341
Norepinephrine 416, 419
Norfluoxetine 355
Normal
benzoic acid 556
phase 657
sera 183
Normalization 313
Normalized peak area 278
Normorphine 224
Nortriptyline 312
Norverapamil 342

N-oxide 62
N-phosphonomethyl
glycine 560
N-tris (hydroxymethyl)
methyl-3-
aminopropanesulfonic
acid 509, 518
Nucelase A variant 178
Nuclear fuel waste 520
Nucleic
acid 140–1, 283, 449–50, 589, 593, 679, 681, 683
Nucleoside 9, 283, 290, 434, 585
Nucleotide sequence 451
Nutrition labeling 595
Nutritional
assessment 430
disorder 430
supplement 372
Nutritive 607
sweetener 607

Obesity 607
Objective
lens 107
zone 25
Octadecylsilica 631
Octanol-water partition
coefficient 632
Octapeptide angiotensin 224
Octopamine 353
Octylglucoside 75
Off rate 275
Off-axis model 97
Off-column 420
detection 84, 86
fluorescence detection 88
Off-line 208, 421
analysis 208
coupling 96
enzymatic reaction
mixture 208
Official AOAC
fluorometric
method 598
Ogston 292
model 292
sieving theory 62
Oligo polysaccharide 617, 620
Oligoadenylate
diasterase 437
synthetase 437
Oligoclonal 434
band 434

724 INDEX

Oligodeoxynucleotide
  isomer 291
Oligomer 555, 566, 630, 665–6, 668
  separation 658
  size 621
Oligomerization 76
Oligonucleotide 63, 122, 223–4, 283, 291, 478, 482, 657, 664
  linker 223
  monomer 122
Oligosaccharide 8, 78, 89, 149, 198–200, 202, 212, 355–6, 592
  mixture 202
O-linked
  glycoprotein 149
N-linked
  oligosaccharide 245
  glycopeptide 202
  oligosaccharide chain 198
On-capillary
  complex formation 510
  complexation 509
  conversion 509
  preconcentration 531, 535
Oncology 219
On-column 84, 91, 112
  absorbance 123
  complexation 517, 520
  derivatization 124, 149, 420
  detection 84, 87
  enrichment 651
  fluorescence
    derivatization 124
  ITP 535
  staining 293
Ondansetron 308
One-spectral-channel
  detection 479
Onized capillary wall
  silanol group 583
On-line
  absorbance 6
  catalytic hydrolysis 213
  chemiluminescence
    detection 89
  chromatographic sample
    concentration 16
  coupling 96
  detection 2, 450, 455
  digestion 22
  enzyme assay 208, 213

immobilized
  enzyme 208, 213
  interface 420–1
  MS 262, 667
  peptide mapping 213
  post-capillary reactor 437
  preconcentration 21, 521
  protease digestion 264
  UV detection 84
On-fly acquisition 682
On-tube detection 682
OPA 112, 368–9
Open-tubular capillary 646
Operator
  efficiency 450
O-Phthaldialdehyde 112, 347
Ophthalmic solution 316
Opiates 224
Optical
  absorption 111
  alignment 84
  axis 107
  boundary 104
  configuration 110
  fiber 87, 109
  filter 109
  isomer 349, 352, 521
  path 84
  resolution 345, 347
  separation 347
  system 107
Optical element 97
Optical-gating injection 22
Optic 86
Optimal
  buffer 611
  condition 72, 611
  pH 45
  resolution 255
  separation condition 517
Optimization 6, 71, 72, 111, 351, 494, 501, 535, 552, 665
  resolution 314
  size 482
Optimized end column
  detection 624
Optimum
  operating condition 521
  pH 438
  position 624
  separation 349
  separation
    parameter 311
  L-menthoxyacetic
    acid 370

Orange 591, 609
  citrus juice by-
    product 598
  juice 520, 590, 596
Organic
  acid 6, 509, 550–1, 554, 583–92, 648–9
  base 550
  hydroxy modifier 632
  ion 499, 535, 550–1
  modifier 356, 368, 372
  phosphate 550, 559
  pollutant 597, 598, 639, 645
  solvent 72
  sulfate 550, 558
Organochlorine 639
Organolead 651
Organomercury 651
  compound 651
  species 651
Organometallic compound 515, 639
Organophosphonic acid 649
Organophosphorus
  compound 560
Organoselenium 651
Organosulfate 559
o-benzylbenzoic acid 533
Orotate 555, 595
Orthogonal
  correlation 54
  geometry 87
Osmotic flow 553, 678
Otosporin 311
Outlet
  buffer vial 118
  end 2, 22
Overdose 392
Overloading 26
Ovine lentivirus 212
Ovomucoid 352
Oxalate 425, 553
Oxazepam 404
Oxazine 632, 750
Oxidation 622
  state 510
Oxide compound 501
Oxidized insulin β-chain 213
Oxomemazine 353
Oxprenolol 351
Oxygen partial pressure 62
Oxyprenolol 397
Oxytocin 255, 263
Oyster sauce 556

P^{53} gene 477
PAA 61
  slab gel 61
PAC 356
Pacific yew 311
Packed
  open-tubular column 597
Packing material 63
PAD 592, 619
PAH gene 477
PAHs 629, 639, 645
Pain 262, 313
Palladium
  decoupler 92
  metal joint 92
p-Aminopyridine 534
Pancreas 255, 264
Pancreatic polypeptide 264
Pantothenate 595
Paper processing 553
PAR, 509, 510, 517
  complex 511
Parabolic flow 2
Paracetomol 313
Parallel
  migration 478, 482
  multi-lane electrophoresis 481
Paraquat 566, 597, 651
Paraxanthine 493
Parenteral solution 520
Particle size 665–6
Particle-capillary
  wall interaction 665–6
Partition 72
  coefficient 7, 29, 31, 647
Partitioning 183
Pasta extract 614
Patent blue 611–12
Path length 104, 138
Pathobiology 418
Pathological
  condition 434
  metabolite 488
Patient 404
  diagnosis 434, 442
  symptom 427
PCP-Cy5.5. 229, 231
PCR 212, 455
  amplification 294, 476–7
  product analysis 143, 293, 466, 477
PCR-amplified
  372 bp DNA sample 477
  DNA fragment 466, 475, 477

rDNA fragment 295
solution 475
PCR-SSCP 477
Pea oligosaccharide 592
Peak
  apex 45
  area 20, 25–6, 402
  area/height 25
  broadening 32, 119, 379, 552
  corrected area 46
  dimension 41
  dispersion 32
  distortion 78
  efficiency 50
  height 20, 444
  homogeneity 308
  identification 41, 616
  identity 309
  impurity 682
  parameter 50
  purity 611, 682
  shape 273, 645
  symmetry 78, 315
  tailing 49, 139
  width 25
Pediatric patient 390
PEG 182, 428, 434, 443, 594, 645
Penciclovir 318, 399
Penicillin 309, 311, 393
Pentafluorobenzoyl 179
Pentanesulfonic acid 310
Pentenyl pyridine 561
Pentobarbital 394
PEO 200, 294, 457
Pepsin 438
Pepsin-catalyzed
  oligopeptide
    cleavage 208
Peptidase 438
Peptide 33, 87, 112, 199, 224, 255, 388, 431, 434, 499, 583–9, 593, 657
  binding membrane 150
  bond 427
  bond scission 247
  hormone 255
  isomer 368
  mapping 123, 200, 213, 264
  substrate 208
Peptide-binding
  membrane 420
Percent
  chiral separation 32
Perdeuterated 556

pyridine 561
Permanent
  non-covalently bonded 57
Permeation 22
Peroxydisulfate 62
Peroxynitrite 244
Peroxyoxalate-
  chemiluminescence 89
Perpendicular 87
Perspex bracket 86
Perylene 630
Pesticide 219, 639, 649
Petroleum ether 312
pH 4, 71, 552
  drift 36
  gradient 73, 123, 141, 183
  shift 71
Pharmaceutical 334, 583–9
  analysis 301, 311, 531
  compound 589
  drug 352
  electrolyte solution 521
  product 595
Pharmacokinetic analysis 22
Pharmacological
  evaluation 301
Phase change 104
PHB ester 610, 614
Phellodendri cortex 315, 597
Phenanthrene 315, 631
Phencyclidine 229–30, 234
Phenobarbital 386, 394, 421
Phenol 73, 503, 550, 553, 557, 632, 639
Phenolic
  acid 557–8
  carboxylic acid 75
  carboxylic derivative 75
  compound 557, 596
  group 227
  hydroxyl group 224
  lignin degradation
    product 557
  sulfonic acid 610
Phenoxy acid
  herbicide 339, 555–6, 557, 650
Phenoxyacetate 556
Phenoxyalkyl
  acid 639, 649–50
  acid herbicide 649–50
Phenoxypropionic acid 650

Phenyl compound 360
Phenylacetate 556
Phenylalanine 124, 156, 360, 425, 442, 488, 593, 609–10
  enantiomer 342
  hydroxylase 477, 488
Phenylalkyl
  alcohol 75, 649
Phenylenediamine 561
  isomer 646
Phenylethanol 361
Phenylketonuria 156, 488, 492
Phenylketonuric 488
  patient 488
Phenylphosphonate 560
Phenylthiohydantoin,
  amino acid 57, 97
Phenylurea 650
Phenytoin 371, 386, 392, 394
Pheochromocytoma cell 150, 420
Pherogram 19, 86, 687
Phosphatase 208
Phosphate 309, 442, 550, 556, 558, 611
  buffer 149, 178, 199, 357, 440, 533, 561, 593–4, 665
Phosphate/borate
  buffer 555–7, 594, 644
Phospholipid 560, 595
Phosphonate 550, 559–60
Phosphonium 550, 561
Phosphopeptide 594
Phosphoramidate 291
Phosphoric 591, 669
Phosphororganic
  compound 649
Phosphorothioate
  oligonucleotide 291
Phosphorous acid 669
Phosphorus-specific flame
  photometric
  detection 560
Phosphotyrosine 245
Photodecomposition 22
Photodiode 86, 97
  array 87, 97, 125, 309
  array detector 86, 610
  array spectral
    analysis 379
  position sensor 97
Photolithographical
  process 482
Photometric method 543

Photomultiplier 87–9, 107, 479, 481
Photon 125
  counting system 89
Photopolymerization 62, 670
Photoresist 122
Photostability 224
Phototube 86
Phthalate 6, 72–3, 553, 591
  buffer 591
  ester 639, 647
  tartrate 522
Phthalic
  acid 533–4, 556, 591
  anhydride 667
Phycobiliprotein 223
Physical
  assumption 33
  parameter 270
Physicochemical
  property 119, 137, 255, 264
Physiological
  concentration 224
Phytate 556, 560
Phytic acid 74, 178, 597
pI 31, 123, 141–2, 200, 222–3, 245, 427, 440, 445, 588, 593
Pigmentation 255
Pilocarpine 316
Pindolol 313
Pine needle 560
Piperidine-dione type
  compound 378
Pituitary gland 255
pKa 29, 310, 314, 361, 397, 402, 511, 551
Placenta 255
Plackett-burman 340
Plano-convex cylindrical
  lens 481
Planorbis corneus 420
Plant
  extract 556
  growth regulator 556
  material 75, 593
Plasma 53–4, 183, 255, 355, 372, 411, 486, 553
  membrane 420
Plastic microwell plate 219
Plasticizer 647
Plate 597
  number 388
Plug 351
  flow 119

PNGase 202
p-Nitrophenyl-2-amino-3-
  hydroxypropanone
  enantiomer 357
Point-mutation
  detection 477–8
Polarity 514, 665
  reversal 678
  switching 140
Polarization 104
Polio virus 455
Pollutant 597, 629, 639, 644
Poly
  (alkyl oxide)
    oligomer 667
  (amidoamine) 76
  (amino acid) polymer 667–8
  (β-malic acid) 667
  (ethylene glycol) 76–7, 141, 466
  (ethylene glycol/oxide) 76
  (ethylene oxide) 76–7, 141, 294, 457
  (methyl methacrylate/
    ethyl acrylate/
    methacrylic acid) 631
  (N-acryloyl-
    aminoethoxyethanol) 77
  (oxyalkylene)
    diamine 112, 666
  (sodium N-undecylenyl-
    L-valinate) 379
  (styrenesulfonate) 77
  (vinyl alcohol) 182
  (vinylchloride) 56
  (9-vinyladenine) 278, 291
  (9-vinyladenine)-
    polyacrylamide
    conjugated gel 291
Polyacrylamide 9, 61, 76, 141, 174, 184, 449–50, 457, 561, 592–3, 666
  electrophoresis 594
  gel 31, 77, 141, 291, 449, 463, 633, 670
  gel electrophoresis 180, 316, 427, 593
  gel polymerization 670
  gel-filled capillary 292, 294
  layer coating 61

matrix 664
polymer 453
sieving medium 482
Polyacrylamide-coated 436
  capillary 222, 436, 487
Polyacrylamide-filled
  capillary 278
Polyacrylate
  polymer 665
Polyamine 150, 418, 499,
    503, 578
Polyamine-coated
  capillary 61
Polyanion 179
Polyaromatic 632
Polybrene 178–9, 352, 361
Polycarboxylic acid 591
Polycation 179
Polycationic
  ammonium compound
    645
  species 179
Polyclonal–monoclonal
    431
Polyclonal 431
  antibody 177, 223,
    229–30
  anti-digoxigenin fab
    fragment 227, 229
  anti-thymocyte equine
    immune globulin
    183
  immunoassay 392
  monoclonal antibody
    transformation
    243–4
Polyclonol–polyclonal
    immunoglobulin 431
Polycyclic
  aromatic amine 501
  aromatic compound 345,
    356
  aromatic hydrocarbon
    88, 111, 501
Polydextran 312
Polydisperse 78
Poly-dT 278
  length isomer 278
Polyelectrolyte 173
Polyether 668
Polyethylene 56, 686
  glycol 418, 428, 460,
    555–6, 558, 566
  glycol-coated
    capillary 63
  glycol 457, 556
  oxide 182, 200, 312, 399,
    450, 566

Polyfluorocarbon 56
Polyglutamate 596
Polyhaloganated
    compound 501
Polyimide 57, 84, 669, 682
  coating 1
Poly-L-aspartic acid 555
Polylysine 665
Polymer 182, 355, 550,
    558, 583, 657
  analysis 671
  chain 457
  composition analysis
    669
  conformation 668
  entanglement 77
  kinetic 665
  oxidation product 665
  separation 657
  sieving 475
  sieving mode 8, 140
  sieving system 681
Polymerase
  chain reaction (PCR)
    212, 291, 466
Polymer-capillary
    interface 458
Polymeric
  additive ficoll-400 664
  buffer additive 664
  carboxylate 555
  cation 555, 558
  impurity 664
  particulate 665
  sieving medium 669
  surfactant 668
Polymerization 61, 181,
    449, 665, 668
  kinetic 670
  scheme 449
  solution 61
Polymorphism 481
Polymyxin 317
Polyol–borate
    complex 592
Polyol 593
Polyoxometalate 669
Polyoxyethylene/propylene
    diamine 566
Polypeptide 31, 255, 317
  protein hormone 255
Polyphenol 557, 596, 598
  oxidase 596
Polyphosphate 597, 649,
    669
Polyphosphonate 597, 649,
    669
Polypropylene 56, 686

capillary 561
Polyquaternary amine
    polymer 178
Polysaccharide 9, 76, 77,
    119, 466, 592, 619
Polystyrene
  carboxylate particle 666
  latex particle 665
  nanoparticle 665
  sulfonate 669
  sulfonate polymer 558
Polysulfonate 558
Polysulfonated
    naphthylurea 399
Polytetrafluoroethylene 56
Polytetrafluoroethylene-
    polyhexafluoro-
    propylene
    copolymer 56
Polyvalent
  hydrophobic cation 552
Polyvinylidene
    difluoride 685
Ponceau 611
Porcine 264
  somatotropin 318
Pork 559
  meat 597
Porous
  gel model 56
  glass 92
Porphyrin
  acid 110
  carboxylate 556–7
Positional
  isomer 556–7, 645
Positively
  charged 509, 666
Post
  capillary reactor 437
  column derivatization
    123
Post-analysis technique
    118
Post-capillary reactor 89
Postcolumn 208
  derivatization 84
  enzymatic assay 124,
    213
  micro-reactor 213
  mixing 213
  radionuclide
    detection 150
Posterior pituitary
    lobe 255
Post-reaction reagent 86
Post-separation 112
Potassium 499, 520, 583

carbonate 442
dichromate 648
sulfate 175
Potato 566, 593
Potentiator 310
Potentiometric
  detection 533
  titration 551
Potentiometry 2, 86, 91, 92
Power
  operation 673
  supply 673
Practolol 347
Praestol 185k 534
Pravastatin sodium 309
Prealbumin 425, 430
Pre-capillary 509
  complexation 509
  derivatization 510
Precipitation 50, 434
Precise quantitation 314
Precision 16, 25, 229,
    308–9, 313–14, 334,
    338, 379, 386, 390,
    401, 403, 425, 444–5,
    551, 667
  data 372
Precolumn 17
  complexation 517
Preconcentration 17, 20,
    21, 63
Prederivatization 112
Predicted optimum
    condition 311
Predictor 438
Pregnancy 265
Preinjection 20
Preparative-scale
    separation 140
Pre post-separation
    sample 456
Prescription 313
Preservative 598, 599, 614,
    616
Pressure 17, 430
  injection 18, 679
  mobilization 184, 681
  mode 19
Pressure-controlled
    Injection 19
Pretreatment 372, 487
Prewashing 57
Primary
  amine 89, 112, 420
  hydroxyl side 345
Primidone 387, 394
Primitive solution 20
Priority 629

phenol 111
pollutant 639
Probe 476
  beam 97
  catalysis 211
Procainamide 397
Processing 609
Product 436
  diffusion 437
Profen 556
Progesterone 265
Prognosis 486
Proinsulin 264
Prolactin 231, 255, 263
Proline 345, 443
Prompt diagnosis 486
Propan-1-ol 355
Propan-2-ol 291, 355, 356
Propanol 590
Propanol-modified-β-CD
    402
Propionate 554
Propranalol 78, 347, 371,
    397
  enantiomer 72
Prostate tumor 399
Protease 436
  enzyme 436
  inhibitor 248
Protease-catalyzed
    hydrolysis of peptide
    211
Protein 87, 140, 173, 199,
    271, 347, 388, 583–9,
    593–5, 657
  adsorption 175
  analysis 173
  conjugate 224
  conjugated 250
  detection 593–4
  gel electrophoresis 588
  hormone 255, 264
  kinase 208
  label 223
  separation 427
  sequencing 685
  succinylation 223
  synthesis 503
Protein/peptide drug 301
Proteinaceous
    compound 593, 598
Protein-carbohydrate 219
Protein-coated 397
Protein-drug 219
  complex 271
Protein-dye 219
Protein-protein 219
Protein-SDS complex 181

Protein-silica interaction
    222
Proteolytic enzyme 425,
    437–8, 445
Protolytic equilibria 552
Pseudo
  epinephrine 341
  ion-exchanger 555
  stationary micellar
    phase 42
Pseudoephedrine 315, 379
Pseudostationary
  hydrophobic micelle
    phase 183
  phase 42, 74, 630
Pseudouridine 442, 503
p-Sulphonate calix 645
PTFE 550
PTH-derivatized D,L-
    amino acid 376
PTH-glycin 97
Pullulan 182
Pulp 553
Pulpwash 591, 598
Pulse laser 125
Pulsed
  amperometric detection
    88, 95, 592, 619, 651
  potential waveform 624
  UV laser 88
Pulsed-field
  CE 119
  condition 669
  gel CE 119
  technique 119
Pump 56
Purification 418
  control 86
Purine 283, 393, 442
  base 442
Purity 308, 309
Putrescine 488, 499, 503
PVA 352
PVP 347
Pyrene 631
Pyrenyldiazomethane 443,
    555
Pyridine 73, 357, 521, 534,
    561, 651
  carboxylate 556
Pyridinecarboxylic acid 79
Pyridylamino 618
Pyrimidine 283, 442
Pyromellitate 72
Pyromellitic acid 533, 558
Pyronin succinimidyl ester
    112
Pyrophosphate 208

INDEX 729

Pyruvate 74, 90, 419

Quadratic regression
    model 314
Quadripole 684
Quality
    control 313, 314, 612
Quanititation 386
Quantifiable assay 231
Quantification 372, 591
Quantify intermediate 273
Quantitation 25, 26, 32,
    50, 138, 142, 227,
    301, 315, 334, 340,
    342, 373, 390, 401–2,
    431, 444
    accuracy 475
    parameter 230
Quantitative
    analysis 16, 25, 41, 310,
        688
    assay 248–9, 314, 455
    CE analysis 342
    data 50, 455
    expression 32
    index 29, 142
    information 29
    reproducibility 20
Quantum yield 224
Quartenary ammonium 3
Quartz 109
    material 108
Quaternary
    alkaloid constituent 315
    alkylamine 561
    amine surfactant 561
    ammonium 76, 519, 651
    phosphonium compound
        566
Quenching 227
Quercitin 596
    glycoside 596
Quinine 560
    sulfate 150
Quinobene 558
Quinolinecarboxaldehyde
    618
Quinolone antibiotic 311
R-136 475
Rabbit skeletal muscle
    tissue extract 208
Racemate 341, 349
Racemethorphan 371, 402
Racemic
    amino acid 369
    amino alcohol 78
    drug 338
    mixture 378

trans-1,2-dihydro-1,2-
    dihydroxy-chrysene
    372
Racemization 368–9
Racemorphan 371, 402
Radial movement 119
Radiation 283
Radioactive
    ATP-G32$_p$ substrate
        208
    DNA precursor 450
Radiodetection 97
Radio-enzymatic
    immunoassay 438
Radioimmunoassay 316,
    417
Radioisotope 173, 208
    assay 208
Radiolabelled
    amino acid 150
    DNA 455
Radiolabelling 455
Radius 3
Rain water 554
Raman
    microprobe spectroscopy
        62
    thermometry 33
Ramp-up voltage control
    120
Randam
    noise 26
Ranitidine 312
Rapidity 84
Rare
    earth element 521–2
    earth metal ion 515, 520
Rat
    brain 156, 416
    brain cortex 418
    cortex microdialysis 416
    gastric ATPase 455
Rate, dissociation 275
Raw data 26
Rb$^+$ 92
Reaction 208
    kinetic 683
    pathway 207
    rate monitoring 334,
        341–2
    tee 89
Reactive dye 224, 559, 648
Reagent consumption 139
Real-time data
    analysis 173
Reanneal 452
Receptor 270, 273
    antagonist 312

molecule 62
    protein 278
Receptor:ligand 219
    complex 273
Recombinant
    biotechnology-derived
        protein 181
    bovine 318
    bovine somatotropin
        183
    chimeric glycoprotein
        317
    erythropoietin 175, 198,
        212
    human erythropoietin
        198
    peptide 208
    protein 198, 202, 317
    therapeutic peptide 224
    therapeutic protein 204
Recondition 680
Recovery 250, 334, 339
    experiment 54
Rectangular
    capillary 63, 66
    channel 125
    column 63
    cross-section 106
    shape 125
    wall 108
Red 609
    He–Ne laser 224
    wine 596
Redox-couple 62
Reduced
    carboxymethylated
        lysozyme 207
Reducing sugar 355
Reduction 622
Redutive amination 202
Reflectance 105
Reflection 105–6
Refractive 84
    index 96, 104–6, 109,
        533, 671
Regeneration 145
Regioselectivity 233
Regression
    line 616, 309
Regulation 610
Regulatory agency 308
Relative
    chiral separation
        factor 32
    molecular mass 245–7
    recovery 54
    standard deviation 293
Relaxation 264

INDEX

time 457
Release transmitter 420
Reliability 609
Renal 430
  clearance 399
  disease 418, 499
  failure 427
  function 397
Rennet
  coagulation 593–4
  whey 593–4
Repeat polymorphism 477
Repeatability 34, 334, 616
Repetition rate 88
Replaceable linear polymer solution 483
Replenishing 680
Replenishment 449, 611, 616
Replicate 227
Reproducibility 2, 16–17, 19, 32–3, 139–40, 142, 174, 313, 315, 334, 386, 388, 402–3, 421, 429, 444–5, 455, 500–1, 551–2, 616, 679
Reproducible injection 679
Reproduction 255
Reptation 292
Research animal 487
Residence time 457
Residual monomer 664, 665, 670
Resolution 74–5, 79, 96, 119, 250, 301–8, 340, 342, 373, 449, 533, 620, 632, 673
  enantiomer 372
  loss 33
Resolving agent 356
Resonance ejection 96
Resorcarene 631
Respiratory syncytial virus 317
Restriction
  endonuclease digestion 453
  fragment length polymorphism 476, 477
  fragment 212
  site 454
Retention
  index 31, 633
  time 139, 597, 616
  window 75

Retinitis pigmentosa 453
Retrieval 682
Reverse
  electroosmosis 179
  migration order 382
  phase 617
  phase liquid chromatography 183, 221, 227
  strand 452
  transcription 475
  transcription-PCR 455
  transport 420
Reversed
  EOF 179, 311, 552
  movement mode method 519
  phase 390, 657
  phase chromatography 183, 369, 607
Reversible polarity 678
Reversion 6
Rhein 597
Rheumatoid
  arthritis 431
  factor 431
Rhodamine 88–90, 108, 125
  acceptor 476
Rhubarb aloe 597
RI 84, 96, 105, 109
  detection 97
  gradient 97, 123
  gradient imaging 123
RIA 417, 500
Riboflavine 62, 109
Ribonuclease 63, 180
Ribonucleotide 533, 649
  monophosphate 597, 669
Ribose 294, 620
Ribosomal RNA gene 295
Rice protein 594
Rifampycin b 78, 349, 353, 355
Right angle 114
Rigid gel 271
Rigidity 379
Rinsing solution 145
Risk arcing 449
River 645
RNA 283, 294, 475
  analysis 291, 294
  reverse transcription 294
Robustness 334
  evaluation 334, 340
Routine
  analysis 478

clinical analysis 430
clinical chemistry laboratory 387
diagnosis 445
rRNA 452
RSD 139, 293, 402
Rt 475
Rt-PCR 455
Ruggedness 139, 342
Running
  buffer 34, 36, 53, 91, 179, 212, 397, 418, 437–8, 517, 610–11
  voltage 518
Run-to-run 37
  peak area 679
  reproducibility 36
  variation 450

S form D-Los Angeles 440
S/N ratio 26
Saccharide 583, 592–3
Saccharine 607, 610, 614
Safety, pharmaceutical 308
Salbutamol 308, 309, 339
  sulphate 308
Salicylate 74, 92, 386, 397, 404, 556
Saline solution 561
Saliva 53, 283, 393, 401, 486, 493
Salivary
  amylase 493
  fluid 493
Salt 534, 678
Sample
  bias 450
  capacity 125
  carry over 49
  clean up 391
  deproteinization 442
  derivatization 618
  extraction 54
  injection 16, 22
  injection length 32
  introduction 139
  manipulation 487
  mass 25
  matrix 53, 138, 211, 411, 425, 444, 678
  preconcentration 20, 84
  preparation 138, 144, 308, 390
  pretreatment 110, 183, 342, 492
  slug 271
  stacking 16, 20
  throughput 125, 681

INDEX 731

tray thermostatting 46
treatment 283, 401
zone 20, 97
Sample-focusing technique 683
Sample-to-sample headspace 679
Sampling procedure 308
Sanger
  method 478
  reaction 294, 480
Savinase 436
SBE-β-CD 345, 353, 360
Scandium(III) 520
Scanning
  detection 682
  diode-array detector 682
Scatchard analysis 273
Scattered light 106
Scattering 87, 105, 107
  effect 87
  light 479
Scavenging 664
Scharlachrot 611
Scintillation 150
SCN⁻ 92, 96
Scrapie prion protein 417
Scute coptis 315
SDS 6, 200, 310, 367, 390, 436, 509, 517, 556, 594, 597, 664
  capillary gel electrophoresis 180
  matrix 667
SDS–CGE 240, 245–7
SDS–PAGE 174, 180
SDS–polyacrylamide 174
SDS–protein complex 9, 31, 76–7, 681
SEC 63
SEC–CZE 63
Second
  fragment 452
  harmonic emission 88
Secondary
  amine 352
  hydroxyl side 345
  structure 291–2, 451, 477, 480
Second-harmonic emission 111
Secretion 264
Sedative 345, 353
Sediment 646
Sedimentation 270, 431
  rate 431
Seed 592, 597

Selectivity 75, 90, 315, 334, 339–40, 352, 416, 452, 487, 494, 501, 528, 552
  mechanism 551
Selenious compound 651
Selenium 499, 651
Self-heating 120
Semiconductor
  laser 88
  lasers/laser diode 88
Sensitive detection 84, 89
Sensitivity 90, 96, 104–5, 111, 338, 402, 404, 452, 494, 532, 535, 647
  limitation 311
Separand 3, 292
  zone detection 87
Separation 4, 32, 96, 108, 111, 314, 334, 349, 372, 390, 429, 482, 552, 610, 649
  buffer 271, 278, 312, 610–11
  capability 137
  capillary 84
  capillary diameter 86
  chemistry 457
  difficulty 119
  efficiency 1, 3, 17, 19, 29, 61, 78, 84, 90, 174, 292, 312, 501, 647, 658
  medium 141, 463
  mode 4
  parameter 308
  process 509
  profile 317
  selectivity 74
  time 26, 673
  tube 106
  voltage 313
Sequence
  analysis 456
  change 452
  isomer 280
Sequence-induced anomalous migration 451
Sequence-specific diagnostic marker 280
Sequencing 110, 450
  accuracy 480
  human genome 480
  reaction 293, 449
Sera 492

Serine 200, 345
Serotonin agonist 339, 369
Serratia macescens nuclease isoform 183
Serum 220, 283, 393, 411, 431, 486, 555
  aspartate aminotransferase 440
  biliprotein species 183
  cicletanine 401
  drug 386
  electrophoresis 430
  lipoprotein 436
  protein 177, 425, 427, 445
  sample 521
Serum-based digoxin 223
Serum-containing medium 252
Serum-free hybridoma culture medium 252
Setup 89
SH residue 62
Shale oil 630
Shape 243
Shear
  force 458
  stress 458
Sheath
  flow connection 118
  flow cuvette 110, 112, 479
  flowcell 107
Sheath-flow
  cuvette 88, 107
  cuvette flow chamber 107
Sheep 264, 594
  liver extract 183
Shell fish toxin 501
Shellfish 310
  poisoning toxin 598
Shielding effect 57
Short tandem repeat 454, 475, 476, 481
Shrinkage 76
Sialic
  acid 198, 199, 204, 212
  acid containing glycopeptide 200
  acid content 184
Sialidase
  activity 211
Siallyation 204
Sialoganglioside 592
Si–C
  bond 174

bond-based vinyl coating 179
Side product 683
Sieving
  matrix 294, 482
  medium 9, 31, 119, 141, 181, 292, 449, 477, 552, 558, 667, 669
  network 457
  polymer buffer 476
  polymer 77, 679
  solution 293
Signal-to-noise ratio 26, 89–90, 96, 338
Silane reagent 181
Silanized column 597
Silanoate 174
Silanol group 2, 56, 78, 173, 180, 181
Silanyl moiety 213
Silica 63
  surface 4, 57, 174, 180, 213, 552
Silica–solution interface 56
Silicone
  wafer 22
  rubber membrane 557
Siloxane bond 61
Siloxy functionality 621
Silver stain 173
Simple
  amine 550
  inclusion 276
Simplicity 342, 390
Simulation 275
  model 275
Simultaneous
  absorbance 114
  analysis 143, 609
  immunoassay 229
Sine-wave signal 119
Single
  antibody 220
  base mutation 456
  cell 411, 418
  human erythrocyte 216
  point mutation 451
  stranded 294
  therapeutic oral dose 372
Single base
  deletion 478
  mismatch DNA 478
  resolution 480
  substitution 477
Single-nucleotide
  mismatch 452

Single-strand
  analysis 452
  conformation polymorphism 451, 476–8
Single-stranded 291
  DNA 62, 480
  oligo-DNA 477
  oligonucleotide 291, 292, 477
  state 680
Site-directed mutagenesis 178
Size 240, 609, 622
  analysis 658
  determination 466
  exclusion 277, 657
  chromatography 63, 270, 614, 657, 667, 669
Size-based
  separation 240
Size-exclusion 617
  chromatography 669
Sizing 455
Skin 262, 431
Skull 255
Slab
  disk gel electrophoresis 76
  format 449
  gel 122, 141–2, 181, 479
  gel electrophoresis 173, 293, 453, 455
  gel SDS–PAGE 245
  gel-based electrophoretic separation 174
Slanted-edge 19
Slide-type injector 19
Slot blot 293
Slow off rate 273
Sludge 639
Small amount injection 139
Small-bore
  capillary 125
Smooth
  muscle 262, 264
  muscle antispasmodic 399
Snake 119
Snap-cooling 477
Soap 669
Sodium 499, 500, 520, 535, 590, 592
  10-undecylenate 76, 630, 647

  acetate 650
  borate 649
  borate buffer 149
  chloride 183
  cholate 313, 556, 557
  cholate micelle 561, 629
  deoxycholate 313, 372–3, 630
  deoxycholate buffer 598
  dodecyl sulfate 6, 180, 200, 240, 367, 390, 509, 517, 556, 596
  dodecylbenzenesulfonate 554
  hydrogencarbonate 611
  hydrogencarbonate buffer 611
  hydroxide 610, 611
  n-Octanesulfonate 519
  phosphate 611
  phosphate buffer 669
  sulfate 594
  sulfobutylether-β-CD 352
  taurodeoxycholate 372, 630
Soft drink 555, 593, 598, 609
Software option 680
Soil 556, 639, 646
  leachate 558
Solid
  laser 88
  phase fluorescent labeling technique 264
  phase-packed column 63
  scintillator 150
  support 221, 229, 234
  surface 118
Solid phase
  extraction 138, 221, 312, 340, 391, 399, 417, 501, 650
  immunoseparation 220
  packed capillary 21
Solubility 311, 392
Soluble
  CD_4 183
  linear polymer 480
Solute 345, 357
  anion 519
  capacity factor 629
  zone 2
Solution viscosity 3, 17–18
Solution-phase, immunochemical kinetic 237

Solvation 552
Solvatochromic parameter 629
Solvent
  evaporation 391
  extraction 138, 391
  gradient 552
  refined coal 630
Solvophobic
  additive 631
  association 629
Somatostatin 263
  analogue 264
Somatotropin 317
Sorbate 555, 558, 591, 597, 598
Sorbic acid 591, 610
  benzoic acid 597
SOS 519
Southern
  analysis 453
  blot 475
  blotting 476
Soybean 560
  genotype 475
Spacer arm 223–4, 233
Spacial width 25
Spaetzle 612, 614
Spatial width 25, 28
Spatial-scanning LIFD 110
Speciation analysis 509, 515
Specific
  amino acid 445
  binding 219
Specificity 211, 308, 404, 647
Spectra library 86
Spectral
  analysis 614
  comparison 611
  identification 611
  information 86
Spectrophotometric
  method 618
  property 270
Spermidine 347, 488, 499, 503
Spermine 488, 499, 503
Spherically porous particle 213
Spontaneous
  extraneous injection 18
  injection 46
Sport-banned drug 488
Sprite 612
Sr 421, 528

S-S bridge 62
SSCP 451, 476, 477, 478
  analysis 452
Stability 110, 301, 342, 509, 610, 671
  constant 521, 522, 523
  testing 334
Stable substrate 211
Stacking 386, 425, 434
  effect 678
Staining 173, 184
  process 142
  step 427
Standard
  deviation 53, 193
  Lac I gene 477
Starburst, dendrimer 76
Starch 591, 607
Stationary
  front 79
  phase 71, 607
Statistical algorithm 33
Steady state 392
Steam distillation 645
Stearic configuration 667
Step gradient filling 62
Stepper motor 118, 125
Stepwise fashion 517
Stereogenic center 352
Stereoisomer 339, 347, 360–1, 372, 388
Stereospecificity 425
Steric
  factor 352
  hindrance 353
Stevia sweetener 598
Stimulant 345
Stoichiometric 227, 273
  amount 220
  study 273
Stoichiometry 278, 310
Stomach 264
Stone formation 442
Storage 609
  condition 308
Straight-edge 19
Stray light 87
Striatal microdialysates 114, 416–17
Striatum 155
Strong
  alkaline buffer 593
  alkaline solution condition 618
  light source 84
Structural 657
  information 84, 95, 96
  sequence 657

specificity 223
Subattomole detection 88
Subcritical fluid extraction 651
Subcutaneous
  injection 312
  tissue 421
Substance P 264
Substituted 629
  base 283
  phenol 557, 644, 645
  purine 283
Substrate 124, 207, 213, 270, 436, 437
Subtle
  difference 137
  factor 314
Subtype 452
Sub-unit 181
Successive run 273
Succinate 6
Succinylation 250
Sudan B black 436
Sugar 111, 202, 589, 591, 624
  acid 555
  juice 553
  moiety 212, 528, 596
Sugarcane cultivar 596
Sulfanilic acid 592
Sulfate 442, 550, 559, 590–1, 665–6
Sulfated disaccharide 559
Sulfobenzoic acid 533
Sulfonamide 394, 550, 559, 597
Sulfonate 224, 550, 554, 558–60
Sulfonated
  azo dye 559
  calixarene 558
  β-CD 357
  polystyrene 119, 669
  reactive dye 558
Sulfonic
  acid 609, 611
Sulfonyl
  urea 95, 550, 559, 597, 650
Sulforhodamine 88
Sulfur
  containing amino acid 442
  containing compound 488
  keratin 404
Sulphametazine 310
Sulphonamide 310

Sumarin 492
Sumatripan 312
Supernatant 397, 434, 438, 444
Support 391
Suppressed
   conductivity
      detection 533, 555, 559, 590, 646
   conductometric CE 91
Suramin 399
Surface charge 212
Surface-bound
   dextran 179
Surface-modified 21
Surface-to-volume ratio 1
Surfactant 6, 71, 74, 183, 312, 367, 368, 372–3, 390, 517, 550, 552, 561, 639, 647
Surfactant/mass ratio 181
Sweetener 598, 607, 610, 612, 614, 616
Sweetening agent 607
Symmetrical square-wave 119
Sympathomimetic 345, 347
   drug 347
Sympathomimetic 345, 347
Syn 646
   anti conformer 646
Synaptic
   vesicle 420
Synchronized
   cyclic CE 122
Synthesis 443
   cationic polymer 665
Synthetic
   chiral surfactant 376
   color 598, 609
   dye 607, 609, 611
   material 657
   polymer 657
   sweetener 607
Systemic lupus
   erythematosus 431

Tablet 313, 612
   capsule 614
Tacrolimus 387
Tagged drug 401
Tamoxifen 311
Tandem
   mass spectrometric
      condition 96
   MS 310
   repeat 454
Tans epimer 316

Tap 509, 518
   water 597, 650
   water analysis 521
Taramasalata 612
Target concentration 372
Tartaric acid 534
Tartrate 528, 554, 590
Tartrazine 610–11, 614–15
Taurine 416, 420
Taurodeoxycholate 369, 372–3, 630
Taxines 311
Taxol 309, 311
Taylor-aris dispersion
   theory 32
Tear 486, 493
   fluid 220
Tem-2-β-lactamase 309
Temperature 4, 33, 72, 291
   change 44
   control 44, 140, 673, 679
   gradient programming 680
Terahydrofolic acid 596
Terbium(III)-acetyl
   acetone 90
Terbutylazine 650
Terephthalic acid 51, 555
Terminator 650
Terminating
   buffer 404
   constituent 12
   electrolyte 535
Tertiary structure 178
Tetraalkylammonium 75
   bromide 399
   salt 518
Tetraantennary
   oligosaccharide 198, 205
Tetraborate buffer 596, 610
Tetrabutylammonium 90, 310, 556
   bromide 519
Tetracycline 311
Tetradecyltrimethyl-
   ammonium
      bromide 519, 553–5
      hydroxide 554
Tetraheptylammonium
   bromide 595
   salt 76
Tetrahydro-cannabinol
   carboxylic acid 404
Tetrahydrofuran 561
Tetraalkyl ammonium 347, 519
Tetramethylrhodamine 222, 294

   isothiocyanate isomer R 90
Tetramethylrhodamine
   isothiocyanate-
   labeled DNA
      fragment 479
Tetramethylrhodamine-
   labeled
      antigen 223
      synthetic disaccharide 211
      trisaccharide 212
Tetranucleotide 477
Tetrapeptide 263
Tetraphenylphosphonium
   ion 566
   salt 519
Tetrasulfonated
   label Cy5.5 231
Textile dye 558, 649
Thalassmia 440, 442
THC-Cooh 230
THC-Cy5 231
Theobromine 393
Theophylline 229, 283, 309, 386, 390, 392–3, 421, 493
Theoretical
   equilibrium model 72
   plate number 86, 122, 137, 550, 597, 673
   plot 522
Therapeutic
   drug 443, 488, 492
   drug level monitoring 219, 301, 342, 386
   screening 486
   treatment 486
   utility 198
Therapy 486, 492
Thermal
   coefficient 33, 453
   cycling 478
   diffusion 1
   dissipation 1
   effect 33, 50, 673, 685
   expansion 120, 611
   fluctuation 34
   gradient 32, 34, 294, 453
   gradient peak
      broadening 479
   noise 97
Thermodynamic 509
   parameter 270
Thermooptical
   absorbance 97
Thermospray
   ionization interface 96

INDEX 735

Thermostable
  DNA ligase 478
Thermostated
  effect 34
THF 561
Thiazole 372
  orange 212, 293
Thin layer chromatography
  308, 607
Thiol 232, 550, 559
  drug 369
  group 222
  reducing agent 180
Three dimensional
  image 125
  data 611
Threonine 149, 345
Throughput 121, 143,
  455–6, 483
Thymidylic acid 291
Thymine 294
Thyroid
  gland 255
  stimulating hormone
    255, 263
Thyrotropin releasing
  hormone 263
Thyroxine 227
Ti:sapphire laser 114
Tight-binding (slow-off)
  system 276
Time
  consuming 141, 388
  window 514
Time-resolved fluorescence
  detection 88
Tin 651
Tinc–torial strength 610
Tissue 411, 417–18, 420–1
  analysis 411
  enzyme level 438
  homogenate 438, 444
  sample 411
Tissue-bound
  cytokine 418
TLC 308, 607
TM-β-CD 355
Toluenesulfonic acid 559,
  560
Toluidine 645
Toothpaste 597, 669
Total ligand 522
Toxicity 371, 492, 639
Toxicological study 403
Toxin 499, 583, 597–8
Transfer RNA 213
Transferrin 184, 425,
  430, 431

Transient
  temperature gradient 33
  thermal gradient 33
Transition 509
  metal ion 73, 510, 515,
    520
  rare earth metal ion 520
  state 207
Translational
  sugar modification 198
Translocation 455
Transpeptidation reaction
  208
Transplantation 488
Transport velocity 78
Transverse flow 63
Trauma 434
Tri-aza aromatic ligand
  compound 373
Triangular shape 45
Triarylmethane
  color 610
  dye 609
Triazine 566, 649
  dye 648
Trichloroacetic acid 442
Tricine 428, 440
Tricine-based buffer 199
Tricyclic 386, 393
  antidepressant 312
Triethylamine 558
Triethylenetetraamine-
  hexaacetic acid 520
Trimellitate 72
Trimethylammonium
  butylsufonate 178
  propylsulfonate 317
Trimetoquinol 373
  hydrochloride 372
Trinitrobenzene
  sulfonate 554
Tri-o-methyl-β-CD 650
Tripeptidase 438
  activity 208
Triplicate 227
Tris 399, 428, 440
  (hydroxymethyl)
    aminomethane 34
  buffer 309
Tris–borate 558, 594
Tris–borate buffer 96, 292,
  664
Tris–borate-EDTA
  buffer 666
Tris–boric acid 594
Tris–SDS mixture 183
Triton 443, 553
Tropolone 534

Trough Level 392
Trypsin 438
Trypsinogen 175
Tryptic
  digest 43, 207, 263, 442
  mapping 202
  peptide 200
Tryptophan 88, 123, 156,
  184, 411, 593
  enantiomer 78, 150
Tubular membrane 22
Tumor
  cell 438, 503
  malignancy 438
  suppresser 477
Tunability 87
Tunable fluorescence
  detection 89
Tungsten 669, 684
  source 682
Turbidity measurement
  666
Turbutaline sulphate 309
Turnover 207, 421
Tween-20 75, 149, 264
Two-dimensional imaging
  system 86
Two-dye labeling 477
Tyrosine 123, 184, 593
Tyrosine-containing
  protein 88

Ubiquitous injection 18
Ucon-coated column 290
Ultracentrifugation 270,
  436
Ultradilute solution 124
Ultrafiltration 421, 442,
  501
  probe 493
Ultra-high-speed analysis
  482
Ultra-high-throughput 482
Ultramicroelectrode 84, 92
Ultramicrosampling 48
Ultrasensitive detection
  111
Ultraviolet 509, 531
  light absorption 104
Unbound
  non-immunoreactive
    compound 219
Uncharged polymer 658
Uncoated fused silica
  capillary 222
Unconjugated
  antibody 247–8
  cytotoxin 247–8

Uncross-linked
    polyacrylamide 9, 666
Undecylenate 630
Undiluted serum 223
Uniconazole 371
Uniformity 56
Unreacted monomer 664–5
Untreated
    fused-silica capillary 175
    silica capillary 665
Uracyl 294
    DNA glycosylase 62
Urea 96, 178, 199, 291, 355, 371, 404, 499, 501, 594
    herbicide 75, 649
Urea-denatured condition 180
Uremia 499
Uremic
    serum 442
    toxin 499
Uric acid 339, 425, 442, 499, 501
Uridine analog 212
Urinary
    5-hydroxyindole acetic acid 442
    metabolite 442
    oxalate 442
    porphyrin 442
    protein 431
Urine 177, 220, 229, 262, 283, 353, 360, 371, 393, 411, 443, 486, 554, 556
    cleanup 443
    drug screening 386
    protein 425
Uronic acid 592, 619
Utility 403
UV 394, 430, 509, 531, 550, 666
    absorbance 220, 345, 588–9
    absorbent 73
    absorbing ions 550
    absorption 104
    chromophore 318, 671
    detection 104, 416
    laser 86, 111, 588–9
    laser induced fluorescence 88
    light 110
    spectrophotometer 386
    spectrum 417

UV/VIS
    absorbance 84, 86
UV-transparent
    hydrophilic polymer 182

V(v) 517
Vacancy peak method 271
Vaccine 317
Vacuum
    injection 18
    suction 17
Validated
    assay 308
    method 308
Validation 139, 248, 301–8, 334, 494
    report 583
Valine 345, 368
    buffer 352
Valproic acid 394
Valveless flow control 123
Van Deemter equation 32
Vanadate 533
Vancomycin 275
Vanillymandelic acid 442
Vaporization
    solvent 95
Variable
    number tandem repeat 475, 477
Variance 310
Variant 442, 452
Variation 184, 456
Varietal origin 594
Various opiates 224
Vasoactive
    intestinal peptide 264, 417
Vasodilator 313
Vasopressin 255, 263
Vegetable 591
Verapamil 314, 342
Vesitone 376
Vinyl
    sulfonate monomer 558, 664
    sulphonate 665
Viral subtype 452
Viscosity 53, 71, 182, 243, 679
    linear polyacrylamide 480
Viscosity-enhancing agent 185
Viscosity-modifying component 679

Viscous
    agent 679
    polymer 9
    sample 18
Visible
    absorbance 588–9
    blue region 88
    laser 111
    range 533
    region 86
Visualization 587
    agent 309
Visualizing agent 520
Vital constituent 609
Vitamin 583, 589, 595
    B12 deficiency 443
    C 595, 598
    C, B1, B3, B6, B12 595
Volatile fatty acid 598
Voltage
    pre-conditioning 44
    ramp 50, 611
Voltammetric
    characterization 416
Volumetric flow 684
Von Willenbrand factor
    gene 475

Waldenstrom's
    macroglobulinemia 431
Wall
    adsorption 45, 78
    coated capillary 669
    effect 403
    interaction 450, 552
Wall-jet
    arrangement 624
    configuration 624
Wall-jet-like
    configuration 95
Warfarin 78, 342, 355, 402
    enantiomer (S/R) 402
Washing
    powder 436
    process 36
Waste water 558
Water 556, 639, 646
    analysis 531
    insoluble polymer 665
    soluble 658, 665
Waveform
    frequency 119
    parameter 624
Wavelength 361, 392
    range 86
    selection 681, 682

Weak complexing
    agent 520, 590
Western blotting 685
Wet chemical method 543
Wheat
    bran 596
    flour 591
    gliadin 594
Whey 183, 555
    protein 178, 594, 598
White noise 86
Whole-column detection 110
Width 482
Wild-type gene 451
Wine 553, 555, 557, 590
Wogonin 314
Wogonin-7-*o*-glucuronide 314
Woolfiber-method 614
Wounding 493

Xanthine 425, 442
Xenon 684
    arc lamp 89
    gun 96
    lamp 86
X-ray crystallography 178
Xylidine 561

Y chromosome-specific
    sequence 212
Yag laser 125, 482
Yellow 609
Yield 683
Yolk 598
Yttrium (III) 520

Z E isomer 357
Z cell 589, 683
Zeta potential 33–4, 78, 119, 174, 552
Zinc 518
    complex 554
Zirconium
    ceramic sample 518
Zolpidem 353
Zone
    broadening 33, 92, 666
    dilution 123
    distortion 184
    electrophoresis 21, 173
    electrophoresis detector 108
    length 309
    mobilization 123, 681
    movement 87
    velocity 25
Zone-broadening 84
Zopiclone 353
Z-shaped cell 66
Zwitterion 3, 71, 78–9, 141, 178, 317, 427
Zwitterionic
    analyte 78
    compound 248
    detergent 263